Drug Delivery Strategies for Poorly Water-Soluble Drugs

Advances in Pharmaceutical Technology

A Wiley Book Series

Series Editors:
Dennis Douroumis, University of Greenwich, UK
Alfred Fahr, Friedrich-Schiller University of Jena, Germany
Jürgen Siepmann, University of Lille, France
Martin Snowden, University of Greenwich, UK

Titles in the Series

Hot-Melt Extrusion: Pharmaceutical Applications
Edited by Dionysios Douroumis

Drug Delivery Strategies for Poorly Water-Soluble Drugs
Edited by Dionysios Douroumis and Alfred Fahr

Forthcoming Title(s):

In Vitro Drug Release Testing of Special Dosage Forms
Edited by Nikoletta Fotaki and Sandra Klein

Drug Delivery Strategies for Poorly Water-Soluble Drugs

Edited by

DENNIS DOUROUMIS

School of Science, University of Greenwich, UK

and

ALFRED FAHR

Friedrich-Schiller University of Jena, Germany

WILEY

A John Wiley & Sons, Ltd., Publication

This edition first published 2013

Registered office
John Wiley & Sons Ltd, The Atrium, Southern Gate, Chichester, West Sussex, PO19 8SQ, United Kingdom

For details of our global editorial offices, for customer services and for information about how to apply for permission to reuse the copyright material in this book please see our website at www.wiley.com.

Library of Congress Cataloging-in-Publication Data

Drug delivery strategies for poorly water-soluble drugs / edited by Dennis Douroumis and Alfred Fahr.
p. ; cm.
Includes bibliographical references and index.
ISBN 978-0-470-71197-2 (cloth)
I. Douroumis, Dennis. II. Fahr, Alfred.
[DNLM: 1. Drug Delivery Systems. 2. Chemistry, Pharmaceutical–methods. 3. Drug Carriers–pharmacology. 4. Hydrophobic and Hydrophilic Interactions. 5. Solubility. QV 785]
615.1–dc23

2012035502

A catalogue record for this book is available from the British Library.

ISBN: 9780470711972

Set in 10/12pt Times by Aptara Inc., New Delhi, India.

Reprinted with corrections January 2014.

Excellence is an art won by training and habituation. We do not act rightly because we have virtue or excellence, but we rather have those because we have acted rightly. We are what we repeatedly do. Excellence, then, is not an act but a habit. (Aristotle, 384–322 BC)

This book is dedicated to my beloved mother Eugenia for her continuous support and unconditional love. It is also dedicated to my brother Bill and sister Angela for their support and patience. Thank you all.

Doubt grows with knowledge.
Johann Wolfgang von Goethe (1749–1832)

I thank my wife for her understanding for spending weekends in my home office for setting and polishing this book. I apologize to my children Fabian and Sophie that their dad was not ready on many weekends for playing and talking. I do hope, they will understand it in the near future somehow.

Contents

List of Contributors

Dr. Naveed Ahmed, University Lyon 1, Villeurbanne, CNRS, UMR-5007, Laboratoire d'Automatique et de Génie des Procédés, France.

Dr. Gavin P. Andrews, The Drug Delivery and Biomaterials Research Group, The School of Pharmacy, Queen's University of Belfast, Northern Ireland.

Dr. Cordin Arpagaus, BÜCHI Labortechnik AG, Flawil, Switzerland.

Dr. Swati Biswas, Center for Pharmaceutical Biotechnology and Nanomedicine, Department of Pharmaceutical Sciences, Northeastern University, Boston, Massachusetts, USA.

Dr. Marcus E. Brewster, Johnson & Johnson Pharmaceutical Research and Development, Beerse, Belgium.

Professor Heike Bunjes, Technische Universität Braunschweig, Institut für Pharmazeutische Technologie, Braunschweig, Germany.

Dr. Dennis Douroumis, School of Science, University of Greenwich, Chatham Maritime, ME4 4TB, Kent, UK.

Professor Abdelhamid Elaissari, University Lyon 1, Villeurbanne, CNRS, UMR-5007, Laboratoire d'Automatique et de Génie des Procédés, France.

Professor Dr. Alfred Fahr, Department of Pharmaceutical Technology, Institute for Pharmacy, Friedrich-Schiller-University, Jena, Germany.

Dr. Dimitrios G. Fatouros, Department of Pharmaceutical Technology, School of Pharmacy, Aristotle University of Thessaloniki, Greece.

Professor Hatem Fessi, University Lyon 1, Villeurbanne, CNRS, UMR-5007, Laboratoire d'Automatique et de Génie des Procédés, France.

Professor Dagmar Fischer, Department of Pharmaceutical Technology, Institute of Pharmacy, Friedrich-Schiller-University, Jena, Germany.

Professor Jonathan Hadgraft, Department of Pharmaceutics, School of Pharmacy, University of London, UK.

Dr. Chiraz Jaafar-Maalej, University Lyon 1, Villeurbanne, CNRS, UMR-5007, Laboratoire d'Automatique et de Génie des Procédés, France.

Professor Narendra K. Jain, Pharmaceutics Research Laboratory, Department of Pharmaceutical Sciences, Gour Central University, Sagar, India.

Dr. David S. Jones, The Drug Delivery and Biomaterials Research Group, The School of Pharmacy, Queen's University of Belfast, Northern Ireland

Dr. Sonja Joseph, Technische Universität Braunschweig, Institut für Pharmazeutische Technologie, Braunschweig, Germany.

Professor Cornelia M. Keck, Department of Applied Logistics and Polymer Sciences, Applied Pharmacy Division, University of Applied Sciences Kaiserslautern, Pirmasens, Germany; Institute of Biosciences (IBS), University Putra Malaysia (UPM), Serdang-Kuala Lumpur, Malaysia; and Department of Pharmaceutics, Biopharmaceutics & NutriCosmetics, Freie Universität Berlin, Berlin, Germany.

Dr. Szymon Kobierski, Department of Pharmaceutics, Biopharmaceutics & NutriCosmetics, Freie Universität Berlin, Berlin, Germany.

Dr. Majella E. Lane, Department of Pharmaceutics, School of Pharmacy, University of London, UK.

Professor Vesa-Pekka Lehto, Department of Applied Physics, University of Eastern Finland, Kuopio, Finland.

Dr. Mathew Leigh, Phares AG, Muttenz, Switzerland.

Dr. Shu Li, The Drug Delivery and Biomaterials Research Group, The School of Pharmacy, Queen's University of Belfast, Northern Ireland.

Professor Thorsteinn Loftsson, Faculty of Pharmacy, University of Iceland, Reykjavik, Iceland.

Dr. Rachmat Mauludin, Department of Pharmaceutics, Universitas Pendidikan, Bandung, Indonesia.

Professor Sylvio May, Department of Physics, North Dakota State University, Fargo, ND, USA

Marco Meuri, BÜCHI Labortechnik AG, Flawil, Switzerland.

Dr. Claudia Elizabeth Mora-Huertas, University Lyon 1, Villeurbanne, CNRS, UMR-5007, Laboratoire d'Automatique et de Génie des Procédés, France.

Dr. Sara Movassaghian, Center for Pharmaceutical Biotechnology and Nanomedicine, Department of Pharmaceutical Sciences, Northeastern University, Boston, Massachusetts, USA; and Department of Pharmaceutics, School of Pharmacy, Shaheed Beheshti University of Medical Sciences, Tehran, Iran.

Professor Rainer H. Müller, Department of Pharmaceutics, Biopharmaceutics & Nutri-Cosmetics, Freie Universität Berlin, Berlin, Germany.

Professor Anette Müllertz, Bioneer: FARMA, Department of Pharmaceutics and Analytical Chemistry, The Faculty of Pharmaceutical Science, University of Copenhagen, Copenhagen, Denmark.

Dr. Joakim Riikonen, Department of Applied Physics, University of Eastern Finland, Kuopio, Finland.

Dr. David Rütti, BÜCHI Labortechnik AG, Flawil, Switzerland.

Assistant Professor Jarno Salonen, Department of Physics and Astronomy, University of Turku, FI-20014 Turku, Finland.

Dr. Hélder A. Santos, Division of Pharmaceutical Technology, Faculty of Pharmacy, University of Helsinki, FI-00014 Helsinki, Finland.

Dr. Stefan Scheler, Sandoz GmbH, Kundl, Austria.

Dr. Hari Singh, Gour Central University, Sagar, India.

Dr. Rakesh K. Tekade, Pharmaceutics Research Laboratory, Department of Pharmaceutical Sciences, Gour Central University, Sagar, India.

Professor Vladimir P. Torchilin, Center for Pharmaceutical Biotechnology and Nanomedicine, Department of Pharmaceutical Sciences, Northeastern University, Boston, Massachusetts, USA.

Dr. Peter van Hoogevest, Phares AG, Muttenz, Switzerland; and Department of Pharmaceutical Sciences, Institute of Pharmaceutical Technology, University of Basel, Switzerland.

Dr. Onkar S. Vaze, Center for Pharmaceutical Biotechnology and Nanomedicine, Department of Pharmaceutical Sciences, Northeastern University, Boston, Massachusetts, USA.

Assistant Professor Xue-Qing Wang, School of Pharmaceutical Sciences, Peking University, Beijing, China.

Professor Qiang Zhang, School of Pharmaceutical Sciences, Peking University, Beijing, China.

Series Preface

The series *Advances in Pharmaceutical Technology* covers the principles, methods and technologies that the pharmaceutical industry use to turn a candidate molecule or new chemical entity into a final drug form and hence a new medicine. The series will explore means of optimizing the therapeutic performance of a drug molecule by designing and manufacturing the best and most innovative of new formulations. The processes associated with the testing of new drugs, the key steps involved in the clinical trials process and the most recent approaches utilized in the manufacture of new medicinal products will all be reported. The focus of the series will very much be on new and emerging technologies and the latest methods used in the drug development process.

The topics covered by the series include:

Formulation: the manufacture of tablets in all forms (caplets, dispersible, fast-melting) will be described, as will capsules, suppositories, solutions, suspensions and emulsions, aerosols and sprays, injections, powders, ointments and creams, sustained release and the latest transdermal products. The developments in engineering associated with fluid, powder and solids handling, solubility enhancement, colloidal systems including the stability of emulsions and suspensions will also be reported within the series. The influence of formulation design on the bioavailability of a drug will be discussed and the importance of formulation with respect to the development of an optimal final new medicinal product will be clearly illustrated.

Drug Delivery: The use of various excipients and their role in drug delivery will be reviewed. Among the topics to be reported and discussed will be a critical appraisal of the current range of modified-release dosage forms currently in use and also those under development. The design and mechanism(s) of controlled release systems including; macromolecular drug delivery, microparticulate controlled drug delivery, the delivery of biopharmaceuticals, delivery vehicles created for gastro-intestinal tract targeted delivery, transdermal delivery and systems designed specifically for drug delivery to the lung will all be reviewed and critically appraised. Further site-specific systems used for the delivery of drugs across the blood–brain barrier including dendrimers, hydrogels and new innovative biomaterials will be reported.

Manufacturing: The key elements of the manufacturing steps involved in the production of new medicines will be explored in this series. The importance of crystallization; batch and continuous processing, seeding; mixing including a description of the key engineering principles relevant to the manufacture of new medicines will all be reviewed and reported. The fundamental processes of quality control including good laboratory practice (GLP), good manufacturing practice (GMP), Quality by Design (QbD), the

Deming Cycle; Regulatory requirements and the design of appropriate robust statistical sampling procedures for the control of raw materials will all be an integral part of this book series.

An evaluation of the current analytical methods used to determine drug stability, the quantitative identification of impurities, contaminants and adulterants in pharmaceutical materials will be described as will the production of therapeutic bio-macromolecules, bacteria, viruses, yeasts, moulds, prions and toxins through chemical synthesis and emerging synthetic/molecular biology techniques. The importance of packaging including the compatibility of materials in contact with drug products and their barrier properties will also be explored.

Advances in Pharmaceutical Technology is intended as a comprehensive one-stop shop for those interested in the development and manufacture of new medicines. The series will appeal to those working in the pharmaceutical and related industries, both large and small, and will also be valuable to those who are studying and learning about the drug development process and the translation of those drugs into new life-saving and life-enriching medicines.

Dennis Douroumis
Alfred Fahr
Jürgen Siepmann
Martin Snowden

Preface

In former times, formulation specialists were not yet exposed to the many problems and subtleties that we face today in producing applicable drugs. In those 'good old days' the best drugs were simply generated using polar media, either by extraction from plants or by synthetic methods. Later, towards the end of the last century, an ever growing number of lipophilic drugs started to appear for oral as well as parenteral administration. Natural substances, like cyclosporine or modifications of aromatic structures that render selected drugs even more lipophilic, began to enter the vanguard of the blockbuster class. Stories about the difficulties of absorption in the gut or the bad pharmacokinetic profiles of these drugs have entered the body of canonical knowledge in many pharmaceutical companies. Desperate attempts to formulate these insoluble drugs – such as 'encapsulating' them in Swiss chocolate to get an oral delivery – are well remembered in the corresponding pharmacists' clubs.

This situation has changed with the advent of the new millennium in two ways:

1. The percentage of new drug molecules that are insoluble in water has risen to about 40% in total; in various therapeutic areas this percentage has even reached 80–90%. Why is this so? Perhaps the best explanation is the invention of the so-called High ThroughPut Screening (HTPS) method. Here, a variety of substances are tested (for cases of more than 100,000 per day, the method is re-named 'ultra-HTPS') for their activity with regard to certain biochemical targets (alternatively a cell, organ, or organism). This often favors the selection of drugs with higher lipophilicity, as most target sites – for example, the active center of an enzyme or a membrane protein – tend to be more accessible to lipophilic drugs, which runs in parallel to non-solubility. Adding to the difficulties for a formulation specialist, oral bioavailability is not among the primary aims of the HTPS procedure.
2. Pharmaceutical scientists have responded to this challenge in the past few decades by developing a variety of formulation principles for these poorly water-soluble drugs. Insoluble drugs should be made dissolvable by physico-chemical or biological means (e.g. transfer to gut cell membranes (p.o.) or lipoproteins (i.v.)) in order to arrive at the pharmacological target in appreciable amounts.

Even though some advertisements of excipient producers do suggest this, there is no ultimate single solution for insoluble drugs (as evidenced by the variety of methods presented in this book). On the other hand, there is the old saying that if there is more than one solution for a problem, there is likely no solution at all.

Therefore, a pharmacist who has to design and develop a formulation for an insoluble drug has to be aware of all the characteristics of the drug, s/he also has to have a profound knowledge of the available and feasible formulation options. To this end, s/he is likely to end up studying the literature in depth, as there are few other resources available that provide comprehensive surveys written by the experts in the field. The present book tries to fill this gap.

The book begins with some theoretical considerations, thereby introducing and discussing basic concepts such as solubility and hydrophobicity, and also provides a modeling framework for nanocarriers and their interactions with drug and the environment (Chapter 1).

Several chapters (2, 3–6, 13, 14) show how cyclodextrins, dendrimers, micelles, liposomes, solid lipid nanoparticles, and polymeric systems can overcome the solubility problem for insoluble drugs by using carrier systems mostly intended for the parenteral route. The carrier systems may be composed of either complex single molecules as hosts (e.g. dendrimers) or an assembly of rather simple molecules (e.g. micelles), or a combination of both. The complex interplay between host and drug often plays a crucial role in the success of such formulations and is extensively discussed in the respective chapters, along with detailed production procedures.

Microemulsion technology serves both the parenteral and the oral administration route for insoluble drugs, as is demonstrated in Chapter 10. Upon the addition of water, anhydrous (micro)emulsions may spontaneously emulsify. This process is used to produce self-emulsifying drug delivery systems (SEDDS) that are mainly used for oral delivery; see the discussion in Chapter 7.

Another feasible approach to improving the solubility of orally administered drugs is the size reduction of solid-state particles, yielding a large specific surface area. In Chapters 8, 9, 16 and 17, the production of nano-sized particles is described, using several different approaches (milling techniques, nanocrystals, nanosuspensions, and spray drying). The amorphous state of, for example, spray-dried particles and nanosuspensions may increase the solubility further and this is discussed thoroughly in the respective chapters.

Hot melt extrusion (solid dispersion technology) is, like the other methods described here, already on the market and attracting ever more attention as a method to enhance the bioavailability of problematic drugs. This is thoroughly described in Chapter 11.

Mesoporous silica nanoparticles (Chapter 15) are an interesting experimental formulation for increasing the solubility of insoluble drugs – they hold promising potential for the future.

Finally, Chapter 12 demonstrates that skin delivery of highly insoluble drugs is equally as challenging as other administration routes.

The different methods described in this book share the underlying goal of improving the solubility and the dissolution rate of poorly water-soluble drugs. We wish to point out that, especially for colloidal systems, these methods can be combined with targeting approaches. Targeting constitutes a fast-growing research field in its own right; its inclusion was outside the scope of the present book.

The interested reader may notice that the chapters integrate with each other. This indeed is the intention of the book as it likely facilitates the decision on which method might be worth trying for a given formulation problem. We emphasize that despite the efforts of all authors – including their careful descriptions, practical tips, and even theoretical considerations – finding the right formulation may in the end still be a matter of educated trial and error. Yet, even in this case, we are confident that this book will speed up the process.

The editors thank all the contributors for their time and effort in composing this compendium, for presenting the current state of the art in formulating insoluble drugs for oral, parenteral and topical administration, and for providing the reader with practical guidelines on how to start a formulation task.

1

Self-Assembled Delivery Vehicles for Poorly Water-Soluble Drugs: Basic Theoretical Considerations and Modeling Concepts

Sylvio May and Alfred Fahr

1.1 Introduction

Poor solubility is a well-recognized property of many drug molecules [1]. Unprotected administration of poorly water-soluble drugs is problematic. Aggregation, precipitation, uncontrolled binding, and direct exposure to a harsh biological environment render this process inefficient. The putative 'solution' of using higher drug concentrations narrows the window between a therapeutic success and unwanted side effects such as locally toxic drug levels. It comes as no surprise that the administration of poorly water-soluble drugs can benefit dramatically from using delivery vehicles. Such vehicles can, in principle, be designed not only to encapsulate a drug and protect it from biological defense mechanisms, but also to release the drug in a controlled manner at the target site and then to be recycled through biodegradation. Different types of delivery vehicles are currently being investigated, including microemulsions [2, 3], gels [4], micelles [5, 6], liposomes [7], polymersomes [8], dendrimers [9], and nanopcrystals [10], or lipid nanoparticles [11]. Notably, most of these are self-assembled structures. Self-assembly is an ubiquitous process in cellular systems, most strikingly perhaps in the cell membrane where a matrix (lipids) contains highly specialized functional units (poorly water-soluble proteins). Functionalization is an advantage that is also increasingly integrated into drug delivery vehicles. As an example

Drug Delivery Strategies for Poorly Water-Soluble Drugs, First Edition. Edited by Dennis Douroumis and Alfred Fahr.

we mention liposomes, which were originally designed as long-circulating transport vehicles for drug molecules [12]. Extending the circulation time by decorating the liposome surface with PEG-chains (stealth liposomes [13]) can be viewed as the first step toward functionalization. Currently designed liposomes raise the concept of functionalization to a new height: they contain targeting ligands and carry out stimuli-sensitive triggering of the drug release [14–16].

Optimizing drug delivery vehicles is promising but also challenging. Self-assembled nanostructures are soft and responsive materials, where entropy becomes an important factor for structure and stability. It is virtually impossible to manipulate one property without affecting others (and sometimes this has drastic implications as one of the authors vividly recalls the disintegration of an entire colloidal formulation upon the replacement of a single -H group by an -OH group in a 1 kDalton drug molecule). Nanocarrier properties are affected by a range of interactions that are well known from colloidal science, including solvation energies, electrostatic and van der Waals interactions, depletion and packing effects, etc. [17, 18]. Appreciation and understanding of these interactions are likely to reflect upon nanocarrier design and optimization. For example, one of the challenges that drug encapsulation in nanocarriers faces is related to the retention of the drug in the carrier. A lipophilic drug does not necessarily remain in a *rigid* lipophilic matrix [19] but is rapidly squeezed out, whereas *soft* structures (like liposomal membranes) tend to increase the residence time in the membrane. What physical mechanisms underlie the ability of soft rather than rigid self-assembled structures to accommodate small lipophilic drugs? And what physical properties determine the release? The latter question relates to the fact that a carrier keeping the drug completely in the interior will ultimately prevent a therapeutic effect. The authors' experience with a liposomal formulation of a peptide showed an increase of lifetime from 3 mins to 24 hours in blood, but there was no pharmacological activity, as the liposomes with the drug inside were eliminated without releasing the drug to blood components or organs. Other practical hurdles are discussed, for example, by van Hoogevest *et al.* [20].

The present chapter presents a conceptual framework for physics-based modeling approaches of self-assembled nanoscaled carrier systems that are associated with lipophilic drugs. Our focus is clearly on the basic physics and underlying concepts [21]. We start with an account of basic thermodynamic relations (Section 1.2) which we subsequently exploit to discuss principles of self-assembly (Section 1.3) and the partitioning of drug molecules into self-assembled carrier systems (Section 1.4). The energetics of individual delivery vehicles depends on a multitude of inter-molecular interactions; of these we discuss electrostatics and the packing of chain-like molecules (Section 1.5). We finally consider kinetic properties of drug transfer from mobile nano-carriers to a target system (Section 1.6). Note that none of the sections aims to give a comprehensive account of the available theoretical concepts (for more comprehensive accounts and discussions of specific applications, see [21–24]). However, for those subjects that we discuss, it is our goal not only to state the final results but also provide some guidance through the physical and mathematical basis of their derivation. We shall focus on simple and generic models, namely those that highlight the underlying physical principles, thereby excluding more advanced theoretical concepts and atomistic simulations. In summary, the present chapter approaches the pharmaceutical scientist who is interested in the process of developing theoretical models for self-assembled delivery vehicles of drug molecules from first principles.

1.2 Brief Reminder of Equilibrium Thermodynamics

We start by briefly recalling some basic concepts of equilibrium thermodynamics and statistical mechanics, thereby focusing on those aspects that will become important in the subsequent parts of this chapter. For each thermodynamic system there exists a thermodynamic potential F that contains all thermodynamic information. This means it encodes for all thermodynamic equations of state (such as the van der Waals equation of state for a van der Waals fluid, etc.). The potential F depends on a number of degrees of freedom that the system possesses. Some of these degrees of freedom can adjust only slowly; they are referred to as constrained variables. The remaining degrees of freedom correspond to unconstrained variables: they can adjust quickly. Equilibrium thermodynamics can only be applied if there exists a time window that separates the constrained from the unconstrained variables. (Note that this statement is very different from the demand of true thermal equilibrium.) Equilibrium thermodynamics – like most other fundamental physical theories – can be formulated as a minimization principle: The thermodynamic potential F adopts its minimum state with respect to all unconstrained variables. This eliminates the unconstrained variables from F. We note that statistical mechanics performs the very same minimization, yet starting from a representation of the potential F that includes all relevant microscopic degrees of freedom. In this entire chapter we treat the temperature T as a fixed constant. The thermodynamic potential – referred to as Helmholtz free energy – can then be expressed as $F = U - TS$ where U and S are the internal energy and entropy of the system under consideration. The free energy $F = U - TS$ can be viewed as a Legendre transformation from given entropy to fixing the temperature of the system by adding a heat bath of temperature T.

It is instructive to actually perform the minimization of F with respect to all its unconstrained variables. To this end, we consider a system that can exist in many different states α. To be specific, the different states α could refer to different spatial arrangements of surfactants and drug molecules that form a mixed surfactant micelle. This is schematically illustrated in Figure 1.1. Obviously, there are many different arrangements of the drug molecules and corresponding conformations of the micelle – so the total number of states may be extremely large. Each state α will be adopted with a probability $P(\alpha)$. These probabilities are normalized according to $\sum_\alpha P(\alpha) = 1$. If each state has an internal energy $u(\alpha)$, we can write the total internal energy as $U = \sum_\alpha u(\alpha)P(\alpha)$.

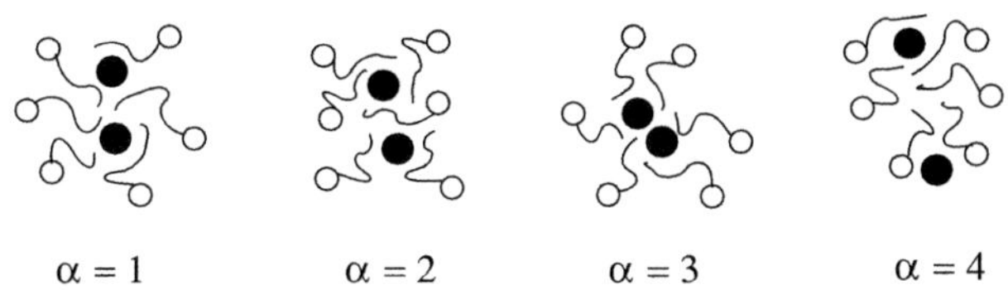

Figure 1.1 *Schematic illustration of a micelle, formed by six single-chained surfactants and two drug molecules (filled circles). Four arbitrarily chosen states α (out of a very large number of possible states) are displayed. Real surfactant micelles are typically composed of 50 to several hundred individual molecules and can contain some tens to 200 of drug molecules [25].*

The entropy can be expressed as a function of $P(\alpha)$ only: $S = -k_B \sum_\alpha P(\alpha) \ln P(\alpha)$. Here, $k_B = 1.38 \times 10^{-23} J/K$ is the Boltzmann constant. We note that only the form $P \ln P$ translates a multiplicativity of the probabilities into an additivity of the entropies; hence, this form of the entropy ensures extensivity of S. The thermodynamic potential thus reads $F = U - TS = \sum_\alpha P(\alpha)[u(\alpha) + k_B T \ln P(\alpha)]$. Clearly, the probabilities $P(\alpha)$ are not constrained; so they correspond to the set of unconstrained degrees of freedom for the present system. Hence, F must adopt a minimum subject to $P(\alpha)$. From the minimization of F we obtain the equilibrium distribution $P(\alpha) = \exp[-u(\alpha)/k_B T]/Q$ where $Q = \sum_\alpha \exp[-u(\alpha)/k_B T]$ ensures proper normalization as stated above. The distribution of probabilities for the different states α is the familiar Boltzmann distribution, and Q is referred to as the partition sum. The partition sum plays a central role as can be seen upon inserting the Boltzmann distribution back into F, yielding $F = -k_B T \ln Q$. Hence calculating Q gives immediate access to the thermodynamic potential F.

As a specific example we consider a fluid that has all its particles (i.e., molecules) confined to the sites of a lattice (this is also known as a lattice gas [26]). The lattice gas is certainly oversimplified but it serves as an instructive model, particularly for a binary fluid with roughly equal-sized solute and solvent molecules. Assume the lattice consists of M sites total; N sites are occupied by solute molecules and $M - N$ sites host solvent molecules. Figure 1.2(a) shows the lattice model. The dimensionality and connectivity of the lattice are irrelevant as long as the molecules do not interact with each other. In this case, the partition sum simply corresponds to the number $Q = M!/[N!(M - N)!]$ of available arrangements of the solute and solvent molecules. The use of Stirling's approximation $\ln x! \approx x \ln x - x$ (which is valid for $x \gg 1$) leads to the thermodynamic potential $F = -k_B T \ln Q = k_B T M[\phi \ln \phi + (1 - \phi) \ln(1 - \phi)]$ where we have defined the mole fraction

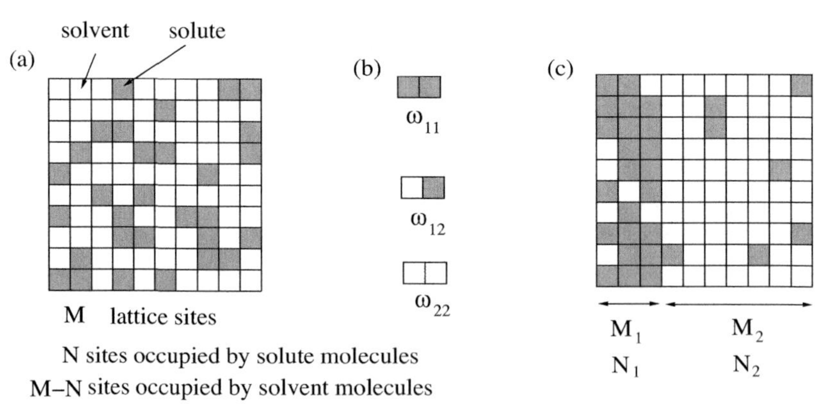

Figure 1.2 (a) *Schematic illustration of the lattice model. A two-dimensional square lattice of $M = 100$ lattice sites is shown, it contains $N = 30$ solute molecules and $M - N = 70$ solvent molecules.* (b) *Inter-molecular interactions can be characterized by a solute-solute (ω_{11}), solute-solvent (ω_{12}), and solvent-solvent (ω_{22}) interaction strength.* (c) *For sufficiently strong interactions the system may phase separate into a solute-rich phase (with N_1 solute molecules in a sublattice of M_1 sites), and a solute-poor phase (with $N_2 = N - N_1$ solute molecules in a sublattice of $M_2 = M - M_1$ sites). Note that M_1, M_2, N_1, N_2 are determined by the common tangent construction; see Equation 1.2.*

$\phi = N/M$ of solute molecules. Dissolving the solute in the solvent leads to a decrease in F which, in principle, can be used as work. This available (free) energy is entirely of entropic origin (i.e., $U = 0$ in $F = U - TS$). We note that for small solute mole fraction, $\phi \ll 1$, the free energy $F = k_B T M\phi(\ln\phi - 1)$ is identical to the free energy of an ideal gas. In fact, for *any* mixed system sufficient dilution leads to an ideal gas.

Let us go one step further and introduce interactions between the molecules. We will do this only on the so-called mean-field level, where correlations are ignored [26]. (Roughly speaking, in mean-field theory the interactions of a given particle are approximated by an effective interaction that only accounts for the net effect but not the details of the environment.) To this end, we consider a lattice of coordination number z. The coordination number simply denotes the number of next neighbors of each given lattice site, i.e., $z = 4$ for the two-dimensional square lattice illustrated in Figure 1.2(a), etc. Denote the (nearest-neighbor) solute-solute, solute-solvent, and solvent-solvent interaction strengths by ω_{11}, ω_{12}, and ω_{22}; see Figure 1.2(b). Neglecting correlations amounts to a random mixing approximation [26] where the total internal energy of the system is $U = Mz[\omega_{11}\phi^2 + 2\omega_{12}\phi(1-\phi) + \omega_{22}(1-\phi)^2 - \phi\omega_{11} - (1-\phi)\omega_{22}]/2$. The first three terms account for the solute-solute, solute-solvent, and solvent-solvent interactions, and the last two terms subtract the fully demixed reference state, one for the solute and the other for the solvent. The total internal energy per lattice site, $U/M = \chi\phi(1-\phi)$, can thus be expressed in terms of a single effective interaction parameter $\chi = z[\omega_{12} - (\omega_{11} + \omega_{22})/2]$, also referred to as the non-ideality parameter. Note that the non-ideality parameter is positive if molecules of the same type effectively attract each other, $(\omega_{11} + \omega_{22})/2 < \omega_{12}$. Based on the interaction term U and the ideal mixing free energy $-TS$, the overall free energy $F = U - TS$ of the interacting lattice gas can be expressed as $F = Mf$ where the free energy per lattice site is

$$f = \chi\phi(1-\phi) + k_B T\,[\phi\ln\phi + (1-\phi)\ln(1-\phi)]\,. \qquad (1.1)$$

The left diagram of Figure 1.3 displays $f = f(\phi)$ for different choices of χ. For sufficiently large χ the curves exhibit a concave region, centered about $\phi = 0.5$. Concavity of $f(\phi)$ is of thermodynamic significance as it implies that the system is *locally* unstable. To explain this instability we consider two equal-sized sublattices with initially the same composition ϕ. If a small number of solute molecules migrate from one to the other sublattice, the total free energy decreases given that the free energy $f(\phi)$ is concave. Mathematically this instability can be expressed as $f(\phi + \Delta\phi) + f(\phi - \Delta\phi) < 2f(\phi)$ for small $\Delta\phi$, or, equivalently, as $f''(\phi) < 0$, where the prime denotes the derivative with respect to the argument ($f'(\phi) = df/d\phi$, etc). Regions with $f''(\phi) > 0$ are locally stable, and regions with $f''(\phi) < 0$ are locally unstable. They are separated by points of inflection where $f''(\phi) = 0$. The points of inflection are marked in Figure 1.3(a) by the symbol $\diamond$ for the curve corresponding to $\chi = 3\,k_B T$). For Equation 1.1 the condition $f''(\phi) < 0$ translates into $k_B T/\chi < 2\phi(1-\phi)$. Concavity is a sufficient but not a necessary condition for thermodynamic instability to occur. A more general condition can be derived by considering two different phases (one of size M_1 with N_1 solute molecules and the other of size $M_2 = M - M_1$ with $N_2 = N - N_1$ solute molecules) with corresponding free energy $F = M_1 f(\phi_1) + (M - M_1) f(\phi_2)$ where $\phi_1 = N_1/M_1$ and $\phi_2 = (N - N_1)/(M - M_1)$ are the mole fractions of solute molecules in the two phases. A lattice that contains two different phases is illustrated in Figure 1.2(c). If the two phases are indeed thermodynamically stable, they must represent a minimum

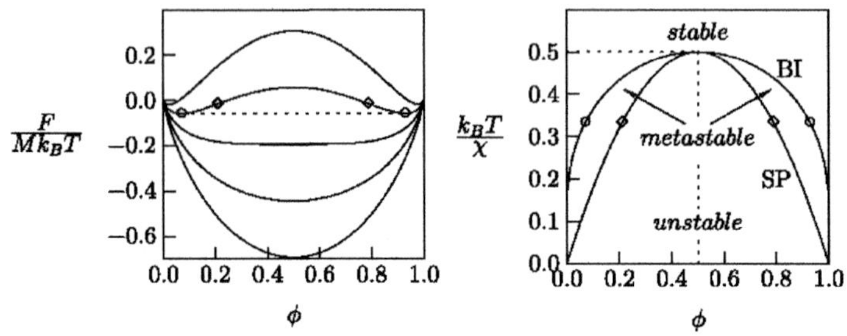

Figure 1.3 Left diagram: *Free energy per lattice site $F/(Mk_BT) = f/k_BT$ according to Equation 1.1 for different choices of the non-ideality parameter, $\chi/k_BT = 0, 1, 2, 3, 4$ (from bottom to top). For $\chi/k_BT = 3$, the points of inflection and the points that give rise to a common tangent are marked by the symbols $\diamond$ and $\circ$, respectively. The common tangent is indicated by the dotted line.* Right diagram: *Phase diagram k_BT/χ versus ϕ, with the spinodal (SP) and binodal (BI) lines indicated. Upon increasing the temperature T the system generally passes from a locally unstable region via a metastable region to a stable single phase system. It thereby crosses first the spinodal and then the binodal line. The symbols $\diamond$ and $\circ$ in the left diagram correspond to the ones in the right diagram (i.e., for $\chi/k_BT = 3$). The dotted lines in the right diagram indicate the critical point $\chi_c/k_BT = 2$ and $\phi_c = 1/2$.*

of F with respect to the two unconstrained parameters N_1 and M_1. This implies the two equations $\partial F/\partial N_1 = f'(\phi_1) - f'(\phi_2) = 0$ and $\partial F/\partial M_1 = f(\phi_1) - f(\phi_2) - \phi_1 f'(\phi_1) + \phi_2 f'(\phi_2) = 0$, giving rise to the common tangent construction [27]

$$f'(\phi_1) = f'(\phi_2) = \frac{f(\phi_2) - f(\phi_1)}{\phi_2 - \phi_1} \tag{1.2}$$

for the two coexisting compositions ϕ_1 and ϕ_2. As long as the system contains coexisting phases, the free energy moves along the common tangent, thus ensuring that the adopted thermodynamic potential does not become concave. The common tangent, which marks the region of global instability, is indicated in the left diagram of Figure 1.3 for the curve corresponding to $\chi = 3\,k_BT$ (see the dotted line between the two $\circ$ symbols). The regions of local and global instability can be mapped into a phase diagram k_BT/χ versus ϕ; see the right diagram of Figure 1.3. The lines separating regions of local and global instability are called spinodal (SP) and binodal (BI), respectively. The region between the spinodal and binodal is metastable. Here, the system can reside in a homogeneous single-phase state or phase separate (which lowers the free energy below that of the single-phase state). The largest k_BT/χ where phase separation occurs is the critical point; that point can generally be found by solving the two equations $f''(\phi) = f'''(\phi) = 0$ with respect to χ and ϕ. For Equation 1.1, this yields the critical point $\chi = \chi_c = 2k_BT$ and corresponding critical composition $\phi = \phi_c = 1/2$.

We note that the free energy in Equation 1.1 is based on a simple lattice model which despite being too simplistic to quantitatively describe real systems captures the principal features that lead to a phase transition. There are plenty of other, more elaborate and realistic, free energy expressions that predict phase diagrams (and corresponding critical points) using the same formalism that yields the phase diagram in Figure 1.3. For example, the van der Waals fluid has a free energy, expressed per lattice site (of unit volume) of $f = a\phi^2 + k_B T\phi \ln[\phi/(1 - b\phi)]$. The two material parameters a and b describe the degree of intermolecular attraction and the molecular eigenvolume; ϕ is the mole fraction of the solute as above. Calculation of the pressure

$$P = -\frac{d[f(\phi)/\phi]}{d(1/\phi)} = \phi\frac{df}{d\phi} - f = \frac{\phi k_B T}{1 - b\phi} + a\phi^2 \tag{1.3}$$

yields the familiar van der Waals equation of state [26]. Upon solving the two equations $f''(\phi) = f'''(\phi) = 0$ we obtain the critical point of the van der Waals fluid, $a = -27b\, k_B T/8$ and $\phi = 1/(3b)$. The mean-field lattice gas model leading to Equation 1.1 can also be extended to a polymer/solvent mixture. Here, the solute (i.e., the polymer) is represented by n (connected) segments, each occupying a single lattice site. Denoting the volume fraction of the polymer by ϕ, the (mean-field) free energy per lattice site becomes

$$f = \chi\phi(1 - \phi) + k_B T\left[\phi \ln \phi + \frac{1}{n}(1 - \phi)\ln(1 - \phi)\right], \tag{1.4}$$

where $\chi/k_B T$ is known as the Flory-Huggins parameter that describes the effective polymer segment-solvent interaction strength [28]. The segment number n has a profound influence on the phase behavior as can be seen from the critical point $\chi_c = (1 + 1/\sqrt{n})^2\, k_B T/2$ and $\phi_c = 1/(1 + \sqrt{n})$. Large n (10^3 is a typical value) drives phase separation already at polymer volume fractions of only a few percent. Flory-Huggins parameters for drug-polymer systems can either be extracted from experiments [29] or predicted via computer simulations [30].

1.3 Principles of Self-Assembly in Dilute Solutions

Self-assembly is a fundamental process with relevance for many of the delivery vehicles that are being used to deliver poorly water-soluble drugs [31]. Among the most relevant examples are micellar structures that are composed of surfactants, lipids, or polymers [24, 32]. Yet, self-assembly is also encountered for hydrophobic drug molecules, such as photosensitizers, when dissolved in an aqueous environment [33]. In this case, the aggregation state can be modulated by adding additional drug-binding proteins [34].

In this section we review the thermodynamic principles behind self-assembly processes [35, 36]. Rather than merely stating the results we aim to make the thermodynamic origin of the principles transparent. We focus on dilute solutions where a number of N chemically identical solute molecules are able to form N_i aggregates of aggregation number i. That is, N_1 is the number of monomers, N_2 the number of dimers, etc. This is illustrated for a specific example in Figure 1.4. As in the previous section, we adopt a lattice description of the system where a solvent molecule and a solute molecule each occupy one single lattice site. The N solute molecules thus self-assemble on a lattice of M sites. Note that using a lattice is convenient and does not limit the generality of the model in the dilute limit (as it did not for the ideal gas). The total volume of the system is then $V = M\nu$, where ν denotes the volume per lattice site. Dilute conditions require $N \ll M$. Here, the system can be

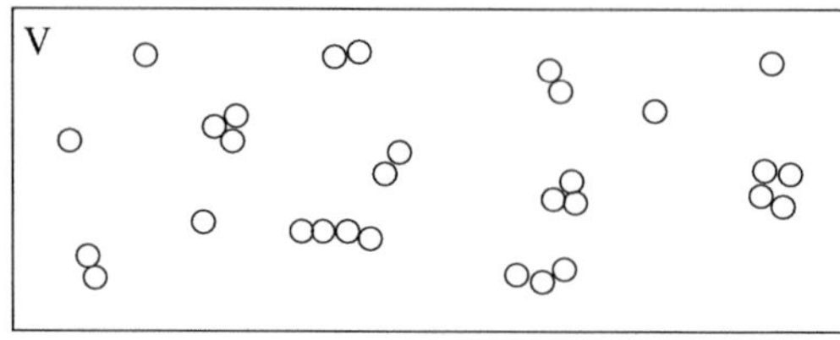

Figure 1.4 *Schematic illustration of $N = 30$ solute molecules (each represented by the symbol $\circ$) that – in this particular snapshot – self-assemble into $N_1 = 5$ monomers, $N_2 = 4$ dimers, $N_3 = 3$ tri-mers, and $N_4 = 2$ four-mers. Note that $\sum_i iN_i = N$.*

described as a mixed ideal gas, where each component is represented by the corresponding N_i aggregates of aggregation number i. The free energy of the system

$$F = \sum_{i=1}^{N} \left[N_i \mu_i^0 + k_B T N_i \left(\ln \frac{N_i}{M} - 1 \right) \right] \tag{1.5}$$

is then simply the sum of the individual ideal gas free energies, each with its own standard chemical potential μ_i^0 per i-mer. The sum runs from $i = 1$ (monomers) to $i = N$ where all solute molecules form one single aggregate. The limit $i = N$ can be relevant. Lipid bilayers, for example, tend to fuse, despite the loss of translational entropy upon the merger of two bilayers into one. We note that the conservation of the total number of solute molecules implies $\sum_{i=1}^{N} iN_i = N$. In the following it is convenient to use the mole fractions $\phi_i = iN_i/M$ of solute molecules that form i-mers. Similarly, we introduce the total mole fraction of solute molecules $\phi = N/M$ in the system. The free energy in Equation 1.5, expressed per lattice site, then reads

$$f = \frac{F}{M} = \sum_{i=1}^{N} \left[\phi_i \tilde{\mu}_i^0 + k_B T \frac{\phi_i}{i} \left(\ln \frac{\phi_i}{i} - 1 \right) \right], \tag{1.6}$$

where we have introduced the standard chemical potential $\tilde{\mu}_i^0 = \mu_i^0/i$ per solute molecule in an i-mer. The distribution ϕ_i constitutes a set of unconstrained degrees of freedom (similar to the probabilities $P(\alpha)$ in the preceding sub-section). Yet, the minimization of f with respect to the ϕ_is must account for the additional conservation $\sum_{i=1}^{N} \phi_i = \phi$ of the total number of solute molecules. A minimization subject to an additional constraint is a problem that occurs frequently in science and for which mathematics offers the method of introducing a so-called Lagrange multiplier [37]. Let us illustrate its use. We define the auxiliary function $\tilde{f} = f - \mu \sum_{i=1}^{N} \phi_i$ which introduces the Lagrange multiplier μ. Note that μ is unspecified as of now but can be determined later so that the additional constraint $\sum_{i=1}^{N} \phi_i = \phi$ becomes satisfied. The introduction of μ allows us to minimize $\tilde{f}$ with respect to *all* ϕ_i. From $\partial \tilde{f}/\partial \phi_i = 0$ we find the equilibrium distribution $\phi_i = i \exp[-(\tilde{\mu}_i^0 - \mu)/k_B T]$. The role of the Lagrange multiplier μ becomes transparent after calculating the chemical potential $\mu_i = \partial F/\partial(iN_i) = \partial f/\partial \phi_i = \tilde{\mu}_i^0 + (k_B T/i) \ln(\phi_i/i)$ of a solute molecule that resides in an i-mer. Comparing this with the equilibrium distribution $\mu = \tilde{\mu}_i^0 + (k_B T/i) \ln(\phi_i/i)$, we conclude $\mu_i = \mu$ for *all* i-mers. Hence, in equilibrium all solute molecules have the same chemical potential, and that chemical potential is equal to the Lagrange multiplier μ. We proceed with two more remarks. First, the relation $\mu_1 = \mu_i$ can be expressed [17] as

$$\frac{(N_i/M)}{(N_1/M)^i} = \frac{\phi_i/i}{\phi_1^i} = e^{-i(\tilde{\mu}_i^0 - \tilde{\mu}_1^0)/k_B T} = K_i. \tag{1.7}$$

This can be interpreted as the familiar mass action law for the association reaction $i\text{A} \overset{K_i}{\rightleftharpoons} \text{A}_i$, where i monomers form one single i-mer; K_i is the corresponding equilibrium constant, and $i(\tilde{\mu}_i^0 - \tilde{\mu}_1^0)$ is the standard change in free energy for the association event. Note that the association reaction does not imply that an i-mer actually forms from i monomers; that process may take a different route such as adding a single monomer and an $(i-1)$-mer. Second, inserting the equilibrium distribution ϕ_i back into the free energy yields, $f = \mu\phi - k_B T \sum_{i=1}^{N} \phi_i/i$. This can be interpreted as the familiar thermodynamic relation for the Helmholtz free energy $F = G - PV$ in terms of the Gibbs free energy $G = \mu N$ and the pressure $P = (\partial F/\partial V)_T$ (where we recall $V = M\nu$). Further analysis of the aggregation behavior requires the specification of the function $\tilde{\mu}_i$. We consider two cases of principal importance: linear growth and cooperative assembly.

1.3.1 Linear Growth

Here aggregates can grow along one dimension only as is the case for biological filaments or worm-like micelles. Each monomer within the aggregate contributes the same (namely μ_b) to the standard chemical potential $\mu_i^0 = i\tilde{\mu}_i^0$ per aggregate. Only the two terminal segments each add an additional contribution $\delta/2$; see the illustration at the top of Figure 1.5. The model for linear growth can thus be written as $\mu_i^0 = i\mu_b + \delta$. It is the excess terminal

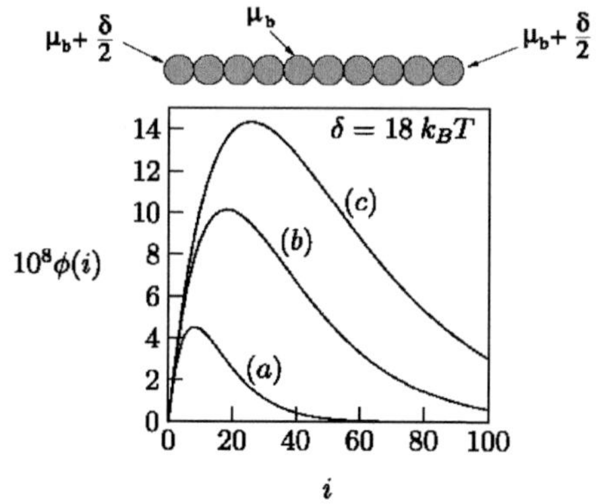

Figure 1.5 Plot of $\phi(i)$ according to Equation 1.8. The excess contribution to the standard chemical potential is $\delta = 18k_BT$ for all curves. The total mole fraction of the solute is $\phi = 10^{-6}$ (a), $\phi = 5 \times 10^{-6}$ (b), and $\phi = 10^{-5}$ (c). Note that for each curve the relation $\int_0^\infty \phi(i)\,di = \phi$ is fulfilled. The top of the diagram shows an illustration of a 10-mer, i.e., a linear aggregate with $i = 10$. Here, μ_b is the standard chemical potential per monomer in the bulk of the aggregate. The two terminal segments each make an additional (i.e., excess) contribution $\delta/2$ to the standard chemical potential $\mu_i^0 = i\mu_b + \delta$.

contribution δ that is gained when two linear aggregates combine into a single one. This gain drives the growth into fewer and longer aggregates, despite the larger entropy that many small aggregates would have. Note that a monomeric unit need not necessarily be identified with a single solute molecule. For example, when modeling worm-like micelles it is convenient to identify a single spherical micelle with a monomeric unit [36]. Upon initiating growth these micelles dynamically combine into elongated (i.e., linearly extended) structures.

Let us investigate the thermodynamics behind self-assembled linear structures. To simplify the calculation we assume $\delta \gg k_BT$ for which the aggregates grow very long so that we can treat $\phi_i \to \phi(i)$ as a function of the continuous variable i and approximate the summation by an integration, $\sum_{i=1}^{N} \to \int_0^\infty di$. From the normalization condition $\phi = \int_0^\infty \phi(i)di$ we then find the chemical potential $\mu = \mu_b - k_BTe^{-\delta/(2k_BT)}/\sqrt{\phi}$. Inserting this into the distribution $\phi(i)$ yields

$$\phi(i) = i \exp\left[-\frac{\delta}{k_BT} - \frac{i}{\sqrt{\phi}}\, e^{-\frac{\delta}{2k_BT}}\right]. \tag{1.8}$$

This distribution adopts a maximum at the aggregation number $i = i_m$ with $i_m = \sqrt{\phi}\, e^{\delta/(2k_BT)}$. To further characterize the distribution, we define the weight-average $\langle Q\rangle = \int_0^\infty \phi(i)\, Q(i)\, di / \int_0^\infty \phi(i)\, di$ of any physical quantity $Q = Q(i)$. The weight average of the size distribution is then $\langle i\rangle = 2i_m$. Similarly, for the standard deviation of the size distribution we obtain $\sigma = \sqrt{\langle (i - \langle i\rangle)^2\rangle} = \sqrt{2}\, i_m$.

Hence, we conclude that the linear growth model leads to a broad equilibrium distribution of aggregate lengths, where the standard deviation of the sizes $\sigma = \langle i\rangle/\sqrt{2}$ is about 70% of the average size distribution $\langle i\rangle$. In Figure 1.5 we display $\phi(i)$ for some examples (all with $\delta = 18k_BT$ but different choices of ϕ). Since μ and $\phi(i)$ are known, we can explicitly calculate the Helmholtz free energy F of the system; this leads to

$$F = N\left[\mu_b - \frac{2k_BT}{\sqrt{\phi}}\, e^{-\frac{\delta}{2k_BT}}\right]. \tag{1.9}$$

For small T all solute molecules assemble into one single aggregate, implying $F = N\mu_b$.

1.3.2 Cooperative Assembly

The second principal scenario that we discuss is when a certain number of solute molecules must come together to form a single aggregate. This is the case, for example, when surfactants start forming micelles or when peptides cooperatively form a pore in a lipid membrane. While the number of surfactants in micelles can certainly vary, we only consider the most simple scenario, where exactly n solute molecules form one single aggregate. To exclude all other aggregation numbers, we chose $\tilde{\mu}_i^0 \to \infty$ (implying $\phi_i = 0$) for all i unless $i = 1$ or $i = n$. The distribution of the remaining monomers and n-mers is then specified by $\phi_1 = e^{-(\tilde{\mu}_1^0 - \mu)/k_BT}$ and $\phi_n = e^{-n(\tilde{\mu}_n^0 - \mu)/k_BT}$, subject to $\phi_1 + \phi_n = \phi$. We thus have three equations for the three unknown quantities ϕ_1, ϕ_n, and μ. These equations can be solved numerically. In Figure 1.6 we show $\phi_1(\phi)$ and $\phi_n(\phi)$ for the two different choices $n = 10$ and $n \to \infty$, both at fixed $\tilde{\mu}_1^0 - \tilde{\mu}_n^0 = 5k_BT$. The limiting behavior for $n \to \infty$ is given by $\phi_1 = \phi$ and $\phi_n = 0$ for $\phi < \phi_{cmc}$, as well as $\phi_1 = \phi_{cmc}$ and $\phi_n = \phi - \phi_{cmc}$ for $\phi > \phi_{cmc}$. The critical value ϕ_{cmc} reflects the so-called critical micelle concentration

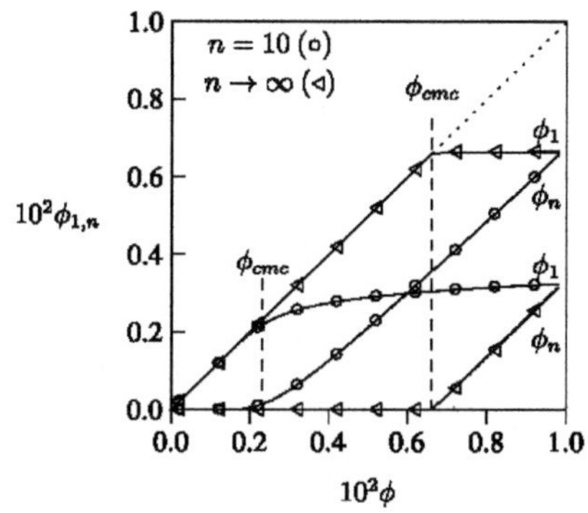

Figure 1.6 *The mole fractions ϕ_1 and ϕ_n as function of the total mole fraction ϕ for cooperative assembly with $n = 10$ (curves marked by the symbol ∘) and $n \to \infty$ (marked by the symbol ◁), both calculated for $\tilde{\mu}_1^0 - \tilde{\mu}_n^0 = 5k_BT$. The corresponding two critical mole fractions ϕ_{cmc} are indicated by the vertical broken lines. Note $\exp[-(\tilde{\mu}_1^0 - \tilde{\mu}_n^0)/k_BT] = 0.673 \times 10^{-2}$; see Equation 1.10. The dotted line marks the total solute mole fraction $\phi = \phi_1 + \phi_n$.*

(CMC). A general definition (one that does not require large n) for ϕ_{cmc} simply identifies the point where 50% of the newly added solute molecules are used to form aggregates (whereas the other 50% remain as monomers) [18]. This can be expressed as $(d\phi_1/d\phi)_{\phi_{cmc}} = 1/2$. Evaluating this expression leads to

$$\phi_{cmc} = (1+n)\, n^{\frac{1+n}{1-n}} \exp\left[-\left(\frac{n}{n-1}\right)\left(\frac{\tilde{\mu}_1^0 - \tilde{\mu}_n^0}{k_BT}\right)\right] \approx \exp\left(-\frac{\tilde{\mu}_1^0 - \tilde{\mu}_n^0}{k_BT}\right), \quad (1.10)$$

where the approximate expression on the right-hand side of Equation 1.10 corresponds to the limit $n \to \infty$. Hence, the CMC can be used to extract the standard free energy of formation $\tilde{\mu}_1^0 - \tilde{\mu}_n^0 = -k_BT \ln \phi_{cmc}$. Finite values of n (see Figure 1.6 for $n = 10$) lead to a smeared-out transition at somewhat smaller ϕ_{cmc} than for $n \to \infty$.

1.4 Solubility and Partitioning of Drugs

1.4.1 Simple Partitioning Equilibria

The basis of designing delivery vehicles for poorly water-soluble drugs is their much higher solubility in apolar solvents as compared to water. We can characterize the thermodynamic partitioning of a solute between two strongly immiscible fluids most simply in the dilute limit. Here, the solute can be treated in each phase as an ideal gas, with corresponding free energy, $F = k_BTN(\ln\phi - 1) + N\mu^0$ where $\phi = N/M$ is the mole fraction of the solute (i.e., N solute molecules are dissolved in a phase that contains M solvent molecules). The chemical potential of the solute is $\mu = dF/dN = \mu^0 + k_BT \ln\phi$. Indexing the two phases '1' and '2', and setting the chemical potentials in the two phases equal, $\mu_1 = \mu_2$, leads to $\phi_2/\phi_1 = \exp[-(\mu_2^0 - \mu_1^0)/k_BT] = K^*$ where K^* is the partition coefficient. Hence, $-k_BT \ln(\phi_2/\phi_1) = \mu_2^0 - \mu_1^0$ directly yields the difference in the standard chemical potential of the solute in the two phases.

As the concentrations grow, deviations from ideal behavior become important. In this case, an activity coefficient $\gamma = \gamma(\phi)$ can be introduced into the expression for the chemical potential through $\mu = \mu^0 + k_B T \ln(\gamma\phi)$. Hence, the partition coefficient

$$K = \frac{\phi_2}{\phi_1} = \frac{\gamma_1}{\gamma_2} e^{-(\mu_2^0 - \mu_1^0)/k_B T} = \frac{\gamma_1(\phi)}{\gamma_2(\phi)} K^* \tag{1.11}$$

now depends on the activity coefficients and thus on the mole fractions of the solute in the two phases. Let us discuss a very simple but illustrative example that is based on the lattice model and corresponding mean-field free energy as introduced in Equation 1.1. Specifically, we consider two separated phases corresponding to M_1 and M_2 lattice sites, filled with two different solvents and, respectively, N_1 and N_2 identical solute molecules. The two phases may be immiscible or may be separated by a semi-permeable barrier (permeable only for the solute). For simplicity we assume $M_1 = M_2 = M/2$ so that the overall mole fraction of the solute becomes $\phi = (N_1 + N_2)/(M_1 + M_2) = (\phi_1 + \phi_2)/2$, where $\phi_1 = N_1/M_1$ and $\phi_2 = N_2/M_2$ are the mole fractions of the solute in the two phases. Let phase '1' be dilute and phase '2' be concentrated. The free energy $f_1 = k_B T \phi_1(\ln \phi_1 - 1) + \mu_1^0 \phi_1$ of the dilute phase can be treated as that of an ideal gas. According to Equation 1.1, we express the free energy of the concentrated phase as $f_2 = k_B T[\phi_2 \ln \phi_2 + (1 - \phi_2)\ln(1 - \phi_2)] + \chi\phi_2(1 - \phi_2)$. The partitioning coefficient is then given by Equation 1.11 with the activity coefficients $\gamma_1 = 1$ and $\gamma_2 = e^{-(2\phi_2\chi/k_B T)}/(1 - \phi_2)$ as well as $K^* = e^{-(\chi - \mu_1^0)/k_B T}$. The relation $\phi_1(\phi)$, together with $\phi_2(\phi) = 2\phi - \phi_1$, is displayed in Figure 1.7 for $K^* = 10$ and three different choices of χ. Note that the tendency of the solute to accumulate in phase '2' is implied by our choice $K^* > 1$. For small ϕ the behavior is ideal and from $\phi_2/\phi_1 = K^*$ and $\phi = (\phi_1 + \phi_2)/2$ we find $\phi_1 = 2\phi/(1 + K^*)$ as well as $\phi_2 = 2\phi K^*/(1 + K^*)$. The ideal behavior is marked in Figure 1.7 by the two dotted lines. For larger ϕ there are deviations from ideal behavior. For $\chi = 0$ (curves *b* and *e* in Figure 1.7) the lattice gas does not exhibit intermolecular interactions. Still, the mole fraction ϕ_2 in the concentrated phase (curve *b*) is smaller than it would be under ideal conditions. The reason is that the

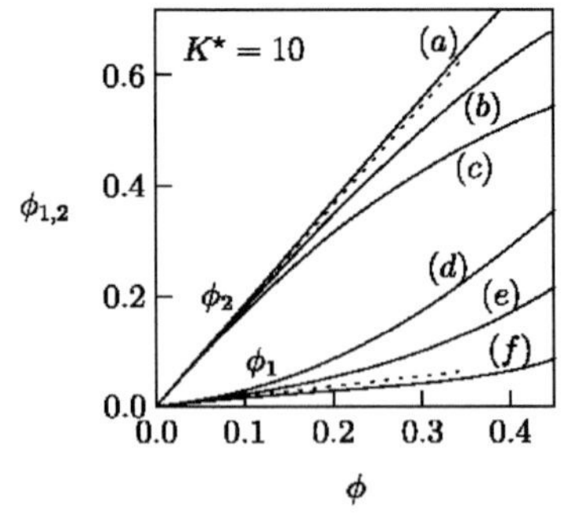

Figure 1.7 *The solute mole fractions $\phi_1(\phi)$ (curves d–f) and $\phi_2(\phi)$ (curves a–c) of the two phases as function of the total mole fraction $\phi = (\phi_1 + \phi_2)/2$, calculated for $\chi = -1\ k_B T$ (curves c and d), $\chi = 0$ (curves b and e), and $\chi = +1\ k_B T$ (curves a and f). Basis of the calculation is Equation 1.11 with $\gamma_1 = 1$, $\gamma_2 = e^{-(2\phi_2\chi/k_B T)}/(1 - \phi_2)$, and $K^* = 10$. The two dotted lines show the ideal cases, $\phi_1 = 2\phi/(1 + K^*)$ and $\phi_2 = 2\phi K^*/(1 + K^*)$, where the activities are $\gamma_1 = \gamma_2 = 1$.*

non-vanishing solute size implied by the lattice model acts as a hard core repulsion and thus tends to displace solute molecules from the concentrated to the dilute phase. This effective repulsion between solute molecules for $\chi = 0$ is manifested by $\gamma_2 = 1/(1 - \phi_2) > 1$ for $\phi_2 > 0$. For $\chi = -1\ k_BT$ (curves *c* and *d*) the additional repulsive interactions between the solvent molecules further enhance the depletion of solvent molecules from the concentrated phase (curve *c*). In contrast, for $\chi = +1\ k_BT$ (curves *a* and *f*) the solvent molecules attract each other. The attraction overcompensates the hard core repulsion of the solute molecules and thus enriches solute molecules in the concentrated phase (curve *a*) as compared to the ideal case. Indeed, for $\chi = +1\ k_BT$ the activity coefficient $\gamma_2 = e^{-(2\phi_2)}/(1 - \phi_2)$ of the solute in the concentrated phase is below that of an ideal gas ($\gamma_2 = 1$) as long as $\phi_2 \lesssim 0.8$.

1.4.2 Partitioning and Micellization

Consider now the case of drug partitioning into mobile carriers, where the carriers are self-assembled structures such as micelles [38] or polymeric aggregates [37]. We have discussed the cooperative self-assembly of solute molecules in Section 1.3.2. The present section adds the presence of drug molecules that due to their low solubility in water exhibit a tendency to be integrated into the self-assembled aggregates. As in Section 1.3.2, we assume that exactly *n* solute molecules (we refer to them as surfactants in the following) cooperatively associate into *n*-mers (i.e., micelles that consist of *n* surfactants). We now add the presence of drug molecules that are poorly soluble in the solvent but can be integrated into the micelles. There will generally be a maximal number of drug molecules that can be incorporated into a single micelle. To keep the present model transparent, we assume the maximal number to be one. That is, each micelle can carry either one single or no drug molecule. A schematic illustration of the system (monomeric surfactants, monomeric drug molecules, empty micelles, and drug-carrying micelles) is displayed in Figure 1.8. We can express the two processes – micellization in the absence of drug and incorporation of a drug molecule into a micelle – as a system of two chemical reactions

$$n\,\mathrm{A} \overset{K_S}{\rightleftharpoons} \mathrm{A}_n, \qquad \mathrm{D} + n\,\mathrm{A} \overset{K_D}{\rightleftharpoons} \mathrm{DA}_n, \tag{1.12}$$

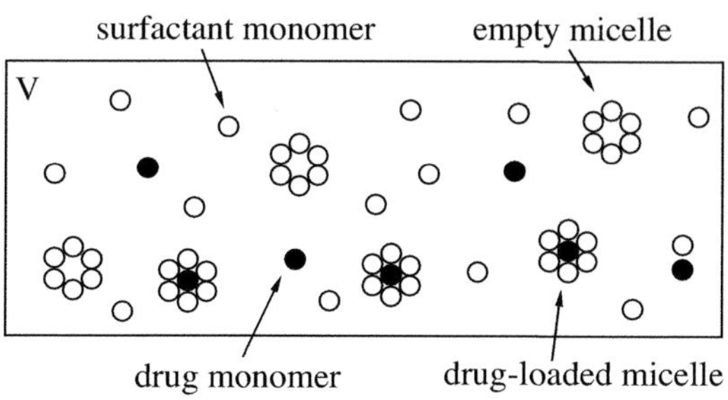

Figure 1.8 *Schematic illustration of surfactants (represented by the symbol ∘) and drug molecules (represented by the symbol •). The surfactants are able to self-assemble into micelles (here displayed with aggregation number n = 6). Each micelle can carry either one or no drug molecule.*

with corresponding equilibrium constants K_S and K_D. We are already familiar from Section 1.3.2 with the reaction on the left-hand side of Equation 1.12. It expresses the association of n monomeric surfactants (each denoted by A) to an empty micelle that consists of n surfactants (denoted by A_n). The reaction on the right-hand side of Equation 1.12 accounts for the assembly of n surfactants and a single drug molecule (denoted by D), resulting in a drug-filled micelle (denoted by DA_n). The mass action law that corresponds to the reaction schemes in Equation 1.12 can be written as

$$\frac{\phi_S^Z}{n(\phi_S^M)^n} = e^{-n(\tilde{\mu}_S^Z - \tilde{\mu}_S^M)/k_BT} = K_S, \qquad \frac{\phi_D^Z}{(\phi_D^M)(\phi_S^M)^n} = e^{-(\tilde{\mu}_D^Z - \tilde{\mu}_D^M - n\tilde{\mu}_S^M)/k_BT} = K_D. \quad (1.13)$$

Here, ϕ_S^Z is the mole fraction of surfactant molecules that are engaged in forming empty micelles. (We again use a lattice description as in Section 1.3.2 where each molecule – solvent, surfactant, and drug – occupies one single lattice site). Similarly, ϕ_S^M denotes the mole fraction of monomeric surfactant molecules, ϕ_D^Z denotes the mole fraction of drug molecules that are incorporated into micelles (which equals the mole fraction of the drug-loaded micelles), and ϕ_D^M denotes the mole fraction of monomeric drug molecules. The equilibrium constants K_S and K_D are related to the standard chemical potentials of a surfactant when monomeric ($\tilde{\mu}_S^M$) or incorporated into an empty micelle ($\tilde{\mu}_S^Z$), as well as to the standard chemical potentials per monomeric drug molecule ($\tilde{\mu}_D^M$) and of a drug-filled micelle ($\tilde{\mu}_D^Z$). More specifically, $\Delta g_{empty} = n(\tilde{\mu}_S^Z - \tilde{\mu}_S^M)$ and $\Delta g_{filled} = \tilde{\mu}_D^Z - \tilde{\mu}_D^M - n\tilde{\mu}_S^M$ represent the standard Gibbs free energies of formation for an empty and drug-filled micelle, respectively. The (experimentally fixed) overall mole fractions of surfactants and drug molecules are

$$\phi_S = \phi_S^Z + \phi_S^M + n\phi_D^Z, \qquad \phi_D = \phi_D^Z + \phi_D^M, \quad (1.14)$$

respectively. The four relations in Equations 1.13 and 1.14 can be used to calculate the four mole fractions $\phi_S^Z, \phi_S^M, \phi_D^Z, \phi_D^M$. Figure 1.9 shows an example, calculated for $n = 10$, $K_S = 9.76 \times 10^{22}$, $K_D = 10^4 \times K_S$. That implies each surfactant gains a standard Gibbs free energy $\tilde{\mu}_S^M - \tilde{\mu}_S^Z = (k_BT/n) \ln K_S = 5.3\, k_BT$ upon incorporation into an empty micelle. For an entire empty micelle this amounts to $\Delta g_{empty} = -53\, k_BT$. Formation of a drug-loaded micelle gains a standard Gibbs free energy $\Delta g_{filled} = -k_BT \ln K_D = -62\, k_BT$. Hence, incorporating a single drug molecule into an initially empty micelle is associated with an energy gain of $(62 - 53)k_BT = 9k_BT$.

The left diagram of Figure 1.9 displays the mole fractions ϕ_S^M (mole fraction of monomeric surfactants), ϕ_S^Z (mole fraction of surfactants in empty micelles), and $n\phi_D^Z$ (mole fraction of surfactants in drug-loaded micelles) as function of ϕ_s for fixed $\phi_D = 0.2 \times 10^{-3}$. The inset shows the corresponding mole fractions ϕ_D^M (mole fraction of monomeric drug) and ϕ_D^Z (mole fraction of drug in micelles). Clearly, for $n = 10$ there is a well-established CMC (roughly at $\phi_S = 2 \times 10^{-3}$), above which micelle formation starts. Immediately above the CMC there is a region (roughly for $2 \times 10^{-3} < \phi_S < 4 \times 10^{-3}$) where there are slightly more drug-filled than empty micelles. Yet, the majority of drug molecules are still monomeric. At $\phi_S = 4 \times 10^{-3}$ (which is twice the CMC), half of the drug molecules reside in micelles and half of the surfactants are engaged in micelle formation. Of all micelles 50% are drug-loaded. This is the optimal case. Further increasing the overall number of

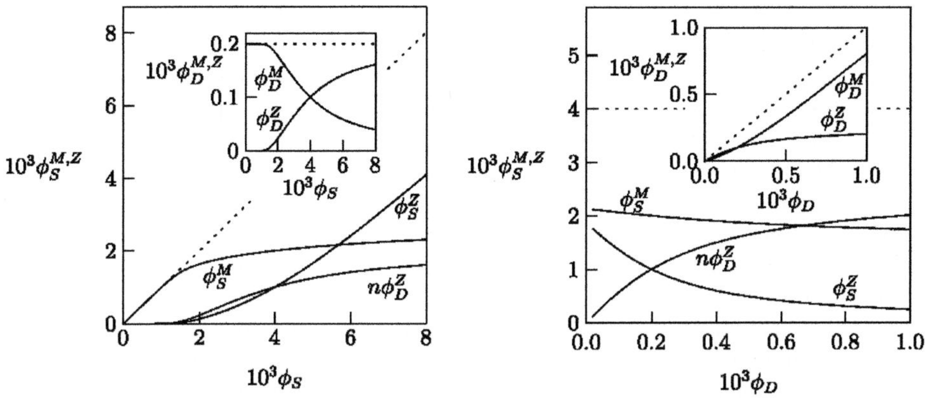

Figure 1.9 Left: *Mole fraction of monomeric surfactants ϕ_S^M, mole fraction of surfactants in empty micelles ϕ_S^Z, and mole fraction of surfactants in drug-loaded micelles $n\phi_D^Z$, displayed as function of the overall mole fraction of surfactants ϕ_S. The inset shows the corresponding mole fraction of monomeric drug ϕ_D^M, and mole fraction of drug in micelles ϕ_D^Z. It is $\phi_D = 0.2 \times 10^{-3}$ in both the main figure and inset.* Right: *Mole fractions ϕ_S^M, ϕ_S^Z, $n\phi_D^Z$, ϕ_D^M, and ϕ_D^Z (the latter two are displayed in the inset) as function of the overall mole fraction of drug molecules ϕ_D, derived for fixed $\phi_S = 4 \times 10^{-3}$. All calculations (in both diagrams) use $n = 10$, $K_S = 9.76 \times 10^{22}$, and $K_D = 10^4 \times K_S$. The dotted lines refer to either ϕ_S or ϕ_D.*

surfactants, ϕ_S, leads to more drug becoming incorporated into micelles but also drastically increases the number of empty micelles.

The right diagram of Figure 1.9 (including the inset) shows the mole fractions ϕ_S^M, ϕ_S^Z, $n\phi_D^Z$, ϕ_D^M, and ϕ_D^Z as function of ϕ_D for fixed $\phi_S = 4 \times 10^{-3}$. As we have discussed above, at $\phi_D = 0.2 \times 10^{-3}$ half of all drug molecules reside in micelles, half of the surfactants form micelles, and half of all micelles contain drug. For $\phi_D < 0.2 \times 10^{-3}$ most micelles are empty and for $\phi_D > 0.2 \times 10^{-3}$ most drug molecules do not reside in micelles. Note also that the total number of micelles is only slightly increased upon increasing ϕ_D. Still, some initially monomeric surfactants become engaged in forming micelles when hydrophobic drug molecules are added to the system.

1.4.3 Hydrophobicity and Ordering of Water

Water solubility is one of the key characteristics in drug design [40]. Yet, what is the physical origin of poor water solubility? Water is a polar molecule that preferentially forms a highly dynamic tetrahedral network of hydrogen bonds between the nonbonding, electron-rich oxygen orbitals and the electron-deficient hydrogen atoms. Hydrogen bonding is sufficiently strong so that in the vicinity of an apolar surface (which is not able to engage in hydrogen bonds) water attempts to maintain or even increase the average number of hydrogen bonds, yet at the cost of being accommodated in a more structured configuration [41,42]. That is, water rearranges in the vicinity of a hydrophobic solute, forming ordered solvation shells (sometimes also referred to as 'clathrates' or 'cages') that essentially preserve their enthalpy, yet at the cost of a reduced entropy. The interaction of hydrophobic solutes with their aqueous environment is thus dominated by entropy. As a rough approximation the free energy of transfer of a hydrophobic solute from an oily phase to water (the so-called solvation free energy is proportional to the solute's water-accessible surface area [43]. The corresponding order of magnitude is equal to the surface tension between water and air ($\approx 17k_BT/\text{nm}^2$). Modeling the underlying physics of solvation has been [44] and continues to be [45,46] an area of active research. We point out that these first-principle models are very different in nature (and must be distinguished) from descriptor-based models such as ASMS (aqueous solubility based on molecular surface), ASM-SAS (aqueous solubility model based on solvent accessible surface areas), and others, which aim to pre-select drugs prior to high throughput screenings [47].

Can we use simple phenomenological models to describe the solvation of solute molecules? A number of approaches are available. Among them are solvation models that incorporate the dipolar properties of water into continuum electrostatics [48,49]. Another, (particularly instructive) approach, the so-called Mercedes-Benz model, has been suggested by Dill and coworkers [50]. Yet, here we suffice to briefly mention the most basic phenomenological model which goes back to Marcelja and Radic [51]. It describes water by a vectorial order parameter $\mathbf{P} = \{P_x, P_y, P_z\}$ and – assuming a small perturbation – employs a quadratic form of the free energy

$$F = C \int_V dv \left[\xi^2(\nabla \cdot \mathbf{P})^2 + \mathbf{P}^2\right]. \tag{1.15}$$

Here, C is a constant and ξ is a characteristic length for the decay of the water perturbation ($\xi \approx 0.2$nm [52]). Minimization of F leads to the vectorial differential equation $\xi^2\nabla^2\mathbf{P} = \mathbf{P}$. For example, assume a single planar surface is located at $z = 0$ and imposes an order parameter $\mathbf{P} = \{0, 0, P_0\}$ at the surface ($z = 0$). The order parameter is then $\mathbf{P} = \{0, 0, P_0 e^{-z/\xi}\}$, and the corresponding free energy per unit area (i.e., the surface tension) becomes $CP_0^2\xi$. Despite being so simple, Marcelja's model correctly predicts the water-mediated repulsion between two identical hydrophilic surfaces as a function of their mutual distance (which has been measured accurately using the surface force apparatus [52]). It also allows for the possibility of attractive water-mediated interactions for hydrophilic surfaces that induce opposite surface orientations of the order parameter. Clearly, the model can serve as a conceptual starting point and reference for more detailed descriptions of water.

1.5 Ways to Model Interactions in Colloidal Systems

In the previous section structural and material properties (partitioning equilibria, CMC, etc) appeared in the form of standard chemical potentials μ_i^0 (see Equation 1.5) and activities γ_i (see Equation 1.11). Their magnitudes result from an often complex interplay of colloidal interactions that include – besides the hydrophobic effect as discussed above – also van der Waals and steric interactions, as well as electrostatic and packing energies of linear chain-like molecules [17,18]. In the following we focus only on the latter two. That is, we briefly discuss methods to model electrostatic interactions and chain-packing energies and what we can learn from such approaches in terms of designing delivery vehicles for poorly water-soluble drug molecules.

1.5.1 Electrostatic Interactions: The Poisson–Boltzmann Model

Although poorly water-soluble drugs are typically uncharged, electrostatic interactions are essential for the stability of drug carriers, partitioning equilibria, and release mechanisms. According to basic electrostatics, the electric potential $\Phi = \Phi(\mathbf{r})$ can be calculated in a medium of uniform dielectric constant ϵ_W (with $\epsilon_W = 80$ in water) from a given local volume charge density $\rho(\mathbf{r})$ at position $\mathbf{r} = \{x, y, z\}$ using Poisson's law $\nabla^2\Phi = -\rho/(\epsilon_W\epsilon_0)$, where $\nabla^2 = \partial^2/\partial x^2 + \partial^2/\partial y^2 + \partial^2/\partial z^2$ is the Laplacian and $\epsilon_0 = 8.85 \times 10^{-12} As/(Vm)$ is the permittivity of free space. It is convenient to switch from Φ and ϵ_W and to the dimensionless potential $\Psi = e\Phi/k_BT$ and Bjerrum length $l_B = e^2/(4\pi\epsilon_W\epsilon_0 k_BT)$, where e denotes the elementary charge. The Bjerrum length is the distance at which two individual elementary charges have an interaction energy of k_BT. At room temperature $l_B \approx 0.7$ nm in water and $l_B \approx 56$ nm in vacuum. In terms of the dimensionless potential and the Bjerrum length, Poisson's equation reads $\nabla^2\Psi = -4\pi l_B\rho/e$. If the potential Ψ is known we can calculate the energy stored in the corresponding electric field through $U = k_BT \int dv\,(\nabla\Psi)^2/(8\pi l_B)$, where $\nabla = (\partial/\partial x, \partial/\partial y, \partial/\partial z)$ denotes the gradient and where the integration runs over all space.

Aqueous solutions usually contain mobile charges such as counterions or salt. Because of their mobility, the positions of these charge carriers are not known. For example, consider the presence of monovalent cations and anions in a large aqueous volume V, with local concentrations $n_+ = n_+(\mathbf{r})$ and $n_- = n_-(\mathbf{r})$, respectively, and bulk concentration n_0 (both bulk concentrations need to be equal to ensure electroneutrality). The local volume charge density is then $\rho = e(n_+ - n_-)$ (and $\rho = 0$ in the bulk). Hence, to solve Poisson's equation for Ψ we first need to know n_+ and n_-. The mobile ions can, as a first approximation (more accurate is the term *mean-field approximation*), be described as an ideal gas with corresponding entropies $S_+ = -k_B \int_V dv\, n_+[\ln(n_+/n_0) - 1]$ and $S_- = -k_B \int_V dv\, n_-[\ln(n_-/n_0) - 1]$, where the integrations run over the aqueous region V that is accessible to the mobile ions. Because the entropies of an ideal gas mixture are additive, the total entropy is $S = S_+ + S_-$, and the total thermodynamic potential of the system $F = U - TS$ becomes

$$\frac{F}{k_BT} = \int\limits_V dv \left[\frac{(\nabla\Psi)^2}{8\pi l_B} + n_+ \ln\frac{n_+}{n_0} - n_+ + n_- \ln\frac{n_-}{n_0} - n_- + 2n_0\right]. \tag{1.16}$$

This expression of the free energy depends on the yet unknown concentrations n_+ and n_- (also Ψ depends on n_+ and n_- through Poisson's equation). In a thermodynamic sense n_+ and n_- constitute two unconstrained degrees of freedom. Hence, in thermal equilibrium F must adopt its minimum with respect to n_+ and n_-. One can show [53] that F becomes minimal for the Boltzmann distributions $n_+ = n_0e^{-\Psi}$ and $n_- = n_0e^{\Psi}$. Inserting these into Poisson's equation yields the Poisson–Boltzmann equation

$$l_D^2\nabla^2\Psi = \sinh\Psi, \tag{1.17}$$

where $l_D = (8\pi l_B n_0)^{-1/2}$ is the Debye screening length. Equation 1.17 is a partial differential equation that usually must be solved numerically within the aqueous region; analytical solutions are available only for a few basic geometries [54]. Boundary conditions for Equation 1.17 reflect the presence of macroions in solution. Typical macroions are bodies of low dielectric constant ϵ_L with surface charge density σ at the surface. Because of $\epsilon_W \gg \epsilon_L$ the boundary condition can, approximately, be written $(\partial\Psi/\partial n)_A = -4\pi l_B\sigma/e$, where the partial derivative of Ψ is taken at the macroion surface A along the surface normal direction, denoted by n and pointing into the aqueous region. Once $\Psi(\mathbf{r})$ is known, the local concentrations n_+ and n_- can be calculated, and the free energy F can be computed according to Equation 1.16. In fact, an equivalent but sometimes more convenient way to compute the free energy is through a so-called *charging process* [18] $F = \int_A do \int_0^\sigma d\bar{\sigma}\, \Phi_0(\bar{\sigma})$, where the equilibrium potential $\Psi_0 = \Psi(x = 0)$ [or, equivalently, $\Phi_0 = \Phi(x = 0)$] needs to be known only at the macroion surface A as function of the surface charge density σ. Increasing σ from zero to its final value and integrating the corresponding surface potential Φ_0 over the macroion surface A then yields the free energy F. A conceptually important limiting case is the Debye-Hückel approximation, which assumes small potential $\Psi \ll 1$ (corresponding to $\Phi \ll 25$ mV) and turns Equation 1.17 into the linear Debye-Hückel equation $l_D^2 \nabla^2 \Psi = \Psi$. In this limit $\Phi_0 \sim \sigma$ and the charging process $\int_0^\sigma d\bar{\sigma}\, \Phi_0(\bar{\sigma}) = \sigma\, \Phi_0(\sigma)/2$ can be carried out. The free energy in the Debye-Hückel limit is thus $F = (1/2) \int_A do\, \sigma\, \Phi_0$, where the integration extends over the macroion surface.

We illustrate the use of the Poisson–Boltzmann model for two examples. The first is a charged lipid layer that we model as a (sufficiently large) single planar surface with charge density σ. For anionic lipids $\sigma < 0$, and for cationic lipids $\sigma > 0$. The lipid layer is in contact with an aqueous region that contains monovalent salt ions of bulk concentration n_0. The potential $\Psi = \Psi(x)$ depends only on the distance x to the surface. It fulfills the Poisson–Boltzmann equation $l_D^2 \Psi''(x) = \sinh \Psi(x)$ with the boundary conditions $\Psi(\infty) = 0$ and $\Psi'(0) = -4\pi l_B \sigma/e$. Here, the second derivative of $\Psi''(x)$ corresponds to the Laplacian (∇^2) in one single dimension, and $\Psi'(0)$ denotes the first derivative of the potential taken at position $x = 0$. A first integration of the Poisson–Boltzmann equation, subject to $\Psi(\infty) = 0$, yields $\Psi'(x) = -(2/l_D) \sinh[\Psi(x)/2]$. At $x = 0$ this equation can be combined with the second boundary condition, implying $\Psi(0) = 2\text{arsinh}(2\pi l_B l_D \sigma/e)$. Yet, this exactly is the surface potential $\Psi_0(\sigma) = \Psi(x = 0, \sigma)$ as a function of the surface charge density that allows us to compute the free energy through the charging process $F = \int_A do \int_0^\sigma d\bar{\sigma}\, \Phi_0(\bar{\sigma}) = k_B T\ A/e \int_0^\sigma d\bar{\sigma}\, \Psi_0(\bar{\sigma})$. Denoting the total lateral area of the lipid layer by A and using the expression of the surface potential, we obtain the free energy per unit area

$$\frac{F}{Ak_BT} = \frac{2}{e} \int_0^\sigma d\bar{\sigma}\ \text{arsinh}\left(2\pi l_B l_D \frac{\bar{\sigma}}{e}\right) = \frac{1}{\pi l_B l_D} \left[1 - \sqrt{1 + p^2} + p\ \text{arsinh}(p)\right], \quad (1.18)$$

where we have defined the dimensionless quantity $p = 2\pi l_B l_D \sigma/e$. Note that positive and negative surface charge density σ corresponds to positive and negative sign of p, respectively. Although the derivation of F did not require calculation of the potential $\Psi(x)$, we can easily do so by performing a second integration of the Poisson–Boltzmann equation, which yields

$$\Psi(x) = 2 \ln \left\{1 + \frac{2}{e^{x/l_D} \coth\left[\text{arsinh}(p)/2\right] - 1}\right\}. \quad (1.19)$$

We also note the linearized Debye-Hückel results, $\Psi(x) = 2p\, e^{-x/l_D}$ and $F/(Ak_BT) = 2\pi l_B l_D\, (\sigma/e)^2$, which follow from Equations 1.18 and 1.19 in the limit of small p. Here, the potential decays exponentially, and we indeed identify l_D as the corresponding screening length. Note that linearization generally overestimates the magnitude of the potential Ψ.

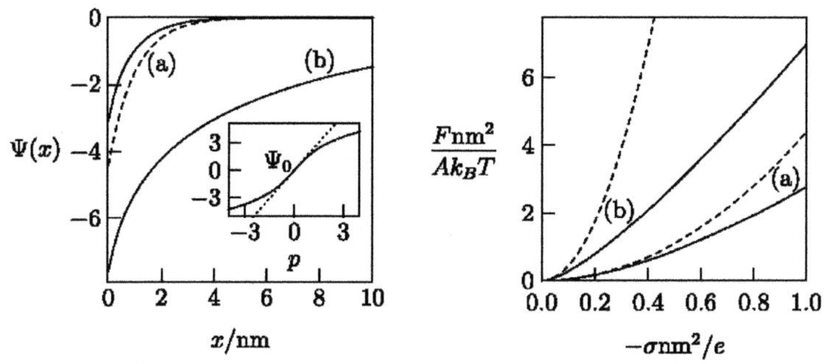

Figure 1.10 *The (dimensionless) potential $\Psi(x)$ as function of x for $\sigma = -0.5e/nm^2$ (left diagram) and the free energy per unit area $F/(Ak_BT)$ as function of $-\sigma/e$ (right diagram). Both diagrams are calculated for $l_B = 0.7$ nm; curves* a *correspond to $l_D = 1$ nm, and curves* b *correspond to $l_D = 10$ nm. Solid and broken lines correspond to the nonlinear (Poisson–Boltzmann) case and linearized (Debye-Hückel) approximation. (The broken line for case (b) in the left diagram is located outside the displayed region; $\Psi(0) = -44$). The inset in the left diagram shows the surface potential $\Psi_0 = \Psi(x = 0)$ as function of $p = 2\pi l_B l_D \sigma/e$ according to Equation 1.19 (solid line) and in the linearized limit (dotted line) where $\Psi_0 = 2p$.*

This can be seen in the inset of Figure 1.10 (left diagram) which shows the surface potential $\Psi_0 = \Psi(x = 0)$ according to Equation 1.19 together with the linearized Debye-Hückel result $\Psi_0 = 2p$. The main diagrams of Figure 1.10 display $\Psi(x)$ from Equation 1.19 and F from Equation 1.18 for two different choices of the Debye screening length, $l_D = 1$ nm and $l_D = 10$ nm. Molarity and Debye screening length are connected through $M = (0.304\ \text{nm}/l_D)^2$, so $l_D = 1$ nm corresponds to physiological conditions, and $l_D = 10$ nm to a 1 mM solution. Surface charge densities of $\sigma \lesssim e/\text{nm}^2$ are typically found for biomacroions such as DNA, lipid membranes, or proteins. Figure 1.10 indicates that the linearized Debye-Hückel limit works reasonably well for $l_D = 1$ nm, but even here there are significant deviations from the prediction of the non-linear model.

Our second example addresses a specific question that is of basic relevance also for the formation of carrier vehicles for drug molecules: What is the energetic cost of incorporating a single charged molecule into a hydrophobic cavity that resides in the aqueous solution? Such a cavity arises, for example, in spherical micelles due to the apolar nature of the surfactant's hydrocarbon chains. For simplicity we model the charged molecule as a sphere of radius R_D with z charges uniformly attached to its surface, implying a charge density $\sigma = ze/(4\pi R_D^2)$. The charged molecule is located at the center of the hydrophobic cavity, which we represent by a sphere of radius R_C. We assume that both the charged molecule and the cavity have the same dielectric constant ϵ_L (with $\epsilon_L \ll \epsilon_W$). We ask the question how much free energy ΔF it would cost to incorporate the charged molecule into the cavity, starting from a separated state as illustrated in Figure 1.11. To make our final result physically insightful, we shall perform the calculation of ΔF within the linear Debye-Hückel approximation. We first calculate the free energy of the charged molecule when residing within the cavity; see the right diagram of Figure 1.11. Because of the spherical symmetry the potential depends only on the radial distance r from the center of the charged sphere. The potential $\Psi_o(r)$ outside the cavity ($R_C < r < \infty$) fulfills the Debye-Hückel

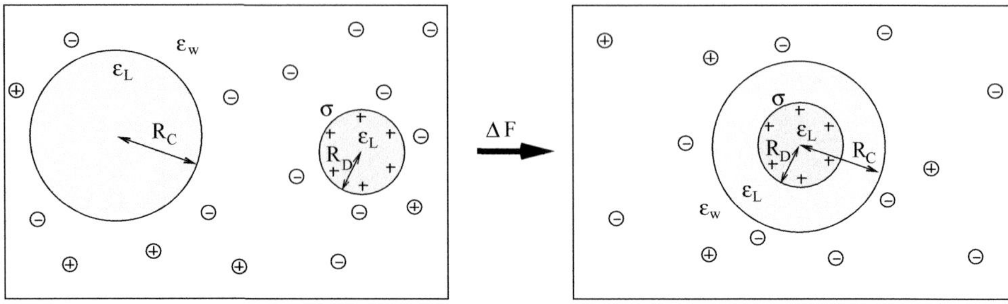

Figure 1.11 *Schematic illustration of a spherical cavity (of radius R_C and dielectric constant ϵ_L) and a charged spherical particle (of radius R_D, dielectric constant ϵ_L, and surface charge density σ). Cavity and particle are initially separated (see the* left diagram*). Inserting the particle into the cavity (see the* right diagram*) incurs a free energy cost ΔF.*

equation $\Psi_o''(r) + (2/r)\Psi_o'(r) = \Psi_o(r)/l_D^2$. Similarly, the potential $\Psi_i(r)$ inside the cavity ($R_D < r < R_C$) fulfills the Laplace equation $\Psi_i''(r) + (2/r)\Psi_i'(r) = 0$. The corresponding boundary conditions are $\Psi_i(R_C) = \Psi_o(R_C)$, $\Psi_o(\infty) = 0$, $\epsilon_W \Psi_o'(R_C) - \epsilon_L \Psi_i'(R_C) = 0$, and $\Psi_i'(R_D) = -l_B z \epsilon_W/(R_D^2 \epsilon_L)$. The latter two account for the change in the dielectric constant at $r = R_C$ and for the presence of the surface charges at $r = R_D$. The solution can be written as

$$\Psi_i(r) = \frac{z l_B}{R_C}\left[\frac{\epsilon_W}{\epsilon_L}\left(\frac{R_C}{r} - 1\right) + \frac{1}{1 + \frac{R_C}{l_D}}\right], \quad \Psi_o(r) = \frac{z}{1 + \frac{R_C}{l_D}} \frac{l_B}{r} e^{\frac{R_C - r}{l_D}}. \tag{1.20}$$

As argued above, the free energy in the Debye-Hückel limit can be calculated according to $F = (1/2)\int_A do\, \sigma \Phi_0$, where the integration runs over the surface $A = 4\pi R_D^2$ of the charged molecule and $\Phi_0 = k_B T \Psi_i(R_D)/e$. Hence,

$$\frac{F}{k_B T} = \frac{z}{2}\Psi_i(R_D) = \frac{z^2 l_B}{2}\left[\frac{\epsilon_W}{\epsilon_L}\left(\frac{1}{R_D} - \frac{1}{R_C}\right) + \frac{1}{R_C\left(1 + \frac{R_C}{l_D}\right)}\right]. \tag{1.21}$$

The first contribution to the free energy in Equation 1.21 is the energy of a spherical capacitor, the second contribution arises due to the diffuse layer of mobile anions in the aqueous solution. In the separated state, see the left diagram of Figure 1.11, potential and thus energy of the cavity vanish, and the energy of the charged particle (which now is immersed in the aqueous solution) can be obtained from Equation 1.21 by replacing $R_C \rightarrow R_D$. Hence, we obtain our final result for the change in free energy

$$\frac{\Delta F}{k_B T} = \frac{z^2 l_B}{2}\left[\frac{\epsilon_W}{\epsilon_L}\left(\frac{1}{R_D} - \frac{1}{R_C}\right) + \frac{1}{R_C\left(1 + \frac{R_C}{l_D}\right)} - \frac{1}{R_D\left(1 + \frac{R_D}{l_D}\right)}\right]. \tag{1.22}$$

Note that ΔF is positive because the energy penalty of the spherical capacitor (the first term in Equation 1.22) overcompensates the energy gain in the diffuse counterion layer. In fact, because of $\epsilon_W \gg \epsilon_L$ the energy cost ΔF is not only dominated by the capacitor energy but also becomes prohibitively large. For example, $z = 5$, $l_B = 0.7$ nm, $l_D = R_D = 1$ nm,

$R_C = 1.5$ nm, $\epsilon_W = 20\epsilon_L$ yield $\Delta F/k_BT = 58 + 2 - 4 = 56$, or more than $10k_BT$ per charge. Although this model is certainly oversimplified, it correctly captures the high energy cost that is associated with inserting small charged molecules from the aqueous environment intro hydrophobic cavities. Similar arguments also rationalize the permeability barrier that prevents the crossing of lipid bilayers by charged molecules [55].

1.5.2 Chain Packing Model

Carriers for poorly water-soluble drug molecules are often composed of chain-like molecules such as surfactants or polymers. This is the case for micelles, liposomes, and polymeric aggregates. Inserting a drug molecule into such an aggregate perturbs the packing properties of the chains. We briefly discuss the corresponding energy penalty that arises from this perturbation. To this end, we focus on a mean-field model for chain packing in amphiphilic aggregates [56]. The model ignores chain-chain correlations and assumes a uniform density of chain segments everywhere in the aggregate. Both are reasonable assumptions well above the main transition temperature [57]. The starting point of the model is an aggregate of given geometry that is composed of a fixed number N of surfactants with their chains residing in the fluid state. Thus, the hydrocarbon chains are flexible and able to adopt a large number of different conformations. In addition, although the chains are anchored to the aggregate's interface A, the local surface density of surfactant headgroups is able to adjust. Yet, because N is given, the average headgroup density $\bar{\sigma} = N/A$ on the interface A is must be fixed. The most relevant case with regard to modeling a delivery vehicle for drug molecules is the additional presence of rigid bodies (i.e., drug molecules) in the hydrophobic core. The left diagram of Figure 1.12 illustrates such a case, where a long cylinder-like inclusion (of radius R) is immersed into the hydrocarbon core of a lipid bilayer with a certain penetration depth p. What is the perturbation energy of the chain packing when inserting the cylinder-like inclusion into the membrane, and how does that energy depend on the penetration depth?

The central quantity of the mean-field chain packing model is the probability $P(\mathbf{r}, \alpha)$ of a given hydrocarbon chain being anchored at position $\mathbf{r}$ on the polar-apolar interface A and residing in a conformation α. We normalize this probability according to $\int_A d^2\mathbf{r} \sum_\alpha P(\mathbf{r}, \alpha) = A$ where the integration $\int_A d^2\mathbf{r}$ extends over the polar-apolar interface of the aggregate (which is represented by an infinitely thin sheet of surface area A) and the summation runs over all possible chain conformations α. For known $P(\mathbf{r}, \alpha)$ we can calculate the local area density of amphiphiles on the surface according to $\sigma(\mathbf{r}) = \bar{\sigma} \sum_\alpha P(\mathbf{r}, \alpha)$, and the total number of amphiphiles in the aggregate is then $N = \int_A d^2\mathbf{r}\sigma(\mathbf{r}) = \bar{\sigma} A$. The conformational free energy of the N chains can be expressed as the sum of the internal chain energy and an entropic contribution

$$F = U - TS = \frac{N}{A} \int_A d^2\mathbf{r} \sum_\alpha P(\mathbf{r}, \alpha) \left[\epsilon(\alpha) + k_BT \ln P(\mathbf{r}, \alpha)\right]. \qquad (1.23)$$

Here, $\epsilon(\alpha)$ denotes the trans-gauge isomerization energy of a chain in conformation α. In thermal equilibrium, F adopts a minimum with respect to the probability distribution $P(\mathbf{r}, \alpha)$. Yet, before the minimization can be carried out, we must account for the additional constraint of constant chain segment density everywhere in the hydrocarbon core (thereby excluding the inserted inclusion). If we denote by $\phi(\alpha, \mathbf{r}', \mathbf{r})$ the number of chain segments

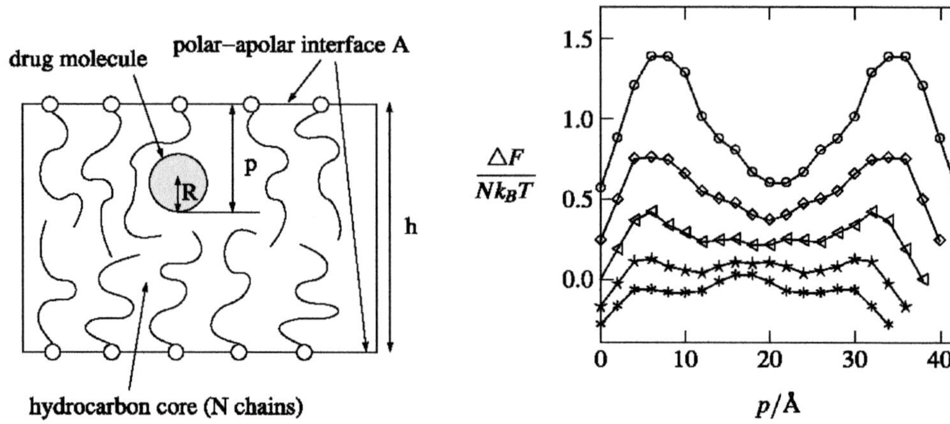

Figure 1.12 Left: *Schematic illustration of a surfactant bilayer with an inserted drug molecule. Shown is a unit cell with N surfactant molecules per drug molecule. The drug molecule is modeled as a long cylinder-like rigid inclusion (shown is the cross-section of the cylinder). The cylinder radius is R, and the hydrocarbon core of the bilayer has thickness h. The penetration depth of the inclusion is p, defined so that for $p = (h/2) + R$ the inclusion resides right in the center of the bilayer. Note that the hydrocarbon chains can adopt many different conformations, subject to being packed uniformly on average and remaining within the hydrocarbon core, excluding the rigid cylinder.* Right: *Change in chain conformational free energy $\Delta F/(Nk_BT)$ induced by inserting the cylinder-like inclusion into the bilayer, calculated for a system of $N = 80$ surfactant chains [of structure $-(CH_2)_{13} - CH_3$] per inclusion and plotted as function of the inclusion's penetration depth p. The different curves correspond to $h = 22$ Å ($*$), $h = 24$ Å ($*$), $h = 26$ Å ($\triangleleft$), $h = 28$ Å ($\diamond$), and $h = 30$ Å ($\circ$). Reproduced with permission from [58].*

at point $\mathbf{r}'$ within the hydrocarbon core contributed from a surfactant that is located at point $\mathbf{r}$ on the interface A and resides in conformation α, we can compute the average segment density at point $\mathbf{r}'$ through

$$\bar{\phi}(\mathbf{r}') = \int_A d^2\mathbf{r} \sum_\alpha P(\mathbf{r}, \alpha)\phi(\alpha, \mathbf{r}', \mathbf{r}). \tag{1.24}$$

The constraint of constant segment density can then be expressed simply as $\bar{\phi}(\mathbf{r}') = \bar{\phi} = const$ (namely, independent of $\mathbf{r}'$) everywhere in the hydrocarbon chain region, excluding the inserted inclusion. Minimization of F with respect to $P(\mathbf{r}, \alpha)$ and subject to Equation 1.24 yields

$$P(\mathbf{r}, \alpha) = \frac{e^{-\frac{1}{k_BT}\left[\epsilon(\alpha) + \int_V d^3\mathbf{r}\lambda(\mathbf{r})\phi(\alpha, \mathbf{r}', \mathbf{r})\right]}}{\frac{1}{A}\int_A d^2\mathbf{r} \sum_\alpha e^{-\frac{1}{k_BT}\left[\epsilon(\alpha) + \int_V d^3\mathbf{r}\lambda(\mathbf{r})\phi(\alpha, \mathbf{r}', \mathbf{r})\right]}}, \tag{1.25}$$

where the function $\lambda(\mathbf{r})$ is a Lagrangian multiplier that ensures the constraint in Equation 1.24 to be fulfilled. In fact, one finds $\lambda(\mathbf{r})$ by inserting $P(\mathbf{r}, \alpha)$ into Equation 1.24 and solving the resulting self-consistency relation numerically. This requires the generation of a representative set of chain conformations at various positions along the interface A, based on a molecular model such as Flory's isomeric state model [28]. Once $\lambda(\mathbf{r})$ is known, the probabilities $P(\mathbf{r}, \alpha)$ can be used to compute the free energy according to Equation 1.23. This can be carried out for different choices of the penetration depth ($F(p)$) or in absence of the cylinder (F_{ref}); the difference $\triangle F(p) = F(p) - F_{ref}$ is then the change in chain conformational free energy induced by inserting the cylinder-like inclusion into the bilayer, at given penetration depth p.

The right diagram of Figure 1.12 displays computational results for the change in chain conformational free energy $\triangle F$ of a planar surfactant bilayer as function of the inclusion's penetration depth p. The conformational free energy profile for translation of the inclusion through the bilayer (i.e., changing the penetration depth p) strongly depends on the thickness of the surfactant layer. For a thick layer (see the upper curve in the right diagram of Figure 1.12) the hydrocarbon chains are significantly stretched. The corresponding large entropic penalty can be relieved by placing the inclusion right in the center of the bilayer. Away from that position the free energy exhibits a large barrier. Hence, in bilayers where strong headgroup attractions lead to small cross-sectional area per surfactant, hydrophobic inclusions tend to be located in the center of the bilayer. In contrast, in thin bilayers (see the lower curve in the right diagram of Figure 1.12) the chains from one monolayer dynamically interdigitate into the apposed monolayer. Here, the center exhibits an energy barrier and is thus an unfavorable location. We emphasize that the right diagram of Figure 1.12 displays the energy per surfactant chain. Because the calculation is based on $N = 80$ chains per inclusion, an energy change of, say, $0.2\, k_BT$ would translate into $40\, k_BT$ per inclusion. Hence, the conformational properties of the chains are expected to provide a significant contribution to defining the position of a hydrophobic inclusion within a surfactant aggregate. These considerations apply to liposomes and similarly to micellar aggregates.

1.6 Kinetics of Drug Transfer from Mobile Nanocarriers

The theoretical concepts we have discussed so far refer exclusively to thermodynamic equilibrium. Yet, the delivery of drugs inevitably involves dynamic properties as well. Dynamic properties determine the kinetics, i.e., the retention properties of the delivery vehicle and the amount of drug molecules transferred from the delivery vehicle to a target system as function of time [59]. Modeling the corresponding transfer properties is often accomplished using methods based on chemical reaction kinetics [60].

In the following we consider a simple and instructive scenario, mobile delivery vehicles and mobile targets that are enclosed in a fixed volume V. We will refer to the delivery vehicles as *donors* and to the target system as *acceptors*. We assume that the numbers of both donors (N_D) and acceptors (N_A) remain unchanged. Hence $N_D + N_A = N$ is a fixed constant. Initially, the donors contain a certain number of drug molecules M (or, equivalently, N_D/M drug molecules per delivery vehicle on average), whereas the acceptors are empty. Over time, some of the drug molecules will migrate from the donors to the acceptors. We can describe the corresponding time dependence by the numbers $M_D = M_D(t)$ and $M_A = M_A(t)$ of drug molecules residing in the donors and acceptors,

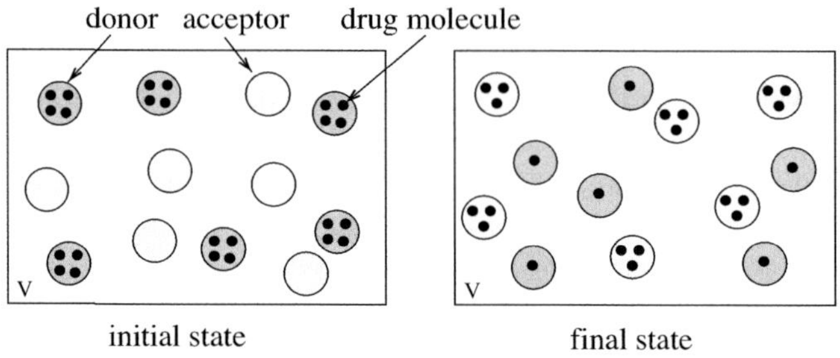

__Figure 1.13__ Schematic illustration of drug transfer from N_d donors (dark shaded) to N_a acceptors (light shaded). Drug molecules can reside in either donors or acceptors; they are represented by the dark bullets. Initially (left diagram), all M drug molecules are located inside the donors. The kinetics of drug transfer can be described by the numbers $M_d(t)$ and $M_a(t)$, of drug molecules in donors and acceptors, respectively. After thermal equilibrium is established (right diagram), the drug molecules are distributed between donors and acceptors with an equilibrium distribution $M_d(t \to \infty) = M_d^{eq}$ and $M_a(t \to \infty) = M_a^{eq}$. The displayed example corresponds to $N_d = N_a = 6$ and $M = M_d^{eq} + M_a^{eq} = 6 + 18 = 24$.

respectively. For poorly water-soluble drugs $M_D(t) + M_A(t) = M$ is a constant, implying that very little drug is dissolved in the aqueous phase. In addition, if the transfer of drug molecules is slow compared to typical diffusion times of donors and acceptors, we may ignore spatial variations of donor and acceptor concentrations. Figure 1.13 shows an illustration of the system. Note that we make a number of significant assumptions (including the fixed volume, the presence of only one type of mobile target system, the slow transfer dynamics of the drugs between donors and acceptors, and the fixed numbers of donors and acceptors). Yet, our system still provides a convenient starting point as it is a generic model for various types of mobile drug delivery vehicles such as liposomes, micelles, and polymeric aggregates.

It is generally useful to distinguish between the dynamics of drug transport within a single delivery vehicle (such as the flip-flop of drug molecules between the two leaflets of a liposomal membrane) and the dynamics of transport from the delivery vehicle to a target site. The latter involves two principal mechanism, namely transfer through collisions and transfer through diffusion [61]. Both are illustrated in Figure 1.14. Note that the collision mechanism requires temporary physical contact between a donor and acceptor (or, similarly,

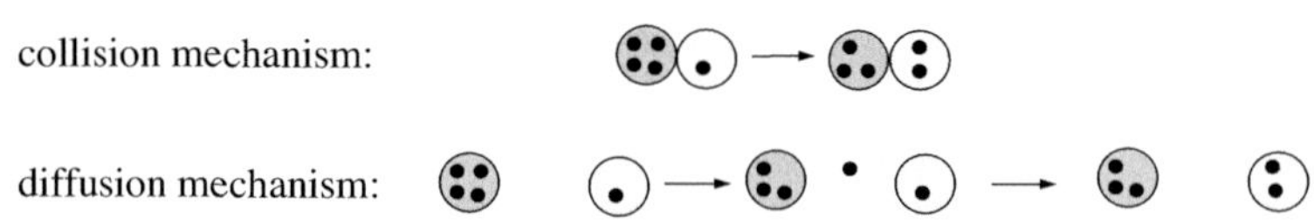

*__Figure 1.14__ Schematic illustration of the collision and diffusion mechanisms for the transfer of drugs from donor (*dark shaded*) to acceptor (*light shaded*). Drug molecules are represented by filled bullets. The collision mechanism involves direct drug transfer upon the collision of donor and acceptor. For the diffusion mechanism, drug molecules are first released into the aqueous phase. In a second step they are taken up by the acceptor.*

donor-donor or acceptor-acceptor), whereas the diffusion mechanism involves drug molecules transferring via the intermediate step of diffusing through the aqueous environment. Both mechanisms have been invoked to explain the transport of lipids and cholesterol [62] between membranes. Let us discuss a model for the kinetics of the collision and diffusion mechanisms.

1.6.1 Collision Mechanism

Because of the need for individual collision between pairs of donor-donor, donor-acceptor or acceptor-acceptor, the kinetics of the collision mechanism is based on a second-order process. Specifically, if only d_i donors that each contain i drug molecules and a_j acceptors that each contain j drug molecules are present, the collision rate between donors and acceptors is proportional to the product $d_i a_j$. If donors and acceptors of given distributions d_i and a_j are present, then all possible combinations of i and j will contribute to the collision rate. Any modeling effort requires knowledge of how the proportionality factor depends on i and j. Typically, it simply reflects the concentration difference. This ignores any non-ideal behavior (such as aggregation of the drug molecules in the donors or acceptors), but it allows us to express the resulting kinetic equations only in terms of $M_d(t)$ and $M_a(t)$. The result [63]

$$\frac{dM_d}{dt} = \frac{K_{coll}}{V}(M_a N_d - M_d N_a + kN_a N_d), \qquad \frac{dM_a}{dt} = \frac{K_{coll}}{V}(M_d N_a - M_a N_d - kN_a N_d), \tag{1.26}$$

contains two constants: a unit transfer rate K_{coll} and the difference k of drug molecules carried on average by each donor and acceptor in thermal equilibrium (i.e., $k = 2$ in the schematic representation of Figure 1.13). The solution of Equations 1.26

$$M_a(t) = M - M_d(t) = \left(1 - e^{-\frac{N}{V}K_{coll}\,t}\right)\frac{N_a}{N}(M - kN_d). \tag{1.27}$$

corresponds to simple exponential behavior. The equilibrium distributions $M_d^{eq} = M_d(t \to \infty)$ and $M_a^{eq} = M_a(t \to \infty)$ indeed fulfill the relation $(M_d^{eq}/N_d) - (M_a^{eq}/N_a) = k$, thus verifying our interpretation of k above. We point out that the rate constant $K = K_{coll}N/V$ of the transfer through collisions depends on the total concentration $N/V = (N_d + N_a)/V$ of donors and acceptors. Hence, larger concentrations N/V increase the transfer speed as collisions become more likely.

Equations 1.26 represent a first-order process. The reason for the microscopic second-order collision mechanism to translate into a first-order process for the overall kinetic behavior is the conservation of the numbers N_D and N_A. That is, collisions do not affect the numbers of 'reactants', namely the donors and acceptors. We can describe the transfer of drug molecules from donors (D) to acceptors (A) by the chemical reaction scheme

$$\mathrm{D} \underset{K_2}{\overset{K_1}{\rightleftharpoons}} \mathrm{A} \tag{1.28}$$

with rate constants $K_1 = (1 - kN_d/M)\,KN_a/N$ and $K_2 = (1 + kN_a/M)\,KN_d/N$. The corresponding equilibrium constant $K_{eq} = K_1/K_2 = (M - kN_d)/(M + kN_a)$ allows us to calculate the standard Gibbs free energy $\Delta g^0 = k_B T\,\ln[(M/N_d + k)/(M/N_a - k)]$. Specifically, if donors and acceptors are chemically similar, $k = 0$ and $\Delta g^0 = k_B T\,\ln(N_d/N_a)$

contains only an entropic component. For $N_d > N_a$ we have $\Delta g^0 > 0$. More drugs will reside in the donors simply because more donors are available. An enthalpic component may arise in Δg^0 through the temperature dependence of k.

1.6.2 Diffusion Mechanism

For the diffusion mechanism the drug molecules pass through the aqueous phase as illustrated in Figure 1.14. If the rates of drug release/uptake of each individual donor and acceptor are strictly proportional to the number of occupied/free binding sites, then the kinetics of the diffusion mechanism can be described according to the chemical reaction scheme

$$\mathrm{D} \underset{K_d^{upt}}{\overset{K_d^{rel}}{\rightleftharpoons}} \mathrm{W} \underset{K_a^{rel}}{\overset{K_a^{upt}}{\rightleftharpoons}} \mathrm{A} \tag{1.29}$$

where the additional water-dissolved state of the drug molecules is denoted by W. Because the drug molecules are poorly water-soluble, it is appropriate to introduce a steady-state approximation $dM_w/dt = 0$, where $M_w(t) \ll M \approx M_a(t) + M_d(t)$ is the number of drug molecules in the aqueous phase. If we also assume equal uptake rates $K_d^{upt} = K_a^{upt}$ of donors and acceptors, we can express the kinetic equations for the diffusion mechanism as

$$\frac{dM_d}{dt} = K_{diff}(M_a N_d - M_d N_a + kN_a N_d), \qquad \frac{dM_a}{dt} = K_{diff}(M_d N_a - M_a N_d - kN_a N_d), \tag{1.30}$$

where we have defined $K_d^{rel} = K_{diff}(1 - kN_d/M)$ and $K_a^{rel} = K_{diff}(1 + kN_a/M)$. Comparison with Equations 1.26 reveals that the diffusion mechanism follows the same first-order kinetics as the collision mechanism, yet with a different rate constant K_{diff}. The total rate constant of the combined collision and diffusion process is thus $K = K_{coll}(N/V) + K_{diff}$. The total rate constant K is a measurable quantity; its dependence on the total concentration of donors and acceptors gives insights into the mechanism of drug transfer. A specific example is the transfer of the drug molecule temoporfin (a second generation photosensitizer) from donor liposomes to acceptor liposomes. Transfer experiment have been conducted [64] using radioactive-labeled temoporfin residing initially in donor liposomes. Addition of acceptor liposomes initiates a transfer process that can be measured as a function of time. To this end, donor and acceptor liposomes carry different amounts of charged lipids and can thus be separated from each other using an ion exchange column. The experimental results for $M_a(t)$ can indeed typically be described by a simple exponential function with inverse rate constants K^{-1} on the order of several hours. Analysis of the concentration dependence $K = K(N/V)$ of the rate constant suggested for unilamellar liposomes (consisting mostly of phospholipids and having a diameter of about 100 nm) that for concentrations of $N/V < 200/\mu\mathrm{m}^3$ the transfer was dominated by the diffusion mechanism whereas for $N/V > 200/\mu\mathrm{m}^3$ the collision mechanism was more prevalent [64].

1.6.3 Internal Kinetics

So far, we have assumed that drug molecules in donors and acceptors can reside in only one single state. This may be appropriate for some cases such as lipid bilayers with all drug molecules buried within the hydrocarbon core; see the upper curve in Figure 1.12. In general, however, there will be a set of either discrete or continuously distributed states in which the drug molecules can reside. The most simple example is a two-state system. Consider, for example, drug molecules that preferentially interact with either the inner or outer leaflet of a liposome; see the lower curve in Figure 1.12. Here, drug molecules initially associated with the inner monolayer need to first flip to the outer monolayer before they can transfer to the target site. Hence a characteristic flip-flop time will enter into the description of the kinetic behavior. A more complex example for the internal dynamics are reservoir systems such as polymeric aggregates [65], hydrogels [66], or microemulsions [67]. Here, the set of accessible states is continuous, and solutions of the diffusion equation determine the availability of the drug molecules at the interface of the carrier with the ambient environment. Different relevant scenarios and theoretical models for these cases have been reviewed recently [23].

Let is briefly discuss the extension of our one-state model for the collision and diffusion kinetics to a two-state model. To make our analysis most instructive, we assume that donors and acceptors are sufficiently large (so that curvature effects can be ignored) and chemically identical liposomes that can host drug molecules in either the inner or outer monolayer; see the illustration on the left-hand side of Figure 1.15. The chemical rection scheme for the transfer of drug molecules from the inner (D^I) to first the outer (D^O) leaflet of the donor liposomes, next to the outer (A^O) to finally to the inner (A^O) leaflets of the acceptor liposomes, can be written as

$$D^I \underset{G}{\overset{G}{\rightleftharpoons}} D^O \underset{K_2}{\overset{K_1}{\rightleftharpoons}} A^O \underset{G}{\overset{G}{\rightleftharpoons}} A^I. \qquad (1.31)$$

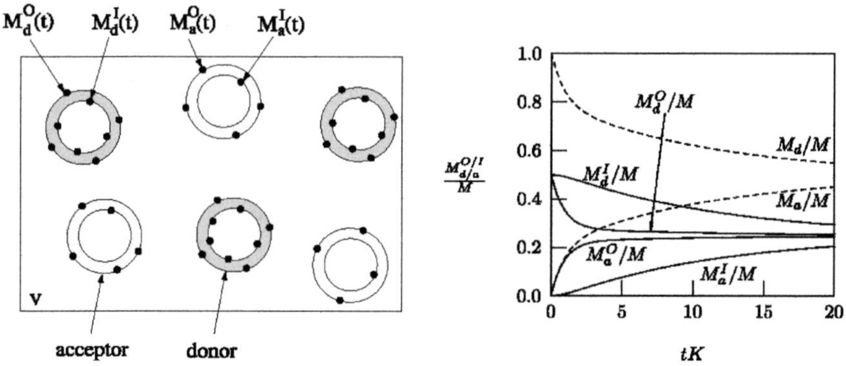

Figure 1.15 Left: *Mixture of N_d donor and N_a acceptor liposomes where the M drug molecules are incorporated either in the inner or outer monolayer. The instantaneous numbers of drug molecules in the inner and outer leaflets of the donor liposomes is denoted by $M_d^I(t)$ and $M_d^O(t)$, respectively. For the acceptor liposomes the corresponding notation is $M_a^I(t)$ and $M_a^O(t)$.* Right: *Plots of $M_d^O(t)$, $M_d^I(t)$, $M_a^O(t)$, and $M_a^I(t)$ according to Eqs. 1.33 for $G/K = 1/10$ and $N_a/N = N_d/N = 0.5$. The broken lines show the biexponential behaviors of the sums $M_d = M_d^O + M_d^I$ and $M_a = M_a^O + M_a^I$. The time is plotted in units of the inverse rate constant K. Note also $\omega_1 = 1.11K$ and $\omega_2 = 0.09K$ are the effective rate constants for the decay. The right diagram is reproduced from [63].*

Here, due to the symmetry of liposomes with regard to the inner and outer leaflet, a single rate constant G fully describes the internal kinetics of the donor and acceptor liposomes. The rate constants for the inter-liposomal transfer are given, as in Equation 1.28 but with $k = 0$ and $K = K_{coll}(N/V) + K_{diff}$, by $K_1 = KN_a/N$ and $K_2 = KN_d/N$. Denote the numbers of drug molecules residing in the inner and outer monolayers of the donor liposomes by $M_d^I(t)$ and $M_d^O(t)$, respectively. For the acceptor liposomes, the corresponding quantities are $M_a^I(t)$ and $M_a^O(t)$. They fulfill the equations

$$\frac{dM_d^O}{dt} = \frac{K}{N}(M_a^O N_d - M_d^O N_a) - G(M_d^O - M_d^I), \qquad \frac{dM_d^I}{dt} = G(M_d^O - M_d^I),$$
$$\frac{dM_a^O}{dt} = \frac{K}{N}(M_d^O N_a - M_a^O N_d) - G(M_a^O - M_a^I), \qquad \frac{dM_a^I}{dt} = G(M_a^O - M_a^I), \qquad (1.32)$$

which depend only on the two rate constants K and G. Initially, all drug molecules are distributed in the donor liposomes, with 50% in each monolayer. The solution of Equations 1.32 is then

$$M_d^I(t) = \frac{M}{2}\left[\frac{N_d}{N} + \frac{N_a}{N}\frac{\omega_2 e^{-\omega_1 t} - \omega_1 e^{-\omega_2 t}}{\omega_2 - \omega_1}\right],$$
$$M_d^O(t) - M_d^I(t) = \frac{M}{2}K\frac{N_a}{N}\frac{e^{-\omega_2 t} - e^{-\omega_1 t}}{\omega_2 - \omega_1},$$
$$M_a^I(t) = \frac{MN_a}{2N}\left[1 - \frac{\omega_2 e^{-\omega_1 t} - \omega_1 e^{-\omega_2 t}}{\omega_2 - \omega_1}\right],$$
$$M_a^O(t) - M_a^I(t) = \frac{M}{2}K\frac{N_a}{N}\frac{e^{-\omega_1 t} - e^{-\omega_2 t}}{\omega_2 - \omega_1}, \qquad (1.33)$$

which describes a biexponential decay with the two effective rate constants ω_1 and ω_2. They follow from G and K through $2G + K = \omega_1 + \omega_2$ and $4G^2 + K^2 = (\omega_2 - \omega_1)^2$. A plot of $M_d^O(t)$, $M_d^I(t)$, $M_a^O(t)$, and $M_a^I(t)$, derived for $G/K = 1/10$ and $N_a/N = N_d/N = 0.5$, is shown in the right diagram of Figure 1.15. In this example, the flip-flop of drug molecules between the liposome leaflets is the rate-limiting step. Hence, the drug molecules in the outer monolayers of donor and acceptor liposomes are quickly exchanged, but it takes much longer to also transfer the drug molecules that initially resided in the inner monolayers of the donor liposomes. It is also interesting to mention two limiting cases. First, for $G = 0$ the drug molecules in the inner leaf do not participate in the transfer, and Equation 1.33 yield $M_d^I(t) = M/2$, $M_a^I(t) = 0$, $M_a^O(t) = M/2 - M_d^O(t) = (1 - e^{-Kt})(MN_a)/(2N)$, which corresponds to simple exponential decay for $M/2$ drug molecules. Second, for $G \to \infty$ the drug molecules exchange (i.e., flip-flop) fast between the liposome leaflets. Equations 1.33 then read $M_a^I(t) = M_a^O(t) = M/2 - M_d^I(t) = M/2 - M_d^O(t) = (1 - e^{-Kt/2})(MN_a)/(2N)$. Here again, we find simple exponential behavior, yet with an apparent rate constant $K/2$ because only the drug molecules in the outer leaflets contribute to the concentration gradients that drive the transport.

We point out that the present two-state model with the identical donor and acceptor liposomes is perhaps the simplest case that leads to biexponential behavior. Different rate constants G_{donor} and $G_{acceptor}$ or the presence of more internal states can be included in a straightforward matter. Also, the two states introduced in our model may not necessarily correspond to the two leaflets of a lipid membrane. They can refer to any states in which the drug is able to reside. For example, for temoporfin in PEGylated liposomes Reshetov *et al.* [68] have suggested that only a fraction of the drug molecules reside inside the lipid bilayer whereas the remaining fraction partitions into the PEG shell. This then leads to a fast and slow component in the transfer process. More generally, biexponential behavior is often observed for the release of drug molecules from nanocarriers; for an account of some experimental systems, see Zeng *et al.* [69].

1.7 Conclusion

We reiterate the scope of our present modeling approach: elucidating basic physical principles that are important for the design of drug delivery vehicles. What contribution can this have to the actual design of specific delivery systems, say, the retention of temoporfin in lipid vesicles?

We first point out that the comprehensive modeling of drug delivery (including that of temoporfin) from basic physical principles is currently an illusion. However, there is an ever increasing body of experimental data that calls for understanding and interpretation rather than mere accumulation and classification. The quest for understanding suggests ultimately envisioning a complete physico-chemical modeling framework of self-associating drug carrier systems, including their interactions with the target system. What would need to be included in this model is the specificity of each drug and drug carrier system, including a molecular-level description of the drug's state inside the carrier and the abundance of potential drug acceptor systems, i.e., in the blood after injecting the carrier. This kind of quantitative modeling would inevitably require large-scale computer simulations, where the complex interactions between carriers, carrier formation, drug encapsulation, and interactions with the target system emerge 'naturally' (i.e., as a result of the simulation). Yet even if an incredibly fast computer were, in principle, able to carry out atomistic simulations of an entire organism, the understanding of the underlying principles will still benefit from the simple kind of models that we have discussed in the present chapter.

Second, some of the models described in the present chapter actually can be applied to specific systems, given that experimental data are present to an extent that allows validation of the model approach. Optimally, modeling and experimental verification proceed in parallel, not independently from each other. For example, the identification of the diffusion and collision mechanism for the transfer of temoporfin from donor to acceptor liposomes (see Section 1.6) requires the measurement of rate constants as a function of the *total* concentration of acceptors and donors (instead of only the concentration of acceptor liposomes). Someone who is testing this theory needs to have this information *prior* to carrying out the kinetic transfer experiments. Conversely, modeling of the transfer process is futile unless guided by a set of well-defined experimental conditions (such as the information that liposome fusion normally can be ignored during a kinetic transfer experiment, that the transfer is typically a first-order process, etc.). In addition, the transfer of measurements have to be done in a system undisturbed by the measurement method. Methods which have been described to investigate the *in vitro* drug release of colloidal drug delivery systems, including sample-and-separate methods, membrane-barrier methods, continuous-flow methods and *in situ* methods [70], are for the majority of the methods not adequate as filtration steps

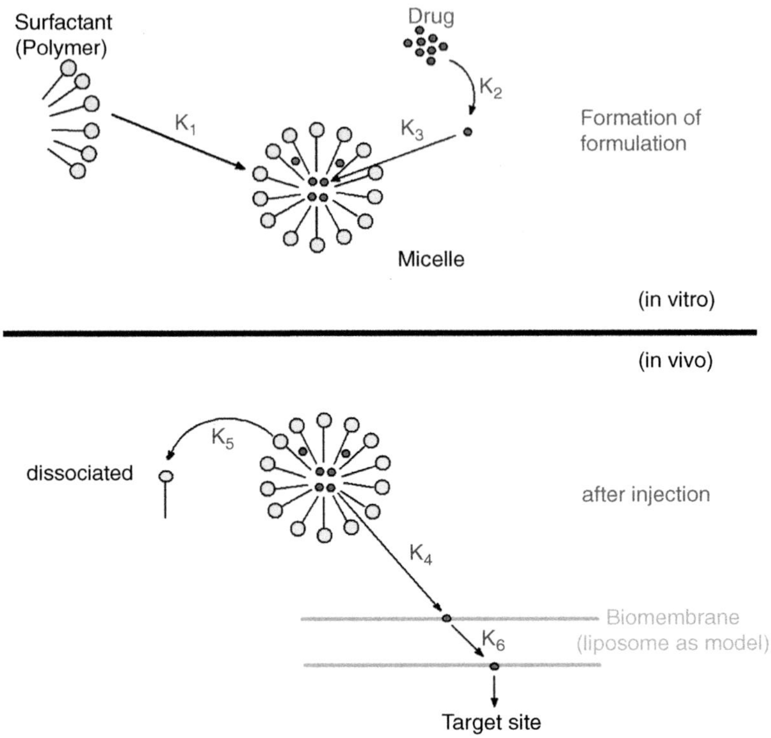

Figure 1.16 Upper part: *Illustration of the* in vitro *formation of a carrier system from surfactants and drug molecules. The drug molecules partition into the micelles or, alternatively, form an aggregate on their own. Different rate constants k_1, k_2 and k_3 are involved in the kinetics of this process.* Lower part: *Illustration of the* in vivo *transfer of drug molecules from a drug-loaded carrier to a biomembrane target. Again, different rate constants k_4, k_5 and k_6 are involved in the kinetics of this process.* (See colour plate section.)

or dialysis membrane barriers introduce a distortion of the real parameters necessary for the modeling process (the described methods are of course usable when comparing and ranking carrier systems). In this respect, fluorescence-based methods (i.e., employing the intrinsic fluorescence of a drug or of the carrier system) offer valuable advantages over invasive methods [19].

The models presented in this chapter may be extended. Recall, for example, the addition of the internal kinetics to the overall transfer kinetics of drugs between mobile nanocarriers (in Section 1.6.3). The final result, see Equations 1.33, depends on the two intrinsic rate constants K and G, corresponding to the transfer between carriers and within each carrier. Already here the final result is of considerable complexity. Yet, many interesting phenomena involve more than two rate constants (and thus even more complexity than Equations 1.33). Figure 1.16 shows two examples, one *in vitro* and one *in vivo*. The first example illustrates the kinetics of forming a carrier system, where surfactants form a micelle and the drug molecules may partition into that micelle or stay in the aqueous phase as monomers or as an aggregate. In contrast to Section 1.4.2, the drug can either partition into the micelle or form an aggregate on its own. Hence, there is a competition to see which one to include in a theoretical model is more relevant and more interesting than our treatment in Section 1.4.2. It is also more complex, but the present chapter has discussed all the tools needed to write down and analyze the kinetic equations. Similarly for the second example, the interaction of a drug-loaded carrier with a biomembrane target, the kinetic equations do not only involve more than two rate constants, but they also require an understanding of the driving forces of the drug molecules to partition between the carrier and the biomembrane. Again, in the present chapter we have introduced some of the basic concepts that contribute to the modeling of these complex systems.

Acknowledgments

We would like to thank Stephan Loew and Alexander Wagner for insightful discussions.

References

1. C. A. Lipinski. Drug-like properties and the causes of poor solubility and poor permeability. *J Pharmacological Toxicological Methods*, **44**(1), 235–249 (2000).
2. M. J. Lawrence, G. D. Rees. Microemulsion-based media as novel drug delivery systems. *Adv Drug Delivery Rev*, **45**(1), 89–121 (2000).
3. C. X. He, Z. G. He, J. Q. Gao. Microemulsions as drug delivery systems to improve the solubility and the bioavailability of poorly water-soluble drugs. *Expert Opinion On Drug Delivery*, **7**(4), 445–460 (2010).
4. B. Jeong, S. W. Kim, Y. H. Bae. Thermosensitive sol-gel reversible hydrogels. *Adv Drug Delivery Rev*, **54**(1), 37–51 (2002).
5. M. C. Jones, J. C. Leroux. Polymeric micelles: a new generation of colloidal drug carriers. *European J Pharmaceutics Biopharmaceutics*, **48**(2), 101–111 (1999).
6. V. P. Torchilin. Micellar nanocarriers: Pharmaceutical perspectives. *Pharmaceutical Research*, **24**(1), 1–16 (2007).
7. J. A. Zasadzinski, B. Wong, N. Forbes, G. Braun, G. H. Wu. Novel methods of enhanced retention in and rapid, targeted release from liposomes. *Current Opinion in Colloid & Interface Science*, **16**(3), 203–214 (2011).
8. D. E. Discher, V. Ortiz, G. Srinivas, M. L. Klein, Y. Kim, C. A. David, S. S. Cai, P. Photos, F. Ahmed. Emerging applications of polymersomes in delivery: From molecular dynamics to shrinkage of tumors. *Progress in Polymer Science*, **32**(8–9), 838–857 (2007).
9. U. Gupta, H. B. Agashe, A. Asthana, N. K. Jain. Dendrimers: Novel polymeric nanoarchitectures for solubility enhancement. *Biomacromolecules*, **7**(3), 649–658 (2006).
10. R. H. Muller, C. M. Keck. Challenges and solutions for the delivery of biotech drugs: a review of drug nanocrystal technology and lipid nanoparticles. *J Biotechnology*, **113**(1–3), 151–170 (2004).
11. H. Bunjes. Lipid nanoparticles for the delivery of poorly water-soluble drugs. *J Pharmacy Pharmacology*, **62**(11), 1637–1645 (2010).
12. G. Poste, D. Papahadjopoulos. Lipid vesicles as carriers for introducing materials into cultured-cells: Influence of vesicle lipid-composition on mechanism(s) of vesicle

incorporation into cells. *Proc National Acad Sciences United States Am*, **73**(5), 1603–1607 (1976).
13. M. L. Immordino, F. Dosio, L. Cattel. Stealth liposomes: review of the basic science, rationale, and clinical applications, existing and potential. *Int J Nanomedicine*, **1**(3), 297–315 (2006).
14. P. Sapra, T. M. Allen. Ligand-targeted liposomal anticancer drugs. *Progress in Lipid Research*, **42**(5), 439–462 (2003).
15. S. Ganta, H. Devalapally, A. Shahiwala, M. Amiji. A review of stimuli-responsive nanocarriers for drug and gene delivery. *J Controlled Release*, **126**(3), 187–204 (2008).
16. R. R. Sawant, V. P. Torchilin. Liposomes as 'smart' pharmaceutical nanocarriers. *Soft Matter*, **6**(17), 4026–4044 (2010).
17. J. N. Israelachvili. *Intermolecular and Surface Forces.* Second edition. New York: Academic Press (1992).
18. D. F. Evans, H. Wennerstrom. *The Colloidal Domain, Where Physics, Chemistry, and Biology Meet.* Second edition. VCH publishers (1994).
19. S. Petersen, A. Fahr, H. Bunjes. Flow cytometry as a new approach to investigate drug transfer between lipid particles. *Mol Pharmaceutics*, **7**(2), 350–363 (2010).
20. P. van Hoogevest, X. Liu, A. Fahr, M. L. S. Leigh. Role of phospholipids in the oral and parenteral delivery of poorly water soluble drugs. *J Drug Delivery Science Technology*, **21**(1), 5–16 (2011).
21. M. Grassi, G. Grassi, R. Lapasin, I. Colombo. *Understanding Drug Release and Absorption Mechanisms: a Physical and Mathematical Approach.* Boca Raton, FL: CRC Press (2006).
22. Martin Malmsten. *Surfactants and Polymers in Drug Delivery*, volume 122 of *Drugs and the Pharmaceutical Sciences.* New York: Marcel Dekker (2002).
23. J. Siepmann, F. Siepmann. Mathematical modeling of drug delivery. *Int J Pharmaceutics*, **364**(2), 328–343 (2008).
24. G. Bonacucina, M. Cespi, M. Misici-Falzi, G. F. Palmieri. Colloidal soft matter as drug delivery system. *J Pharmaceutical Sciences*, **98**(1), 1–42 (2009).
25. P. A. Kralchevsky, K. D. Danov, N. D. Denkov. Chemical physics of colloid systems and interfaces. In K. S. Birdi, (Ed.) *Handbook of Surface and Colloid Chemistry*, 2 edition, chapter 5. Boca Raton, FL: CRS Press, 137–344 (2003).
26. H. T. Davis. *Statistical Mechanics of Phases, Interfaces, and Thin Films.* New York: VCH publishers (1996).
27. R. DeHoff. *Thermodynamics in Materials Science.* Second edition. Boca Raton, FL: CRC Press (2006).
28. P. J. Flory. *Statistical Mechanics of Chain Molecules.* New York: Wiley-Interscience (1969).
29. D. X. Lin, Y. B. Huang. A thermal analysis method to predict the complete phase diagram of drug-polymer solid dispersions. *Int J Pharmaceutics*, **399**(1–2), 109–115 (2010).
30. S. Patel, A. Lavasanifar, P. Choi. Application of molecular dynamics simulation to predict the compatability between water-insoluble drugs and self-associating poly(ethylene oxide)-b-poly(epsilon-caprolactone) block copolymers. *Biomacromolecules*, **9**(11), 3014–3023 (2008).

31. M. C. Branco, J. P. Schneider. Self-assembling materials for therapeutic delivery. *Acta Biomaterialia*, **5**(3), 817–831 (2009).
32. A. Mahmud, X. B. Xiong, H. M. Aliabadi, A. Lavasanifar. Polymeric micelles for drug targeting. *J Drug Targeting*, **15**(9), 553–584 (2007).
33. R. W. Redmond, E. J. Land, T. G. Truscott. Aggregation effects on the photophysical properties of porphyrins in relation to mechanisms involved in photodynamic therapy. *Adv in Experimental Medicine Biol*, **193**, 293–302 (1985).
34. S. Sasnouski, V. Zorin, I. Khludeyev, M. A. D'Hallewin, F. Guillemin, L. Bezdetnaya. Investigation of Foscan (R) interactions with plasma proteins. *Biochimica Et Biophysica Acta-General Subjects*, **1725**(3), 394–402 (2005).
35. J. N. Israelachvili, D. J. Mitchell, B. W. Ninham. Theory of self-assembly of hydrocarbon amphiphiles into micelles and bilayers. *J Chem Society, Faraday Transactions Ii*, **72**, 1525–1568 (1976).
36. A. Ben-Shaul, W. M. Gelbart. Statistical thermodynamics of amphiphile self-assembly: Structure and phase transitions in micellar solutions. In W. M. Gelbart, A. Ben-Shaul, D. Roux, (Eds.) *Micelles, Membranes, Microemulsions, and Monolayers*, first edition, section 1. New York: Springer, 359–402 (1994).
37. K. A. Dill, S. Bromberg. *Molecular Driving Forces: Statistical Thermodynamics in Chemistry & Biology.* First edition. New York: Garland Publishing (2003).
38. A. Muellertz, A. Ogbonna, S. Ren, T. Rades. New perspectives on lipid and surfactant based drug delivery systems for oral delivery of poorly soluble drugs. *J Pharmacy Pharmacology*, **62**(11), 1622–1636 (2010).
39. K. Kataoka, A. Harada, Y. Nagasaki. Block copolymer micelles for drug delivery: design, characterization and biological significance. *Adv Drug Delivery Rev*, **47**(1), 113–131 (2001).
40. S. N. Bhattachar, L.A. Deschenes, J. A. Wesley. Solubility: it's not just for physical chemists. *Drug Discovery Today*, **11**(21–22), 1012–1018 (2006).
41. P. R. Bergethon. *The Physical Basis of Biochemistry: The Foundations of Molecular Biophysics.* First edition. New York: Springer Verlag (1998).
42. D. Chandler. Interfaces and the driving force of hydrophobic assembly. *Nature*, **437**(7059), 640–647 (2005).
43. K. A. Sharp, A. Nicholls, R. F. Fine, B. Honig. Reconciling the magnitude of the microscopic and macroscopic hydrophobic effects. *Science*, **252**(5002), 106–109 (1991).
44. D. Sitkoff, K. A. Sharp, B. Honig. Accurate calculation of hydration free-energies using macroscopic solvent models. *J Phys Chem*, **98**(7), 1978–1988 (1994).
45. H. Tjong, H. X. Zhou. Prediction of protein solubility from calculation of transfer free energy. *Biophys J*, **95**(6), 2601–2609 (2008).
46. Y. Shi, C. J. Wu, J. W. Ponder, P. Y. Ren. Multipole electrostatics in hydration free energy calculations. *J Computational Chem*, **32**(5), 967–977 (2011).
47. J. M. Wang, T. J. Hou. Recent advances on aqueous solubility prediction. *Combinatorial Chem & High Throughput Screening*, **14**(5), 328–338 (2011).
48. M. Manciu, E. Ruckenstein. The polarization model for hydration/double layer interactions: the role of the electrolyte ions. *Adv in Colloid Interface Science*, **112**(1–3), 109–128 (2004).

49. C. Azuara, H. Orland, M. Bon, P. Koehl, M. Delarue. Incorporating dipolar solvents with variable density in Poisson–Boltzmann electrostatics. *Biophys J*, **95**(12), 5587–5605 (2008).
50. K. A. T. Silverstein, A. D. J. Haymet, K. A. Dill. A simple model of water and the hydrophobic effect. *J Am Chem Soc*, **120**(13), 3166–3175 (1998).
51. S. Marcelja, N. Radic. Repulsion of interfaces due to boundary water. *Chem Phys Lett*, **42**(1), 129–130 (1976).
52. J. N. Israelachvili, H. Wennerstrom. Entropic forces between amphiphilic surfaces in liquids. *J Phys Chem*, **96**(2), 520–531 (1992).
53. E. S. Reiner, C. J. Radke. Variational approach to the electrostatic free-energy in charged colloidal suspensions: General theory for open systems. *J Chem Society, Faraday Transactions*, **86**(23), 3901–3912 (1990).
54. D. Andelman. Electrostatic properties of membranes: The Poisson–Boltzmann theory. In R. Lipowsky, E. Sackmann, (Eds.) *Structure and Dynamics of Membranes*, volume 1, second edition, section 12. Amsterdam: Elsevier, 603–642 (1995).
55. A. Kessel, B. Musafia, N. Ben-Tal. Continuum solvent model studies of the interactions of an anticonvulsant drug with a lipid bilayer. *Biophys J*, **80**(6), 2536–2545 (2001).
56. A. Ben-Shaul, I. Szleifer, W. M. Gelbart. Chain organization and thermodynamics in micelles and bilayers. i. theory. *J Chem Phys*, **83**, 3597–3611 (1985).
57. D. Harries, A. BenShaul. Conformational chain statistics in a model lipid bilayer: Comparison between mean field and Monte Carlo calculations. *J Chem Phys*, **106**(4), 1609–1619 (1997).
58. A. Zemel, A. Ben-Shaul, S. May. Membrane perturbation induced by interfacially adsorbed peptides. *Biophys J*, **86**(6), 3607–3619 (2004).
59. A. Fahr, P. Van Hoogevest, J. Kuntsche, M. L. S. Leigh. Lipophilic drug transfer between liposomal and biological membranes: What does it mean for parenteral and oral drug delivery? *J Liposome Research*, **16**(3), 281–301 (2006).
60. S. Balaz. Lipophilicity in trans-bilayer transport and subcellular pharmacokinetics. *Perspectives in Drug Discovery Design*, **19**(1), 157–177 (2000).
61. J. D. Jones, T. E. Thompson. Spontaneous phosphatidylcholine transfer by collision between vesicles at high lipid-concentration. *Biochemistry*, **28**(1), 129–134 (1989).
62. L. R. McLean, M. C. Phillips. Mechanism of cholesterol and phosphatidylcholine exchange or transfer between unilamellar vesicles. *Biochemistry*, **20**(10), 2893–2900 (1981).
63. S. Loew, A. Fahr, S. May. Modeling the release kinetics of poorly water-soluble drug molecules from liposomal nanocarriers. *J Drug Delivery*, **376548**(2011).
64. H. Hefesha, S. Loew, X. L. Liu, S. May, A. Fahr. Transfer mechanism of temoporfin between liposomal membranes. *J Controlled Release*, **150**(3), 279–286 (2011).
65. L. Masaro, X. X. Zhu. Physical models of diffusion for polymer solutions, gels and solids. *Progress in Polymer Science*, **24**(5), 731–775 (1999).
66. N. A. Peppas, A. R. Khare. Preparation, structure and diffusional behavior of hydrogels in controlled-release. *Adv Drug Delivery Rev*, **11**(1-2), 1–35 (1993).
67. S. Tenjarla. Microemulsions: An overview and pharmaceutical applications. *Critical Rev in Therapeutic Drug Carrier Systems*, **16**(5), 461–521 (1999).
68. V. Reshetov, D. Kachatkou, T. Shmigol, V. Zorin, M.A. D'Hallewin, F. Guillemin, L. Bezdetnaya. Redistribution of meta-tetra(hydroxyphenyl)chlorin (m-THPC) from

conventional and PEGylated liposomes to biological substrates. *Photochemical & Photobiological Sciences*, **10**(6), 911–919 (2011).
69. L. Zeng, L. An, X. Wu. Modeling drug-carrier interaction in the drug release from nanocarriers. *J Drug Delivery*, **370308**(2011).
70. C. Washington. Drug release from microparticulate systems (microencapsulations methods and industrial applications). In S. Benita, (Ed.) *Drugs and the Pharmaceutical Sciences 73*. New York: Marcel Dekker, 155–181 (1996).

2

Liposomes as Intravenous Solubilizers for Poorly Water-Soluble Drugs

Peter van Hoogevest, Mathew Leigh and Alfred Fahr

2.1 Introduction

Liposomes were discovered by Alec Bangham in 1961 by dispersing phospholipids in aqueous media and making the microscopic observation that cell-like structures were formed [1–3]. Initially, they were mainly studied as model systems for biological membranes, because the basic phospholipid bilayer structure of phospholipid molecules was quite similar to the phospholipid bilayer structure of biological membranes.

Based on the number of membranes per particle and their average size, liposomes are classified as Small Unilamellar Vesicles (SUVs) (diameter size range ca. 15–100 nm), Large Unilamellar Vesicles (LUVs) (size range ca. 100 nm–1000 nm), and Multilamellar Vesicles (MLVs) (ca. 1–5 μm).

In the following decade liposomes were also increasingly considered for drug delivery and, more specifically, for drug targeting. Due to their physico-chemical composition, liposomes can carry water-soluble drugs in their interior water phase, lipophilic drugs in the fatty acid domain of the liposomal membrane and amphiphilic drugs at the membrane water interphase (Figure 2.1).

Originally, there were hopes that liposomes could act as intravenous drug carriers ('targeted missiles') which due to addition of specific ligands (e.g. antibodies) to the liposomal membrane would specifically interact with very specific disease targets (e.g. antigens on the surface of tumour cells) and discharge their cargo at the site of action, thus yielding highly efficacious therapies with reduced systemic toxicity. The emphasis was on the exploration of combinations of water soluble cytostatics which have a narrow therapeutic index and liposomes for improving cancer treatment.

Drug Delivery Strategies for Poorly Water-Soluble Drugs, First Edition. Edited by Dennis Douroumis and Alfred Fahr.

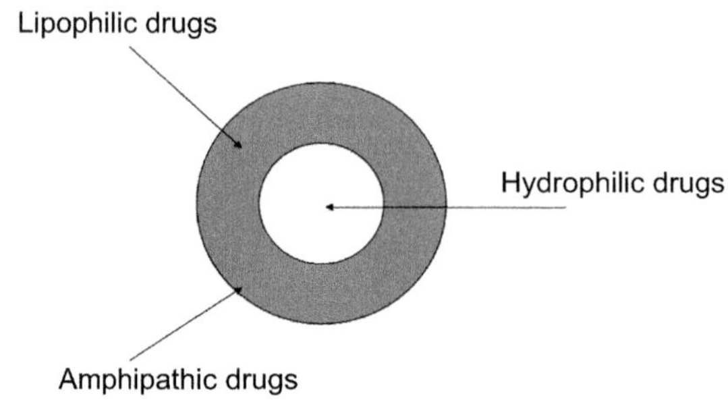

Figure 2.1 *Incorporation of drugs in liposomes (cross-section).*

It appeared that this concept was not easy to realize for various biological reasons: the liposomes could not reach the tumour target because they were taken up by macrophages situated in the liver and the spleen (Reticuloendothelial System (RES) or Mononuclear Phagocytotic System (MPS)), the existence of tumour heterogeneity, the lack of stability in the blood circulation, giving rise to leakage of the entrapped water-soluble compound. In addition, the liposomes themselves were not stable after long-term storage, the phospholipids used had variable quality and the characterization and quality control of the liposomes were relatively demanding, and large-scale production methods were challenging.

After this initial disappointment, in particular, small pharmaceutical companies and dedicated university institutes did overcome all the hurdles related to the use of liposomes in the field of drug delivery. By the addition of pegylated phospholipids, liposomes could escape from the MPS system and reach the tumour tissue [4]; the disadvantage of uptake by MPS cells could be turned into an advantage by selectively activating macrophages using MTP-PE (muramytripeptide-phosphatidylethanolamine) liposomes to become tumouricidal [5–8]; the stability in the blood circulation could be greatly enhanced by the use of cholesterol in combination with higher melting (gel state) phospholipids [9]. Even the original idea of a targeted liposome by binding antibodies to the liposomes to make immunoliposomes is still being dynamically pursued [10, 11].

Also on the technical front, breathtaking progress has been made. Nowadays, parenteral quality phospholipids, which are adequately characterized, are available from natural or synthetic sources; long-term storage can be achieved by freeze-drying in the presence of disaccharides [12] or the selection of optimal pH conditions of aqueous liposomes dispersions [13–17]; the availability of e.g. photon correlation spectroscopy [18] or cryo Transmission Electron Microscopy (cryoTEM) [18] to assess the particle size of the liposomes, phospholipid-specific HPLC methods [19], and TLC [20], high field P-NMR [21], etc. to assess the quality of phospholipids and liposomes makes a robust QC feasible; large-scale sterile production using lyophilization from solvent [22], dialysis methods [23], and high shear homogenizers are nowadays possible [24, 25].

The evidence that indeed these problems have been overcome to the satisfaction of scrutinizing regulatory authorities is the existence of at least 15 approved (either in the EU or/and the USA) parenteral liposome products on the market (Table 2.1).

The list of products in Table 2.1 demonstrates the usefulness of liposomes for various drug delivery concepts and that they are being used to create a superior line extension or

Table 2.1 *Marketed parenteral liposomal products status, 2011.*

Product	Drug substance
Abelcet*	Amphotericin B
AmbiSome*	Amphotericin B
DaunoXome	Daunorubicin citrate
Definity*	Octafluoropropane
DepoCyte	Cytarabine
Depodure	Morphine HCl
Doxil/Caelyx	Doxorubicin
EndoTAG-1*	Paclitaxel
Epaxal® (Hep A) and Inflexal® (Influenza) Vaccines	Antigens
Mepact*	MTP-PE
Myocet	Doxorubicin
Octocog alpha	Recombinant human factor VIII
Regulon	Cisplatin
Sonovue® (SFl6) ('micro-bubbles')	Contrast agent
Visudyne*	Benzoporphyrin

Note: *use liposomes as solubilizer.

to formulate New Chemical Entities (NCEs). Analyses of the originators of the products also clearly show that major pharmaceutical companies (with one exception: Ciba-Geigy for muramyltripeptide-phosphatidylethanolamine (MTP-PE) liposomes [22]) are currently not involved in the initial research and development of these products.

Six of the presented products use liposomes as solubilizer for poorly water-soluble drugs, with aqueous solubility lower than ca. 10–100 μg/ml. In the literature, the lipid formulations for amphotericin B, MTP-PE and Visudyne (benzoprophyrin) are best described. Amphotericin B is in the original Fungizone product formulated with sodium desoxycholate to solubilize the water-insoluble polyene antibiotic amphotericin B. The liposomal formulation and lipid complex formulation (AmbiSome and Abelcet, respectively) of amphotericin B avoid the use of the haemolytic solubilizer desoxycholate and use liposomes or lipid complex to solubilize the drug and to achieve a higher therapeutic index compared to the desoxycholate formulation [26, 27].

MTP-PE (muramyltripeptide-phosphatidylethanolamine) is an intravenous immunomodulator designed to activate macrophages to become tumouricidal [28]. MTP-PE on its own forms micelles and shows an increased intravenous toxicity compared to MTP-PE incorporated in multilamellar liposomes. The linking of the phosphatidylethanolamine (PE) to the muramyltripeptide ensures the association/solubilization of the drug molecule with the liposomal membrane. Large liposomes comprising POPC (1-palmitoyl, 2-oleoyl phosphatidylcholine) and DOPS (1,2 dioleoyl phosphatidylserine) are preferred as carriers because of the fast uptake by macrophages.

Benzoporphyrin (Verteporfin) is a photosensitizing drug for photodynamic therapy (PDT) activated by low-intensity, non-heat-generating light of 689 nm wavelength. Activation generates cytotoxic free oxygen radicals. The specificity and uptake of benzoporphyrin for target cells with a high expression of low density lipoprotein (LDL) receptors, such as tumour and neovascular endothelial cells, are enhanced by the use of a liposome, which acts as a solubilizer for the compound, and as donor reservoir for rapid transfer and uptake by plasma LDL. Verteporfin therapy (at light doses < 150 J/cm) selectively damages neovascular endothelial cells leading to thrombus formation and specific occlusion of choroidal neovascular vessels in subfoveal lesions in patients with age-related macular degeneration (AMD) [29–33].

Since there is an increasing number of NCEs coming from Drug Discovery which are poorly water-soluble, and the number of formulation approaches to solubilize drugs to enable intravenous administration are limited, it is of interest to compare in more detail in the following section of this chapter the pros and cons of other formulation options, compared to liposomes as an intravenous solubilizer.

2.2 Intravenous Administration of Poorly Water-Soluble Compounds (PWSC)

At the preclinical stage, intravenous testing of the efficacy, distribution, metabolism and elimination and toxicity of NCEs in comparison with oral testing of the NCEs is mandatory. The initial profiling of NCEs through the intravenous route eliminates the need for extensive oral formulation screens to obtain formulations with adequate oral absorption of the NCEs. It avoids the necessity for unnecessary extensive oral formulation screen efforts being invested in compounds before it is even known whether the compounds are biologically efficacious and/or toxic. In addition, since the uncertainty about the degree of oral absorption is eliminated, a straightforward head-to-head comparison of the therapeutic index and level of efficacy of series of compounds on a mg/kg basis can be made. Finally, the intravenous testing also allows the assessment of the degree of oral absorption by comparison of the area under the curve (AUC) after intravenous administration compared to the AUC after oral testing.

If the NCEs are intended to indicate diseases, which is exclusively linked to oral administration, then oral formulation screens have to be performed on the NCE candidates with the best therapeutic index, after the intravenous screen. For the cancer indication, parenteral screening of NCEs is the first choice [34].

A further preference for first screening and profiling compounds using the intravenous route is related to the relatively simple formulation vehicles (which is basically restricted to vehicles wherein the NCEs is solubilized) which can be used compared to the oral route. Besides the solubilized physical form of the compound, oral administration can also be performed with the formulations containing the solid crystalline and solid amorphous form of the NCEs. The use of solid compounds requires measurement of the particle size and control of the crystal form and knowledge of the tendency for polymorphism.

In the case of pH-sensitive compounds, which may have the tendency to form a salt, also the salt (the counter ion) needs to be selected and the tendency for polymorphism of the salt needs to be checked before the salt can be selected as the solid form for oral administration. As derived from these complications, it is easy to see that the attempts to find an adequate oral vehicle, which is needed to assess the prospects of obtaining a reasonable degree of oral absorption, are quite considerable, compared to finding a vehicle for intravenous administration.

Although parenteral administration is mainly restricted to formulations wherein the compound is solubilized and these vehicle options seemingly appear quite simple, there are still some pitfalls and drawbacks which the formulation scientist, when designing the vehicle, and the biologist who is using the vehicle, should be aware of.

In general, poorly water-soluble compounds can be solubilized for intravenous administration using: (1) water miscible solvents; (2) pH adjustment; (3) surfactants; (4) cyclodextrins; (5) mixed micelles; (6) oil-in-water emulsions; and (7) liposomes. As pointed out in a recent review [35], every formulation option has its pros and cons and a parallel screen to select the best option rather than a sequential screen (giving higher preference to seemingly simple, but potentially toxic formulation options such as organic solvent water mixtures and pH adjustment) should be considered. The intravenous formulation options for PWSC can be divided into two subgroups. The first is water miscible/dilutable and loses its solubilizing capacity relatively rapidly upon dilution/injection into the blood circulation; to this group belong water-miscible solvents, pH adjustment, surfactants and cyclodextrins; the second group does not immediately lose its solubilizing capacity upon dilution/injection into the blood circulation because the macroscopic structure of the solubilizing entity does not dissolve/change immediately; to this group belong oil-in-water emulsions and liposomes. Mixed micelles comprising glycocholate and lecithin [35] may convert upon dilution/injection into the blood circulation immediately into liposomes and the resulting liposomes may act as a more consistent solubilizing vehicle. Therefore, the mixed micelles may belong to both groups.

The final selection of the composition of the vehicles is determined by the purpose of use. Higher concentrations of potentially toxic excipients or deviations from physiological conditions (pH, tonicity) will be more acceptable for one-off bolus injections compared to chronic repetitive administration of infusions. Still, for proper administration of the compounds, precipitation upon dilution in the blood circulation should be avoided at all times. Lack of knowledge of the degree of precipitation after intravenous administration is a poor research and development practice. When the development goal is an injectable formulation for clinical research, such deficits are unacceptable.

2.2.1 Solubilizing Vehicles with Precipitation Risk upon Dilution

2.2.1.1 Co-Solvents

The use of solvent/water mixtures is limited to relatively low concentrations of a few solvents like PEG 300 and PEG 400, propylene glycol and ethanol [35, 37]. Considering the high water content in these formulations, it is questionable whether a meaningful concentration (e.g. to enable toxicity testing of poorly water-soluble compound with a water solubility lower than 100 μg/ml) can be dissolved in these mixtures. The (old) drugs in products on the market using this approach still have a reasonable degree of water solubility which reduces the risk of precipitation upon dilution in the blood circulation. In addition to precipitation risks, the toxicity of the used solvents and influence of the excipients on the metabolism of the compounds (as part of pharmacokinetic studies) may be a reason for concern [37].

2.2.1.2 pH Adjustment

When PWSCs have deprotonatable or protonatable groups in their molecular structure, such compounds can be dissolved by adjusting the pH. Neutral PWSCs are excluded from this approach. The use of a non-physiological pH may give rise to distortion of the blood buffer and cause phlebitis and pain at the injection site. Just as with the co-solvent approach, there is a risk of precipitation of the PWSC in the blood circulation, thus falsifying the PK results and causing a risk of embolism. It is claimed in the literature that at Pfizer the majority of NCEs can be injected by means of pH adjustment [38], This may be true, but this also indicates that careful *in vitro* assessments of precipitation tendencies, risk of phlebitis and assessment of experience of pain (by patients) need to be performed before this formulation approach can be considered for further development. The authors of the publication did not mention the percentage of drugs still using pH adjustment, which enter further development stages. In addition, there is no mention of whether the formulations were suitable for repetitive dosing as part of e.g. chronic tox testing in animals. Also, information on the solubility of the compound at neutral pH is missing. Finally, too acidic and alkaline pH may have a detrimental influence on the (long-term) stability of the compound. When for the reasons mentioned above a reformulation has to be made, the preclinical and clinical assessment may have to be partially or completely repeated.

2.2.1.3 Surfactants

The use of surfactants like Tween 80 (polysorbate 80), Cremophor El (macrogolglycerol ricinoleate, polyoxyl 35 castor oil), and Solutol HS-15 (polyethyleneglycol 660-12-hydroxystearate) may also be considered [35,37]. Cremophor EL still enjoys some popularity in solubilizing compounds, in spite of its potential to cause anaphylactic reactions (like Tween 80 and Solutol HS-15 to a lower extent). The serious drawbacks of the intravenous use of Cremophor have been reviewed extensively [34,39,40].

It was pointed out that the introduction of paclitaxel on the market as a cancer drug (Taxol) was delayed for ten years because of the absence of a suitable formulation. The developed solubilizing vehicles using Cremophor appear to have side effects which are quite serious: anaphylaxis and severe hypersensitivity reactions characterized by dyspnea and hypotension requiring treatment, angioedema, and generalized urticaria have occurred in 2–4% of patients receiving Taxol in clinical trials. Fatal reactions have occurred in patients despite premedication. All patients should be pre-treated with corticosteroids, diphenhydramine, and H2 antagonists [41]. A further issue is that Cremophor may persist at high doses as micelles in the blood circulation after intravenous administration [39]. Under these circumstances Cremophor may influence pharmacokinetics and the pharmcodynamics of drugs. For these reasons, high dose surfactants, which do not readily disperse/dissociate in the blood, may also be classified in the group of solubilizers which do not lose their solubilization capacity upon dilution. The authors consider the Cremophor vehicles to be hardly sufficient and claim that there is a serious need to optimize the formulation for poorly water-soluble cancer drugs [39].

In spite of this serious criticism by oncologists, the pharmaceutical industry continues to use this vehicle. Recently an epothilone type of anti-cancer drug of BMS (Ixempra) has been approved using Cremophor EL as solubilizer [42]. Typically, antihistamines and steroids have to be co-administered to eliminate the risk of anaphylactic reactions. Since the interaction of the drug with antihistamines and steroids was not specifically investigated, it may be assumed that throughout the whole preclinical and clinical development these drugs have been co-administered. The formulation is further characterized by a limited stability. After constituting Ixempra (ixabepilone), the constituted solution should be further diluted with infusion fluid as soon as possible, but may be stored in the vial (not the syringe) for a maximum of 1 hour at room temperature and room light. Once diluted with infusion fluid, the solution is stable at room temperature and room light for a maximum of 6 hours. Administration of diluted IXEMPRA (ixabepilone) must be completed within this 6-hour period. The infusion solution must be administered through an appropriate in-line filter with a microporous membrane of 0.2 to 1.2 μm. DEHP-free infusion containers, and administration sets must be used.

The drug product is indicated for patients for whom Taxol treatment fails. Simultaneously, patients who are allergic to Cremophor in the Taxol formulation are excluded. This means that because of the use of Cremophor, patients are unable to receive the best therapy for their disease. This is a rather unethical situation which warrants the reformulation of the epitholone product and the replacement of the Cremophor vehicle by other possibly better vehicles.

It can be assumed that regulatory authorities have approved the drug product because of its clinical efficacy but certainly not because of its robust formulation, which could be the subject of considerable improvement. An approach, as suggested in the patent literature [43, 44], to use liposomes for epothilones may be a viable alternative to serve as an alternative solubilizer and to possibly increase the therapeutic index of the epothilones.

Solutol HS 15 will not be an attractive alternative to the formulating NCEs of companies striving for world-wide development as long as this excipient is not being used in a marketed product in the USA and the US FDA will consider this surfactant, consequently, as a novel excipient in need of an extensive tox data base.

2.2.1.4 Cyclodextrins

At present, two cyclodextrins are accepted for human intravenous use: Hydroxypropyl-beta-cyclodextrin and sulfobutyl-beta-cyclodextrin sodium salt [45].

The toxicity of these cyclodextrines is well documented [46]. However, only a few intravenous products are on the market with these cyclodextrins [35]. The commercial use of these excipients is complicated by the existence of a few patents covering the use of these cyclodexrins in formulations [35]. One-to-one molar complexes of cyclodextrin and drugs are very rare. In practice, compounds may require a large excess of cyclodextrin compared to the drug to solubilize the drug. A corresponding weak binding constant may result after dilution and the corresponding dissociation of the complex in precipitation of the drug [35]. Also in the kidney where the cyclodextrins are excreted, a PWSC may precipitate because of the reduced concentration of the solubilizer in the kidney.

2.2.2 Solubilizing Vehicles Maintaining Solubilization Capacity upon Dilution

2.2.2.1 Oil-in-Water Emulsions

Oil-in-water emulsions may be used to solubilize poorly water-soluble compounds [37, 47]. In contrast to the other category of solubilizers, the oil particles have some stability in the blood circulation and the drug dissolved in the oil phase may be transferred from the oil droplets to other lipid domains in the blood circulation [47]. There will be no immediate risk of precipitation. Intrinsic to this approach, intact oil particles loaded with the drug may be taken up by the macrophages. The lack of toxicity of oil-in-water emulsions is underscored by the use of these emulsions in large volumes as vehicles for parenteral nutrition.

This approach is restricted to compounds which have a reasonably high solubility in the oil phase of the emulsions to achieve therapeutic or (in case of preclinical safety assessments) toxic levels. Unfortunately, it appears that many research NCEs do not have a high solubility in oils (triglycerides). The drug containing an oil-in-water emulsion must be stable in liquid form and also be sterilizable preferably by rotating autoclaving. The presence of the drug at high concentrations may influence the physical stability of the emulsion. An *in situ* loading of oil-in-water emulsions with the drug may be possible [48] but due to the milky appearance, visual monitoring of the absence of drug crystals and successful loading of the oil droplets with the PWSC are not possible.

2.2.2.2 *Mixed Micelles*

Mixed micelles comprising sodium glycocholate and soy bean lecithin are being used to solubilize vitamin K and other lipophilic vitamins for intravenous use [49]. The mixed micelles possess a low parenteral toxicity in animal studies [50]. It is not very well documented whether and to what degree liposomes may be formed upon dilution/injection into the blood circulation. Hoffmann-La Roche had a mixed micellar formulation on the market for diazepam to replace the organic solvent-based formulation. Later, the mixed micellar formulation was replaced by the original formulation.

2.2.2.3 *Liposomes*

As mentioned in the Introduction, liposomes are being used as solubilizer for poorly water-soluble compounds in marketed products. Depending on the stability properties of the associated drug, marketed liposomal products are supplied in liquid or freeze-dried form. Liposomes belong to the class of solubilizers which do not show immediately (i.e. within one minute) upon dilution/injection a change of morphology or disintegration, thus maintaining an intact lipidic domain of the liposomal membrane and the solubilization capacity of the liposomes. The associated compound may transfer (with a kinetic specific for a compound and composition of the membrane) to other lipid domains/components in the blood circulation (e.g. albumin, lipoproteins and red blood cell membranes) [50].This is explained in detail below.

The standard laboratory preparation of liposomes, starting from the selection of the phospholipids, making a mixed film of phospholipids and the drug by organic solvent evaporation, drying the film in vacuum, dispersing the film, applying energy to downsize the liposomes (extrusion or high pressure homogenization), sterile filtration [24], followed by characterizing the resulting liposomes, all requires rather sophisticated equipment (e.g. photon correlation spectroscopy) and is certainly not within the scope of the laboratory capabilities of a biology-orientated scientist interested in formulating poorly water-soluble drugs for preclinical experiments. Instead of this, simple formulations such as aqueous solutions of surfactants, co-solvents, cyclodextrins, etc. [35, 37], to which the drug can be added as a single preparation step and where the degree of solubilization of the drug can be visually inspected, are preferred. It is remarkable to note that in practice the clear solutions obtained are considered to be perfectly suitable for intravenous administration without checking for any precipitation tendency after dilution in the blood circulation, by performing serial dilution experiments with e.g. buffer, serum or plasma.

The advent of instantaneous solubilization using a liposome solubilizer [52] comprising a prefabricated liposome dispersion which is as clear as solutions, allowing the same visual inspection to check for adequate solubilization, places the liposomes in the same simple, single-step preparation category as the other solubilizers. In addition, since, upon intravenous administration, the liposomes do not disintegrate, the solubilizing capacity is maintained and precipitation of the liposomally associated drug is prevented. The characteristics of the liposome solubilizer and release mechanisms from the liposome are described in more detail below.

2.2.3 Mechanistic Release Aspects/Transfer Liposomal PWSC

It is a paradigm in pharmaceutical sciences that a drug should be unbound (i.e. a so-called 'free drug') to any other component before it can unfold its action by interaction with the target receptor. This way of thinking was stimulated by findings of some 25 to 30 years ago that relatively water-soluble sulphonamides and antibiotics, when strongly bound to albumin through electrostatic and lipophilic interactions, lost their activity [53].

After the advent of poorly water-soluble compounds, for which the only possibility of being transported by the blood is by binding to lipophilic domains or particles (blood cells, lipoproteins, albumin [51, 54], the issue of binding possibly influencing efficacy is of even more interest. Since these compounds need to be administered with the aid of a solubilizer, the binding of the drug to the solubilizer and the related equilibrium binding constant versus the equilibrium binding constant of the drug and the lipid blood domain to which the drug will bind, will determine which fraction of the administered dose will be bound to the solubilizer, the lipid domains in the blood and eventually the target receptor.

These fractions can be assessed at a true equilibrium situation. In nature, there is, however, not an equilibrium situation, but a continuously changing highly dynamic environment. When liposomes loaded with drug are injected, the first process which will take place is a release of the compound from the liposomes, which is transferred/distributed to other lipid domains to reach the equilibrium. In other words, also the kinetics at which the equilibrium will be reached is of paramount importance to assess whether 'binding' indeed means a loss of activity or not, and to what extent the drug is still liposomally bound.

From a regulatory perspective, the release of the drug from the liposome (and any other intravenous formulation type (including micelles) which has some affinity for the drug) is a chemical manufacturing and control (CMC) aspect. The further distribution from the receiving (acceptor) lipid domains in the blood to organs, receptors, etc. is a biological property of the drug. The release should be reproducible from batch to batch during the assigned shelf life. In order to prove this, the formulation should be adequately characterized and an *in vitro* drug release method, which is able to monitor a change of (key) characteristics of the formulation essential for determining the release rate is recommended.

Using poorly water-soluble drugs-associated liposomes with aqueous solubilities lower than 1 μg/ml and binding for 98% to lipoproteins and/or albumin and/or red blood cells, it is clearly proven that neither the binding to liposomes nor the binding to blood proteins prevent these drugs from becoming active. Examples are amphotericin B, benzoporphyrin and paclitaxel [31, 55, 56].

This may be explained by the fact that these drugs, although they bind to these domains strongly (under equilibrium conditions), they are not irreversibly bound, but can shuttle between lipid domains and in this way distribute along a concentration gradient to all available lipid domains in the blood and are efficiently transported by these domains to their site of action.

2.2.3.1 *Transfer (Release) Kinetics*

In Vitro The transfer (release) kinetics of lipophilic drugs from liposomes to other lipidic domains can be assessed *in vitro*. These assays comprise two populations of liposomes: a donor liposome population containing the drug; and an acceptor liposome population which is free of the drug. The acceptor liposome population serves as model for lipid domains in the blood. Upon mixing the two populations, the populations need to be separated at predetermined time points and the increasing drug content or decreasing drug content in acceptor or donor population, respectively, is determined by a drug-specific analytical method. The separation can be achieved by using, for example, ion exchange chromatography columns and neutral and negatively charged liposomes as donor/acceptor pair [54], analogous to the assay proposed by McLean and Phillips in 1981. This consists of the insertion of the drug into negatively charged liposomes using standard techniques. Neutral liposomes (mostly PC liposomes) are used as an acceptor medium at an excess of 10 by weight in relation to the donor liposomes. After mixing the two liposome populations together, samples were processed at defined time points over an ion-exchange column, which allows only the neutral (acceptor) liposomes to be eluted [54]. Centrifugation and the use of SUV and MLV populations which differ largely in centrifugation kinetics may also be used as part of an interliposomal transfer assay [57]. The possible cross-contamination of the two populations can be corrected by including non-exchangeable markers.

The *in vitro* measurement of drug transfer rates between liposomes and e.g. liposomes and lipoproteins, albumin or red blood cells is only possible when there is sufficient difference in e.g. either chromatographic or centrifugation characteristics of the two populations. As an alternative to chromatographic and centrifugation techniques, equilibrium dialysis may be considered to monitor the release of drug from the liposomes as well. Possibly, analytical methods like field flow fractionation may be useful as well.

If the drug possesses fluorescent properties, the transfer between two liposome populations may be followed without separation, when the compound could be present in the donor population at high concentrations which gives rise to quenching of the fluorescence signal and dequenching upon dilution. Upon addition of an acceptor liposome population, the fluorescent compound distributes to the acceptor population and the fluorescence signal of the mixtures increases. This principle was applied to the fluorescent drug benzoporphyrin [58]. Transfer kinetics between liposomes containing a fluorescent drug at dequenching concentration and plasma or serum is also possible. The quantification of the change of fluorescence signal does, however, requires knowledge of the quantum yield of the fluorescent compound when bound to the various lipid domains and, without this knowledge, is very difficult to quantify accurately.

When the concentration of the drug is not high enough to cause quenching, a change of the fluorescence signal when the drug moves from the liposome to e.g. albumin may be an option to monitor the transfer [59]. Another option which may not require separation of the donor liposomes and released compound, is an immunoassay, which uses antibodies which can discriminate between the liposomal and released compound. The assay can then also be used to assess the ratio of released and bound drug in blood samples [60].

The transfer properties of a compound are expressed in $t_{1/2}$, which is the time period needed to achieve 50% of the equilibrium level when the drug is equilibrated over the donor and acceptor population. Another way to express the degree of transfer is to plot the mass transfer of the liposomally administered drug dose to the acceptor domains. This may be more relevant to the *in vivo* situation because a 1:1 to 1:10 weight lipid ratio of donor to acceptor arbitrarily used in *in vitro* assays may be dependent on the lipid dose, 1:100 or more *in vivo*. This means that, at a 100-fold excess of acceptor, in spite of, for instance, a long transfer $t_{1/2}$ of e.g. 30 min, within 3 min, x% of the administered dose can be transferred to the acceptor domains.

Cyclosporin A Cyclosporin A (CyA) was investigated thoroughly for encapsulation in liposomes. Several studies were concerned with the binding and release of CyA to liposomal membranes and the pharmacokinetic impact [54, 61, 62].

When using the assay with ion exchange chromatography and analysis of the amount of drug in the acceptor liposomes by HPLC, or most easily by radioactive tracers, which were available in the case of CyA, the interliposomal transfer rate of CyA could be determined (Figure 2.2).

It can easily be seen that despite its high lipophilicity, CyA exchanges between its binding sites with a half-time of ca. 4.5 min. In agreement with previous studies on the spontaneous cholesterol transfer between liposomal membranes [63, 64], cholesterol exchanges more slowly compared with CyA. As a control for the assay, the radioactive lipid marker, cholesteryloleoylether exchanges even more slowly, as one would expect from a lipid marker, which is claimed to be non-exchangeable. A parallel experiment using radiolabelled lecithin showed that under the experimental conditions used, the lecithin did not co-migrate with CyA.

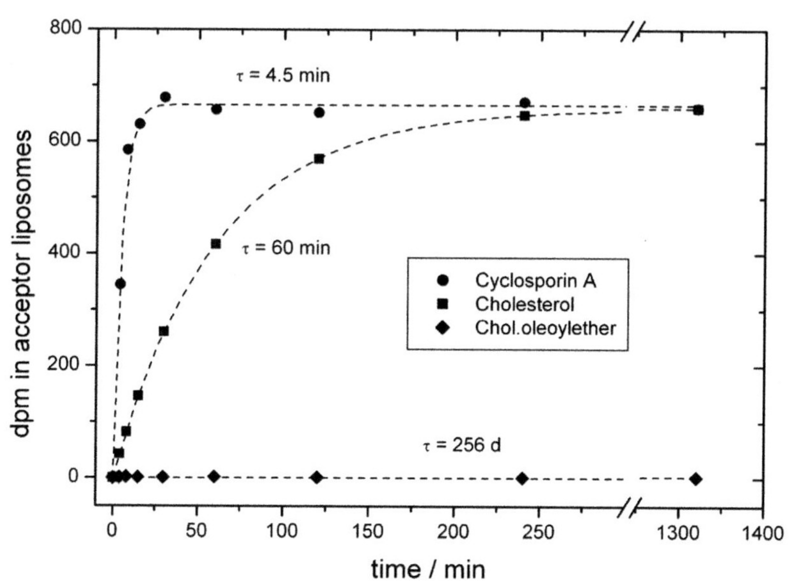

Figure 2.2 *Interliposomal transfer rate of CyA.*

These *in vitro* experiments were supported by pharmacokinetic experiments, which showed that liposomal CyA intravenous pharmacokinetics are (at the studied lipid and Cremophor doses) not significantly different from a classical detergent-based (Cremophor®) formulation [61].

An exception was found for very high dosages of liposomes, which may act as a competing binding reservoir for CyA with physiological binding places in the blood (lipoproteins, erythrocytes, etc.). Alternatively, the cells of the MPS in the liver and the spleen may be saturated and the elimination of the liposomes (carrying the CyA) from the circulation by these cells is delayed.

Paclitaxel Paclitaxel is a natural product with antitumour activity. Natural paclitaxel is extracted from *Taxus chinensis* and purified. Liposomal formulations are described to lower toxicity and to change body distribution [65]. Paclitaxel transfer studies were also performed, similar to the one described for CyA above [66]. The findings were also similar to those using CyA. In a time frame of minutes, paclitaxel seems to exchange between membrane binding sites.

Amphotericin B Amphotericin B is known to transfer rapidly (i.e. within seconds) between membranes [67, 68]. The transfer kinetics depend on the physical state (i.e., gel state or liquid crystalline state) of the donor membrane and the acceptor membrane. Transfer is rapid (within seconds) between the gel state, DPPC-containing liposomes and also between liquid crystalline, egg-PC-liposomes, however, no transfer was observed at room temperature from DPPC liposomes to egg-PC liposomes. However, when the temperature was increased to above the phase transition temperature of DPPC to 48°C (at which the DPPC is in the liquid crystalline state), transfer occurred rapidly.

The role of liposome composition (i.e. the influence of the presence of negatively charged phospholipids like DMPG) and temperature in the distribution of amphotericin B with serum lipoproteins were determined [69]. The authors conclude that lipid charge and temperature play a role in amphotericin B distribution into serum lipoproteins. Furthermore, amphotericin B and DMPG may co-transfer as an intact drug-lipid complex to serum lipoproteins. In addition, phospholipid transfer proteins may play a role in (*in vivo*) plasma distribution of amphotericin B [70].

In vivo Any *in vitro* result needs to be scrutinized for its relevance for the *in vivo* situation. The release or transfer of a lipophilic compound from a liposome to lipidic domains present in blood, because of its complexity, is difficult to assess in detail.

When isotopes are being used in animal experiments, the ratio of e.g. ^{14}C labelled drug to ^{3}H labelled liposomes may give some information. In clinical research, such radiolabelled compounds cannot be used. In the literature only one example is described where an immunoassay was used to discriminate between free MTP-PE and liposomal MTP-PE [60].

When such assays are not available, density gradient centrifugation [70] or gel electrophoresis [71,72] have been used to assess the degree of drug release from the liposomes to lipoproteins. The time period between sampling and getting assay results is rather long, which means that no time course can be made. However, when a liposomal drug is infused over very long periods and equilibrium levels are reached, density centrifugation to find out the distribution of the drug within the blood pool may very well be meaningful [70].

Because of these analytical difficulties, the *in vivo* transfer characteristics of liposomal drugs can, most of the time, only be indirectly studied. As part of Phase I clinical research, measurement of the total drug (free plus liposomal) is common, because of the absence of a more sophisticated assay.

How far the observed *in vitro* half-times for e.g. cyclosporine and paclitaxel truly reflect the transfer rates *in vivo* situation still needs to be proven. An indication that *in vitro* half-times of minutes may be *in vivo* only seconds was obtained from experiments with benzoporphyrin liposomes. When using an interliposomal transfer assay with SUV – MLV at a 1.1 lipid weight ratio of donor and acceptor assay, a $t_{1/2}$ of about 60 min was observed (pers. comm. H.J. Schwebel). In contrast, when the benzoporphyrin liposomes were mixed with plasma, an immediate change from quenched benzoporphyrin to 80% of the unquenched fluorescence signal in Triton-X100 (the benzoporphyrin is then completely mono-molecularly dissolved) was observed [58].

An explanation for this discrepancy may be that the *in vivo* situation is quite different compared to an interliposomal transfer assay condition in many aspects: the ratio of donor to acceptor will be different dependent on which acceptor is considered; the MLV (diameter of 1 μm) as acceptor possesses a surface area which is many thousand-fold lower than, for instance, lipoproteins (diameter ca 50 nm), the affinity of the drug to lipoproteins may be higher compared to liposomes; when the drug transfers to albumin, albumin is also a particle with much lower diameter compared to the acceptor MLV; in addition, albumin may bind to the liposome surface (leakage of sucrose from phosphatidylcholine liposomes induced by interaction with serum albumin [73], so encouraging the drug transfer; finally blood comprises phospholipid exchange proteins which may play a role in the plasma distribution and accelerate the transfer process [74–80]. Further research in this area is required to elucidate the *in vivo* transfer rates in more detail and to bridge *in vitro* and *in vivo* results.

2.2.3.2 Thermodynamic Perspective

The lipophilicity of a drug determines the partition equilibrium between an aqueous and an oily phase. The more lipophilic the drug is, the further the partition equilibrium is shifted to the oily phase. The oily phase can be an alkane phase or – in order to better represent the amphipathic nature of a lipid bilayer – consist of an alcohol such as octanol. Regarding its lipophilicity, the lipid chain region of a lipid bilayer is comparable to an oily phase. There is, however, a significant difference: The structural properties of the oil phase are uniform, isotropic, and homogeneous. In contrast, a lipid bilayer is a thin, self-assembled film with strongly non-homogeneous properties. The lipid chains within a membrane – despite being in the fluid-like state – are orientated in an ordered manner and thus render the bilayer anisotropic. The packing properties of the hydrocarbon chains in a lipid bilayer are significantly different from those in an alkane phase. The free energy of the transfer process related to this difference in packing properties is described in Equation 2.1:

$$\Delta F = \Delta F_{\text{sol}} + \Delta F_{\text{lip}} \tag{2.1}$$

ΔF is the difference of the free energy of a drug when it transfers from the water phase into the lipid bilayer or with the opposite sign for ΔF, when it transfers from lipid layer to the water phase. The equation is composed of two major contributions (for a discussion of other relevant contributions, see [81]; and for a detailed discussion of the so-called cratic entropy, see [82]. The first contribution is the solvation free energy, ΔF_{sol}, which accounts for changes in electrostatic and hydrophobic interactions of membrane-drug associations and constitutes the classical hydrophobic effect. The second contribution, ΔF_{lip}, is characteristic of a lipid bilayer and arises from the drug-induced structural membrane perturbation. The presence of the contribution ΔF_{lip} is a consequence of the above-mentioned anisotropic orientation of the lipid chains within the lipid bilayer.

Based on these theoretical thermodynamic considerations, in spite of high solubility in an (isotropic) oily phase, it may be speculated that drugs may be repelled/excluded from the (anisotropic) lipid bilayer, because of packing defects caused by the incorporation of the drug into the fatty acid chain region. The magnitude of this tendency is governed by the individual properties of the drug molecule such as its size, shape, orientation, and hydrophobic moment. As a result, an equilibrium exists between the drug dissolved in the membrane and the drug dissolved (to a much lower extent) in the water phase. Upon addition of an acceptor membrane, a flux of the drug, through the water phase, from the donor to the acceptor membrane will be initiated. Alternatively, when the transfer occurs through the collision of vesicles (see below), the drug moves directly along its concentration gradient from the donor lipid domain to the acceptor lipid domain (possibly through an aqueous boundary layer) at the moment of collision.

We shall illustrate the influence of the characteristic packing properties of the hydrocarbon chains in a lipid bilayer on the location and orientation of a drug in a membrane using a simple generic model. In this model we represent a rigid, bulky, entirely lipophilic drug as a stiff cylinder. We choose the cylinder length, $L \approx 1.6$ nm, to correspond roughly to the length of a lipid tail, and its radius, $R \approx 0.6$ nm, to account for the volume of a typical drug. CyA is an example whose spatial dimensions roughly fit with our choice of L and R [62]. In the following example, it is argued that our simple lipophilic model drug is interacting with the membrane in a way that does not agree with the simplified assumption of the drug being buried deeply within the membrane hydrocarbon core and being safely shielded by the lipid headgroups from any contact with the aqueous phase.

Let us compare two different drug-membrane association states: Consider first the long axis of the cylindrical model drug being parallel to the bilayer midplane as illustrated schematically in Figure 2.3(a).

For this drug orientation, we expect the solvation free energy, ΔF_{sol}, to be strongly negative because after transfer from water into the hydrocarbon core of the membrane, no

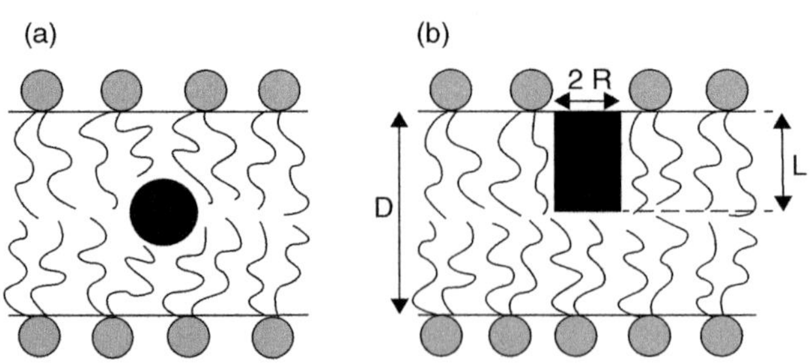

Figure 2.3 (a) and (b), Drug membrane association states.

part of the drug is exposed to the polar aqueous milieu. However, the presence of the drug induces a substantial structural membrane perturbation. This perturbation results from the rigidity of the cylinder which prevents the penetration of the lipid hydrocarbon chains into the drug interior and thus restricts the number of potentially accessible chain conformations. In order to get an impression of the corresponding cost of free energy, we refer to the results of a recently presented statistical-thermodynamic mean-field theory of chain packing [83] which resulted in $\Delta F_{\text{lip}}/L = 3.2$ kJ/mole · Å.

With the cylinder length, L = 16 Å, the perturbation free energy of the membrane amounts to $\Delta F_{\text{lip}} \approx 51.2$ kJ/mole. In contrast to this, the same chain packing theory predicts $\Delta F_{\text{lip}} = 12.4$ kJ/mole [84] if the cylinder is oriented in parallel to the fatty acid chains of the lipid bilayer, see Figure 2.3(b). The different magnitudes for ΔF_{lip} in (A) and (B) reflect a different loss of chain conformational freedom. The membrane-buried cylinder in (A) mainly restricts the terminal region of the hydrocarbon chains in both lipid monolayers. In the absence of the cylinder, this region is particularly 'fluid' as indicated by measurements of the lipid chain order parameter [85]. Inserting a rigid lipophilic inclusion (i.e. the drug) within the hydrocarbon region strongly decreases the number of chain conformations that the lipids in the immediate vicinity of the drug can adopt and thus entails a large entropic penalty. Hence, ΔF_{lip} is large. In contrast to (A), the conformational restrictions in case (B) are much smaller. Here, the perturbation of the drug-containing lipid monolayer results mainly from weak motional limitations of a few lipid chains in the immediate vicinity of the drug molecule. The opposite lipid monolayer remains nearly unperturbed.

The calculated value of $\Delta F_{\text{lip}} = 12.4$ kJ/mole (see Figure 2.3(b)) can be explained by considering a simple model which has recently been found to be qualitatively in agreement with mean-field chain packing calculations [86]. This so-called 'director model' is based on the representation of a single hydrocarbon chain by a rigid fluctuating director. To be more specific, a director is defined as the end-to-end vector of a hydrocarbon chain, pointing from the glycerol backbone to the terminal methyl segment. Different chain conformations correspond to different director orientations. Within the director model it is assumed that all possible director orientations within the hydrocarbon core occur with equal probability; yet, the director cannot penetrate into the model drug. Based on a simple statistical model [86], one obtains the perturbation free energy induced by a monolayer spanning cylinder (see Figure 2.3):

$$\Delta F_{\text{lip}} = (1 - \ln 2)\frac{\pi DR}{a_0} k_B T \tag{2.2}$$

Here $a_0 = 30$ Å2 denotes the cross-sectional area of a single, fluid-like lipid chain that resides in the membrane (the typical cross-sectional area per double-chained lipid in a lipid bilayer is about 60 Å2), kB is the Boltzmann constant and T is the absolute temperature. Inserting furthermore $D = 30$ Å, and $R = 6$ Å into Equation 2.2 leads to $\Delta F_{\text{lip}} = 14.9$ kJ/mole which is close to the value of $\Delta F_{\text{lip}} = 12.4$ kJ/mole, calculated on the basis of the chain packing theory. The director model suggests that ΔF_{lip} results mainly from the entropic loss due to conformational restrictions of the drug-facing lipid chains. Of course, interactions between the drug and the lipid head groups, which are not accounted for by the director model, may further affect ΔF_{lip}.

2.2.3.3 Modelling

Recently the transfer of poorly water-soluble drugs between liposomes has been mathematically modelled [87], and the authors found that apart from liposomes, the scope of the model includes other types of small and mobile pharmaceutical nanocarriers, such as micelles, colloids, and nanoparticles. Starting from a microscopic distribution function of drug molecules among donor and acceptor liposomes, we have specified the conditions that lead to an apparent first-order kinetic behavior. These include low drug loading of the liposomes, strict proportionality of all rate constants to drug concentrations, no aggregation phenomena of drugs within liposomes, and no overlap of the intraliposomal flip-flop kinetics. Systems that do not fulfil these conditions do not, generally, exhibit apparent first-order kinetics. Instead the behavior may become biexponential or sigmoidal. High drug loading may preserve the first-order kinetics but with increased apparent rate constant.

2.2.4 *In Vivo* Consequences

As pointed out above, poorly water-soluble compounds can be solubilized by two types of formulation principles: the first loses its solubilization potential in a relatively short time frame (ca < 1 min.) upon dilution in the blood circulation (examples: formulations based on detergents, co-solvents); the second maintains its solubilization capacity because the solubilizing entity does not disintegrate rapidly (examples: oil-in-water emulsions and liposomes). This classification does not mean that a grey zone may exist between the two classes. Detergent micelles may be quite resistant in the blood circulation, especially at high dose, as exemplified by Cremophor EL [38]. On the other hand, liposomes may be destabilized by adding relatively high amounts of detergents like lysolecithin and oleic acid [88].

2.2.4.1 Type of Drug

When using *in vitro* interliposomal transfer assays, significant differences in transfer rates (expressed as the time ($t_{1/2}$) required to transfer half of the equilibrium concentration at $t_{infinite}$ form donor to acceptor population) between lipophilic compounds/drugs can be found. For instance, amphotericin B and benzoporphyrin transfer in seconds, cyclosporine and paclitaxel in minutes; compounds with similar structures as membrane components (i.e., phospholipids and cholesterol) transfer in 18–36 hours). However, the half-times of transfer *in vivo* may be significantly faster because of accelerating factors not present in the *in vitro* assay. At high lipid dose the excess of transfer accelerating factors will be lower and the transfer rates found in the interliposomal assay with the various drugs may be more relevant.

In any case it is clear that the binding of a lipophilic drug to the lipid domain of the liposomal bilayer does not give rise to an irreversible binding of the drug to the liposomes. The more the drug has a shape like a phospholipid molecules or has properties like cholesterol which has a high affinity for the phospholipid bilayer, the slower the transfer. Many poorly water-soluble drugs can be associated with liposomes up to 10 mole %. The molecular structure of these drugs are mostly far from phospholipid molecules. Such drugs, however, still prefer to squeeze themselves between the non-isotropic bilayer structures simply because these compounds dislike the environing water phase wherein the liposomes are dispersed more than they dislike the lipid domain of the liposomal membrane.

In the presence of acceptor domains, the drug will therefore transfer quickly to the other acceptor domain to lower their concentration in the donor membrane and decrease the thermodynamically unfavourable situation.

2.2.4.2 Lipid Dose

As derived from the pre-clinical studies on rats [54] using cyclosporine liposomes as the model drug, the influence of the lipid dose on the pharmacokinetics and circulation time of cyclosporine associated with liposomes can be easily understood.

At a relatively low lipid dose of $\leq$40 mg/kg and a dose of $\leq$1 mg/kg cyclosporine, the cyclosporine liposomes showed the same blood elimination curves of cyclosporine, irrespective whether the liposomes comprised a lipid composition showing a long circulating or a faster eliminating composition (DPPC) was used. Obviously, at this lipid dose the fate of the liposomes is not related to the fate of the cyclosporine. Therefore, the cyclosporine must have been released quickly from the liposomes and the pool of acceptor lipid domains must have been large enough to take up the cyclosporine dose.

At higher lipid doses (ca. >40 mg/kg), the cells of the monophagocytotic systems (the macrophages mainly present in the spleen and liver) will increasingly saturate and the elimination of the liposomes from the circulation is slowed down. As a result, the lipid pool related to the liposomes in the blood circulation will increase and the ratio of liposome lipid pool to acceptor lipid pools will increase as well. The co-administration of cyclosporine in the liposome then results in a higher (total) cyclosporine concentration (AUC) of cyclosporine. In other words, at higher phospholipid doses, the AUC of the liposomal lipid and the AUC of the cyclosporine increase simultaneously. Under these conditions, the drug will probably mainly be bound to the liposomes.

The binding of a drug to liposomes, however, does not deactivate the activity of the drug but may reduce its activity and a higher dose of the liposomal drug may have to be administered to achieve the same effect compared to the drug administered in e.g. detergents. For instance, AmbiSome (liposomal amphotericin B) is used at 3–5 mg/kg/day whereas Fungizone (amphotericin B solubilized with cholate) is used at 0.7 mg/kg/day for the treatment of Cryptococcal Meningitis [89].

For cytostatics, the liposomal formulations have in general a lower toxicity [90] compared to Cremophor formulations and the Maximal Tolerated Dose is increased as well. The liposomal cytostatic can/should be administered at a higher dose which may show a higher efficacy and yet an acceptable low degree of toxicity.

Since at these higher lipid doses the drug is kept in the blood circulation for a longer period, the kinetics and degree at which the drug reaches tox-sensitive organs may be slowed down as well and overall or organ-specific toxicity may be reduced, because the sensitive tissues are not confronted with a peak concentration and have more time to cope with the dose over a longer period. In addition, during the time when the drug is still associated with the liposomes, the drug may reach other organs or even tumours, giving rise to more efficacy and/or to a changed toxicity profile. There are also indications that the longer residence time of the drug may also be partially caused by a release of the drug from the macrophages into the blood circulation which previously phagocytosed the liposomal drug [90, 91].

The lipid dose also influences the disintegration kinetics of the liposomes in the blood and the concomitant release of drug. At a high lipid dose, the ratio of phospholipid concentration versus the concentration of destabilizing blood components like albumin and HDL able to extract phospholipid from the liposomal membrane is high, and the liposomes are relatively intact.

At lower lipid doses, the ratio of phospholipid concentration versus the concentration of destabilizing blood components is low, and the drug release from the liposomes will be the sum of the drug transfer from liposomes which are still intact and the accelerated transfer caused by the disintegration of the liposomes.

2.2.4.3 Liposome Type and Composition

As demonstrated by the model studies with cyclosporine [54], at lower lipid doses the release of the drug is not dictated by the lipid composition but by the transfer kinetics of the drug. The studies were, however, performed with liposomes which contain either gel state phospholipids or liquid crystalline state phospholipids plus cholesterol which are relatively stable in blood. No studies were performed with liposomes containing exclusively phospholipids which are in the liquid crystalline state and therefore relatively unstable in blood. Faster disintegrating liposomes comprising phospholipids which are at 37°C in the liquid crystalline state may therefore have faster-releasing properties when the bilayer structure is essential to capture the drug and when the drug has even slow transferring properties.

More stable liposomes with longer circulation times which comprise gel state phospholipids (DPPC or DSPC or hydrogenated phospholipids) and pegylated phospholipids will, at higher lipid dose, maintain the higher ratio of liposomal phospholipids versus blood lipids for a longer time and may further keep the drug longer in the blood compartment compared to less stable liposomes which comprise phospholipids in the liquid crystalline state [92].

If solubilization is required and/or a longer circulation time, small liposomes are preferred, such as SUVs of 20–50 nm diameter, compared to larger LUVs and MLVs which are faster taken up by the macrophages. The smaller size not only makes the liposome less attractive for the macrophages but also offers a much larger surface area of interaction with the acceptor lipid domains in the blood circulation, thus facilitating the transfer of the drug to these domains. Also this 'rule of thumb' should not be too generalized, because, for example, in the case of intravenous administration of large amphotericin B liposomes, it is claimed, that after fast uptake by the macrophages, the macrophages do not completely digest the liposomes in the drug but are also able to excrete the drug slowly into the blood circulation [91,93].

The fate of the poorly water-soluble drugs after administration in liposomes and the sort of interactions the drug may have with blood components and macrophages are depicted in Figure 2.4.

Figure 2.4 clearly shows that the distribution of the drugs among the lipid domains is given by the individual dissociation and bindings constants of the complexes of the drug with the individual lipid domains. These general considerations are true for any solubilizing formulations using a solubilizer for which the drug has a certain affinity [94]. At the beginning of the intravenous dosing, there will be a tendency for the poorly water-soluble drug to distribute over the available lipid domains in the blood pool. The kinetics at which this happens will be dependent on the transfer properties. Since the liposomes may stay intact, the uptake and possible release of the drug by these cells may be another factor influencing the overall pharmacokinetics of the drug.

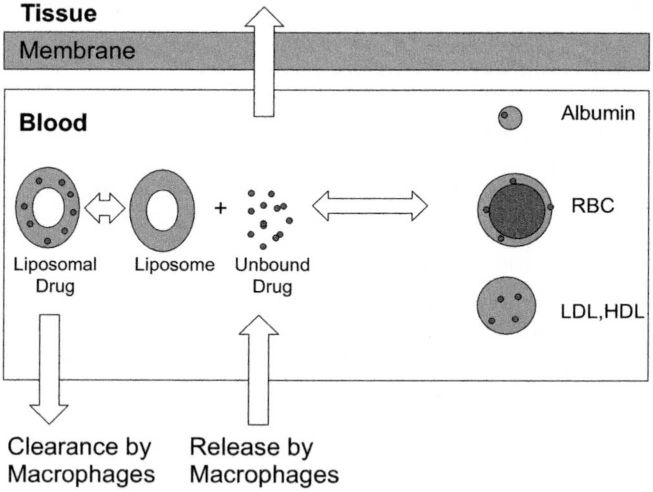

Figure 2.4 *Interaction of poorly water-soluble drugs with lipid domains in the blood pool after intravenous administration in liposomal form.*

Derived from Figure 2.4, the influence of the liposomal lipid dose and the ratio of the liposomal lipid dose versus the competing 'lipid' dose in the lipid domains of the blood pool can now be easily understood (Table 2.2).

How much a higher lipid dose may influence the therapeutic index of a drug will be drug-specific. When drugs are toxic to peripheral tissues because of high peak concentrations,

Table 2.2 *Correlation between ratio of concentration of liposomal lipid and blood pool lipid domains and liposomal drug fate after intravenous administration.*

Ratio of liposomal lipid dose to concentration of lipid domains in blood	Effect
HIGH	Disintegration/destabilization decreased MPS/RES cells more saturated Longer circulation time of liposomes Reduced mass transfer from liposome to other lipid domains in the blood Longer circulation time of drug in blood pool in liposomal form Slower flux of drug into peripheral organs by either increase of drug conc. in blood compartment or/and slow release from RES Possible reduction of toxicity
LOW	Disintegration/destabilization increased MPS/RES cells not saturated Shorter circulation time of liposomes Increased mass transfer from liposome to other lipid domains in the blood Shorter circulation time of drug in blood pool Faster flux of drug into peripheral organs

maintenance of the drug dose and slower distribution into the peripheral tissue obtained after administering a high lipid dose may be beneficial. When the efficacy of the drug is related to high peak levels, the approach of using high lipid dose may be disadvantageous.

2.2.5 Preclinical Parenteral Assessment Liposomal PWSC

2.2.5.1 Technical Aspects

The use of liposomes as solubilizers for poorly water-soluble drugs in the early research stage requires the preparation of these formulations containing the drug at lab scale. At this stage only minimal amounts of drugs (2–10 mg) may be available. Taking classical preparation methods into consideration [24], first a mixed film of phospholipids and drug has to be prepared by evaporation of an organic solvent solution. This has to be done at several drug to lipid ratios because the maximal incorporation degree of the drug into the liposomal membrane has be assessed. After careful removal of the residual solvent under high vacuum for at least 18 hours, the dry film of drug and lipid is hydrated to form multilamellar vesicles. At this stage, the multilamellar vesicle dispersion can be inspected for the presence for drug crystals and maximal drug to lipid ratio can be selected. Since multilamellar vesicles are less preferred (see above; in addition, sterile filtration through 0.22 μm pore size filter should be possible), the liposomes have to be downsized to SUVs. In the case of small amounts, only a sonicator may be used. Sonication with a titanium probe sonicator has to be performed under nitrogen and in a cooling bath to prevent lipid (and drug) degradation. After that, the abrasion of the titanium probe in the form of black metal particles needs to be removed by low speed centrifugation or by filtration. In order to control the preparation, drug and lipid content determinations (to determine the yield) have to be made and particle size needs to be assessed.

It goes without saying that such procedures are prohibitive regarding any broad screening efforts in early research settings. Preference will be for single-step preparation methods by adding the drug to clear detergent or cyclodextrin solutions, organic solvent solutions or applying pH adjustment to make, in a very simple way, a solution of the drug. The amount of drug dissolved can be very simply visually checked.

After the advent of a straightforward instant solubilization method using Phares Liposome Solubilizer [52], liposomal formulations can be considered for broad screening of discovery compounds when minute amounts of the drug are available. Prefabricated, fully characterized sterile, isotonic, isohydric liposome dispersions, if necessary, produced under cGMP, with particle size 20–80 nm in diameter and optically clear appearance, supplied in various vial sizes, containing 100 mg/ml of pharmaceutically acceptable phospholipids (Soy PC and DMPG- Na) of pharmaceutical grade, are mixed with a solution of the drug in a pharmaceutically acceptable solvent (PEG300-400, ethanol, propyleneglycol or mixtures thereof). The volume of solvent in the final mixture should not exceed 10% for reasons of solvent toxicity and disruption of the liposomal membrane as solubilizing domain. The final injections are transparent dispersions containing the drug fully solubilized in the membrane of the liposomes. The increase of solubility by the liposomes can be easily checked by comparison with a mixture of the same solution of solvent and addition to water. Also, the maximal drug load in the liposomes can be easily checked by titrating in increasing volumes of the drug solution or adding the same amount of solution in solvent with increasing concentrations of the test compound. If necessary, the amount of drug that is solubilized in the liposome membrane can be more accurately assessed by analysis of the filtrate obtained after filtration of the drug-loaded liposomes through a 0.22 μm pore size membrane. Additionally, the transmission characteristics of the dispersion before filtration and the corresponding size of the lipid particles are good indicators of solubilization.

The loading method appears to be highly reproducible and robust [52]. The addition rate of the drug solution to the liposomes may be critical. Drugs with low lipid solubility and slow transferring properties may require slow addition rates to avoid oversaturation and precipitation of the drug. In contrast, drugs with high lipid solubility and fast transfer kinetics may be added very quickly to the liposomes without any risk of precipitation of the drug.

To guard against injection of unwanted large particles, the optical clarity of the lipid dispersion before and after mixing is an excellent guide prior to injection. External factors such as temperature and the duration of mixing did not affect spontaneous solubilization of diazepam and cyclosporine A in the lipid particles. Both the ease of observing the clarity visually and the reproducibility of the method, under a variety of conditions which are typical of routine use in the laboratory and clinic, make the method suitable for broader use. These practical considerations do not apply to injectable oil-in-water emulsions which have a milky appearance, because the oil droplets have a diameter of more than 200 nm. The simple option of observing the drug precipitation visually in such formulations is not possible. Therefore extemporaneous loading of lipophilic drugs in oil-in-water systems may not be advisable [48].

The instant solubilization method has several practical advantages compared to other methods of solubilizing lipophilic compounds. Provided that the drug is reasonably soluble in pharmaceutically acceptable solvents to achieve therapeutic doses, the amount of organic solvent required can be reduced to less than 10% because the lipophilic domain of the particles serves as an alternative solubilizing phase for the drug. Due to the association of the drug with the lipophilic domain, drug precipitation at the injection site and in the blood compartment is prevented upon dilution with the infusion medium, blood and tissue fluids.

Although one-step mixing procedures to load liposomes with drugs using two liquid media are preferred, because of convenience, other alternative methods do exist to load prefabricated liposomes with poorly water-soluble drugs. The methods may avoid the use of solvents at all. In the case of drugs with increased solubility at $>$pH 8–9 or pH $<$5 and very low solubility at ca. pH 7, they may be mixed in acidic or basic solution with the liposomes followed by neutralization of the mixture. Also the drug may be added as a solid substance to the liposomes, followed by spontaneous dissolving of the drug in the liposomes. The rate at which the drug dissolves in the liposomes will depend on the lipophilicity and transfer properties, particle size, and physical form (amorphous or crystalline) of the drug and the temperature at which the drug is incubated with the liposomes. Also, co-homogenization of the drug with liposomes under high pressure may serve the purpose of associating the drug with the liposomes. Simultaneously, the drug and the liposomes are reduced in size under high pressure and increased temperature and pressure conditions, which may force the drug into the liposomes. The choice of the final loading method may also be related with the stability of the drug related to the time period needed to load the drug and the used loading conditions.

The method using a liquid solution of the drug in water-miscible, pharmaceutically acceptable, solvents is viable for evaluating the preclinical pharmacological profile of lipophilic compounds in DMSO solutions typically emanating from High Throughput Screening in Drug Discovery. The DMSO solution of the drug may be added to an excess of phospholipid dispersion and the resulting solubilized drug may be administered parenterally as well as orally. The availability of a simple-to-use, intravenous formulation at a very early pre-clinical stage, in which the drug is consistently and completely solubilized, is useful for rapidly determining the absolute bioavailability and drug efficacy after intravenous administration.

Recently, it is found that the Instantaneous Loading method could be made even more attractive by using a suitable cryo/lyoprotectant [95–98], in the freezing of liposomes in the prefabricated liposomal formulation, enabling the freezing and long-term storage of the drug-unloaded and drug-loaded liposomes at −18°C. In this way stability testing is simplified to long-term testing under frozen conditions and 24-hour stability testing at 2–8°C, RT and 40°C of thawed samples, and the general risk of development interruptions due to lack of stability is eliminated. The freezing/thawing procedures appear to be performed at much higher phospholipid concentrations compared to freeze drying without a change in particle size characteristics. In the case of further development towards a marketed form (and if the drug, because of instability issues, does not allow storage for 2 years at 2–8°C), the frozen liposomes may be converted to freeze-dried liposomes using the same cryo/lyoprotectant.

2.2.5.2 *Toxicological Aspects*

In general, liposomes are well tolerated after intravenous administration. Depending on the dosing schedule and animal species, maximal tolerated doses may go up to 5 g phospholipid/kg in case of single dose and 100 mg/kg (dogs) to 1 g/kg (rats) for repetitive dosing [49]. The LD_{50} in mice for liposomes comprising eggPC and Chol 80/20 molar ratio was found to be ca 13.5 g lipid/kg [99]. Repetitive dosing of liposomes may give rise to splenomegaly. The uptake of liposomes by macrophages of the spleen interferes with the degradation of erythrocytes in this organ. This results in the occurrence of macrocytic anaemia which is reversible most of the time.

Liposomal dispersions need to be administered slowly in diluted form, to avoid the occurrence of embolism [49] and possible immunological reactions (see below). In a recent publication, the possible immunological side effects of intravenous liposomes were reviewed [100]. It is claimed that liposomes can also cause anaphylactic side effects like e.g. Cremophor. Warnings in the package leaflets of liposomal drugs are given to illustrate the statement that the liposomes cause side effects. In the list of characteristics of liposomes which especially may cause side effects, the presence of gel state phospholipid in Doxil as well as Ambisome as the possible reason for the increased incidence of immunological side effects was not mentioned. Gel state bilayer membranes do not occur in mammals and may be responsible for the observed effects. Furthermore, it is mentioned that the anaphylactic side effect after administration of Visudyne occurs in less than 1% of the patient population [101]. This low incidence is obtained without premedication with antihistamines or steroids. In sharp contrast, Cremophor is even showing these side effects in combination with premedication with antihistamines and steroids.

The absence of placebo data as a control experiment, to study whether the liposomes on their own are responsible for the side effects, or a statement that the antigenicity of the encapsulated drug may be increased by the encapsulation in liposomes, are further research topics. It is further unclear whether the lipophilic drug or hydrophilic drug combinations with liposomes are equally causing the same side effects. Also the absence of immunological side effects when administering emulsions for parenteral nutrition has not been mentioned. The emulsion particles possess a surface area of monolayered (egg) phosphatidylcholine, which, due to the grade of purity, has a negative charge, and in spite of this characteristic and the use of this type of formulation by patients in need of parenteral nutrition at high doses for very long periods, the frequency of anaphylactic-like side reactions is less than 1% [102].

We therefore conclude that when using liquid crystalline phospholipids (neutral as well as in combination with negatively charged phospholipids) and the precautions mentioned in the chapter (start with low infusion rate), that liposome solubilizers are a solubilizer with much less (immune) toxicity than Cremophor and are to be preferred (assuming at least the same solubilization potential) compared to Cremophor.

2.3 Conclusion

Nowadays liposomes are still considered to be somewhat exotic formulations. The reasons for this may be manifold. The main reason may be a lack of knowledge of the existing literature on liposomes and a lack of awareness of the existence of many liposomal products on the market. Another reason may be that the preparation/manufacturing of liposomes, and the concomitant selection of the phospholipid do require some in-depth knowledge of the related pharmaceutical technology and physicochemical properties of phospholipids.

Also within the field of liposomology, the use of liposomes for targeting tumour cells and the increase of the therapeutic index of cytostatics by using long circulating liposomes as exemplified by daunorubicin (DaunoXome) and doxorubicin (Doxil) are celebrated as a major technological and clinical achievement. These achievements are distracting from a more straightforward use of liposomes, namely to use them as a solubilizer for poorly water-soluble drugs.

Finally, the prices of liposomal products are excessive compared to non-liposomal alternatives. When NCEs are formulated with liposomes, such a price comparison does not exist. In the case of generic drugs like amphotericin B and doxorubicin, the price increase of the liposomal product must be compensated for by a better therapeutic profile and/or toxicity profile. Since the original non-liposomal products are still on the market, it seems that many MDs and health insurance representatives have another opinion of the usefulness/superiority of the particular liposomal line extensions.

Also in the case of the use of liposomes as a solubilizer, not only pharmaceutical benefits such as the prevention of precipitation, low toxicity, the technical accessibility of the liposomes and the corresponding price, for an increasingly cost-sensitive market, will play a role in the decision-making process on which the formulation approach will be selected. This decision may also be significantly influenced by the expected intravenous dose of the drug and the possibility of administering the drug in larger volumes at lower concentration (which would allow the use of a solubilization method with lower maximal concentration) or at higher concentration with lower (bolus) volume. At the preclinical tox testing stage when very high drug concentrations need to be tested with a low toxicity vehicle background, liposomes may very well be a useful tool to assess the properties of NCEs.

The advent of a prefabricated liposome solubilizer in combination with a straightforward loading method is a significant step in establishing liposomes as a solubilizer for poorly water-soluble compounds in the pharmaceutical industry. The method will have to compete, however, with lower cost excipient approaches. When NCEs really need liposomes as a solubilizer, because there is no viable formulation alternative (examples: MTP-PE and benzoporphyrin), these considerations do not play a role.

In any case, the characterization of liposomal solubilizers regarding *in vitro* and *in vivo* release/dissolution methods, including proper separation/isolation methods of liposomes, will be of paramount importance to develop such dosage forms. Also a deeper understanding of the stability of the liposomes in blood circulation and the concomitant release of the poorly water-soluble drug, its lipid dose dependency, the interaction with other lipid domains and the uptake and release kinetics from MPS/RES cells will be needed to enable interpretation of the obtained PK, efficacy and toxicity results.

This seems to be highly complicated and demanding, but it should be realized that these requirements are general requirements for any other solubilizer as well and they are not specific to liposomes. The only point of difference from non-particulate solubilizers is, however, that the uptake by cells of the RES/MPS and following release of drug by these cells may influence the pharmacokinetics and body distribution of the drug.

Considering the progress made in the understanding of the mechanism underlying the transfer of poorly water-soluble compounds from liposomes to other lipid domains and the advent of a standardized liposome solubilizer enabling further exploration of liposomal solubilization with NCEs, it can be expected that this type of solubilization will find its place beside the established formulation technologies.

References

1. A.D. Bangham, R.W. Horne, Negative staining of phospholipids and their structural modification by surface-active agents as observed in the electron microscope. *J Mol Biol*, **8**, 660–668 (1964).
2. R.W. Horne, A D. Bangham, V.P. Whittaker, Negatively stained lipoprotein membranes. *Nature*, **200**, 1340 (1963).
3. A.D. Bangham, R.W. Horne, A.M. Glauert, J.T. Dingle, J.A. Lucy, Action of saponin on biological cell membranes, *Nature*, **196**, 952–955 (1962).
4. M.L. Immordino, F. Dosio, L. Cattel, Stealth liposomes: Review of the basic science, rationale, and clinical applications, existing and potential, *Int J Nanomedicine*, **1**(3), 297–315, Review (2006).
5. P.M. Anderson, M. Tomaras, K. McConnell, Mifamurtide in osteosarcoma: A practical review, *Drugs Today* (Barc), **46**(5), 327–337 (2010).
6. Anon, Mifamurtide: CGP 19835, CGP 19835A, L-MTP-PE, liposomal MTP-PE, MLV 19835A, MTP-PE, muramyltripeptide phosphatidylethanolamine. [No authors listed.] *Drugs R D*, **9**(2), 131–135 (2008).
7. K. Mori, K. Ando, D. Heymann, Liposomal muramyl tripeptide phosphatidyl ethanolamine: A safe and effective agent against osteosarcoma pulmonary metastases. *Expert Rev Anticancer Ther*, **8**(2), 151–159, Review (2008).

8. A. Nardin, M.L. Lefebvre, K. Labroquère, O. Faure, J.P.L. Abastado, Liposomal muramyl tripeptide phosphatidyl ethanolamine: Targeting and activating macrophages for adjuvant treatment of osteosarcoma, *Curr Cancer Drug Targets*, **6**(2), 123–133 (2006).
9. E.A. Forssen, The design and development of DaunoXome for solid tumor targeting *in vivo*, *Adv Drug Del Rev*, **24**, 133–150 (1997).
10. A.S. Manjappa, K.R. Chaudhari, M.P. Venkataraju, P. Dantuluri, B. Nanda, C. Sidda, K.K. Sawant, R.S. Murthy, Antibody derivatization and conjugation strategies: Application in preparation of stealth immunoliposome to target chemotherapeutics to tumor, *J Control Release*, **150**(1), 2–22 (2011).
11. A. Schnyder, A. Huwyler, Drug transport to brain with targeted liposomes, *J NeuroRx*, **2**(1), 99–107 (2005).
12. C. Chen, D. Han, C. Cai, X. Tang, An overview of liposome lyophilization and its future potential, *J Control Release*, **142**(3), 299–311 (2010).
13. N.J. Zuidam, D.J. Crommelin, Chemical hydrolysis of phospholipids, *J Pharm Sci*, **84**(9), 1113–1119 (1995).
14. M. Grit, D.J. Crommelin, Chemical stability of liposomes: Implications for their physical stability, *Chem Phys Lipids*, **64**(1–3), 3–18. Review (1993).
15. M. Grit, N.J. Zuidam, W.J. Underberg, D.J. Crommelin, Hydrolysis of partially saturated egg phosphatidylcholine in aqueous liposome dispersions and the effect of cholesterol incorporation on hydrolysis kinetics, *J Pharm Pharmacol*, **45**(6), 490–495 (1993).
16. M. Grit, W.J. Underberg, D.J. Crommelin, Hydrolysis of saturated soybean phosphatidylcholine in aqueous liposome dispersions, *J Pharm Sci*, **82**(4), 362–366 (1993).
17. M. Grit, D.J.A. Crommelin, The effect of surface charge on the hydrolysis kinetics of partially hydrogenated egg phosphatidylcholine and egg phosphatidylglycerol in aqueous liposome dispersions, *Biochim Biophys Acta*, **1167**(1), 49–55 (1993).
18. M.M. Kamal, D. Mills, M. Grzybek, J. Howard, Measurement of the membrane curvature preference of phospholipids reveals only weak coupling between lipid shape and leaflet curvature, *Proceedings of the National Academy of Sciences*, **106**, 22245 (2009).
19. G.M. Patton, S.J. Robins, Separation and quantitation of phospholipid classes by HPLC. *Methods Mol Biol*, **110**, 193–215 (1998).
20. B. Fuchs, R. Süss, K. Teuber, M. Eibisch, J. Schiller, Lipid analysis by thin-layer chromatography: A review of the current state, *J Chromatogr A*, **1218**(19), 2754–2774, Epub 2010 Dec 3. Review (2011).
21. J. Schiller, K. Arnold, Application of high resolution 31P NMR spectroscopy to the characterization of the phospholipid composition of tissues and body fluids: A methodological review, *Med Sci Monit*, **8**(11), 205–222 (2002).
22. P. van Hoogevest, P. Fankhauser, An industrial liposomal dosage form for muramyl-tripeptide-phosphatidyl ethanolamine (MTP-PE). In: *Liposomes in the Therapy of Infectious Diseases and Cancer*, I.J. Fidler, G. Lopez-Berestein (eds.), New York: Alan R. Liss Inc., 1989, p. 453.

23. U. Isele, P. van Hoogevest, R. Hilfiker, H.G. Capraro, K. Schieweck, H. Leuenberger, Large-scale production of liposomes containing monomeric zinc phthalocyanine by controlled dilution of organic solvents, *J Pharm Sci*, **83**, 1608–1616 (1994).
24. R.R.C. New (Ed.), *Liposomes: A Practical Approach*, Oxford: Oxford University Press (1990).
25. R. Barnadas-Rodríguez, M. Sabés, Factors involved in the production of liposomes with a high-pressure homogenizer, *Int J Pharm*, **213**(1–2), 175–186 (2001).
26. R Janknegt, S. de Marie, I.A.J.M. Bakker-Woudenberg, D.J.A. Crommelin, Liposomal and lipid formulations of amphotericin B, *Clin Pharmacokinet*, **23**(4), 279–291 (1992).
27. R.M. Fielding, A.W. Singer, L.H. Wang, S. Babbar, L.S.S. Guo, Relationship of pharmacokinetics and drug distribution in tissue to increased safety of amphotericin B colloidal dispersion in dogs, *Antimicrob Agents and Chemother*, **26**(2), 299–307 (1992).
28. G. Schumann, P. Van Hoogevest, P. Fankhauser, A. Probst, A. Peil, M. Court, J/C/ Schaffner, M. Fischer, T. Skripsky, P. Graepel, Comparison of free and liposomal MTP-PE: Pharmacological and toxicological aspects. In: *Liposomes in the Therapy of Infectious Diseases and Cancer*, I.J. Fidler, G. Lopez-Berestein (eds.), New York: Alan R. Liss Inc. (1989).
29. B.I. Gaynes, R.G. Fiscella, Safety of verteporfin for treatment of subfoveal choroidal neovascular membranes associated with age-related macular degeneration, *Expert Opin Drug Saf*, **3**(4), 345–361 (2004).
30. K.J. Messmer, S.R. Abel,Verteporfin for age-related macular degeneration, *Ann Pharmacother*, **35**(12), 1593–1598 (2001).
31. S.B. Brown, K.J. Mellish, Verteporfin: A milestone in opthalmology and photodynamic therapy, *Expert Opin Pharmacother*, **2**(2), 351–361 (2001).
32. L.J. Scott, K.L. Goa, Verteporfin, *Drugs Aging*, **16**(2), 139–146; discussion 147–148 (2000).
33. L. Wu, R.P. Murphy, Photodynamic therapy: A new approach to the treatment of choroidal neovascularization secondary to age-related macular degeneration, *Curr Opin Ophthalmol*, **10**(3), 217–220 (1999).
34. H.J. Gelderblom, K. Verweij, A. Nooter, A .Sparreboom, Cremophor EL: The drawbacks and advantages of vehicle selection for drug formulation, *European Journal of Cancer*, **37**, 1590–1598 (2001).
35. P. van Hoogevest, X. Liu, A. Fahr, Drug delivery strategies for poorly water-soluble drugs: the industrial perspective, *Expert Opinion*, **8**(11), 1481–1500 (2011).
36. C. Gähwiller, C. von Planta, D. Schmidt, H.Z. Steffen, Size, structure, and dynamics of bile salt/lecithin mixed micelles (author's trans.) *Naturforsch C*, **32**(9–10), 748–755 (1977). In German.
37. R.G. Strickley, Solubilizing excipients in oral and injectable formulations. *Pharm Res*, **21**, 201–230 (2004).
38. Y.-C. Lee, P.D. Zocharski, B. Samas, An intravenous formulation decision tree for discovery compound formulation development, *International Journal of Pharmaceutics*, **253**, 111–119 (2003).
39. A. Sparreboom, L. van Zuylen, E. Brouwer, W.J. Loos, P. de Bruijn, H. Gelderblom, M. Pillay, K. Nooter, G. Stoter, J. Verweij, Cremophor EL-mediated alteration of

paclitaxel distribution in human blood clinical pharmacokinetic implications, *Cancer Res* **59**, 1454 (1999).
40. A. Sparreboom, O. van Tellingen, W.J. Nooijen, J.H. Beijnen, Nonlinear pharmacokinetics of paclitaxel in mice results from the pharmaceutical vehicle Cremophor EL, *Cancer Res*, **56**, 2112 (1996).
41. See http://www.rxlist.com/taxol-drug.htm.
42. See http://www.rxlist.com/ixempra-drug.htm.
43. Epothilone Compositions United States Patent Application 20090191264.
44. Epothilone compositions European Patent Application EP1198225 Sonntag, Jean-Claude (FR), Wartmann, Markus (CH), Altmann, Karl-Heinz (CH).
45. M.E. Brewster, T. Loftsson, Cyclodextrins as pharmaceutical solubilizers, *Adv Drug Deliv Rev*, **59**, 645–666 (2007).
46. S. Gould, R.C. Scott, 2-Hydroxypropyl-beta-cyclodextrin (HP-beta-CD): A toxicology review, *Food Chem Toxicol*, **43**, 1451–1459 (2005).
47. H. Bunjes, Lipid nanoparticles for the delivery of poorly water-soluble drugs, *J Pharm Pharmacol*, **62**(11), 1637–1645. doi: 10.1111/j.2042-7158.2010.01024.x (2010).
48. W. Sila-on, N. Vardhanabhuti, B. Ongpipattanakul, P. Kulvanich, Influence of incorporation methods on partitioning behavior of lipophilic drugs into various phases of a parenteral lipid emulsion, *AAPS Pharm Sci Tech*, **9**(2) (2008).
49. P. van Hoogevest, X. Liu, A. Fahr, M.L.S. Leigh, Role of phospholipids in the oral and parenteral delivery of poorly water soluble drugs. *J Drug Del Sci Tech*, **21**(1), 5–16 (2011).
50. K. Teelmann, B. Schlappi, M. Schupbach, A. Kistler, Preclinical safety evaluation of intravenously administered mixed micelles, *Arzneimittelforschung*, **34**, 1517–1523, (1984).
51. A. Fahr, P. van Hoogevest, S. May, N.S. Bergstrand, M.L. Leigh, Transfer of lipophilic drugs between liposomal membranes and biological interfaces: consequences for drug delivery, *Eur J Pharm Sci*, **26**(3–4), 251–265 (2005).
52. P. Van Hoogevest, V. Rogue, M. Brumec, H. Schwebel, J.L. Grunkemeyer, M.L. Leigh, Instant solubilization of poorly water-soluble drugs by in-situ loading of aqueous phospholipid dispersions suitable for parenteral administration, PDA, *J Pharm Sci Technol*, **60**, 366–377 (2006).
53. H.A. Aaron, *J Pharmacol Exp Ther* **129**(3), 282–290, 1960.
54. A. Fahr, J. Seelig, Liposomal formulations of Cyclosporin A: A biophysical approach to pharmacokinetics and pharmacodynamics, *Crit Rev Ther Drug Carrier Syst*, **18**(2), 141–172 (2001).
55. F. Lanternier, O. Lortholary, Liposomal amphotericin B: What is its role in 2008? *Clin Microbiol Infect*, **14**, Suppl. 4, 71–83 (2008).
56. S. Singh, A.K. Dash, Paclitaxel in cancer treatment: Perspectives and prospects of its delivery challenges, *Crit Rev Ther Drug Carrier Syst*, **26**(4), 333–37 (2009).
57. H.J. Schwebel, A. Fahr, P. van Hoogevest, Exploration of an interliposomal transfer assay of poorly water soluble drugs, in preparation.
58. R.K. Chowdary, I. Shariff, D. Dolphin, Drug release characteristics of lipid based benzoporphyrin derivative, *J Pharm Pharm Sci*, **6**, 13–19 (2003).

59. I.W. Schmider, A. Fahr, H.E. Blum, G.J. Kurz, Transport of heptafluorostearate across model membranes: Membrane transport of long-chain fatty acid anions, *Lipid Res*, **41**(5), 775–787 (2000).
60. B. Gay, J.M. Cardot, C. Schnell, P. van Hoogevest, D. Gygax, Comparative pharmacokinetics of free muramyl tripeptide phosphatidyl ethanolamine (MTP-PE) and liposomal MTP-PE. *J Pharm Sci*, **82**(10), 997–1001 (1993).
61. A. Fahr, M. Holz, G. Fricker, Liposomal formulations of cyclosporin A, *Pharmaceut. Res.* **12**, 1189–1198 (1995).
62. A. Fahr, G. Reiter, Biophysical characterization of liposomal delivery systems for lipophilic drugs: cyclosporin A as an example, *Cellular & Molecular Biology Letters*, **4**, 611–623 (1999).
63. Y. Barenholz, T.E. Thompson, Fraction of cholesterol undergoing spontaneous exchange between small unilamellar phosphatidylcholine vesicles. *Biochemistry*, **25**, 6701–6705 (1986).
64. L.K. Bar, Y. Barenholz, T.E. Thompson, Dependence on phospholipid composition of the fraction of cholesterol undergoing spontaneous exchange between small unilamellar vesicles, *Biochemistry* **26**, 5460–5465 (1987).
65. A. Cabanes, K.E. Briggs, P.C. Gokhale, J.A. Treat, A. Rahman, Comparative *in vivo* studies with paclitaxel and liposome-encapsulated paclitaxel, *Int J Oncol*, **12**, 1035–1040 (1998).
66. A. Fahr, R. Reszka, R. Schaufelberger, unpublished results.
67. P. van Hoogevest, B. de Kruijff, Effect of amphotericin B on cholesterol-containing liposomes of egg phosphatidylcholine and didocosenoyl phosphatidylcholine: A refinement of the model for the formation of pores by amphotericin B in membranes, *Biochem Biophys Acta* **511**, 397–407 (1978).
68. J. Bolard, Interaction of polyene antibiotics with membrane lipids: physicochemical studies of the molecular basis of selectivity. *Drugs Exp Clin Re*. **12**, 613–6181 (1986).
69. K.M. Wasan, G.A. Brazeau, A. Keyhani, A.C. Hayman, G. Lopez-Berestein, Roles of liposome composition and temperature in distribution of amphotericin B in serum lipoproteins. *Antimicrobial Agents and Chemotherapy*, **37**, 246–250 (1993).
70. R.M. Fielding, D.E. Dressler, J.W. Lee, D.N. Buell, T.J. Walsh, Plasma protein binding of amphotericin B and pharmacokinetics of bound versus unbound amphotericin B after administration of intravenous liposomal amphotericin B (AmBisome) and amphotericin B deoxycholate. Bekersky I, *Antimicrob Agents Chemother*, **46**(3), 834–840 (2002).
71. P.C.N. Rensen, W.G. Love, P.W. Taylor, *In vitro* interaction of zinc-phthalocyanine-containing and plasma lipoproteins, liposomes ref ZnPc and benzoporphyrin, *J Photochem and Photobiol B: Biol*, **26**, 29–35 (1994).
72. A.J. Versluis, P.C.N. Rensen, M.E. Kuipers, W.G. Love, P.W. Taylor, Interaction between zinc(II)-phthalocyanme-containing liposomes and human low density lipoprotein, *Photochem Photobiol B: Biol*, **23**, 141–148 (1994).
73. J. Zborowski, F. Roerdink, G. Scherphof, General Subjects, *Biochimica et Biophysica Acta (BBA)*, **497**(1), 183–191 (1977).
74. N. Patankar, K.M. Wasan, Role of phospholipid transfer protein on the plasma distribution of amphotericin B following the incubation of different amphotericin B formulations, *Pharm Res*, **23**(5), 1020–1024 (2006).

75. J. Seki, S. Sonoke, A. Saheki, T. Koike, H. Fukui, M. Doi, T. Mayumi, Lipid transfer protein transports compounds from lipid nanoparticles to plasma lipoproteins, *International Journal of Pharmaceutics*, **275**(1–2), 239–248 (2004).
76. K.M. Wasan, M. Ramaswamy, W. Wong, P. Haydn Pritchard, Lipid transfer protein in facilitated transfer of cyclosporine from low- to high-density lipoproteins is only partially dependent on its cholesteryl ester transfer activity, *JPET*, **284**(2), 599–605 (1998).
77. K.M. Wasan, Modifications in plasma lipoprotein concentration and lipid composition regulate the biological activity of hydrophobic drugs, *Journal of Pharmacological and Toxicological Methods*, **36**(1), 1–11 (1996).
78. K.M. Wasan, S.M. Cassidy, Role of plasma lipoproteins in modifying the biological activity of hydrophobic drugs, *Journal of Pharmaceutical Sciences*, **87**(4), 411–424 (1999).
79. M. Kwong, K.M. Wasan, Cholesteryl ester transfer protein facilitates the movement of water-insoluble drugs between lipoproteins: A novel biological function for a well-characterized lipid transfer protein, *Biochemical Pharmacology*, **64**(12), 1669–1675 (2002).
80. K.M. Wasan, D. Brocks, S. D. Lee, K. Sachs-Barrable, S.J. Thornton, Impact of lipoproteins on the biological activity and disposition of hydrophobic drugs: Implications for drug discovery, *Nature Reviews Drug Discovery*, **7**, 84–99 (2008).
81. N. Ben-Tal, A. Ben-Shaul, A. Nicholls,B. Honig, Free-energy determinants of a-helix insertion into lipid bilayers. *Biohys J*, **70**, 1803–1812 (1996).
82. H.S. Chan, K.A. Dill, Solvation: how to obtain microscopic energies from partitioning and solvation experiments. *Ann Rev BiophysBiomol Struct*, **26**, 425–459 (1997).
83. A. Zemel, A. Ben-Shaul, S. May, Membrane perturbation induced by interfacially adsorbed peptides. *Biophys J*, **86**, 3607–3619 (2004).
84. A. Zemel, A. Ben-Shaul, S. May, Perturbation of a lipid membrane by amphipathic peptides and its role in pore formation. *Eur Biophys J*, **34**, 230–243 (2005)
85. M. Lafleur, P.R. Cullis, M. Bloom, Modulation of the order profile in the La-phase. *Eur. Biophys J*, **19**, 55–62 (1990).
86. S. May, A. Ben-Shaul, A molecular model for lipid-mediated interaction between proteins in membranes. *Phys Chem Chem Phys*, **2**, 4494–4502(2000).
87. S. Loew, A. Fahr, S May, Modelling the release kinetics of release of poorly water soluble drug molecules from liposomal nanocarriers, *J Drug Del*, (2011), article id 376548.
88. M. Madörin, P. van Hoogevest, R. Hilfiker, H. Leuenberger, The use of fluorescence resonance energy transfer to study the disintegration kinetics of liposomes containing lysolecithin and oleic acid in rat plasma, *Pharm Res*, **17**(9), 1118–1123 (2000).
89. See http://www.rxlist.com/ambisome-drug.htm.
90. Fetterly and Straubinger, *AAPS PharmSci*, **5**(4), article 32, (2003), available at: http:/aapspharmsci.org.
91. P. Legrand, A. Vertut-Doi, J. Bolard, Comparative internalization and recycling of different amphotericin B formulations by a macrophage-like cell line, *J Antimicrob Chemother*, **37**(3), 519–533 (1996).

92. P. Crosasso, M. Ceruti, P. Brusa, S. Arpicco, F. Dosio, L. Catte, Preparation, characterization and properties of sterically stabilized paclitaxel-containing liposomes, *Journal of Controlled Release*, **63**(1–2), 19–30 (2000).
93. J. Brajtburg, J. Bolard, Carrier effects on biological activity of amphotericin B, *Clinical Microbiology Reviews*, **9**(4), 512–531 (1996).
94. B. Bittner, R.J. Mountfield, Intravenous administration of poorly soluble new drug entities in early drug discovery: The potential impact of formulation on pharmacokinetic parameters, *Cur Opin in Drug Disc & Dev*, **5**(1), 59–72 (2002).
95. G.J. Fransen, P.J.M. Salemink, D.J.A. Crommelin, Critical parameters, *Int J Pharm.* **33**, 27–35 (1986).
96. K. Miyajima, K .Tomita, M. Nagakaki, Effect of saccharides on the freezing and thawing of liposome dispersion, *Chem Pharm Bull*, **34**(7), 2689–2697 (1986).
97. Y. Oezer, H. Talsma, D.J.A. Crommelin, A.A. Hincal, Influence of freezing and freeze drying on the suitability of liposomes dispersed in aqueous media, *Acta Pharm Technol*, **34**(3), 129–139 (1989).
98. N.J. Zuidam, S.S.L. Lee, D.J.A. Crommelin Gamma irradiation of non-frozen, frozen, freeze dried liposomes, *Pharm Res*, **12**(11), 1761–1768 (1995).
99. T. Fritz, E. Unger, S. Wilson-Sanders, Q.F. Ahkong, C. Tilcock, Detailed toxicity studies of liposomal gadolinium, *DTPA*, **26**(11), 960–968 (1991).
100. J. Szebeni, F. Muggia, A. Gabizon, Y. Barenholz, Activation of complement by therapeutic liposomes and other lipid excipient based therapeutic products: Prediction and prevention, *Advanced Drug Delivery Reviews*, doi: 10.1016/j.addr.2011.06.017 (2011).
101. See http://www.rxlist.com/visudyne-drug.htm.
102. See http://www.rxlist.com/intralipid-20-drug.htm.

3

Drug Solubilization and Stabilization by Cyclodextrin Drug Carriers

Thorsteinn Loftsson and Marcus E. Brewster

3.1 Introduction

In 1891, a French scientist called A. Villiers isolated crystalline dextrins from a bacterial digest which he named 'cellulosine' [1]. Twelve years later an Austrian microbiologist called Franz Schardinger described crystalline compounds A and B which he had isolated from a bacterial digest of potato starch [2]. He was only able to isolate a very small amount of compound A but significantly more of compound B which he identified as Villiers' 'cellulosine' [3]. Later Schardinger named the compounds α-dextrin and β-dextrin. Now these compounds are commonly called α-cyclodextrin (αCD) and β-cyclodextrin (βCD) or less commonly cyclomaltohexaose and cyclomaltoheptaose, respectively. For the next eight years, Schardinger continued his studies of cyclodextrins and during that time laid the foundation for cyclodextrin chemistry. γ-Cyclodextrin (γCD) was discovered in 1935 by Freudenberg and Jacobi [4] but the existence of larger cyclodextrins was not discovered until much later [5–9]. The first cyclodextrin-related patent entitled 'Verfahren zur Herstellung von Einschlußverbindungen physiologisch wirksamer organischer Verbindungen' was issued in 1953 [10] and in 1954 Friedrich Cramer published a book entitled *Einschlussverbindungen* in which he described all the basic structural and physicochemical characteristics of αCD, βCD and γCD, including their ability to form inclusion complexes, their ability to solubilize poorly soluble compounds and their ability to stabilize chemically labile compounds [11]. Although cyclodextrins were discovered over 100 years ago and although their effects on drug solubility and chemical stability have been well known for over 60 years, today they are still regarded as novel enabling pharmaceutical excipients.

Drug Delivery Strategies for Poorly Water-Soluble Drugs, First Edition. Edited by Dennis Douroumis and Alfred Fahr.

This is partly due to their slow commoditization related to large-scale production challenges. Bacterial treatment of starch produces crude mixtures of αCD, βCD and γCD together with linear and branched dextrins that were historically difficult to purify. The biotechnological advances that occurred in the 1970s led to dramatic improvements in the cyclodextrin production methodologies that made their industrial exploitation possible. In 1970, βCD was only available as a rare fine chemical at a price of about US$2000 per kg. Today the annual βCD production is more than 10,000 metric tons and the bulk price is about US$5 per kg. Various cyclodextrin derivatives have been synthesized, such as 2-hydroxypropyl-β-cyclodextrin (HPβCD) and sulfobutyl ether β-cyclodextrin sodium salt (SBEβCD). In 1976, the world first pharmaceutical product containing a cyclodextrin, prostaglandin E2/βCD sublingual tablets (Prostarmon E™, Ono, Japan), was marketed in Japan. About 12 years later, piroxicam/βCD tablets (Brexin®, Chiesi Farmaceutici, Italy) were marketed in Europe, and in 1997 itraconazole/HPβCD oral solution (Sporanox®, Janssen Pharmaceutica, Belgium) was marketed in the US. Worldwide, about 35 different drugs are now marketed as cyclodextrin complexes [2, 12]. This chapter will focus on the physiochemical and biological characteristics of cyclodextrins and their role as drug solubilizers and stabilizers.

3.2 Structure and Physiochemical Properties

Cyclodextrins are cyclic oligosaccharides formed by 6 (αCD), 7 (βCD) or 8 (γCD) (α-1,4)-linked D-glucopyranose units (Table 3.1). Cyclodextrins with more than eight glucopyranose units do exist but are of limited interest as drug solubilizers and stabilizers and will not be explored in this chapter. Cyclodextrin manufacturing is a three-step process. First, cyclodextrin glycosyltransferase (CGTase) is manufactured by bacterial fermentation. Second, the enzyme is used to produce cyclodextrin from starch (Figure 3.1). An aqueous starch slurry is liquefied by the addition of α-amylase, or by heating, and then CGTase is added for the enzymatic degradation of the linear dextrins to form crude mixture of cyclodextrins. In order to produce, for example, predominantly βCD, the enzymatic degradation can be performed in the presence of solvents such as toluene or p-xylene. These solvents form poorly soluble complexes with βCD that are separated from the crude mixture as precipitates. Other solvents can be used for the production of predominantly αCD or γCD. In the final step, the solvent is removed from the precipitate by boiling the complex in an aqueous solution followed by distillation. The cyclodextrin is then recovered by precipitation or crystallization.

Due to the chair structure of the glucopyranose units, the CD molecule is cone-shaped with the secondary hydroxy groups extending from the wider rim of the torus and the primary hydroxy groups from the narrower rim (Table 3.1). The result is a very hydrophilic molecule ($\log K_{\text{octanol/water}}$ between −13 and −17) with a central cavity that has a lipophilicity comparable to about a 60–70% aqueous ethanolic solution. Although the natural αCD, βCD and γCD and their complexes are hydrophilic, their aqueous solubility is somewhat limited, especially that of βCD. This is believed to be due to their relatively high crystal lattice energy as seen by their relatively high melting point. Random substitution of

Table 3.1 *Structure and physicochemical properties of the natural αCD, βCD and γCD.*

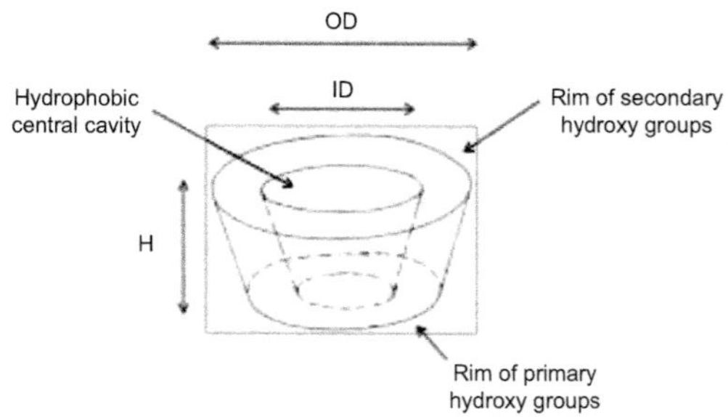

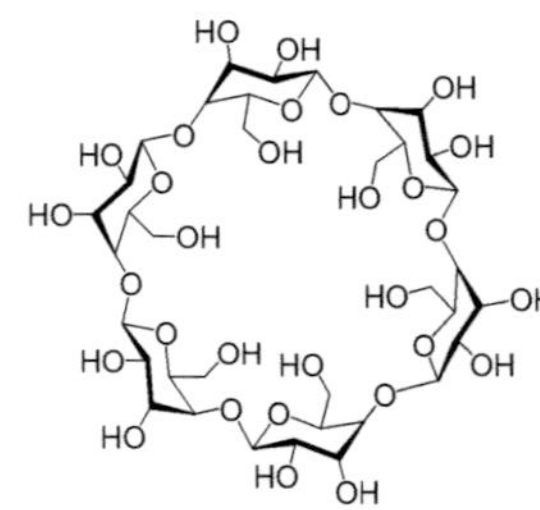

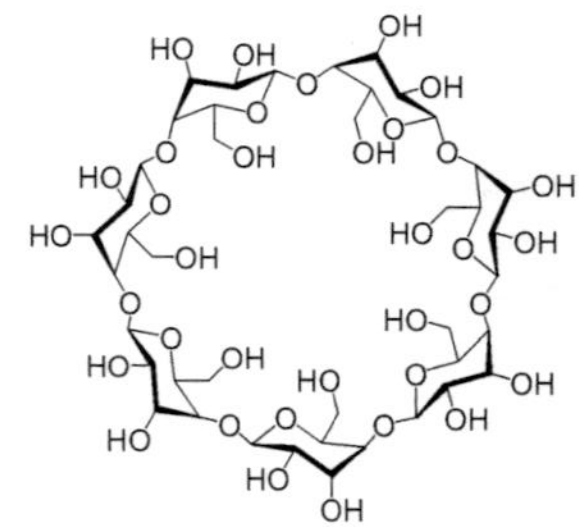

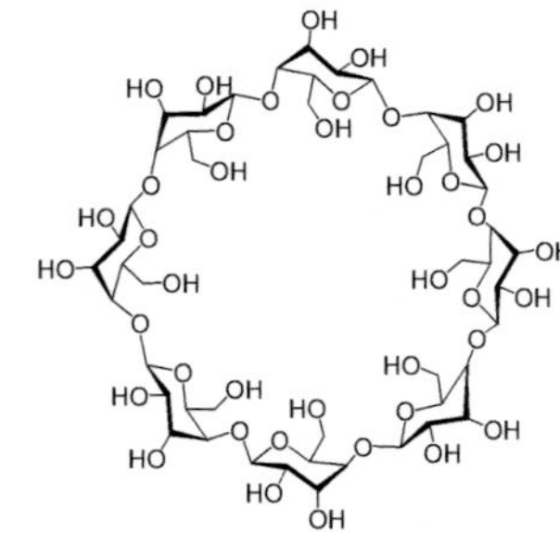

Property	α-Cyclodextrin	β-Cyclodextrin	γ-Cyclodextrin
Synonyms:	Cyclo-α-(1→4)-D-hexaglucopyranoside Cyclomaltohexaose Cyclohexaamylose Alfadex	Cyclo-α-(1→4)-D-heptaglucopyranoside Cyclomaltoheptose Cycloheptaamylose Betadex	Cyclo-α-(1→4)-D-octaglucopyranoside Cyclomaltooctaose Cyclooctaamylose Gammadex
Molecular weight of anhydrous compound (Dalton):	972.84	1134.98	1297.12
Number of glucopyranose units:	6	7	8
Melting temperature range:[a]	255–260	255–265	240–245
Crystal water content (% w/w):[a]	10.2	13.0–15.0	8–18
Approximate dimensions (nm):[a]			
height (H):	0.78	0.78	0.78
inner diameter (ID):	0.50	0.62	0.80
outer diameter (OD):	1.46	1.54	1.75
Solubility in water at 25°C (mg/ml):[b]	129.5 ± 0.7	18.4 ± 0.2	249.2 ± 0.2
Calculated LogK(octanol/water) at 25°C:[c]	−13	−14	−17
$K_{1:1}$ (population mean ± SD, 25°C)[d]	130 ± 8	490 ± 8	350 ± 9

Notes:
[a]From [125].
[b]From [126].
[c]From SciFinder, ACS, USA (scifinder.cas.org) and ChemExper Chemical Directory (www.chemexper.com).
[d]From [127, 128].

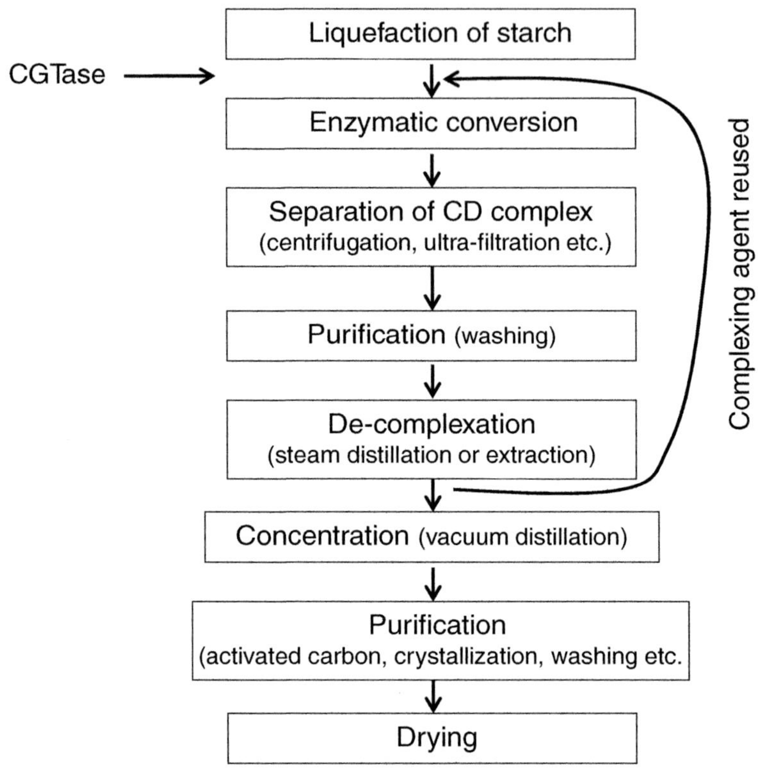

Figure 3.1 *Schematic description of cyclodextrin production.*

the hydroxy groups, even by somewhat lipophilic methoxy functions, will result in amorphous CD derivatives of relatively high aqueous solubility (Table 3.2). Through complex formation, these amorphous CD derivatives can convert poorly soluble crystalline drugs into water-soluble amorphous drug/CD complexes. Commercially available CD derivatives of pharmaceutical interest (Table 3.2) include the hydroxypropyl derivatives of αCD, βCD and γCD (HPαCD, HPβCD and HPγCD), sulfobutylether βCD sodium salt (SBEβCD), the randomly methylated βCD (RMβCD) and the so-called branched CDs such as maltosyl βCD (MβCD). The physicochemical properties of the CD derivatives, including their aqueous solubility and ability to form drug/CD complexes, depend not only on the structure of the attached substituents but also on their number and location within the CD molecule. The molar degree of substitution (MS) is defined as the average number of substituents per one glucopyranose repeat unit (Table 3.2). For methylated CDs, the MS has a value between 0 (no substitution) and 3 when all hydroxy groups have been methylated. In other cases, such as in hydroxypropylation of CDs, the hydroxy groups on the substituents can also be substituted to form oligomeric or polymeric side chains. In such cases the MS value can be greater than 3. For optimal solubility and complexation efficiency, the MS value is, in general, kept somewhat low. Fully substituted CDs can possess both poor aqueous solubility

Table 3.2 *Some cyclodextrin derivatives of pharmaceutical interest.*

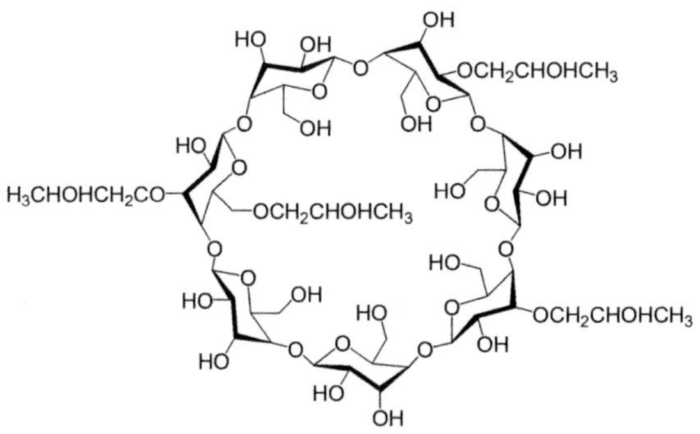

Rim of primary hydroxy groups

Cyclodextrin	Trade name and manufacturer	MS	MW (Da)	Solubility (mg/ml)	$LogK_{o/w}$[a]
2-Hydroxypropyl-α-cyclodextrin (HPαCD)	Cavasol® W6 HP (Wacker, Germany)	0.65	1199	>500	< −10
2-Hydroxypropyl-β-cyclodextrin (HPβCD)	Cavasol® W7 HP (Wacker, Germany)	0.65	1400	>600	−11
(Hydroxypropylbetadex, Ph.Eur., USP/NF)	Kleptose® HPB (Roquette, France)	0.65	1400	>600	−11
Sulfobutylether β-cyclodextrin Na (SBEβCD) (Betadex Sulfobutyl Ether Sodium, USP/NF)	Captisol® (CyDex Pharmaceuticals, USA)	0.9	2163	>500	< −10
Randomly methylated β-cyclodextrin (RMβCD)	Cavasol® W7 M (Wacker, Germany)	1.8	1312	>600	−6
	Kleptose® Crysmeb (Roquette, France)	0.57	1191	200	
Maltosyl β-cyclodextrin (MβCD)	(Ensuiko Sugar Refining Co., Japan)	0.14	1459	>500	< −10
2-Hydroxypropyl-γ-cyclodextrin (HPγCD)	Cavasol® W8 HP (Wacker, Germany)	0.6	1576	>600	−13

Note:
[a] From SciFinder, ACS, USA (scifinder.cas.org) and ChemExper Chemical Directory (www.chemexper.com).

and less than an optimal ability to form complexes due to substituent obstruction of the central cavity. In carbohydrate chemistry, degree of substitution (DS) is defined as the number of hydroxyl groups per anhydroglucose repeat unit that have been substituted and, thus, the maximum obtainable DS is 3. By contrast, in CD chemistry, DS frequently refers to the average number of substituents per CD molecule [13].

CDs possess many of the same physicochemical and biological characteristics as the analogous water-soluble linear dextrins. However, due to their cyclic structure, they are more resistant to both enzymatic and non-enzymatic hydrolysis than the linear sugars [14]. In the solid state, CDs are at least as stable as sucrose or starch and can be stored for several years at room temperature without any detectable degradation [15.] Non-enzymatic degradation of CDs in aqueous solutions is specific acid catalyzed hydrolysis of the α-acetal linkages to form glucose, maltose and non-cyclic oligosaccharides [16]. The half-life ($t_{1/2}$) for the ring-opening of βCD was determined to be about 15 hours at 70°C and pH 1.1 [16]. The CD derivatives are hydrolyzed at about the same rate, ring-opening being the dominant degradation pathway. In aqueous solutions CDs are essentially chemically stable under neutral and basic conditions. CDs are resistant to β-amylases that hydrolyze starch from the non-reducing end, but are slowly hydrolyzed by α-amylases that hydrolyze starch from within the carbohydrate chain. The hydrolytic rate depends on the ring size and fraction of free CD. The mechanism by which CDs resist hydrolysis includes burying all bridge oxygens within the central cavity and, thus, free CD is hydrolyzed more rapidly than CD bound in an inclusion complex with the rate of hydrolysis increasing with increasing cavity size [17]. For example, αCD and βCD are essentially stable towards α-amylase in saliva while γCD is rapidly digested by salivary and pancreatic α-amylase [18,19]. All the natural CDs and their previously mentioned derivatives are susceptible to bacterial digestion in the gastrointestinal tract [20–26]. After oral administration, γCD is almost completely digested in the gastrointestinal tract while both αCD and βCD are, to a large extent, digested by bacteria in the colon. αCD is digested more slowly than βCD.

3.3 Cyclodextrin Complexes and Phase Solubility Diagrams

The main characteristic of CDs with respect to their use as functional pharmaceutical excipients is their ability to form inclusion complexes with many lipophilic, poorly water-soluble compounds (Figure 3.2). Inclusion complexes are formed when a small lipophilic molecule, or a lipophilic moiety of a larger molecule (a guest) enters the central cavity of a CD molecule (a host). The two most important properties of the complexes are their stoichiometry and the numerical values of their stability constants. If m drug molecules (D) associate with n CD molecules (CD) to form a complex (D_m/CD_n), the following overall equilibrium is attained:

$$m \cdot \mathrm{D} + n \cdot \mathrm{CD} \overset{K_{m:n}}{\rightleftarrows} \mathrm{D}_m/\mathrm{CD}_n \tag{3.1}$$

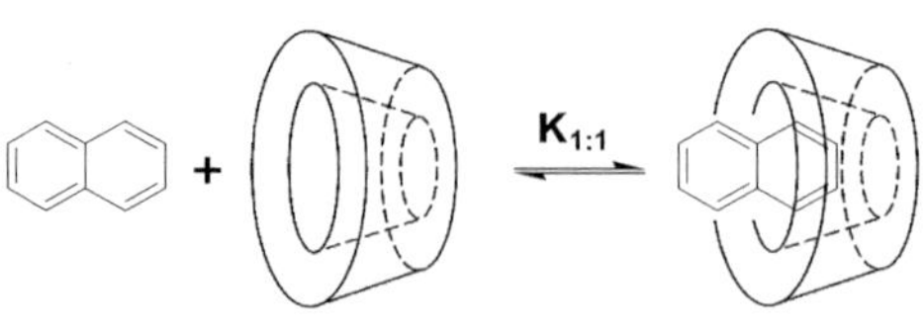

Figure 3.2 *Formation of a guest/host inclusion complex. In this case one guest molecule (m = 1) forms a complex with one host molecule (n = 1).*

where $K_{m:n}$ is the stability constant of the drug/CD complex. The stoichiometry of drug/CD complexes and the numerical values of their stability constants are often obtained from phase-solubility diagrams where the drug solubility is monitored as a function of total CD added to the complexation media (Figure 3.2) [27–29]. If an A_L-type (i.e., linear) phase-solubility diagram is obtained, then it can be assumed that the complex is first order with respect to CD ($n = 1$ in Equation 3.1) and first or higher order with respect to the drug ($m \geq 1$). In this case the total concentration of dissolved drug (S_{tot}) is equal to the sum of intrinsic drug solubility (S_0) and the concentration of drug in the complex ($m \cdot [D_m/CD]$):

$$S_{tot} = S_0 + m \cdot [D_m/CD] \tag{3.2}$$

The most common type of D/CD complexes in dilute aqueous solutions are 1:1 D/CD complexes (Figure 3.2). In this case, the slope of the linear phase-solubility diagram is less than unity and Equation 3.3 can be applied to calculate stability constant ($K_{1:1}$):

$$K_{1:1} = \frac{\text{Slope}}{S_0 \cdot (1-\text{Slope})} \tag{3.3}$$

Positive deviation from linearity (A_P-type phase-solubility diagram) suggests formation of higher-order complex with respect to CD. The stoichiometry of the system can be obtained by curve fitting with a quadratic model. A good fit to this model could suggest the formation of a 1:2 drug/CD complex:

$$S_{tot} = S_0 + K_{1:1} \cdot S_0 \cdot [CD] + K_{1:1} \cdot K_{1:2} \cdot S_0[CD]^2 \tag{3.4}$$

where [CD] represents the concentration of free CD. A third-order model is suggestive of a 1:3 complex, etc. [29, 30]. Here consecutive complexation is assumed where, for example, the 1:2 complex is formed when one additional CD molecule forms a complex with an existing 1:1 complex. It is very important to remember that this curve fitting technique does not indicate whether a given drug forms an inclusion complex with CD, but only how CD influences the drug solubility. Phase-solubility studies are performed in aqueous solutions saturated with the drug where formation of higher-order complex aggregates is more likely than in diluted unsaturated solutions. The natural CD and their derivatives, as well as their complexes, are all known to form aggregates, especially at elevated drug and CD concentrations [31–36]. Formation of non-inclusion complexes and CD aggregates contribute to the overall drug solubilization in aqueous CD solutions [34–46]. A_N-type profiles have been explained by changes in the complexation media and self-association of CD molecules and/or their complexes at higher CD concentrations. A-type phase-solubility diagrams are commonly observed in complexation media containing the water-soluble CD derivatives, i.e. HPβCD, SBEβCD and HPγCD.

B-type phase-solubility diagrams (Figure 3.3) suggest formation of poorly soluble complexes and they are commonly observed in aqueous complexation media containing the natural αCD, βCD and γCD. B_S-type phase-solubility diagrams are thought to be formed when the drug/CD complex has limited solubility in the complexation medium with the profile plateau indicating the total drug solubility, i.e. the intrinsic drug solubility (S_0) plus the drug solubility in the form of CD complexes. The ascending part of the profile can be treated mathematically as an A-type diagram and the previously described techniques used to gain information on the apparent stoichiometry of the complex. The decrease of total drug solubility at higher CD concentrations which is manifested in the B-type profile has

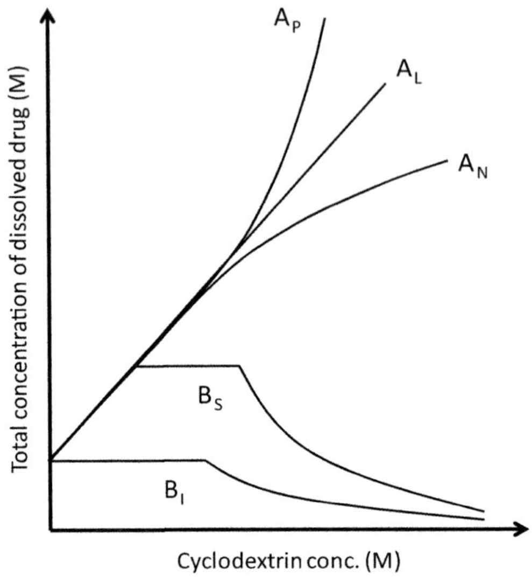

Figure 3.3 *Different types of phase-solubility diagrams, i.e. plots of total drug solubility (S_{tot}) vs total amount of CD present in the complexation media. The Higuchi and Connors complex classification [27].*

been explained by completion of the available drug in the complexation media. However, this descent is frequently observed when excess drug is available and, thus, these stoichiometric explanations may be inadequate. B_i-type profiles are similar to B_S-type except that the drug/CD complexes formed are insoluble in the complexation media. However, phase-solubility studies are generally performed in drug-saturated media, most commonly drug-saturated aqueous CD solutions, and such solutions are non-ideal so that they do not necessarily obey the Higuchi and Connors' system for classifying molecular complexes [27].

According to the phase-solubility technique, the intrinsic solubility (S_0) should be identical to the Y-intercept value of the phase-solubility diagram. However, this is rarely the case for poorly soluble drugs. Thus, complexation efficacy (CE) is frequently a better measure for comparison of solubilization effects of different CDs [47]. If the slope of a linear phase-solubility diagram is less than unity, the CE can be calculated from the following equation (Table 3.3):

$$\mathrm{CE} = S_0 \cdot K_{1:1} = \frac{[\mathrm{D/CD}]}{[\mathrm{CD}]} = \frac{\mathrm{Slope}}{1-\mathrm{Slope}} \tag{3.5}$$

where [D/CD] is the concentration of dissolved complex, [CD] is the concentration of dissolved free cyclodextrin and Slope is the slope of the linear phase-solubility profile. The complexation efficiency can be used to calculate the D:CD ratio, which can be correlated to the expected increase in formulation bulk:

$$\mathrm{D:CD} = 1 : \left(1 + \frac{1}{\mathrm{CE}}\right) \tag{3.6}$$

Table 3.3 *Phase-solubility of propofol in aqueous HPβCD, RMβCD and SBEβCD solutions at room temperature (22–23°C). The table shows the slopes of the A_L-type diagrams, the Y-intercept, the correlation coefficient, and the calculated values of the stability coefficient (Equation 3.3), complexation efficiency (CE, Equation 3.5) and the formulation bulk (Equation 3.7). The molar ratio is the drug:CD molar ratio in the aqueous solution (Equation 3.6).*

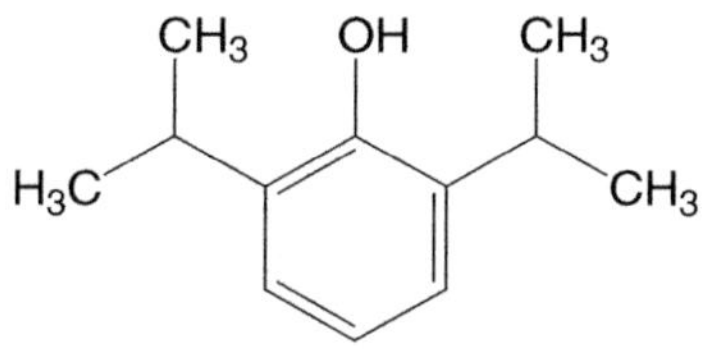

Solubility in water at 23°C (S_0): 0.16 mg/ml or $9.0 \cdot 10^{-4}$ M
Molecular weight: 178 Da
Log $P_{(octanol/water)}$ = 3.8
Protein binding about 98%
Dose = 10 mg

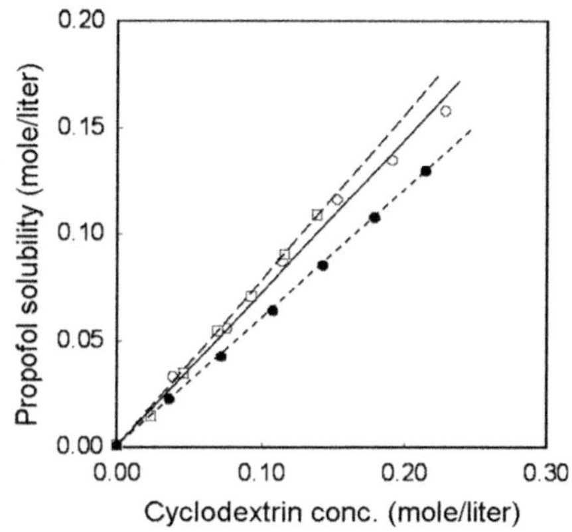

Cyclodextrin		Slope	Intercept (M)	Corr.	$K_{1:1}$ (M^{-1})	CE	Molar ratio	Formulation bulk (mg)
HPβCD	●	0.59994	$6 \cdot 10^{-4}$	1.000	1,700	1.50	1:1.7	130
RMβCD	○	0.68974	$50 \cdot 10^{-4}$	0.998	2,500	2.22	1:1.5	110
SBEβCD	□	0.78836	$-9 \cdot 10^{-4}$	0.999	4,100	3.73	1:1.3	150

Equation 3.7 shows the relationship between the increase in formulation bulk and molecular weights of the cyclodextrin (MW_{CD}) and the drug (MW_{Drug}), and the value of CE:

$$\text{Relative increase in the formulation bulk} = \frac{MW_{CD}}{MW_{Drug}} \cdot \left(1 + \frac{1}{CE}\right) \tag{3.7}$$

The formulation bulk associated with a CD complex can be found by multiplying the number obtained from Equation 3.7 with the drug dose (Table 3.3). The MWs of the natural CDs and some common CD derivatives are listed in Tables 3.1 and 3.2, respectively. The formulation bulk will increase with increasing molecular weight of the CD used (MW_{CD}) and decrease with increasing CE. For example, in the case of propofol while SBEβCD (MW 2163 Da) has the highest CE the formulation bulk is lowest for RMβCD (MW 1312 Da). All things being equal, the CD derivatives will result in a greater increase in the formulation bulk than their parent CDs. Although the aqueous solubility of drug complexes of the parent αCD, βCD and γCD is in general much lower than those of their derivatives, their solubilities are frequently sufficient to prevent the dissolution rate limited drug absorption from a solid dosage form in the gastrointestinal tract.

Drug/CD 1:1 complexes are most common in dilute aqueous solutions and the values of the $K_{1:1}$ stability constants are usually between 10 and 2000 M^{-1}. Values much greater than 5000 M^{-1} are very rarely observed. There are, however, two known exceptions. Sugammadex (Bridion®, Merck, USA) is a γCD derivative that was designed to specifically bind the neuromuscular blocking agent rocuronium. The binding constant of the rocuronium/sugammadex complex has been determined to be 1.8×10^7 M^{-1} [48]. Another example is complexation of SBEβCD with some adamantane ozonide antimalarial drug candidates where the binding constants are about 10^6 M^{-1} [49]. The very snug fit of the adamantane ozonide structure within the CD cavity resulted in this unusually high binding constant.

3.4 Cyclodextrin Complexes

Although CDs are able to form non-inclusion complexes and complex aggregates, it is believed that most frequently individual CD molecules and CD complexes predominate in dilute aqueous solutions [36]. Methods for determining the value of stability constant of the guest molecule (i.e. drug)-CD complexes are based on titrating a certain chemical or physical property of the guest molecule with the CD molecule in relatively dilute solutions and then analyzing the concentration dependencies. Additive properties of the drug molecule that can be addressed in this manner include aqueous solubility (i.e. the previously mentioned phase-solubility relationships), chemical reactivity, molar absorptivity and other spectrophotometric properties, NMR chemical shifts and other spectroscopic properties, pKa values and HPLC retention time. It must, however, be remembered that most pharmaceutical solutions are concentrated, non-ideal systems where both free CD molecules and their complexes interact with each other as well as with other pharmaceutical excipients. Therefore, studies of dilute aqueous CD solutions cannot always be used to explain phenomenon observed in more concentrated CD solutions and pharmaceutical formulations without further validation.

3.4.1 Self-Assembly of Cyclodextrins and their Complexes

It is known that CDs and their complexes self-associate to form aggregates [33, 36, 42, 45, 50–56]. The natural αCD, βCD and γCD self-associate in dilute aqueous solutions but the degree of aggregation is very low or less than 0.1% of the total amount of dissolved CD [33, 53]. Likewise, the more water-soluble CD derivatives, such as HPβCD, RMβCD, SBEβCD and HPγCD, have a relatively low tendency to aggregate in dilute aqueous solutions. Formation of a water-soluble inclusion complex with a lipophilic poorly soluble drug increases, in general, the aggregate formation. The degree of aggregation depends on the availability of guest/CD complexes in the aqueous solution, increasing with increasing CE. Thus addition of excipients that decrease the CE or increasing the temperature will reduce the aggregation. The aggregate size depends on the properties of both the guest and the CD molecules [45, 46].

3.4.2 Thermodynamic and Driving Forces for Complexation

The van't Hoff equation is used to explain the effects of temperature on equilibrium constants such as stability constants of cyclodextrin complexes ($K_{1:1}$):

$$\ln K_{1:1} = \text{Intercept} - \frac{\Delta \text{H}}{\text{R}} \cdot \frac{1}{\text{T}} \tag{3.8}$$

where ΔH is the standard enthalpy change, R is the gas constant and T is the temperature in Kelvin. A plot of T^{-1} versus $\ln K_{1:1}$ will give a straight line with slope equal to −(ΔH/R). The standard free energy change (ΔG) for equilibrium processes is calculated from Equation 3.8:

$$\Delta G = -R \cdot T \cdot \ln K_{1:1} \tag{3.9}$$

And then finally the standard free entropy change from the ΔG = ΔH − T · ΔS or:

$$-\Delta S = \frac{\Delta G - \Delta H}{T} \tag{3.10}$$

The thermodynamic parameters (i.e. ΔH and ΔS) for several series of drugs and other compounds have been determined and analyzed [57–61]. Thermodynamic parameters of several other drugs are listed in Table 3.4. The complex formation is almost always associated with a relatively large negative ΔH and ΔS which can either be positive or negative. Also, the complex formation is largely independent of the chemical properties of the guest (i.e. drug) molecules. The association of binding constants with substrate polarizability suggests that van der Waal's forces are important in the complex formation [62]. Hydrophobic interactions are associated with a slightly positive ΔH value and a large positive ΔS value and, therefore, classical hydrophobic interactions are entropy-driven, suggesting that they are not involved with cyclodextrin complexation since, as indicated, these are enthalpically driven processes. Furthermore, for a series of guests there tends to be a linear relationship between enthalpy and entropy, with increasing enthalpy related with less negative entropy values and this effect, termed compensation, is often correlated with solvent restructuring as a driving force in complex formation. Thus, it was thought that the main driving force for the complex formation was the release of enthalpy-rich water from the cyclodextrin cavity. The water molecules located inside the cavity cannot satisfy their hydrogen bonding potentials and therefore they are of higher enthalpy. The energy of the system is lowered when these enthalpy-rich water molecules are replaced by suitable guest molecules which are less polar than water. However, recent studies have indicated that the enthalpy and entropy changes of the complexation are not good criteria to judge whether a particular driving force is present or important [63]. It is now thought that van der Waals interaction and hydrophobic interaction constitute the major driving forces for cyclodextrin complexation, whereas electrostatic interaction and hydrogen bonding can significantly affect the conformation of particular inclusion complex [63].

3.5 Effects on Drug Stability

CDs can enhance both chemical and physical stability of drugs and enhance their shelf-life. CDs reduce the rate of drug hydrolysis, oxidation, steric rearrangement, racemization, isomerization and polymerization as well as enzymatic decomposition of drugs [29, 64]. CD complexes are always under kinetic equilibrium in aqueous solutions, i.e., the molecules forming a complex are in equilibrium with free molecules in the solution, where the rate of formation and dissociation of the complex is diffusion controlled. In kinetic studies the stability constant ($K_{m:n}$ in Equation 3.1) can be determined from the stabilizing or destabilizing effects of CDs on a given drug. If a CD has, for example, a stabilizing effect on a drug molecule in an aqueous solution, then the rate of disappearance of the drug will

Table 3.4 *The standard enthalpy change (ΔH) and the standard entropy change (ΔS) of the stability constant ($K_{1:1}$) for some drug · cyclodextrin complexes.*

Cyclodextrin[a]	Drug	pH	ΔH (kJ/mol)	ΔS (J/mol/K)	References
HPαCD	Hydrocortisone	–	−32	−70	129
βCD	Phenytoin, unionized	7	−38	−67	130
	Phenytoin, ionized	7	−21	−21	130
βCD	Naproxen	–	−13	18	131
βCD	Adenine arabinoside	7	−28	−64	132
βCD	Adenosine	7	−21	−53	132
βCD	Ibuprofen (pKa 5.2)	2	−29	15	133
		4	−32	4	133
		5	−29	3	133
		6	−17	34	133
βCD	Diazepam (pKa 3.3)	2	−0.2	70	133
		3	−3.3	69	133
		4	−17	22	133
		6	−18	19	133
βCD	Hydrochlorothiazide	5	−40	62	133
	(pKa 8.8 and 10.4)	8	−39	59	133
		9	−42	70	133
AβCD	Naproxen (pKa 4.2)	6	−5.9	51	134
HPβCD	Aspirin (pKa 3.5)	1	−68	−170	135
HPβCD	Acetazolamide	–	−18	−26	129
HPβCD	17β-Estradiol	–	−71	−150	129
HPβCD	Hydrocortisone	–	−20	−6	129
HPβCD	Methyl acetylsalicylate	1	−55	−130	135
HPβCD	Methyl salicylate	1	−63	−140	135
HPβCD	Naproxen	6	−12	23	136
SBEβCD	Hydrocortisone	–	−18	5	137
SBEβCD	Naproxen	–	−25	7	137
SBEβCD	Prednisolone	–	−15	11	137
SBEβCD	Progesterone	–	−31	−19	137
SBEβCD	Testosterone	–	−27	−26	137
AγCD	Naproxen	6	−3.6	44	134
HPγCD	Aspirin	1	−28	−56	135
HPγCD	Methyl acetylsalicylate	1	−75	−190	135
HPγCD	Methyl salicylate	1	−73	−180	135

Note:
[a]HPαCD: 2-hydroxypropyl-α-cyclodextrin; βCD: β-cyclodextrin; AβCD: acetyl-β-cyclodextrin; HPβCD: 2-hydroxypropyl-β-cyclodextrin; SBEβCD: sulfobutyl ether β-cyclodextrin; AγCD: acetyl-γ-cyclodextrin; HPγCD: 2-hydroxypropyl-γ-cyclodextrin.

decrease when the CD-to-drug concentration ratio is increased. The observed first-order rate constant (k_{obs}) for a degradation of a drug in an aqueous CD solution is the weighted average of the first-order rate constant for the degradation of the free drug (k_f) and the first-order rate constant for the degradation of the drug within the CD complex (k_c):

$$k_{obs} = k_f \cdot f_f + k_c \cdot f_c \tag{3.11}$$

where f_f is the fraction of free drug and f_c is the fraction of drug in complex. If we assume that only 1:1 drug-CD complex (D/CD) is being formed and that the total drug concentration ($[\mathrm{D}]_\mathrm{T}$) is the sum of the concentration of free drug ([D]) and the concentration of the complex ([D/CD]) the following equations are obtained:

$$[\mathrm{D}]_\mathrm{T} = [\mathrm{D}] + [\mathrm{D/CD}] \tag{3.12}$$

$$[\mathrm{CD}]_\mathrm{T} = [\mathrm{CD}] + [\mathrm{D/CD}] \tag{3.13}$$

$$K_{1:1} = \frac{[\mathrm{D/CD}]}{[\mathrm{D}] \cdot [\mathrm{CD}]} \tag{3.14}$$

$$f_f = \frac{[\mathrm{D}]}{[\mathrm{D}] + [\mathrm{D/CD}]} = \frac{1}{1 + K_{1:1} \cdot [\mathrm{CD}]} \tag{3.15}$$

$$f_c = 1 - f_f = \frac{K_{1:1} \cdot [\mathrm{CD}]}{1 + K_{1:1} \cdot [\mathrm{CD}]} \tag{3.16}$$

$$k_\mathrm{obs} = \frac{k_f + k_\mathrm{c} \cdot K_{1:1} \cdot [\mathrm{CD}]}{1 + K_{1:1} \cdot [\mathrm{CD}]} \tag{3.17}$$

$$-\frac{\mathrm{d}[\mathrm{D}]_\mathrm{T}}{\mathrm{dt}} = k_\mathrm{obs} \cdot [\mathrm{D}]_\mathrm{T} = \left(\frac{k_f + k_c \cdot K_{1:1} \cdot [\mathrm{CD}]}{1 + K_{1:1} \cdot [\mathrm{CD}]}\right) \cdot [\mathrm{D}]_\mathrm{T} \tag{3.18}$$

If the total CD concentration is much greater than the total drug concentration ($[\mathrm{CD}]_\mathrm{T} \geq 10 \cdot [\mathrm{D}]_\mathrm{T}$) then it can be assumed that $[\mathrm{CD}] \approx [\mathrm{CD}]_\mathrm{T}$:

$$k_{obs} = \frac{k_f + k_c \cdot K_{1:1} \cdot [\mathrm{CD}]_\mathrm{T}}{1 + K_{1:1} \cdot [\mathrm{CD}]_\mathrm{T}} \tag{3.19}$$

Equation 3.19 can then be rearranged into several different formats including those of the Lineweaver-Burk plot of $(k_f - k_\mathrm{obs})^{-1}$ versus $([\mathrm{CD}]_\mathrm{T})^{-1}$ that gives a straight line where k_c can be obtained from the intercept and $\mathrm{K}_{1:1}$ from the slope:

$$\frac{1}{k_f - k_\mathrm{obs}} = \frac{1}{K_{1:1} \cdot (k_f - k_c)} \cdot \frac{1}{[\mathrm{CD}]_\mathrm{T}} + \frac{1}{(k_f - k_c)} \tag{3.20}$$

Alternatively, k_c and $K_{1:1}$ can be obtained by non-linear fitting of k_obs according to Equation 3.19.

Chlorambucil is a nitrogen mustard-type anticancer drug and is as such susceptible to hydrolysis in aqueous solutions (Table 3.5). SBEβCD is negatively charged, while both HPβCD and RMβCD are uncharged. At pH 7.35 chlorambucil is a monovalent anion and, thus, the value of the stability constant ($K_{1:1}$) is affected by the charge of the cyclodextrin molecule. The anionic cyclodextrin, SBEβCD, is repelled from the negatively charged chlorambucil and, thus the value of $K_{1:1}$ is decreased by a factor of about 3 compared to the uncharged cyclodextrins. The k_f/k_c ratio is on the other hand influenced by the substituents on the cyclodextrin molecules where the large charged groups on the SBEβCD molecules provide better protection of the included drug molecule than the small uncharged groups on the HPβCD and RMβCD molecules. Since the observed hydrolytic degradation rate constants are pH-sensitive, and since the complexation will change with ionization stage of the drug, the values of k_f, k_c and $K_{1:1}$ are all pH-dependent.

Table 3.5 *The value of the observed first-order rate constants for the hydrolysis of chlorambucil at pH 7.35 and 40°C.*

k_f (min^{-1})	Cyclodextrin[a]	k_c (min^{-1})	k_f/k_c	$K_{1:1}$ (M^{-1})
$6.5 \pm 0.1 \cdot 10^{-2}$	SBEβCD	$1.0 \pm 0.4 \cdot 10^{-3}$	65	1400 ± 50
	HPβCD	$3.6 \pm 1.0 \cdot 10^{-3}$	18	3400 ± 720
	RMβCD	$3.3 \pm 0.6 \cdot 10^{-3}$	20	3550 ± 530

Note:
[a]SBEβCD: sulfobutyl ether-β-cyclodextrin; HPβCD: 2-hydroxypropyl-β-cyclodextrin; RMβCD: randomly methylated β-cyclodextrin.

3.6 Cyclodextrins and Drug Permeation through Biological Membranes

Water molecules are bound to the skin surface as well as within the outermost layer of the skin, the stratum corneum, but the resulting unstirred water layer (UWL) is relatively thin and permeable [65]. Mucosal epithelium (mucosa), on the other hand, contains mucosal cells that secretes mucus, a gel-like fluid containing mainly water (~95%) and mucin [66]. Mucin forms hydrogen bonds with the surrounding water molecules, enhancing cluster formation and, consequently, decreased water mobility. This leads to up to 10^5-fold enhancement in the thickness of the UWL in, for example, the gastrointestinal tract, the respiratory tract, the ocular-rhino-otolaryngeal tracts and the reproductive tract [67] The tear film on the eye surface is about 8 μm thick but the thickness of the gastrointestinal mucus layer can be about 100 μm. Under unstirred *in vitro* conditions the thickness of the UWL can be much greater, even in the absence of mucus [68–71]. Passive drug permeation through multilayer barriers, such as through the UWL and lipophilic epithelium, is described as a series of additive resistances analogous to electric circuits [72–75] Assuming independent and additive resistances of the individual layers, the total resistance (R_T) of a simple membrane (Figure 3.4) can be defined as:

$$J = P_T \cdot C_V = R_T^{-1} \cdot C_V = (R_D + R_M)^{-1} \cdot C_V = \left(\frac{1}{P_D} + \frac{1}{P_M}\right)^{-1} \cdot C_V = \left(\frac{P_D \cdot P_M}{P_D + P_M}\right) \cdot C_V \tag{3.21}$$

where J is the flux of the drug through the membrane, P_T is the overall permeability coefficient, C_V is the drug concentration in the vehicle (i.e. donor phase), R_D and R_M, and P_D and P_M are the resistances and permeability coefficients in the UWL at the donor side and within the membrane, respectively [75, 76] The relationship between the permeation coefficient (P) and the diffusion coefficient (D) is given by Equation 3.22:

$$P = \frac{D \cdot K}{h} \tag{3.22}$$

where h is the thickness (h_D, or h_M) and K is the partition coefficient between the aqueous phase and the membrane. For P_D the value of K is unity. Finally, D can be estimated from the Stokes-Einstein equation:

$$D \approx \frac{R \cdot T}{6\pi \cdot \eta \cdot r \cdot N} \tag{3.23}$$

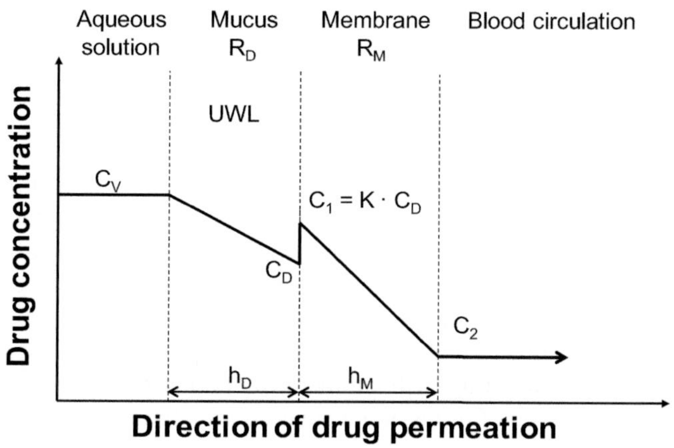

Figure 3.4 *Drug permeation through a simple two-layer barrier consisting of an unstirred water layer (UWL) and a lipophilic biological membrane. The aqueous solution is the vehicle containing the dissolved drug,* R_D, h_D, R_M *and* h_M *are the resistance and the thickness of the UWL (D) and the membrane (M), respectively.* C_V *is the drug concentration in the vehicle,* C_D *is the drug concentration in the UWL immediate to the membrane surface,* C_1 *and* C_2 *are the drug concentrations within the membrane at the outer and inner surface, respectively,* D_D *and* D_M *are the drug diffusion constants in the UWL and the membrane, respectively, and K is the drug partition coefficient between the UWL and the membrane.*

where R is the molar gas constant, T is the absolute temperature, η is the apparent viscosity within the UWL or the lipophilic membrane, r is the radius of the permeating drug molecule and N is Avogadro's number. Thus, the diffusion constant within the UWL (D_D) will decrease with increasing viscosity of the layer as well as with increasing molecular weight of the drug. For example, small lipophilic drug molecules frequently possess a large permeability coefficient through the lipophilic membrane (i.e. large P_M value) and, thus, may be able to permeate lipophilic membrane much faster than they can be transported through the UWL. Under such conditions, diffusion through the UWL becomes the rate-limiting step in the absorption process. Presence of mucin in the mucus layer not only increases the thickness (h) of this UWL but also its viscosity (η), both of which will increase its resistance (R_D) with a consequent decrease in permeability. Other surface structures, such as villi and microvilli, can also increase h and η of the UWL. Studies have shown that drug diffusion through mucus is up to 100-times slower than through pure water [77]. In the preceding discussion it has been assumed that the drug permeated through both the UWL and the membrane is via passive diffusion. It is, however, well known that although permeation through the UWL is always passive, the drug permeation through biological membranes can either be passive, facilitated or active. No matter how the drug is transported through the membrane, the drug molecules will always have to permeate the UWL to reach the membrane surface. Thus, Equation 3.22 can also be applied to describe the effects of the UWL on active transport of drugs through biomembranes.

There are numerous studies on the effects of cyclodextrins on drug permeation through biological membranes. The following are some general observations [75, 76, 78]. First, CDs are unable to enhance drug delivery from non-aqueous dermal vehicles through biomembranes or when an aqueous phase is not in contact with the skin surface (i.e. no enhancement if no UWL is present at the membrane surface). For example, CDs enhance dermal delivery of drugs from o/w creams but not w/o creams [79, 80]. Second, hydrophilic CDs are only able to enhance drug delivery through biological membranes if the UWL is the main barrier or contributes significantly to the overall permeation barrier of the membrane. In other words, CDs only enhance permeation when R_D is about equal or greater than R_M [75]. Third, CDs and CD complexes are unable to permeate biological membranes as indicated by their very poor oral bioavailability. Thus, drug molecules have to be released from the complexes before they can permeate the membranes. Fourth, CDs do not, in general, enhance membrane permeation of hydrophilic water-soluble drugs. Fifth, excess CD, more than is needed to solubilize the drug, will hamper drug permeation. Sixth, CDs are able to prevent enzymatic degradation of drugs at the aqueous membrane exterior.

3.7 Drug Solubilization in Pharmaceutical Formulations

The most common pharmaceutical application of CDs is their use as solubilizers of lipophilic drugs in aqueous media. The increased apparent solubility can enable solution-based dosage forms such as parenteral IV (intravenous) formulation and oral liquids. In addition, increasing the apparent solubility of a drug can, through the Noyes-Whitney equation, increase the drug dissolution rate and, for compounds whose oral bioavailability is limited by solubility or dissolution rate, can act to increase oral bioavailability. The use of CDs is often preferred to organic solvents from both a toxicological perspective but also from a mechanistic point of view. Based on an inclusion complex formation, CDs frequently solubilize compounds as a linear function of their concentration (i.e. display A_L-type phase-solubility diagrams). This means that as a solution is administered, both the drug and CD concentration are reduced in a linear manner suggesting that precipitation is, at least theoretically, not likely after either oral or IV dosing. Organic solvents, on the other hand, solubilize solutes as a log function of their concentration as described by the extended Hildebrand equation [81, 82]. This log-linear relationship means that as an organic solvent is introduced into an aqueous environment, the solubilizing power of the formulation is exponentially lost and precipitation can occur. Depending on the nature of this precipitation, the drug can be effectively removed from the equilibria required for absorption from the gastrointestinal tract. For parenteral liquids, precipitation can occur at the site of injection or at other loci in the body.

In most cases, only limited amounts of CDs can be included in drug formulations. According to Equation 3.7, the increase in the formulation weight is proportional to the molecular weight of the CD and inversely proportional to the value of the CE. The mean CE (CE $\pm$ standard deviation) of 23 different drugs (MW 356 $\pm$ 205 Da) with HPβCD (MW 1400 Da) was determined to be 0.57 $\pm$ 0.75 in pure water or aqueous buffer solutions (Table 3.6) [47]. This indicates that less than two out of every five CD molecules is forming a pharmaceutically relevant complex with the drug. Only 5 out of 23 drugs had CE greater than unity and 8 had CE of 0.1 or lower. HPβCD complexation of a drug with MW 356 Da and CE 0.57 will result in an eleven-fold increase in the formulation bulk and if the CE is

Table 3.6 *The solubility of the unionized drug in pure water or aqueous buffer solution, the slope of the phase-solubility diagram of the unionized drug in aqueous 2-hydroxypropyl-β-cyclodextrin (molecular weight 1400 Da; HPβCD) solution or aqueous randomly methylated β-cyclodextrin (molecular weight 1312 Da; RMβCD) solution, the correlation of the linear slope (Corr.), and the complexation efficiency (CE) calculated from the slope according to Equation 3.5, at ambient temperature.*

Drug	Molecular weight (Da)	Cyclodextrin	Solubility (mg/ml)	Slope	Corr.	CE
Acetazolamide	222.3	HPβCD	0.64	0.197	0.995	0.246
Alfaxalone	332.5	HPβCD	0.00	0.553	0.995	1.24
Calcipotriol	412.6	RMβCD	0.00	0.278	1.000	0.385
Carbamazepine	236.3	HPβCD	0.26	0.404	0.991	0.679
Cyclosporine A	1202.6	HPβCD	0.01	0.004	0.978	0.004
Dexamethasone	392.5	HPβCD	0.16	0.246	1.000	0.326
Dextromethorphan	271.4	RMβCD	0.09	0.663	0.998	1.96
Diethylstilbestrol	268.4	HPβCD	0.00	0.739	0.997	2.82
Ergotamine	581.7	HPβCD	0.00	0.001	0.967	0.001
Estradiol	272.4	HPβCD	0.09	0.243	0.998	0.322
Finasteride	372.6	RMβCD	0.04	0.458	0.994	0.844
Flunitrazepam	313.3	HPβCD	0.00	0.012	0.998	0.012
Hydrocortisone	362.5	HPβCD	0.42	0.667	1.000	2.00
Ketoprofen	254.3	HPβCD	0.01	0.601	0.997	1.51
Methazolamide	236.3	HPβCD	0.70	0.092	0.999	0.101
Miconazole	416.1	HPβCD	0.09	0.052	0.993	0.055
Naproxen	230.3	HPβCD	0.12	0.282	0.998	0.393
Omeprazole	345.4	HPβCD	0.00	0.004	0.974	0.004
Oxazepam	286.7	RMβCD	0.05	0.140	0.968	0.163
Prazepam	324.8	HPβCD	0.00	0.018	0.995	0.018
Pregnenolone	316.5	HPβCD	0.03	0.110	0.999	0.123
Progesterone	314.5	HPβCD	0.00	0.240	0.998	0.315
Propofol	178.3	HPβCD	0.16	0.602	1.000	1.51
Sulfamethoxazole	253.3	HPβCD	0.39	0.359	0.998	0.561
Tamoxifen	371.5	HPβCD	0.00	0.004	1.000	0.004
Terfenadine	471.7	HPβCD	0.00	0.165	0.988	0.197
Triclosan	289.5	HPβCD	0.00	0.391	0.998	0.643

Source: [47].

0.1 then an over 40-fold increase in the formulation bulk will be observed. Several methods can be applied to increase the CE and several are listed in Table 3.7 [83]. It should also be emphasized that numerous pharmaceutical excipients have been found to reduce the CE and, thus, it is important to determine the CE in an aqueous environment that resembles the final formulation as closely as possible.

The following is a brief review of the usage of CDs in pharmaceutical formulations with the main emphasis on their effects on the aqueous solubility of drugs and drug permeation through biological membranes.

Table 3.7 *Methods that have been applied to enhance the complexation efficacy (CE).*

Effect	Consequences	References
Drug ionization	Unionized drugs do usually form more stable complexes than their ionic counterparts. However, ionization of a drug increases its apparent intrinsic solubility (i.e. increasing apparent S_0) that can result in enhanced complexation	138–141
Salt formation	It is sometimes possible to enhance the apparent intrinsic solubility of a drug through salt formation (i.e. increasing apparent S_0)	142–145
Acid/base ternary complexes	It has been shown that certain organic hydroxy acids (such as citric acid) and certain organic bases are able to enhance the complexation efficiency by the formation of ternary drug/CD/acid or base complexes (i.e. increasing the apparent S_0 and/or $K_{1:1}$)	146–150
Polymer complexes	Water-soluble polymers form a ternary complex with drug/CD complexes increasing the observed stability constant of the drug/CD complex ($K_{1:1}$)	151, 152
Metal complexes	Many drugs are able to form somewhat water-soluble metal complexes without decreasing the drugs' ability to form complexes with CDs (i.e. increasing apparent S_0). Thus, the complexation efficiency can be enhanced by the formation of drug:metal ion:CD complexes	153
Co-solvents	Addition of co-solvents to the complexation media can increase the apparent intrinsic solubility (S_0) of the drug that can lead to enhanced CE	154, 155
Ion pairing	Ion pairing of positively charged compounds with negatively charged CDs enhances the complexation efficiency (i.e. increasing the apparent $K_{1:1}$)	156
Combination of two or more methods	Frequently the complexation efficiency can be enhanced even further by combining two or more of the above-mentioned methods. For example, drug ionization and the polymer method, or solubilization of the CD aggregates by adding both polymers and cations or anions to the aqueous complexation medium	144, 153, 157

3.7.1 Oral Drug Delivery

In general, the effects of CDs on drug bioavailability can be classified according to the biopharmaceutics classification system (BCS, see Table 3.8). The oral bioavailability of hydrophilic CDs is normally less than 3% and thus intact drug/CD complexes are unable to permeate the lipophilic gastrointestinal (GI) epithelium. Hence, CDs and especially

Table 3.8 *The biopharmaceutics classification system (BCS) and the effect of drug/CD complexation on drug bioavailability from instant release solid dosage forms after oral administration. The flux of drug from the GI tract (J) through the mucosal membrane is proportional to the drug's permeability constant (P) and the concentration of dissolved drug (C_{Aq}) or its aqueous solubility.*

$J = P \cdot C_{Aq}$

Parameters	Class I	Class II	Class III	Class IV
Aqueous solubility (C_{Aq})[a]	High	Poor	High	Poor
Permeability (P)[b]	High	High	Poor	Poor
Absorption rate control	Gastric emptying	Dissolution	Permeability	Dissolution and permeability
Examples of drugs and relative bioavailability ($F_{rel} = [AUC]_{CD}/[AUC]_{no\ CD}$)	Some NSAIDs	4-Biphenylacetic acid (2.9) Carbamazepine (5.5) Digoxin (5.4) Fenbufen (5.5) Glibenclamide (6.2) Ketoprofen (2.9) Miconazole (2.3) Nifedipine (2.9) Phenytoin (5.0) Piroxicam (1.4) Tolbutamide (81.5)	Acyclovir (1.1) Diphenhydramine HCl (0.9)	Cyclosporin A (4.7) Saquinavir (9.3)

Notes:
[a]Solubility (S) of the drug dose (D) in aqueous solution (high: D:S $\leq$ 250 ml; poor: D:S > 250 ml).
[b]Permeability of a drug through a lipophilic biological membrane.
Source: Modified from [158].

the natural CDs and their hydroxypropyl derivatives (HPαCD, HPβCD and HPγCD) and sulfobutylether β-cyclodextrin (SBEβCD), and their complexes belong to BCS Class III. However, it has been shown that the rates for formation and dissociation of drug/CD complexes are very close to the diffusion controlled limits with complexes being constantly formed and dissociated [84]. Consequently, the presence of water-soluble drug/CD complexes right at the hydrated epithelial surface can increase the availability of dissolved drug molecules, especially of lipophilic drugs with poor aqueous solubility [85]. The dissolved drug/CD complexes will increase the drug concentration gradient over the UWL (i.e. the aqueous GI mucus), enhancing drug permeation through the UWL to the surface of the lipophilic epithelium where the complex will dissociate. Thus the best results are obtained when the oral absorption is dissolution limited, i.e. with Class II and Class IV drugs, but less effect is observed with drugs that belong to Class I, or with drugs belonging to Class III where low bioavailability is due to the inability of the dissolved drug molecules to permeate the mucosal epithelium (Table 3.8). Also, since excess CD will hamper drug partition from the aqueous exterior into the lipophilic epithelium, excess CD, more than is needed to solubilize the drug, will slow down drug absorption from the GI tract [86]. Likewise, CD complexation of water-soluble drugs (e.g. BCS Class I and Class III drugs) can delay their absorption from the GI tract. It must, however, be remembered that CDs can have other effects, such as drug stabilization, and generation and stabilization of supersaturated drug solutions, that are difficult to account for and, thus, these general guidelines might not apply to all drugs under all conditions.

3.7.2 Sublingual, Buccal, Nasal, Pulmonary, Rectal and Vaginal Drug Delivery

Some examples of sublingual, buccal, nasal, pulmonary, rectal and vaginal drug delivery from formulations containing CDs are shown in Table 3.9. For instant release oral dosage forms, the drug:solubility ratio (D:S) should preferably be less than 250 ml. For many other types of delivery forms, the D:S ratio should be much less or from about 5 ml for buccal drug delivery to 0.3 ml for nasal drug delivery. Thus, solubility requirements can much stricter for non-oral formulations. Enzymatic activity at these mucosal surfaces can be quite high and, thus, the observed permeation enhancements are sometimes due to the prevention of enzymatic degradation of the active compound. This is especially true in the

Table 3.9 *Examples of cyclodextrin containing formulations for buccal, sublingual, nasal, pulmonary, rectal and vaginal drug delivery.*

Drug	Cyclodextrin	D:S ratio (ml)[a]	Reference
Buccal and sublingual drug delivery:		≤5	
Atenolol	βCD, MβCD, RMβCD		159, 160
Cyclosporin A	αCD		161
Clomipramide	HPβCD		162
Danazol	HPβCD, SBEβCD		163, 164
17β-Estradiol	HPβCD		165–167
Nasal drug delivery:		≤0.3	
17β-Estradiol	DMβCD		168
Midazolam	SBEβCD		169, 170
Pulmonary drug delivery:		≤0.5	
Beclomethasone	γCD		171
Budesonide	γCD		172
Cyclosporin A	HPαCD		173
Itraconazole	HPβCD		174
Rectal drug delivery:		≤3	
Edaravone	HPβCD		90
Flurbiprofen	HPβCD		91
Vaginal drug delivery:		≤3	
Itraconazole	HPβCD		92
Natamycin	γCD		93

Note:
[a] Dose:Solubility ratio (D:S) indicating the maximum amount of water in ml needed to solubilize one drug dose.

case of proteins and peptide drugs [20]. Hydrophilic CDs, such as HPβCD, have been used successfully in a number of sublingual and buccal solid dosage forms of lipophilic poorly water-soluble drugs. The hydrophilic CDs enhance drug dissolution in the aqueous saliva and deliver the drug molecules to the lipophilic membrane. In nasal delivery CDs are used to solubilize the drug in the aqueous nasal spray. Only about 100–150 μl of the solution can be sprayed into each nostril and, thus, the D:S ratio is less than 0.2–0.3 ml. CDs are more readily absorbed from the lungs than from other routes of delivery but CDs that are considered safe for parenteral administration are considered safe for pulmonary drug delivery [20, 87–89]. Drug/HPβCD complexes have been included in aqueous solutions and *in situ* forming hydrogels (liquid suppositories) for rectal drug delivery [90, 91], and vaginal cream [92] and gel [93].

3.7.3 Ophthalmic Drug Delivery

The aqueous tear fluid and mucus layer on the eye surface form an UWL, i.e. an aqueous diffusion barrier for topical drug delivery to the eye. Furthermore, the constant secretion of the tear fluid and its draining from the eye surface severely limit the drug contact time with the eye surface. After eye drop administration and the initial drainage the precorneal half-life of topically applied drugs is between 1 and 3 min [78, 94]. Thus, increasing the aqueous solubility of poorly soluble drugs and increasing their contact time with the eye surface through CD complexation and enhanced mucoadhesion enhances the ophthalmic bioavailability of topically applied drugs [95]. Furthermore, CD complexation can reduce local drug irritation in the eye [96–98].

3.7.4 Dermal and Transdermal Drug Delivery

Depending on the vehicle composition, CDs can either increase or decrease dermal and transdermal drug delivery [75]. The main barrier to percutaneous absorption is the outermost layer of the skin, the *stratum corneum*. Only 0.02% of HPβCD is able to permeate through intact hairless mouse skin over a 24 h period while 24% of HPβCD was able to permeate the skin when the stratum corneum had been removed [99]. Even somewhat lipophilic cyclodextrin derivatives, like methylated βCD, do not easily permeate intact skin. CDs do not enhance percutaneous drug permeation if the UWL does not contribute to the skin barrier. Thus, CDs do not enhance dermal and transdermal drug delivery from ointments, w/o creams and other drug formulations where the thickness UWL is negligible but can have an effect on drug delivery from hydrogels, o/w creams and other drug formulations where the formulation induces UWL formation at the skin surface [75]. Furthermore, CDs have been shown to deliver drugs to the hair follicles and other openings on the skin surface [100]. Excess CD, more than is needed to solubilize the drug and deliver it through the UWL, will reduce the percutaneous drug permeation. In this way, CDs have been added to sunscreen formulations and mosquito repellents to reduce systemic absorption of the active ingredients into the systemic blood circulation and to enhance their duration [101–104].

3.7.5 Injectable Formulations

Injectable formulations of lipophilic and poorly water-soluble drugs frequently consist of mixtures of water, organic solvents and surfactants. Such solvent systems frequently result

Table 3.10 *The composition of the current diazepam formulation (Valium®) and a HPβCD (Cavasol) formulation.*

	Current formulation	HPβCD formulation
Diazepam	5 mg	5 mg
Propylene glycol	40%	
Ethyl alcohol	10%	
Sodium benzoate/benzoic acid	5%	
Benzyl alcohol	1.5%	
HPβCD sodium salt		6%
Sodium chloride		0.6%
Water for injection	∼43%	∼93%

Source: [106, 158].

in drug precipitation, hemolysis, inflammation and pain on injection [105]. In some cases organic solvents and surfactants can be replaced by CDs. For example, the diazepam parenteral solutions containing propylene glycol, ethyl alcohol and benzoate and only about 43% water can be replaced by a CD formulation that contains over 90% water (Table 3.10). In addition, the diazepam formulation containing CDs is isotonic while the formulation containing organic solvent is very hypertonic (osmolarity 7775 mOsm/kg) [106]. Also, milky white parenteral solutions (e.g. propofol in o/w emulsion) can be replaced by clear colorless drug solutions (Table 3.11). On parenteral administration, especially after intravenous injection, the drug is both rapidly and efficiently released from the CD complex upon dilution and protein binding of the drug [21, 107, 108]. Studies have indicated that parenteral doses of CD have to be very high to have a detectable effect on drug protein binding and drug pharmacokinetics [108–116]. There are, however, a couple of exceptions such as the sugammadex binding of rocuronium [48, 117] and SBEβCD binding of some ozonide antimalarial drug candidates [49, 118]. In both cases the binding constant is $\geq 10^6\ M^{-1}$ or sufficiently large to efficiently bind the drugs in blood plasma and clear them with the urine as hydrophilic drug/CD complexes.

Table 3.11 *The composition of the current propofol formulation (Diprivan®) and the SBEβCD (Captisol) formulation [106, 175]. The two formulations give identical pharmacodynamic and pharmacokinetic profile in pigs [112].*

	Oil-in-water emulsion	SBEβCD formulation
Propofol	10 mg	10 mg
Soybean oil	100 mg	
Glycerol	22.5 mg	
Egg lecitin	12 mg	
Disodium edetate	0.05 mg	
NaOH	ad pH 12	
SBEβCD sodium salt		220 mg
Water for injection	ad 1.00 ml	ad 1.00 ml

3.8 Toxicology and Pharmacokinetics

Most CDs of current pharmaceutical interest (Tables 3.1 and 3.2) are hydrophilic and, due to their bacterial digestion origins, of high molecular weight (from almost 1000 to over 2000 Da), a large number of hydrogen donors and acceptors, and high hydrophilicity ($\text{Log}K_{o/w}$ between −6 and −17), thus, their oral bioavailability is generally below 3% (Table 3.12) [75]. The oral bioavailability of HPβCD in humans is between 0.5 to 3.3% with 50 to 65% of the oral dose excreted intact in the feces with the remainder mainly being metabolized by bacteria in the colon. Absorbed CD is rapidly excreted with the urine. Toxicological studies have demonstrated that orally administered CDs of pharmaceutical interest are generally nontoxic due to lack of absorption from the gastrointestinal tract [12, 20]. However, RMβCD is somewhat more lipophilic ($\text{Log}K_{o/w} = -6$) and has fewer hydrogen bond donors than the other CDs. Consequently its oral bioavailability is slightly higher or up to 15% in rats [22]. Currently, oral administration of methylated βCDs is limited by their potential toxicity. Oral administration of αCD (up to 6 grams per day) is well tolerated and is not associated with any observable harmful effects [119, 120]. The same applies to βCD [121] (up to 0.5 grams per day), γCD [19] (up to 10 grams per day), HPβCD [122] and SBEβCD [12, 21, 75]. The main side effects of oral administration of high doses of these CDs are similar to those related to poorly digestible carbohydrates and include diarrhea, flatulence and soft stools. αCD, βCD and HPβCD can all be found is various oral drug products and all three parent cyclodextrins (i.e. αCD, βCD and γCD) are being used in dietary supplement products. The maximum CD dose that can be found in oral dietary or drug products is shown in Table 3.12.

Table 3.12 *Some pharmacokinetic data on common CD and CD derivatives.*

		After iv injection to rats[a,b]			After iv injection to humans[c]		Max. dosage in marketed drug products[d,e]	
Cyclodextrin	Rats f_{oral} (%)	V_D (L/kg)	$t_{1/2}$ (h)	$f_{urine\ unch.}$ (%)	V_D[a] (L/kg)	$t_{1/2}$[a] (h)	Oral (mg/day)	iv (mg/day)
αCD	2–3		0.4	~90			6,000[e]	1
βCD	~0.6	0.2	0.4	~90			500[e]	No use[a]
HPβCD	≤3	0.2	0.4	~90	0.2	1.9	8,000	16,000
SBEβCD	1.6	0.3	0.3	≥90	0.2	1.4		14,000
RMβCD	0.5–15	~2.5	0.3	~95				No use[a]
γCD	<0.1		0.3	~90			10,000[e]	~50
HPγCD								~50

Notes:

[a] f_{oral} = fraction absorbed intact after oral administration (i.e. oral bioavailability); V_D = volume of distribution; $t_{1/2}$ = biological half-life; $f_{urine\ unch.}$ = fraction excreted unchanged with urine; No use: not for parenteral usage.

[b] *Source:* [20, 22–25, 176].

[c] *Source:* [21, 26].

[d] *Source:* [12].

[e] As dietary supplement.

Parenteral administration of CDs can be somewhat more limited. The hemolytic effects of CDs on human erythrocytes in phosphate buffered saline are in the order of methylated βCDs > βCD > HPβCD > αCD > γCD > HPγCD > SBEβCD [20, 123, 124]. There appears to be a correlation between the hemolytic activity and the ability of the CDs to bind or extract cholesterol from the membranes [20]. This *in vitro* cellular lysis study, as well as other comparable *in vitro* studies using intestinal cells, E. coli, human skin fibroblasts and liposomes, do not necessarily translate to *in vivo* toxicity but rather provide a method to classify CDs according to their potential to destabilize or disrupt cellular membranes [123]. βCD cannot be given parenterally due to its low aqueous solubility and related adverse effects (e.g. nephrotoxicity). These and other *in vitro* and *in vivo* studies have shown that the methylated βCDs and βCD cannot be used in parenteral formulations while HPβCD, αCD, γCD, HPγCD and SBEβCD can all be found in marketed parenteral formulations. Due to their favorable toxicological profile CDs are frequently preferred to organic solvents during *in vitro*/*in vivo* evaluation of new chemical entities.

HPβCD has a small volume of distribution ($V_D \approx 0.2$ liter/kg) and short half-life ($t_{1/2} \approx 1.7$ h), and is mainly excreted unchanged with the urine after parenteral administration to humans (Table 3.12) [12, 21, 26, 75]. In man, there is a linear relationship between the parenterally administered HPβCD dose and the area under the plasma concentration – time curve (AUC) indicating non-saturable excretion mechanism even at high doses. No side effects were observed after parenteral administration of up to 24 grams of HPβCD daily (12 g bid) for 15 days. The pharmacokinetics of SBEβCD is very similar to that of HPβCD [21]. The total plasma clearance of both HPβCD and SBEβCD is similar to the glomerular filtration rate and since CDs are predominately eliminated unchanged with urine, their elimination half-life ($t_{1/2}$) will increase with impaired or reduced kidney function. However, in individuals with normal kidney function, about 90% of parenterally administered CD will be excreted within 6 hours from the administration and about 99% within 12 hours. Thus, administration of CD containing drug formulations will result in negligible accumulation of CD in individuals with normal kidney function.

3.9 Regulatory Issues

Cyclodextrins are pharmaceutical excipients which, unlike the active pharmaceutical ingredient (API), have no regulatory status in and of themselves. In the US, their use is generally supported by widespread experience with these materials based on (1) being recognized as 'generally regarded as safe (GRAS)' by the FDA and other agencies (USA 21CFR182, 184 and 186), (2) being approved as a food additive by the FDA (21CFR171) or other agencies, and/or (3) being present in an approved drug product. In this instance, the use of the excipient, without additional data, is limited to circumstances where the same material is used in the same administration route, at the same level of exposure, for the same duration and in the same patient population. The FDA maintains a list of 'Inactive Pharmaceutical Ingredients' which contains the product in which they are formulated, their route and the amount in the dosage form [177]. Novel excipients are those that either have not been used before or have not been used by the suggested route of administration. Novel excipients require additional testing for NDA approval at a level similar to that of the drug product.

The nature and extent of toxicology needed to use an excipient are included in *Guidance for Industry – Nonclinical Studies for the Safety Evaluation of Pharmaceutical Excipients* [178]. The regulatory status of CDs is reasonable well established. βCD is listed in a number of pharmacopoeia sources including the US Pharmacopoeia/National Formulary (USP/NF), the European Pharmacopoeia (Ph.Eur.) and the Japanese Pharmaceutical Codex (JPC). αCD is similarly listed in the Ph.Eur., USP/NF and JPC and γCD is referenced in the USP/NF and JPC and will soon be included in the Ph.Eur. A monograph for HPβCD is available in the USP and Ph.Eur. A monograph for SBEβCD is available in the USP. Other derivatives are not yet compendial but efforts are underway for their inclusion. αCD, βCD and γCD were also introduced into the generally regarded as safe (GRAS) list of the FDA for use as a food additive in 2004, 2001 and 2000, respectively, and HPβCD is cited in the FDA's list of Inactive Pharmaceutical Ingredients.

3.10 Conclusion

Cyclodextrins are value-adding functions excipients which have garnered significamt interest and are widely used in pharmaceutical and nutriceutical products. These materials can contribute to the optimal biopharmaceutical performance of a medicine and in some cases are enabling, meaning that the drug product would not be possible without the technology. It has been estimated that products enabled by chemically modified cyclodextrins account for sales of almost US$2 billion (2008). While all compounds will not necessarily benefit from the application of cyclodextrins, the approach is simple to test. Based on this simplicity as well as the low toxicological potential of these materials, they are often used in early drug screening and early clinical assessments and figure highly in formulation decision trees. While solubilization is, to date, the most often cited reason for using cyclodextrins, their ability to impact stability as well as other aspects of a drug suggest a significant potential for the derivatives in the future. Thus, assessments of cyclodextrin complexation and aggregation point to the possibility of increased delivery through biological membranes as well as drug products with improved shelf-lives. Cyclodextrins have also gained interest as components of siRNA delivery vectors as well as in numerous industrial applications.

References

1. A. Villiers, Sur la fermentation de la fécule par l'action du ferment butyrique. *Compt Rend Acad Sci*, **112**, 536–538 (1891).
2. T. Loftsson, D. Duchêne, Cyclodextrins and their pharmaceutical applications. *Int J Pharm*, **329**, 1–11 (2007).
3. F. Schardinger, Über Thermophile Bakterien aus verschiedenen Speisen und Milch, sowie über einige Umsetzungsprodukte derselben in kohlenhydrathältigen Nährlösungen, darunter krystallisierte Polysaccharide (Dextrine) aus Stärke. *Z Untersuch Nahr u Genussm*, **6**, 865–880 (1903).
4. K. Freudenberg, R. Jacobi, Über Schardinger Dextrine aus Stärke. *Liebigs Ann Chem*, **518**, 102–108 (1935).

5. K. Freudenberg, F. Cramer, Die Konstitution der Schardinger-Dextrine a, b und g. *Z Naturforsch*, **3b**, 464 (1948).
6. D. French, A.O. Pulley, J.A. Effenberger, M.A. Rougvie, M. Abdullah, Studies on the Schardinger dextrins. XII. The molecular size and structure of the d-, e-, z-, and h-dextrins. *Arch Biochem Biophys*, **111**, 153–160 (1965).
7. K.L. Larsen, Large cyclodextrins. *J Incl Phenom Macroc Chem*, **43**, 1–13 (2002).
8. H. Ueda, T. Endo, H. Nagase, S. Kobayashi, T. Nagai, Isolation, purification, and characterization of cyclomaltodecaose. in *Proceedings of the Eighth International Symposium on Cyclodextrins*, Szejtli, J., Szente, L. (eds). Dordrecht: Kluwer Academic Publishers, pp. 17–20 (1996).
9. H. Ueda, T. Endo, Large-ring cyclodextrins, in *Cyclodextrins and their Complexes: Chemistry, Analytical Methods, Applications*, Dodziuk, H. (ed.), Weinheim: Wiley-VCH Verlag, pp. 370–380 (2006).
10. K. Freudenberg, F. Cramer, H. Plieninger, *Verfahren zur Herstellung von Einschlußverbindungen physiologisch wirksamer organischer Verbindungen.* Knoll A.-G. Chemische Fabriken (1953).
11. F. Cramer, *Einschlussverbindungen*. Berlin: Springer-Verlag (1954).
12. T. Loftsson, M.E. Brewster, Pharmaceutical applications of cyclodextrins: Basic science and product development. *J Pharm Pharmacol*, **62**, 1607–1621 (2010).
13. J. Blanchard, S. Proniuk, Some important considerations in the use of cyclodextrins. *Pharm Res*, **16**, 1796–1798 (1999).
14. K.H. Frömming, J. Szejtli, *Cyclodextrins in Pharmacy*. Dordrecht: Kluwer Academic Publishers (1994).
15. J. Szejtli, *Cyclodextrin Technology*. Dordrecht: Kluwer Academic Publishers (1988).
16. F. Hirayama, M. Yamamoto, K. Uekama, Acid-catalyzed hydrolysis of maltosyl-β-cyclodextrin. *J Pharm Sci*, **81**, 913–916 (1992).
17. S. Buedenbender, G.E. Schulz, Structural base for enzymatic cyclodextrin hydrolysis. *J Mol Biol*, **385**, 606–617 (2009).
18. J. Szejtli, The metabolism, toxicity and biological effects of cyclodextrins, in Duchêne, D. (ed.), *Cyclodextrins and Their Uses*. Paris: Editions de Santé, pp. 173–212 (1987).
19. I.C. Munro, P.M. Newberne, R.R. Young, A. Bär, Safety assessment of γ-cyclodextrin. *Regul Toxicol Pharmacol*, **39**(Suppl. 1), S3–S13 (2004).
20. T. Irie, K. Uekama, Pharmaceutical applications of cyclodextrins. III. Toxicological issues and safety evaluation. *J Pharm Sci*, **86**(2), 147–162 (1997).
21. V.J. Stella, Q. He, Cyclodextrins. *Tox Pathol*, **36**, 30–42 (2008).
22. G. Antlsperger, G. Schmid, Toxicological comparison of cyclodextrins. In Szejtli, J., Szente, L., editors. *Proceedings of the Eighth International Symposium on Cyclodextrins.* Budapest, Hungary, March 30–April 2, 1996, ed., Dordrecht: Kluwer Academic Publishers, pp. 149–155 (1996).
23. G. Antlsperger, New aspects in cyclodextrin toxicology. In Hedges, A.R. (ed.), *Minutes of the Sixth International Symposium on Cyclodextrins*, Paris: Editions de Santé. pp. 277–283 (1992).
24. A.T.H.J. De Bie, B. van Ommen, A. Bär, Disposition of 14C-γ-cyclodextrin in germ-free and conventional rats. *Regul Toxicol Pharmacol*, **27**, 150–158 (1998).

25. B. Van Ommen, A.T.H.J. De Bie, A. Bär, Disposition of 14C-α-cyclodextrin in germ-free and conventional rats. *Regul Toxicol Pharmacol*, **39**, S57–S66 (2004).
26. H. Zhou, M. Goldman, J. Wu, R. Woestenborghs, A.E. Hassell, P. Lee, A. Baruch, L. Pesco-Koplowitz, J. Borum, L.J. Wheat, A pharmacokinetic study of intravenous intraconazole followed by oral administration of intraconazole capsules in patients with advanced human immunodeficiency virus infection. *J Clin Pharmacol*, **38**, 593–602 (1998).
27. T. Higuchi, K.A. Connors, Phase-solubility techniques, *Adv Anal Chem Instrum*, **4**, 117–212 (1965).
28. A.J. Repta, Alteration of apparent solubility through complexation. In Yalkowski, S.H. (ed.) *Techniques of Solubilization of Drugs*. New York: Marcel Dekker, pp. 135–157 (1985).
29. M.E. Brewster, T. Loftsson, Cyclodextrins as pharmaceutical solubilizers. *Adv Drug Deliv Rev*, **59**, 645–666 (2007).
30. M.E. Brewster, T. Loftsson, Complexation: Use of cyclodextrins to improve pharmaceutical properties of intramuscular formulations. In Gupta, P.K., Brazeau, G.A. (eds), *Injectable Drug Development Techniques to Reduce Pain and Irritation*. Denver: Interpharm Press, pp. 307–336 (1999).
31. A. Wu, X. Shen, Y. He, Investigation of γ-cyclodextrin nanotube induced by *N,N′*-diphenylbenzidine molecule. *J Coll Interf Sci*, **297**, 525–533 (2006).
32. A. Wu, X. Shen, Y. He, Micrometer-sized rodlike structure formed by the secondary assembly of cyclodextrin nanotube. *J Coll Interf Sci*, **302**, 87–89 (2006).
33. M. Bonini, S. Rossi, G. Karlsson, M. Almgren, P. Lo Nostro, P. Baglioni, Self-assembly of β-cyclodextrin in water. Part 1: Cryo-TEM and dynamic and static light scattering. *Langmuir*, **22**, 1478–1484 (2006).
34. P. Jansook, S.V. Kurkov, T. Loftsson, Cyclodextrins as solubilizers: formation of complex aggregates. *J Pharm Sci*, **99**, 719–729 (2010).
35. P. Jansook, M.D. Moya-Ortega, T. Loftsson, Effect of self-aggregation of γ-cyclodextrin on drug solubilization. *J Incl Phenom Macroc Chem*, **68**, 229–236 (2010).
36. M. Messner, S.V. Kurkov, P. Jansook, T. Loftsson, Self-assembled cyclodextrin aggregates and nanoparticles. *Int J Pharm*, **387**, 199–208 (2010).
37. M. Kajtár, M. Vikmon, E. Morlin, J. Szejtli, Aggregation of amphotericin B in the presence of g-cyclodextrin. *Biopolymers*, **28**, 1585–1596 (1989).
38. A.W. Coleman, I. Nicolis, N. Keller, J.P. Dalbiez, Aggregation of cyclodextrins: an explanation of the abnormal solubility of β-cyclodextrin. *J Incl Phenom Macroc Chem*, **13**, 139–143 (1992).
39. S.A. Andronati, Y.E. Shapiro, L.N. Yakubovskaya, V.Y. Gorbatyuk, K.S. Andronati, S.P. Krasnoschekaya, Inclusion compounds of psychotropic agents and cyclodextrins. *J Incl Phenom Macroc Chem*, **24**, 175–186 (1996).
40. V. Gabelica, N. Galic, E. De Pauw, On the specificity of cyclodextrin complexes detected by electrospray mass spectrometry. *J Am Soc Mass Spectrom*, **13**, 946–953 (2002).
41. I. Correia, N. Bezzenine, N. Ronzani, N. Platzer, J.-C. Beloeil, B.-T. Doan, Study of inclusion complexes of acridine with b- and (2,6-di-*O*-methyl)-b-cyclodextrin by

use of solubility diagrams and NMR spectroscopy. *J Phys Org Chem*, **15**, 647–659 (2002).
42. T. Loftsson, M. Másson, M.E. Brewster, Self-association of cyclodextrins and cyclodextrin complexes. *J Pharm Sci*, **93**, 1091–1099 (2004).
43. Z. Bikádi, R. Kurdi, S. Balogh, J. Szemán, E. Hazai, Aggregation of cyclodextrins as an important factor to determine their complexation behavior. *Chem Biodiv*, **3**, 1266–1276 (2006).
44. P. Yan, J.-N. Tang, J.-X. Xiao, Interaction between β-cyclodextrin and mixed cationic-anionic surfactants (2): aggregation behavior. *J Dispers Sci Technol*, **28**, 623–626 (2007).
45. M. Messner, S.V. Kurkov, M.E. Brewster, P. Jansook, T. Loftsson, Self-assembly of cyclodextrin complexes: Aggregation of hydrocortisone/cyclodextrin complexes. *Int. J. Pharm*, **407**, 174–183 (2011).
46. M. Messner, S.V. Kurkov, R. Flavià-Piera, M.E. Brewster, T. Loftsson, Self-assembly of cyclodextrin complexes: The effect of the guest molecule (manuscript) (2011).
47. T. Loftsson, D. Hreinsdóttir, M. Másson, Evaluation of cyclodextrin solubilization of drugs. *Int J Pharm*, **302**, 18–28 (2005).
48. J.M. Adam, D.J. Bennett, A. Bom, J.K. Clark, H. Feilden, E.J. Hutchinson, R. Palin, A. Prosser, D.C. Rees, G.M. Rosair, D. Stevenson, M.-Q. Zhang, Cyclodextrin-derived host molecules as reversal agents for the neuromuscular blocker rocuronium bromide: synthesis and structure-activity relationships. *J Med Chem*, **45**, 1806–1816 (2002).
49. C.S. Perry, S.A. Charman, R.J. Prankerd, F.C.K. Chiu, M.J. Scanlon, D. Chalmers, W.N. Charman, The binding interaction of synthetic ozonide antimalarials with natural and modified β-cyclodextrins. *J Pharm Sci*, **95**, 146–158 (2006).
50. L. Szente, J. Szejtli, G.L. Kis, Spontaneous opalescence of aqueous γ-cyclodextrin solutions: complex formation or self-aggregation. *J Pharm Sci*, **87**, 778–781 (1998).
51. L. Szente, J. Szejtli, Highly soluble cyclodextrin derivatives: chemistry, properties, and trends in development. *Adv Drug Deliv Rev*, **36**, 17–28 (1999).
52. G. González-Gaitano, P. Rodríguez, J.R. Isasi, M. Fuentes, G. Tardajos, M. Sánchez, The aggregation of cyclodextrins as studied by photon correlation spectroscopy. *J Incl Phenom Macrocycl Chem*, **44**, 101–105 (2002).
53. W. He, P. Fu, X.H. Shen, H.C. Gao, Cyclodextrin-based aggregates and characterization by microscopy. *Micron*, **39**, 495–516 (2008).
54. T. Loftsson, A. Magnúsdóttir, M. Másson, J.F. Sigurjónsdóttir, Self-association and cyclodextrin solubilization of drugs. *J Pharm Sci*, **91**, 2307–2316 (2002).
55. A. Magnusdottir, M. Másson, T. Loftsson, Self association and cyclodextrin solubilization of NSAIDs. *J Incl Phenom Macroc Chem*, **44**, 213–218 (2002).
56. H.H. Sigurdsson, A. Magnusdottir, M. Masson, T. Loftsson. The effects of cyclodextrins on hydrocortisone permeability through semi-permeable membranes. *J Incl Phenom Macrocycl Chem*, **44**, 163–167 (2002).
57. P.D. Ross, M.V. Rekharsky, Thermodynamics of hydrogen bond and hydrophobic interactions in cyclodextrin complexes. *Biophys J*, **71**, 2144–2154 (1996).
58. R.I. Gelb, L.M. Schwartz, B. Cardelino, H.S. Fuhrman, R.F. Johnson, D.F. Laufer, Binding mechanisms in cyclohexaamylose complexes. *J Am Chem Soc*, **103**, 1750–1757 (1981).

59. R.I. Gelb, L.M. Schwartz, M. Radeos, D.A. Laufer, Cycloamylose complexation of organic anions. *J Phys Chem*, **87**, 3349–3354 (1982).
60. M.V. Rekharsky, R.N. Goldberg, F.P. Schwarz, Y.B. Tewari, P.D. Ross, Y. Yamashoji, Y. Inoue, Thermodynamic and nuclear magnetic resonance study of the interactions of α- and β-cyclodextrin with model substances: phenethylamine, ephedrines, and related substances. *J Am Chem Soc*, **145**, 8830–8840 (1995).
61. M.V. Rekharsky, F.P. Schwarz, Y.B. Tewari, R.N. Goldberg, M. Tanaka, Y. Yamashoji, Thermodynamic and NMR study of the inclusions of cyclodextrins with cyclohexane derivatives. *J Phys Chem*, **98**, 4098–4103 (1994).
62. W. Saenger, Cyclodextrin inclusion compounds in research and industry. *Angew Chem Int Ed Engl*, **19**, 344–362 (1980).
63. L. Liu, Q.-X. Guo, The driving forces in the inclusion complexation of cyclodextrins. *J Incl Phenom Macroc Chem*, **42**, 1–14 (2002).
64. T. Loftsson, Effects of cyclodextrins on chemical stability of drugs in aqueous solutions. *Drug Stability*, **1**, 22–33 (1995).
65. P.W. Wertz, Stratum corneum lipids and water. *Exog Dermatol*, **3**, 53–56 (2004).
66. R. Bansil, B.S. Turner, Mucin structure, aggregation, physiological functions and biomedical applications. *Curr Opin Colloid Interface Sci*, **11**, 164–170 (2006).
67. H. Lennernäs, Human intestinal permeability. *J Pharm Sci*, **87**, 403–410 (1998).
68. J.P. Karlsson, P. Artursson, A method for the determination of cellular permeability coefficients and aqueous boundary layer thickness in monolayers of intestinal epithelia (Caco-2) cells grown in permeable filter chambers. *Int J Pharm*, **71**, 55–64 (1991).
69. K.A. Youdim, A. Avdeef, N.J. Abbott, *In vitro* trans-monolayer permeability calculations: often forgotten assumptions. *Drug Discov Today* **8**, 997–1003 (2003).
70. A. Avdeef, P.E. Nielsen, O. Tsinman, PAMPA: A drug absorption in vitro model 11. Matching the in vivo unstirred water layer thickness by individual-well stirring in microtitre plates. *Eur J Pharm Sci*, **22**, 365–374 (2004).
71. M.E. Brewster, M. Noppe, J. Peeters, T. Loftsson, Effect of the unstirred water layer on permeability enhancement by hydrophilic cyclodextrins. *Int J Pharm*, **342**, 250–253 (2007).
72. T. Higuchi, Physical chemical analysis of percutaneous absorption process from creams and ointments. *J Soc Cosmet Chem*, **11**(2), 85–97 (1960).
73. G.L. Flynn, O.S. Carpender, S.H. Yalkowsky, Total mathematical resolution of diffusion layer control of barrier flux. *J Pharm Sci*, **61**, 312–314 (1972).
74. G.L. Flynn, S.H. Yalkowsky, Correlation and prediction of mass transport across membranes I: influence of alkyl chain length on flux-determining properties of barrier and diffusant. *J Pharm Sci*, **61**, 838–852 (1972).
75. T. Loftsson, M.E. Brewster, Pharmaceutical applications of cyclodextrins: effects on drug permeation through biological membranes. *J Pharm Pharmacol*, **63**, 1119–1135 (2011).
76. T. Loftsson, S.B. Vogensen, M.E. Brewster, F. Konráðsdóttir, Effects of cyclodextrins on drug delivery through biological membranes. *J Pharm Sci*, **96**, 2532–2546 (2007).
77. K. Khanvilkar, M.D. Donovan, D.R. Flanagan, Drug transfer through mucus. *Adv Drug Deliv Rev*, **48**, 173–193 (2001).

78. T. Loftsson, H.H. Sigurðsson, F. Konráðsdóttir, S. Gísladóttir, P. Jansook, E. Stefánsson, Topical drug delivery to the posterior segment of the eye: anatomical and physiological considerations. *Pharmazie*, **63**, 171–179 (2008).
79. A. Preiss, W. Mehnert, K.H. Frömming, In-vitro hydrocortisone release from ointments in presence of cyclodextrins. *Pharmazie*, **49**, 902–906 (1994).
80. A. Preiss, W. Mehnert, K.-H. Frömming, Penetration of hydrocortisone into excised human skin under the influence of cyclodextrins. *Pharmazie*, **50**, 121–126 (1995).
81. J. Pitha, T. Hoshino, J. Torres-Labandeira, T. Irie, Preparation of drug: hydroxypropylcyclodextrin complexes by method using ethanol or aqueous ammonium hydroxide as co-solubilizers. *Int J Pharm*, **80**, 253–258 (1992).
82. J. Pitha, T. Hoshino, Effects of ethanol on formation of inclusion complexes of hydroxypropylcyclodextrins with testosterone or with methyl organe. *Int J Pharm*, **80**, 243–251 (1992).
83. T. Loftsson, M. Másson, J.F. Sigurjónsdóttir, Methods to enhance the complexation efficiency of cyclodextrins. *STP Pharma Sci*, **9**, 237–242 (1999).
84. V.J. Stella, R.A. Rajewski, Cyclodextrins: their future in drug formulation and delivery. *Pharm Res*, **14**(5), 556–567 (1997).
85. T. Loftsson, F. Konráðsdóttir, M. Másson, Influence of aqueous diffusion layer on passive drug diffusion from aqueous cyclodextrin solutions through biological membranes. *Pharmazie*, **61**, 83–89 (2006).
86. J.D. Artiss, K. Brogan, M. Brucal, M. Moghaddam, K.-L.C. Jen, The effects of a new soluble dietary fiber on weight gain and selected blood parameters in rats. *Metab Clin Exp*, **55**, 195–202 (2006).
87. H.M. Cabral Marques, J. Hadgraft, I.W. Kellaway, G. Taylor, Studies of cyclodextrin inclusion complexes. III. The pulmonary absorption of β-, DM-β- and HP-β-cyclodextrins in rabbits. *Int J Pharm*, **77**, 297–302 (1991).
88. A. Hussain, J.J. Arnold, M.A. Khan, F. Ahsan, Absorption enhancers in pulmonary protein delivery. *J Control Rel*, **94**, 15–24 (2004).
89. B. Evrard, P. Bertholet, M. Gueders, M-P. Flament, G. Piel, L. Delattre, A. Gayot, P. Leterme, J.-M. Foidart, D. Cataldo, Cyclodextrins as a potential carrier in drug nebulization. *J Control Rel*, **96**, 403–410 (2004).
90. T. Sato, K. Mizuno, F. Ishii, A novel administration route of edaravone – II: mucosal absorption of edaravone from edaravone/hydroxypropyl-β-cyclodextrin complex solution including L-cysteine and sodium hydrogen sulfite. *Pharmacology*, **85**, 88–94 (2010).
91. J.K. Kim, M.S. Kim, J.S. Park, C.K. Kim, Thermo-reversible flurbiprofen liquid suppository with HP-β-CD as a solubility enhancer: improvement of rectal bioavailability. *J Incl Phenom Macroc Chem*, **64**, 265–272 (2009).
92. M. Francois, E. Snoeckx, P. Putteman, F. Wouters, E. De Proost, U. Delaet, J. Peeters, M.E. Brewster, A mucoadhesive, cyclodextrin-based vaginal cream formulation of itraconazole. *AAPS PharmSci*, **5**(1), Article Number: 5 www.aapsj.org/view.asp?art=ps050105 (2003).
93. E. Cevher, D. Şensoy, M. Zloh, L. Mülazimoğlu, Preparation and characterisation of natamycin: γ-cyclodextrin inclusion complex and its evaluation in vaginal mucoadhesive formulations. *J Pharm Sci*, **97**, 4319–4335 (2008).

94. M.F. Sugrue, The pharmacology of antiglaucoma drugs. *Pharm Ther*, **43**, 91–138 (1989).
95. P. Jansook, E. Stefánsson, M. Thorsteinsdóttir, B.B. Sigurdsson, S.S. Kristjánsdóttir, J.F. Bas, H.H. Sigurdsson, T. Loftsson, Cyclodextrin solubilization of carbonic anhydrase inhibitor drugs: formulation of dorzolamide eye drop microparticle suspension. *Eur J Pharm Biopharm*, **76**, 208–214 (2010).
96. T. Järvinen, K. Järvinen, A. Urtti, D. Thompson, V.J. Stella, Sulfobutyl ether β-cyclodextrin (SBE-β-CyD) in eye drops improves the tolerability of a topically applied pilocarpine prodrug in rabbits. *J Ocul Pharmacol Ther*, **11**, 95–106 (1995).
97. P. Jarho, K. Järvinen, A. Urtti, V.J. Stella, T. Järvinen, Modified β-cyclodextrin (SBE7-β-CyD) with viscous vehicle improves the ocular delivery and tolerability of pilocarpine prodrug in rabbits. *J Pharm Pharmacol*, **48**, 264–270 (1996).
98. P. Suhonen, T. Jarvinen, K. Lehmussaari, T. Reunamaki, A. Urtti, Ocular absorption and irritation of pilocarpine prodrug is modified with buffer, polymer, and cyclodextrin in the eyedrop. *Pharmaceutical Research*, **12**, 529–533 (1995).
99. M. Tanaka, Y. Iwata, Y. Kouzuki, K. Taniguchi, H. Matsuda, H. Arima, S. Tsuchiya, Effect of 2-hydroxypropyl-β-cyclodextrin on percutaneous absorption of methyl paraben. *J Pharm Pharmacol*, **47**, 897–900 (1995).
100. F. Konrádsdóttir, H. Ogmundsdóttir, V. Sigurdsson, T. Loftsson, Drug targeting to the hair follicles: a cyclodextrin based drug delivery. *AAPS PharmSciTech*, **10**, 266–269 (2009).
101. L.A. Felton, C.J. Wiley, D.A. Godwin, Influence of hydroxypropyl-β-cyclodextrin on transdermal permeation and skin accumulation of oxybenzone. *Drug Devel Ind Pharm*, **28**, 1117–1124 (2002).
102. J. Yang, C.J. Wiley, D.A. Godwin, L.A. Felton, Influence of hydroxypropyl-β-cyclodextrin on transdermal penetration and photostability of avobenzone. *Eur J Pharm Biopharm*, **69**, 605–612 (2008).
103. L.A. Felton, C.J. Wiley, D.A. Godwin, Influence of cyclodextrin complexation on the in vivo photoprotective effects of oxybenzone. *Drug Devel Ind Pharm*, **30**, 95–102 (2004).
104. V. Sarveiya, J.F. Templeton, H.A.E. Benson, Inclusion complexation of the sunscreen 2-hydroxy-4-methoxy benzophenone (oxybenzone) with hydroxypropyl-β-cyclodextrin: effect on membrane diffusion. *J Incl Phenom Macroc Chem*, **49**, 275–281 (2004).
105. R.G. Strickley, Solubilizing excipients in oral and injectable formulations. *Pharm Res*, **21**, 201–230 (2004).
106. L.A. Trissel (ed.), *Handbook on Injectable Drugs*. 12th edn. Bethesda, MD: American Society of Health-System Pharmacists (2003).
107. V.J. Stella, V.M. Rao, E.A. Zannou, V. Zia, Mechanism of drug release from cyclodextrin complexes, *Adv Drug Deliv Rev*, **36**, 3–1 (1999).
108. S.V. Kurkov, T. Loftsson, M. Messner, D. Madden, Parenteral delivery of HPβCD: effects on drug-HSA binding. *AAPS PharmSciTech*, **11**, 1152–1158 (2010).
109. G. Piel, B. Evrard, T. Van Hees, L. Delattre, Comparison of the IV pharmacokinetics in sheep of miconazole–cyclodextrin solutions and a micellar solution. *Int J Pharm*, **180**, 41–45 (1999).

110. J. Liu, L. Qiu, J. Gao, Y. Jin, Preparation, characterization and in vivo evaluation of formulation of baicalein with hydroxypropyl-β-cyclodextrin. *Int J Pharm*, **312**, 137–143 (2006).
111. B.W. Muller, E. Alberts, Effect of hydrotropic substances on the complexation of sprangly soluble drugs with cyclodextrin derivatives and the influence of cyclodextrin complexation on the pharmacokinetics of the drugs. *J Pharm Sci*, **80**, 599–604 (1991).
112. T.D. Egan, S.E. Kern, K.B. Johnson, N.L. Pace, The pahrmacokinetics and pharmacodynamics of propofol in modified cyclodextrin formulation (Captisol®) versus propofol in a lipid formulation (Diprivan®): an electroencephalographic and hemodynamic study in a porcine model. *Anesth Analg*, **97**, 72–79 (2003).
113. T.R. Buggins, P.A. Dickinson, G. Taylor, The effects of pharmaceutical excipients on drug disposition. *Adv Drug Deliv Rev*, **59**, 1482–1503 (2007).
114. V.J. Stella, H.K. Lee, D.O. Thompson, The effect of SBE4-β-CD on i.m. prednisolone pharmacokinetics and tissue damage in rabbits: comparison to a co-solvent solution and a water-soluble prodrug. *Int J Pharm*, **120**, 197–204 (1995).
115. M.P. McIntosh, N. Schwarting, R.A. Rajewski, In vitro and in vivo evaluation of a sulfobutyl ether β-cyclodextrin enabled etomidate formulation. *J Pharm Sci*, **93**, 2585–2594 (2004).
116. Z. Zuo, Y.K. Tam, J. Diakur, L.I. Wiebe, Hydroxypropyl-b-cyclodextrin-flutamide inclusion complex. II. Oral and intravenous pharmacokinetics of flutamide in the rat. *J Pharm Pharmaceut Sci*, **5**, 292–298 (2002).
117. M. Welliver, New drug sugammadex: A selective relaxant binding agent. *AANA Journal*, **74**, 357–363 (2006).
118. S.A. Charman, C.S. Perry, F.C.K. Chiu, K.A. McIntosh, R.J. Prankerd, W.N. Charman, Alteration of the intravenous pharmacokinetics of a synthetic ozonide antimalarial in the presence of a modified cyclodextrin. *J Pharm Sci*, **95**, 256–267 (2006).
119. B.A.R. Lina, A. Bär, Subchronic oral toxicity studies with a-cyclodextrin in rats. *Regul Toxicol Pharmacol*, **39**, S14–S26 (2004).
120. B.A.R. Lina, A. Bär, Subchronic (13-week) oral toxicity study of a-cyclodextrin in dogs, *Regul Toxicol Pharmacol*, **39**, S27–S33 (2004).
121. M.E. Bellringer, T.G. Smith, R. Read, C. Gopinath, P. Oliver, β-Cyclodextrin: 52-week toxicity studies in the rat and dog, *Food Chem Toxicol*, **33**, 367–376 (1995).
122. S. Gould, R.C. Scott, 2-Hydroxypropyl-b-cyclodextrin (HP-b-CD): a toxicology review. *Food Chem Toxicol*, **43**, 1451–1459 (2005).
123. D.O. Thompson, Cyclodextrins-enabling excipients: their present and future use in pharmaceuticals. *Crit Rev Ther Drug Carrier Syst*, **14**, 1–104 (1997).
124. K. Uekama, Design and evaluation of cyclodextrin-based drug formulation. *Chem Pharm Bull*, **52**, 900–915 (2004).
125. H. Dodziuk (ed.), *Cyclodextrins and their Complexes*. Weinheim: Wiley-VCH Verlag (2006).
126. E. Sabadini, T. Cosgrove, F.C. Egídio, Solubility of cyclomaltooligosaccharides (cyclodextrins) in H_2O and D_2O: a comparative study. *Carbohydr Res*, **341**, 270–274 (2006).
127. K.A. Connors, Population characteristics of cyclodextrin complex stabilities in aqueous solution. *J Pharm Sci*, **84**, 843–848 (1995).

128. K.A. Connors, The stability of cyclodextrin complexes in solution. *Chem Rev*, **97**, 1325–1357 (1997).
129. T. Loftsson, H. Fridriksdottir, A.M. Sigurdardottir, H. Ueda, The effect of water-soluble polymers on drug-cyclodextrin complexation. *Int J Pharm*, **110**, 169–177 (1994).
130. F.A. Menard, M.G. Dedhiya, C.T. Rhodes, Studies of the effect of pH, temperature and ring size on the complexation of phenytoin with cyclodextrins. *Pharm Acta Helv*, **63**, 303–308 (1988).
131. G.P. Bettinetti, P. Mura, A. Liguori, G. Bramanti, F. Giordano, Solubilization and interaction of naproxen with cyclodextrinsin aqueous solution and in the solid state. *Farmaco*, **44**, 195–213 (1989).
132. T.-X. Xiang, B.D. Anderson, Inclusion complexes of purine nucleosides with cyclodextrins. 2. Investigation of inclusion complex-geometry and cavity microenvironment. *Int J Pharm*, **59**, 45–55 (1990).
133. F.A. Menard, M.G. Dedhiya, C.T. Rhodes, Physicochemical aspects of the complexation of some drugs with cyclodextrins. *Drug Devel Ind Pharm*, **16**, 91–113 (1990).
134. G. Bettinetti, P. Mura, M.T. Faucci, M. Sorrent, M. Setti, Interaction of naproxen with noncrystalline acetyl β- and acetyl γ-cyclodextrins in the solid and liquid state. *Eur J Pharm Sci*, **15**, 21–29 (2002).
135. T. Loftsson, B.J. Ólafsdóttir, H. Fridriksdóttir, S. Jónsdóttir, Cyclodextrin complexation of NSAIDS: physicochemical characteristics. *Eur J Pharm Sci*, **1**, 95–101 (1993).
136. P. Mura, M.T. Faucci, G.P. Bettinetti, The influence of polyvinylpyrrolidone on naproxen complexation with hydroxypropyl-β-cyclodextrin. *Eur J Pharm Sci*, **13**, 187–194 (2001).
137. V. Zia, R.A. Rajewski, V.J. Stella, Thermodynamics of binding of neutral molecules to sulfobutyl ether b-cyclodextrins (SBE-β-CDs): the effect of total degree of substitution. *Pharm Res*, **17**, 936–941 (2000).
138. T. Loftsson, N. Bodor, Effects of 2-hydroxypropyl-β-cyclodextrin on the aqueous solubility of drugs and transdermal delivery of 17β-estradiol. *Acta Pharm Nord*, **1**, 185–194 (1989).
139. M.A. Hussain, R.C. Diluccio, M.B. Maurin, Complexation of moricizine with nicotinamide and evaluation of the complexation constants by various methods. *J Pharm Sci*, **82**, 77–79 (1993).
140. R. Krishnamoorthy, A.K. Mitra, Prodrugs for nasal drug delivery. *Adv Drug Deliv Rev*, **29**, 135–146 (1998).
141. P. Li, E. Tabibi, S.H. Yalkowsky, Combined effect of complexation and pH on solubilization. *J Pharm Sci*, **87**, 1535–1537 (1998).
142. G. Piel, B. Evrard, M. Fillet, G. Llabres, L. Delattre, Development of a non-surfactant parenteral formulation of miconazole by the use of cyclodextrins. *Int J Pharm*, **169**, 15–22 (1998).
143. E. Redenti, L. Szente, J. Szejtli, Cyclodextrin complexes of salts of acidic drugs. Thermodynamic properties, structural features, and pharmaceutical applications. *J Pharm Sci*, **90**, 979–986 (2001).
144. G. Granero, C. Granero, M. Longhi, The effect of pH and triethanolamine on sulfisoxazole complexation with hydroxypropyl-β-cyclodextrin. *Eur J Pharm Sci*, **20**, 285–293 (2003).

145. T. Loftsson, H.H. Sigurðsson, M. Másson, N. Schipper, Preparation of solid drug/cyclodextrin complexes of acidic and basic drugs. *Pharmazie*, **59**, 25–29 (2004).
146. A.Y. Tinwalla, B.L. Hoesterey, T.-X. Xiang, K. Lim, B.D. Anderson, Solubilization of thiazolobenzimidazole using a combination of pH adjustment and complexation with 2-hydroxypropyl-β-cyclodextrin. *Pharm Res*, **10**, 1136–1143 (1993).
147. A. Selva, E. Redenti, P. Ventura, M. Zanol, B. Casetta, Study of β-cyclodextrin-ketakonazole-tartaric acid multicomponent non-covalent association by positive and netagive ionspray mass spectrometry. *J Mass Sepectrom*, **33**, 729–734 (1998).
148. E. Fenyvesi, M. Vikmon, J. Szeman, E. Redenti, M. Delcanale, P. Ventura, J. Szejtli, Interaction of hydoxy acids with β-cyclodextrin. *J Incl Phenom Macroc Chem*, **33**, 339–344 (1999).
149. E. Redenti, L. Szente, J. Szejtli, Drug/cyclodextrin/hydroxy acid multicomponent systems. Properties and pharmaceutical applications. *J Pharm Sci*, **89**, 1–8 (2000).
150. P. Mura, F. Maestrelli, M. Cirri, Ternary systems of naproxen with hydroxypropyl-β-cyclodextrin and aminoacids. *Int J Pharm*, **260**, 293–302 (2003).
151. T. Loftsson, Increasing the cyclodextrin complexation of drugs and drug biovailability through addition of water-soluble polymers. *Pharmazie*, **53**, 733–740 (1998).
152. T. Loftsson, M. Másson, The effects of water-soluble polymers on cyclodextrins and cyclodextrin solubilization of drugs. *J Drug Del Sci Tech*, **14**, 35–43 (2004).
153. T. Yamakawa, S. Nishimura, Liquid formulation of a novel non-fluorinated topical quinolone, T-3912, utilizing the synergic solubilizing effect of the combined use of magnesium ions and hydroxypropyl-β-cyclodextrin. *J Control Rel*, **86**, 101–113 (2003).
154. T. Furuta, H. Yoshii, A. Miyamoto, A. Yasunishi, H. Hirano, Effects of water of inclusion complexes of d-limonene and cyclodextrins, *Supramolecular Chemistry*, **1**, 321–325 (1993).
155. P. Li, L. Zhao, S.H. Yalkowsky, Combined effect of cosolvent and cyclodextrin on solubilization of nonpolar drugs. *J Pharm Sci*, **88**, 1107–1111 (1999).
156. V. Zia, R.A. Rajewski, V.J. Stella, Effect of cyclodextrin charge on complexation of neutral and charged substrates: comparison of (SBE)7M-β-CD to HP-β-CD. *Pharm Res*, **18**, 667–673 (2001).
157. T. Loftsson, Í.B. Össurardóttir, M. Duan, N. Zhao, T. Thorsteinsson, M. Másson, Cyclodextrin solubilization of the antibacterial agents triclosan and triclocarban: effect of ionization and polymers. *J Incl Phenom Macroc Chem*, **52**, 109–117 (2005).
158. T. Loftsson, P. Jarho, M. Másson, T. Järvinen, Cyclodextrins in drug delivery. *Expert Opin Drug Deliv*, **2**, 335–351 (2005).
159. A. Figueiras, J. Hombach, F. Veiga, A. Bernkop-Schnürch, In vitro evaluation of natural and methylated cyclodextrins as buccal permeation enhancing system for omeprazole delivery. *Eur J Pharm Biopharm*, **71**, 339–345 (2009).
160. M. Jug, M. Bećirević-Laćan, S. Bengez, Novel cyclodextrin-based film formulation intended for buccal delivery of atenolol. *Drug Devel Ind Pharm*, **35**, 796–807 (2009).
161. J. Mannila, K. Jarvinen, J. Holappa, L. Matilainen, S. Auriola, P. Jarho, Cyclodextrins and chitosan derivatives in sublingual delivery of low solubility peptides: A study using cyclosporin A, alpha-cyclodextrin and quaternary chitosan N-betainate. *Int J Pharm*, **381**, 19–24 (2010).

162. S.D. Yoo, B.M. Yoon, H.S. Lee, K.C. Lee, Increased bioavailability of clomipramine after sublingual administration in rats. *J Pharm Sci*, **88**, 1119–1121 (1999).
163. S.I.F. Badawy, M.M. Ghorab, C.M. Adeyeye, Bioavailability of danazol-hydroxypropyl-β-cyclodextrin complex by different routes of administration, *Int J Pharm*, **145**, 137–143 (1996).
164. A.C. Jain, B.J. Aungst, M.C. Adeyeye, Development and in vivo evaluation of buccal tablets prepared using danazol-sulfobutylether 7 β-cyclodextrin (SBE 7) complexes, *J Pharm Sci*, **91**, 1659–1668 (2002).
165. T.J. Hoon, M.Y. Dawood, F.S. Khan-Dawood, J. Ramos, R.L. Batenhorst, Bioequivalence of 17β-estradiol hydroxypropyl-β-cyclodextrin complex in postmenopausal women. *J Clin Pharmacology*, **33**, 1116–1121 (1993).
166. H. Fridriksdóttir, T. Loftsson, J.A. Gudmundsson, G.J. Bjarnason, M. Kjeld, T. Thorsteinsson, Design and in vivo testing of 17β-estradiol-HPβCD sublingual tablets, *Pharmazie*, **51**, 39–42 (1996).
167. T. Loftsson, J.A. Guðmundsson, R.O. Árnadóttir, H. Friðriksdóttir, Sublingual delivery of 17β-estradiol from cyclodextrin containing tablets. *Pharmazie*, **58**, 358–359 (2003).
168. W.A.J.J. Hermens, M.J.M. Deurloo, S.G. Romeijn, J.C. Verhoef, F.W.H.M. Mercus, Nasal absorption enhancement of 17β-estradiol by dimethyl-β-cyclodextrin in rabbits and rats. *Pharm Res*, **7**, 500–503 (1990).
169. T. Loftsson, H. Gudmundsdottir, J.F. Sigurjonsdottir, H.H. Sigurdsson, S.D. Sigfusson, M. Masson, E. Stefansson, Cyclodextrin solubilization of benzodiazepines: formulation of midazolam nasal spray. *Int J Pharm*, **212**, 29–40 (2001).
170. O. Dale, T. Nilsen, T. Loftsson, H.H. Tønnesen, P. Klepstad, S. Kaasa, T. Holand, P.G. Djupesland, Intranasal midazolam: a comparison of two delivery devices in human volunteers. *J Pharm Pharmacol*, **58**, 1311–1318 (2006).
171. H. Cabral-Marques, R. Almeida, Optimization of spray-drying process variables for dry powder inhalation (DPI) formulations of corticosteroid/cyclodextrin inclusion complexes. *Eur J Pharm Biopharm*, **73**, 121–129 (2009).
172. T. Kinnarinen, P. Jarho, K. Järvinen, T. Järvinen, Pulmonary deposition of a budesonide/γ-cyclodextrin complex in vitro, *J Control Rel*, **90**, 197–205 (2003).
173. L. Matilainen, K. Jarvinen, T. Toropainen, E. Nasi, S. Auriola, T. Jarvinen, P. Jarho, In vitro evaluation of cyclodextrin complexation on pulmonary deposition of a peptide, cyclosporin, *A. Int J Pharm*, **318**, 41–48 (2006).
174. W. Yang, K.T. Chow, B. Lang, N.P. Wiederhold, K.P. Johnston, R.O. Williams, In vitro characterization and pharmacokinetics in mice following pulmonary delivery of itraconazole as cyclodextrin solubilized solution, *Eur J Pharm Sci*, **39**, 336–347 (2010).
175. D.O. Thompson, G.L. Mosher, *Formulations Containing propofol and a Sulfoalkyl Ether Cyclodextrin*. ed., US: Cydex, InC. (2006).
176. M.E. Davis, M.E. Brewster, Cyclodextrin-based pharmaceutics: Past, present and future. *Nature Rev*, **3**, 1023–1035 (2004).
177. FDA, Inactive Pharmaceutical Ingredients (2006).
178. FDA, Guidance for Industry: Nonclinical Studies for the Safety Evaluation of Pharmaceutical Excipients (2005).

4

Solid Lipid Nanoparticles for Drug Delivery

Sonja Joseph and Heike Bunjes

4.1 Introduction

Since many drugs with poor water solubility have lipophilic properties, lipid-based delivery systems are an interesting formulation option for such substances. Colloidal lipid carriers offer the advantage of being, in principle, applicable by virtually any route of administration, nonparenteral as well as parenteral, including i.v. administration. Compared to other types of nanoparticulate carriers such as, for example, colloidal fat emulsions, liposomes or mixed phospholipid micelles, solid lipid nanoparticles represent a relatively new type of carrier system that has not yet led to a commercial drug product. The first preliminary attempt at preparing solid lipid nanoparticles was reported in 1985 [1] and the first successful approaches were described around 1990 [2–4]. Since then, colloidal lipid suspensions have been under intensive investigation as a new type of carrier system, in particular for the parenteral administration of poorly water-soluble, lipophilic drugs.

Solid lipid nanoparticles consist of a core of solid lipids which is surrounded by a shell of stabilizing agents and are usually dispersed in an aqueous medium. According to their classification as a colloidal system, their mean particle size is usually in the range between below 100 and several hundreds of nanometers. In the majority of cases, glycerides, fatty acids and fatty alcohols are used as matrix components in the preparation of solid lipid nanoparticles. As these substances are components of physiological lipids, they are expected to be well tolerated and are classified as 'generally recognized as safe' (GRAS) by the FDA [5]. Matrix lipids based on waxes and solid paraffins are also well known in the preparation of biocompatible colloidal lipid dispersions but may not or may not be as easily biodegradable as the physiological lipids. The choice of matrix material thus

Drug Delivery Strategies for Poorly Water-Soluble Drugs, First Edition. Edited by Dennis Douroumis and Alfred Fahr.

depends a lot on the intended type of administration of the formulation of interest. This also applies to the type of emulsifier which has to ensure a good stability of the formulation as well as a good physiological compatibility. The potential toxicity of the components has to be taken into particular consideration with regard to intravenous administration. There are only a few approved emulsifying agents available for parenteral administration, including phospholipids (e.g. lecithin), sodium oleate, sodium glycocholate, poloxamer 188, polysorbate 80 and sorbitan trioleate [6].

Solid lipid nanoparticles were developed in order to improve the drug carrier properties of lipid carriers with a liquid or liquid crystalline matrix. In particular, they can be regarded as a further development of colloidal fat emulsions which are being used in parenteral nutrition (e.g. Intralipid®, Lipofundin®) and as drug carrier systems (e.g. Disoprivan®, Etomidat-®Lipuro, Lipotalon®, Stesolid®). The high mobility of drug molecules in liquid emulsion droplets usually causes a burst release of drug from the carrier in biological fluids. It was hypothesized that the exchange of the liquid with a solid particle core might reduce the mobility of incorporated drug molecules and might thus provide the potential to sustain and/or control drug release. Furthermore, a better physical and chemical stability was expected as a result of the solid physical state of the dispersed lipid phase. Also some advantages concerning the ease of surface modification enabling modification of pharmacokinetics and potential drug targeting were proposed [3, 7–9]. Not all of these expectations have proven realistic (as will be discussed later in this chapter) but solid lipid nanoparticles still remain an interesting formulation option.

This chapter first introduces and compares the different methods to prepare solid lipid nanoparticles followed by a short discussion of their general physicochemical properties. Afterwards, the interaction of these carriers with lipophilic drugs is considered and the final section of the chapter deals with the administration aspects of these particles with special regard to the oral and parenteral way of administration, as these are the most challenging when dealing with poorly water-soluble substances.

4.2 Preparation Procedures for Solid Lipid Nanoparticles

A broad variety of preparation techniques for solid lipid nanoparticles have been successfully developed. These techniques can be roughly divided into those involving top-down processes such as dispersion of the lipid phase (melt dispersion, cold homogenization, precipitation from solvent-in-water emulsions) and bottom-up procedures usually involving some kind of precipitation of the lipid particles from homogeneous systems (warm microemulsions, solutions in water-miscible organic solvents). Typically, top-down processes require rather high energy input whereas bottom-up procedures are lower-energy processes. In the following, the general principles of the different procedures are described, concluding with a comparison and a discussion of their use on different scales.

4.2.1 Melt Dispersion Processes

As most of the lipid matrix materials of interest have a melting point below 100°C, solid lipid nanoparticles can be conveniently prepared by dispersing the liquid melt into fine droplets in a hot, surfactant-containing aqueous phase (Figure 4.1). Except for the elevated

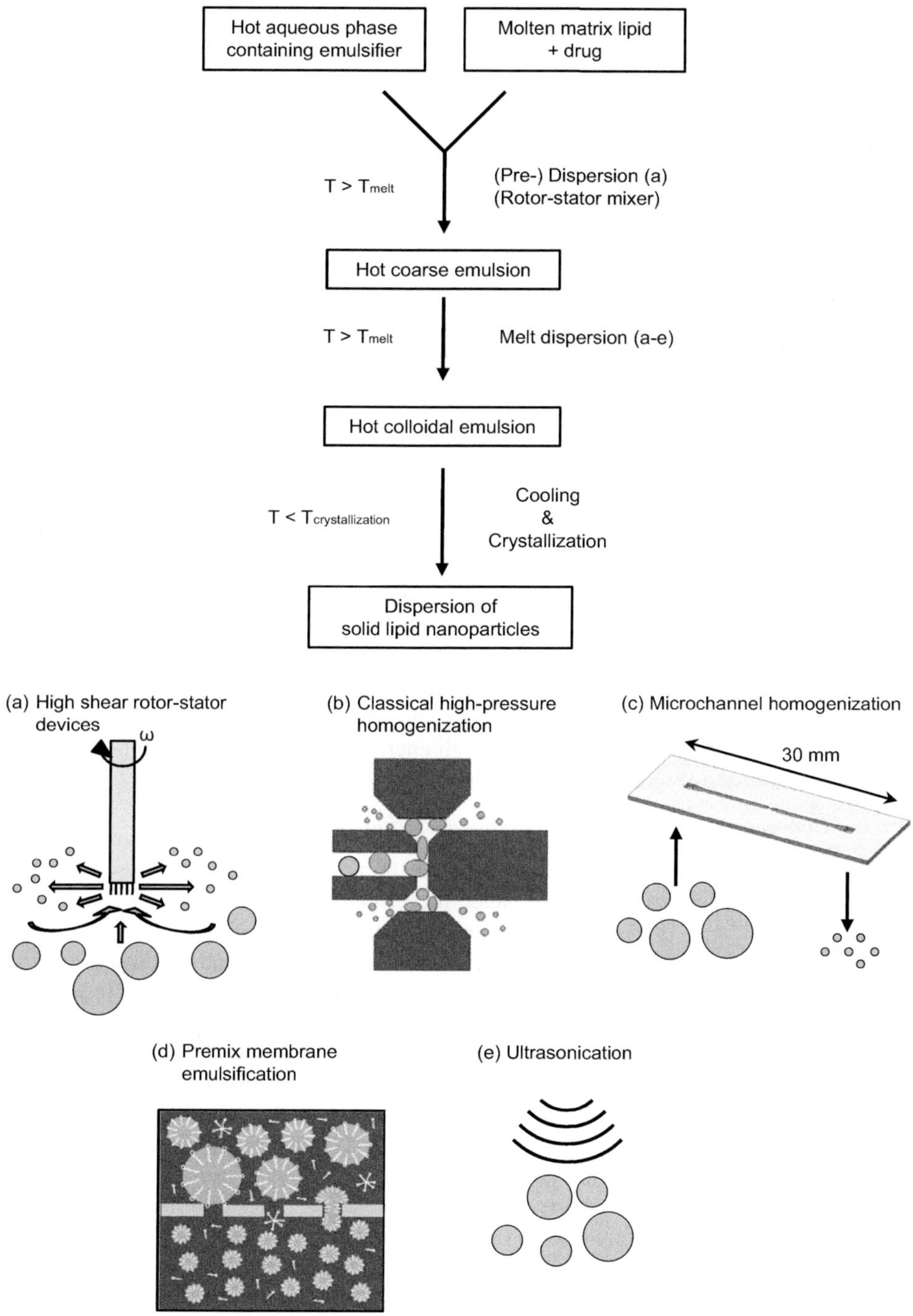

Figure 4.1 *General processes of melt homogenization for the preparation of solid lipid nanoparticles and variations of the procedure. (a–c, e) Reproduced from [18] with permission from Springer Science and Business Media. (d) Reproduced from [30] with permission from John Wiley and Sons.*

temperature required, the droplet generation process is completely analogous to that during the preparation of submicron lipid emulsions. This means that all the technologies established for emulsion preparation can be used [10]. A broad variety of matrix lipids can be employed in melt dispersion processes. Classically, they are of a more nonpolar nature (such as triglycerides, waxes or paraffins) but the processing of more polar matrices such as fatty acids, fatty alcohols or partial glycerides, sometimes in combination with nonpolar ones, has also been described. Lipid soluble drugs are usually processed by dissolving them in the melt prior to droplet dispersion; this process may be supported by using a solution of the drug in a volatile organic solvent. The emulsification process is carried out at temperatures above the melting point of the lipid droplets to retain the liquid state. To solidify the particles after generation of the colloidal droplets, a cooling stage is required. The conditions of the cooling process need to be carefully chosen in order to ensure complete crystallization of the lipids. Some lipid matrix materials (e.g. monoacid triglycerides like trimyristin or trilaurin) exhibit extended supercooling in the colloidal state and require adequate thermal treatment to be crystallized [7, 11]. Usually, the cooling process is carried out batch-wise (e.g. in small vials positioned in a thermostat or in a stirred tank) but the use of micro heat exchangers has also been explored in an attempt to provide precisely controlled cooling conditions and to allow continuous processing [12]. The physical transformations that take place during crystallization of the particles and potentially following polymorphic transitions of the lipids may impose particular requirements for the formulations of the nanoparticle dispersion. These aspects will be discussed in Section 4.3.

Melt dispersion of the matrix lipids usually is carried out as a two-step process involving a predispersion step (e.g. by high-shear rotor-stator vortexing) followed by disruption of the predispersed lipid phase into colloidal particles (Figure 4.1). A variety of different technical principles, commonly employing high energy input, can be used to perform the second dispersion step (Figure 4.1a–e) which will be outlined in the following.

4.2.1.1 Conventional High Pressure Homogenization

Homogenization with conventional high pressure homogenizers has been used for the preparation of solid lipid nanoparticles since they were first described in the beginning of the 1990s [3, 13]. It is currently the most frequently employed preparation process even though special manufacturing equipment is required. Corresponding devices of different designs and sizes are commercially available from several manufacturers (e.g. APV, Avestin, Niro Soavi, Microfluidics). In most cases, high pressure homogenizers push the predispersion through a homogenization valve, consisting of a narrow gap in the range of a few microns (Figure 4.1(b)). Very high shear and expansion forces in laminar flows, inertia forces through energy dissipating eddies in turbulent flows as well as cavitation forces are responsible for the droplet disruption during the process. Similar processes occur in the microfluidizer which uses narrow channels to achieve droplet break-up [14]. Typically applied homogenization pressures are in the range of several hundred to about 1500–2000 bar, and volume fractions of lipid phase of up to more than 10% can easily be processed.

Process parameters such as, e.g. the homogenization pressure and the number of homogenization cycles as well as the formulation parameters such as the type of matrix lipid and emulsifier as well as their concentration ratio, the volume fraction of disperse phase and the quality of the predispersion, influence the mean particle size and the particle size distribution, which are typically in the range of 50–400 nm. Usually, the particle size decreases with increasing pressure and cycle number reflecting a higher energy input and a more intense droplet break-up. Multiple passes are often preferable to achieve narrower particle size distributions. Smaller particles were also observed with a higher ratio of emulsifier to matrix lipid as well as higher temperatures [15, 16]. When optimized formulations are processed under adequate conditions, small-size dispersions with narrow size distribution and good storage stability can be obtained.

4.2.1.2 Melt Dispersion in Microchannels

As a variation of the conventional high pressure homogenization process, the hot, coarse dispersion of molten lipids can be forced with high pressures through microchannel systems yielding hot nanoemulsions (Figure 4.1c). Customized micro structures of varying geometric principles have been established for the preparation of nanoemulsions [17, 18] and further investigated for application in melt homogenization by a single passage [19]. The particle size and the particle size distribution of solid lipid nanoparticles obtained after microchannel homogenization in stainless steel microsystems depend on the homogenization pressure as well as on the geometric structure of the microchannels. The preparation of narrowly distributed solid lipid nanoparticles in a single passage requires pressures up to 1500 bar.

4.2.1.3 Membrane Techniques

Membrane structures can also be used to disrupt the droplets in the coarse pre-emulsion. The so-called premix membrane emulsification method developed in the late 1990s [20] is related to the extrusion technique for the preparation of unilamellar liposomes with controlled particle size [21]. A predispersed emulsion is extruded several times through the pores of a membrane yielding smaller emulsion droplets (Figure 4.1d). In order to prepare uniform particle size products through premix membrane emulsification, the porous membrane must have a narrow pore size distribution and be strong enough not to deform or compact, even if pressure is applied [22]. Premix membrane emulsification has been used for the preparation of simple [23] and multiple emulsions [24], solid lipid microcapsules [25], polymer micro- and nanoparticles [26, 27] as well as low density lipoprotein (LDL) analogues [28, 29]. Recently, this technique was also established for the preparation of solid lipid nanoparticles [30, 31]. Very narrow particle size distributions of the solid lipid nanoparticles were observed in particular when using the membranes in a commercial pump-driven extruder. The process could, however, also be performed in a small-scale ($\leq$1 ml), hand-held extruder (as typically used for the preparation of liposomes) which allowed the rapid preparation of a large number of samples within a short time. The processable concentrations of the lipid phase were in the range of those found in high pressure homogenization. The mean particle sizes of the lipid nanoparticle dispersions depended mainly on the membrane pore size and the number of extrusion cycles. Ratios of average pore size to average particle size in the range of 1:0.5 to 1:2.5 were obtained [30]. Most experiments were performed with track etched polycarbonate membranes but other types of membrane filters, as well as Shirasu Porous Glass (SPG) membranes could also be used [32, 33].

A different approach that does not rely on the preparation of a coarse pre-emulsion is used in the classical or 'direct' membrane emulsification process (Figure 4.2). The liquid lipid phase is forced by low pressures (e.g. 0.15 to 10 bar, depending on the membrane pore size) through the pores of a membrane into the emulsifier-containing aqueous continuous phase.

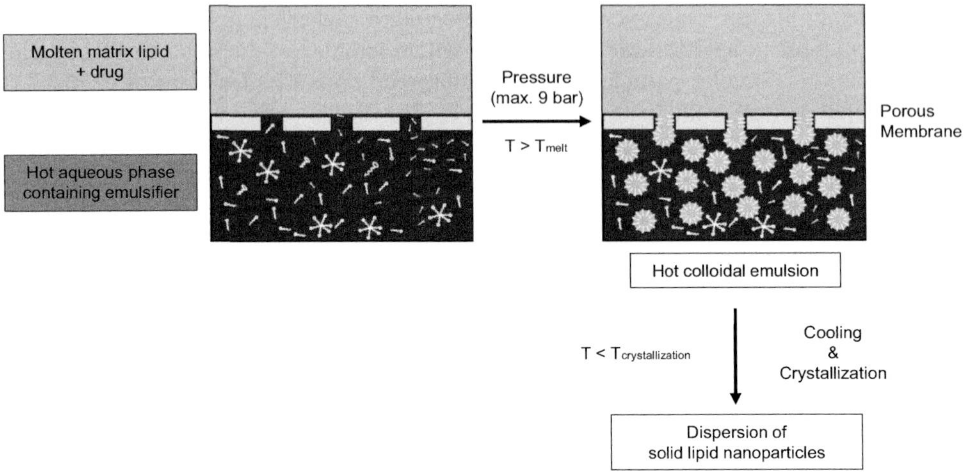

Figure 4.2 *Direct membrane emulsification for the formation of solid lipid nanoparticles.*

Mechanically robust membranes like glass or ceramic membranes are usually required for this process [22, 34, 35]. Lipid droplets grow at the pore openings at the membrane surface and are stabilized by emulsifiers from the continuous phase. When the droplets reach a certain size, they are detached from the membrane by the stream of the aqueous phase which is recirculated over the membrane surface [22, 34]. The resulting particle size is primarily controlled by the type and pore size of the membrane and the particle size distributions are very narrow. For emulsions prepared with Shirasu Porous Glass membranes, typical ratios of average pore size to average particle size were obtained in the range from 1:2 to 1:10 [36, 37]. Although this process has mainly been used to prepare micron-sized drug delivery systems [22, 38, 39], there are also a few reports on the preparation of solid lipid nanoparticles [40–43]. In these investigations, however, the self-emulsifying matrix lipid Gelucire 44/14 was used, which is not likely to form common solid lipid nanoparticles and which does not require a special dispersion process. The processing of lipid materials containing Compritol 888 or Precirol ATO 5 stabilized with poloxamer 188 did not yield fine colloidal dispersions but lipid particles in the upper nanometer range with high instability on storage were obtained [43, 44]. In principle, however, the preparation of stable, colloidal and narrowly distributed lipid dispersions on the basis of nonpolar lipids can be achieved when membranes with small pore diameters (100 nm) are used [32]. In most cases, the particles were, however, rather large as a result of the typical pore size to particle size ratios observed in this process. This and the long process times which often still lead to low concentrations of disperse lipid phase make this technique less promising for routine use in solid lipid nanoparticle preparation.

4.2.1.4 Ultrasonication

In several studies, probe ultrasonication has been reported for the preparation of solid lipid nanoparticles by melt dispersion. In this process, acoustic cavitation is the driving force of droplet disruption [45] (Figure 4.1e). Upon collapse of the cavitation bubbles, high energy is transferred into the dispersion. In the resulting hot spots, temperatures of approximately 5000°C and pressures of about 500 bar have been determined [46]. The particle size and the particle size distribution of the ultrasonicated dispersions are influenced by the energy performance of the probe as well as by the composition of the formulation [47, 48]. For solid lipid nanoparticle formulations, the formation of particles in the adequate size range has been reported [3, 49] but the dispersion quality may be compromised due to the presence of submicron-sized particles, resulting in instabilities during storage [3,50,51]. Moreover, particularly during long process times, the use of ultrasonication carries the risk of metal contamination from the probe [52].

4.2.1.5 Dispersion by High Shear Rotor-Stator Devices

Rotor-stator vortexing (e.g. by UltraTurrax®), which is commonly employed in the predispersion process, has also been used for the main homogenization stage [53, 54] (Figure 4.1a). Here, the droplet disruption depends on high shear forces resulting from the high energy input. As the energy input is typically lower than for most of the techniques described above, it is, however, rather difficult to produce homogeneous colloidal dispersions of solid lipid particles with this method. The use of optimized combinations of formulation and preparation conditions is crucial to achieve dispersions of acceptable quality [54]. Often, rather coarse and heterogeneous lipid dispersions are obtained (e.g. see [55]).

4.2.2 Other Top-Down Processes

4.2.2.1 Precipitation from Solvent-in-Water Emulsions

Instead of as a melt, the lipid can also be emulsified as solution in non-water-miscible organic solvents (Figure 4.3a). The matrix lipid and the pharmaceutically active ingredients are dissolved in a water-immiscible organic solvent and emulsified in the aqueous phase containing emulsifier. Typically, the mixture is processed in a high-pressure homogenizer yielding a dispersion of nanodroplets by high energy input. Afterwards, the organic solvent is removed by evaporation under reduced pressure to precipitate the solid lipid nanoparticles. This process does not require any heat treatment or low melting point of the lipid matrix materials, and particle sizes far below 100 nm have been achieved [4, 56]. However, triglyceride suspensions obtained by this method tended to be physically less stable than dispersions prepared by melt homogenization [56]. Moreover, residual amounts of organic solvent may lead to toxicological problems during the administration of these colloidal drug carrier systems, which is the major drawback of this preparation procedure.

In a modification of the original procedure, Trotta *et al.* used partially water soluble solvents with low toxicity instead of water-immiscible organic solvents [57]. The dispersion process was accomplished with rotor-stator equipment. Upon dilution of the resulting emulsion with a large amount of water, colloidal lipid particles precipitated from the organic phase as a result of solvent extraction into the aqueous phase. As this process is dilution-based, relatively low concentrations of particles are obtained. In a further variation, the matrix lipid and drug were dissolved in diethyl ether and injected into a warm (40°C) agitated, emulsifier-containing aqueous phase. Due to its low boiling point, the diethyl ether evaporated and thus the lipid precipitated from the dispersed droplets [58].

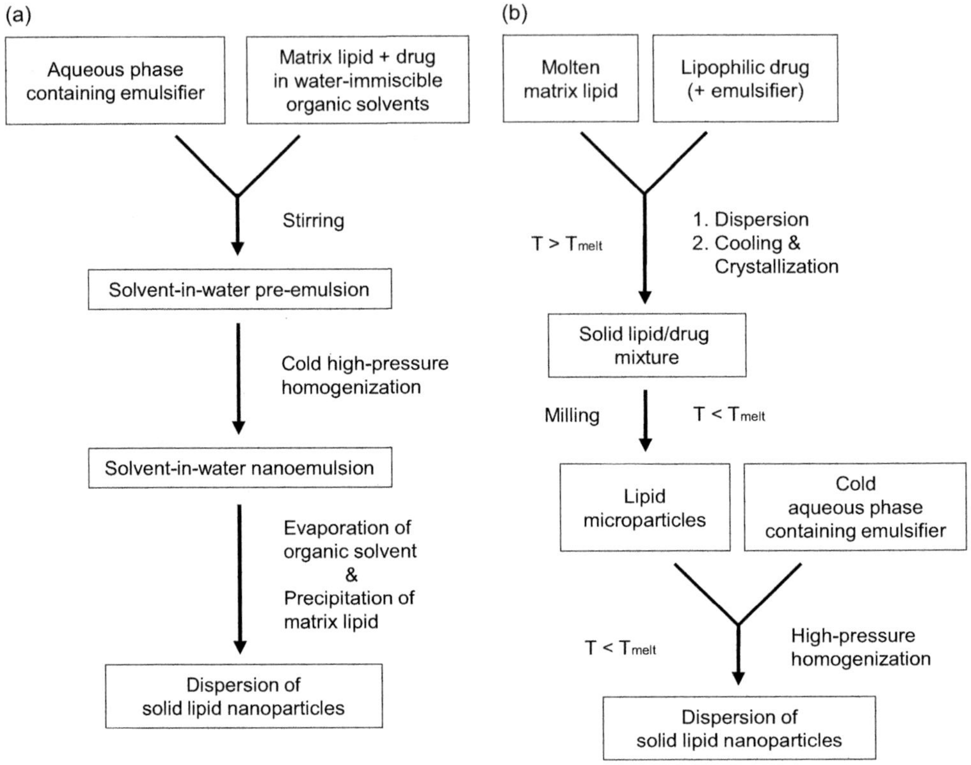

Figure 4.3 *Precipitation from solvent-in-water emulsions (a) and cold homogenization (b) for the preparation of solid lipid nanoparticles.*

4.2.2.2 *Cold Homogenization*

A further possibility for the preparation of solid lipid particles is the high-pressure homogenization of the lipid matrix in the solid state. A coarse dispersion of powdered lipids in the emulsifier-containing aqueous phase is processed below the melting point of the matrix lipid (Figure 4.3b). For drug incorporation, the matrix lipid is melted and the drug is dissolved or dispersed – sometimes with the aid of organic solvents. The solidified mixture is milled into microparticles in liquid nitrogen or using dry ice cooling [59–61]. Also this technique allows the processing of the matrix lipid materials with a melting point above 100°C, which cannot be used in melt homogenization, e.g. cholesterol (melting point ~148°C) [60]. However, larger particles and rather broad particle size distributions often containing particles in the upper nanometer or even in the micrometer range were observed in spite of comparatively harsh homogenization conditions. Typically, higher pressures (1000–1500 bar) and more homogenization cycles (30–40) are required than for melt homogenization [59, 62]. Due to the high energy input, heating of the dispersion may occur. Friedrich *et al.* measured product temperatures in the range of the melting point of the matrix lipid after processing at room temperature and observed increasing homogenization efficiency due to the partial melting of the lipids [61].

4.2.3 Precipitation from Homogeneous Systems

Solid lipid nanoparticles can also be precipitated from homogeneous solutions or colloidal systems. Such processes usually do not require the use of high energy input and can in most cases be performed with conventional laboratory equipment. Often, rather small particles can be obtained but – since supersaturation phenomena are commonly involved – it is sometimes not easy to prevent them from increasing in size.

4.2.3.1 Precipitation from Warm Microemulsions

The solidification of lipid nanoparticles upon precipitation from warm microemulsions introduced by Gasco *et al.* was the first preparation method published for colloidal solid lipid dispersions [2]. This preparation principle is still widely used due to the simple process sequence. In the original process, fatty acids (e.g. stearic acid) were used as matrix constituents but the process was later extended also to other lipids [63, 64]. The molten matrix lipid (containing the drug) is mixed by mechanical stirring with the hot aqueous phase composed of water, emulsifier and cosurfactant. An optically transparent, homogeneous colloidal system is formed spontaneously in the heat and is subsequently diluted with cold water (2–3°C) to precipitate solid lipid nanoparticles (Figure 4.4). Rather high dilution ratios are often required to obtain small-size dispersions, thus leading to low lipid concentrations. Moreover, a high concentration of surface active additives (typical examples are phospholipids and bile salts), is usually employed to form the initial microemulsion. In consequence, the process was enhanced continuously: ultrafiltration, dialysis and centrifugation were used to concentrate the diluted nanoparticulate systems after precipitation as

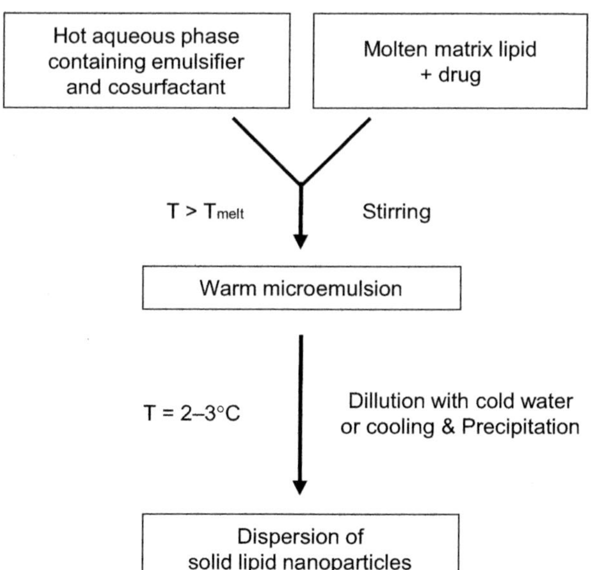

Figure 4.4 *Precipitation from oil-in-water microemulsions for the preparation of solid lipid nanoparticles.*

well as to remove most of the cosurfactants [65–67]. Freeze drying is typically performed to avoid particle growth in aqueous systems upon storage [68].

Systems with modified compositions (usually based on a mixture of fatty alcohols and nonionic surfactants) allow direct cooling of the warm microemulsion to form solid lipid nanoparticles under stirring [69–75]. Lipophilic drugs are processed with the lipophilic components of the microemulsion. The size of the resulting nanoparticle dispersion may be affected by drug loading and the composition of the microemulsion may require some adjustment to yield high-quality solid lipid nanoparticle dispersions [73]. Although dilution of the microemulsion during precipitation is not required, these dispersions typically contain matrix lipid concentrations below 1%. Upon storage, the aqueous suspensions tend to display particle growth [70, 71, 76] and thus some kind of stabilization, e.g. by freeze drying, appears to be necessary.

4.2.3.2 Precipitation from Water-Miscible Organic Solvents

The precipitation of nanoparticulate solid material from water-miscible organic solvents such as acetone or ethanol (Figure 4.5) is becoming increasingly popular. A solution of lipid matrix material, drug and sometimes stabilizers is injected into an agitated, emulsifier-containing aqueous phase. The whole process can be carried out at room temperature [77] or at increased temperatures of the aqueous or organic phase, or both [66, 78, 79]. The use of hot organic solutions may increase the solubility in the organic solvent. To improve dispersion of the lipidic ingredients, ultrasonication or a combination of heat treatment and ultrasonication or vortexing during the solvent injection has been used [80–82]. In

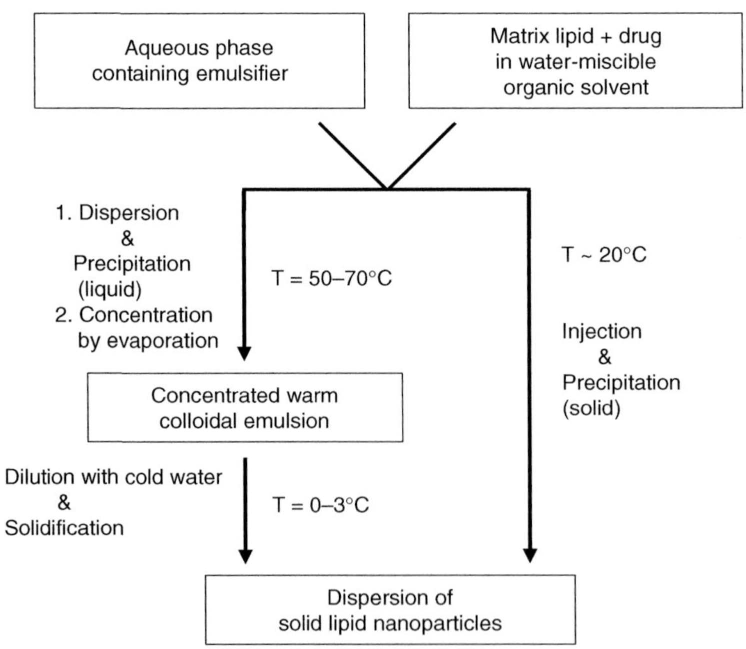

Figure 4.5 *Precipitation from organic solution for the preparation of solid lipid nanoparticles.*

most cases, organic solvents are later removed from the dispersion, e.g. by evaporation at increased temperature and/or reduced pressure [78, 83, 84] or by centrifugation and subsequent redispersion of the concentrated solid lipid nanoparticles [79, 85]. Removal of organic solvents reduces toxicological problems and increases the colloidal stability of the dispersions.

While the above-mentioned approaches use batch-wise processing, microchannel techniques allow the precipitation of lipid nanoparticles from organic solutions in a continuous process. Microchannel assemblies of different geometry (e.g. a co-flowing assembly with inner and outer capillaries or cross-shaped channels) were employed for the preparation of solid lipid nanoparticles [86–88]. The solution of solid lipids in water-miscible solvents and the emulsifier-containing aqueous phase are injected simultaneously into the different channels of the microsystem by separate syringe pumps. When the phases are combined at the junction of the channels, the solvent starts to diffuse into the aqueous phase during the passage along the main, common channel. As a result of supersaturation of the lipid in the aqueous phase, solid lipid nanoparticles are formed.

4.2.4 Comparison of the Formulation Procedures and Scale-Up Feasibility

An ideal preparation procedure would lead to the reproducible formation of lipid nanoparticle dispersions with narrow particle size distributions and would provide good control over the particle size range obtained. The particle size properties should be retained during storage without further processing steps which would make the process more tedious and expensive. In any case, the preparation method has to be compatible with the physicochemical properties of the drug to be incorporated. For example, a heat-sensitive active pharmaceutical ingredient should be processed without or with only minimal heat exposure. Generally, the whole procedure should be gentle without too much stress on the formulation and its ingredients. The process would be easily scalable and would not require the use of sophisticated equipment. Clearly, the processes described above can only partially meet these criteria.

The conventional high-pressure melt homogenization process appears particularly promising with regard to commercial production as it uses well-established instrumentation that is available for very different scales. Several studies have proven the scaling-up potential of this manufacturing process [89, 90]. The particle size parameters can be reliably controlled by the homogenization conditions and storage-stable aqueous dispersions are obtained with appropriate formulations. This also applies to the more recent modifications such as the premix membrane or high pressure microchannel emulsification. Concerning potentially negative influences on the formulation, heat and shear stress are parameters that cannot be avoided in the conventional melt homogenization process. For most lipophilic drugs, shear stress should not pose much of a problem. If so, the membrane techniques which involve lower energy densities [91] and typically lead to dispersions with very well-controlled particle size distributions may be an interesting alternative.

Thermolabile ingredients may, however, suffer from the heat inevitably present during melt-dispersion processes. The high energy input upon pressurization combined with rather poor energy utilization in the high-pressure homogenization process leads to a temperature increase in the equipment dependent on the homogenization pressure (e.g. 15°C per cycle at 1250 bar [92], 2.5°C per 100 bar for water [93]) and the emulsion composition [94]. From this point of view, membrane techniques may be advantageous [91]. While the additional evolution of thermal energy in the homogenization equipment can be controlled by adequate thermostatization, the heat required to prepare the (drug-containing) lipid melt and to keep the formulation above the melting temperature of the lipid during processing cannot be avoided. In production, very rapid processing might reduce heat exposure. Although the cold homogenization process reduces heat exposure and thus problems with temperature-induced drug degradation [59] and may additionally circumvent problems with the crystallization step of melt-homogenized nanodispersions [15, 16], it does not appear to be a particularly promising alternative with regard to a good control over the dispersion properties. The only possible way to strictly avoid working at elevated temperatures in homogenization-based procedures is the processing of the lipid in organic solution. However, the organic solvents need to be removed completely from the formulation afterwards, which requires additional process steps.

The absence of heat during the entire process makes precipitation methods involving water-miscible organic solvents interesting for the formulation of heat-sensitive drugs. Furthermore, these methods allow the processing of high-melting lipids which cannot be used in melt homogenization. However, the organic solvents have to be removed from the final dispersion in order to ensure the physicochemical stability of the compounds and physiological compatibility. Moreover, the resulting nanoparticle dispersions may have a rather broad particle size distribution, resulting in physical instabilities of the drug carrier formulations. Potentially, current developments in the use of micromixing equipment may solve this challenge by providing better control over the precipitation conditions. Still, such procedures will probably always require concentration steps as precipitation is typically carried out at rather low concentrations.

Precipitation from warm microemulsions does not require specialized equipment and is thus widely being used in laboratory set-ups. Its potential for commercial large-scale production appears to be somewhat limited as it combines rather than solves many of the problematic issues discussed above. First of all, microemulsion formulation requires a delicate choice of excipients and their ratios. Quite a high concentration of surfactants is usually necessary to form the initial, homogeneous colloidal system. Also, heat is required for processing although the heat exposure of drugs can probably be minimized by adding them only shortly before the cooling step. The physical stability of the precipitated particles is limited (as is the concentration of lipid phase) and further process steps (washing, drying) appear to be inevitable. The washing step applied in the original method to remove the co-surfactants used for microemulsion formation leads to a loss of cost-intensive surfactants (e.g. bile salts) and can also reduce the drug content in the formulation. Nevertheless, some investigations on the scaling-up possibilities of microemulsion-derived solid lipid nanoparticle dispersions have been performed [95–97].

In contrast to the scale-up feasibilities of the preparation techniques, scaling-down possibilities have only scarcely been discussed. Small-volume processes may be advantageous for formulation screening where the possibility of preparing a large number of samples within a short time is required. Moreover, they allow the handling of scarce, expensive or highly potent substances (which require effective containment). For example, the microemulsion-based process has often been used on a comparatively low scale. Melt-dispersion processes with small-scale lab homogenizers typically require at least several milliliters up to more than 100 ml and are difficult to perform on smaller scales. Ultrasonication may be considered an alternative but it may lead to radical formation, resulting in oxidation and hydrolysis of sensitive substances [98, 99]. The best approach for a small-scale screening set-up would be with processes that allow the transfer of process parameters from the laboratory to the production scale. Membrane as well as high-pressure microchannel emulsification have recently been investigated in this regard [19, 30].

In conclusion, the advantages and disadvantages of the different manufacturing procedures have to be carefully taken into consideration when choosing an appropriate particle preparation technique. Important aspects include the physicochemical properties and stability of formulation components (matrix lipids, emulsifiers, drugs) to be incorporated, the desired particle size properties, the concentration and stability of the colloidal formulation, the preparation scale (and/or the possibility of scaling up or scaling down the process) as well as the available equipment.

4.2.5 Further Processing of Solid Lipid Nanoparticle Suspensions

Depending on the intended route of administration, there are several further processing approaches of interest for solid lipid nanoparticle formulations. With regard to parenteral or ocular administration, sterilization is an important issue. Surprisingly, there is still little information on the effect of sterilization on the properties of solid lipid nanoparticles. It has been reported that the stability of melt-homogenized systems upon autoclaving depends on the nature of the lipid/surfactant combination and that optimized systems can be autoclaved at 121°C [13]. Care has to be taken upon steam sterilization of purely sterically stabilized systems which may not withstand the heat treatment [100]. The combination of nonionic with ionic stabilizers as, for example, used in [101] might be an approach to overcome this problem.

In order to increase the chemical and physical stability of the particles as well as to prepare solid dosage forms, drying processes such as lyophilization or spray drying have been employed. Freeze drying requires the addition of a cryoprotectant to enhance redispersion and to reduce particle growth. In spite of cryoprotective additives, redispersion of the lyophilizates may not be possible without additional energy input, e.g. by ultrasonication [13, 102]. Moreover, significant particle growth was observed for melt-homogenized formulations based on glycerides as matrix lipids after freeze drying, even when using trehalose or sucrose as cryoprotectant [102, 103]. Optimization of the freezing program may help to preserve the dispersed state of the formulation upon freeze drying [104]. Concerning long-term storage, potential alterations of the quality of lyophilizates need to be taken into consideration [102, 105]. For stearic acid-based dispersions obtained from the microemulsion process, an evaporative drying process has been developed which eliminates freezing as a potential source of agglomeration [106, 107].

Also spray drying of solid lipid nanoparticle dispersions requires the addition of additives such as sugar to enable redispersion. Optimized formulations obtained under adequate process conditions may display similar particle size characteristics before and after the spray drying procedure [108, 109]. Direct incorporation of melt-homogenized solid lipid nanoparticles into pellet formulations by using the aqueous dispersion as granulation liquid has also been described [110].

Compared with the transformation into solid dosage forms, the incorporation of solid lipid nanoparticle suspensions into semisolid dosage forms is much more straightforward.

As solid lipid nanoparticles are under intensive investigation with regard to dermal administration, the processing into creams or hydrogels has often been described [111–114].

4.3 Structural Parameters and Their Influence on Product Quality and Pharmaceutical Performance

Solid lipid nanoparticles as drug carrier systems have to be based on physiologically compatible excipients. Many lipid matrix materials (e.g. glycerides, fatty acids and waxes) and emulsifiers (such as lecithin, bile salts, polysorbates or poloxamers) meet this demand but the actual choice of excipients will also have to consider the specific requirements of the corresponding administration route. For example, the choice of excipients for parenteral dosage forms is more limited than that for oral or dermal dosage forms. The composition of the formulation as well as the preparation procedure may have pronounced effects on the physicochemical properties of solid lipid nanoparticle dispersions. Particle size and shape, the composition and stability of the colloidal structures involved as well as issues related to the crystalline state of the nanoparticles are of major interest in this regard and are thus the focus of current research.

4.3.1 Particle Size and Size Distribution

Particle size and size distribution are natural quality-determining parameters of colloidal dispersions for drug delivery. Not only with regard to quality assurance, these parameters should be well defined and reproducibly be obtained by the chosen preparation procedure. In particular with regard to intravenous administration, the importance of particle size on the biodistribution and the clearance of colloidal particles is well known. Smaller particles are better able to reach a target tissue via the circulation [115, 116]. Moreover, attention needs to be paid to the particle size distribution in order to minimize the risk of embolism and other side effects due to the presence of large particles [117]. Also with respect to other administration routes, particle size effects have been reported [118–120]. The influence of different preparation procedures on the achievable particle sizes and the particle size stability has already been outlined above. Moreover, the choice of the emulsifier and its ratio to the matrix lipid are very important in the development of high-quality solid lipid nanoparticle dispersions. Relatively high emulsifier concentrations are required to stabilize the surface of small colloidal particles. Moreover, for preparation procedures that involve morphological alterations of the particles during the process, an initial excess of emulsifier may be required to prevent problems with particle size stability (cf. Sections 4.3.3, 4.3.4). On the other hand, a distinct excess of emulsifier is usually undesirable because it may lead to side effects as has been observed for parenteral fat emulsions [121].

The particle size and the size distribution of colloidal lipid drug carrier dispersions are routinely determined by light scattering methods (e.g. photon correlation spectroscopy (PCS)) or laser diffraction equipped with an adequate submicron instrumentation (e.g. polarization intensity differential scattering technology) to confirm the desired colloidal state and to monitor its stability during storage and further processing steps. The confirmation of the colloidal state can also be achieved by microscopic techniques (e.g. electron and atomic force microscopy), but the possibility of substantial changes in the ultrastructure due to the preparation steps required for these techniques needs to be taken into consideration. Moreover, such methods are usually of limited applicability in the investigation of the particle size distribution. Unfortunately, solid lipid nanoparticle dispersions are often not checked for the absence of microparticles. As a specialized technique for the colloidal particle size range, PCS is not very reliable in detecting microparticulate fractions (although hints towards the presence of microparticulate contaminations may be obtained with modern instrumentation, depending on the measurement conditions). Even with laser diffraction, it may be difficult to observe very small amounts of micron-sized particles. Simple optical microscopy can be very helpful in assuring the absence of at least larger fractions of microparticles. Single particle counters such as single particle optical sensing (SPOS) may be used to determine the absolute number of particles in the micrometer range. SPOS has already been developed into a USP method to determine the number of large particles in parenteral fat emulsions [122]. As also recommended for other types of colloidal particles and routinely performed with colloidal fat emulsions [122, 123], a combination of at least two methods is highly advisable to obtain reliable information on the particle size parameters of solid lipid nanoparticle dispersions.

4.3.2 Surface Properties

The characteristics of the particle surface of colloidal dispersions are important in terms of their colloidal stability and *in vivo* behavior. The large specific surface area of the solid lipid nanoparticles needs to be stabilized in aqueous dispersions. This can be achieved by using amphiphilic substances due to electrostatic or steric repulsion or to a combination of both principles. The repulsion forces are imposed by the surfactant molecules on the particle surface and their interaction with the surrounding medium. The use of ionic surfactants results in a positive or negative zeta potential and thus induces electrostatic stabilization. Most surfactants employed in the preparation of solid lipid nanoparticles result in a negative surface charge. However, cationically charged particles have also been investigated as transfection agents for nucleic acids (RNA and DNA) and as protein carriers [124–127]. Particles prepared with nonionic and macromolecular surfactants are stabilized by steric repulsion forces. Additionally, such particles often bear negative surface charges as evident from zeta potential measurements [128]. Steric stabilization agents often contain polyethylene glycol chains resulting in a the hydrophilic cladding of the lipophilic particle core. The polyethylene glycol chains are responsible for the reduced phagocytic uptake of the particles and thus for longer circulation times [129, 130]. Further modifications of the particle surface are directed towards active drug targeting, i.e. a tissue-specific delivery induced by specific ‘homing devices’ on the particle surface. For example, surface modification with galactoside led to an increased accumulation of solid lipid nanoparticles in the liver after intravenous administration [131].

4.3.3 Solid State Properties of Solid Lipid Nanoparticles

In principle, the lipids serving as a matrix in solid lipid nanoparticles retain their material properties when dispersed into the colloidal state but there may be some modifications due to the small particle size. As many of the preparation procedures commonly applied rely on the processing of the matrix lipids above their melting temperatures, issues related to the crystallization of the nanoparticles are of special importance. Differential scanning calorimetry (DSC) and X-ray diffraction are the methods of choice to study phenomena of crystallinity and polymorphism in colloidal lipid dispersions [132].

As the solid state of the particles is one of the main features of solid lipid nanoparticles, its presence has to be ensured in the formulation. Typically, lipid nanoparticles display a lower crystallization tendency than the bulk lipids and may thus not readily crystallize after cooling below their melting temperature. Supercooling (i.e. the temperature difference between the melting and the crystallization temperature) is differently pronounced in different materials ranging, e.g. from less than 10°C for cetyl palmitate or glyceryl behenate [133] and up to more than 40°C for triglycerides [7, 11]. Nanoparticles prepared from comparatively low melting lipids such as certain hard fats or short-chain monoacid triglycerides (e.g. trimyristin, trilaurin) thus may not crystallize when simply cooled to room temperature after processing in the heat. Instead emulsions of supercooled, liquid particles may be formed that can display a surprisingly high long-term stability of the supercooled droplets (up to several years) [7, 11, 134]. Such dispersions require specific processing such as cooling to very low temperatures or isostatic high pressure treatment to reliably transform the particles into the solid state [11, 135, 136]. In such cases the required crystallization conditions need to be carefully established during the development process. This can easily be done by DSC investigations on the aqueous dispersions. At least, all new formulations need to be checked for their crystallization behavior since the crystallization temperature can be modified by the presence of certain emulsifiers or drugs. For example, the incorporation of larger amounts of ubidecarenone or vitamin E led to a distinct decrease in the crystallization temperature [133, 137]. On the other hand, emulsifiers with long, saturated alkyl chains, including saturated phospholipids are able to increase the crystallization temperature of some matrix materials [135, 138–140].

Although usually less affected than the crystallization temperature, some peculiarities of the melting behavior are also commonly observed. Frequently, the nanoparticles display lower melting temperatures than the bulk material. This common feature of small colloidal particles is expressed in the Gibbs–Thomson equation that predicts an inverse relation between the particle radius and the melting temperature. Clear correlations between the particle size and the melting temperature of nanoparticles prepared from different triglycerides could indeed be established. For small particles (e.g. $<$150–200 nm) prepared from monoacid triglycerides, a very specific melting behavior could be observed [141]. DSC heating curves of the corresponding nanoparticle dispersions display a series of well-defined melting events (Figure 4.6) that correspond to the melting of nanoparticles with different thicknesses [142, 143]. More to be expected, the melting temperature of solid lipid nanoparticles is influenced by the presence of incorporated foreign substances such as additional matrix materials or drugs. Eutectic behavior was, for example, observed for nanoparticles prepared from binary mixtures of triglycerides [11]. The presence of larger amounts of incorporated active substances may decrease the melting temperature of solid lipid nanoparticles distinctly [133, 137, 144]. Eutectic melting was also observed for the mixture of matrix components in fatty alcohol-based nanoparticles prepared from warm microemulsions and there may be some correlation of this phenomenon with the increased particle growth rate of these particles observed with increasing temperature [76].

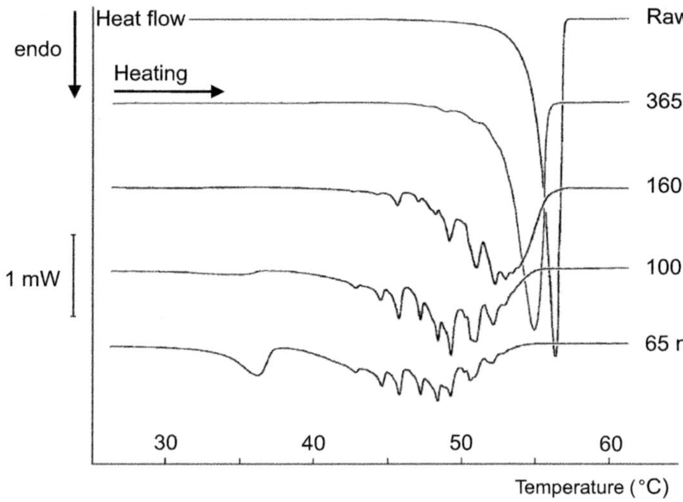

Figure 4.6 *Differential scanning calorimetry heating curves (0.04°C/min) displaying the melting behavior of tyloxapol-stabilized trimyristin dispersions with different particle sizes (z-average values and corresponding polydispersity indices are shown at the respective curve). The raw material was dispersed in an aqueous phase containing 1% tyloxapol. The curves are displaced along the ordinate. Adapted from [141] with permission from the American Chemical Society.*

Many matrix lipids used in the preparation of solid lipid nanoparticles can occur in various crystalline modifications, for example triglycerides or fatty acids [95, 145]. For example, triglycerides usually crystallize in the metastable α-modification from the melt and undergo a time- and temperature-dependent polymorphic transition into the stable β-modification upon storage of the colloidal dispersions [146]. The different polymorphic forms can be identified by X-ray diffraction and often also by DSC analysis. As melting temperatures observed for solid lipid nanoparticles deviate from that of the bulk material, and DSC curves of particles with more complex matrix composition are sometimes difficult to interpret, it is often advisable to perform investigations with both methods to avoid erroneous interpretation. The kinetics of the transitions between the different polymorphic forms depends on the excipients incorporated in the colloidal dispersion, in particular on the matrix lipid and the emulsifier. Shorter chain triglycerides (e.g. trilaurin (C12)) transform more quickly from the metastable modification into the stable one than triglycerides with longer chains (e.g. tristearin (C18)) [11]. Some emulsifiers, like poly(vinyl alcohol), sodium glycocholate or long-chain hydrogenated phospholipids induce a very slow transition rate whereas many typical nonionic surfactants and their combination with unsaturated phospholipids lead to a rather rapid transformation [135, 138, 140, 147, 148]. Particles stabilized with poly(vinyl alcohol) or the ionic emulsifier glycocholate may assume an atypical, 'nonlamellar' structure of the α-form (Figure 4.7a) [147, 148] that might be responsible for their extended stability in the α-modification. Generally, the polymorphic transitions

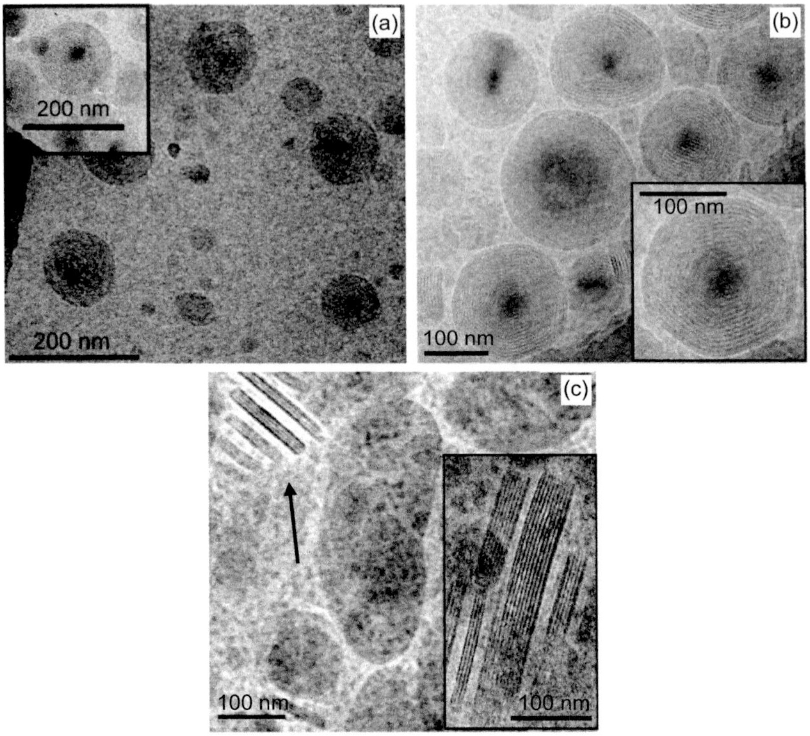

Figure 4.7 *Cryogenic transmission electron micrographs of solid tristearin nanoparticles in the metastable 'nonlamellar' (a) and 'lamellar' (b) α-modification (spheroidal particles) and in the stable β-modification (platelet-shaped particles) (c). The anisometric particles shown in (c) appear as elongated irregular ellipsoidal structures when viewed from the top and as sharply edged 'needles' when viewed from the side. The arrow indicates a stack of particles. The insets show no distinct internal structure (a) or a clear lamellar order (b, c) of the molecular glyceride layers within single particles. (a) Adapted from [148] with permission from the American Chemical Society. (b, c) Adapted from [151] with permission from the American Chemical Society.*

of the matrix lipids proceed faster in lipid nanoparticles than in the corresponding bulk material [7, 134, 146] and are also more rapid in smaller than in larger nanoparticles [141]. For drug-loaded solid lipid nanoparticles, faster as well as slower transformation rates than in the unloaded counterparts have been reported [137, 144, 149].

The profound alterations of lipid packing during the polymorphic transition result in the transformation of the inner structure of lipid particles which may have an influence on the drug-loading capacity. Drug expulsion was observed for different particle compositions upon transformation into more stable forms [7, 150]. Polymorphic transitions may also lead to an increase in particle anisometry (Figure 4.7) and, hence, to an increase in specific surface area [147, 151]. Several studies revealed instabilities of the colloidal state of glyceride nanoparticle dispersions as a result of ongoing changes in the crystalline structure [152,153]. Slowly changing particle properties during storage are unacceptable in pharmaceutical formulations. Since transformation into more stable polymorphs may proceed over very long periods of time depending on the type of dispersion and on the storage conditions [135, 148, 151], ways have to be found to either completely stabilize metastable structures within the particles or – as a more realistic approach – to transform the particles into their stable form directly upon production. For dispersions that are critical with regard to the formation of metastable forms, highly controlled ways of thermal treatment such as, e.g. with micro heat exchangers [12], will probably have to be developed to ensure a precisely controlled and constant product quality.

4.3.4 Particle Morphology and Overall Structure of the Dispersions

Knowledge of the organization of colloidal structures within dispersions is an important basis for the interpretation of the behavior of the dispersion. Different (electron) microscopic methods can be used to image the lipid particles and to visualize additional colloidal structures such as, e.g. vesicles and micelles [154, 155]. These additional components of the dispersions can be formed by self-assembly of an excess of emulsifier molecules in the aqueous phase. Such colloidal structures may be a further site of drug localization and may thus modify the intended behavior of the dispersions, e.g. with regard to drug biodistribution. In spite of this potential importance, relatively few studies have hitherto addressed this question directly. The presence of liposomes was clearly shown for dispersions stabilized with phospholipid/bile salt combinations [134, 156]. In tripalmitin nanodispersions stabilized with high amounts of a phospholipid/PEG-stearate mixture, evidence of the formation of mixed micelles was found in Nuclear Magnetic Resonance (NMR) studies [157]. On the other hand, small angle X-ray scattering and electron microscopic investigation did not give indications of the presence of surfactant aggregates in the aqueous phase of hard fat dispersions stabilized with a mixture of phospholipid and nonionic surfactant [158].

For solid nanoparticles containing glycerides in the stable β-modification, a platelet-like shape is usually observed (Figure 4.7c) [147, 151, 156, 158–160]. Particles in the metastable α-modification can assume a more isometric shape up to the formation of spherical particles (Figure 4.7a, b) [147, 148, 151]. The particle shape also depends on the emulsifier used to stabilize the particles. For example, β-form triglyceride nanoparticles stabilized with poly(vinyl alcohol) are less anisometric than those prepared with more conventional surfactants and surfactant combinations like polysorbate 80 or phospholipid/bile salt combinations [148, 161]. Also rheological investigations pointed to a dependence of particle shape on surfactant composition. Moreover, they revealed that the anisometry of triglyceride nanoparticles increases with the chain length of the triglyceride [162]. A slightly anisometric shape was also reported for fatty alcohol-based solid lipid nanoparticles prepared from microemulsion templates [76].

Investigations by freeze-fracture electron microscopy and – under favorable conditions – by cryoelectron microscopy also allow insights into the internal structure of the nanoparticles. Thus, it has become clear that the single molecular layers in glyceride nanoparticles are orientated in parallel to the large planes of the single particles [132, 141, 151]. On the

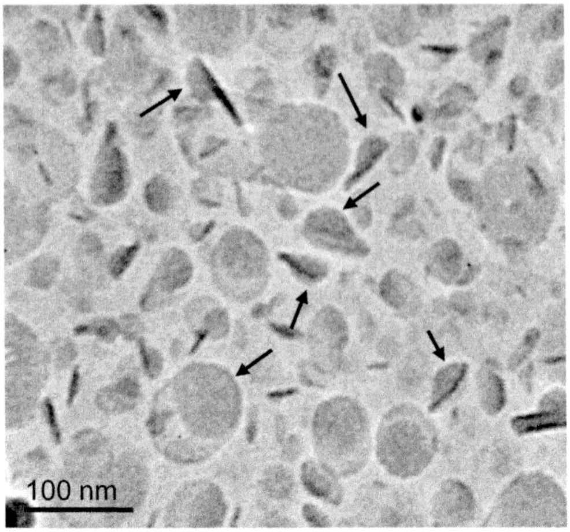

Figure 4.8 *Cryogenic transmission electron micrograph of ubidecarenone loaded (50%) solid tripalmitin nanoparticles obtained by conventional high pressure homogenization (arrows mark the droplets of separated liquid drug phase within the single lipid particles). Adapted from [137] with permission from Springer Science and Business Media.*

other hand, there were no indications of any lamellarity for the triglyceride molecules in the uncommon 'nonlamellar' α-modification (Figure 4.7a) [147, 148]. The detailed structure of this crystalline form still remains to be investigated. For triglyceride particles containing large amounts of the lipophilic drug ubidecarenone, a phase separation within the single nanoparticles was observed [137]. As shown in Figure 4.8, the drug formed a separate phase appearing as a small droplet on top of the solid platelets. The liquid, supercooled nature of these ubidecarenone compartments was confirmed by NMR spectroscopy [7]. Similar structures have been reported for particles containing liquid oils or tocotrienols in addition to the solid matrix [133, 159, 160].

Solid lipid nanoparticles are able to form different types of superstructure in the aqueous dispersion medium. Above a critical concentration, platelet-like triglyceride nanoparticles can self-assemble into stacks (Figure 4.7c) that may extend over long distances throughout the dispersion. Stack formation leads to an increase in dispersion viscosity but is reversible upon dilution [163, 164].

The formation of semi-solid gels has also been observed frequently in dispersions of solid lipid nanoparticles. An increase in viscosity can simply be the result of increasing concentration of anisometric lipid particles leading to particle interactions [162, 165–167]. Transformation of liquid into semi-solid systems has, however, also been observed at comparatively low lipid concentrations (e.g. 5–10% matrix lipid concentration). It often occurs in dispersions stabilized only with phospholipids upon crystallization of the nanodroplets into crystalline particles after melt homogenization. In this case, gelation results from the insufficient stabilization of the nanoparticles when they suddenly transform from spherical emulsion droplets into platelet-like crystalline structure which leads to irreversible particle aggregation. The addition of highly mobile emulsifiers, such as e.g. glycocholate, can effectively prevent this phenomenon [3, 146, 156]. Gelation can, however, also occur in dispersions containing other stabilization systems where it is less well understood. For example, an often reversible time- and temperature-dependent process of gelation was observed upon storage of colloidal triglyceride dispersions containing a stabilizer regime of lecithin and non-ionic surfactants (e.g. Cremophor EL, tyloxapol and polysorbates) [138, 147, 158]. High temperatures, exposure to light, mechanical stress and high ionic strengths were identified as factors promoting gelation processes in poloxamer-stabilized dispersions [168, 169]. As outlined above, these processes were attributed to changes in crystalline structure [152]. For polysorbate 20 stabilized tripalmitin dispersions, gel formation could clearly be related to the polymorphic transition from the α- to the β-modification and a low concentration of emulsifier which could not stabilize the larger surface area of the anisometric particles in the more stable modification [153]. The addition of excess surfactant could prevent the aggregation and thus gel formation.

4.4 Incorporation of Poorly Soluble Drugs and *In Vitro* Release

4.4.1 Drug Incorporation

Solid lipid nanoparticle dispersions can be loaded with a broad variety of poorly soluble drugs, for example, those belonging into the groups of cytostatics, immunosuppressants, corticoids and lipophilic vitamins (Table 4.1). With the exception of a few specific cases, the loading capacity of the lipid nanoparticles (mass of incorporated drug related to the matrix lipid or to the total content of dispersed material) is usually in the range of below 5–10%. This is not unexpected considering the crystalline nature of the matrix lipids in the solid particles, leaving only very limited space for the incorporation of foreign substances.

The drug-loading capacity of such dispersions can be determined by incorporating increasing concentrations of drug into the suspensions until phase separation of drug from the particles or other instability phenomena (e.g. particle growth or increase in viscosity) are observed [7, 60, 170, 171]. Using this procedure the drug-loaded lipid nanodispersions need to be investigated carefully for drug precipitation after preparation, e.g. by light microscopy. Thereby, it needs to be taken into consideration that precipitated drug substance is not always easy to detect, in particular when precipitates are present only at low concentrations or if very tiny crystals or liquid precipitates are formed. Moreover, it should be considered that signs of drug overloading may not be evident directly after preparation but that drug expulsion may continue over extended periods of time [7]. Thus, alternative methods, like the loading of preformed dispersions are currently under evaluation for their suitability to determine the loading capacity in a more reliable and rapid way [172].

The incorporation of drugs may result in alterations of the physicochemical behavior of the nanoparticles, e.g. regarding their melting and crystallization properties as well as the time-course of polymorphic transitions. Although sometimes detectable already at low fractions of incorporated drugs, such effects become particularly prominent in the rare cases of rather high drug loads. For example, a drastic decrease in melting and crystallization

Table 4.1 *Examples of poorly water soluble drugs loaded into solid lipid nanoparticle dispersions.*

Class	Examples	References
Anticancer drugs	Camptothecin	[174,229]
	Docetaxel	[241]
	Etoposide	[109]
	Idarubicin	[197,240]
	Paclitaxel	[58,63,72,78,239,242,243]
	Tamoxifen	[244]
Antiviral compounds	Delavirdine	[245,246]
	Lopinavir	[222]
	Saquinavir	[245,246]
Immunosuppressants	Cyclosporine	[144,188,230]
Glucocorticoids	Betamethasone valerate	[178]
	Clobetasol propionate	[79]
	Dexamethasone palmitate	[73]
	Prednicarbate	[178]
	Prednisolone	[60]
Lipophilic vitamins and related substances	β-carotene	[54,185]
	Menadione	[7,11]
	Retinoids	[7,103,150,170,171,247]
Miscellaneous	Bromocriptine	[160]
	Cloricromene	[66]
	Clozapine	[100]
	Curcuminoids	[80,182]
	Diazepam	[7,68,148,248]
	Enrofloxacine	[249]
	Estradiol	[250]
	Etomidate	[60,102]
	Indomethacin	[251]
	Lovastatin	[223]
	Nevirapine	[252]
	Nifedipine	[253]
	Nitrendipine	[226]
	Quercetin	[227]
	Riluzole	[254]
	Risperidone	[255]
	Simvastatin	[85]
	Tetracaine	[60,102]
	Triptolide	[256]
	Vinpocetine	[189]
	Ubidecarenone	[7,137,146,187]

Source: Expanded from [9].

temperature as well as an increase in the rate of polymorphic transitions resulted from incorporation of high amounts of ubidecarenone into triglyceride nanoparticles [7, 137]. Upon loading of high concentrations of cyclosporine into monostearate nanoparticles, also a decrease in melting temperature but a retardation of the transition into more stable modifications was observed [144].

Although such distinct effects on the physicochemical behavior of the nanoparticles point to strong interaction of the drug molecules with the matrix of the nanoparticles, there is still not much knowledge concerning the exact localization of drug molecules incorporated into solid lipid nanoparticle dispersions. Beside the particle core, the particle surface may act as an additional site for drug attachment and some of the drug may also reside in the aqueous phase of the dispersion, either in aqueous solution or associated with additional colloidal structures formed by excess stabilizer molecules. Studies on colloidal lipid emulsions have shown that incorporated active substances may be distributed between all corresponding potential sites of localization [173]. Drug distribution between the lipid particles and the aqueous phase has also been observed for solid lipid nanoparticle suspensions [60, 78, 100, 103, 174]. Comparatively few studies have hitherto addressed the question of the exact site of drug localization within single solid lipid nanoparticles in spite of the presumably high relevance for the pharmaceutical performance of the dispersions (e.g. with regard to chemical protection of the drug and drug release). This topic is not easy to address experimentally. Nevertheless, there is increasing evidence that often at least a large fraction of drug is localized at the surface of the particles instead of being incorporated in the particle core [60, 103, 137, 175–181]. A preferential surface localization may be due to a high crystalline order of the matrix lipids, a certain surface activity of the incorporated drug substance but may also result from the fact that there is often not much core volume in solid lipid nanoparticles. For example, the platelet-shaped particles typically found in glyceride dispersions may be very thin, with their thickness corresponding only to that of a few molecular layers of glyceride (Figures 4.7c, 4.8).

Several spectroscopic techniques were employed to elucidate drug–particle interactions. Nuclear Magnetic Resonance (NMR) spectroscopy has mainly been used to basically confirm the association of the drug with the particles via the resulting reduction in molecular mobility [7]. First studies with Raman spectroscopy on drug-loaded solid lipid nanoparticle dispersions have also been reported [182] but the potential of both techniques remains to be further exploited. More detailed investigations have been performed with Electron Spin Resonance (ESR) and fluorescence spectroscopy [175–180, 182, 183]. Using spin probes or fluorescent model substances, respectively, these techniques distinguish between more polar and more non-polar environments of the incorporated marker molecules as well as reveal the local viscosity. ESR measurements addressing the chemical accessibility of the marker molecules for reactants dissolved in the aqueous phase can provide additional information on drug localization [175, 176, 183]. With systematic alterations of the physicochemical properties of the marker substances, basic knowledge on the means of interaction with the particles can be obtained. While these techniques are extremely useful for mechanistic investigations, 'real' drug substances cannot, in most cases, be detected and the concentration of incorporated marker substances has often to be kept very low for experimental reasons. Information on the localization of, in particular, glucocorticoid drugs has been obtained with parelectric spectroscopy [178, 184] but this specialized technique is not readily available for most investigators.

Beside its relevance regarding the controlled release potential, the topic of drug localization is also important with respect to the chemical stability of incorporated substances. While localization within a solid particle core should lead to considerable protection against chemical or enzymatic attack, surface localization would render the incorporated substance more susceptible. In fact, there is not much evidence for a special chemical protection of active substances by incorporating them into solid lipid nanoparticles, e.g. as compared with the liquid core of emulsion droplets. Observation of the degradation behavior of incorporated drugs has been used to draw conclusions on their localization within the dispersions in several studies. For example, the rapid degradation of dexamethasone palmitate incorporated into fatty alcohol nanoparticles was assumed to result from the insertion of this substance into the surface layer of the nanoparticles [73, 75]. The degradation rates of retinoids and β-carotene were used to consider their distribution behavior in solid lipid nanoparticle dispersions based on different matrix lipids, or prepared with different surfactants, respectively. While the study with retinoids always revealed a higher degradation rate for retinol incorporated into solid lipid nanoparticle dispersions in comparison to loaded emulsions [170], the situation for carotene was more complex [185]. This was attributed to the occurrence of different polymorphic forms in the solid particles under investigation. Eicosane nanoparticles in the solid state were not significantly better at protecting an incorporated fluorescence dye from degradation by reactive oxygen species than the corresponding emulsion of a supercooled melt. Using micron-sized eicosane particles as a model, significant redistribution of fluorescent dyes from the core into the surface region of the particles upon their solidification could be shown by fluorescence microscopy, which would explain the low protective activity of the solid particles [186].

4.4.2 *In Vitro* Drug Release

One important aspect during the development of solid lipid nanoparticles as drug carriers was to achieve sustained or controlled release of poorly water-soluble substances upon systemic administration [13, 187]. Sustained release would also be the prerequisite for intravenous drug targeting. So far, it could not, however, be proven that these particles do indeed provide a universal platform for controlled drug release. Already the finding that even lipophilic drugs tend to be localized to a large extent at the surface of the single particles raises doubts about the controlled release potential of such formulations. On the other hand, there are numerous reports of long-term drug release from solid lipid nanoparticles under conditions aimed at simulating the situation upon oral or intravenous administration, see e.g. [100, 174, 188, 189]. The determination of drug release from colloidal carriers under sink conditions is not, however, an easy task and inadequate methods are, unfortunately, very often applied to study the release from colloidal lipid carrier systems. By analogy with the determination of drug release from, for example solid oral dosage forms, a large volume of release medium needs to be usually provided in order to trigger the drug release from the particles. Solubility-enhancing substances such as surfactants may be added to enhance the transfer of the poorly soluble drugs into the aqueous medium [78, 79]. As the solid lipid particles are in most cases not degraded within the time frame of the experiment, they remain in the release medium and a distribution of drug between carrier particles and release medium instead of a complete release is observed. The direct analytical determination of released drug in the aqueous phase is often impaired by the presence of the colloidal particles. Unless specialized analytical techniques that are sensitive only to the drug present in the aqueous phase are used, the release medium needs to be separated from the particles. This can, for example, be achieved by (ultra)filtration or (ultra)centrifugation but care has to be taken that the separation technique applied provides sufficiently high time resolution, as the drug release from colloidal lipid particles can be extremely rapid [190–193]. Experiments performed under adequate, 'sink-like' conditions indicate that this often also applies to solid lipid nanoparticles [13, 60, 66, 148]. Under these conditions the drug release seems to be mainly controlled by drug partition processes and not be much influenced by diffusion processes through the solid lipid matrix. Comparative studies of the release behavior of solid lipid nanoparticles and corresponding lipid nanoemulsions revealed that the rate and extent of release from the solid particles were comparable or even higher than that from the emulsion systems. In order to investigate also the release behavior of extremely lipophilic compounds, these experiments were not performed in an aqueous release medium but by using lipid emulsion droplets as acceptor compartments for the drug. The release behavior depended much more on the lipophilicity of the incorporated model substance than on the nature of the lipid core (Figure 4.9). Transfer to the acceptor droplets was usually completed within minutes and was only retarded for a very lipophilic compound or in highly diluted systems [149]. By comparing the release of dexamethasone and its lipophilic ester prodrug from solid lipid nanoparticles, a much higher retention of dexamethasone palmitate by the particles was observed even when human plasma was used as a release medium [73, 75]. For drug-loaded lipid emulsion droplets, it has been proposed that only extremely lipophilic drugs (log$P \geq 9$) may remain associated with the particles within the circulation [194–196]. Similar relations seem to apply for solid lipid nanoparticle formulations. The studies with dexamethasone palmitate-loaded solid lipid nanoparticles also pointed out that the release of this drug depends on the enzymatic activity of the release medium. Release media with high esterase activity (e.g. rodent plasma) led to rapid cleavage of the ester bond and thus to a rapid release of dexamethasone from the particles [181].

Even if intended to mimic release conditions after intravenous or oral administration, many drug release studies reported in the literature do not, however, bring the particles into direct contact with the release medium. Very often, such experiments rely on set-ups where the lipid nanoparticle dispersion is separated from the release medium by a dialysis membrane [65, 78, 100, 174, 188, 189, 197, 198]. This approach leads to distorted release profiles since the rate of drug transfer is determined by the concentration gradient over the dialysis membrane and is not much influenced by the true rate of drug release from the particles. The concentration gradient over the membrane is typically very small for lipophilic drugs which tend to be distributed towards the lipid particles in the membrane-confined donor compartment. Therefore, the drug crosses the membrane only very slowly and an artificially extended release is observed [192, 199, 200]. Of course, the appropriate experimental set-up depends on the intended mode of administration of the formulation of interest. Local (e.g. dermal or ocular) administration is usually not associated with a high dilution of the formulation and release experiments using a confined donor compartment separated from the acceptor compartment by natural or artificial membranes may be perfectly adequate [150, 201].

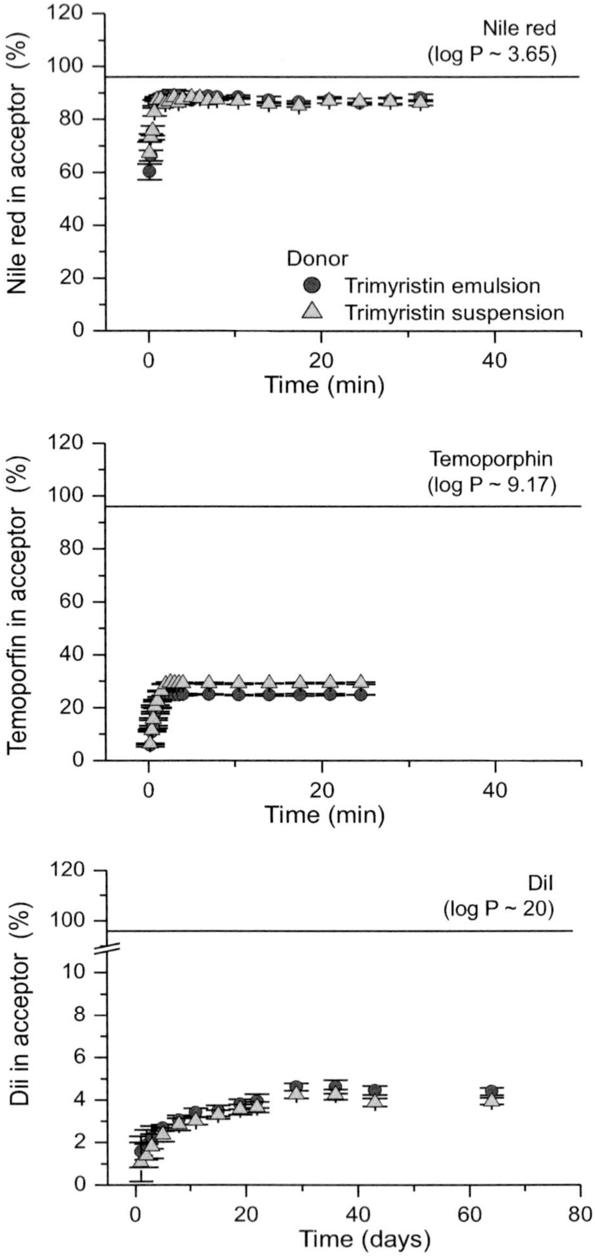

Figure 4.9 *Time course of transfer of different lipophilic drugs from trimyristin nanoemulsions and solid lipid nanoparticles (suspensions) into micron-sized emulsion droplets as lipophilic acceptor compartments. For the transfer studies 1 part (lipid mass) of donor particles was mixed with 22 parts of acceptor droplets. The horizontal line indicates the theoretical acceptor concentration upon complete equilibration between donor and acceptor.*
Source: Data from [149].

4.5 Safety Aspects, Toxicity and Pharmacokinetic Profiles

4.5.1 *In Vitro* Behavior and Toxicity Studies

As they are mainly based on lipid compounds, solid lipid nanoparticles are generally assumed to be physiologically compatible and well tolerable *in vivo*. However, some of the formulations under investigation may contain components that have not been fully evaluated for all the potential ways of administration (e.g. certain surfactants but also solid lipids in intravenous formulations). Thus, the assumption of good biocompatibility needs to be carefully proven before such dispersions can actually be considered for human use. Although several studies have addressed this question, in particular by *in vitro* investigation on the cell compatibility, much further work is still required to develop a realistic view on this topic. The situation is complicated by the highly variable composition of the formulations that have been described and by different requirements that have to be met, depending on the route of administration. Generally it appears reasonable to assume that excipients that have already been used in registered formulations for a given route of administration should also be applicable in solid lipid nanoparticle formulations. However, the specific combination of formulation ingredients and the small size of solid lipid nanoparticles may add new aspects that need to be considered.

The *in vitro* degradability of glyceride and wax nanoparticles by lipase/colipase has already been studied in an early development stage of solid lipid nanoparticles. The degradation rate depended on the type of lipid matrix and on the type of surfactant used for stabilization. Compared to stabilization with lecithin and cholate, the degradation of sterically stabilized particles (e.g. employing poloxamer or polysorbate) was slower, presumably due to steric hindrance. At comparatively low degradation rates, an increasing degradation rate with decreasing particle size could be observed. Also a reduced crystallinity appeared to accelerate degradation [202–204].

The *in vitro* stability of solid glyceride nanoparticles in blood or plasma has not yet received much attention. Tentatively, some analogy with intravenous fat emulsions may be assumed. These dispersions have been studied in detail with regard to their metabolism and are enzymatically degraded by lipoprotein lipases [196]. However, the metabolism is influenced by the composition of the formulation [195] and it may be expected that the effects found for the enteral way of degradation (e.g. decrease in degradation rate for highly crystalline particles or those carrying steric stabilizers) will apply in a similar way for intravenous degradation. A preliminary investigation with human plasma indicated that the particle size of cetyl palmitate nanoparticles decreased more rapidly than that of glyceryl behenate nanoparticles, pointing to a faster degradation of the wax nanoparticles [205]. The fatty alcohol component of solid lipid nanoparticles based on mixtures of fatty alcohol and nonionic surfactant can be oxidized by alcohol dehydrogenase, as has been observed in *in vitro* studies [206]. Such particles were shown not to be hemolytic or thrombogenic at clinically relevant concentrations [207]. The tendency towards particle growth in aqueous suspension that has frequently been reported for this type of particle [70, 71, 73, 208] may, however, require some attention with regard to safety issues in parenteral delivery. A very low hemolytic potential was also found for triglyceride nanoparticles stabilized with lecithin and the nonionic surfactant polyethylene glycol 660-12-hydroxystearate [209].

The *in vitro* compatibility and interaction of solid lipid nanoparticle dispersions with cells have been studied with a broad variety of formulations and with different cell types. Most of the current research focuses on drug-loaded systems and information on the influence of the carrier itself and its components is thus not easy to extract. For this purpose, comparative studies with unloaded formulations of varying composition and structure are very helpful. When comparing results from different studies, it has to be borne in mind that the interaction behavior may be quite specific for a given nanoparticle dispersion/cell line combination and may also depend on the experimental conditions.

Early investigations in this regard were directed towards the comparison of solid lipid and polymeric nanoparticles. The phagocytic uptake of poloxamer- or poloxamine-stabilized glyceride nanoparticles by human granulocytes was significantly lower than that of polystyrene nanoparticles [210, 211]. The cytotoxicity of solid lipid nanoparticles was found to be much lower than that of polylactide/glycolide and butylcyanoacrylate nanoparticles [211,212]. Exploiting the pronounced supercooling effect of trimyristin nanoparticles, a direct comparison of corresponding nanoemulsion and solid lipid nanoparticle formulations revealed a lower compatibility of solid lipid nanoparticles with mouse fibroblasts than of the lipid emulsion particles. The effect appeared to be more related to the physical state of the nanoparticles than to the differences in particle shape [161].

The effect of solid lipid nanoparticles on the cell viability greatly depends on the type of surfactant used for dispersion preparation [67, 161, 212–216]. In particular, the presence of excess, not particle-bound, surfactant in the aqueous phase has a negative influence on cell compatibility which can be counteracted by removing free surfactant from the formulation [67, 212]. The presence of cationic components as matrix or surfactant components seems to be rather critical with respect to cell viability with single-chain cationic surfactants being more toxic than double-chain cationic lipids [67, 213, 216, 217]. In comparison to glyceride and wax nanoparticles, stearic acid nanoparticles have been found to be distinctly more toxic in several cell models [216, 217]. Comparative investigations performed on different glyceride as well as glyceride and cholesterol ester nanoparticles pointed to no or a rather marginal influence of the type of matrix lipid on cell compatibility [161, 212].

There are also some studies that are even closer to the *in vivo* situation. For example, studies on bovine brain microvessel endothelial cells combined with an *in situ* rat brain perfusion technique were employed to study the effect of emulsifying wax nanoparticles on blood–brain barrier integrity. Even at high nanoparticle concentrations, no significant alterations in cerebral perfusion, sucrose permeability or thiourea permeation were observed [208]. With regard to inhalative administration, detailed studies with *in vitro* and *ex vivo* cell models of the lung environment were performed to identify promising conditions for an *in vivo* study. Indeed, the results could be successfully translated into a dose at which solid lipid nanoparticle formulations could be administered in a mouse model without causing an inflammatory response in the animals [218, 219].

An *in vivo* toxicity study performed with glyceride and wax nanoparticles in mice did not lead to acute neurotoxicity after intravenous administration even of extremely high doses. Glyceryl behenate nanoparticles accumulated, however, in the liver and spleen resulting in pathological changes in these organs. The effects were dose-dependent and reversible to a certain extent. For cetyl palmitate nanoparticles no adverse effects were observed. The differences were attributed to the faster degradation of cetyl palmitate nanoparticles in plasma [205].

4.5.2 Bioavailability and Pharmacokinetics

4.5.2.1 Oral Drug Delivery

Solid lipid particles may be an interesting carrier for drugs that are poorly absorbed after peroral administration due to solubility limitations in the gastrointestinal tract. Association with the carrier particles may prevent precipitation of such substances during intestinal passage and thus foster their absorption. If mechanisms of lymphatic absorption can be exploited first pass metabolism may be circumvented since the mesenteric lymph is directly drained into the systemic blood circulation without passing the liver. Lymphatic uptake can, for example, take place by transcytosis of particle-associated drug, such as via the M-cells of the Peyer's Patches in the gut tissue [220] or by interaction with the lipid digestion process, resulting in secretion of drug-carrying lipoproteins [221].

An improvement of systemic bioavailability after peroral or duodenal administration of solid lipid nanoparticle formulations has been reported, for example, for lopinavir [222], lovastatin [223], clozapine [224], nitrendipine [225,226], quercetin [227], puerarin [82] and cryptotanshinone [228]. Suspensions of the active substances served as control samples in these studies. However, a higher bioavailability of camptothecin [229] and vinpocetine [189] compared to solutions of the respective drugs has also been reported. A comparison with another lipid formulation has hitherto only been carried out with cyclosporine-loaded solid lipid nanoparticles. In a pig study, the bioavailability from the solid lipid nanoparticle formulation was slightly lower than that of the microemulsion formulation Sandimmun® Neoral but avoided the high peak blood levels of the commercial formulation which has been related to adverse effects of this product [230].

By comparing plasma and lymph concentrations of the drug, Aji Alex *et al.* found evidence of an improvement of lymphatic uptake when lopinavir was administered as a solid lipid nanoparticle formulation in comparison to a drug suspension [222]. Other authors also reported indications of lymphatic uptake from solid lipid nanoparticle formulations [231,232] but more investigations into this direction will be required to elucidate the details of this interesting delivery option.

4.5.2.2 Parenteral Drug Delivery

Like other types of colloidal lipid carrier systems, solid lipid nanoparticles are an interesting option to solubilize lipophilic drugs in order to render them applicable by the parenteral, in particular, the intravenous route. Provided that the incorporated drug remains appropriately associated with the particles, such carrier systems can also be used to modulate the biodistribution of active substances. The pharmacokinetics of intravenously administered colloidal particles differs from that of plain solutions and depends on different parameters, including particle size and surface structure [115, 116]. After intravenous injection, colloidal particles are usually cleared by cells of the reticuloendothelial system (RES), with the phagocytic uptake being mediated by the adsorption of plasma proteins. RES uptake results in predominant accumulation of the particles in liver and spleen. Surface modification of the particles, for example, by the adsorption of steric stabilizers like PEG-containing surface active agents, may help to avoid RES clearance, thus resulting in a longer circulation time. Long-circulating particles can be directed to, for example, tumor sites either by exploiting passive accumulation due to the enhanced permeation and retention (EPR) effect [233] or by using specific targeting moieties attached to the colloidal drug carrier [234]. For uptake into the tumor tissue, a small particle size is required to allow transport through pores in the endothelium and in the interstitial space. On the other hand, preferential phagocytic uptake may be utilized to target intracellular disorders, such as leishmaniasis [235, 236].

These effects can also be exploited with solid lipid nanoparticle formulations. For fluorescently-labeled stearic acid nanoparticles containing either dipalmitoyl phosphatidylethanolamine-PEG 2000 or stearic acid-PEG 2000 in their stabilizer mixture, a much lower *in vitro* uptake into murine macrophages was observed compared to nanoparticles stabilized only with a phospholipid/bile salt mixture [237]. In cetyl palmitate-based nanoparticles prepared with nonionic surfactants, the addition of a PEG 6000 surface modifier reduced the uptake into murine macrophages in a concentration-dependent manner [73]. Intravenous administration of stearic acid nanoparticles carrying doxorubicin as a lipophilic ion pair to rats markedly prolonged the circulation time of doxorubicin compared to a doxorubicin solution. PEGylation of the nanoparticles resulted in higher doxorubicin plasma concentrations 12 to 24 hours after administration than using non-PEGylated nanoparticles [129]. In a rabbit study, the AUC of doxorubicin after intravenous administration of stearic acid nanoparticles increased with the concentration of stearic acid-PEG 2000 used for the surface modification of the nanoparticles [130]. Investigations with polyethylene glycol 2000 conjugated, radioactively labeled stearic acid nanoparticles in comparison to non-PEGylated nanoparticles revealed a much lower accumulation of the sterically stabilized particles in the RES organs liver and lung 60 minutes after intravenous injection to rats [238]. After intravenous administration of paclitaxel-loaded stearic acid nanoparticles surface-modified with poloxamer 188 or polyoxyethylene 20 stearyl ether to mice, higher plasma levels of the drug could be observed several hours after injection than for a paclitaxel solution [78]. On the other hand, a biodistribution study after intravenous administration of emulsifying wax-based, paclitaxel-loaded nanoparticles to mice did not reveal significant differences in organ distribution compared to the free drug, except for the lung and brain, where the drug concentration was lower when given in the nanoparticulate formulation. Tentatively, this result was attributed to the presence of a considerable fraction of the drug in the free form also in the nanoparticle dispersions [239].

However, several studies do indicate the ability of solid lipid nanoparticles to enhance drug delivery to the brain. For example, the transport of radioactively labeled solid lipid nanoparticles generated from microemulsion precursors across the blood–brain barrier (BBB) was demonstrated in an *in situ* rat brain perfusion model [198]. Control investigations with different *in vitro* and *in vivo* blood–brain barrier models indicated that primary BBB parameters were not affected by the presence of such nanoparticles [208]. In a drug-loaded dispersion, brain delivery of camptothecin was significantly increased after incorporation into stearic acid nanoparticles compared to a camptothecin solution [174]. Also doxurubicin could be delivered into the brain when incorporated as a lipophilic ion pair into stearic acid nanoparticles containing a sufficient concentration of stearic acid-PEG 2000 as surface modifier [130]. Brain delivery after incorporation into solid lipid nanoparticles was not only observed upon parenteral use but was also found after oral or duodenal administration, for example, for idarubicin (as lipophilic ion pair) and campthotecin [229, 240].

As with other types of colloidal carrier systems, the main focus of the *in vivo* use of solid lipid nanoparticles is their application in tumor therapy with special regard to tumor-specific delivery. Thus, many studies are concerned with the incorporation of anti-tumor drugs like taxans or camptothecin (Table 4.1). *In vitro*, paclitaxel-loaded emulsifying wax-based solid lipid nanoparticles had a much higher cytotoxic effect on multidrug-resistant colorectal adenocarcinoma cells than had free paclitaxel [72]. This effect was attributed to the circumvention of p-glycoprotein mediated drug efflux by association of paclitaxel with the nanoparticles as the cells were shown to take up such nanoparticles [239]. In a mouse model, paclitaxel-loaded solid lipid nanoparticles were slightly more efficient in reducing tumor growth upon intratumoral administration than free paclitaxel [239]. Solid lipid nanoparticle formulations loaded with docetaxel could inhibit tumor growth in hepatoma-bearing mice much better than the formulation with free drug (Taxotere®). Galactosylation of the nanoparticle surface in order to target the particles to the tumor cells led to particularly effective formulations. The good efficacy of the lipid nanoparticle formulations may result from a better drug accumulation in the tumor that was observed in biodistribution studies. Moreover, cell culture studies indicated that the nanoparticles could be taken up by the hepatoma cells, in particular when carrying the galactose surface modification. Studies in healthy mice confirmed that the nanoparticle formulations were well tolerated by the animals. As a further benefit with regard to therapeutic use, the tolerated dose of docetaxel was much higher when incorporated into solid lipid nanoparticles than that of the free drug [241].

4.6 Conclusion

Solid lipid nanoparticles are an interesting formulation option for lipophilic drugs. Many preparation techniques are available that tailor the manufacturing process to the requirements of the drug and formulation of interest. Compared to other types of lipid formulations, such as, for example, lipid emulsions, the solid state of the nanoparticles adds several specificities to the behavior of the dispersions that have to be carefully taken into consideration when developing such formulations. Although much research has already been done to elucidate the physicochemical aspects of the dispersions, not all details have yet been fully clarified. One important aspect that still deserves more attention in the future is the exact interaction of the incorporated drug with the carrier particles. Moreover, there is only limited knowledge on the influence of structural parameters of the dispersions on their *in vivo* performance. Comparative studies with other types of colloidal lipid carrier systems would be extremely helpful to create a rational basis for choosing the best formulation option for a given drug delivery problem. There are some indications that solid lipid nanoparticles interact in a different way with biological systems than do liquid ones. Thus, it is an important and fascinating future task to explore where the specificities of the solid particle state (or of the materials that lead to this state) can most favorably be exploited for drug delivery purposes.

References

1. P. Speiser, Lipidnanopellets als Trägersystem für Arzneimittel zur peroralen Anwendung. EP 0167825 (1986).
2. M.R. Gasco, S. Morel, Lipospheres from microemulsions. *Farmaco*, **45**, 1127–1128 (1990).

3. B. Siekmann, K. Westesen, Submicron-sized parenteral carrier systems based on solid lipids. *Pharm Pharmacol Lett*, 123–126 (1992).
4. B. Sjöström, B. Bergenståhl, Preparation of submicron drug particles in lecithin-stabilized o/w emulsions. I. Model studies of the precipitation of cholesteryl acetate. *Int J Pharm*, **88**, 53–62 (1992).
5. FDA, Code of Federal Regulations Title 21.
6. S. Nema, R.J. Brendel, Excipients and their role in approved injectable products: Current usage and future directions. *PDA J Pharm Sci and Tech*, **65**, 287–332 (2011).
7. K. Westesen, H. Bunjes, M.H.J. Koch, Physicochemical characterization of lipid nanoparticles and evaluation of their drug loading capacity and sustained release potential. *J Control Release*, **48**, 223–236 (1997).
8. K. Westesen, Novel lipid-based colloidal dispersions as potential drug administration systems – expectations and reality. *Colloid Polym Sci*, **278**, 608–618 (2000).
9. H. Bunjes, Lipid nanoparticles for the delivery of poorly water-soluble drugs. *J Pharm Pharmacol*, **62**, 1637–1645 (2010).
10. S. Tamilvanan, S.R. Senthilkumar, R. Baskar, T.R. Sekharan, Manufacturing techniques and excipients used during the formulation of oil-in-water type nanosized emulsions for medical applications. *J Excipients Food Chemicals*, **1**, 11–29 (2010).
11. H. Bunjes, K. Westesen, M.H.J. Koch, Crystallization tendency and polymorphic transitions in triglyceride nanoparticles. *Int J Pharm*, **129**, 159–173 (1996).
12. K. Jasch, N. Barth, S. Fehr, H. Bunjes, W. Augustin, S. Scholl, A microfluidic approach for a continuous crystallization of drug carrier nanoparticles. *Chem Eng Technol*, **32**, 1806–1814 (2009).
13. R.H. Müller, W. Mehnert, J.-S. Lucks, C. Schwarz, A. zur Mühlen, H. Weyhers, C. Freitas, D. Rühl, Solid lipid nanoparticles (SLN): An alternative colloidal carrier system for controlled drug delivery. *Eur J Pharm Biopharm*, **41**, 62–69 (1995).
14. B. Freudig, S. Tesch, Tropfenzerkleinerung in Hochdruckhomogenisatoren, in *Emulgiertechnik: Grundlagen, Verfahren und Anwendungen*, H. Schubert (ed.), 1st edn. Hamburg: Behr, pp. 343–368 (2005).
15. H. Bunjes, B. Siekmann, Manufacture, characterization and applications of solid lipid nanoparticles as drug delivery systems, in *Microencapsulation: Methods and Industrial Applications*, S. Benita (ed.), 2nd edn. New York: Taylor & Francis, pp. 213–268 (2006).
16. W. Mehnert, K. Mäder, Solid lipid nanoparticles: Production, characterization and applications. *Adv Drug Deliv Rev*, **47**, 165–196 (2001).
17. T. Gothsch, J.H. Finke, S. Beinert, C. Lesche, J. Schur, S. Büttgenbach, C. Müller-Goymann, A. Kwade, Effect of microchannel geometry on high-pressure dispersion and emulsification. *Chem Eng Technol*, **34**, 335–343 (2011).
18. C. Richter, T. Krah, S. Büttgenbach, Novel 3D manufacturing method combining microelectrial discharge machining and electrochemical polishing. *Microsyst Technol*, **18**, 1109–1118 (2012).
19. J.H. Finke, J. Schur, C. Richter, T. Gothsch, A. Kwade, S. Büttgenbach, C.C. Müller-Goymann, The influence of customized geometries and process parameters on nanoemulsion and solid lipid nanoparticle production in microsystems, *Chem Eng J*, **209**, 126–137 (2012).

20. K. Suzuki, I. Shuto, Y. Hagura, Characteristics of the membrane emulsification method combined with preliminary emulsification for preparing corn oil-in-water emulsions. *Food Sci Technol Int*, **2**, 43–47 (1996).
21. R.C. MacDonald, R.I. MacDonald, B.P.M. Menco, K. Takeshita, N.K. Subbarao, L.-R. Hu, Small-volume extrusion apparatus for preparation of large, unilamellar vesicles. *Biochim Biophys Acta Biomembr*, **1061**, 297–303 (1991).
22. T. Nakashima, M. Shimizu, M. Kukizaki, Membrane emulsification by microporous glass, *Key Eng Mater*, **61-62**, 513–516 (1991).
23. J. Drew, A. Liodakis, R. Chan, H. Du, M. Sadek, R. Brownlee, W.H. Sawyer, Preparation of lipid emulsions by pressure extrusion, *Biochem Int*, **22**, 983–992 (1990).
24. G.T. Vladisavljević, M. Shimizu, T. Nakashima, Preparation of monodisperse multiple emulsions at high production rates by multi-stage premix membrane emulsification, *J Membr Sci*, **244**, 97–106 (2004).
25. M. Kukizaki, Preparation of solid lipid microcapsules via solid-in-oil-in-water dispersions by premix membrane emulsification, *Chem Eng J*, **151**, 387–396 (2009).
26. H. Sawalha, N. Purwanti, A. Rinzema, K. Schroën, R. Boom, Polylactide microspheres prepared by premix membrane emulsification: Effects of solvent removal rate, *J Membr Sci*, **310**, 484–493 (2008).
27. Q. Wei, W. Wei, B. Lai, L.Y. Wang, Y.X. Wang, Z.G. Su, G.H. Ma, Uniform-sized PLA nanoparticles: Preparation by premix membrane emulsification, *Int J Pharm*, **359**, 294–297 (2008).
28. M. Nikanjam, E.A. Blakely, K.A. Bjornstad, X. Shu, T.F. Budinger, T.M. Forte, Synthetic nano-low density lipoprotein as targeted drug delivery vehicle for glioblastoma multiforme, *Int J Pharm*, **328**, 86–94 (2007).
29. M.D. Owens, G.W. Halbert, Production and characterization of protein-free analogues of low density lipoprotein, *Eur J Pharm Biopharm*, **41**, 120–126 (1995).
30. S. Joseph, H. Bunjes, Preparation of nanoemulsions and solid lipid nanoparticles by premix membrane emulsification, *J Pharm Sci*, **101**, 2479–2489 (2012).
31. S. Fehr, H. Bunjes, Preparation of solid lipid nanoparticles by small- and large-volume extrusion, *Proceedings of the 7th World Meeting Pharm Biopharm Pharm Technol, Valetta, Malta, 2010.*
32. S. Joseph, H. Bunjes, Evaluation of Shirasu Porous Glass (SPG) membrane emulsification techniques for the preparation of colloidal lipid dispersions as drug carrier systems, submitted for publication.
33. S. Joseph, H. Bunjes, Influence of membrane structure on the preparation of colloidal lipid dispersions by premix membrane emulsification, *Int J Pharm*, **446**, 59–62 (2013).
34. V. Schröder, O. Behrend, H. Schubert, Effect of dynamic interfacial tension on the emulsification process using microporous, ceramic membranes, *J Colloid Interf Sci*, **202**, 334–340 (1998).
35. S. Berot, S. Giraudet, A. Riaublanc, M. Anton, Y. Popineau, Key factors in membrane emulsification, *Chem Eng Res Des*, **81**, 1077–1082 (2003).
36. S.M. Joscelyne, G. Trägårdh, Membrane emulsification: A literature review, *J Membr Sci*, **169**, 107–117 (2000).
37. C. Charcosset, Preparation of emulsions and particles by membrane emulsification for the food processing industry, *J Food Eng*, **92**, 241–249 (2009).

38. S. Higashi, M. Shimizu, T. Nakashima, K. Iwata, F. Uchiyama, S. Tateno, S. Tamura, T. Setoguchi, Arterial-injection chemotherapy for hepatocellular carcinoma using monodispersed poppy-seed oil microdroplets containing fine aqueous vesicles of epirubicin: Initial medical application of a membrane-emulsification technique, *Cancer*, **75**, 1245–1254 (1995).
39. S. Omi, K. Katami, A. Yamamoto, M. Iso, Synthesis of polymeric microspheres employing SPG emulsification technique, *J. Appl. Polymer Sci.*, **51**, 1–11 (1994).
40. C. Charcosset, A. El-Harati, H. Fessi, Preparation of solid lipid nanoparticles using a membrane contactor, *J Control Release*, **108**, 112–120 (2005).
41. A.A. El-Harati, C. Charcosset, H. Fessi, Influence of the formulation for solid lipid nanoparticles prepared with a membrane contactor, *Pharm Dev Technol*, **11**, 153–157 (2006).
42. P. Czermak, T. Steinle, M. Ebrahimi, T. Schmidts, F. Runkel, Membrane-assisted production of S1P loaded SLNs for the treatment of acne vulgaris, *Desalination*, **250**, 1132–1135 (2010).
43. C. D'oria, C. Charcosset, A.A. Barresi, H. Fessi, Preparation of solid lipid particles by membrane emulsification-Influence of process parameters, *Colloids Surf, A*, **338**, 114–118 (2009).
44. Y. Li, H. Fessi, C. Charcosset, Preparation of indomethacin-loaded lipid particles by membrane emulsification, *Adv Sci Lett*, **4**, 591–595 (2011).
45. O. Behrend, Emulgieren mit Ultraschall, in *Emulgiertechnik: Grundlagen, Verfahren und Anwendungen*, H. Schubert (Ed.), 1st ed., Hamburg: Behr, 2005, pp. 299–342.
46. T. Leighton, The principles of cavitation, in *Ultrasound in food processing*, M.J.W. Povey, T.J. Mason (Eds.), 1st ed., London: Blackie Academic & Professional, 1998, pp. 55–65.
47. T.S.H. Leong, T.J. Wooster, S.E. Kentish, M. Ashokkumar, Minimising oil droplet size using ultrasonic emulsification, *Ultrason Sonochem*, **16**, 721–727 (2009).
48. S. Kentish, T.J. Wooster, M. Ashokkumar, S. Balachandran, R. Mawson, L. Simons, The use of ultrasonics for nanoemulsion preparation, *Innovative Food Sci Emerg Technol*, **9**, 170–175 (2008).
49. S.-H. Hsu, S.A. Al-Suwayeh, C.-F. Hung, C.-Y. Tsai, J.-Y. Fang, Mechanistic studies of the skin delivery of lipid colloid systems with different oil/fatty ester ratios for both lipophilic and hydrophilic drugs, *Curr Nanosci*, **7**, 200–209 (2011).
50. G. Abdelbary, R.H. Fahmy, Diazepam-loaded solid lipid nanoparticles: Design and characterization, *AAPS PharmSciTech*, **10**, 211–219 (2009).
51. M.-J. Tsai, Y.-B. Huang, P.-C. Wu, Y.-S. Fu, Y.-R. Kao, J.-Y. Fang, Y.-H. Tsai, Oral apomorphine delivery from solid lipid nanoparticles with different monostearate emulsifiers: Pharmacokinetic and behavioral evaluations, *J Pharm Sci*, **100**, 547–557 (2011).
52. S. Freitas, G. Hielscher, H.P. Merkle, B. Gander, Continuous contact- and contamination-free ultrasonic emulsification: A useful tool for pharmaceutical development and production, *Ultrason Sonochem*, **13**, 76–85 (2006).
53. P. Ahlin, J. Kristl, J. Šmid-Korbar, Optimization of procedure parameters and physical stability of solid lipid nanoparticles in dispersions, *Acta Pharm*, **48**, 259–267 (1998).

54. M.D. Triplett II, J.F. Rathman, Optimization of β-carotene loaded solid lipid nanoparticles preparation using a high shear homogenization technique, *J Nanoparticle Res*, **11**, 601–614 (2009).
55. D. Doroud, A. Vatanara, F. Zahedifard, E. Gholami, R. Vahabpour, A.R. Najafabadi, S. Rafati, Cationic solid lipid nanoparticles loaded by cysteine proteinase genes as a novel anti-leishmaniasis DNA vaccine delivery system: Characterization and in vitro evaluations, *J Pharm Pharm Sci*, **13**, 320–335 (2010).
56. B. Siekmann, K. Westesen, Investigations on solid lipid nanoparticles prepared by precipitation in o/w emulsions, *Eur J Pharm Biopharm*, **42**, 104–109 (1996).
57. M. Trotta, F. Debernardi, O. Caputo, Preparation of solid lipid nanoparticles by a solvent emulsification-diffusion technique, *Int J Pharm*, **257**, 153–160 (2003).
58. D. Pandita, A. Ahuja, T. Velpandian, V. Lather, T. Dutta, R.K. Khar, Characterization and in vitro assessment of paclitaxel loaded lipid nanoparticles formulated using modified solvent injection technique, *Pharmazie*, **64**, 301–310 (2009).
59. A.J. Almeida, S. Runge, R.H. Müller, Peptide-loaded solid lipid nanoparticles (SLN): Influence of production parameters, *Int J Pharm*, **149**, 255–265 (1997).
60. A. zur Mühlen, C. Schwarz, W. Mehnert, Solid lipid nanoparticles (SLN) for controlled drug delivery: Drug release and release mechanism, *Eur J Pharm Biopharm*, **45**, 149–155 (1998).
61. I. Friedrich, C.C. Müller-Goymann, Characterization of solidified reverse micellar solutions (SRMS) and production development of SRMS-based nanosuspensions, *Eur J Pharm Biopharm*, **56**, 111–119 (2003).
62. A. zur Mühlen, E. zur Mühlen, H. Niehus, W. Mehnert, Atomic force microscopy studies of Solid Lipid Nanoparticles, *Pharm Res*, **13**, 1411–1416 (1996).
63. R. Cavalli, O. Caputo, M.R. Gasco, Preparation and characterization of solid lipid nanospheres containing paclitaxel, *Eur J Pharm Sci*, **10**, 305–309 (2000).
64. E. Ugazio, E. Marengo, C. Pellizzaro, D. Coradini, E. Peira, M.G. Daidone, M.R. Gasco, The effect of formulation and concentration of cholesteryl butyrate solid lipid nanospheres (SLN) on NIH-H460 cell proliferation, *Eur J Pharm Biopharm*, **52**, 197–202 (2001).
65. R. Cavalli, A. Bargoni, V. Podio, E. Muntoni, G.P. Zara, M.R. Gasco, Duodenal administration of solid lipid nanoparticles loaded with different percentages of tobramycin, *J Pharm Sci*, **92**, 1085–1094 (2003).
66. M.L. Bondì, G. Fontana, B. Carlisi, G. Giammona, Preparation and characterization of solid lipid nanoparticles containing cloricromene, *Drug Deliv*, **10**, 245–250 (2003).
67. A.V. Heydenreich, R. Westmeier, N. Pedersen, H.S. Poulsen, H.G. Kristensen, Preparation and purification of cationic solid lipid nanospheres - Effects on particle size, physical stability and cell toxicity, *Int J Pharm*, **254**, 83–87 (2003).
68. R. Cavalli, O. Caputo, M.E. Carlotti, M. Trotta, C. Scarnecchia, M.R. Gasco, Sterilization and freeze-drying of drug-free and drug-loaded solid lipid nanoparticles, *Int J Pharm*, **148**, 47–54 (1997).
69. M.O. Oyewumi, R.J. Mumper, Gadolinium-loaded nanoparticles engineered from microemulsion templates, *Drug Dev Ind Pharm*, **28**, 317–328 (2002).
70. Z. Cui, R.J. Mumper, Genetic immunization using nanoparticles engineered from microemulsion precursors, *Pharm Res*, **19**, 939–946 (2002).

71. C.H. Hsu, Z. Cui, R.J. Mumper, M. Jay, Preparation and characterization of novel coenzyme Q10 nanoparticles engineered from microemulsion precursors, *AAPS PharmSciTech*, **4**, E32 (2003).
72. J.M. Koziara, P.R. Lockman, D.D. Allen, R.J. Mumper, Paclitaxel nanoparticles for the potential treatment of brain tumors, *J Control Release*, **99**, 259–269 (2004).
73. X. Lu, M.D. Howard, M. Mazik, J. Eldridge, J.J. Rinehart, M. Jay, M. Leggas, Nanoparticles containing anti-inflammatory agents as chemotherapy adjuvants: Optimization and in vitro characterization, *AAPS J*, **10**, 133–140 (2008).
74. X. Dong, C.A. Mattingly, M.T. Tseng, M.J. Cho, Y. Liu, V.R. Adams, R.J. Mumper, Doxorubicin and paclitaxel-loaded lipid-based nanoparticles overcome multidrug resistance by inhibiting P-glycoprotein and depleting ATP, *Cancer Res*, **69**, 3918–3926 (2009).
75. J.-K. Kim, M.D. Howard, T.D. Dziubla, J.J. Rinehart, M. Jay, X. Lu, Uniformity of drug payload and its effect on stability of solid lipid nanoparticles containing an ester prodrug, *ACS Nano*, **5**, 209–216 (2011).
76. M.D. Howard, X. Lu, J.J. Rinehart, M. Jay, T.D. Dziubla, Physicochemical characterization of nanotemplate engineered solid lipid nanoparticles, *Langmuir*, **27**, 1964–1971 (2011).
77. M.A. Schubert, C.C. Müller-Goymann, Solvent injection as a new approach for manufacturing lipid nanoparticles – Evaluation of the method and process parameters, *Eur J Pharm Biopharm*, **55**, 125–131 (2003).
78. D.-B. Chen, T.-Z. Yang, W.-L. Lu, Q. Zhang, In vitro and in vivo study of two types of long-circulating solid lipid nanoparticles containing paclitaxel, *Chem Pharm Bull*, **49**, 1444–1447 (2001).
79. F.Q. Hu, H. Yuan, H.H. Zhang, M. Fang, Preparation of solid lipid nanoparticles with clobetasol propionate by a novel solvent diffusion method in aqueous system and physicochemical characterization, *Int J Pharm*, **239**, 121–128 (2002).
80. D. Chirio, M. Gallarate, M. Trotta, M.E. Carlotti, E.C. Gaudino, G. Cravotto, Influence of α- and γ-cyclodextrin lipophilic derivatives on curcumin-loaded SLN, *J Incl Phenom Macrocyclic Chem*, **65**, 391–402 (2009).
81. S. Jain, P. Khare, A. Gulbake, D. Bansal, S.K. Jain, Design and development of solid lipid nanoparticles for topical delivery of an anti-fungal agent, *Drug Deliv*, **17**, 443–451 (2010).
82. C.-F. Luo, M. Yuan, M.-S. Chen, S.-M. Liu, L. Zhu, B.-Y. Huang, X.-W. Liu, W. Xiong, Pharmacokinetics, tissue distribution and relative bioavailability of puerarin solid lipid nanoparticles following oral administration, *Int J Pharm*, **410**, 138–144 (2011).
83. T. Wang, N. Wang, Y. Zhang, W. Shen, X. Gao, T. Li, Solvent injection-lyophilization of tert-butyl alcohol/water cosolvent systems for the preparation of drug-loaded solid lipid nanoparticles, *Colloids Surf B*, **79**, 254–261 (2010).
84. M.K. Rawat, A. Jain, S. Singh, Studies on binary lipid matrix based solid lipid nanoparticles of repaglinide: In vitro and in vivo evaluation, *J Pharm Sci*, **100**, 2366–2378 (2011).
85. M. Shah, K. Chuttani, A.K. Mishra, K. Pathak, Oral solid compritol 888 ATO nanosuspension of simvastatin: Optimization and biodistribution studies, *Drug Dev Ind Pharm*, **37**, 526–537 (2011).

86. S.-H. Zhang, S.-C. Shen, Z. Chen, J.-X. Yun, K.-J. Yao, B.-B. Chen, J.-Z. Chen, Preparation of solid lipid nanoparticles in co-flowing microchannels, *Chem Eng J*, **144**, 324–328 (2008).
87. S.-H. Zhang, J. Yun, S. Shen, Z. Chen, K.J. Yao, J. Chen, B. Chen, Formation of solid lipid nanoparticles in a microchannel system with a cross-shaped junction, *Chem Eng Sci*, **63**, 5600–5605 (2008).
88. J. Yun, S. Zhang, S. Shen, Z. Chen, K. Yao, J. Chen, Continuous production of solid lipid nanoparticles by liquid flow-focusing and gas displacing method in microchannels, *Chem Eng Sci*, **64**, 4115–4122 (2009).
89. A. Dingler, S. Gohla, Production of solid lipid nanoparticles (SLN): Scaling up feasibilities, *J Microencapsulation*, **19**, 11–16 (2002).
90. V. Jenning, A. Lippacher, S.H. Gohla, Medium scale production of solid lipid nanoparticles (SLN) by high pressure homogenization, *J Microencapsulation*, **19**, 1–10 (2002).
91. A. Nazir, K. Schroën, R. Boom, Premix emulsification: A review, *J Membr Sci*, **362**, 1–11 (2010).
92. N.B. Raikar, S.R. Bhatia, M.F. Malone, D.J. McClements, M.A. Henson, Predicting the effect of the homogenization pressure on emulsion drop-size distributions, *Ind Eng Chem Res*, **50**, 6089–6100 (2011).
93. S. Jahnke, The theory of high-pressure homogenization, in *Emulsions and nanosuspensions for the formulation of poorly soluble drugs*, R.H. Müller (Ed.), Stuttgart: Medpharm Scientific Publ, 1998, pp. 117–200.
94. M. Cortés-Muñoz, D. Chevalier-Lucia, E. Dumay, Characteristics of submicron emulsions prepared by ultra-high pressure homogenisation: Effect of chilled or frozen storage, *Food Hydrocolloids*, **23**, 640–654 (2009).
95. R. Cavalli, E. Marengo, L. Rodriguez, M.R. Gasco, Effects of some experimental factors on the production process of solid lipid nanoparticles, *Eur J Pharm Biopharm*, **42**, 110–115 (1996).
96. E. Marengo, R. Cavalli, O. Caputo, L. Rodriguez, M.R. Gasco, Scale-up of the preparation process of solid lipid nanospheres. Part I, *Int J Pharm*, **205**, 3–13 (2000).
97. R.J. Mumper, Z. Cui, M.O. Oyewumi, Nanotemplate engineering of cell specific nanoparticles, *J Disp Sci Technol*, **24**, 569–588 (2003).
98. H.O. Hauser, The effect of ultrasonic irradiation on the chemical structure of egg lecithin, *Biochem Biophys Res Commun*, **45**, 1049–1055 (1971).
99. F. Chemat, I. Grondin, P. Costes, L. Moutoussamy, A.S.C. Sing, J. Smadja, High power ultrasound effects on lipid oxidation of refined sunflower oil, *Ultrason Sonochem*, **11**, 281–285 (2004).
100. V. Venkateswarlu, K. Manjunath, Preparation, characterization and in vitro release kinetics of clozapine solid lipid nanoparticles, *J Control Release*, **95**, 627–638 (2004).
101. R. Lopes, C.V. Eleutério, L.M.D. Gonalves, M.E.M. Cruz, A.J. Almeida, Lipid nanoparticles containing oryzalin for the treatment of leishmaniasis, *Eur J Pharm Sci*, **45**, 442–450 (2012).
102. C. Schwarz, W. Mehnert, Freeze-drying of drug-free and drug-loaded solid lipid nanoparticles (SLN), *Int J Pharm*, **157**, 171–179 (1997).

103. S.-J. Lim, C.-K. Kim, Formulation parameters determining the physicochemical characteristics of solid lipid nanoparticles loaded with all-trans retinoic acid, *Int J Pharm*, **243**, 135–146 (2002).
104. E. Zimmermann, R.H. Müller, K. Mäder, Influence of different parameters on reconstitution of lyophilized SLN, *Int JPharm*, **196**, 211–213 (2000).
105. B. Siekmann, K. Westesen, Melt-homogenized solid lipid nanoparticles stabilized by the nonionic surfactant tyloxapol. II. Physicochemical characterization and lyophilisation, *Pharm Pharmacol Lett*, **3**, 225–228 (1994).
106. R. Cavalli, M. Gasco, A. Barresi, G. Rovero, Evaporative drying of aqueous dispersions of solid lipid nanoparticles, *Drug Dev Ind Pharm*, **27**, 919–924 (2001).
107. E. Marengo, R. Cavalli, G. Rovero, M.R. Gasco, Scale-up and optimization of an evaporative drying process applied to aqueous dispersions of solid lipid nanoparticles, *Pharm Dev Technol*, **8**, 299–309 (2003).
108. C. Freitas, R.H. Müller, Spray-drying of solid lipid nanoparticles (SLN^{TM}), *Eur J Pharm Biopharm*, **46**, 145–151 (1998).
109. L. Harivardhan Reddy, R. Sharma, K. Chuttani, A. Mishra, R. Murthy, Influence of administration route on tumor uptake and biodistribution of etoposide loaded solid lipid nanoparticles in Dalton's lymphoma tumor bearing mice, *J Control Release*, **105**, 185–198 (2005).
110. J.F. Pinto, R.H. Müller, Pellets as carriers of solid lipid nanoparticles (SLN) for oral administration of drugs, *Pharmazie*, **54**, 506–509 (1999).
111. T. de Vringer, H.A.G. de Ronde, Preparation and structure of a water-in-oil cream containing lipid nanoparticles, *J Pharm Sci*, **84**, 466–472 (1995).
112. A. Dingler, R.P. Blum, H. Niehus, R.H. Müller, S. Gohla, Solid lipid nanoparticles (SLN^{TM}/Lipopearls^{TM}) – A pharmaceutical and cosmetic carrier for the application of vitamin E in dermal products, *J Microencapsulation*, **16**, 751–767 (1999).
113. V. Jenning, A. Gysler, M. Schäfer-Korting, S.H. Gohla, Vitamin A loaded solid lipid nanoparticles for topical use: Occlusive properties and drug targeting to the upper skin, *Eur J Pharm Biopharm*, **49**, 211–218 (2000).
114. S.A. Wissing, R.H. Müller, The influence of solid lipid nanoparticles on skin hydration and viscoelasticity: In vivo study, *Eur J Pharm Biopharm*, **56**, 67–72 (2003).
115. S.M. Moghimi, A.C. Hunter, J.C. Murray, Long-circulating and target-specific nanoparticles: Theory to practice, *Pharmacol Rev*, **53**, 283–318 (2001).
116. S.M. Moghimi, A.C. Hunter, T.L. Andresen, Factors controlling nanoparticle pharmacokinetics: An integrated analysis and perspective, *Ann Rev Pharmacol Toxicol*, **52**, 481–503 (2012).
117. D.F. Driscoll, Lipid injectable emulsions: 2006, *Nutr Clin Prac*, **21**, 381–386 (2006).
118. A.T. Florence, N. Hussain, Transcytosis of nanoparticle and dendrimer delivery systems: Evolving vistas, *Adv Drug Delivery Rev*, **50**, S69 (2001).
119. S. Hasani, Y. Pellequer, A. Lamprecht, Selective adhesion of nanoparticles to inflamed tissue in gastric ulcers, *Pharm Res*, **26**, 1149–1154 (2009).
120. F. Knorr, J. Lademann, A. Patzelt, W. Sterry, U. Blume-Peytavi, A. Vogt, Follicular transport route: Research progress and future perspectives, *Eur J Pharm Biopharm*, **71**, 173–180 (2009).
121. Y.A. Carpentier, I.E. Dupont, Advances in intravenous lipid emulsions, *World J Surg*, **24**, 1493–1497 (2000).

122. D.F. Driscoll, Lipid injectable emulsions: Pharmacopeial and safety issues, *Pharm Res*, **23**, 1959–1969 (2006).
123. M. Gaumet, A. Vargas, R. Gurny, F. Delie, Nanoparticles for drug delivery: The need for precision in reporting particle size parameters, *Eur J Pharm Biopharm*, **69**, 1–9 (2008).
124. C. Olbrich, U. Bakowsky, C.-M. Lehr, R.H. Müller, C. Kneuer, Cationic solid-lipid nanoparticles can efficiently bind and transfect plasmid DNA, *J Control Release*, **77**, 345–355 (2001).
125. Z. Cui, R.J. Mumper, Intranasal administration of plasmid DNA-coated nanoparticles results in enhanced immune responses, *J Pharm Pharmacol*, **54**, 1195–1203 (2002).
126. Z. Cui, R.J. Mumper, Coating of cationized protein on engineered nanoparticles results in enhanced immune responses, *Int J Pharm*, **238**, 229–239 (2002).
127. G. Montana, M.L. Bondì, R. Carrotta, P. Picone, E.F. Craparo, P.L. San Biagio, G. Giammona, M. Di Carlo, Employment of cationic solid-lipid nanoparticles as RNA carriers, *Bioconj Chem*, **18**, 302–308 (2007).
128. E. Zimmermann, R.H. Müller, Electrolyte- and pH-stabilities of aqueous solid lipid nanoparticle (SLN^{TM}) dispersions in artificial gastrointestinal media, *Eur J Pharm-Biopharm*, **52**, 203–210 (2001).
129. A. Fundarò, R. Cavalli, A. Bargoni, D. Vighetto, G.P. Zara, M.R. Gasco, Non-stealth and stealth solid lipid nanoparticles (SLN) carrying doxorubicin: Pharmacokinetics and tissue distribution after i.v. administration to rats, *Pharmacol Res*, **42**, 337–343 (2000).
130. G.P. Zara, R. Cavalli, A. Bargoni, A. Fundaro, D. Vighetto, M.R. Gasco, Intravenous administration to rabbits of non-stealth and stealth doxorubicin-loaded solid lipid nanoparticles at increasing concentrations of stealth agent: Pharmacokinetics and distribution of doxorubicin in brain and other tissues, *J. Drug Targeting*, **10**, 327–335 (2002).
131. W. Lu, L.C. He, C.H. Wang, Y.H. Li, S.Q. Zhang, The use of solid lipid nanoparticles to target a lipophilic molecule to the liver after intravenous administration to mice, *Int J Biol Macromol*, **43**, 320–324 (2008).
132. H. Bunjes, T. Unruh, Characterization of lipid nanoparticles by differential scanning calorimetry, X-ray and neutron scattering, *Adv Drug Del Rev*, **59**, 379–402 (2007).
133. H. Ali, K. El-Sayed, P.W. Sylvester, S. Nazzal, Molecular interaction and localization of tocotrienol-rich fraction (TRF) within the matrices of lipid nanoparticles: Evidence studies by Differential Scanning Calorimetry (DSC) and Proton Nuclear Magnetic Resonance spectroscopy (^{1}H NMR), *Colloids Surf B*, **77**, 286–297 (2010).
134. K. Westesen, M. Drechsler, H. Bunjes, Colloidal dispersions based on solid lipids, in *Food Colloids: Fundamentals of Formulation*, E. Dickinson, R. Miller (Eds.), Cambridge: CUP, 2001, pp. 103–115.
135. H. Bunjes, M.H.J. Koch, Saturated phospholipids promote crystallization but slow down polymorphic transitions in triglyceride nanoparticles, *J Control Release*, **107**, 229–243 (2005).
136. C. Blümer, K. Mäder, Isostatic ultra-high-pressure effects on supercooled melts in colloidal triglyceride dispersions, *Pharm Res*, **22**, 1708–1715 (2005).

137. H. Bunjes, M. Drechsler, M.H.J. Koch, K. Westesen, Incorporation of the model drug ubidecarenone into solid lipid nanoparticles, *Pharm Res*, **18**, 287–293 (2001).
138. H. Bunjes, M.H.J. Koch, K. Westesen, Effects of surfactants on the crystallization and polymorphism of lipid nanoparticles, *Prog Colloid Polym Sci*, **121**, 7–10 (2002).
139. T. Awad, K. Sato, Fat crystallization in O/W emulsions controlled by hydrophobic emulsifier additives, *in Physical Properties of Lipids*, A. Marangoni, S. Narine (Eds.), New York: Marcel Dekker, 2002, pp. 37–62.
140. T. Sonoda, Y. Takata, S. Ueno, K. Sato, Effects of emulsifiers on crystallization behavior of lipid crystals in nanometer-size oil-in-water emulsion droplets, *Cryst Growth Design*, **6**, 306–312 (2006).
141. H. Bunjes, M.H.J. Koch, K. Westesen, Effect of particle size on colloidal solid triglycerides, *Langmuir*, **16**, 5234–5241 (2000).
142. T. Unruh, H. Bunjes, K. Westesen, M.H.J. Koch, Observation of size-dependent melting in lipid nanoparticles, *J Phys Chem B*, **103**, 10373–10377 (1999).
143. T. Unruh, H. Bunjes, K. Westesen, M.H.J. Koch, Investigations on the melting behaviour of triglyceride nanoparticles, *Colloid Polym Sci*, **279**, 398–403 (2001).
144. R.H. Müller, S.A. Runge, V. Ravelli, A.F. Thünemann, W. Mehnert, E.B. Souto, Cyclosporine-loaded solid lipid nanoparticles (SLN®): Drug-lipid physicochemical interactions and characterization of drug incorporation, *Eur J Pharm Biopharm*, **68**, 535–544 (2008).
145. J.W. Hagemann, J.A. Rothfus, Polymorphism and transformation energetics of saturated monoacid triglycerides from differential scanning calorimetry and theoretical modeling, *J Am Oil Chem Soc*, **60**, 1123–1131 (1983).
146. B. Siekmann, K. Westesen, Thermoanalysis of the recrystallization process of melt-homogenized glyceride nanoparticles, *Colloids Surf B*, **3**, 159–175 (1994).
147. H. Bunjes, M.H.J. Koch, K. Westesen, Influence of emulsifiers on the crystallization of solid lipid nanoparticles, *J Pharm Sci*, **92**, 1509–1520 (2003).
148. K.M. Rosenblatt, H. Bunjes, Poly(vinyl alcohol) as emulsifier stabilizes solid triglyceride drug carrier nanoparticles in the α-modification, *Mol Pharmaceutics*, **6**, 105–120 (2009).
149. S. Petersen, A. Fahr, H. Bunjes, Flow cytometry as a new approach to investigate drug transfer between lipid particles, *Mol Pharmaceutics*, **7**, 350–363 (2010).
150. V. Jenning, M. Schäfer-Korting, S. Gohla, Vitamin A-loaded solid lipid nanoparticles for topical use: Drug release properties, *J Control Release*, **66**, 115–126 (2000).
151. H. Bunjes, F. Steiniger, W. Richter, Visualizing the structure of triglyceride nanoparticles in different crystal modifications, *Langmuir*, **23**, 4005–4011 (2007).
152. C. Freitas, R.H. Müller, Correlation between long-term stability of solid lipid nanoparticles (SLN™) and crystallinity of the lipid phase, *Eur J Pharm Biopharm*, **47**, 125–132 (1999).
153. T. Helgason, T.S. Awad, K. Kristbergsson, D.J. McClements, J. Weiss, Influence of polymorphic transformations on gelation of tripalmitin solid lipid nanoparticle suspensions, *J Am Oil Chem Soc*, **85**, 501–511 (2008).
154. H. Bunjes, Characterization of solid lipid nano- and microparticles, in *Lipospheres in Drug Targets and Delivery*, C. Nastruzzi (Ed.), Boca Raton, FL: CRC Press, 2005, pp. 41–66.

155. J. Kuntsche, J.C. Horst, H. Bunjes, Cryogenic transmission electron microscopy (cryo-TEM) for studying the morphology of colloidal drug delivery systems, *Int J Pharm*, **417**, 120–137 (2011).
156. K. Westesen, B. Siekmann, Investigation of the gel formation of phospholipid-stabilized solid lipid nanoparticles, *Int J Pharm*, **151**, 35–45 (1997).
157. M. Garcia-Fuentes, D. Torres, M. Martín-Pastor, M.J. Alonso, Application of NMR spectroscopy to the characterization of PEG-stabilized lipid nanoparticles, *Langmuir*, **20**, 8839–8845 (2004).
158. M.A. Schubert, M. Harms, C.C. Müller-Goymann, Structural investigations on lipid nanoparticles containing high amounts of lecithin, *Eur J Pharm Sci*, **27**, 226–236 (2006).
159. K. Jores, W. Mehnert, M. Drechsler, H. Bunjes, C. Johann, K. Mäder, Investigations on the structure of solid lipid nanoparticles (SLN) and oil-loaded solid lipid nanoparticles by photon correlation spectroscopy, field-flow fractionation and transmission electron microscopy, *J Control Release*, **95**, 217–227 (2004).
160. E. Esposito, M. Fantin, M. Marti, M. Drechsler, L. Paccamiccio, P. Mariani, E. Sivieri, F. Lain, E. Menegatti, M. Morari, R. Cortesi, Solid lipid nanoparticles as delivery systems for bromocriptine, *Pharm Res*, **25**, 1521–1530 (2008).
161. S. Petersen, F. Steiniger, D. Fischer, A. Fahr, H. Bunjes, The physical state of lipid nanoparticles influences their effect on in vitro cell viability, *Eur J Pharm Biopharm*, **79**, 150–161 (2011).
162. A. Illing, T. Unruh, Investigation on the flow behavior of dispersions of solid triglyceride nanoparticles, *Int J Pharm*, **284**, 123–131 (2004).
163. T. Unruh, K. Westesen, P. Bösecke, P. Lindner, M.H.J. Koch, Self-assembly of triglyceride nanocrystals in suspension, *Langmuir*, **18**, 1796–1800 (2002).
164. A. Illing, T. Unruh, M.H.J. Koch, Investigation on particle self-assembly in solid lipid-based colloidal drug carrier systems, *Pharm Res*, **21**, 592–597 (2004).
165. A. Lippacher, R.H. Müller, K. Mäder, Investigation on the viscoelastic properties of lipid based colloidal drug carriers, *Int J Pharm*, **196**, 227–230 (2000).
166. A. Lippacher, R.H. Müller, K. Mäder, Preparation of semisolid drug carriers for topical application based on solid lipid nanoparticles, *Int J Pharm*, **214**, 9–12 (2001).
167. A. Lippacher, R.H. Müller, K. Mäder, Semisolid SLN™ dispersions for topical application: Influence of formulation and production parameters on viscoelastic properties, *Eur J Pharm Biopharm*, **53**, 155–160 (2002).
168. C. Freitas, R.H. Müller, Effect of light and temperature on zeta potential and physical stability in solid lipid nanoparticle (SLN®) dispersions, *Int J Pharm*, **168**, 221–229 (1998).
169. C. Freitas, R.H. Müller, Stability determination of solid lipid nanoparticles (SLN®) in aqueous dispersion after addition of electrolyte, *J Microencapsulation*, **16**, 59–71 (1999).
170. V. Jenning, S.H. Gohla, Encapsulation of retinoids in solid lipid nanoparticles (SLN®), *J Microencapsulation*, **18**, 149–158 (2001).
171. S. Das, W.K. Ng, P. Kanaujia, S. Kim, R.B.H. Tan, Formulation design, preparation and physicochemical characterizations of solid lipid nanoparticles containing a hydrophobic drug: Effects of process variables, *Colloids Surf B*, **88**, 483–489 (2011).

172. E. Kupetz, L. Preu, C. Kunick, H. Bunjes, Parenteral formulation of an antileishmanial drug candidate – Tackling poor solubility, chemical instability, and polymorphism, *Eur J Pharm Biopharm*, **85**, 511–520 (2013).
173. J. Han, C. Washington, Partition of antimicrobial additives in an intravenous emulsion and their effect on emulsion physical stability, *Int J Pharm*, **288**, 263–271 (2005).
174. S.C. Yang, L.F. Lu, Y. Cai, J.B. Zhu, B.W. Liang, C.Z. Yang, Body distribution in mice of intravenously injected camptothecin solid lipid nanoparticles and targeting effect on brain, *J Control Release*, **59**, 299–307 (1999).
175. P. Ahlin, J. Kristl, M. Šentjurc, J. Štrancar, S. Pečar, Influence of spin probe structure on its distribution in SLN dispersions, *Int J Pharm*, **196**, 241–244 (2000).
176. K. Jores, W. Mehnert, K. Mäder, Physicochemical investigations on solid lipid nanoparticles and on oil-loaded solid lipid nanoparticles: A nuclear magnetic resonance and electron spin resonance study, *Pharm Res*, **20**, 1274–1283 (2003).
177. K. Jores, A. Haberland, S. Wartewig, K. Mäder, W. Mehnert, Solid Lipid Nanoparticles (SLN) and oil-loaded SLN studied by spectrofluorometry and raman spectroscopy, *Pharm Res*, **22**, 1887–1897 (2005).
178. R. Sivaramakrishnan, C. Nakamura, W. Mehnert, H.C. Korting, K.D. Kramer, M. Schäfer-Korting, Glucocorticoid entrapment into lipid carriers: Characterisation by parelectric spectroscopy and influence on dermal uptake, *J Control Release*, **97**, 493–502 (2004).
179. S. Lombardi Borgia, M. Regehly, R. Sivaramakrishnan, W. Mehnert, H.C. Korting, K. Danker, B. Röder, K.D. Kramer, M. Schäfer-Korting, Lipid nanoparticles for skin penetration enhancement: Correlation to drug localization within the particle matrix as determined by fluorescence and parelectric spectroscopy, *J Control Release*, **110**, 151–163 (2005).
180. C. Braem, T. Blaschke, G. Panek-Minkin, W. Herrmann, P. Schlupp, T. Paepenmüller, C. Müller-Goyman, W. Mehnert, R. Bittl, M. Schäfer-Korting, K.D. Kramer, Interaction of drug molecules with carrier systems as studied by parelectric spectroscopy and electron spin resonance, *J Control Release*, **119**, 128–135 (2007).
181. X. Lu, M.D. Howard, D.R. Talbert, J.J. Rinehart, P.M. Potter, M. Jay, M. Leggas, Nanoparticles containing anti-inflammatory agents as chemotherapy adjuvants II: Role of plasma esterases in drug release, *AAPS J*, **11**, 120–122 (2009).
182. A. Noack, G. Hause, K. Mäder, Physicochemical characterization of curcuminoid-loaded solid lipid nanoparticles, *Int J Pharm*, **423**, 440–451 (2012).
183. P. Ahlin, J. Kristl, S. Pecar, J. Štrancar, M. Šenturc, The effect of lipophilicity of spin-labeled compounds on their distribution in solid lipid nanoparticle dispersions studied by electron paramagnetic resonance, *J Pharm Sci*, **92**, 58–66 (2003).
184. T. Blaschke, L. Kankate, K.D. Kramer, Structure and dynamics of drug-carrier systems as studied by parelectric spectroscopy, *Adv Drug Deliv Rev* **59**, 403–410 (2007).
185. T. Helgason, T.S. Awad, K. Kristbergsson, E.A. Decker, D.J. McClements, J. Weiss, Impact of surfactant properties on oxidative stability of β-carotene encapsulated within solid lipid nanoparticles, *J Agricult Food Chem*, **57**, 8033–8040 (2009).
186. R.V. Tikekar, N. Nitin, Effect of physical state (solid vs. liquid) of lipid core on the rate of transport of oxygen and free radicals in solid lipid nanoparticles and emulsion, *Soft Matter*, **7**, 8149–8157 (2011).

187. K. Westesen, B. Siekmann, M.H.J. Koch, Investigations on the physical state of lipid nanoparticles by synchrotron radiation X-ray diffraction, *Int J Pharm*, **93**, 189–199 (1993).
188. E. Ugazio, R. Cavalli, M.R. Gasco, Incorporation of cyclosporin A in solid lipid nanoparticles (SLN), *Int J Pharm*, **241**, 341–344 (2002).
189. Y. Luo, D. Chen, L. Ren, X. Zhao, J. Qin, Solid lipid nanoparticles for enhancing vinpocetine's oral bioavailability, *J Control Release*, **114**, 53–59 (2006).
190. B. Magenheim, M.Y. Levy, S. Benita, A new in vitro technique for the evaluation of drug release profile from colloidal carriers: Ultrafiltration technique at low pressure, *Int J Pharm*, **94**, 115–123 (1993).
191. C. Washington, K. Evans, Release rate measurements of model hydrophobic solutes from submicron triglyceride emulsions, *J Control Release*, **33**, 383–390 (1995).
192. B.J. Boyd, Characterisation of drug release from cubosomes using the pressure ultrafiltration method, *Int J Pharm*, **260**, 239–247 (2003).
193. K.M. Rosenblatt, D. Douroumis, H. Bunjes, Drug release from differently structured monoolein/poloxamer nanodispersions studied with differential pulse polarography and ultrafiltration at low pressure, *J Pharm Sci*, **96**, 1564–1575 (2007).
194. T. Takino, K. Konishi, Y. Takakura, M. Hashida, Long circulating emulsion carrier systems for highly lipophilic drugs, *Biol Pharm Bull*, **17**, 121–125 (1994).
195. S. Kawakami, F. Yamashita, M. Hashida, Disposition characteristics of emulsions and incorporated drugs after systemic or local injection, *Adv Drug Deliv Rev*, **45**, 77–88 (2000).
196. K. Hippalgaonkar, S. Majumdar, V. Kansara, Injectable lipid emulsions-advancements, opportunities and challenges, *AAPS PharmSciTech*, **11**, 1526–1540 (2010).
197. R. Cavalli, O. Caputo, M.R. Gasco, Solid lipospheres of doxorubicin and idarubicin, *Int J Pharm*, **89**, R9 (1993).
198. J.M. Koziara, P.R. Lockman, D.D. Allen, R.J. Mumper, In situ blood-brain barrier transport of nanoparticles, *Pharm Res*, **20**, 1772–1778 (2003).
199. C. Washington, Evaluation of non-sink dialysis methods for the measurement of drug release from colloids: effects of drug partition, *Int J Pharm*, **56**, 71–74 (1989).
200. C. Washington, Drug release from microdisperse systems: A critical review, *Int J Pharm*, **58**, 1–12 (1990).
201. Z. Mei, H. Chen, T. Weng, Y. Yang, X. Yang, Solid lipid nanoparticle and microemulsion for topical delivery of triptolide, *Eur J Pharm Biopharm*, **56**, 189–196 (2003).
202. R.H. Müller, D. Rühl, S.A. Runge, Biodegradation of solid lipid nanoparticles as a function of lipase incubation time, *Int J Pharm*, **144**, 115–121 (1996).
203. C. Olbrich, O. Kayser, R.H. Müller, Enzymatic degradation of Dynasan 114 SLN – Effect of surfactants and particle size, *J Nanopart Res*, **4**, 121–129 (2002).
204. C. Olbrich, O. Kayser, R.H. Müller, Lipase degradation of Dynasan 114 and 116 solid lipid nanoparticles (SLN): Effect of surfactants, storage time and crystallinity, *Int J Pharm*, **237**, 119–128 (2002).
205. H. Weyhers, S. Ehlers, H. Hahn, E.B. Souto, R.H. Müller, Solid lipid nanoparticles (SLN) – Effects of lipid composition on in vitro degradation and in vivo toxicity, *Pharmazie*, **61**, 539–544 (2006).

206. X. Dong, R.J. Mumper, The metabolism of fatty alcohols in lipid nanoparticles by alcohol dehydrogenase, *Drug Dev Ind Pharm*, **32**, 973–980 (2006).
207. J.M. Koziara, J.J. Oh, W.S. Akers, S.P. Ferraris, R.J. Mumper, Blood compatibility of cetyl alcohol/polysorbate-based nanoparticles, *Pharm Res*, **22**, 1821–1828 (2005).
208. P.R. Lockman, J. Koziara, K.E. Roder, J. Paulson, T.J. Abbruscato, R.J. Mumper, D.D. Allen, In vivo and in vitro assessment of baseline blood-brain barrier parameters in the presence of novel nanoparticles, *Pharm Res*, **20**, 705–713 (2003).
209. M.A. Schubert, C.C. Müller-Goymann, Characterisation of surface-modified solid lipid nanoparticles (SLN): Influence of lecithin and nonionic emulsifier, *Eur J Pharm Biopharm*, **61**, 77–86 (2005).
210. R.H. Müller, S. Maaßen, H. Weyhers, W. Mehnert, Phagocytic uptake and cytotoxicity of solid lipid nanoparticles (SLN) sterically stabilized with poloxamine 908 and poloxamer 407, *J Drug Targeting*, **4**, 161–170 (1996).
211. R.H. Müller, S. Maaßen, C. Schwarz, W. Mehnert, Solid lipid nanoparticles (SLN) as potential carrier for human use: Interaction with human granulocytes, *J Control Release*, **47**, 261–269 (1997).
212. R.H. Müller, D. Rühl, S. Runge, K. Schulze-Forster, W. Mehnert, Cytotoxicity of solid lipid nanoparticles as a function of the lipid matrix and the surfactant, *Pharm Res*, **14**, 458–462 (1997).
213. N. Schöler, C. Olbrich, K. Tabatt, R.H. Müller, H. Hahn, O. Liesenfeld, Surfactant, but not the size of solid lipid nanoparticles (SLN) influences viability and cytokine production of macrophages, *Int J Pharm*, **221**, 57–67 (2001).
214. J. Kristl, K. Teskac, M. Milek, I. Mlinaric-Rascan, Surface active stabilizer Tyloxapol in colloidal dispersions exerts cytostatic effects and apoptotic dismissal of cells, *Toxicol Appl Pharmacol*, **232**, 218–225 (2008).
215. K. Tabatt, M. Sameti, C. Olbrich, R.H. Müller, C.-M. Lehr, Effect of cationic lipid and matrix lipid composition on solid lipid nanoparticle-mediated gene transfer, *Eur J Pharm Biopharm*, **57**, 155–162 (2004).
216. W. Weyenberg, P. Filev, D. van den Plas, J. Vandervoort, K. de Smet, P. Sollie, A. Ludwig, Cytotoxicity of submicron emulsions and solid lipid nanoparticles for dermal application, *Int J Pharm*, **337**, 291–298 (2007).
217. N. Schöler, H. Hahn, R.H. Müller, O. Liesenfeld, Effect of lipid matrix and size of solid lipid nanoparticles (SLN) on the viability and cytokine production of macrophages, *Int J Pharm*, **231**, 167–176 (2002).
218. M. Nassimi, C. Schleh, H.-D. Lauenstein, R. Hussein, K. Lübbers, G. Pohlmann, S. Switalla, K. Sewald, M. Müller, N. Krug, C.C. Müller-Goymann, A. Braun, Low cytotoxicity of solid lipid nanoparticles in in vitro and ex vivo lung models, *Inhal Toxicol*, **21**, 104–109 (2009).
219. M. Nassimi, C. Schleh, H.D. Lauenstein, R. Hussein, H.G. Hoymann, W. Koch, G. Pohlmann, N. Krug, K. Sewald, S. Rittinghausen, A. Braun, C. Müller-Goymann, A toxicological evaluation of inhaled solid lipid nanoparticles used as a potential drug delivery system for the lung, *Eur J Pharm Biopharm*, **75**, 107–116 (2010).
220. A.T. Florence, The oral absorption of micro- and nanoparticulates: Neither exceptional nor unusual, *Pharm Res*, **14**, 259–266 (1997).

221. C.J.H. Porter, N.L. Trevaskis, W.N. Charman, Lipids and lipid-based formulations: Optimizing the oral delivery of lipophilic drugs, *Nature Rev Drug Discov*, **6**, 231–248 (2007).
222. M.R. Aji Alex, A.J. Chacko, S. Jose, E.B. Souto, Lopinavir loaded solid lipid nanoparticles (SLN) for intestinal lymphatic targeting, *Eur J Pharm Sci*, **42**, 11–18 (2011).
223. G. Suresh, K. Manjunath, V. Venkateswarlu, V. Satyanarayana, Preparation, characterization, and in vitro and in vivo evaluation of lovastatin solid lipid nanoparticles, *AAPS PharmSciTech*, **8**, E162-E170 (2007).
224. K. Manjunath, V. Venkateswarlu, Pharmacokinetics, tissue distribution and bioavailability of clozapine solid lipid nanoparticles after intravenous and intraduodenal administration, *J Control Release*, **107**, 215–228 (2005).
225. K. Manjunath, V. Venkateswarlu, Pharmacokinetics, tissue distribution and bioavailability of nitrendipine solid lipid nanoparticles after intravenous and intraduodenal administration, *J Drug Target*, **14**, 632–645 (2006).
226. V.V. Kumar, D. Chandrasekar, S. Ramakrishna, V. Kishan, Y.M. Rao, P.V. Diwan, Development and evaluation of nitrendipine loaded solid lipid nanoparticles: Influence of wax and glyceride lipids on plasma pharmacokinetics, *Int J Pharm*, **335**, 167–175 (2007).
227. H. Li, X. Zhao, Y. Ma, G. Zhai, L. Li, H. Lou, Enhancement of gastrointestinal absorption of quercetin by solid lipid nanoparticles, *J Control Release*, **133**, 238–244 (2009).
228. L. Hu, Q. Xing, J. Meng, C. Shang, Preparation and enhanced oral bioavailability of cryptotanshinone-loaded solid lipid nanoparticles, *AAPS PharmSciTech*, **11**, 582–587 (2010).
229. S. Yang, J. Zhu, Y. Lu, B. Liang, C. Yang, Body distribution of camptothecin solid lipid nanoparticles after oral administration, *Pharm Res*, **16**, 751–757 (1999).
230. R.H. Müller, S. Runge, V. Ravelli, W. Mehnert, A.F. Thünemann, E.B. Souto, Oral bioavailability of cyclosporine: Solid lipid nanoparticles (SLN®) versus drug nanocrystals, *Int J Pharm*, **317**, 82–89 (2006).
231. A. Bargoni, R. Cavalli, O. Caputo, A. Fundarò, M.R. Gasco, G.P. Zara, Solid lipid nanoparticles in lymph and plasma after duodenal administration to rats, *Pharm Res*, **15**, 745–750 (1998).
232. H. Yuan, J. Chen, Y.-Z. Du, F.-Q. Hu, S. Zeng, H.-L. Zhao, Studies on oral absorption of stearic acid SLN by a novel fluorometric method, *Colloids Surf B*, **58**, 157–164 (2007).
233. J. Fang, H. Nakamura, H. Maeda, The EPR effect: Unique features of tumor blood vessels for drug delivery, factors involved, and limitations and augmentation of the effect, *Adv Drug Deliv Rev*, **63**, 136–151 (2011).
234. M.A. Phillips, M.L. Gran, N.A. Peppas, Targeted nanodelivery of drugs and diagnostics, *Nano Today*, **5**, 143–159 (2010).
235. C.R. Alving, Macrophages as targets for delivery of liposome-encapsulated antimicrobial agents, *Adv Drug Deliv Rev*, **2**, 107–128 (1988).
236. A.K. Agrawal, C.M. Gupta, Tuftsin-bearing liposomes in treatment of macrophage-based infections, *Adv Drug Deliv Rev*, **41**, 135–146 (2000).

237. C. Bocca, O. Caputo, R. Cavalli, L. Gabriel, A. Miglietta, M.R. Gasco, Phagocytic uptake of fluorescent stealth and non-stealth solid lipid nanoparticles, *Int J Pharm*, **175**, 185–193 (1998).
238. V. Podio, G.P. Zara, M. Carazzone, R. Cavalli, M.R. Gasco, Biodistribution of stealth and non-stealth solid lipid nanospheres after intravenous administration to rats, *J Pharm Pharmacol*, **52**, 1057–1063 (2000).
239. J.M. Koziara, T.R. Whisman, M.T. Tseng, R.J. Mumper, In-vivo efficacy of novel paclitaxel nanoparticles in paclitaxel-resistant human colorectal tumors, *J Control Release*, **112**, 312–319 (2006).
240. G.P. Zara, A. Bargoni, R. Cavalli, A. Fundarò, D. Vighetto, M.R. Gasco, Pharmacokinetics and tissue distribution of idarubicin-loaded solid lipid nanoparticles after duodenal administration to rats, *J Pharm Sci*, **91**, 1324–1333 (2002).
241. Z. Xu, L. Chen, W. Gu, Y. Gao, L. Lin, Z. Zhang, Y. Xi, Y. Li, The performance of docetaxel-loaded solid lipid nanoparticles targeted to hepatocellular carcinoma, *Biomaterials*, **30**, 226–232 (2009).
242. L. Serpe, M.G. Catalano, R. Cavalli, E. Ugazio, O. Bosco, R. Canaparo, E. Muntoni, R. Frairia, M.R. Gasco, M. Eandi, G.P. Zara, Cytotoxicity of anticancer drugs incorporated in solid lipid nanoparticles on HT-29 colorectal cancer cell line, *Eur J Pharm Biopharm*, **58**, 673–680 (2004).
243. M.-K. Lee, S.-J. Lim, C.-K. Kim, Preparation, characterization and in vitro cytotoxicity of paclitaxel-loaded sterically stabilized solid lipid nanoparticles, *Biomaterials*, **28**, 2137–2146 (2007).
244. G. Fontana, L. Maniscalco, D. Schillaci, G. Cavallaro, G. Giammona, Solid lipid nanoparticles containing tamoxifen characterization and in vitro antitumoral activity, *Drug Deliv*, **12**, 385–392 (2005).
245. Y.-C. Kuo, F.-L. Su, Transport of stavudine, delavirdine, and saquinavir across the blood-brain barrier by polybutylcyanoacrylate, methylmethacrylate-sulfopropylmethacrylate, and solid lipid nanoparticles, *Int J Pharm*, **340**, 143–152 (2007).
246. Y.-C. Kuo, C.-Y. Chung, Solid lipid nanoparticles comprising internal Compritol 888 ATO, tripalmitin and cacao butter for encapsulating and releasing stavudine, delavirdine and saquinavir, *Colloids Surf B*, **88**, 682–690 (2011).
247. L. Hu, X. Tang, F. Cui, Solid lipid nanoparticles (SLNs) to improve oral bioavailability of poorly soluble drugs, *J Pharm Pharmacol*, **56**, 1527–1535 (2004).
248. M. Sznitowska, M. Gajewska, S. Janicki, A. Radwanska, G. Lukowski, Bioavailability of diazepam from aqueous-organic solution, submicron emulsion and solid lipid nanoparticles after rectal administration in rabbits, *Eur J Pharm Biopharm*, **52**, 159–163 (2001).
249. S. Xie, L. Zhu, Z. Dong, X. Wang, Y. Wang, X. Li, W. Zhou, Preparation, characterization and pharmacokinetics of enrofloxacin-loaded solid lipid nanoparticles: Influences of fatty acids, *Colloids Surf B*, **83**, 382–387 (2011).
250. I. Friedrich, S. Reichl, C.C. Müller-Goymann, Drug release and permeation studies of nanosuspensions based on solidified reverse micellar solutions (SRMS), *Int J Pharm*, **305**, 167–175 (2005).

251. F. Castelli, C. Puglia, M.G. Sarpietro, L. Rizza, F. Bonina, Characterization of indomethacin-loaded lipid nanoparticles by differential scanning calorimetry, *Int J Pharm*, **304**, 231–238 (2005).
252. Y.-C. Kuo, J.-F. Chung, Physicochemical properties of nevirapine-loaded solid lipid nanoparticles and nanostructured lipid carriers, *Colloids Surf B*, **83**, 299–306 (2011).
253. R. Cavalli, D. Aquilano, M.E. Carlotti, M.R. Gasco, Study by X-ray powder diffraction and differential scanning calorimetry to two model drugs, phenothiazine and nifedipine, incorporated into lipid nanoparticles, *Eur J Pharm Biopharm*, **41**, 329–333 (1995).
254. M.L. Bondì, E.F. Craparo, G. Giammona, F. Drago, Brain-targeted solid lipid nanoparticles containing riluzole: Preparation, characterization and biodistribution, *Nanomedicine*, **5**, 25–32 (2010).
255. A.C. Silva, E. González-Mira, M.L. García, M.A. Egea, J. Fonseca, R. Silva, D. Santos, E.B. Souto, D. Ferreira, Preparation, characterization and biocompatibility studies on risperidone-loaded solid lipid nanoparticles (SLN): High pressure homogenization versus ultrasound, *Colloids Surf B: Biointerfaces*, **86**, 158–165 (2011).
256. Z. Mei, X. Li, Q. Wu, S. Hu, X. Yang, The research on the anti-inflammatory activity and hepatotoxicity of triptolide-loaded solid lipid nanoparticle, *Pharmacol Res*, **51**, 345–351 (2005).

5

Polymeric Drug Delivery Systems for Encapsulating Hydrophobic Drugs

Naveed Ahmed, Claudia Elizabeth Mora-Huertas, Chiraz Jaafar-Maalej, Hatem Fessi and Abdelhamid Elaissari

5.1 Introduction

The application of nanotechnology in the biomedical field is amazing. In the past two decades, it has resulted in enormous work leading to hundreds of articles with new ideas, trends and methods for the development of a suitable drug delivery system. Of these, nanoparticles for drug delivery and targeting has attracted most researchers' attention. This can be considered the result of the advantages of nanoparticles over micro-particles and liposomes in terms of their small size, which provides a high surface-to-volume ratio and solves the solubility problem of hydrophobic active molecules, resulting in better efficiency, precise delivery and safety for organisms.

Different researchers have defined nanoparticles depending upon their size, and normally these are particles ranging in size from 1–500 nm. It should be noted that nanoparticles is a collective term used for nanospheres and nanocapsules in biomedical nanotechnology. The nanoparticles based on polymers are well investigated without the differentiation of the natural and synthetic polymers but the important point to be considered is the biodegradability and biocompatibility of the polymers. Polymeric nanoparticles can be prepared by using the preformed polymers or by direct polymerization of the monomers, resulting in nanoparticles using polymerization techniques. Developments in the polymeric nanoparticle field have enabled us to divide the preparation methods of nanoparticles into two main divisions: via preformed polymers and by polymerization techniques. However, in this chapter, focus will be on the use of preformed polymers for the preparation of nanoparticles. Knowledge

Drug Delivery Strategies for Poorly Water-Soluble Drugs, First Edition. Edited by Dennis Douroumis and Alfred Fahr.

of safety and the biocompatibility of polymers used for nanoparticles preparation will be helpful in the choice of a suitable process, keeping in mind the preparation method as well as *in vivo* and *in vitro* applications of the final particles.

Preformed polymers are used for the encapsulation of both hydrophilic and hydrophobic drugs. Researchers have devised specific techniques for both types but here the discussion will focus on hydrophobic drugs. The different methods used for the hydrophobic molecules include nanoprecipitation, emulsification, polymer coating, polymersomes, supercritical fluid and layer-by-layer methods. Generally, an aqueous and an organic phase are required for the preparation of nanoparticles, further selection of a suitable method should be on the basis of the drug nature, the selected polymer and the solvent to be used. The study of the different characteristics of nanoparticles prepared by any method can help to compile a suitable data sheet and can give an idea of success in the preparation of conceived nanoparticles. Nevertheless, as mentioned above, it is a dynamic research topic, so it has resulted in the preparation of more intelligent particles and specifically has led to the concept of finding multiple advantages, as can be seen by the recent developments in the preparation of theranostics targeted for therapy and diagnosis.

In this chapter, first of all the safety and biocompatibility of the polymers used for nanoparticles preparation, then the techniques used for preparation and, finally, the properties of the nanoparticles will be discussed.

5.2 Safety and Biocompatibility of Polymers

Polymeric nanoparticles are widely used in medical and pharmaceutical applications. They exhibit several desirable properties: safety, biocompatibility and biodegradability are the most important ones that nanoparticles must meet for *in vivo* applications [1].

Delivered nanoparticles are always in direct and sustained contact with biological tissues, and some of them degrade *in situ*. Thus, a fundamental understanding of the *in vivo* degradation criteria and mechanisms, as well as an appreciation of the cellular and tissue responses which determine the biocompatibility of polymeric nanoparticles, are of great interest and are vital components in the design and development of biocompatible vectors [2]. To achieve this, both polymer materials and degradation derivatives must be devoid of toxicity.

As shown in Table 5.1, among the synthetic polymers, biodegradable aliphatic polyesters such as poly (ε-caprolactone) (PCL), poly (lactic acid) (PLA), poly (glycolic acid) (PGA), and their copolymers have received considerable attention [3]. Owing to the high degree of crystallinity and hydrophobicity, PCL degrades rather slowly and is less biocompatible with soft tissue, which restricts its further clinical applications. Therefore, modification is proposed; Poly (ethylene glycol) (PEG), appreciated for its hydrophilicity, non-toxicity, absence of antigenicity and immunogenicity, has been selected to be attached to PCL. The preformed PCL–PEG copolymers have improved the hydrophilicity, biodegradability and mechanical properties of PCL, leading to wider applications [3]. Due to their biodegradability, good biocompatibility, excellent shaping and modelling properties, both PLA and poly lactic-glycolic acid (PLGA) have been widely applied in the biomedical field [4].

Table 5.1 *Examples of polymers used in the nanoparticles preparation via different methods.*

Polymer	Reference
Nanoprecipitation	
PACA	13
Methacrylic acid copolymers: Eudragit® S90, S100, RS, L100-55	14,17
PCL Mw 10 kDa, 14 kDa, 40 kDa, 42,5 kDa, 64 kDa, 65 kDa, 80 kDa, 100 kDa	14, 18,33
PLA Mw 9 kDa, 42 kDa, 75 kDa, 88 kDa, 100 kDa, 110 kDa, 200 kDa.	20–22, 24, 25, 34–44
PLGA 50:50 Mw 40 kDa, 50–75 kDa; 75:25 Mw 75–120 kDa	20, 21, 24, 25, 45–47
PCL-PEG 40–5 kDa, 45–5 kDa	20, 21, 24, 48
PLA-PEG 45–5 kDa, 90% PLA Mw 45 kDa – 10% PEG Mw 20 kDa, 90% PLA Mw 49 kDa – 10% PEG Mw 5 kDa	20, 21, 24, 43, 44
PLGA-PEG 45–5 kDa, 45–20 kDa,	20,21, 24
PCLLA	49
Poly [H_2NPEGCA-co-HDCA]	13
Poly (N-acryloylamide)	16
Poly (vinylene)	16
PVM/MA	50
Gliadin	51, 52
Cellulose acetate	53
Emulsification-diffusion	
Methacrylic acid copolymer	54–57
PCL Mw 10 kDa, 14 kDa, 40 kDa, 42,5 kDa, 60 kDa, 80 kDa	54, 55, 58–66
PLA	54, 55, 67–71
PLGA 50:50, 65:35, 70:30, 75:25, 85:15 Mw 14.5 kDa, 45 kDa, 85 kDa, 137 kDa, 213 kDa	54, 56,69, 72–79
PHBHV Mw 23 kDa, 300 kDa	32, 33
Modified starch	80
PLGA-mPEG	81
Emulsification-coacervation	
Sodium alginate Mw 75 - 120 kDa	82, 83
Swine skin gelatin type II	84, 85
Poly (1,4 butadiene) (PB)-block-polystyrene (PS)-block-poly(ethylene oxide) (PEO) triblock terpolymer Mn 76 - 86 kDa	86
Emulsification-evaporation	
PLA	84,87, 88
PCL Mw 10 kDa	89
PLGA 75:25 Mw 10 and 92 kDa	90
PLGA 50:50 Mw 7, 14, 24, 48, 63 and 85 kDa; 65:35 Mw 97 and 114 kDa; 85:15 Mw 87 kDa)	78, 88
Dextran derivatives	91

(continued)

Table 5.1 (Continued)

Polymer	Reference
Double emulsification	
PLGA Mw 34 kDa	92
Polymersome preparation	
PCL-PEG	93–97
PLA-PEG	93, 94, 96–99
PB-PEG	93, 97, 98,100
Poly(FA/SC)-co-PEG	101
PEG-co-poly(FA/SC)-co-PEG	101
PTMC-PEG	96
PMPC-PDPA	102
PEO-PDPA	102
PB-b-PGA	103
CMPC	104
Supercritical fluid technology	
PLA Mw 10, 100 kDa	105, 106
Polymer coating	
Chitosan oligomers or medium molecular weight chitosan	18, 107
PCL Mw 40 kDa poly-L-lysine	18
PLA Mw 28 kDa/PLGA chitosan Mw Mn >50 kDa	108
Sodium alginate Mw 80 – 120 kDa – chitosan Mn 41 and 72 kDa	109
Layer-by-layer	
Chitosan Mw 250 kDa – sodium alginate Mw 70 kDa	110
PDDA Mw 200 kDa – sodium alginate Mw 70 kDa	110
Gelatin Mw 500 kDa – sodium alginate Mw 70 kDa	110
PAH Mw 15 kDa – PSS Mw 70 kDa	111, 112
PDDA – PSS	112
Chitosan Mw 400 g/mol – Lambda-carrageenan	113

Note: PACA: Poly (alkylcyanoacrylate) derivate; PCL: Poly (e-caprolactone); PLA: Poly (lactide); PLGA: Poly (lactide-co-glycolide); PCL-PEG: copolymer of poly (e-caprolactone) and poly (ethylene glycol); PLA-PEG: copolymer of poly (lactide) and poly (ethylene glycol); PGLA-PEG: copolymer of poly (lactide-co-glycolide) and poly (ethylene glycol); PCLLA: copolymer of e-caprolactone and L-lactide; poly (H2NPEGCA-co-HDCA): Poly [aminopoly(ethylene glycol)cyanoacrylate-co-hexadecyl cyanoacrylate]; PHBHV: poly (3-hydroxybutyrate-co-hydroxyvalerate); PB-PEG: copolymer of poly (butadiene) and poly (ethylene glycol); PEG-co-poly (FA/SC)-co-PEG: copolymer of poly (ethylene glycol) and poly (fumaric acid-co-sebacoyl chloride); PTMC-PEG: copolymer of poly(trimethylene carbonate) and poly (ethylene glycol); PMPC-PDPA: block copolymer (2-(methacryloyloxy)ethyl phosphorylcholine)-poly(2-(diisopropylamino)ethyl methacrylate); PEO-PDPA: poly (ethylene oxide)-poly (2-diisopropylamino)ethyl methacrylate; PB-b-PGA: poly (butadiene)-b-poly(g-L-glutamic acid); CMPC: cholesterol-end-capped poly (2-methacryloyloxyethyl phosphorylcholine); PDDA: poly(dimethyldiallylamide ammonium chloride); PAH: poly (allylamine hydrochloride); PSS: sodium poly (styrene sulphonate).

Furthermore, poly glycolic acid (PGA) and poly methyl methacrylate (PMMA), typically included in biodegradable nanoparticles formulation, have frequently been used to improve bioavailability and reduce the patients' risks of toxicity [5]. On the other hand, chitosan has already attracted increasing attention as a polymeric drug carrier because of its desirable properties such as biocompatibility, biodegradability, and biological activities [6, 7]. In addition to this, it has been reported that chitosan metabolic derivatives are biodegradable, biocompatible and poorly immunogenic.

The interactions between polymers and cells are known primarily to depend on the physical and chemical properties of the biomaterial surfaces, such as the microstructure, the surface wettability and the hydrophobic to hydrophilic ratio. The influence of the surface hydrophilicity or hydrophobicity on biocompatibility of polymers has been reported in many studies [8, 9]. Recent data highlight the impact of surface characteristics on nanomaterials' biocompatibility and point to the inadequacy of the current size-dependent mechanistic paradigms [10]. Bulk and surface modifications have been used to improve the compatibility of materials with the surrounding tissues as well as the interactions between two different materials [11].

Generally, the evaluation of a formulation's biocompatibility progresses through *in vitro* and *in vivo* phases. In many cases, the types of experiments involved are quite similar. Nonetheless, the specifics of the evaluation depend on the nature of the problems that can be anticipated given the compounds involved [12].

In vitro studies provide a rough assessment of the relevant cell-type survival in the presence of an investigated polymer and its derivatives. Numerous tests such as an MTT cell proliferation assay, measures of DNA synthesis and cell proliferation, and dye-based cell membrane integrity tests can be applied. For further investigation, the mechanism of toxicity can be addressed by looking at cell death or tissue injury markers [12].

In vivo studies, an essential tool for biocompatibility understanding, can vary considerably, depending on the context. A common assay for injected or implanted nanoparticles is hematoxylin and eosin staining of tissue and cell sections. These are generally excellent for looking at gross tissue morphology and tissue reaction, including inflammation [12]. There is a range of specialized stains, as well as immune-histo-chemical methods, for detecting specific biological changes. Epon-embedded sections or electron microscopy may allow detection of more subtle signs of tissue injury [12].

For further progress, it is important to address appropriately the nanoparticles' safety profile, including possible immunogenicity and inflammation reactions [114]. Given the increasing novelty of materials proposed for second-generation polymer therapeutics' development (with the increasing complexity of conjugate composition), and the growing debate as to the safety of nanomedicines, the need for the development of an appropriate regulatory framework is at the forefront of the scientific discussion. The adequacy of the current tests and models used to define safety is also constantly being reviewed [115].

In spite of numerous advantages, cytotoxic effects have been reported for polyester polymers, induced by the particles' nanometric size range which is conducive to internalization by cells (macrophages), and degradation inside the cell [116]. Furthermore, high tissue accumulation of non-biodegradable nanoparticles has caused toxicity problems. The toxicity and safety of non-biodegradable nanoparticles are subject to future research and more studies on the toxicological effects of nanoparticles on humans need to be done [117].

5.3 Encapsulation Techniques of Hydrophobic Drugs

5.3.1 The Nanoprecipitation Method

The nanoprecipitation method can also be called a solvent displacement or interfacial deposition method. But Reis *et al.* [118] described minute differences between solvent displacement and interfacial deposition, explaining as first forms the nanospheres or nanocapsules while the later is restricted only to nanocapsules. Nevertheless, both are based on the spontaneous emulsification of the organic phase into the aqueous phase. According to Fessi *et al.* [119], for the synthesis of nanoparticles, solvent and non-solvent phases are essential. This division of solvent or non-solvent phases is based on the dissolution of the polymer and can be termed the organic and aqueous phases respectively. This is because solvent mostly is an organic medium for the dissolution of polymer or active substances and the non-solvent phase consists of water.

The solvent phase consists of a single or a mixture of solvents (e.g. ethanol, acetone or dioxane) and in the same way the non-solvent phase can be a single or a mixture of non-solvents. Hence keeping in mind the miscibility, solubility and stability, two aqueous phases and two organic phases can be used. The solvent phase can contain the active substance to be encapsulated, the stabilizing agent and the polymer that can be natural, synthetic or semi-synthetic, while the non-solvent phase can contain one or more stabilizing agents.

The resulting particles will be used for the drug delivery purposes. Therefore, the selection of polymer should be based on biodegradability and biocompatibility. The most common polymers include poly e-caprolactone (PCL), poly (lactic acid) (PLA), poly (lactic-co-glycolic acid) (PLGA), Eudragit and poly (alkylcyanoacrylate) (PACA), etc. According to Khoee and Yaghoobian [120], synthetic polymers show greater reproducibility and higher purity over natural polymers and can be modified according to requirements, for example, to avoid recognition of nanocapsules by the mononuclear phagocyte system, some polymers can be copolymerized with PEG [43]. This discussion does not imply that biodegradable nanoparticles are only prepared using preformed polymers; it is possible with monomers or macromonomers passing through polycondensation reactions [121, 122]. To prepare nanocapsules via a nanoprecipitation method ranging in size from 150–200 nm, the composition is suggested in Table 5.2. However, in practical research, only a few of these are in use and among them different capric/caprylic triglyceride types, with the ability to solubilize most of the active substances, are frequently used as the oily phase. For w/o surfactants, sorbitan esters and phospholipids are considered. As a polymer solvent, acetone is mostly preferred while solvents such as ethanol are used for the dissolution of the active molecule or oil. The non-solvent can contain only water with a surfactant or a buffer solution. In preparation, the organic phase is added slowly by stirring into the aqueous phase and a colloidal suspension containing the nanocapsules is obtained. This simple method is presented in Figure 5.1.

Other techniques for nanoparticle preparation, such as the dialysis method, have been proposed, considering the fundamental principles of nanoparticle formation using the nanoprecipitation method [123–125].

The different parameters affecting the process are the stirring speed, the method and rate of organic phase addition to the aqueous phase, and the organic to aqueous phase ratio. However, the nature and the concentration of the substance to be encapsulated are

Table 5.2 *Different compositions used for nanoparticles preparation via different methods.*

	Nano precipitation[*1]	Emulsification-diffusion[*1]	Emulsification-coacervation[*2]	Emulsification-evaporation[*3]	Double emulsification[*4]	Polymersome preparation[*5]	Supercritical fluid[*6]	Polymer-coating[*7]	Layer-by-layer[*8]
Active substance (g)	0.1–0.25	0.1–0.5	≈0.07	≈ 0.00024	≈ 0.01	≈0.45	1–1.5	≈ 0.1	≈ 0.095
Polymer (g)	0.5–1.25	1.0–2.0	≈0.7	≈ 0.4	0.5–2	≈ 10	1–1.5	≈ 2	≈ 0.005
Oil (g)	2.5–5.0	2.5–5	N.A.	N.A.	N.A.	N.A.	N.A.	≈ 5	N.A.
Lipophilic surfactant (g)	0.5–1.25	N.A.	N.A.	N.A.	N.A.	N.A.	N.A.	≈ 1	N.A.
Salt (g)	N.A.	N.A.	≈10	N.A.	N.A.	N.A.	N.A.	N.A.	N.A.
Cross-linking agent (g)	N.A.	N.A.	≈0.7	N.A.	N.A.	N.A.	N.A.	N.A.	N.A.
Oxidizing agent (g)	N.A.	N.A.	≈4	N.A.	N.A.	N.A.	N.A.	N.A.	N.A.
Stabilizing agent (g)	1–2.5	8–20	N.A.	0.4–1.6	≈0.6	≈ 2.5	N.A.	≈ 1	N.A.
Water (q.s.; ml)	100	100	≈100	≈ 100	≈100	≈ 100	N.A.	≈ 100	N.A.

Notes:

*Estimated composition for 100 ml of nanoparticle dispersion after the total evaporation of the organic solvents and before any additional purification step. N.A.: non-applicable.

*1 Based on [161, 162].

*2 Based on [84].

*3 Based on [152].

*4 Based on [163].

*5 Based on [95].

*6 Based on [164].

*7 Based on [18].

*8 Based on [110].

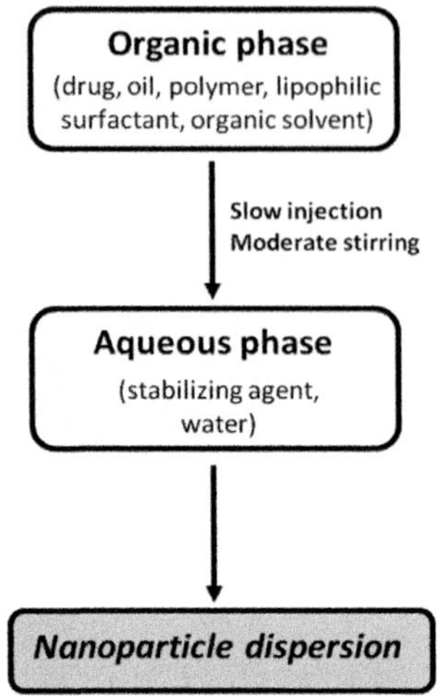

Figure 5.1 *The nanoprecipitation method.*

external to the process parameters that affect the characteristics of the resulting particles [126–129]. For a detailed understanding of the mechanism involved in the preparation of nanoparticles via the solvent displacement technique, some researchers have explained two types of effects: the Maragoni effect [130, 132], and the classical precipitation or ouzo effect [133, 134]. It can be called classical precipitation when nanospheres are prepared, and the ouzo effect refers to nanocapsule preparation. Most studies are based on an understanding of the physico-chemical principles and specially emphasize the type of solvent, the molecular weight of the polymer, its concentration, viscosity and the mixing speed of the polymer solution in the non-solvent phase and the interfacial tension of the non-solvent phase [135].

Cyclosporin A encapsulation with an entrapment efficiency of 98% is an excellent illustration of the interfacial deposition method application [136]. Spironolactone nanocapsules with an entrapment efficiency of more than 90% have been successfully prepared by Blouza *et al.* [137].

5.3.2 The Emulsification Methods

Some widely used techniques for nanoparticle preparation are the emulsification-based methods, which require the formation of a nanoemulsion before the particle formation. They include the techniques of emulsification-diffusion, emulsification-evaporation, emulsification-coacervation and double emulsification.

5.3.2.1 The Emulsification-Diffusion Method

The emulsification-diffusion method is one of the most commonly used methods for polymeric nanoparticles preparation. In a broader sense, this technique is a combination of the emulsification process and the diffusion process. This technique was utilized by Leroux *et al.* [54] for nanosphere preparation and then by Quintanar-Guerrero *et al.* [67] to obtain nanocapsules. Generally, both the organic and aqueous phases are necessary but Quintanar *et al.* [55, 138] suggested three phases are required, considering dilution by

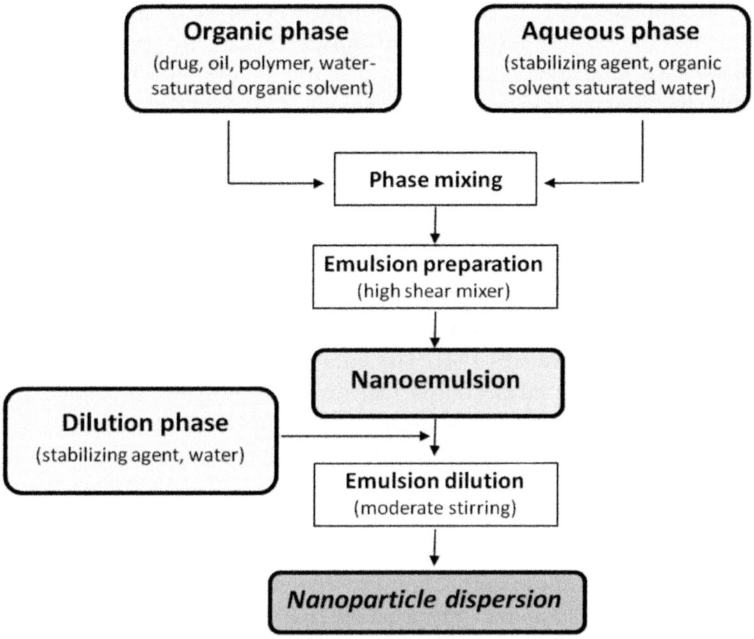

Figure 5.2 *The emulsion-diffusion method.*

water as a third phase, hence, these are termed the organic, aqueous and dilution phases. The organic solvent should be partially miscible with water and capable of dissolving the polymer, active substance, oil or any other compound (excipient) in the preparation. The aqueous phase normally contains a stabilizing agent, and, for dilution, water is used. Hence, preparation of o/w emulsion followed by dilution induces diffusion of the organic solvent, resulting in nanoparticles formation. Benzyl alcohol was used by Leroux *et al.* [54], propylene carbonate by Quintanar-Guerrero *et al.* [55] and ethyl acetate was considered best by Moinard-Chécot *et al.* [65]. A general recipe for nanocapsules preparation by the emulsification-diffusion method, with the resulting size between 150–200 nm is shown in Figure 5.2.

Solvents such as benzyl alcohol, propylene carbonate and ethyl acetate are used along with dichloromethane (DCM). However, poly vinyl alcohol (PVA) is the common choice of the stabilizing agent although poloxamer and ionic surfactants have also been used. Most of the polymers used are based on different capric/caprylic triglyceride types including PLA, PCL, eudragit, but poly (hydroxybutyrate-co-hydroxyvalerate) (PHBHV) is also considered.

To prepare nanoparticles using this technique, the water-saturated organic phase (dissolving all the required compounds) is first emulsified in an organic solvent-saturated aqueous phase containing a stabilizing agent. This emulsification process is accompanied by high energy stirring. Then more water (the dilution phase) is poured into this emulsion to induce the diffusion of the solvent in the external phase that results in the formation of nanoparticles. Finally, according to the boiling point of solvent, it is eliminated by distillation, filtration, or by evaporation in some cases.

According to Guinebretière [139] and Moinard-Chécot *et al.* [65], in the emulsion formation process, the shear rate, the oil to polymer ratio, the polymer concentration in the organic phase, and the chemical composition of the solvent phase are the key factors affecting the final particle size. This technique is used for drug-loaded nanoparticles preparation and some efficiently encapsulated drugs applying this method include coumarin [140], cyclosporine A [141] and indocyanine [142]. Plasmid DNA was also encapsulated by this technique [143].

5.3.2.2 The Emulsification-Coacervation Method

This method uses the naturally occurring polymers, i.e., gelatine and sodium alginate, although some synthetic polymers can also be used, such as carboxy methyl cellulose (CMC) and polystyrene derivatives [54]. The emulsion-coacervation method is based on emulsion formation and then coacervation that results in polymer precipitation as a thin layer on the template forming the nanocapsules. Then stability is achieved by the addition of a cross-linking agent or by altering the temperature or the pH, which results in physical intermolecular or covalent cross-linking. The active agent is dissolved in a solvent and then a poor solvent/non-solvent is added to induce coacervation based on the phase separation [144]. In the o/w emulsification step, the organic phase contains oil, an active substance and, if necessary, some other organic material (e.g. an active substance solvent) while the external aqueous phase consists of polymer and a stabilizing agent in water. The coacervation process can be done by electrolyte addition [83,109], temperature modification [86], or by the addition of a dehydrating agent [84], and is finalized by a cross-linking step to create a stable aqueous dispersion of the nanocapsules. Nonetheless, the work of Lutter *et al.* [86] is normally used as the principal emulsion-coacervation technique. The general concept of this method is explained in Figure 5.3.

The coacervation step is the most important one in this technique and is investigated in detail to establish the pH effects,the electrolyte effects and the change in polymer type such as cellulose acetate trimellitate [145], ethylcellulose [146,147] and polyvinyl chloride [148]. This type of research leads to the emulsion-coacervation technique having nanocapsules sized in nanometres. The use of a template and a technique to form a concentrated, colloid-rich dispersed phase to cover the template can be advantageous to form nanoparticles. The polymers are precipitated on the core material and these core nanoparticles are then dispersed in another solvent, mostly an aqueous one. According to Gander *et al.* [149], the polymer functional groups are solvated by the water molecules. Coacervating agents are added to avoid solvation, inducing a thin solvated shell and allowing the attraction among functional groups to form a network with increasing polymer concentration.

As discussed above, temperature modifications, electrolytes, or dehydrating agents are used for the completion of the coacervation process, so in the case of temperature modification to induce polymer precipitation, the Flory-Huggins theory on interaction parameter χ is important. According to this theory, if the interaction parameter value is less than a specific value, i.e., the critical value χ_c, the polymer will dissolve in solvent and this χ_c value depends on the degree of polymerization of polymer. Similarly, for electrolytes' efficiency in the desolvation process, the Hofmeister or lyotropic series is followed, which is responsible for arranging the ions according to their capacity to immobilize water molecules in solvation. Yin *et al.* [150] performed a practical study on polymer coacervation behavior with respect to the lyotropic series. The electrolytes used in the polymer desolvation can

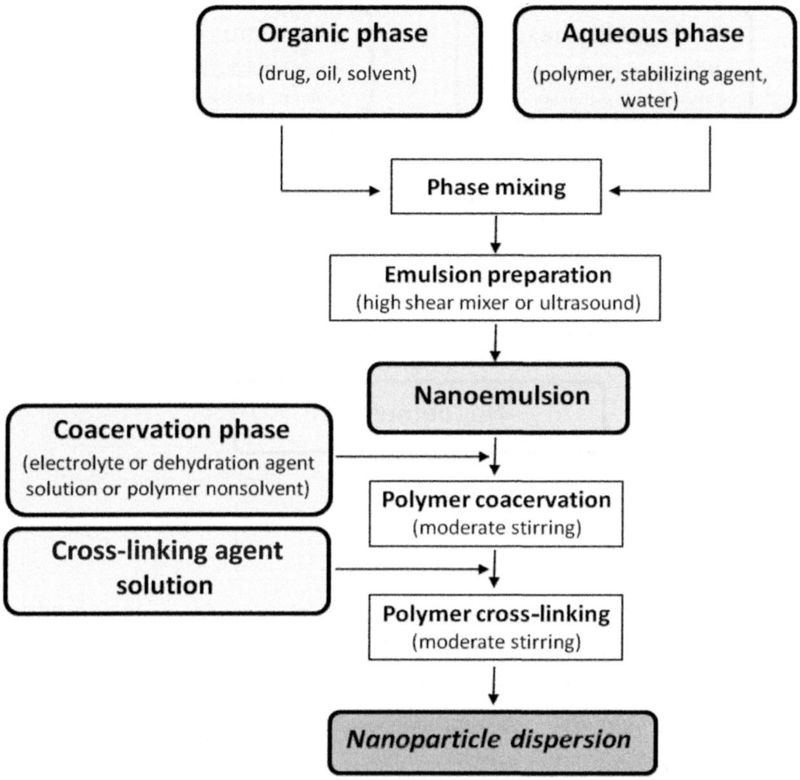

Figure 5.3 *The emulsion-coacervation method.*

be designed to salt out. In the case of the dehydrating agent, the polymer concentration increase is due to the solvent-solvation competition process that leads to the desolvation of the polymer chains, resulting in phase separation. However, other factors such as the pH change or the addition of other incompatible materials can also be used to reduce polymer solvation.

Capsaicin, a poorly soluble active agent, obtained from red hot peppers, was successfully nanoencapsulated by Xing *et al.* using the coacervation method [151].

5.3.2.3 The Emulsion-Evaporation Method

The emulsion-evaporation method is another widely used technique for nanoparticles preparation, particularly by Pisani *et al.* [152]. It is also known as the solvent evaporation method. It is based on the nanoemulsion preparation followed by evaporation of the organic solvent, leading to the point of insolubility and the precipitation of the polymer-engaging active material. The aqueous phase outside behaves as the dispersion medium, therefore, there is involvement of the two phases; aqueous and organic. For evaporation, simple evaporation by stirring or evaporation under reduced pressure may be used, depending on the nature of the solvent. The factors affecting the size of the particles can be the stirring time and

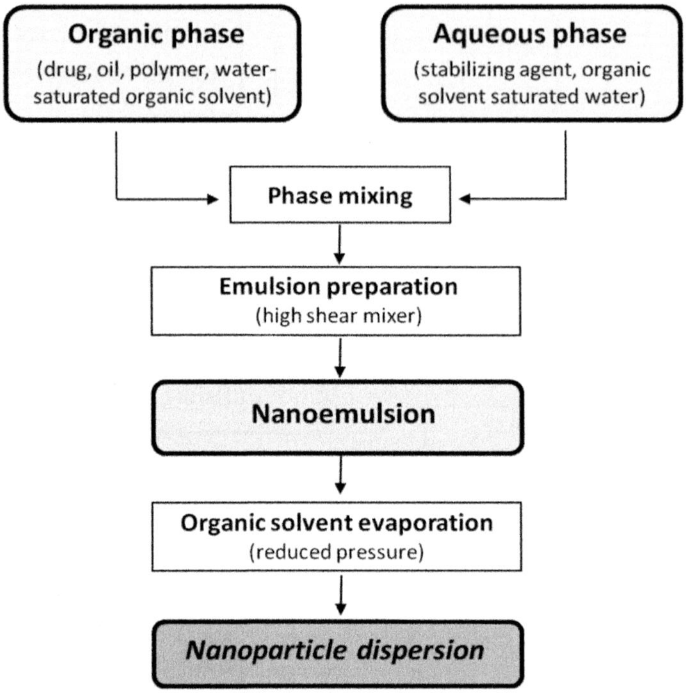

Figure 5.4 *The emulsion-evaporation method.*

the stirring speed for the emulsification, and the nature and concentration of the stabilizing agent. A simple schematic representation is given here in Figure 5.4.

The main drawback of this technique is the susceptibility of the formation of different interfaces in the organic and aqueous phases, which can result in the restriction of the solvent diffusion in the presence of the polymer, hence creating a difficulty in nanocapsules formation during the rapid solvent removal via evaporation [152]. However, Moinard-Chécot *et al.* [65] noted that resistance in evaporation can be explained as mechanical stress inside the emulsion due to the formation of gas bubbles. Nanoparticles of docetaxel were prepared by Quaglia *et al.* [153] using the emulsion-evaporation method.

5.3.2.4 The Multiple Emulsions Method

Multiple emulsions are the emulsification of already prepared emulsions. However, in this multiple emulsion technique, the double emulsion technique (Figure 5.5) is praiseworthy, allowing the encapsulation of both types of materials: hydrophilic and lipophilic. According to researchers [154,155], double emulsion is considered the complex heterodisperse system and is called the 'emulsions of emulsions'. On the basis of two phases, aqueous and organic, the double emulsion can be classified into two categories: oil-water-oil emulsion (o/w/o) and water/oil/water (w/o/w) emulsion. For the encapsulation of hydrophobic drugs, both categories are used, depending upon the nature of the drug and the choice of starting materials. Of these double emulsion methods, the most common is the double emulsion

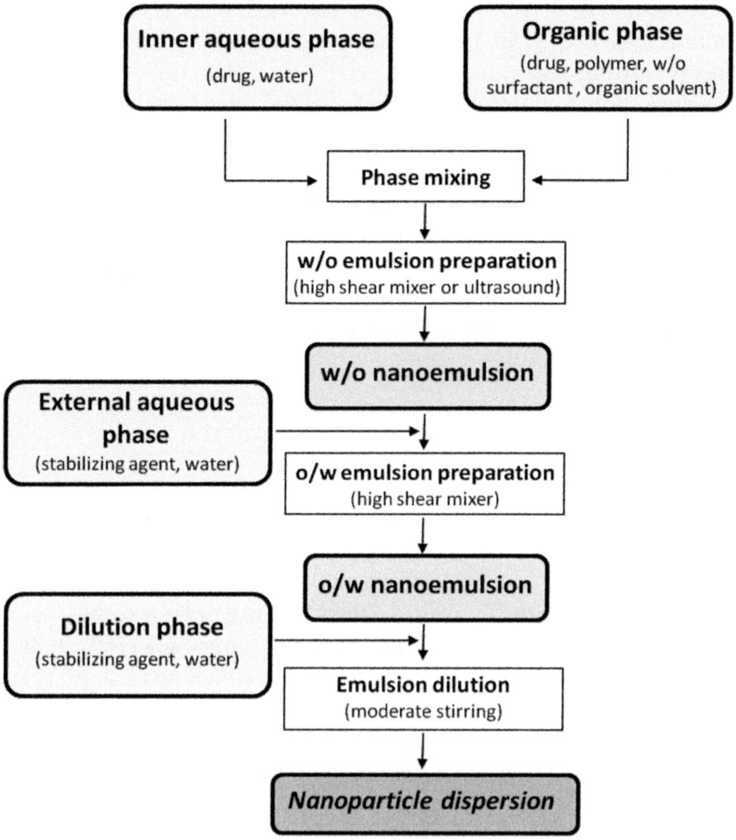

Figure 5.5 *The double emulsion method.*

evaporation method, in which a primary emulsion is prepared that is emulsified again by another phase to prepare the secondary emulsion, and after this evaporation takes place. For example, in the case of w/o/w emulsion, a primary emulsion is prepared by emulsifying the aqueous phase in the organic phase, this primary emulsion is again emulsified in the aqueous phase to form w/o/w double emulsion and finally the organic phase is evaporated. Hence, the contents of the organic phase (mostly polymer) act as the film between the two aqueous phases in the particles, and that film can also act as a barrier for release of the drug from the particles [120].

Like other encapsulation methods, the polymer choice accounts for its biodegradability and biocompatibility profile. These emulsions are normally prepared with two stabilizing agents, a hydrophobic and hydrophilic one, but in some cases, a single stabilizing agent is sufficient, normally in the second emulsification phase. For evaporation, simple evaporation by stirring or evaporation under reduced pressure is used, depending on the nature of the solvent. Different experimental variables affecting the physico-chemical characteristics of the final particles are: the ratio between phase volumes, the stirring speed, the stirring time, the nature and concentration of both polymer and stabilizing agents, the preparation temperature, the method of solvent evaporation, etc. This method was earlier used for the preparation of microparticles [156, 157] but later was applied to nanoparticles preparation using high energy stirring techniques [158–160]. The advantage of this technique is its ability to encapsulate both hydrophilic and hydrophobic drugs, thus other constituents like polymer, solvent and stabilizing agents, etc. can be selected accordingly, depending upon the nature of the drug.

Different polymers used in this technique include PLA, PCL and PLGA, and as stabilizing agents, polysorbates and PVA are studied. In the case of solvents, ethyl acetate, DCM and chloroform are the common choices. However, ethyl acetate is preferred because of its less toxicological effects and smaller particle size as compared to that of DCM and chloroform. Table 5.2 shows the different compositions for nanocapsules preparation via the double emulsion method.

5.3.3 Polymersome Preparation

Another method for nanoparticles preparation is the polymersome preparation for drug delivery and diagnostic purposes. Polymersome is a kind of artificial vesicle, enclosing a solution, made of amphiphilic block copolymers and forming a vesicle membrane that recalls the structure of lipids in cell membranes [94]. The amphiphilic block copolymers have been widely studied for drug delivery systems because of their self-assembling ability in aqueous solutions, and polymersome is developed on same idea [165]. Polymersome is important as artificial cells, having a large compartment which gives stability properties, a tunable membrane and the encapsulation of biofunctional compounds [94].

The copolymers used are poly (2-methylaxone) derivatives, PEG (polyethylene glycol) or PEO (polyethylene oxide) derivatives and cholesterol derivatives along with some new polypeptides [99,104,166,167]. Nevertheless, other polymers applied in polymersome formulation include diblock copolymers, such as, poly(ethyl ethylene)-*b*-poly(ethylene glycol) (PEE-PEG), poly(lactic acid)-*b*-poly(ethylene glycol) (PLA-PEG), poly(ε-caprolactone)-*b*-poly(ethylene glycol) (PCL-PEG), poly(*N*-isopropylacrylamide)-*b*-poly(ethylene glycol) (PNIPAAM-PEG), poly(styrene-*b*-acrylic acid) (PS-PAA), poly(2-vinylpyridine)-*b*-poly(ethylene glycol) (PV2P-PEG), and polybutadiene-*b*-poly(glutamic acid) (PBD-PGIA), and also triblock copolymers including poly(ethylene glycol)-*b*-poly(propylene sulfide)-*b*-poly(ethylene glycol) (PEG-PPS-PEG). Graft copolymers and branched polymers have also been described for polymersome formulation [168].

The techniques used for polymersome preparation are similar to that of liposomes in which different structures of the block copolymers are used. For the methods applied, Kita *et al.* [169] described two main divisions, i.e., the solvent-accompanied method and the solvent-free method. The solvent-free method is without any organic solvent, hence for the preparation amphiphile in dry state contacts the aqueous medium and is hydrated subsequently. In the solvent-accompanied method, block copolymer is dissolved in a specific organic solvent and mixed with water, then removal of solvent is done by evaporation or any other technique. As it is difficult to remove the solvent completely, there is the possibility that residual solvent can cause problems in biological and pharmaceutical applications. A number of factors, including both the structure of amphiphile (the chemical constitution and the relative lengths of the individual blocks) and the properties of the solution (such as the concentration, the pH, the temperature and the solvent) can control the size and shape of

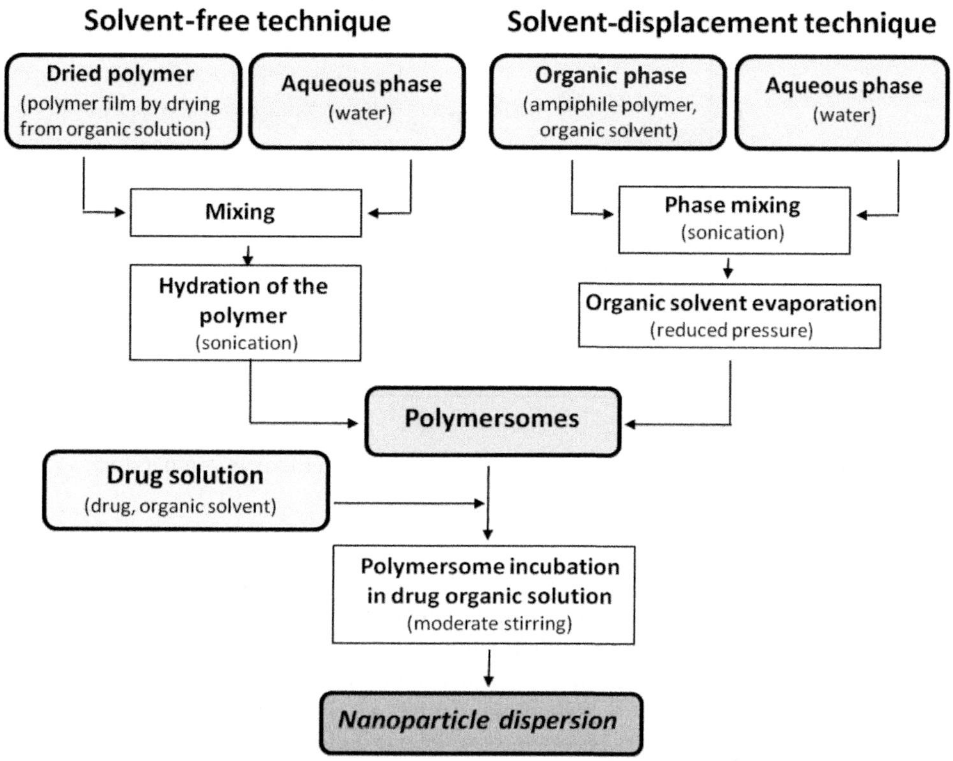

Figure 5.6 *The polymersome preparation.*

nanocapsules in this process [169]. Figure 5.6 shows the polymersome preparation method, including both solvent-free and solvent-displacement techniques.

In general, the active substance encapsulation in polymer vesicles is done by using techniques based on incubation, and the choice of the polymer is based on the nature of the active molecule. However, both hydrophilic and lipophilic molecules can be encapsulated. According to Ahmed *et al.* [98], polymersome can be used in the controlled release of different drugs against tumors, offering enhanced permeability, retention effect and high drug-loading efficiency as compared to liposomal formulation. The encapsulated active molecules are mostly anticancer agents, e.g. paclitaxel [170, 171]. The combination of doxorubicin (a hydrophilic active molecule) and paclitaxel into a polymersome exhibited greater effect in reducing tumor growth due to the increased synergistic effect and less cardiac cytotoxicity [155]. Other examples include the development of biotin functionalized leuko-polymersome [172] to monitor or treat inflammation, cancer and cardiovascular disease, and Tat peptide-conjugated polymersomes [173] for cellular trafficking. Early developments in polymersome application were focused on parenteral and oral administration with controlled release of active molecules. However, recently the drug targeting applications by ligand incorporation within the membranes, biosensing applications and other routes of administration are being explored [174, 175].

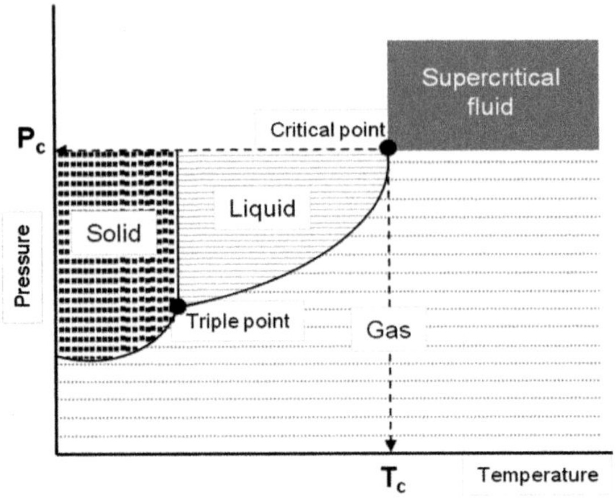

Figure 5.7 Phase diagram of supercritical point via pressure vs. temperature.

5.3.4 Supercritical Fluid Technology

Nanosized particles can be prepared by using the supercritical fluid technique. Any liquid or gas can be used as a supercritical fluid (SCF), although the most commonly used are carbon dioxide and water. SCF can be defined as any substance above its critical temperature and pressure (i.e., $T > T_c$ and $P > P_c$) where no distinct gas and liquid phases exist and the physical properties of SCF exist between those of gas and liquid. The phase diagram of SCF is given for a better understanding of the phenomenon (Figure 5.7).

SCFs are used as organic solvent substitutes in the laboratory as well as on an industrial scale, as they are harmless to the environment. Like other fields, SCF-based technology is a new potential technology for nanoparticle preparation and has been adopted by many researchers on nanoparticles involved in drug delivery purposes. These methods are preferred because of environment-friendly solvents and the rapid unistep process. Nanoparticles can be obtained without traces of organic solvents, hence with high purity. SCF technology is a good choice compared to existing nanoparticles preparation methods, as it provides an opportunity to control the particle size and morphology in a more efficient manner [175–177].

The preparation of nanoparticles using supercritical technology can be further divided on the basis of the method used and this has been done by various authors [178–180], but the important study is by Mishima [175] who divided all the processes into four general methods, depending on whether SCF was used as a solvent, a solute or an antisolvent.

1. Rapid expansion of supercritical solution (RESS).
2. Particles from gas-saturated solution (PGSS) process.
3. The gas antisolvent (GAS) process.
4. The supercritical antisolvent (SAS) process and its various modifications.

Explanation of all the processes will not be given here, only the general technique will be discussed. As a summary of the literature [175, 178–181], a mixture of polymer, drug and other excipients (if present in the recipe) are put in a vessel and SCF is passed through this vessel to absorb/dissolve the drug and pressure treatment (expansion of the solution) is

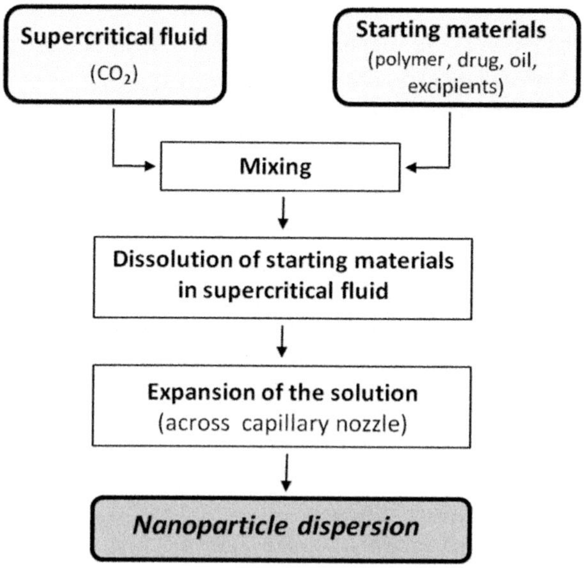

Figure 5.8 *The supercritical fluid technology technique.*

carried out. Then nanoparticles are obtained in a vessel as precipitate after passing the solution through a capillary tube. The general scheme of this process is explained in Figure 5.8.

The characteristics of the nanoparticles prepared by the supercritical fluid technology can be influenced by the solute properties (drug, polymer and other excipients, if present), the parameters used in the technique (temperature and pressure applied, solute and solvent flow rate, nozzle geometry) and, most importantly, the nature and type of SCF used. The disadvantages of SCF technology are the need for special equipment requirements which is difficult for every laboratory, the involvement of processes like extraction, partitioning and expansion of solvents that may cause porosity, resulting in the quick release of drugs and low encapsulation efficiency [175], and last but not least, the ability of SCF to dissolve only a small number of hydrophobic drugs and polymers [182].

5.3.5 The Polymer-Coating Method

Polymer coating is a relatively new technique. Actually coating is the deposition of a new layer on the particles and in the case of nanoparticles for biological applications, this is used to acquire stability, good solubility or specific functionality for good targeting. In the surface engineering of these nanoparticles, both inorganic and polymeric materials [183, 184] are used. Polymers used for coating purposes can easily be classified as natural and synthetic ones, although some polymeric surfactants are also found for coating but here the focus is on the polymer used in the coating technique. The natural polymers used are gelatine, chitosan and dextrane, etc. [185–187] while synthetic polymers include poly (lactic-*co*-glycolic acid) (PLGA), poly (vinyl alcohol) (PVA), poly (vinylpyrrolidone) (PVP), poly (ethylene-*co*-vinyl acetate), and poly (ethylene glycol) (PEG) [188, 189].

The mechanism of deposition of this new layer on the particle surface can be considered as a result of van der Waals forces, hydrogen binding, or other forces for adsorption of the polymer on the particle surface, depending on the nature of the surface and the polymer. Different propositions are given to prepare nanoparticles using the polymer coating technique. According to Calvo *et al.* [18], this can be done by putting non-coated nanoparticles in a stirred polymeric dispersion for a specific time and hence a thin layer of polymer is obtained. This can be done by a small modification of the traditional nanoparticle preparation methods (e.g. nanoprecipitation) by adding a polymer to the external aqueous medium, leading to charged particles' precipitation and diffusion of the solvent [18, 108]. However, Prego *et al.* [107] mentioned the preparation of a nanoemulsion template and the addition of the polymer in a continuous phase for its precipitation in droplets, and then solvent diffusion and polymer coacervation as shown in Figure 5.9.

It is impossible to present the general criteria and starting materials for preparing nanoparticles using this technique although some references on nanocapsules preparation by different polymer coating methods are given in Table 5.3.

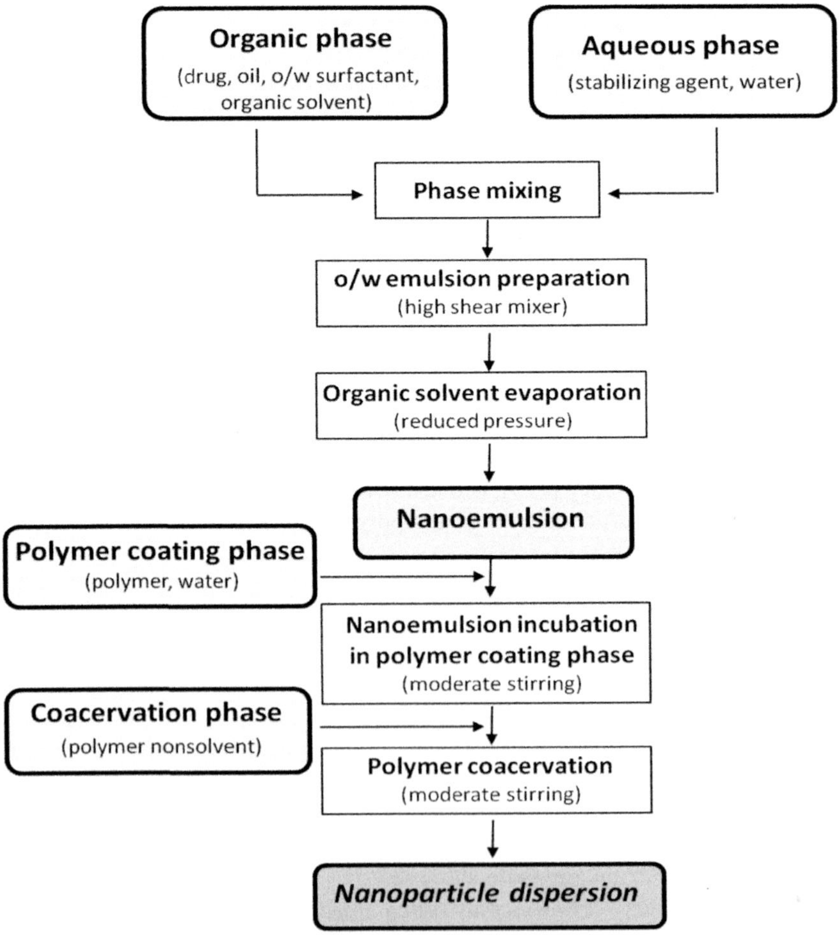

Figure 5.9 *The polymer coating method.*

Table 5.3 *Examples of mean size and zeta potential of nanoparticles prepared by different methods.*

Active ingredient	Nanoparticle type	Mean diameter range (nm)	Zeta potential (mV)	Reference
Nanoprecipitation				
All-trans-retinoic acid	Nanosphere	70–460	nr.	253
Atovaquone	Nanocapsule	200–270	nr.	25,37
Benzathine penicillin G	Nanocapsule	130–230	nr.	45
Cyclosporin A	Nanosphere	110–215	nr.	254
Dexamethasone	Nanocapsule	270–320	nr.	34
Diclofenac	Nanocapsule	120–420	nr.	14, 35
	Nanosphere	85–195	nr.	14
Flubiprofen	Nanosphere	150–290	−25 to −30	255
5-Fluorouracil	Nanosphere	75–255	nr.	256
Gemcitabine derivatives	Nanocapsule	180–300	nr.	13
	Nanosphere	150–190	+25 to +35	
Griseofulvine	Nanocapsule	380–400	nr.	27
	Nanosphere	250–325	nr.	
Ibuprofen	Nanosphere	105–145	nr.	57
Indomethacin	Nanocapsule	120–450	−35 to −50	18, 22, 23,28, 34, 36
	Nanosphere	170–180	nr.	22
Indomethacin ethyl ester	Nanocapsule	230–295	−45	28, 30
Isradipine	Nanosphere	200–210	−20 to −35	257
Melatonin	Nanocapsule	185–295	−30	15
Oridonin	Nanosphere	105–195	~ -15	258
Paclitaxel	Nanosphere	115–160	−20 to −35	259
Primaquine	Nanosphere	150–165	−35 to +1.5	260
Primidone	Nanocapsule	222	nr.	19
Progesterone	Nanocapsule	225–275	nr.	34
Rifabutine	Nanocapsule	200–500	nr.	37
Spironolactone	Nanocapsule	310–410	nr.	29
Tacrolimus	Nanocapsule	350–400	nr.	17
Tamoxifen	Nanosphere	180–800	−15 to 800	261
Taxol	Nanocapsule	240–280	nr.	34
^{99m}Tc-HMPAO complex	Nanocapsule	240–360	−60	44
Tretinoin	Nanocapsule	220–230	−5 to −7	31
Usnic acid	Nanocapsule	140–290	−30	262
Vitamin E	Nanocapsule	260	nr.	26
Vitamin K	Nanocapsule	240–310	nr.	34
Xanthone	Nanocapsule	265–300	−40	46

(continued)

Table 5.3 (Continued)

Active ingredient	Nanoparticle type	Mean diameter range (nm)	Zeta potential (mV)	Reference
Emulsification-diffusion				
17-b-estradiol benzoate	Nanosphere	75–350	nr.	72
Chlorambucil	Nanocapsule	330–335	nr.	131
Clofibrate	Nanocapsule	365–375	nr.	
Cyclosporine	Nanosphere	60–270	nr.	77
Enalaprilat	Nanosphere	180–615	−30 to −60	56
Estradiol	Nanocapsule	335–345	nr.	131
	Nanosphere	95–650	+70 to +95	75, 78, 79,
Eugenol	Nanocapsule	320–325	−6	66
Hinokitol	Nanocapsule	225–600	nr.	64
Ibuprofen	Nanosphere	310–430	nr.	57
Indomethacin	Nanocapsule	310–520	nr.	59, 131, 139
4-Nitroanisole	Nanocapsule	550–600	nr.	68
Progesterone	Nanocapsule	505–515	nr.	131
Vitamin E	Nanocapsule	355–365	nr.	
Tacrolimus	Nanosphere	215–220	−20 to −30	81
p-THPP	Nanosphere	90–160	−4 to −8	69
Triclosan	Nanosphere	175–450	nr.	71
Emulsification-coacervation				
$HAuCl_4$	Nanocapsule	185–219	nr.	86
Hydrocortisone	Nanosphere	150–230	−4 to −12	92
Triamcinolone acetonide	Nanocapsule	300	nr.	84
Turmeric oil	Nanocapsule	195–470	−17	83
Double emulsification				
Triptorelin	Nanosphere	350–600	nr.	163
Polymersome preparation				
NC-1900	Nanocapsule	90–110	−21	95
Paclitaxel	Nanocapsule	80–120	nr.	98
Supercritical fluid technology				
Cefurexime axetil	Nanosphere	<1000	nr.	164
Dexamethasone acetate	Nanosphere	500–1000	nr.	175
Polymer coating				
Indomethacin	Nanocapsule	245–255	+30	18
Salmon calcitonin	Nanocapsule	255–270	+35	107
Layer-by-layer				
Artemisinin	Nanocapsule	800	−30	110
Tamoxifen	Nanocapsule	100–120	−20 to 50	112

Note: nr.: Non-reported data; ^{99m}Tc-HMPAO complex: Technetium-99m D,L-hexamethyl propyleneamine oxime; HAuCl4: Hydrogen tetrachloroaureate; NC-1990: Vasopressin fragment analog; p-THPP: Mesotetra(p-hydroxyphenyl)porphyrin.

5.3.6 The Layer-by-Layer Method

The idea of nanoscale objects assembling by electrostatic attraction in a layer-by-layer fashion was first coined by Iler [190], however, work by the following researchers [191–196], especially the work done by Sukhorukov *et al.* [197] for biotechnological applications significantly progressed this technique. Almost every type of macromolecular species, including inorganic molecular clusters [198], nanoparticles [199], nanotubes and nanowires [200, 201], nanoplates [202], organic dyes [203], organic nanocrystals [204], dendrimers [205], porphyrin [206], polysaccharides [207, 208], polypeptides [209], nucleic acids and DNA [210], proteins [211, 212], and viruses [213] have been successfully used as assembly components [214]. Remarkable versatility has further led to a number of novel designs and applications, such as superhydrophobic surfaces [215, 216], chemical sensors and semi-permeable membranes [214, 217–219], drug and biomolecules delivery systems [214, 220, 221], memory devices [222] optically active and responsive films [223–225], cell and protein adhesion resistant coatings [214, 226], fuel cells and photovoltaic materials [227], biomimetic and bio-responsive coatings [228], semiconductors [229, 230], catalysts [231, 232], magnetic devices [233, 234], and many more [214, 235].

The mechanism of this newer technique for nanocapsules preparation is described as an irreversible electrostatic attraction, resulting in polyelectrolyte adsorption which needs molecules with a sufficient number of charged groups in order to allow a good adsorption on the opposite surface. The alternation of the negative and positive surface charges results in the formation of a continuous assembly between opposite charged materials, providing freedom in a number of layers and their sequence. However, the method is similar to polymer coating except it requires multiple coatings on the template. According to Radtchenko *et al.* [236], adsorption of the polymer layer on the template is done by its incubation in polymer solution or by decreasing the polymer solubility using a miscible solvent.

According to Chen *et al.* [110], like inorganic particles, the solid form of an active agent to encapsulate can act as a template for the layer-by-layer deposition of the polymer. Adsorption can occur by subsequent core dissolution and then hollow nanocapsules can be loaded with the required substance [237–242]. The general mechanism is explained in Figure 5.10.

Radtchenko *et al.* [236] mentioned penetration of molecules in the polyelectrolyte multilayers; small solutes can easily penetrate, but for large macromolecules it is difficult task. Large macromolecules will be present only inside capsules, resulting in a difference of physicochemical properties between the bulk and capsule interior; a polarity gradient will be established across the capsule wall, a useful step for the precipitation of hydrophobic molecules. On the same lines, Antipov *et al.* [238] and Ai and Gao [240] explained the pH effect on the permeability properties of hollow polyelectrolyte nanocapsules and the temperature effect on the reversible behavior of the nanocapsules. Preetz *et al.* [113] have mentioned that the nanocapsules having an oily core covered with polyelectrolytes were designed for the incorporation of lipophilic drugs and prepared by high pressure homogenization. The different polyelectrolyte layers were deposited as in the complex coacervation process and then nanocapsules dispersion was treated by homogenization and centrifugation finally. The advantage over other common methods mentioned by Preetz *et al.* [113] is the deletion of the separation step.

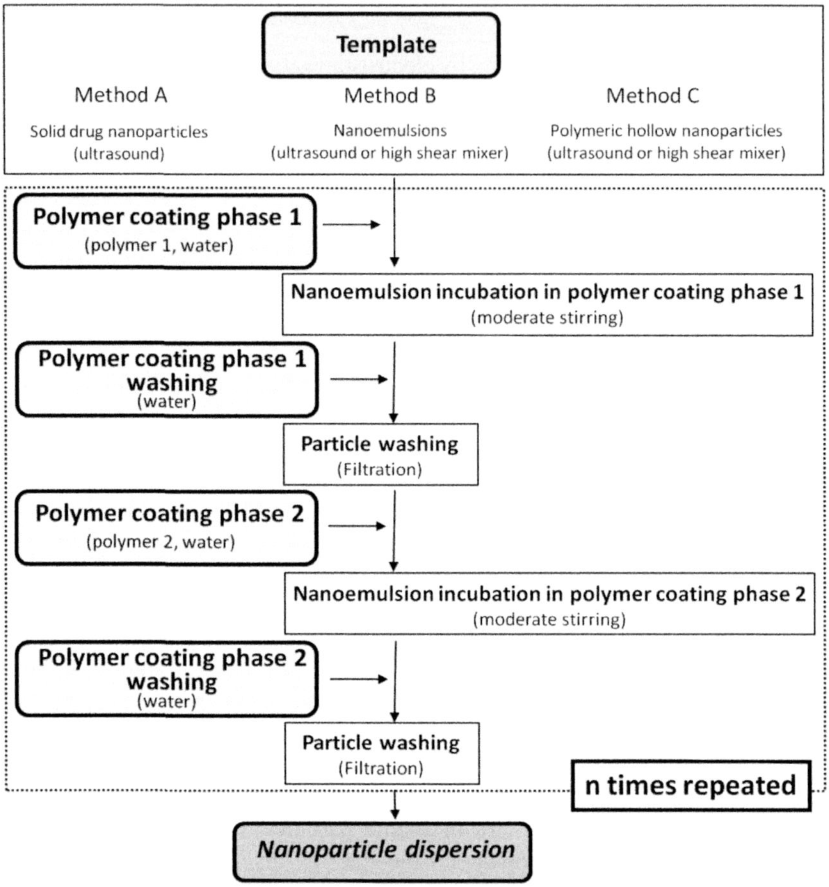

Figure 5.10 *Schematic representation of the polymer coating method with the template.*

Small particles are obtained by ultrasonic sonication of the suspension to decrease the size of the individual drug particles, which can be further stabilized by coating, using ultrasonication and hence thin polyelectrolytes are assembled on their surfaces [112]. The other difficulties include formation of contraion aggregates, separation of free polyelectrolytes from the particles before the next layer's deposit, and polyelectrolyte bridging during centrifugation. According to Sukhorukov *et al.* [197], a close particle–particle encounter can result in unnecessary interaction with films, leading to aggregate formation and film destruction.

As mentioned above, the essential point about this method is the choice of polymers having a sufficient number of charged groups, utilizing polyanions e.g. poly (styrene sulfonate) (PSS), sodium alginate, poly (acrylic acid), dextran sulfate, carboxymethyl cellulose, hyaluronic acid, gelatin A, chondroitin heparin and polycations, e.g. polylysine, chitosan, gelatin B, poly (allylamine) (PAA) poly (ethyleneimine) (PEI), amino-dextran and protamine sulphate [112, 243, 244]. Examples of hydrophobic drugs encapsulated include paclitaxel tamoxifen, and curcumin [112, 245–247].

5.4 Behavior of Nanoparticles as Drug Delivery Systems

This section discusses some nanoparticles' properties including size, zeta potential, drug encapsulation efficiency, release of the active ingredients and performance related to the above-mentioned preparation methods.

5.4.1 Mean Size

The particle size is an important characteristic of any particulate system, especially one used in pharmaceutical applications. The importance increases in view of *in vitro* and *in vivo* applications. Table 5.3 shows some examples of the size of particles prepared via the different methods discussed. According to Hans *et al.* [248], the molecular weight and concentration of the polymer have effects on the size of the nanoparticles but targeted size depends on the purpose of the delivery system, e.g. if the goal of administration is rapid dissolution or arterial uptake, then approximately 100 nm particles are suitable but if the purpose is targeting the mononuclear phagocytic system (MPS) or prolonged dissolution, then 800 nm particles are preferred, as the drug release kinetics depends on mean size and size distribution [249]. In different nanoparticles preparation techniques, several factors can be modified to obtain the required nanoparticle size as well. The most important of these factors are the nature and concentration of the polymers and surfactants, the solvent choice, the stirring rate (when emulsification is involved), and the compatibility and ratio of external/internal phase, but there are contradictory views on the nature and concentration of the drug to be encapsulated. Different methods used for size measurement include dynamic light scattering (DLS), transmission electron microscopy (TEM), scanning electron microscopy (SEM) and atomic force microscopy (AFM).

5.4.2 Zeta Potential

Zeta potential gives an idea of the charge of the particle; the higher the zeta potential, the higher the value will be the charge on the particle surface. Zeta potential is important in characterizing nanoparticles because the stability, adhesion and rheological properties of colloids depend on their zeta potential value [250] and indirectly on the repulsive and attractive forces that a colloidal particle experiences. It gives an idea about the potential at the diffused double layer [251] but is unable to explain completely the phenomenon of coagulation. However, it is considered important by researchers to characterize the nanoparticles, and more descriptive behavior on coagulation is related to critical coagulation concentration (c.c.c.) along with a consideration of slow coagulation regions [252].

In nanocapsule preparation, the zeta potential depends on the nature of the polymer and the stabilizing agent on the medium pH values. Hence, we are unable to explain any general trend for the zeta potential of nanoparticles prepared by the above-mentioned methods because of the unknown pH and salinity of the solution. Some examples of the zeta potential of particles are included in Table 5.3. Generally, in water and other polar solvents, the adsorption of anionic surfactants induces a negative zeta potential, and the adsorption of cationic surfactants induces positive values, while in hydrocarbons and other non-polar solvents, the case is reversed and anionic surfactants induce a positive zeta potential, and cationic surfactants induce a negative zeta potential [263]. According to Mora-Huertas *et al.* [162], zeta potential values are dependent on the adsorption of the stabilizing agent on the

Table 5.4 *General tendencies in drug encapsulation by different methods of nanoparticle preparation.*

	Encapsulation efficiency (%)	
Nanoparticle preparation method	Nanocapsules	Nanospheres
Nanoprecipitation	50–100	20–100
Emulsification-diffusion	30–100	20–85
Emulsification-coacervation	~45	~40
Double emulsification	n/a	480
Polymersome preparation	~20	–
Supercritical fluid technology	–	n/a
Polymer coating	40–90	n/a
Layer-by-layer	85–100	n/a

Note: *n/a = not available. It is not possible to prepare nanoparticles by this method.

nanocapsule surfaces and hence can be expressed as non-ionic stabilizing agents, which give negative zeta potential values with polyester polymers/methacrylate derivatives and positive values with cationic polymers. They further explained that negative zeta potential values are obtained by using negative charged stabilizing agents, and positive values are obtained by using positive charged stabilizing agents.

5.4.3 Encapsulation Efficiency

Encapsulation efficiency is one of the important criteria for the characterization of the drug delivery systems. It evidences the success of a selected technique for the encapsulation of the active ingredient. There are various factors associated with both the preparation and the starting materials used for preparing nanoparticles that can affect the drug entrapment efficiency. As can be seen, drug encapsulation data is variable, even when using the same methods. Then it further depends on the operational conditions involved, such as the volume of the organic solvent [19], the volume of the aqueous phase [29] or the active substance concentration [46] and the type of nanoparticles prepared. The encapsulation efficiency of nanocapsules and nanospheres obtained by the above-mentioned different preparation method is given in Table 5.4.

However, taking particular examples of nanoparticles prepared via the nanoprecipitation technique and the emulsification-diffusion method, it seems that nanocapsules render the most efficient encapsulation rates, as mentioned in Table 5.5.

Regardless of the technique and operational conditions used, molecule properties are relevant for the successful encapsulation of active substances. They include drug solubility in aqueous and organic media, the drug stability and the drug–polymer interactions [178]. As mentioned above, nanoparticle preparation methods often require organic and aqueous phases. Usually hydrophobic drugs and the polymeric matrix are dissolved in organic solvents making it essential to determine the drug solubility in different organic solvents in order to select the best option for the optimal drug quantities to be processed. On the other hand, considering the particularities of the method and the mechanism of nanoparticle formation, it could be advisable to guarantee the active molecule in its non-dissociate state

Table 5.5 *Drug encapsulation by using nanoprecipitation and emulsification-diffusion methods for the preparation of nanocapsules and nanospheres.*

Active molecule	Encapsulation efficiency (%)	Reference	Active molecule	Encapsulation efficiency (%)	Reference
Nanocapsules			Nanospheres		
Nanoprecipitation					
Indomethacin	100	34	Indomethacin	100	22
Indomethacin	100	28	Primaquine	85–94	260
Atovaquone	100	25	Cyclosporin A	82–98	264
Atovaquone	100	37	Cyclosporin A	90–98	254
Rifabutine	93	37	Pentamidine	40–76	26
Tretinoin	100	31	Isradipine	74–97	257
Primidone	75–97	19	Nimodipine	20–90	49
Spironolactone	76–96	29	Rolipram	2–20	89
Griseofulvine	78–99	27	Tyrphostin AG-1295	70	127
Melatonin	50	15	Paclitaxel	15–100	259
Diclofenac	97–100	14	Acyclovir	1–3.5	266
Diclofenac	100	35	Diclofenac	100	14
Xanthone	85–89	46	Ibuprofen	40–50	57
All-trans retinoic acid	68–97	253	Xanthone	26–40	46
			Griseofulvin	78–98	27
			Docetaxel	10–23	267
			Flurbiprofen	74–97	255
			Haloperidol	0.25–4	88
			Coenzyme Q10	49–72	268
			Silymarin	20–62	269
			Carvedilol	41–56	270
			Phenobarbital	94	90
Emulsification-diffusion					
Indomethacin	100	131	17-β-estradiol benzoate	67	72
Vitamin E	92		Enalaprilat	24–46	56
Progesterone	99		Ibuprofen	62–86	57
Estradiol	52		Estradiol	46–73	75
Chlorambucil	32		Cyclosporine	16–23	77
Hinokitol	92–94	64	Estradiol	35–68	78
			Estradiol	48–95	79
			Tacrolimus	50–60	81

in order to attain the highest encapsulation efficiency via the manipulation of aqueous phase properties (e.g. by the pH adjustment). Likewise, nanoparticles are dispersed in aqueous media which, depending on its composition (e.g. containing surfactant agents) or its pH, could cause drug leakage from the nanoparticles during storage. As a particular case, it is interesting to highlight the effect of pH on the encapsulation efficiency of triptorelin by using the double emulsion evaporation method [163]. Here, the highest encapsulation efficiency is achieved when the isoelectric point reaches the internal and external aqueous phases.

Drug stability in the operating conditions also governs its encapsulation. Then, emulsification conditions in some cases use high shear forces rendering high local temperatures, which could induce the degradation of temperature-sensitive drugs. On the other hand, high drug–polymer interactions could be considered when a polymer is chosen, mainly because the chemical affinity could govern the drug trapping. It is expected that a good interaction between the active molecule and the polymeric matrix will allow a drug partition favoring its encapsulation.

5.4.4 Drug Release Properties

The active substance release from the nanoparticles is an important parameter to consider in their development as drug carriers, being the basic point for their *in vivo* performance.

In vitro release of the active substance depends on the concentration and physicochemical properties of the drug, biodegradability, the nature and concentration of the polymer, the nature of the solvent and its compatibility with other starting materials, the size of the nanocapsules, the conditions of preparation and also the conditions of *in vitro* testing, e.g. temperature, pH, presence of surface-active agents, etc.

Unfortunately, drug release data reported by the different research works on nanoparticle preparation is limited and it is not possible to make a comparative study using the same active molecule. Moreover, the same active molecule is encapsulated by different methods; the experimental conditions used by one research team to investigate *in vitro* drug release are different from others. Considering the former's drawbacks which limit the discussion, but as an attempt to provide an overall view of the drug release from nanoparticles prepared using the different methods, Figure 5.11 compiles the behaviors of typical examples. It is obvious that immediately or extended drug release profiles can be obtained. Nanoparticles prepared via the supercritical fluid technology lead to the drug availability in the dissolution medium during the first 15 min. of beginning the test. On the contrary, polymersomes and nanoparticles prepared by using double emulsification seem to release the active ingredient very slowly (50% of the molecule are released after one day of beginning the dissolution study). Active substances encapsulated using the other nanoparticle preparation methods exhibit release behaviors situated between these ranges.

5.4.5 General Performance of the Nanoparticles

Nanoparticles have been developed for better delivery of drugs and for targeting specific cells in the body to overcome the problems related to the conventional systems. The most important problem was the insolubility of drugs in water, but detailed research on nanoencapsulation has been for other purposes, such as sustained drug delivery, targeting of specific organs, overcoming multidrug resistance, achieving higher bioavailability, minimizing side effects, protecting the drugs from the specific body defense mechanisms and, most importantly, the required therapeutic activity. For example, in cancer research, use of nanoparticles can overcome the drug resistance at the tumor and cellular level, to enhance the pharmacokinetic properties of anti-cancer agents and, most importantly, to reduce the toxic effects to normal cells, these targets are not easy to achieve because of

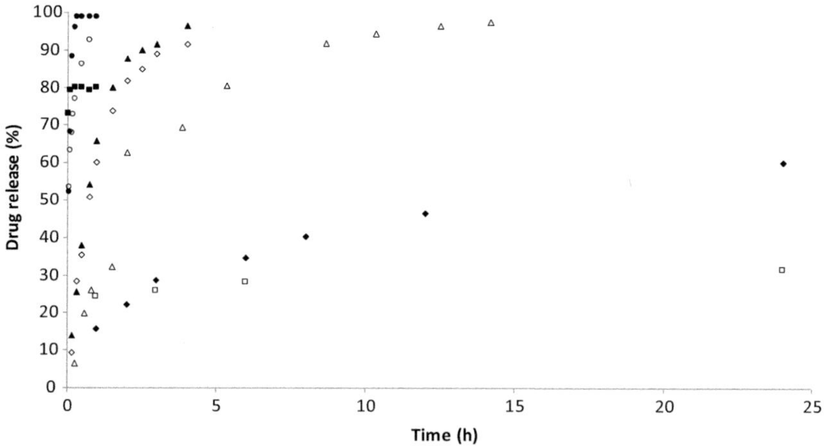

Figure 5.11 *Drug release behaviors of nanoparticles prepared by different methods. ◇ NP: Indomethacin nanocapsules by nanoprecipitation [18]; ■ ED: Indomethacin nanocapsules by emulsification-diffusion [139]; ○ EC: Triamcinolone acetonide nanocapsules by emulsification-coacervation [84]; □ DE: Triptorelin nanospheres by double emulsification [163]; ◆ Pms: NC-1900 nanocapsules by polymersome preparation [95]; • SCF: Cefuroxime axetil nanospheres by supercritical fluid technology [164]; ▲ PC: Indomethacin nanocapsules by polymer coating [18]; △ LbL: Artemisnin nanocapsules by layer-by-layer [110].*

the many influencing factors such as preparation difficulties, *in vitro* and *in vivo* analysis problems, and the interaction of the body with the contents of the nanocapsules including drug, polymer, and other ingredients, etc, although there has been some success, and in time, the techniques will improve with the discovery of new approaches, e.g. Soma *et al.* [272] report encapsulation of doxorubicin, a water-soluble molecule with cyclosporine A to reduce the side effects of both drugs while increasing their therapeutic efficacy, giving a new idea of co-encapsulation.

Keeping in mind all the above mechanisms and requirements, researchers have been working hard on *in vitro* and *in vivo* modeling, for more than a decade to find a good solution. The delay in success or less success in achieving these targets can be attributed to the limitations of the nanoparticles. A review of the success and results of researchers' trials is shown in Table 5.6.

5.5 Conclusion

The primary goal of nanoencapsulation is to solve the insolubility problem of different active ingredients, especially hydrophobic drugs, which can be accomplished by the nanoparticles preparation providing increased surface area and especially an increased surface to volume ratio. The other factors responsible for the choice of nanocapsules include the improvement in the pharmacokinetic and pharmacodynamic profiles of the drugs (the most important factor is the controlled release properties of the drugs) and the decrease in the side effects, especially on the neighboring normal cells of the target. The main focus

Table 5.6 In vitro *and* in vivo *performance of polymeric nanoparticles prepared by several methods.*

Active	Preparation method	Test	Conclusion	Reference
Atovaquone	Nanoprecipitation	Antiparasitic activity.	Increasing of the therapeutic effect.	37
Indomethacin	Nanoprecipitation	Pharmacokinetic study and potential irritant effect on the rectal mucosa in rabbits.	Improvement of the extravascular distribution. A limited protective effect on the rectal mucosa was evidenced.	36
Indomethacin	Nanoprecipitation	Active ocular distribution and acute ocular tolerance studies in rabbits.	Good ocular tolerancy and ocular bioavailability.	18
Indomethacin ethyl ester	Nanoprecipitation	Antiedematogenic activity study in rats.	Nanoencapsulation was not able to target the prodrug to the site of action and the antiedematogenic effect observed was due to metabolite effect.	30
Melatonin	Nanoprecipitation	Acute antioxidant effect of intraperitoneal administration in mice.	Improvement of the drug antioxidant effect.	15
Muramyltripeptide cholesterol (MTP-Chol)	Nanoprecipitation	Immunomodulating capacity towards a mouse macrophage cell line *in vitro*.	MTP-chol included within biodegradable polymeric nanoparticles activates mouse macrophages.	38
Oridonin	Nanoprecipitation	Pharmacokinetics and tissue distribution studies.	Drug could obtain prolonged circulation time and accumulate in liver, spleen and lung.	258
4-(N)-Stearoylgemcitabine	Nanoprecipitation	Cytotoxic activity on human cancer cell lines.	Active incorporation in nanocapsules did not change the IC_{50} compared to the free active.	13
Tacrolimus	Nanoprecipitation	Pharmacokinetic study in rats and pigs.	Nanocapsules yielded significantly higher drug levels than an active emulsion, resulting in a more enhanced bioavailability.	17

Usnic acid	Nanoprecipitation	Antitumor activity in Sarcoma 180-bearing mice and subchronic toxicity in healthy animals.	Nanoencapsulation was able to maintain and improve the usnic acid antitumor activity and considerably reduce the hepato-toxicity of this drug.	47
Salmon calcitonin	Polymer coating	Hypocalcemic effect in rats.	Pulsatile pharmacological profile and enhancement of the hypocalcemic effect when compared to the peptide solution.	107
Tamoxifen	Nanoprecipitation	Cell uptake and distribution studies using breast cancer cells.	Drug vectorization could serve as a useful form of targeted drug delivery system towards breast cancer.	273
Docetaxel	Nanoprecipitation	Interaction with biomembrane models.	Drug was not able to diffuse into the external medium through the polymer matrix, then longer times are needed to obtain complete drug release.	267
Estradiol	Emulsification-diffusion	Drug release behavior in rats.	Nanoparticles can be effective in improving the drug oral bioavailability.	78
Hinikitol	Emulsification-diffusion	Permeation study in hairless mice.	The active nanoencapsulated was more skin-permeable than active in propyleneglycol.	64
Tacrolimus	Emulsification-diffusion	Drug targeting to the lymphatic system.	Nanoparticles accumulate in the target lymph site during continuous systemic circulation. They easily pass through the walls of lymphatic capillaries acting as 'passive targeting' drug delivery system.	81
Triclosan	Emulsification-diffusion	Study in dogs with induced periodontal defects.	Nanoparticles penetrate through the junctional epithelium, and then they could help decrease gingival inflamation.	71
Paclitaxel	Polymersome	Antitumor activity in human breast carcinoma xenograft model and *in situ* apoptosis in isolated tumor tissue.	Polymersomes persist in the circulation and accumulate drug in solid tumors following a pattern of controlled release.	98
Paclitaxel	Layer-by-layer	Cytotoxic activity.	Targeted particles increase cytotoxicity of tumor cells.	112

of the earlier research on nanotechnology application in the biomedical field includes the work on antineoplastics, anti-inflammatory drugs, antigens, hormones, vitamins, antivirals, antibacterials, antifungals, antipneumocystics and diuretics. The methods of nanocapsules preparation are almost always based on the use of preformed polymers. The properties of the polymer (e.g. type, molecular weight, solubility, biodegradability, biocompatibility, etc.) and the properties of the active molecules (e.g. nature, molecular weight, charge, solubility, etc.) play an important role in the selection of the appropriate technique of nanocapsules preparation using preformed polymers. The different researchers have proved that the nanoencapsulation of hydrophobic drugs can be achieved via nanoprecipitation, emulsification methods, supercritical fluid technology, polymersome, polymer coating and layer-by-layer methods, according to the specific requirements and with modifications of the technique. The required properties can be enhanced encapsulation, controlled release, improved bioavailability, targeting a specific site (tissues or cells), reduced toxicological effects and suitability of a specific route of administration. However, there are factors to be considered that can significantly affect each method, resulting in effects on the physico-chemical properties of the nanocapsules, the most important being size.

Research progress has opened up new horizons rich with new ideas to obtain many requirements/advantages at the same time from nanocapsules. The preparation of hybrid compounds using a metal and the active ingredients leads to several advantages, especially use of the nanoparticles as theranostic, i.e., having both diagnostic and therapeutic properties. The metals used normally include gadolinium, gold, silver, and iron, etc. For example, the use of such kind of theranostic agents for cancer includes diagnostic purposes, mostly imaging of cancerous cells, e.g. MRI (magnetic resonance imaging). In this case, iron oxide nanoparticles are encapsulated with active ingredients to give the therapeutic and diagnostic advantages. Further research will lead to the creation of intelligent/smart particles, e.g. thermo-sensitive particles which release the active ingredients at a specific temperature.

After the above discussion it can simply be concluded that the preparation of nanoparticles and their use in drug delivery systems are a milestone in the biomedical field and also result in setting new trends for researchers.

References

1. C. Chauvierre, L. Leclerc, D. Labarre, M. Appel, M. C. Marden, P. Couvreur, C. Vauthier, Enhancing the tolerance of poly (isobutylcyanoacrylate) nanoparticles with a modular surface design. *Int J Pharm*, **338**, 327–332 (2007).
2. E. Fournier, C. Passirani, C. N. Montero-Menei, J. P. Benoit, Biocompatibility of implantable synthetic polymeric drug carriers: Focus on brain biocompatibility. *Biomaterials*, **24**, 3311–3331 (2003).
3. X. V. Wei, C. Y. Gong, M. Gou, S. Fu, Q. Guo, S. Shi, F. Luo, G. Guo, L. Qiu, Z. Qian, Biodegradable poly (ε-caprolactone)–poly (ethylene glycol) copolymers as drug delivery system. *Int J Pharm*, **381**, 11–18 (2009).
4. M. S. Shive, J. M. Anderson, Biodegradation and biocompatibility of PLA and PLGA microspheres. *Adv Drug Deliv Rev*, **28**, 5–24 (1997).

5. K. Avnesh, Y. Sudesh Kumar, C. Subhash, Y. Biodegradable polymeric nanoparticles based drug delivery systems. *Colloids and Surfaces B: Biointerfaces*, **75**, 1–18 (2010).
6. S. Hirano, Chitin and chitosan as novel biotechnological materials, *Poly Int*, **48**, 732–734 (1999).
7. A. Doustgani, E. V. Farahani, M. Imani, A.H. Doulabi, Dexamethasone Sodium Phosphate Release from Chitosan Nanoparticles Prepared by Ionic Gelation Method. *J. Colloid Sci. Biotechnol.* **1**, 42–50 (2012).
8. D. Fallahi, H. Mirzadeh, M. T. Khorasani, Physical, mechanical, and biocompatibility evaluation of three different types of silicone rubber. *Journal of Applied Polymer Science*, **88**, 2522–2529 (2003).
9. Hong Chena, Lin Yuana, Wei Songa, Zhongkui Wub, Dan Lia. Biocompatible polymer materials: Role of protein–surface interactions. *Progress in Polymer Science*, **33**, 1059–1087 (2008).
10. S. Stern, S. McNeil, Nanotechnology safety concerns revisited. *Toxico. Sci*, **101**, 4–21(2008).
11. S. M. Rezaei, Z. A. Mohd Ishak, The biocompatibility and hydrophilicity evaluation of collagen grafted poly(dimethylsiloxane) and poly (2-hydroxyethylmethacrylate) blends. *Polymer Testing*, **30**, 69–7 (2011).
12. D. S. Kohanea, R. Langer, Biocompatibility and drug delivery systems. *Chem Sci*, **1**, 441–446 (2010).
13. B. Stella, S. Arpicco, F. Rocco, V. Marsaud, J. M. Renoir, L. Cattel, P. Couvreur, Encapsulation of gemcitabine lipophilic derivatives into polycyanoacrylate nanospheres and nanocapsules. *Int. J Pharm*, **344**, 71–77 (2007).
14. S. R. Schaffazick, A. R. Pohlmann, T. Dalla-Costa, S. S. Guterres, Freeze-drying polymeric colloidal suspensions: nanocapsules, nanospheres and nanodispersion. A comparative study. *Eur. J Pharm Biopharm*, **56**, 501–505 (2003).
15. M. S. Khan, G. D. Vishakante, A. Bathool, Development and characterization of Brimonidine Tartrate loaded eudragit nanosuspensions for ocular drug delivery. *J. Colloid Sci. Biotechnol.* **1**, 122–128 (2012).
16. A. Jäger, V. Stefani, S. S. Guterres, A. R. Pohlmann, Physico-chemical characterization of nanocapsule polymeric wall using fluorescent benzazole probes. *Int. J Pharm*, **338**, 297–305 (2007).
17. T. Nassar, A. Rom, A. Nyska, S. Benita, Novel double coated nanocapsules for intestinal delivery and enhanced oral bioavailability of tacrolimus a P-gp substrate drug. *J Control. Rel*, **133**, 77–84 (2009).
18. P. Calvo, J. L. Vila-Jato, M. J. Alonso, Evaluation of cationic polymer-coated nanocapsules as ocular drug carriers. *Int. J Pharm*, **153**, 41–50 (1997).
19. V. Ferranti, H. Marchais, C. Chabenat, A. M. Orecchioni, O. Lafont, Primidone-loaded poly-e-caprolactone nanocapsules: incorporation efficiency and *in vitro* release profiles. *Int. J Pharm*, **193**, 107–111 (1999).
20. V. C. Furtado, P. Legrand, A. Gulik, O. Bourdon, R. Gref, D. Labarre, G. Barratt, Relationship between complement activation, cellular uptake and surface physicochemical aspects of novel PEG-modified nanocapsules. *Biomaterials*, **22**, 2967–2979 (2001).
21. V. C. Furtado, P. Legrand, J. L. Morgat, M. Vert, E. Mysiakine, R. Gref, J. P. Devissaguet, G. Barratt, Biodistribution of long-circulating PEG-grafted

nanocapsules in mice: Effects of PEG chain length and density. *Pharm Res*, **18**, 1411–1419 (2001).

22. A. R. Pohlmann, V. Weiss, O. Mertins, N. Pesce da Silveria, S. S. Guterres, Spray-dried indomethacin-loaded polyester nanocapsules and nanospheres: development, stability evaluation and nanostructure models. *Eur. J Pharm Sci*, **16**, 305–312 (2002).
23. A. R. Pohlmann, G. Mezzalira, C. de Garcia Venturini, L. Cruz, A. Bernardi, E. Jäger, A. M. O. Battastini, N. Pesce da Silveira, Determining the simultaneous presence of drug nanocrystals in drug-loaded polymeric nanocapsule aqueous suspensions: A relation between light scattering and drug content. *Int. J Pharm*, **359**, 288–293 (2008).
24. T. Ameller, V. Marsaud, P. Legrand, R. Gref, G. Barrat, J. M. Renoir. Polyester-Poly (ethylene glycol) nanoparticles loaded with the pure antiestrogen RU 58668: Physicochemical and opsonization properties. *Pharm Res*, **20**, 1063–1070 (2003).
25. E. Cauchetier, M. Deniau, H. Fessi, A. Astier, M. Paul, Atovaquone-loaded nanocapsules: influence of the nature of the polymer on their *in vitro* characteristics. *Int. J Pharm*, **250**, 273–281 (2003).
26. C. Charcosset, H. Fessi, Preparation of nanoparticles with a membrane contactor. *J Membr. Sci*, **266**, 115–120 (2005).
27. Z. Zili, S. Sfar, H. Fessi, Preparation and characterization of poly-e-caprolactone nanoparticles containing griseofulvin. *Int. J Pharm*, **294**, 261–267 (2005).
28. L. Cruz, L. U. Soares, T. D. Costa, G. Mezzalira, N. P. da Silveira, S. S. Guterres, A. R. Pohlmann, Diffusion and mathematical modeling of release profiles from nanocarriers. *Int. J Pharm*, **313**, 198–205 (2006).
29. I. Limayem, C. Charcosset, S. Sfar, H. Fessi, Preparation and characterization of spironolactone-loaded nanocapsules for paediatric use. *Int. J Pharm*, **325**, 124–131 (2006).
30. V. B. Cattani, A. R. Pohlmann, T. D. Costa, Pharmacokinetic evaluation of indomethacin ethyl ester-loaded nanoencapsules. *Int. J Pharm*, **363**, 214–216 (2008).
31. A. F. Ourique, A. R. Pohlmann, S. S. Guterres, R. C. R. Beck, Tretionoin-loaded nanocapsules: Preparation, physicochemical characterization, and photostability study. *Int. J Pharm*, **352**, 1–4 (2008).
32. F. S. Poletto, E. Jäger, L. Cruz, A. R. Pohlmann, S. S. Guterres, The effect of polymeric wall on the permeability of drug-loaded nanocapsules. *Mater. Sci Eng. C*, **28**, 472–478 (2008).
33. F. S. Poletto, L. A. Fiel, B. Donida, M. I. Ré, S. S. Guterres, A. R. Pohlmann, Controlling the size of poly (hydroxybutyrate-co-hydroxyvalerate) nanoparticles prepared by emulsification-diffusion technique using ethanol as surface agent. *Colloid Surf. A*, **324**, 105–112 (2008).
34. H. Fessi, F. Puisieux, J. P. Devissaguet, N. Ammoury, S. Benita, Nanocapsule formation by interfacial polymer deposition following solvent displacement. *Int. J Pharm*, **55**, R1–R4 (1989).
35. S. S. Guterres, H. Fessi, G. Barrat, J. P. Devissaguet, F. Puisieux, Poly (DL-lactide) nanocapsules containing diclofenac: I. Formulation and stability study. *Int. J Pharm*, **113**, 57–63 (1995).

36. F. Fawaz, F. Bonini, M. Guyot, A. M. Lagueny, H. Fessi, J. P. Devissaguet, Disposition and protective effect against irritation after intravenous and rectal administration of indomethacin loaded nanocapsules to rabbits. *Int. J Pharm*, **133**, 107–115 (1996).
37. F. Dalençon, Y. Amjaud, C. Lafforgue, F. Derouin, H. Fessi, Atovaquone and rifabutine-loaded nanocapsules: formulation studies. *Int. J Pharm*, **153**, 127–130 (1997).
38. I. Seyler, M. Appel, J. P. Devissaguet, P. Legrand, G. Barratt, Macrophage activation by a lipophilic derivative of muramyldipeptide within nanocapsules: Investigation of the mechanism of drug delivery. *J Nanopart Res*, **1**, 91–97 (1999).
39. A. Rübe, G. Hause, K. Mäder, J. Kohlbrecher, Core-shell structure of miglyol/poly (D, L-lactide)/poloxamer nanocapsules studied by small angle neutron scattering. *J Control. Rel*, **107**, 244–252 (2005).
40. P. Tewa-Tagne, S. Briançon, H. Fessi, Spray-drying microparticles containing polymeric nanocapsules: Formulation aspects, liquid phase interactions and particles characteristics. *Int. J Pharm*, **325**, 63–74 (2006).
41. P. Tewa-Tagne, G. Degobert, S. Briançon, C. Bordes, J. Y. Gauvrit, P. Lanteri, H. Fessi, Spray-drying nanocapsules in presence of colloidal silica as drying auxiliary agent: Formulation and process variables optimization using experimental designs. *Pharm Res*, **24**, 650–660 (2007).
42. P. Tewa-Tagne, S. Briançon, H. Fessi, Preparation of redispersible dry nanocapsules by means of spray-drying: Development and characterisation. *Eur. J Pharm Sci*, **30**, 124–135 (2007).
43. D. Nogueira de Assis, V. C. Furtado, J. M. Carneiro, M. Spangler, V. Nascimento, Release profiles and morphological characterization by atomic force microscopic and photon correlation spectroscopy of 99mTechnetium-fluconazole nanocapsules. *Int. J Pharm*, **349**, 152–160 (2008).
44. M. A. Pereira, V. C. Furtado, J. M. Carneiro, M. Spangler, G. Andrade, V. Nascimento, PLA-PEG nanocapsules radiolabelled with 99mTechnetium-HMPAO: Release properties and physicochemical characterization by atomic force microscopy and photon correlation spectroscopy. *Eur. J Pharm Sci*, **33**, 42–51 (2008).
45. N. S. Santos-Magalhães, A. Pontes, V. M. Pereira, M. N. P. Caetano, Colloidal carriers for benzathine penicillin G: Nanoemulsions and nanocapsules. *Int. J Pharm*, **208**, 71–80 (2000).
46. M. Texeira, M. J. Alonso, M. M. Pinto, C. Barbosa, Development and characterization of PLGA nanospheres and nanocapsules containing xanthone and 3-methoxyxanthone. *Eur. J Pharm Biopharm*, **59**, 491–500 (2005).
47. N. Pereira, S. Carneiro, M. Silvestre, N. Teles, J. Figuereido da Silva, C. M. Machado, E. C. Pereira, N. H. da Silva, N. K. Honda, N. S. Santos-Magalhães, Nanoencapsulation of usnic acid: An attempt to improve antitumor activity and reduce hepatoxicity. *Eur. J Pharm Biopharm*, **64**, 154–160 (2006).
48. J. Cheng, B. A. Teply, I. Sherifi, J. Sung, G. Luther, F. X. Gu, E. Levy, A. F. Radovic, R. Langer, O. C. Farokhzad. Formulation of functionalized PLGA-PEG nanoparticles for *in vivo* targeted drug delivery. *Biomaterials*, **28**, 869–876 (2007).
49. H. Ge, Y. Hu, S. Yang, X. Jiang, C. Yang. Preparation, characterization and drug release behaviors of drug-loaded e-caprolactone/L-lactide copolymer nanoparticles. *J Appl Polym Sci*, **75**, 874–882 (2000).

50. P. Arbós, M. Wirth, M. A. Arangoa, F. Gabor, J. M. Irache. Gantrez® AN as a new polymer for the preparation of ligand-nanoparticle conjugates. *J Control Release*, **83**, 321–330 (2002).
51. M. A. Arangoa, M. A. Campanero, M. J. Renedo, G. Ponchel, J. M. Irache. Gliadin nanoparticles as carriers for the oral administration of lipophilic drugs. Relationships between bioadhesion and pharmacokinetics. *Pharm Res*, **18**, 1521–1527 (2001).
52. I. Ezpeleta, J. M. Irache, S. Stainmesse, C. Chabenat, J. Gueguen, Y. Popineau, A. M. Orecchioni. Gliadin nanoparticles for the controlled release of all-trans-retinoic acid. *Int J Pharm*, **131**, 191–200 (1996).
53. M. R. Kulterer, M. Reischl, V. E. Reichel, S. Hribemik, M. Wu, S. Köstler, R. Kargl, V. Ribitsch. Nanoprecipitation of cellulose acetate using solvent/nonsolvent mixtures as dispersive media. *Colloids Surf A.: Physicochem Eng Asp*, **375**, 23–29 (2011).
54. J. C. Leroux, E. Allemann, E. Doelker, R. Gurney, New approach for the preparation of nanoparticles by an emulsification-diffusion method. *Eur. J. Pharm. Biopharm*, **41**, 14–18 (1995).
55. D. Quintanar-Guerrero, E. Allémann, E. Doelker, H. Fessi, Preparation and characterization of nanocapsules from preformed polymers by a new process based on emulsification-diffusion technique. *Pharm Res*, **15**(7), 1056–1062 (1998).
56. P. Ahlin, J. Kristl, A. Kristl. P. Vrečer, Investigation of polymeric nanoparticles as carriers of enalaprilat for oral administration. *Int J Pharm*, **239**, 113–120 (2002).
57. S. Galindo, F. Puel, S. Briançon, E. Allémann, E. Doelker, H. Fessi, Comparative scale-up of three methods for producing ibuprofen-loaded nanoparticles. *Eur. J Pharm Sci*, **25**, 357–367 (2005).
58. S. Guinebretière, S. Briançon, H. Fessi, V. S Teodorescu, M. G. Blanchin, Nanocapsules of biodegradable polymers: preparation and characterization by direct high resolution electron microscopy. *Materials Science and Engineering C*, **21**, 137–142 (2002).
59. I. Limayem, C. Charcosset, H. Fessi, Purification of nanoparticle suspensions by a concentration/diafiltration process. *Separ. Purif. Tech*, **38**, 1–9 (2004).
60. W. Abdelwahed, G. Degobert, H. Fessi, A pilot study of freeze drying of poly(epsilon-caprolactone) nanocapsules stabilized by poly(vinyl alcohol): Formulation and process optimization. *Int. J Pharm*, **309**, 178–188 (2006).
61. W. Abdelwahed, G. Degobert, H. Fessi, Freeze-drying of nanocapsules: Impact of annealing on the drying process. *Int J Pharm*, **324**, 74–82 (2006).
62. W. Abdelwahed, G. Degobert, H. Fessi, Investigation of nanocapsules stabilization by amorphous excipients during freeze-drying and storage. *Eur J Pharm Biopharm*, **63**, 87–94 (2006).
63. J. Gang, S. B. Park, W. Hyung, E. H. Choi, J. Wen, H. S. Kim, Y. G. Shul, S. Haam, S. Y. Song, Magnetic poly-e-caprolactone nanoparticles containing Fe3O4 and gemcitabine enhance anti-tumor effect in pancreatic cancer xenograft mouse model. *J Drug Target*, **15**, 445–453 (2007).
64. H. H. Joo, H. Y. Lee, Y. S. Guan, J. C. Kim, Colloidal stability and *in vitro* permeation study of poly (e-caprolactone) nanocapsules containing hinokitiol. *J Ind Eng Chem*, **14**, 608–613 (2008).

65. M. Moinard-Checot, Y. Chevalier, S. Briancon, L. Beney, H. Fessi, Mechanism of nanocapsules formation by the emulsion–diffusion process. *J Colloid Interface Sci*, **317**, 458–68 (2008).
66. M. J. Choi, A. Soottutantawat, O. Nuchuchua, S. G. Min, U. Ruktanonchai, Physical and light oxidative properties of eugenol encapsulated by molecular inclusion and emulsion-diffusion method. *Food Res. Int*, **42**, 148–156 (2009).
67. D. Quintanar-Guerrero, H. Fessi, E. Allémann, E. Doelker. Influence of stabilizing agents and preparative variables on the formation of poly (D, L-lactic acid) nanoparticles by an emulsification-diffusion technique. *International Journal of Pharmaceutics*, **143**(2), 133–141 (1996).
68. M. S. Romero-Cano, B. Vincent, Controlled release of 4-nitroanisole from poly (lactic acid) nanoparticles. *J Control Rel*, **82**, 127–135 (2002).
69. Y. N. Konan, R. Cerny, J. Favet, M. Berton, R. Gurny, E. Allémann, Preparation and characterization of sterile sub-200 nm meso-tetra(4-hydroxyphenyl) porphyrin-loaded nanoparticles for photodynamic therapy. *Eur J Pharm Biopharm*, **55**, 115–124 (2003).
70. T. Trimaille, C. Pichot, A. Elaissari, H. Fessi, S. Briançon, T. Delair, Poly(D,L-lactic acid) nanoparticle preparation and colloidal characterization. *Colloid Polym Sci*, **281**, 1184–1190 (2003).
71. E. Piñón, A. Ganem, V. Alonso, D. Quintanar. Preparation and characterization of triclosan nanoparticles for periodontal treatment. *Int J Pharm*, **294**, 217–232 (2005).
72. H. Y. Kwon, J. Y. Lee, S. W. Choi, Y. Jang, J. H. Kim, Preparation of PLGA nanoparticles containing estrogen by emulsification-diffusion method. *Colloid Surf A*, **182**, 123–130 (2001).
73. Sung-Wook Choi, Hye-Young Kwon, Woo-Sik Kim, Jung-Hyun Kim. Thermodynamic parameters on poly (D, L-lactide-*co*-glycolide) particle size in emulsification–diffusion process. *Colloids Surf A.: Physicochem Eng Asp*, **201**(1–3), 283–289 (2002).
74. M. N. V. Ravi, U. Bakowsky, C. M. Lehr, Preparation and characterization of cationic PLGA nanospheres as DNA carriers. *Biomaterials*, **25**, 1771–1777 (2004).
75. S. Hariharan, V. Bhardwaj, I. Bala, J. Sitterberg, U. Bakowsky, M. N. V. Ravi Kumar, Design of estradiol loaded PLGA nanoparticulate formulations: A potential oral delivery system for hormone therapy. *Pharm Res*, **23**, 184 (2006).
76. K. C. Song, H. S. Lee, I. Y. Choung, K. I. Cho, Y. Ahn, E. J Choi, The effect of type of organic phase solvents on the particle size of poly (D, L-lactide-co-glycolide) nanoparticles. *Colloid Surf A*, **276**, 162–167 (2006).
77. J. L. Italia, D. K. Bhatt, V. Bhardwaj, K. Tikoo, M. N. V. Ravi, PLGA nanoparticles for oral delivery of cyclosporine: nephrotoxicity and pharmacokinetic studies in comparison to sandimmune Neoral®. *J Control Release*, **119**, 197–206 (2007).
78. G. Mittal, D. K. Sahana, V. Bhardwaj, M. N. V. Ravi Kumar. Estradiol loaded PLGA nanoparticles for oral administration: Effect of polymer molecular weight and copolymer composition on release behavior *in vitro* and *in vivo*. *J Control Release*, **119**, 77–85 (2007).
79. D. K. Sahana, G. Mittal, V. Bhardwaj, M. N. V. Ravi Kumar, PLGA nanoparticles for oral delivery of hydrophobic drugs: influence of organic solvent on nanoparticle

formation and release behavior *in vitro* and *in vivo* using estradiol as a model drug. *J Pharm Sci*, **97**, 1530–1542 (2008).

80. M. J. Santander, T. Stauner, B. Loretz, J. L. Ortega, D. Bastos, G. Wenz, U. F. Schaefer, C. M. Lehr. Nanoparticles made from novel starch derivatives for transdermal drug delivery. *J Control Release*, **141**, 85–92 (2010).
81. S. B. Shin, H. Y. Cho, D. D. Kim, H. G. Choi, Y. B. Lee. Preparation and evaluation of tacrolimus-loaded nanoparticles for lymphatic delivery. *European Journal of Pharmaceutics and Biopharmaceutics*, **74**, 164–171 (2010).
82. M. Rajaonarivony, C. Vauthier, G. Couarraze, F. Puisieux, P. Couvreur. Development of a new drug carrier made from alginate. *J Pharm Sci*, **82**, 912–917 (1993).
83. P. Lertsutthiwong, K. Noomun, N. Jongaroonngamsang, P. Rojsitthisak, U. Nimmannit, Preparation of alginate nanocapsules containing turmeric oil. *Carbohydr Polymer*, **74**, 209–214 (2008).
84. H. J. Krause, P. Rohdewald, Preparation of gelatin nanocapsules and their pharmaceutical characterization. *Pharm Res*, **5**, 239–243. (1985).
85. W. Fraunhofer, G. Winter, C. Coester. Asymmetrical flow field-flow fractionation and multiangle light scattering for analysis of gelatin nanoparticle drug carrier systems. *Anal Chem*, **76**, 1909–1920 (2004).
86. S. Lutter, J. Koetz, B. Tiersch, A. Boschetti de Fierro, V. Abetz, Formation of gold nanoparticles in triblock terpolymer-modified inverse microemulsions. *Colloid Surf A*, **329**, 160–176 (2008).
87. P. Chiewpattanakul, R. Covis, R. Vanderesse, B. Thanomsub, E. Marie, A. Durand. Design of polymeric nanoparticles for the encapsulation of monoaylglycerol. *Colloid Polym Sci*, **288**, 959–967 (2010).
88. A. Budhian, S. J. Siegel, K. I. Winey. Haloperidol-loaded PLGA nanoparticles: Systematic study of particle size and drug content. *Int J Pharm*, **336**, 367–375 (2007).
89. A. Lamprecht, N. Ubrich, H. Yamamoto, U. Schäfer, H. Takeuchi, C. M. Lehr, P. Maincent, Y. Kawashima. Design of rolipram-loaded nanoparticles: comparison of two preparation methods. *J Control Release*, **71**, 297–306 (2001).
90. J. M. Barichello, M. Morishita, K. Takayama, T. Nagai. Encapsulation of hydrophilic and lipophilic drugs in PLGA nanoparticles by the nanoprecipitation method. *Drug Dev Ind Pharm*, **25**, 471–476 (1999).
91. A. Aumelas, A. Serrero, A. Durand, E. Dellacherie, M. Leonard. Nanoparticles of hydrophobically modified dextrans as potencial drug carrier Systems. *Colloid Surf B Biointerfaces*, **59**, 74–80 (2007).
92. J. Vandervoort, A. Ludwig. Biocompatible stabilizers in the preparation of PLGA nanoparticles: a factorial design study. *Int J Pharm*, **238**, 77–92 (2002).
93. F. Ahmed, D. Discher, Self-porating polymersomes of PEG-PLA and PEG-PCL: hydrolysis-triggered controlled release vesicles. *J Control Release*, **96**, 37–53 (2004).
94. F. Meng. G. H. M Engbers, J. Feijen, Biodegradable polymersomes as a basis for artificial cells: encapsulation, release and targeting. *J Control Rel*, **101**, 187–198 (2005).
95. Z. Pang, W. Lu, H. Gao, K. Hu, J. Chen, C. Zhang, X. Gao, X. Jiang, C. Zhu, Preparation and brain delivery property of biodegradable polymersomes conjugated with OX26. *J Control Release*, **128**, 120–127 (2008).
96. W. Zhou, J. Feijen, Biodegradable polymersomes for controlled drug release. *J Control Release*, **132**, 19–36 (2008).

97. Y. Kim, M. Tewari, J. D. Pajerowski, S. Cai, S. Sen, J. Williams, S. Sirsi, G. Lutz, Discher, D. E., Polymersome delivery of siRNA and antisense oligonucleotides *J Control Rel*, **134**, 132–140 (2009).
98. F. Ahmed, R. I. Pakunlu, A. Brannan, F. Bates, T. Minko, D. E. Discher, Biodegradable polymersomes loaded with both paclitaxel and doxorubicin permeate and shrink tumors, inducing apoptosis in proportion to accumulated drug. *J Control Rel*, **116**, 150–158 (2006).
99. J. Zhou, L. Wang, X. Dong, T. Chen, Q. Yang, C. Chen X. Chen. Preparation of organic/inorganic hybrid nanoballs using aggregates of PTMSPMA-b-PSMA-Fc-PSMA-b-PTMSPMA block copolymers as precursors. *Nanotechnology*, **17**, 2745 (2006).
100. P. J. Photos, L. Bacakova, B. Discher, F. S. Bates, D. E. Discher, Polymer vesicles *in vivo*: correlations with PEG molecular weight. *J Control Rel*, **90**, 323–334 (2003).
101. F. Najafi, M. N. Sarbolouki, Biodegradable micelles/polymersomes from fumaric/sebacic acids and poly(ethylene glycol). *Biomaterials*, **24**, 1175–1182 (2003).
102. V. Hearnden, H. Lomas, S. MacNeil, M. Thornhill, C. Mirdoch, A. Lewis, J. Madsen, A. Blanazs, S. Armes, G. Battaglia, Diffusion studies of nanometer polymersomes across tissue engineered human oral mucosa. *Pharm Res*, **26**, 1718–1728 (2009).
103. F. Chécot, S. Lecommandoux, H. A. Klok, Y. Gnanou. From supramolecular polymersomes to stimuli-responsive nano-capsules based on poly (diene-b-peptide) diblock copolymers. *Eur Phys J E*, **10**, 25–35 (2003).
104. J. P. Xu, J. Ji, W. D. Chen, J. C. Shen, Novel biomimetic polymersomes as polymer therapeutics for drug delivery. *J Control Rel*, **107**, 502–512 (2005).
105. J. W. Tom, P. G. Debenedtti, Precipitation of poly(L-lactic acid) and composite poly(L-lactic acid)-pyrene particles by rapid expansion of supercritical solutions. *J Supercrit Fluids*, **7**, 9–29 (1994).
106. D. J. Jarmer, C. S. Lengsfeld, T. W. Randolph, Manipulation of particle size distribution of poly(L-lactic acid) nanoparticles with a jet-swirl nozzle during precipitation with a compressed antisolvent. *J Supercrit Fluids*, **27**, 317–336 (2003).
107. C. Prego, M. Fabre, D. Torres, M. J. Alonso, Efficacy and mechanism of action of chitosan nanocapsules for oral peptide delivery. *Pharm Res*, **23**, 549–556. (2006).
108. A. Vila, A. Sánchez, M. Tobío, P. Calvo, M. J. Alonso, Design of biodegradable particles for protein delivery. *J Control Rel*, **78**, 15–24. (2002).
109. P. Lertsutthiwong, P. Rojsitthisak, U. Nimmannit, Preparation of turmeric oil-loaded chitosan-alginate biopolymeric nanocapsules *Mater Sci Eng C*, (2008).
110. Y. Chen, X. Lin, H. Park, R. Greever, Study of artemisinin nanocapsules as anticancer drug delivery systems. *Nanomedicine: NBM*, **5**(3), 316–322 (2009).
111. S. Krol, A. Diaspro, O. Cavalleri, D. Cavanna, P. Ballario, B. Grimaldi, P. Filetici, P. Ornaghi, A. Gliozzi, Nanocapsules: A novel tool for medicine and science. In *Frontiers of Multifunctional Integrated Nanosystems*, E. Buzaneva, P. Scharff (eds). Dordrecht: Academic Publishers, pp. 439–446 (2004).
112. A. Agarwal, Y. Lvov, R. Sawant, V. Torchilin, Stable nanocolloids of poorly soluble drugs with high drug content prepared using the combination of sonication and layer-by-layer technology. *J Control Rel*, **128**, 255–260 (2008).
113. C. Preetz, A. Rübe, I. Reiche, G. Hause, K. Mäder, Preparation and characterization of biocompatible oil-loaded polyelectrolyte nanocapsules. *Nanomedicine: NMB*, **4**, 106–114 (2008).

114. E. Rytting, M. Bur, R. Cartier, T. Bouyssou, X. Wang, M. Krüger, M. Lehr, T. Kissel, *In vitro* and *in vivo* performance of biocompatible negatively-charged salbutamol-loaded nanoparticles. *J Cont Rel*, **141**, 101–107 (2010).
115. R. Gaspar, R. Duncan, Polymeric carriers: Preclinical safety and the regulatory implications for design and development of polymer therapeutics. *Adv Drug Deliv Rev*, **61**, 1220–1231 (2009).
116. M. Rawat, D. Singh, S. Saraf, Nanocarriers: Promising vehicle for bioactive drugs. *Biol Pharm Bull*, **29**, 1790–1798 (2006).
117. M. Nahar, T. Dutta, S. Murugesan, A. Aathana, D. Mishra, V. Rajkumar, M. Tare, S. Saraf, J. Narendra Kumar, Functional polymeric nanoparticles: An efficient and promising tool for active delivery of bioactives. *Crit Rev Therap Drug Carr Sys*, **23**, 259–318 (2006).
118. C. P. Reis, R. J. Neufeld, A. J. Ribeiro, F. Veiga, Nanoencapsulation 1, Methods for the preparation of drug loaded polymeric nanoparticles. *Nanomedicine: Nanotechnology, Biology, and Medicine*, **2**, 8–21 (2006).
119. H. Fessi, F. Puisieux, J. P. Devissaguet, Procédé de préparation de systèmes colloïdaux dispersibles d'une substance sous forme de nanocapsules. *European Patent* 274961 A1, 20 July (1988).
120. S. Khoee, M.Yaghoobian, An investigation into the role of surfactants in controlling particle size of polymeric nanocapsules containing penicillin-G in double emulsion. *Eur J Med Chem*, **44**(6), 2392–2399 (2009).
121. S. Sakuma, M. Hayashi, M. Akashi, Design of nanoparticles composed of graft copolymers for oral peptide delivery. *Adv Drug Deliv Rev*, **47**, 21–37 (2001).
122. N. Behan, C. Birkinshaw, N. Clarke, Poly *n*-butyl cyanoacrylate nanoparticles: a mechanistic study of polymerization and particle formation. *Biomaterials*, **22**, 1335–44 (2001).
123. S. W. Choi, J. H. Kim. Design of surface-modified poly(D,L-lactide-co-glycolide) nanoparticles for targeted drug delivery to bone. *J Control Release*, **122**, 24–30 (2007).
124. H. J. Jeon, Y. I. Jeong, M. K. Jang, Y. H. Park, J. W. Nah, Effect of solvent on the preparation of surfactant-free poly(D,L-lactide-co-glycolide) nanoparticles and norfloxacin release characteristics. *Int J Pharm*, **207**, 99–108 (2000).
125. T. Akagi, T. Kaneko, T. Kida, M. Akashi. Preparation and characterization of biodegradable nanoparticles based on poly(g-glutamic acid) with L-phenylalanine as a protein carrier. *J Control Release*, **108**, 226–236 (2005).
126. E. Plasari, P. H. Grisoni, J. Villermaux, Influence of process parameters on the precipitation of organic nanoparticles by drowning-out. *Chem Eng Res Des*, **75**, 237–244 (1997).
127. M. Chorny, I. Fishbein, H. D. Danenberg, G. Golomb, Lipophilic drug loaded nanospheres prepared by nanoprecipitation: effect of formulation variables on size; drug recovery and release kinetics. *J Control Rel*, **83**, 389–400 (2002).
128. P. Legrand, S. Lesieur, A. Bochot, R. Gref, W. Raatjes, G. Barratt, C. Vauthier, Influence of polymer behavior in organic solution on the production of polylactide nanoparticles by nanoprecipitation. *Int J Pharm*, **344**, 33–43 (2007).
129. F. Lince, D. L. Marchisio, A. A. Barresi, Strategies to control the particle size distribution of poly-e-caprolactone nanoparticles for pharmaceutical applications. *J Colloid Interf Sci*, **322**, 505–515 (2008).

130. C. V. Sternling, L. E. Scriven, Interfacial turbulence: hydrodynamic instability and the Marangoni effect. *A.I.Ch.E Journal*, **5**, 514–523 (1959).
131. D. Quintanar, E. Allémann, H. Fessi, E. Doelker, Preparation techniques and mechanisms of formation of biodegradable nanoparticles from preformed polymers. *Drug Dev Ind Pharm*, **24**, 1113–1128. (1998).
132. S. Galindo-Rodriguez, E. Allemann, H. Fessi, E. Doelker, Physicochemical parameters associated with nanoparticle formation in the salting-out, emulsification-diffusion and nanoprecipitation methods. *Pharm Res*, **21**, 1428–1439 (2004).
133. S. A. Vitale, J. L. Katz, Liquid droplet dispersions formed by homogeneous liquid-liquid nucleation: "The ouzo effect", *Langmuir*, **19**(10), 4105–4110 (2003).
134. F. Ganachaud, J. L. Katz, Nanoparticles and nanocapsules created using the ouzo effect: Spontaneous emulsification as an alternative to ultrasonic and high-shear devices. *Chem Phys Chem*, **6**(2), 209–216 (2005).
135. Moritz Beck-Broichsitter, Erik Rytting, Tobias Lebhardt, Xiaoying Wang, Thomas Kissel. Preparation of nanoparticles by solvent displacement for drug delivery: A shift in the "ouzo region" upon drug loading. *Eur J Pharm Biopharm*, **41**, 244–253 (2010).
136. E. Allémann, J. C Leroux, R. Gurny, Polymeric nano-microparticles for the oral delivery of peptides and peptidomimetics. *Adv Drug Deliv Rev*, **34**,171–89 (1998).
137. E. Blouza *et al.* Preparation and characterization of spironolactone-loaded nanocapsules for paediatric use. *International Journal of Pharmaceutics*, **325**, 1-2, 124–131 (2006).
138. D. Quintanar, H. Fessi, E. Doelker, E. Alleman, Method for preparing vesicular nanocapsules. US Patent 6884438, 26 April (2005).
139. S. Guinebretière, Nanocapsules par emulsion-diffusion de solvant: Obtention, caractérisation et mécanisme de formation. Ph.D Thesis, Université Claude Bernard-Lyon 1, France (2001).
140. W. Lu, Y. Zhang, Y.-Z. Tan, K.-L. Hu, X.-G. Jiang, S. K. Fu, Cationic albumin-conjugated pegylated nanoparticles as novel drug carrier for brain delivery. *J Control Release*, **107**, 428–48 (2005).
141. M. H. El-Shabouri Positively charged nanoparticles for improving the oral bioavailability of cyclosporin-A. *Int J Pharm*, **249** (2002).
142. V. Saxena, M. Sadoqi, J. Shao, Indocyanine green-loaded biodegradable nanoparticles: preparation, physicochemical characterization and *in vitro* release. *Int J Pharm*, **278**, 293–301 (2004).
143. C. Perez, A. Sanchez, D. Putnam, D. Ting, R. Langer, M. J. Alonso, Poly (lactic acid)-poly (ethylene glycol) nanoparticles as new carriers for the delivery of plasmid DNA, *J Control Release*, **75**, 211–224 (2001).
144. Tomoko Nii, Fumiyoshi Ishii. The novel liposome preparation methods based on in-water drying and phase separation: microencapsulation vesicle method and coacervation method. *Advances in Planar Lipid Bilayers and Liposomes*, **5**, 41–61 (2006).
145. S. P. Sanghvi, J. G. Nairn, Phase diagram studies for microencapsulation of pharmaceuticals using cellulose acetate trimellitate, *J Pharm Sci*, **80**, 394–398 (1991).
146. W. Feinstein, J. J. Sciarra, Development and evaluation of a dexamethasone timed-release aerosol formulation, *J Pharm Sci*, **64**, 408–414 (1975).
147. J. R. Nixon, M. R. Maleka, The preparation and characterization of ethyl cellulose-walled theophylline microcapsules, *J Microencapsul*, **1**, 53–64 (1984).

148. S. K. Das, S. Palchowdhury, Design and *in vitro* evaluation of polyvinyl chloride microcapsules containing sulphamethoxazole, *J Microencapsul*, **6**, 53–58 (1989).
149. B. Gander, M. J. Blanco-Príeto, C. Thomasin, Ch. Wandrey, D. Hunkeler, Coacervation/Phase separation. In *Encyclopedia of Pharmaceutical Technology*. J. Swarbrick, J. C. Boylan (eds). New York: Marcel Dekker, pp. 481–496 (2002).
150. W. Yin, H. Zhang, L. Huang, K. Nishinari, Effects of the lyotropic series salts on the gelation of konjac glucomannan in aqueous solutions. *Carbohydr Polymer*, **74**, 68–78 (2008).
151. Fubao Xing, Guoxiang Cheng, Kejing Yi, Linrong Ma. Nanoencapsulation of capsaicin by complex coacervation of gelatin, acacia, and tannins. *Journal of Applied Polymer Science 96*, **6**, 2225–2229 (2005).
152. E. Pisani, E. Fattal, J. Paris, C. Ringard, V. Rosilio, N. Tsapis, Surfactant dependent morphology of polymeric capsules of perfluorooctyl bromide: Influence of polymer adsorption at the dichloromethane-water interface. *J Colloid Interf Sci*, **326**, 66–71 (2008).
153. Fabiana Quaglia, Luisanna Ostacolo, Antonino Mazzaglia, Valentina Villari, Daniela Zaccaria and Maria T. Sciortino. The intracellular effects of non-ionic amphiphilic cyclodextrin nanoparticles in the delivery of anticancer drugs. *Biomaterials*, **30**(3) 374–382 (2009).
154. N. Garti, Double emulsions–scope, limitations and new achievements. *Colloid Surf A*, **123–124**, 233–246 (1997).
155. D. Georgieva, V. Schmitt, F. Leal-Calderon, D. Langevin, On the possible role of surface elasticity in emulsion stability, *Langmuir*, **25**, 5565–5573 (2009).
156. R. Alex, R. Bodmeier, Encapsulation of water-soluble drugs by a modified solvent evaporation method. I. Effect of process and formulation variables on drug entrapment. *J Microencapsul*, **7**, 347–355 (1989).
157. S. Cohen, T. Yoshioka, M. Lucarelli, L. H. Hwang, R. Langer, Controlled delivery systems for proteins based on poly (lactic: glycolic acid) microspheres. *Pharm Res*, **8**, 713–720 (1991).
158. M. D. Blanco, M. J. Alonso, Development and characterization of protein-loaded poly (lactide-co-glycolide) nanospheres. *Eur. J Pharm Biopharm*, **43**, 287–294 (1997).
159. D. Lemoine, V. Preat, Polymeric nanoparticles as delivery system for influenza virus glycoproteins. *J Control. Rel*, **54**, 15–27 (1998).
160. M. Tobıo, R. Gref, A. Sanchez, R. Langer, M. J. Alonso, Stealth PLA-PEG nanoparticles as protein carriers for nasal administration: *Pharm Res*, **15**, 270–275 (1998).
161. C. E. Mora-Huertas, H. Fessi, A. Elaissari. Polymer based nanocapsules for drug delivery. *Int J Pharm*, **385**, 113–142 (2010).
162. C. E. Mora, H. Fessi, A. Elaissari, Influence of process and formulation parameters on the formation of submicron particles by solvent displacement and emulsification-diffusion methods. Critical comparison. *Adv Colloid Interf Sci*, **163**, 90–122 (2011).
163. S. Nicoli, P. Santia, P. Couvreurb, G. Couarrazeb, P. Colomboa, E. Fattal, Design of triptorelin loaded nanospheres for transdermal iontophoretic administration. *Int J Pharm*, **214**, 31–35 (2001).
164. Seoung Wook Jun, Min-Soo Kim, Guk Hyun Jo, Sibeum Lee, Jong Soo Woo, Jeong-Sook Park, Sung-Joo Hwang. Cefuroxime axetil solid dispersions prepared using

solution enhanced dispersion by supercritical fluids. *Journal of Pharmacy and Pharmacology*, **57**(12), 1529–1537 (2005).
165. D. E. Discher, A. Eisenberg, Polymer vesicles. *Science*, **297**(5583) 967–973 (2002).
166. C. Nardin, T. Hirt, J. Leukel, W. Meier, Polymerized ABA triblock copolymer vesicles. *Langmuir*, **16**, 1035–1041 (2000).
167. F. Chécot, S. Lecommandoux, H. A. Klok and Y. Gnanou. From supramolecular polymersomes to stimuli-responsive nano-capsules based on poly (diene-b-peptide) diblock copolymers. *The European Physical Journal E: Soft Matter and Biological Physics*, **10**(1), 25–35 (2003).
168. F. Meng, C. Hiemstra, G. M. H. Engbers, J. Feijen, Biodegradable polymersomes. *Macromolecules*, **36**, 3004–3006 (2003).
169. K. Kita-Tokarczyk, J. Grumelard, T. Haefele, W. Meier, Block copolymer vesicles–using concepts from polymer chemistry to mimic biomembranes. *Polymer*, **46**, 3540–3563 (2005).
170. S. Li, B. Byrne, J. Welsh, A. F. Palmer, Self-assembled poly (butadiene)-β-poly (ethylene oxide) polymersomes as paclitaxel carriers. *Biotechnol Prog*, **23**, 278–285 (2007).
171. D. H. Levine, P. P. Ghoroghchian, J. Freudenberg, G. Zhang, M. J. Therien, M. I. Greene, D. A. Hammer, R. Murali, Polymersomes: A new multi-functional tool for cancer diagnosis and therapy. *Methods*, **46**, 25–32 (2008).
172. D. A. Hammer, G. P. Robbins, J. B. Haun, J. J. Lin, W. Qi, L. A. Smith, P. P. Ghoroghchian, M. J. Therien, F. S. Bates, Leuko-polymersomes. *Faraday Discuss*, **139**, 129–141; discussion 213–228, 419–420 (2008).
173. N. A. Christian, M. C. Milone, S. S. Ranka, G. Li, P. R. Frail, K. P. Davis, F. S. Bates, M. J. Therien, P. P. Ghoroghchian, C. H. June, D. A. Hammer. Tat-functionalized near-infrared emissive polymersomes for dendritic cell labeling. *Bioconjug Chem*, **18**, 31–40 (2007).
174. R. Rastogi, S. Anand, V. Koul, Flexible polymerosomes: An alternative vehicle for topical delivery. *Colloids and Surfaces B: Biointerfaces*, **72** :161–166 (2009).
175. K. Mishima, Biodegradable particle formation for drug and gene delivery using supercritical fluid and dense gas. *Adv Drug Deliv Rev*, **60**, 411–32 (2008).
176. E. Reverchon, Supercritical antisolvent precipitation of micro-and nano-particles. *J Supercrit Fluids*, **15**, 1–21 (1999).
177. A. Shariati, C. J. Peters, Recent developments in particle design using supercritical fluids. *Curr Opin Solid State Mater Sci*, **7**, 371–83 (2003).
178. C. Wischke, S. P. Schwendeman. Principles of encapsulating hydrophobic drugs in PLA/PLGA microparticles. *International Journal of Pharmaceutics*, **364**, 298–327 (2008).
179. J. P. Rao, K. E. Geckeler, Polymer nanoparticles: Preparation techniques and size-control parameters. *Prog Polym Sci*, (2011).
180. B. Mishra, Bhavesh B. Patel, Sanjay Tiwari, Colloidal nanocarriers: A review on formulation technology, types and applications toward targeted drug delivery. *Nanomedicine Nanotechnology Biology and Medicine* **6**(1), 9–24 (2010).
181. H. K. Chan, P. C. L. Kwok, Production methods for nanodrug particles using the bottom-up approach, *Adv Drug Deliv. Rev.* (2011).

182. A. Vatanara, A. R. Najafabadi, M. Khajeh, Y. Yamini, Solubility of some inhaled glucocorticoids in supercritical carbon dioxide. *J Supercrit. Fluid*, **33**, 21–25 (2005).
183. L. M. Liz-Marzán, P. V. Kamat. Nanoparticles with polymeric surfaces: fillers and models for ultrasoft colloids. In *Nanoscale Materials*, Boston: Kluwer Academic Publishers (2003).
184. J. Yu, C-W. Lee, S-S. Im, J-S. Lee, Structure and magnetic properties of SiO_2 coated Fe_2O_3 nanoparticles synthesized by chemical vapor condensation process, *Rev Adv Mater Sci*, **4**, 55–59 (2003).
185. H. G. Schwick, K. Heide, Immunochemistry and immunology of collagen and gelatin, *Bibliotheca Haematol*, **3**, 111–125 (1969).
186. S. P. Massia, J. Stark, D. S. Letbetter, Surface immobilized dextran limits cell adhesion and spreading, *Biomaterials*, **21**, 2253–2261(2000).
187. J. K. Li, N. Wang, X. E. Wu, A novel biodegradable system based on gelatin nanoparticles and poly(lactic-co-glycolic acid) microspheres for protein and peptide drug delivery, *J Pharm Sci*, **86**(8), 891–895 (1997).
188. E. Miller, N. A. Peppas, D. N. Winslow, Morphological changes of ethylene/vinyl acetate based on controlled delivery systems during release of water-soluble solutes, *J Memb Sci*, **14**, 79–92 (1983).
189. X. Zhao, J. M. Harris, Novel degradable poly (ethylene glycol) hydrogels for controlled release of protein, *J Pharm Sci*, **87**(11), 1450–1458 (1998).
190. R. K. Iler, Multilayers of colloidal particles, *J Colloid Interface Sci*, **21**, 569–594 (1966).
191. G. Decher, J. D. Hong, Buildup of ultrathin multilayer films by a self-assembly process: I. consecutive adsorption of anionic and cationic bipolar amphiphiles. *Makromol. Chem. Macromol. Symp*, **46**, 321–327 (1991).
192. G. Decher, J. D. Hong, J. Schmitt. Buildup of ultrathin multilayer films by a self-assembly process: III. Consecutively alternating adsorption of anionic and cationic polyelectrolytes on charged surfaces. *Thin Solid Films*, **210–211**, 831–835 (1992).
193. G. Decher, J. Schmitt. Fine-Tuning of the film thickness of ultrathin multilayer films composed of consecutively alternating layers of anionic and cationic polyelectrolytes. *Colloid Polymer Sci*, **89**, 160–164 (1992).
194. Y. Lvov, K. Ariga, I. Ichinose, T. Kunitake. Assembly of multi-component protein films by means of electrostatic layer by layer adsorption. *J Am. Chem. Soc*, **117**(22), 6117–6123 (1995).
195. N. A. Kotov, I. Dekany, J. H. Fendler. Layer-by-layer self-assembly of polyelectrolyte-semiconductor nanoparticle composite films. *The Journal of Physical Chemistry*, **99**(35), 13065–13069 (1995).
196. K. Ariga, Y. Lvov, T. Kunitake. Assembling alternate dye—polyion molecular films by electrostatic layer-by-layer adsorption. *Journal of the American Chemical Society*, **119**(9), 2224–2231 (1997).
197. G. B. Sukhorukov, E. Donath, H. Lichtenfeld, E. Knippel, M. Knippel, A. Budde, E. Möhwald. Layer-by-layer self assembly of polyelectrolytes on colloidal particles. *Coll. Surf A*, **137**, 253–265 (1998).
198. D. Ingersoll, P. J. Kulesza, L. R. Faulkner, Polyoxometallate-based layered composite films on electrode. *J. Electrochem Soc*, **141**, 140–147 (1994).

199. N. A. Kotov, I. Dekany, J. H. Fendler, Layer-by-layer self-assembly of polyelectrolyte-semiconductor nanoparticle composite films. *J Phys. Chem*, **99**, 13065–13069 (1995).
200. A. A. Mamedov, N. A. Kotov, M. Prato, D. M. Guldi, J. P. Wicksted, A. Hirsch, Molecular design of strong single-wall carbon nanotube/polyelectrolyte multilayer composites. *Nat Mater*, **1**, 190 (2002).
201. C. Y. Jiang, H. Y. Ko, V. V. Tsukruk, Strain-sensitive Raman modes of carbon nanotubes in deflecting freely suspended nanomembranes. *Adv Mater.*, **17**, 2127–2131 (2005).
202. S. W. Keller, H. N. Kim, T. E. Mallouk, Layer-by-layer assembly of intercalation compounds and heterostructures on surfaces: toward molecular "beaker" epitaxy. *J Am Chem Soc*, **116**, 8817–8818 (1994).
203. T. M. Cooper, A. L. Campbell, R. L. Crane, Formation of polypeptide-dye multilayers by electrostatic self-assembly technique. *Langmuir*, **11**, 2713–2718 (1995).
204. P. Podsiadlo, L. Sui, Y. Elkasabi, P. Burgardt, J. Lee, A. Miryala, W. Kusumaatmaja, M. R. Carman, M. Shtein, J. Kieffer, J. Lahann, N. A. Kotov, Layer-by-layer assembled films of cellulose nanowires with antireflective properties. *Langmuir*, **33**, 7901–7906 (2007).
205. J. A. He, R. Valluzzi, K. Yang, T. Dolukhanyan, C. M. Sung, J. Kumar, S. K. Tripathy, L. Samuelson, L. Balogh, D. A. Tomalia. Electrostatic multilayer deposition of a gold–dendrimer nanocomposite. *Chem Mater*, **11**, 3268–3274 (1999).
206. K. Araki, M. J. Wagner, M. S. Wrighton, Layer-by-layer growth of electrostatically assembled multilayer porphyrin films. *Langmuir*, **12**, 5393–5398 (1996).
207. Y. Lvov, M. Onda, K. Ariga, T. Kunitake, Ultrathin films of charged polysaccharides assembled alternately with linear polyions. *J Biomater Sci-Polym Ed*, **9**, 345–355 (1998).
208. L. Richert, P. Lavalle, D. Vautier, B. Senger, J. F. Stoltz, P. Schaaf, J. C. Voegel, C. Picart, Cell interactions with polyelectrolyte multilayer films. *Biomacromolecules*, **3**, 1170–1178 (2002).
209. F. Boulmedais, V. Ball, P. Schwinte, B. Frisch, P. Schaaf, J. C. Voegel, Buildup of exponentially growing multilayer polypeptide films with internal secondary structure. *Langmuir*, **19**, 440–445 (2003).
210. Y. Lvov, G. Decher, G. Sukhorukov, Assembly of thin films by means of successive deposition of alternate layers of DNA and poly(allylamine). *Macromolecules*, **26**, 5396–5399 (1993).
211. J. D. Hong, K. Lowack, J. Schmitt, G. Decher, Prog. layer-by-layer deposited multilayer assemblies of polyelectrolytes and proteins: From ultrathin films to protein arrays. *Colloid Polym Sci*, **93**, 98–102 (1993).
212. Y. Lvov, K. Ariga, T. Kunitake. Layer-by-layer assembly of alternate protein/polyion ultrathin films. *Chem Lett*, 23,12, 2323 (1994).
213. P. J. Yoo, K. T. Nam, J. F. Qi, S. K. Lee, J. Park, A. M. Belcher, P. T. Hammond. Spontaneous assembly of viruses on multilayered polymer surfaces. *Nat Mater*, **5**, 234 (2006).
214. Z. Y. Tang, Y. Wang, P. Podsiadlo, N. A. Kotov, Biomedical applications of layer-by-layer assembly: from biomimetics to tissue engineering. *Adv Mater*, **18**, 3203–3224 (2006).

215. L. Zhai, F. C. Cebeci, R. E. Cohen, M. F. Rubner, Stable superhydrophobic coatings from polyelectrolyte multilayers. *Nano Lett*, **4**, 1349–1353 (2004).
216. L. Zhai, M. C. Berg, F. C. Cebeci, Y. Kim, J. M. Milwid, M. F. Rubner, R. E. Cohen, Patterned superhydrophobic surfaces: toward a synthetic mimic of the Namib Desert Beetle. *Nano Lett*, **6**, 1213–1217 (2006).
217. D. L. Ellis, M. R. Zakin, L. S. Bernstein, M. F. Rubner, Conductive polymer films as ultrasensitive chemical sensors for hydrazine and monomethylhydrazine vapor. *Anal Chem*, **68**, 817–822 (1996).
218. C. A. Constantine, S. V. Mello, A. Dupont, X. H. Cao, D. Santos, O. N. Oliveira, F. T. Strixino, E. C. Pereira, T. C. Cheng, J. J. Defrank, R. M. Leblanc, Layer-by-layer self-assembled chitosan/poly(thiophene-3-acetic acid) and organophosphorus hydrolase multilayer. *J Am Chem Soc*, **125**, 1805 (2003).
219. D. S. Koktysh, X. R. Liang, B. G. Yun, I. Pastoriza-Santos, R. L. Matts, M. Giersig, C. Serra-Rodriguez, L. M. Liz-Marzan, N. A. Kotov, Biomaterials by design: Layer-by-layer assembled ion-selective and biocompatible films of TiO2 nanoshells for neurochemical monitoring. *Adv Funct Mater*, **12**, 255 (2002).
220. K. C. Wood, H. F. Chuang, R. D. Batten, D. M. Lynn, P. T. Hammond. Controlling interlayer diffusion to achieve sustained, multiagent delivery from layer-by-layer thin films. *Proc Natl Acad Sci U.S.A.*, **103**, 10207–10212 (2006).
221. C. M. Jewell, J. Zhang, N. J. Fredin, D. M. Lynn, Multilayered polyelectrolyte films promote the direct and localized delivery of DNA to cells. *Journal of Controlled Release*, **106**, 214–223 (2005).
222. J. S. Lee, J. Cho, C. Lee, I. Kim, J. Park, Y. M. Kim, H. Shin, J. Lee, F. Caruso. Layer-by-layer assembled charge-trap memory devices with adjustable electronic properties. *Nat Nanotechnol*, **2**, 790–795 (2007).
223. J. Hiller, J. D. Mendelsohn, M. F. Rubner. Reversibly erasable nanoporous anti-reflection coatings from polyelectrolyte multilayers. *Nat Mater*, **1**, 59–63 (2002).
224. D. M. DeLongchamp, P. T. Hammond. High contrast electrochromism and controllable dissolution of assembled Prussian blue polymer nanocomposites. *Adv Funct Mater*, **14**, 224–232 (2004).
225. I. Moriguchi, J. H. Fendler. Characterization of electrochromic properties of ultrathin film self-assembled from poly (diallyldimethylammonium) chloride and sodium decatungstate *Chem Mater*, **10**, 2205–2211 (1998).
226. R. Heuberger, G. Sukhorukov, J. Voros, M. Textor, H. Mohwald. Biofunctional polyelectrolyte multilayers and microcapsules: control of non-specific and bio-specific protein adsorption. *Adv Funct Mater*, **15**, 357–366 (2005).
227. H. Tokuhisa, P. T. Hammond, Solid-state photovoltaic thin films using TiO2, organic dyes, and layer-by-layer polyelectrolyte nanocomposites. *Adv Funct Mater*, **13**, 831–839 (2003).
228. J. Zhang, B. Senger, D. Vautier, C. Picart, P. Schaaf, J. C. Voegel, P. Lavalle. Natural polyelectrolyte films based on layer-by layer deposition of collagen and hyaluronic acid. *Biomaterials*, **26**, 3353–3361 (2005).
229. A. A. Mamedov, A. Belov, M. Giersig, N. N. Mamedova, N. A. Kotov. ChemInform abstract: nanorainbows: graded semiconductor films from quantum dots. *J Am Chem Soc*, **123**, 7738–7739 (2001).

230. D. Y. Wang, A. L. Rogach, F. Caruso. Semiconductor quantum dot-labeled microsphere bioconjugates prepared by stepwise self-assembly. *Nano Lett*, **2**, 857–861 (2002).
231. J. Y. Liu, L. Cheng, Y. H. Song, B. F. Liu, S. J. Dong. Simple preparation method of multilayer polymer films containing PD nanoparticles. *Langmuir*, **17**, 6747–6750 (2001).
232. Y. Shen, J. Y. Liu, J. G. Jiang, B. F. Liu, S. J. Dong. Fabrication of a metalloporphyrin–polyoxometalate hybrid film by a layer-by-layer method and its catalysis for hydrogen evolution and dioxygen reduction. *J Phys Chem B*, **107**, 9744–9748 (2003).
233. A. J. Nolte, M. F. Rubner, R. E. Cohen. Creating effective refractive index gradients within polyelectrolyte multilayer films: Molecularly assembled rugate filters. *Langmuir*, **20**, 3304–3310 (2004).
234. A. A. Mamedov, N. A. Kotov. Free-standing layer-by-layer assembled films of magnetite nanoparticles. *Langmuir*, **16**, 5530–5533 (2000).
235. P. T. Hammond. Form and function in multilayer assembly: New applications at the nanoscale. *Adv Mater*, **16**, 1271–1293 (2004).
236. I. L. Radtchenko, G. B. Sukhorukov, H. Möhwald, Incorporation of macromolecules into polyelectrolyte micro- and nanocapsules via surface controlled precipitation on colloidal particles. *Colloid Surf A*, **202**, 127–133 (2002).
237. I. L. Radtchenko, G. B. Sukhorukov, H. Möhwald, A novel method for encapsulation of poorly water-soluble drugs: precipitation in polyelectrolyte multilayer shells. *Int J Pharm*, **242**, 219–223 (2002).
238. A. A. Antipov, G. B. Sukhorukov, S. Leporatti, I. L. Radtchenko, E. Donath, H. Möhwald, Polyelectrolyte multilayer capsule permeability control. *Colloid Surf A*, **198–200**, 535–541 (2002).
239. J. Fans, J. J. Bozzola, Y. Gao, Encapsulation of uranyl acetate molecules using hollow polymer templates. *J Colloid Interf Sci*, **254**, 108–112 (2002).
240. H. Ai, J. Gao, Size-controlled polyelectrolyte nanocapsules via layer-by-layer self-assembly. *J Mater Sci*, **39**, 1429–1432 (2004).
241. S. Krol, A. Diaspro, O. Cavalleri, D. Cavanna, P. Ballario, B. Grimaldi, P. Filetici, P. Ornaghi, A. Gliozzi, Nanocapsules: A novel tool for medicine and science. In *Frontiers of Multifunctional Integrated Nanosystems*. E. Buzaneva, P. Scharff (eds). Dordrecht: Academic Publishers, pp. 439–446 (2004).
242. J. Cui, D. Fan, J. Hao, Magnetic {Mo72Fe30}-embedded hybrid nanocapsules. *J Colloid Interf Sci*, **330**, 488–492 (2009).
243. T. Crouzier, T. Boudou, C. Picart, Polysaccharide-based polyelectrolyte multilayers, *Curr Opin Colloid Interface Sci*, **15**, 417–426 (2010).
244. F. J. Pavinatto, L. Caseli, O. N. Oliveira, Jr, Chitosan in nanostructured thin films, *Biomacromolecules*, **11**, 1897-1190 (2010).
245. Y. Lvov, A. Agarwal, R. Sawant, V. Torchilin, Sonication assisted polyelectrolyte nanoencapsulation of paclitaxel and tamoxifen, *Pharma Focus Asia*, **7**, 36–40 (2008).
246. N. Pargaonkar, Y. Lvov, N. Li, J. Steenekamp, M. M. de Villiers, Controlled release of dexamethasone from microcapsules produced by polyelectrolyte layer-by-layer nanoassembly, *Pharm Res*, **22**, 826–835 (2005).

247. Z. Zheng, X. Zhang, D. Carbo, C. Clark, C-A. Nathan, Y. Lvov, Sonication assisted synthesis of polyelectrolyte-coated curcumin nanoparticles, *Langmuir*, **26**, 7679–2681 (2010).
248. M. L. Hans, A. M. Lowman. Biodegradable nanoparticles for drug delivery and targeting. *Current Opinion in Solid State and Materials Science*, **6**, 319–327 (2002).
249. Z. M. Miao, S. X. Cheng, X. Z. Zhang, R. X. Zhuo, Study on drug release behaviors of poly-alpha,beta-[*N*-(2-hydroxyethyl)-l-aspartamide]-g-poly(epsiloncaprolactone) nano- and microparticles, *Biomacromolecules*, **7**, 2020–2026 (2006).
250. M. Kosmulski, P. Próchniak, J. B. Rosenholm. Solvents, in which ionic surfactants do not affect the zeta potential. *Journal of Colloid and Interface Science*, **342**(1), 110–113 (2010).
251. E. Ofir, Y. Oren, A. Adin. Electrofloculation: The effect of zeta-potential on particle size. *Desalination*, **204**, 33–38 (2007).
252. S. Rossi, P. F. Luckham, T. F. Tadros. Influence of non-ionic polymers on the rheological behavior of Na^+-montmorillonite clay suspensions. Part II. Homopolymer ethyleneoxide and polypropylene oxide–polyethylene oxide ABA copolymers. *C. S. A: Physiochem Eng Aspects*, **215**(1–3), 1–10 (2003).
253. Y.-I. Jeong, M.-K. Kang, H.-S. Sun, S.-S. Kang, H.-W. Kim, K.-S. Moon, K.-J. Lee, S.-H. Kim, S. Jung. All-trans-retinoic acid release from core-shell type nanoparticles of poly (ε-caprolactone)/poly (ethylene glycol) diblock copolymer. *International Journal of Pharmaceutics*, **273**(1–2), 95–107 (2004).
254. J. Molpeceres, M. Guzmán, P. Bustamante, M. del Rosario Aberturas. Exothermic-endothermic heat of solution shift of cyclosporin a related to poloxamer 188 behavior in aqueous solutions. *International Journal of Pharmaceutics* **130**(1), 75–81 (1996).
255. E. Vega, M. A. Egea, O. Valls, M. Espina, M. L. García. Flurbiprofen loaded biodegradable nanoparticles for ophthalmic administration. *Journal of Pharmaceutical Sciences*, **95**(11), 2393–2405 (2006).
256. A. Bozkir, O. M. Saka, Formulation and investigation of 5-FU nanoparticles with factorial design-based studies. *Farmaco*, **60**(10), 840–846 (2005).
257. M. Leroueil-Le Verger, L. Fluckiger, Y.-I. Kim, M. Hoffman, P. Maincent. Preparation and characterization of nanoparticles containing an antihypertensive agent. *European Journal of Pharmaceutics and Biopharmaceutics*, **46**, **2**, 137–143 (1998).
258. J. Xing, D. Zhang, T. Tan. Studies on the oridonin-loaded poly (D, L-lactic acid) nanoparticles *in vitro* and *in vivo*. *International Journal of Biological Macromolecules*, **40**(2), 153–158 (2007).
259. C. Fonseca, S. Simonesa, B. Rogerio Gaspar, Paclitaxel-loaded PLGA nanoparticles: Preparation, physicochemical characterization and *in vitro* anti-tumoral activity. *Journal of Controlled Release*, **83**, 273–286 (2002).
260. J. M. Rodrigues Jr, H. Fessi, C. Bories, F. Puisieux, J.-ph Devissaguet. Primaquine-loaded poly (lactide) nanoparticles: physicochemical study and acute tolerance in mice. *International Journal of Pharmaceutics*, **126**, 1–2, 29 December, 253–260 (1995).
261. D. Shenoy, S. Little, R. Langer, and M. Amiji. Poly (ethylene oxide)-modified poly (β-amino ester) nanoparticles as a pH-sensitive system for tumor-targeted delivery of hydrophobic drugs: Part I. *In vitro* evaluations. *Mol Pharm*, **2**(5), 357–366 (2005).
262. Noemia Pereira Da Silva Santos, Silene Carneiro Nascimento, Marcela Silvestre Outtes Wanderley, Nicodemos Teles Pontes-Filho, José Figueiredo Da Silva, Eugênia

Cristina Pereira, Nicácio Henrique Da Silva, Neli Kika Honda, Nereide Stela Santos-Magalhães. Nanoencapsulation of usnic acid: An attempt to improve antitumour activity and reduce hepatotoxicity. *European Journal of Pharmaceutics and Biopharmaceutics*, **64**(2), 154–160 (2006).

263. M. Chacón, L. Berges, J. Molpeceres, M. R. Aberturas, M. Guzman. Optimized preparation of poly D, L (lactic-glycolic) microspheres and nanoparticles for oral administration. *International Journal of Pharmaceutics*, **141**(1–2), 6, 81–91 (1996).
264. P. C. van der Hoeven, J. Lyklema. Electrostatic stabilization in non-aqueous media. *Adv Colloid Interface Sci*, **42**, 205–277 (1992).
265. M. Paul, H. Fessi, A. Laatiris, Y. Boulard, R. Durand, M. Deniau, A. Astier. Pentamidine-loaded poly (D, L-lactide) nanoparticles: Physicochemical properties and stability work. *International Journal of Pharmaceutics*, **159**(2), 223–232 (1997).
266. C. Giannavola, Claudio Bucolo, Adriana Maltese, Donatella Paolino, Maria Angela Vandelli, Giovanni Puglisi, Vincent H. L. Lee and Massimo Fresta. Influence of preparation conditions on acyclovir-loaded poly-*d,l*-lactic acid nanospheres and effect of peg coating on ocular drug bioavailability. *Pharmaceutical Research*, **20**(4), 584–590 (2006).
267. C. A. Musumeci, I. Ventura, B. Giannone, L. Ruozi, R. Montenegro, G. Pignatello, Puglisi. PLA/PLGA nanoparticles for sustained release of docetaxel. *Int J Pharm*, **325**(1–2), 172–179 (2006).
268. J. Barrett, J. Nehilla, M. Bergkvist, K. C. Popat, T. A. Desai, Purified and surfactant-free coenzyme Q10-loaded biodegradable nanoparticles. *International Journal of Pharmaceutics*, **348**, 107–114 (2008).
269. S. A. Guhagarkar, V. C. Malshe, P. V. Devarajan, Nanoparticles of polyethylene sebacate: A new biodegradable polymer. *AAPS Pharm Sci Technol*, **10**, 935–942 (2009).
270. N. Jawahar, D. Nagasamy Venkatesh, R. Sureshkumar, V. Senthil, G. N. K. Ganesh, P. Vinoth, S. Sood and M. K. Samanta. Development and charecterization of PLGA-nanoparticles containing carvedilol. *J Pharm Sci & Res.*, **1**(3), 123–128 (2009).
271. I. Brigger, C. Dubernet, P. Couvreur: Nanoparticles in cancer therapy and diagnosis. *Advanced Drug Delivery Reviews*, **54**, 631–651 (2002).
272. C. E. Soma, C. Dubernet, D. Bentolila, S. Benita, P. Couvreur, Reversion of multidrug resistance by co-encapsulation of doxorubicin and cyclosporin A in polyalkylcyanoacrylate nanoparticles. *Biomaterials*, **21**, 1–7 (2000).
273. J. S. Chawla, M. M. Amiji. Biodegradable poly (e-caprolactone) nanoparticles for tumor-targeted delivery of tamoxifen. *International Journal of Pharmaceutics*, **249**, 127–138 (2002).

6

Polymeric Drug Delivery Systems for Encapsulating Hydrophobic Drugs

Dagmar Fischer

6.1 Introduction

Low drug solubility and stability in the physiological environment constitute the main hurdle in attaining appropriate bioavailability. About 40% of all active substances identified by combinatorial screening methods are difficult to formulate by conventional approaches and therefore it is difficult to control their distribution in the body due to their lack of water solubility. Their application requires more sophisticated drug delivery systems aimed at optimizing solubility, pharmacokinetics and release characteristics [1, 2].

Several polymer-based technologies are intended to optimize the technological aspects of hydrophobic drugs. Polymeric materials present a promising approach to the development of new drug delivery systems for the controlled release of poorly soluble drugs due to their favourable, adjustable characteristics and possibilities for further modification [3, 4]. However, the selection and design of a new polymer are a challenging task due to the inherent diversity of structures which requires a detailed understanding of the relationship between the polymer structure and the physico-chemical and biological properties of the polymer [5, 6].

Colloidal carriers based on polymers have largely influenced the controlled drug delivery concept [7]. Micro- and nanotechnology have given the opportunity of controlling and manipulating structures and creating novel architectures. Innovative systems in the micro- and nanometer range include micro- and nanoparticles which can be subdivided into two groups: spheres and capsules [4]. Whereas spheres contain drugs homogeneously dispersed in a polymer matrix, capsules demonstrate a core-shell structure with an active pharmaceutical ingredient distributed in the core that controls the leaching and protects the

Drug Delivery Strategies for Poorly Water-Soluble Drugs, First Edition. Edited by Dennis Douroumis and Alfred Fahr.

encapsulated drug from the external environment. The use of micro- and nanoparticulate systems with different hard-to-formulate drugs has been realized through several concepts. The increase in total surface area in the micro- and nanoscale enhances not only the drug entrapment efficiency but also the dissolution rate of hydrophobic drugs and improves the pharmacokinetic properties [8]. These sophisticated polymer-based drug formulation technologies provide the pharmaceutical industry with strategies to solve some of the limitations associated with hydrophobic active ingredients in therapy and enhance their therapeutic efficiency.

6.2 Drug Encapsulation by Monomer Polymerization

Many different technologies have been developed to prepare particulate drug delivery systems on the basis of polymers [9, 10]. They can be classified into two important groups according to whether the formulation is based on a polymerization reaction or is achieved by direct formulation from preformed synthetic or natural polymers and by desolvation of macromolecules [11]. The use of preformed polymers has already been discussed in Chapter 5. The *in situ* polymerization methods which are characterized by synthesis of the polymer during particle formation will be discussed in the following text. For pharmaceutical applications it can be further distinguished into emulsion polymerization, interfacial polymerization and interfacial condensation. Two different types of emulsion polymerization, i.e., organic and aqueous, depending on the type of the continuous phase, were described in the literature [10, 11]. Figure 6.1 gives an overview of the different *in situ* polymerization techniques and typically used polymers. *In situ* polymerization techniques are a special category of free radical polymerization which encompasses bulk, solution, suspension and emulsion polymerization and is defined as a method of polymerization by which a polymer is formed from successive addition of free radical building blocks.

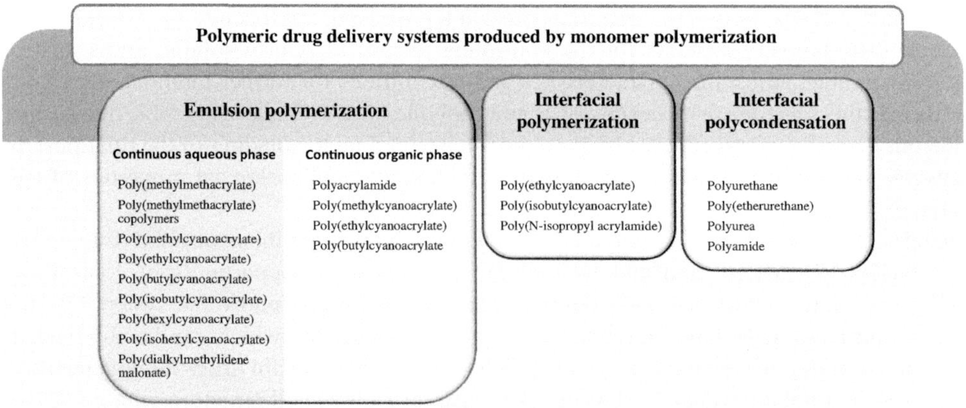

Figure 6.1 *Overview over different in situ polymerization techniques often used in pharmaceutical applications and representative polymers.*

For the formation of nanosized particles by *in situ* polymerization two steps are necessary: first, the preparation of an emulsified system and, second, the formation of the particles by polymerization of the monomers which is the step that gave the method its name [12]. Only in a few cases do both events happen in parallel. Suitable emulsified systems in pharmaceutical applications can be conventional emulsions, mini-emulsions, and micro-emulsions [12, 13]. Particle-forming polymerization processes are heterogeneous polymerizations, i.e., two-phase systems in which the starting monomer is in the form of a fine dispersion in an immiscible liquid. The encapsulation of drugs by *in situ* polymerization techniques can only be undertaken in the presence of active pharmaceutical ingredients that withstand the polymerization and do not react with the monomers. Otherwise the drug has to be adsorbed into the particles after particle formation [2, 3].

6.2.1 Emulsion Polymerization

Emulsion polymerization was first commercialized in the early 1930s and boomed during World War II to produce synthetic rubbers. Since then the method has been extended to a variety of pharmaceutical applications and is today one of the fastest methods for the preparation of micro- and nanoparticles, easily scalable to large dimensions [10, 12, 14]. By definition, emulsion polymerization is a type of heterogeneous free radical polymerization process which involves the emulsification of a relatively hydrophobic monomer in water by an emulsifier (surfactant), followed by the initiation reaction with an initiator. The term 'emulsion polymerization' is misleading since polymerization takes place in so-called latex particles that form spontaneously in the first step of the process, and not in emulsion droplets [14]. Emulsion polymerization can also be accomplished in an organic phase which has been adapted, for instance, for poly(alkyl cyanoacrylate) nanoparticles.

The theory behind the polymerization process was successfully explained by Smith and Ewart [15–17], as well as Harkins [18], and modified by Gardon [19, 20] in the 1940s, based on studies with polystyrene. They described the heterogenous polymerization system as a very complex process consisting of the three steps: (i) nucleation; (ii) growth; and (iii) stabilization of the polymer particles, all of them controlled by free radical polymerization mechanisms in combination with several colloidal phenomena. Although this three-step system is not suitable for all preparations, the Smith-Ewart theory is a useful way of understanding the process. Figure 6.2 shows the emulsion polymerization process which will be described in the following.

Monomer, water, surfactant and initiator are the typical components of an emulsion polymerization system. In classical emulsion polymerization, the monomer is insoluble or hardly soluble in the outer phase whereas the initiator should be soluble in the medium but not in the monomer [10, 14]. The monomer is dispersed and emulsified in a solution of o/w surfactant in water and forms emulsified monomer droplets (about 1–10 μm in diameter, 10^{12}–10^{14} dm^{-3}) [14]. When the concentration of surfactant in the aqueous phase is above its critical micelle concentration (CMC), monomer-swollen micelles (about 5–10 nm in diameter depending on the type and concentration of emulsifier, 10^{19}–10^{21} dm^{-3}) are formed as second species [14, 21]. Additionally, a low amount of free monomer can be molecularly dissolved in the aqueous phase. Most of the monomer molecules accumulate in the monomer droplets that act as giant reservoirs [22]. The volume ratio of monomer phase

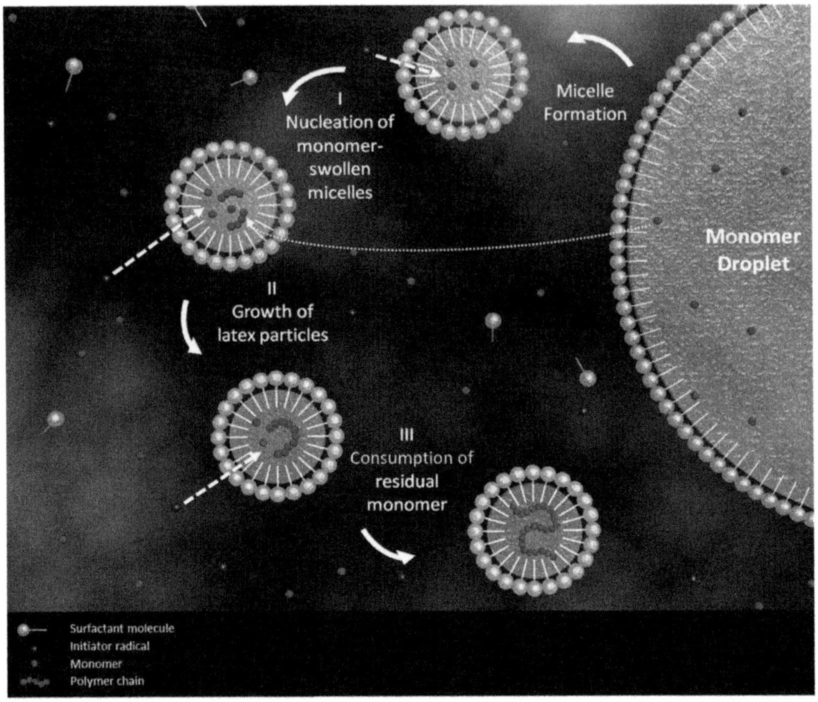

Figure 6.2 *The emulsion polymerization process [27].* (See colour plate section.)

to medium is usually in the range of 0.1–0.5. Polymerizations described in the literature are to be performed at 40–80°C [21].

Polymerization starts with the addition of a water-soluble initiator to the water phase by either free radical or ion formation (Figure 6.3). The initial location of the polymerization is the medium outside the droplets and micelles. Radicals interact with the free monomer in the water phase, polymerizing very quickly up to a critical chain length correlated with an increase in hydrophobicity of the resulting oligomers. The hydrophobic character promotes their uptake into the monomer-swollen micelles followed by a reaction with the monomers therein. The radicals preferentially react with monomers in micelles due to their larger total surface area compared to that of the fewer and larger monomer droplets [21, 22]. The subsequent polymerization of monomers in micelles forms submicron latex particles (about 0.05–1 μm in diameter, 10^{16}–10^{18} dm^{-3}) [14]. The micelles turn into particle nuclei which acquire more reactants from monomer droplets and other micelles for their growth. The absorbance of surfactant by the growing particles reduces the surfactant concentration below the CMC, followed by the dissolution of the inactive micelles. Colloidal stabilization of the growing micelles is additionally accomplished by a surfactant which desorbs from the monomer droplets [10, 14]. About one of every 10^2–10^3 micelles can be transferred into latex particles with typical sizes in the range from 50–300 nm [21]. The so-called particle nucleation stage (the *Smith-Edward Interval I*) ends after exhaustion of the micelles with about 2–15% of total polymer conversion [14, 21]. Details of particle formation and

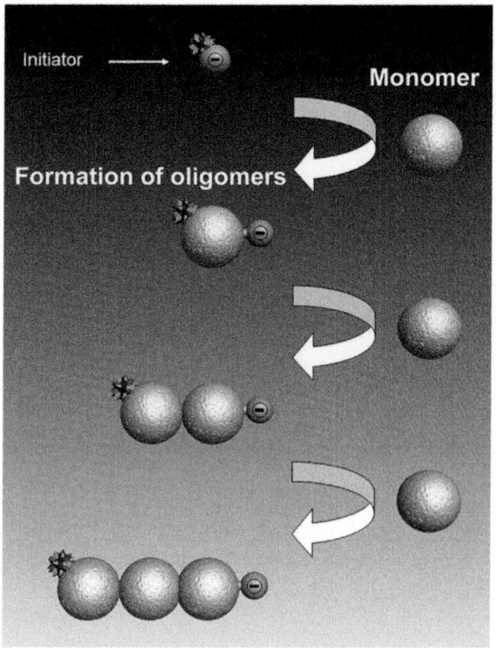

Figure 6.3 *Initiation and propagation step after initiator activation in emulsion polymerization [modified according to 13, 22].*

growth characteristics are strongly dependent on the nature of the polymerization mixture [12, 14, 21].

In the *Smith-Edward Interval II*, the number of latex particles and the polymerization rate remain relatively constant (Figure 6.4). The reaction is mainly localized in monomer-swollen particles whereas monomer droplets supply the growing particles with monomer and surfactant molecules. Stage II is completed when the monomer droplets disappear, with polymer conversion rates up to about 60% [13, 14]. The degree of polymerization can be controlled by the number of latex particles and therefore also by the concentration of the emulsifier.

In *stage III* the number of particles remains constant but the polymerization rate decreases because the latex particles become monomer-starved, i.e. the concentration of monomer in the reaction loci decreases (Figure 6.4). For an application in pharmaceuticals, it is crucial to minimize the residual monomer in the latex products to avoid toxicological hazards to the patients [14, 21].

The following advantages were reported for the emulsion polymerization process in pharmaceutical applications [23, 24]:

- Organic solvents that could be harmful to health can be avoided and replaced by solvents with low toxic potential.
- Under optimized conditions the final product is ready to use. Further processing is not necessary.

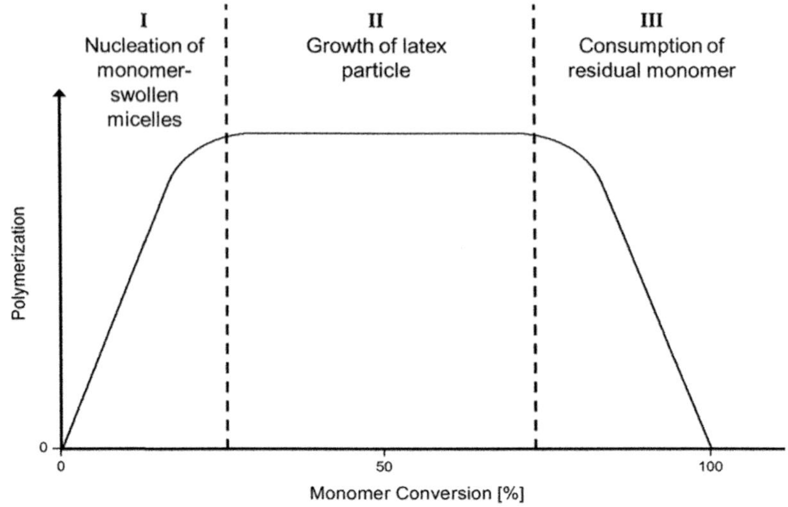

Figure 6.4 *Polymerization rate and monomer conversion in three Smith-Edward Intervals during emulsion polymerization [modified according to 14].*

- The viscosity of the reaction medium is comparable to water and not dependent on the molecular weight of the resulting polymer which facilitates the stirring process especially in large-scale production.
- The polymerization is faster than bulk or solution polymerization at the same temperature and greater average molar masses can be achieved.
- The temperature of the reaction can be controlled easily since the water phase allows a good heat transfer.

Nevertheless, some disadvantages of the emulsion polymerization can also be identified [23,24]:

- Surfactant and other excipients can contaminate the final product and are difficult to remove. Especially water removal is an energy-intensive process during the drying of the final product.
- Emulsion polymerizations are usually designed to operate at a high conversion of monomer to polymer. This can result in significant chain transfer to polymer.
- Generally, the method cannot be used for condensation, ionic or Ziegler-Natta polymerization, although some exceptions are known.

Typical size ranges of droplets produced by the conventional emulsion polymerization as described above are >1 μm [25,26]. A broad range of low water soluble monomers such as acrylics, methacrylics, styrene and vinyl acetate have been used for this application. Inverse emulsions containing a continuous organic phase with aqueous droplets stabilized by stabilizers with HLB (hydrophilic lipophilic balance) values <7 (Span, Tween, Pluronic) and non-ionic surfactants were discussed in only a few reports in pharmacy, mostly for poly(acrylamide) products [27].

Table 6.1 *Comparison of the characteristics of macroemulsion, miniemulsion and microemulsion for* in situ *polymerization [12].*

	Macro-emulsion	Mini-emulsion	Micro-emulsion
Size range	1–10 μm Large polydispersity	50–500 nm Low polydispersity	About 10 nm Low polydispersity
Stability	Seconds to months Non-thermodynamically stable	Hours to months Non-thermodynamically stable	Thermodynamically stable
Surfactant concentration	< 5%	< 5%	10%
Equipment	Simple mechanical stirrer optional: High pressure homogenizator Ultrasound Colloidal mills	High pressure homogenizator Ultrasound Colloidal mills	Spontaneous formation Simple mechanical stirrer

Additionally, mini-emulsion and micro-emulsion techniques are available which have particle nucleation and growth mechanisms as well as kinetics quite different from the classical emulsions. Table 6.1 shows the differences of the systems compared to macro-emulsions. Mini-emulsions with typical droplet sizes of about 50–500 nm are formed by homogenization using high shear forces [25,26]. They are stabilized by surfactants and often an additional hydrophobic, low molecular weight co-stabilizer such as hexadecane or cetyl alcohol to retard the diffusion of monomer molecules from smaller to larger droplets. They demonstrate a kinetic stability over hours to months [27, 28]. The major difference from macro-emulsions is the so-called 'droplet nucleation' [15, 29, 30]. Because of the smaller size and higher surface area of the monomer droplets in mini-emulsions compared to classical emulsions, they are competitive to the micellar and homogeneous nucleation reaction resulting in the entry of radicals directly into the droplets followed by polymerization in the droplets [31]. The same types of monomers with limited water solubility as described for the macro-emulsion can be used also in mini-emulsion polymerization. Mini-emulsion techniques have been demonstrated to be advantageous in the use of hydrophobic monomers and encapsulation of water-insoluble drugs such as indomethacin [32]. To produce non-ionic nanoparticles suited for the encapsulation of hydrophobic drugs, the mini-emulsion polymerization of vinyl acetate with oligocaprolactone macromonomers, benzyl benzoate, or triglycerides from fatty acids as biocompatible hydrophobes and nonionic surfactants has been carried out [33].

Micro-emulsions are one-phase, translucent, thermodynamically stable systems and form spontaneously from mixtures of water and monomer in the presence of a large amount of surfactant (ca. 10%, for comparison in macro-emulsions ca. 1–3%) and often a co-surfactant [13, 27]. The high surfactant concentration has to be taken into consideration for toxicological evaluations. The most often reported stabilization system consists of the anionic surfactant sodium dodecyl sulphate (SDS) and the co-surfactant 1-pentanol [13]. One-pentanol decreases the electrostatic repulsion between the anionic SDS and enhances the membrane flexibility. The same polymers are used as for the other types of emulsions discussed above. Droplet sizes from 10 nm to 100 nm can be obtained which have an extremely large interfacial area (10^5 m^5 dm^{-3}) [13, 27].

In the three different emulsion techniques, drugs can be incorporated during the polymerization step or can be adsorbed afterwards to the surface of the particles [34]. Depending on the monomer type, some drugs are able to initiate the polymerization step by itself, resulting in the covalent attachment of drug molecules to the particles. As an example, photosensitizers such as naphthalocyanines can initiate the polymerization of alkylcyanoacrylates [35]. The addition of cyclodextrins to the polymerization medium can promote the encapsulation of poorly water-soluble drugs [36].

6.2.2 Interfacial Polymerization

Interfacial polymerizations are reactions in which two reactive monomers, dissolved in different immiscible solvents, react at the interface between the two solvents [10, 12, 34]. An ethanolic phase containing an oil or an immiscible organic solvent, the lipophilic monomer and the drug is dispersed in an aqueous external phase containing a surfactant. The ethanol rapidly diffuses out of the organic phase giving tiny oil droplets. Unstirred interfacial polymerization produces films or membranes whereas agitated interfacial polymerization forms micro- and nanoparticles. While nanospheres can be produced by the polymerization of different types of monomers in emulsion as described above, defined core-shell-structured nanocapsules can be obtained by interfacial polymerization techniques. However, up to now, only a few polymers were found to be suitable to produce nanoparticles with this technique for pharmaceutical applications [34].

The major advantage of the interfacial polymerization is that the polymer is very rapidly formed *in situ* by following the contours of the inner phase of an o/w or w/o emulsion. Disadvantages are: (i) the possible occurrence of unwanted reactions between drug and monomer because highly reactive monomers are used; and (ii) the requirement of organic solvents for external phases which requires time-consuming and extensive washing steps of the final product [10].

For the encapsulation of hydrophobic drugs, the following standard technique has been described, but modifications are possible. The lipophilic drug is dissolved together with the oil and the lipophilic monomer in a water-miscible organic solvent (e.g. ethanol). Afterwards the organic phase is injected by needle under intensive stirring into the aqueous phase consisting of a hydrophilic surfactant in water, possibly supplemented with ethanol or acetone. After initiation of the polymerization, a milky suspension of capsules is formed. After removal of the organic phase by vacuum evaporation, particles can be purified, e.g. by repeated centrifugation steps [10, 34]. Lipophilic drugs were described as encapsulated by dissolution in the organic phase. Encapsulation efficiencies of poorly soluble drugs are related to their solubility in the oily core and can be in the range of 50% and 30% for larger and smaller capsules, respectively [37]. For instance, carbamazepine, phenytoin, ethosuccimide, iomustine and indomethacin were successfully encapsulated drugs [38–43].

Mineral and vegetable oils as well as compounds like ethyl oleate, benzyl benzoate, Miglyol and Lipiodol are suitable for the formation of the oily core [34,44,45]. Low toxicity and high chemical stability, compatibility with the polymer shell and a high solubility of the selected drug are the parameters for the selection of the oily core materials. Typical surfactants employed in interfacial polymerizations for the handling of lipophilic drugs are Tweens, Triton X-100 and Poloxamers [34,46]. Hollow microspheres made of poly(N-isopropylacrylamide) (PNIPAM) were reported to be formed by dissolution in an aqueous phase, emulsification with toluene to an o/w emulsion in the presence of a low HLB surfactant (Span 80), followed by initiation by a redox initiation system in the water phase [47].

6.2.3 Interfacial Polycondensation

Interfacial polycondensation is characterized by the *in situ* synthesis of a polymer film at the interface between two immiscible phases, resulting from the condensation of two monomers, each dissolved in one phase [10, 12, 48]. The principles of interfacial polycondensation and parameters to control polymer formation at the interface were first described by Morgan and collegues [49, 50]. Although solid particles as well as hollow capsules can be prepared, this technique has mainly been employed for the preparation of core-shell capsules in pharmaceutical applications. Given that the oligomer molecules during the first stage of polymerization are highly soluble in the droplets, spheres can be formed. In the case that the oligomer molecules are highly insoluble in the droplets, they precipitate at the interface and preferentially form capsules [48].

The classical interfacial polycondensation leads to particles with a mean size diameter of 10–200 μm. For the preparation of nanoparticles, a modified method was developed combining interfacial polycondensation with spontaneous emulsification [51–53]. The emulsification step produces a dispersion of oil droplets in the continuous aqueous phase with a surface, where subsequently the polycondensation takes place, forming a thin, flexible polymer wall at the interface under optimized conditions. The volume ratio of the droplet phase to the outer phase is usually adjusted to ratios of about 1:5 to 1:20 [48]. The process of capsule formulation can be categorized into three steps: (i) initial period of polycondensation; (ii) formation of the primary membrane around the droplets; and (iii) subsequent growth of the membrane to the final capsule shell [48]. Some of the factors controlling the process of capsule formation in interfacial polycondensation reactions are listed in Table 6.2. The partition coefficient of the monomers in the two-phase system determines the initial location of the polycondensation. Solubility and swelling power of aqueous and organic phases control the formation of the primary membranes around the droplets. The higher the solvency of the medium for the polymer, the thicker and less porous the

Table 6.2 *Important factors controlling the course of capsule formation in interfacial polymerization [48]*

- Concentration and partition coefficients of monomers
- Molar mass, crystallizability and chemical structure of the polymer
- Volume ratio of the two phases
- Solvency of both phases for the polymer
- Polymerization rate
- Reaction temperature
- Type of stirrer and stirring rate
- Type and concentration of formulation components like stabilizers, buffer, drugs, etc.

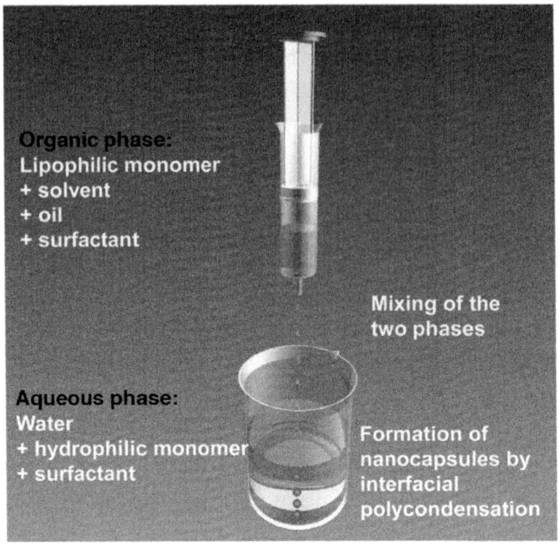

Figure 6.5 *Preparation of nanocapsules by interfacial polycondensation.*

membrane is expected to be. Solubility characteristics of both phases can be substantially modified by additives such as monomers, stabilizers, salts, etc. After formation of the first membrane, the reaction can proceed only if one monomer can permeate the film. This is influenced by the chemical structure of the membrane as well as solvent effects [48,54,55].

For the delivery of hydrophobic drugs, capsules comprising an oily core and a polymeric envelope can be produced by this technique (Figure 6.5). The organic phase is typically composed of a water-miscible solvent, the lipophilic monomer, oil and a lipophilic surfactant while the aqueous phase contains water, the hydrophilic monomer and a hydrophilic surfactant [56]. The characteristics and properties of the polymer shell can be varied by the selection of the type of monomers. Frequently used monomers in pharmaceutical applications are the water-soluble diol or diamine and the oil-soluble di(acid chloride) or diisocyanate forming polyamides, polyesters, polyurea, polyurethanes, or poly(ether urethane) [10,56]. The possibility of modulating the thickness of the polymer wall by simple variation of the monomer concentration and the low concentration of surfactants necessary for production make this technique interesting for nanoparticle production. The thickness of the polymer wall was found to be mainly dependent on the concentration of the lipophilic monomer but independent from that of the hydrophilic monomer [12,55]. Changing the amount of monomer to be added during the preparation process had no particular influence on size. Nevertheless, limitations like the necessity of a purification step at the end of preparation as well as the fact that the method is well adapted only for the encapsulation of lipophilic drugs have to be taken into consideration [12,46,57].

The method to prepare oil-containing nanocapsules by this technique is well established in pharmacy for the encapsulation of lipophilic drugs. As a typical example for lipophilic drugs, alpha-tocopherol was described as efficiently encapsulated with yields up to 65–92% by interfacial polycondensation in several studies [51,53,55,57]. An organic phase consisting of the solvent acetone, the lipophilic monomer isophorone diisocyanate, oil and lipophilic surfactant was injected in an aqueous phase containing hydrophilic diol monomers (1,2-ethanediol, 1,4-butanediol, 1,6-hexanediol) and a hydrophilic emulsifier. The polymer wall of the resulting poly(urethane) nanocapsules, especially that based on the 1,6-hexanediol monomer, showed good stability against aggregation over a period of six months and protected alpha-tocopherol against damage caused by oxygen, temperature or UV irradiation better than a nanoemulsion prepared under the same conditions [57]. Substituting the conventional small monomers like ethanediol, butanediol or hexanediol by poly(ethylene glycol) additionally improved the yield of drug encapsulation.

6.3 Polymeric Nanospheres and Nanocapsules Produced by Polymerization

Nanoparticulate formulations offer several advantages over larger particles due to their increase in total surface area to volume ratio. Higher dissolution rates (according to the Noyes-Whitney equation) and increased saturation solubilities (according to the Kelvin- and Ostwald-Freundlich equation) are parameters which have to be taken into consideration as advantages for the delivery of low soluble drugs [8, 10]. However, up to now only a few monomers were found to be suitable to form nanoparticles for *in vivo* applications [8].

Emulsion polymerization to produce nanoparticles in an outer organic phase up to now has been less important because of the toxic hazards of organic solvents and the other polymerization components of the formulation recipe [10]. Typically poly(acryl amide) nanospheres were produced by this technique, but have the disadvantage of nonbiodegradability [58, 59]. The most abundant literature is related to poly(alkylcyanoacrylates) such as poly(methylmethacrylate), poly(ethylcyanoacrylate) and poly(butylcyanoacrylate) nanoparticles which were produced by dispersion via surfactants in less toxic solvents like cyclohexane, n-pentane and toluene as the organic phase [60]. Examples of encapsulated drugs are timolol [61], triamcinolone [52], and pilocarpine [61].

With the continuous aqueous phase method, the monomer methylmethacrylate (MMA) was used, for example, for encapsulation of an influenza antigen [62] and doxorubicin [63] with high entrapment efficiencies. Copolymerization of methylmethacrylate with hydroxypropylmethacrylate, methacrylic acid or ethylene glycol dimethacrylate was shown to modify particle hydrophilicity and body distribution [64]. Although drugs can be entrapped with high efficiency in PMMA, the missing biodegradability of the polymers and the necessity of a chemical or physical (e.g. gamma-irradiation) initiation have stimulated the search for alternatives. Again, poly(alkylcyanoacrylate) polymers served as an alternative with advantages such as anionic polymerization that performs at room temperature without the need for critical initiators, and biodegradability [10].

Also in interfacial polymerization, alkylcyanoacrylate is typically used to form oil- and water-containing capsules [10]. Not only hydrophilic drugs like doxorubicin, ampicillin, insulin, and monoclonal antibodies were encapsulated with high efficiency [10], but also more lipophilic drugs such as progesterone [65] and triamcinolon diacetate [66]. They have to be dissolved in a solvent like ether or in a liquid surfactant before addition to the aqueous phase.

Many references describe the use of alkylcyanoacrylates in *in situ* polymerization for the encapsulation of small or large and hydrophilic or lipophilic drugs forming all kinds of particles such as spheres, oil- or water-containing capsules [10,67,68]. Poly(alkylcanoacrylates) are usually formed by anionic polymerization initiated by nucleophilic groups of molecules in the polymerization recipe or spontaneously initiated by hydroxyl groups of water [67]. Acidic conditions can reduce the rate of anionic polymerization and avoid the production of polymer aggregates. Very strong acidic conditions are required to initiate redox radical polymerization.

The ethyl-, isobutyl- and isohexyl-derivatives of poly(alkylcyanoacrylates) were used for the production of nanoparticles by interfacial polymerization to encapsulate drugs like calcitonin [59], indomethacin [41], octreotide [69], darodipin [70] and phthalocyanines [71]. They were entrapped in the polymers by dissolution of the monomer and drug in a mixture of oil and absolute ethanol followed by extrusion of the mixture into an aqueous solution containing ethanol or acetone and surfactant. After initiation by ions present in the water, polymerization of the cyanoacrylate started and nanocapsules were formed.

Alternative polymer examples are poly(dialkylmethylidene malonate)s in the form of the ethyl derivatives as primaquine nanocarrier using emulsion polymerization [72]. Nanoparticles can be prepared by interfacial polycondensation of lipophilic monomers such as phthaloyldichloride or the hydrophilic monomer diethylenetriamine in the presence and absence of surfactant.

6.4 Formulation Components

For the *in situ* polymerizations performed as described above the most important formulation components will be discussed.

Typical monomers and comonomers are shown in Figure 6.1 and can also be used in the copolymerization processes [13, 21, 22]. They should be poorly soluble in water which greatly influences polymerization and copolymerization kinetics, have a liquid or gaseous consistency and react by radical polymerization. Highly water-soluble monomers do not form particles, tend to partition in the aqueous phase and not in the polymer particle, and react as in solution polymerization. They will not be incorporated as readily in the polymer chain as monomers with lower aqueous solubility. Insoluble or solid monomers are difficult to disperse in water [10, 12]. Ethylene and olefins, mostly used as minor comonomers in emulsion polymerization, for example, in vinyl acetate copolymers, must be polymerized at high pressures (up to 800 bar). Small amounts of acrylic acid or other ionizable monomers are sometimes used to accomplish colloidal stability. Factors for the selection of monomers are: (i) the performance requirements for the intended applications (e.g. degradability, biocompatibility, release, etc.); (ii) characteristics of the active pharmaceutical ingredient; (iii) and costs. Comonomers can be selected to modify characteristics of the monomers such as mechanical properties, and glass transition temperature [10].

Most initiators used in emulsion polymerization are water-soluble or at least partially water-soluble [10, 21]. Triggers for polymer growth include thermal and redox reactions, high-energy radiation, UV light, or hydroxyl ions. Water-solubility enables the partitioning into the water phase and consequently, the particle generation as described above. Typical water-soluble initiators are ammonium or potassium persulfate, hydrogen peroxide, sulfite, and 2,2′-azobis(amidinopropane) dihydrochloride. Partially water-soluble initiators are succinic acid peroxide, t-butyl-hydroperoxide and azo derivates such as 4,4′-azobis(4-cyanopentanoic acid). Organic peroxides and hydroperoxides are used only in a small number of publications. In redox initiation, at least one component, i.e., the oxidant or the reducing agent (or both) must be water-soluble. The main advantage of redox systems is the independence on temperature with the performance at temperatures lower than 50°C still resulting in acceptable initiation rates. Redox systems typically consist of an oxidant (e.g. persulfate, tert-butyl hydrogen peroxide) combined with a reducing agent (e.g. glucose, sulfite, sodium formaldehyde sulfoxylate) and a redox catalyst (e.g. ferrous ion) [24, 73]. Selection of the surfactant is based on its partitioning between the oil and aqueous phase and its half-life [10].

With regard to the aqueous phase, the following aspects should be taken into consideration. Water is preferred because it can extract high heat energy during polymerization due to its great heat capacity [27]. High quality, deionized water should be used to avoid ions in the water affecting the initiation process or the emulsifier. The addition of inorganic electrolytes may influence the rate of polymerization and stability of emulsion. Alkali metal salts can be complexed and inactivated by ethylenediamine tetraacetic acid (EDTA). Changes of the pH in the aqueous phase can be avoided by bicarbonate, phosphate or citrate buffers. For instance, persulfate initiators decompose to form radicals which will lower the pH. Microbial contamination can be avoided by the addition of preservatives especially in products sold as liquid dispersions [21, 24, 27, 73].

Solvents form the basis for the organic phase in polymerization reactions. When choosing, the possible risk to human health should be taken into consideration. According to the classification of the International Conference of Harmonization (ICH), three different classes can be defined: class 1 (solvents to be avoided); class 2 (solvents to be limited); and class 3 (solvents with low toxic potential). Additionally, the pressure for environmental protection has increased the trend towards the development of water-based polymers although up to now they are often inferior to the solvent-based polymers [10, 27].

One of the key components in the polymerization process is the selection of an adequate surfactant. These stabilizers achieve colloidal stability by electrostatic or steric stabilization methods. The surfactant should guarantee a high rate of polymerization, reduce coagulation during the polymerization process, minimize microbial contamination and fouling especially in large-scale systems, prevent high viscosities otherwise leading to poor heat transfer, and improve tensile strength, gloss and water absorption of the final product. Particle sizes can be decreased and particle numbers increased with higher surfactant concentrations in the initial reaction. To avoid effects on these particle parameters, the addition of surfactant can be delayed after completion of the nucleation to improve particle stability [21, 24, 74].

Nonionic, anionic and cationic surfactants have been used, although anionic surfactants are the most favoured. Surfactants with a low critical micelle concentration (CMC) are preferred, since the polymerization rate shows an increase when the surfactant concentration is above the CMC. In general, a low surfactant concentration is advantageous for economic reasons and due to the often observed adverse effect of the surfactant on the physical properties of the formed polymer.

Most preferred are anionic surfactants usually applied at concentrations of 0.2–3% based on the amount of water. They include sulfates and sulfonates like sodium lauryl sulfate or sodium dodecylbenzene sulfonate, or fatty acid soaps (sodium and potassium stearate, palmitate or laurate). For polymerizations under acidic conditions sulfates and sulfonates are used where fatty acids are unstable. Cationic surfactants (e.g. cetyltrimethylammonium bromide, dodecylammonium chloride) have not been described as first choice for the polymerization in pharmaceutical applications due to their high costs because of the high pressure hydrogenation reaction to be carried out during synthesis, the inefficient emulsifying properties compared to the anionic surfactants or adverse effects on initiator decompositions. Mixtures of cationic and anionic surfactants are not useful since they may form insoluble salts [10, 21, 24, 74]. To avoid hazards to health and the environment, new green surfactants are now under investigation [75].

Nonionic stabilizers (e.g. poly(vinyl alcohol), dextrin, hydroxyethyl cellulose, poly (ethylene oxide)) do not typically form micelles and do not lower the surface tension. They seem to act by physical absorption to or chemical incorporation into the polymer particles and stabilize them via steric interference with the van der Waals attraction forces between the particles. Mechanical shear stability, freeze-thaw stability, and additional electrolyte tolerance are additional functions of the non-ionic stabilizers. The nonionic systems are used in a range from 2–10%, in most cases in a mixture with anionic surfactants since the efficiency to produce stable emulsions alone is quite low compared to others [21, 24, 27].

After particle formation the final product can optionally be supplemented with additional surfactants as post-stabilizers to improve freeze-thaw stability or electrolyte tolerance, antioxidants, UV stabilizers, antimicrobial agents to inhibit microbial or fungal growth or modifiers of rheology [27].

6.5 Control of Particle Morphology

Particle morphology with regard to shape, size, size distribution and surface characteristics plays an important role in the biodistribution of particles in the body, cell uptake and intracellular distribution [4]. Due to the higher surface area leading to faster solubilization rates nano-sized structures usually show higher plasma concentrations and area under the curve values [8]. Wall thickness and porosity determine drug release and drug protection against the environment [55]. Bouchemal *et al.* developed a model for polyamide nanocapsules to predict the thickness of the capsule wall from monomer concentration [46, 55, 57]. The thickness of the polymer wall seems to depend on the concentration of the lipophilic monomer whereas it is independent of the concentration of the hydrophilic monomer [55]. For polyurea and polyamide nanocapsules prepared by interfacial polycondensation, it could be shown that the chemical structure of the monomer is responsible for obtaining different polymeric walls [76]. Low crystallinity is favourable to the growth of the nanocapsules and to the stability of the final suspension.

Using *in situ* polymerization techniques, nanospheres and nanocapsules can be formed depending on the selected process as described above. Nanospheres are characterized by matrices, i.e., the mass is solid whereas nanocapsules have a reservoir form with a solid material shell surrounding a liquid or semisolid core. One of the advantages of capsules over spheres is their lower polymer content and higher loading capacity for lipophilic drugs. Most preparations are spherical although nanospheres with shapes different from spheres were also described. Fully water-soluble solvents (acetone, acetonitrile) have the advantage of forming well-defined capsules whereas water-miscible solvents (ethanol, n-butanol, isopropanol) carry the risk of sphere formation [8, 46].

Particle sizes of *in situ* polymerization products are influenced by the fraction of monomer molecularly dissolved in the continuous phase, initiator and emulsifier concentration as well as the polymerization temperature [21, 22]. Therefore, it has to be taken into consideration that the sizes of monomer droplets and micelles do not represent the size of the final product [21]. For poly(vinyl toluene) particles produced by emulsion polymerization, the particle size increased with increasing temperature which is in line with the dependency of size on nucleation rate [77]. Salt concentration determines the viscosity and ionic strength which also influences the nucleation [21]. The polydispersity of size distribution was found to correlate with the length of nucleation time relative to the entire reaction. Multimodal distributions are often the result of secondary nucleation [55]. Correlations between concentration and molar mass of the monomers and properties of the polymeric wall such as thickness, porosity, polymer swelling, flexibility and storability have been described by several authors [21, 55].

The particular morphology is determined by kinetic (rate of phase separation versus rate of polymerization) and thermodynamic (equilibrium) considerations [21]. It is well known that the properties of the final product largely depend on the particular particle morphology, i.e., the spatial arrangement of the two polymer phases within the particles [78–81]. This arrangement can be controlled by either thermodynamic or kinetic factors. By thermodynamic control the particle morphology development is driven by a minimization of the interfacial free energy resulting in the so-called equilibrium morphology [77, 82]. However, high viscosity in the particles can inhibit the rearrangement of the two polymer phases and prevent the formation of the equilibrium morphology. In these cases the morphology is kinetically controlled, which can be described as the ability of second stage polymer radicals to penetrate into the interior of seed particles [79, 83]. Using water-soluble initiators a slow diffusion inhibits the entry of the particles and leads to polymerization, preferentially at the outer shell, resulting in the formation of core-shell particles [79]. Possibilities to influence the penetration process include a change of the seed polymer nature (e.g. Tg) or radicals, use of chain transfer agents, or anchorage of the second stage polymer chains to the particle surface, for example, by ionic end groups [82, 83]. Ionic initiators increase the possibility of obtaining a core-shell particle by decreasing the penetration of radicals [82]. Chain transfer agents were found to increase radical penetration correlated with deviations from the core-shell structure, but only at high concentrations of the agent [82]. The described effects are often only observed under very specific conditions and compositions of the formulation recipes [82].

6.6 Toxicity and *In Vivo* Performance

Polymers to be selected for *in situ* polymerization of particles need to fulfil several requirements for *in vivo* applications such as biocompatibility, non-immunogenicity, biodegradability, elimination from the body without undesired accumulation, and the possibility of being formulated to accomplish the desired drug release profiles [4].

The colloidal properties of micro- and nanoparticles for drug release are of great importance from both the academic and the industrial point of view [3]. However, since nanoparticles are defined as an intermediate state of matter between bulk and molecular material, many of their properties are dependent on the particle size. Consequently, not only the toxicological hazards related to the components of the recipe such as surfactants, monomers or initiators which could be found in the product as impurities have to be taken into consideration, as discussed above, but also the toxicological aspects related to the nano-size. Careful evaluation of any new particle and its components intended for clinical use should be based upon sufficient data from systematic *in vitro* and *in vivo* testing to ensure that the delivery systems are safe for *in vivo* use.

Regarding the polymer material, based on many reports, the toxicity is thought to be influenced by characteristics such as: (i) molecular weight and polydispersity of the carrier material; (ii) charged functionalities; (iii) charge density; (iv) structure and sequence as well as (v) architecture and conformational flexibility of the molecules [4]. Furthermore, the conditions used to perform the experiments such as particle concentration, exposure time, and the selection of the cell types all play an important role and have to be taken into consideration, especially when different publications are compared. Molar mass of the carrier materials of the particles plays an important role since these materials can be released after degradation or dissociation of the particles and come into contact with tissues, cells and fluids of the body [84–86]. Circulation in the bloodstream, half-life time in the blood as well as rate and mechanism of polymer clearance are highly dependent on molar mass. Non-ionic and ionic polymers with a molar mass below the renal threshold barrier are usually cleared relatively quickly from the bloodstream and excreted by the kidneys without significant extravasation into tissues. Surface charge was reported to be an important parameter for cytotoxicity, haemocompatibility, complement activation, body distribution and internalization into cells [87, 88]. Cationic polymers showed intensive electrostatic interactions with plasma proteins, cell membranes and connective tissue, leading to unwanted rapid elimination from the blood. Negatively charged polymers are known to be cleared from the blood more slowly than cationic systems, while non-ionic particles exhibit increased circulation half-lives. Consequently, extended blood circulation could be achieved by using macromolecules with a hydrophilic, non-charged or slightly anionic structure and sufficiently large size. A strong correlation between the density of accessible positive surface charges and the extent of complement activation was reported [87, 89, 90].

The use of a nanotechnological carrier in therapeutic and diagnostic medicine is currently one of the most active and perspective research area. Synthetic biodegradable polymers like poly(alkyl cyanoacrylates) (PACA) were preferred compared to nonbiodegradable polymers such as poly(alkyl methacrylates), poly(acryl amides) or polystryrole since they can be repeatedly applied without the risk of accumulation in the body [10]. Poly(alkyl cyanoacryles) are one of the most potential synthetic materials for *in situ* polymerizations in drug delivery due to their biocompatible, biodegradable and low-toxic charcteristics. They are suitable especially for the delivery of hydrophobic drugs due to their inherent hydrophobic nature. Several formulations are under investigation including cancer agents, antibiotics, and antiviral agents [91].

6.7 Scale-Up Considerations

In the area of conflict between the reduction of margins, increasing competition and public sensitivity to environmental hazards, industrial producers have to achieve a safe, environmentally-friendly and efficient production of particles of high quality [22]. The *in situ* polymerizations produce 'products by process' whose main properties are determined

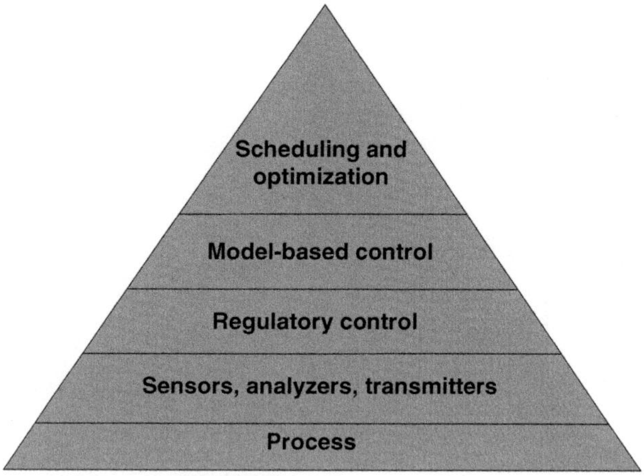

Figure 6.6 *The process control hierarchy in polymerization reactors as postulated by Richards and Congalidis [modified according to 94].*

during the formation process [21]. Consequently, critical points are the knowledge and controllability of process variables since not only the recipe but also the preparation techniques can influence the final product properties. The control of polymerization reactors and their safe operation, selection of process parameters and specifications of the final product are therefore challenging due to the complexity of the physical processes and kinetics during production [92]. For successful application and modelling of process control in *in situ* polymerization, a hierarchical approach was postulated by Richards and Congalidis [93] (Figure 6.6) based on (i) process knowledge; (ii) appropriate sensors; transmitters and analyzers as basic control systems to (iii) regulate pressure, temperature, level and flow; (iv) model-based control algorithms; and (v) on-line scheduling and optimization strategies [93, 94].

Three different types of reactors (so-called 'feed strategies') such as batch, semi-batch and continuous tank reactors were described for the large-scale production of *in situ* polymerization, especially for emulsion polymerization (Figure 6.7) [13, 27, 95]. Continuous working systems are employed for large-scale production whereas batch and semi-batch processes are typically used for low volume, high value products [13]. The choice depends on the properties desired in the final polymer or dispersion and on the economics of the product.

The quite simple batch approach is used on the laboratory scale to investigate new formulations, understand reaction mechanisms and kinetics and to generate information for the scale-up [27, 96]. All ingredients are added into a stirred reactor and heated with no material added or removed during polymerization. The technique is limited due to the small surface-to-volume ratio of the reactor, resulting in a low heat transfer capacity and a difficult control of reaction temperature [13]. Since there is little control beyond recipe, temperature and agitation, batch-to-batch reproducibility is difficult to accomplish.

Semi-batch or semi-continuous processes are characterized by a seeding step and a programmed feeding of monomer to the reactor [27, 97, 98]. During seeding only a fraction of the emulsion is charged in the reactor and initiator is added to start polymerization. For

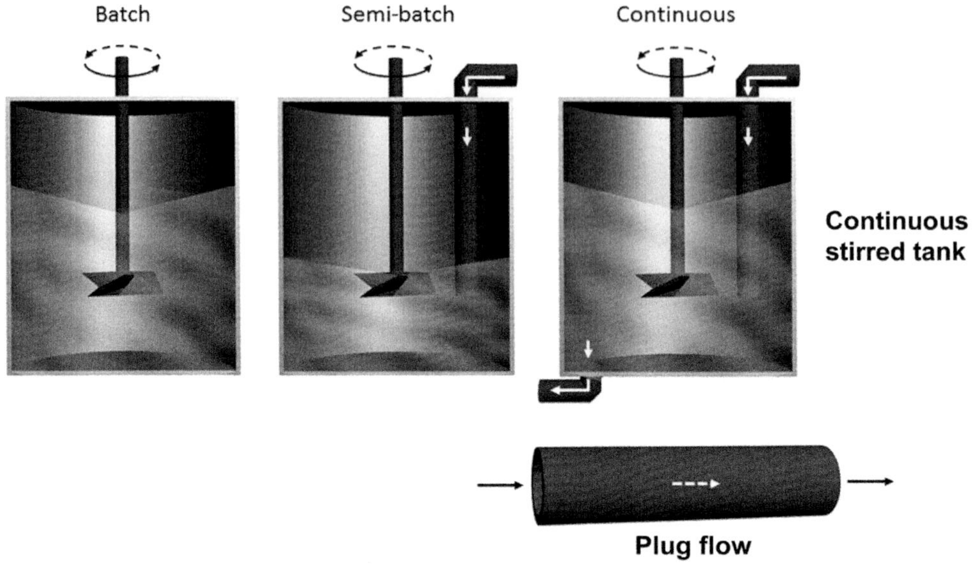

Figure 6.7 *Type of polymerization reactors for production of large-scale latex products [13].*

feeding, additional monomer is continuously added to supply the growing particles. This enables a starve-fed reaction to insure a good distribution of monomers into the polymer chain [99, 100]. Control of the rate of monomer addition accomplishes the control of the rate of heat generation. Continuous reactors can be designed as tubular plug flow reactors, continuous stirred tank reactors or a combination of both [13, 27, 101, 102]. The monomer emulsion is fed at a constant rate to the reactor for polymerization, and in parallel the final product is continuously removed from the reactor. Reactor conditions do not change during the entire process so that in consequence the final product is consistent and batch-to-batch variability is low. Heat removal can be effectively controlled. Plug flow reactors are not as flexible as continuous stirred tanks in terms of processing and production capabilities and are therefore not often used in industry [27].

Several key parameters were identified which could influence the properties of the end-products in large-scale synthesis. Due to the reduction of residence time of the growing particles in the reactor in the order continuous > semi-batch > batch reactor, the breadth of particle size distribution decreased in the same order [13]. The rate of polymerization followed the ranking batch > semi-batch > continuous system [13]. In a dynamic model for semi-batch reactors, both particle growth and particle size distribution were found to be affected by the monomer feed rate [99, 100]. The molecular weight distribution was a function of reactor temperature. Several authors have investigated the initiation of emulsion polymerization by ultrasound, for instance, for methyl methyacrylate, butyl acrylate, and styrene [103–105]. This technique demonstrated some advantages compared to the conventional emulsification process such as narrow particle size distribution, higher polymerization rate and monomer conversion already with a lower amount of initiator. The distribution of the initiator was found to be faster due to mechanical shearing and increased the speed of formation of radical initiators.

6.8 Conclusion

The development and optimization of *in situ* polymerization techniques have been driven by the need for safe applications with less toxic reagents, environmental considerations and simplification of the processes, thus allowing an economic upscale and the improvement of yield and drug entrapment efficiency. This is particularly true for the delivery of hydrophobic active pharmaceutical ingredients, since an efficient entrapment with high biological activity of the drug is of great importance for industrial applications. In this chapter, it was shown that several such *in situ* polymerization techniques are available to form polymeric spheres and capsules on the micro- and nano-scale, depending on the physico-chemical characteristics of drugs and polymers. Nevertheless, several problems remain, such as efficient purification of the particles, efficient and reliable residual solvent analysis, long-term stability of the active components, and highly reproducible release characteristics. Despite these technological challenges, micro- and nanoparticles have today shown great promise in the development of hydrophobic drug administration.

Acknowledgements

A very special thank you goes to Hannes Ahrem for the excellent graphics and Stefan Grund for his excellent support in the preparation of the manuscript.

References

1. P. van Hoogevest, X. Liu, A. Fahr, Drug delivery strategies for poorly water-soluble drugs: the industrial perspective. *Expert Opinion on Drug Delivery*, **8**(11), 1481–1500 (2011).
2. E.M.G. Merisko-Liversidge, G. Liversidge, Drug nanoparticles: Formulating poorly water-soluble compounds. *Toxicologic Pathology*, **36**(1), 43–48 (2008).
3. V.K. Gupta, *et al.*, Nanoparticle formulation for hydrophilic and hydrophobic drugs. *International Journal of Pharmaceutical Sciences and Research*, **1**(2), 163–169 (2010).
4. S. Grund, M. Bauer, D. Fischer, Polymers in drug delivery: State of the art and future trends. *Advanced Engineering Materials*, **13**(3), B61–B87 (2011).
5. T.M. Allen, P.R. Cullis, Drug delivery systems: Entering the mainstream. *Science*, **303**(5665), 1818–1822 (2004).
6. O. Pillai, R. Panchagnula, Polymers in drug delivery. *Current Opinion in Chemical Biology*, **5**(4), 447–451 (2001).
7. A.S. Hoffman, The origins and evolution of 'controlled' drug delivery systems. *Journal of Controlled Release*, **132**(3), 153–163 (2008).
8. A. Sosnik, A.M. Carcaboso, D.A. Chiappetta, Polymeric nanocarriers: New endeavors for the optimization of the technological aspects of drugs. *Recent Patents on Biomedical Engineering*, **1**(1), 43–59 (2008).
9. J. Coelho, *et al.*, Drug delivery systems: Advanced technologies potentially applicable in personalized treatments. *The EPMA Journal*, **1**(1), 164–209 (2010).

10. C. Pinto Reis, *et al.*, Nanoencapsulation I. Methods for preparation of drug-loaded polymeric nanoparticles. *Nanomedicine: Nanotechnology, Biology and Medicine*, **2**(1), 8–21 (2006).
11. P. Couvreur, C. Dubernet, F. Puisieux, *Controlled Drug Delivery with Nanoparticles: Current Possibilities and Future Trends*. Vol. 41. Amsterdam: Elsevier (1995).
12. C. Vauthier, K. Bouchemal, Methods for the preparation and manufacture of polymeric nanoparticles. *Pharm Res*, **26**(5), 1025–1058 (2009).
13. C.S. Chern, *Principles and Applications of Emulsion Polymerization*. New York: Wiley (2008).
14. C.S. Chern, Emulsion polymerization mechanisms and kinetics. *Progress in Polymer Science*, **31**(5), 443–486 (2006).
15. W.V. Smith, R.H. Ewart, Kinetics of emulsion polymerization *Journal of Chemical Physics*, **16**(6), 592–599 (1948).
16. W.V. Smith, The kinetics of styrene emulsion polymerization. *Journal of the American Chemical Society*, **70**(11), 3695–3702 (1948).
17. W.V. Smith, Chain initiation in styrene emulsion polymerization. *Journal of the American Chemical Society*, **71**(12), 4077–4082 (1949).
18. W.D. Harkins, A general theory of the mechanism of emulsion polymerization. *Journal of the American Chemical Society*, **69**(6), 1428–1444 (1947).
19. J.L. Gardon, Emulsion polymerization. I. Recalculation and extension of the Smith-Ewart theory. *Journal of Polymer Science Part A-1: Polymer Chemistry*, **6**(3), 623–641 (1968).
20. J.L. Gardon, Emulsion polymerization. II. Review of experimental data in the context of the revised Smith-Ewart theory. *Journal of Polymer Science Part A-1: Polymer Chemistry*, **6**(3), 643–664 (1968).
21. R. Arshady, Suspension, emulsion, and dispersion polymerization: A methodological survey. *Colloid & Polymer Science*, **270**(8), 717–732 (1992).
22. J.M. Asua, Emulsion polymerization: From fundamental mechanisms to process developments. *Journal of Polymer Science Part A: Polymer Chemistry*, **42**(5), 1025–1041 (2004).
23. P.R. Lockman, *et al.*, Nanoparticle technology for drug delivery across the blood–brain barrier. *Drug Development and Industrial Pharmacy*, **28**(1), 1–13 (2002).
24. G. Odian, *Principles of Polymerization*. New York: Wiley (2004).
25. K. Landfester, Polyreactions in miniemulsions. *Macromolecular Rapid Communications*, **22**(12), 896–936 (2001).
26. J. Qiu, B. Charleux, K. Matyjaszewski, Controlled/living radical polymerization in aqueous media: Homogeneous and heterogeneous systems. *Progress in Polymer Science*, **26**(10), 2083–2134 (2001).
27. C.D. Anderson, E.S. Daniels, Emulsion polymerisation and latex applications. *Rapra Reviews Reports*. Vol. 14., Shropshire: iSmithers Publishing (2003).
28. C. Vauthier, K. Bouchemal, Methods for the preparation and manufacture of polymeric nanoparticles. *Pharmaceutical Research*, **26**(5), 1025–1058 (2009).
29. R.M. Fitch, C.H. Tsai, Particle formation in polymer colloids, III: Prediction of the number of particles by a homogeneous nucleation theroy. *Polymer Colloids*, ed. R.M. Fitch. New York: Plenum Press (1971).

30. W.D. Harkins, A general theory of the reaction loci in emulsion polymerization *Journal of Chemical Physics*, **13**(9), 381 (1945).
31. I. Capek, C.-S. Chern, Radical polymerization in direct mini-emulsion systems, In *New Polymerization Techniques and Synthetic Methodologies*, Berlin: Springer Verlag, pp. 101–165 (2001).
32. I. Rajot, *et al.*, Nonionic nanoparticles by miniemulsion polymerization of vinyl acetate with oligocaprolactone macromonomer or miglyol as hydrophobe application to the encapsulation of indomethacin. *Macromolecules*, **36**(20), 7484–7490 (2003).
33. K. Landfester, Synthesis of colloidal particles in miniemulsions, in *Annual Review of Materials Research*. Palo Alto, CA: Annual Reviews, pp. 231–279 (2006).
34. V.P. Torchilin, *Nanoparticulates as Drug Carriers*. London: Imperial College Press (2006).
35. A. Labib, *et al.*, Biodegradable nanospheres containing phthalocyanines and naphthalocyanines for targeted photodynamic tumor therapy. *Pharmaceutical Research*, **8**(8), 1027–1031 (1991).
36. H. Boudad, *et al.*, Combined hydroxypropyl-beta-cyclodextrin and poly(alkylcyanoacrylate) nanoparticles intended for oral administration of saquinavir. *International Journal of Pharmaceutics*, **218**(1–2), 113–124 (2001).
37. E. Allémann, J. Leroux, R. Gurny, Polymeric nano- and microparticles for the oral delivery of peptides and peptidomimetics. *Advanced Drug Delivery Reviews*, **34**(2–3), 171–189 (1998).
38. M. Fresta, *et al.*, Preparation and characterization of polyethyl-2-cyanoacrylate nanocapsules containing antiepileptic drugs. *Biomaterials*, **17**(8), 751–758 (1996).
39. L. Marchal-Heussler, *et al.*, Colloidal drug delivery systems for the eye: A comparison of the efficacy of three different polymers: PolYisobutylcyanoacrylate, polylactic-co-glycolic acid, polyepsilon-caprolacton. *STP Pharma Sciences*, **2**, 98–104 (1992).
40. S.D. Desai, J. Blanchard, Pluronic® F127-based ocular delivery system containing biodegradable polyisobutylcyanoacrylate nanocapsules of pilocarpine. *Drug Delivery*, **7**(4), 201–207 (2000).
41. A. Gürsoy, *et al.*, Evaluation of indomethacin nanocapsules for their physical stability and inhibitory activity on inflammation and platelet aggregation. *International Journal of Pharmaceutics*, **52**(2), 101–108 (1989).
42. N. Ammoury, *et al.*, Physicochemical characterization of polymeric nanocapsules and in vitro release evaluation of indomethacin as a model drug. *STP Pharma Sciences*, **5**, 642 (1989).
43. M.S. El-Samaligy, *et al.*, Ocular disposition of nanoencapsulated acyclovir and ganciclovir via intravitreal injection in rabbit's eye. *Drug Delivery*, **3**(2), 93–97 (1996).
44. F. Chouinard, *et al.*, Preparation and purification of polyisohexylcyanoacrylate nanocapsules. *International Journal of Pharmaceutics*, **72**(3), 211–217 (1991).
45. N. Al Khouri Fallouh, *et al.*, Development of a new process for the manufacture of polyisobutylcyanoacrylate nanocapsules. *International Journal of Pharmaceutics*, **28**(2–3), 125–132 (1986).
46. G. Puglisi, *et al.*, Influence of the preparation conditions on poly(ethylcyanoacrylate) nanocapsule formation. *International Journal of Pharmaceutics*, **125**(2), 283–287 (1995).

47. Q. Sun, Y. Deng, In situ synthesis of temperature-sensitive hollow microspheres via interfacial polymerization. *Journal of the American Chemical Society*, **127**(23), 8274–8275 (2005).
48. R. Arshady, Preparation of microspheres and microcapsules by interfacial polycondensation techniques. *Journal of Microencapsulation*, **6**(1), 13–28 (1989).
49. E.L. Wittbecker, P.W. Morgan, Interfacial polycondensation. *I. Journal of Polymer Science*, **40**(137), 289–297 (1959).
50. P.W. Morgan, S.L. Kwolek, Interfacial polycondensation. II. Fundamentals of polymer formation at liquid interfaces. *Journal of Polymer Science*, **40**(137), 299–327 (1959).
51. K. Bouchemal, *et al.*, Simultaneous emulsification and interfacial polycondensation for the preparation of colloidal suspensions of nanocapsules. *Materials Science and Engineering: C*, **26**(2–3), 472–480 (2006).
52. H-J.Krause, A. Schwarz, P. Rohdewald, Interfacial polymerization, a useful method for the preparation of polymethylcyanoacrylate nanoparticles. *Drug Development and Industrial Pharmacy*, **12**(4), 527–552 (1986).
53. K.Bouchemal, *et al.*, Synthesis and characterization of polyurethane and poly(ether urethane) nanocapsules using a new technique of interfacial polycondensation combined to spontaneous emulsification. *International Journal of Pharmaceutics*, **269**(1), 89–100 (2004).
54. E. Jabbari, Characterization of microcapsules prepared by interfacial polycondensation of methylene bis(phenyl isocyanate) with hexamethylene diamine. *Iranian Polymer Journal*, **10**(1), 33–43 (2001).
55. K. Bouchemal, *et al.*, Polyamides nanocapsules: Modeling and wall thickness estimation. *AIChE Journal*, **52**(6), 2161–2170 (2006).
56. L. Torini, J.F. Argillier, N. Zydowicz, Interfacial polycondensation encapsulation in miniemulsion. *Macromolecules*, **38**(8), 3225–3236 (2005).
57. K. Bouchemal, *et al.*, Stability studies on colloidal suspensions of polyurethane nanocapsules. *Journal of Nanoscience and Nanotechnology*, **6**(9–10), 3187–3192 (2006).
58. B.I. Ekman, I. Sjöholm, Improved stability of proteins immobilized in microparticles prepared by a modified emulsion polymerization technique. *Journal of Pharmaceutical Sciences*, **67**(5), 693–696 (1978).
59. P.J. Lowe, C.S. Temple, Calcitonin and insulin in isobutylcyanoacrylate nanocapsules: protection against proteases and effect on intestinal absorption in rats. *Journal of Pharmacy and Pharmacology*, **46**(7), 547–552 (1994).
60. M.S. El-Samaligy, P. Rohdewald, H.A. Mahmoud, Polyalkyl cyanoacrylate nanocapsules. *Journal of Pharmacy and Pharmacology*, **38**(3), 216–218 (1986).
61. T. Harmia-Pulkkinen, A. Tuomi, E. Kristoffersson, Manufacture of polyalkylcyanoacrylate nanoparticles with pilocarpine and timolol by micelle polymerization: Factors influencing particle formation. *Journal of Microencapsulation*, **6**(1), 87–93 (1989).
62. J.P. Kreuter, P. Speiser, New adjuvants on a polymethylmethacrylate base. *Infection and Immunity*, **13**(1), 204–210 (1976).
63. A. Rolland, *et al.*, Purification et propriétés physico-chimiques des suspensions de nanoparticules de polymère. *Journal de Pharmacie de Belgique*, **41**, 94–105 (1986).

64. E. Allémann, R. Gurny, E. Doekler, Drug-loaded nanoparticles-preparation methods and drug targeting issues. *European Journal of Pharmaceutics and Biopharmaceutics*, **39**, 173–191 (1993).
65. V.H.K. Li, *et al.*, Ocular drug delivery of progesterone using nanoparticles. *Journal of Microencapsulation*, **3**(3), 213–218 (1986).
66. M.S. El-Samaligy, P. Rohdewald, Triamcinolone diacetate nanoparticles: a sustained release drug delivery system suitable for parental administration. *Pharmaceutica Acta Helvetiae*, **57**, 201–204 (1982).
67. J. Kreuter, Nanoparticle: Preparation and applications. In *Microcapsules and Nanoparticles in Medicine and Pharmacy*, M. Donbrow (ed.). Boca Raton, FL: CRC Press (1992).
68. L. Grislain, *et al.*, Pharmacokinetics and distribution of a biodegradable drug-carrier. *International Journal of Pharmaceutics*, **15**(3), 335–345 (1983).
69. C. Damgé, *et al.*, Poly(alkyl cyanoacrylate) nanocapsules as a delivery system in the rat for octreotide, a long-acting somatostatin analogue. *Journal of Pharmacy and Pharmacology*, **49**(10), 949–954 (1997)
70. B. Hubert, *et al.*, The preparation and acute antihypertensive effects of a nanocapsular form of darodipine, a dihydropyridine calcium entry blocker. *Pharmaceutical Research*, **8**(6), 734–738 (1991).
71. V. Lenaerts, *et al.*, Nanocapsules with a reduced liver uptake: Targeting of phthalocyanines to EMT-6 mouse mammary tumour in vivo. *European Journal of Pharmaceutics and Biopharmaceutics*, **41**, 38–43 (1995).
72. F. Lescure, *et al.*, Synthesis and evaluation of a new biodegradable monomer. *Proceedings of International Symposium on Controlled Release Bioactive Materials*, **18**, 325–326 (1991).
73. Emulsion polymerization. 2012: http://en.wikipedia.org.
74. J-L. Salager, *Surfactants: Types and Uses.* FIRP Booklet # 300A (1999).
75. A.M. Fernandez, *et al.*, New green surfactants for emulsion polymerization. *Progress in Organic Coatings*, **53**(4), 246–255 (2005).
76. I. Montasser, S. Briançon, H. Fessi, The effect of monomers on the formulation of polymeric nanocapsules based on polyureas and polyamides. *International Journal of Pharmaceutics*, **335**(1–2), 176–179 (2007).
77. P. Bagchi, B.V. Gray, S.M. Birnbaum, Preparation of model poly(vinyl toluene) lattices and characterization of their surface charge by titration and electrophoresis. *Journal of Colloid and Interface Science*, **69**(3), 502–552 (1979).
78. V. Dimonie, *et al.*, Control of particle morphology. In *Emulsion Polymerization and Emulsion Polymers*, P.A. Lovell, M.S. El-Aasser (eds). New York: John Wiley and Sons (1997).
79. J. Stubbs, *et al.*, Non-equilibrium particle morphology development in seeded emulsion polymerization. 1: Penetration of monomer and radicals as a function of monomer feed rate during second stage polymerization. *Colloids and Surfaces A: Physicochemical and Engineering Aspects*, **153**(1–3), 255–270 (1999).
80. D.C. Sundberg, *et al.*, Morphology development of polymeric microparticles in aqueous dispersions. I. Thermodynamic considerations. *Journal of Applied Polymer Science*, **41**(7–8), 1425–1442 (1990).

81. J.A. Waters, Predicting the surface morphology of composite latex particles. *Colloids and Surfaces A: Physicochemical and Engineering Aspects*, **83**(2), 167–174 (1994).
82. J Stubbs, D. Sundberg, Fundamental studies on morphology control for latex systems with application to waterborne coatings: The effect of polymer radical mobility in latex particles during polymerization. *Journal of Coatings Technology*, **75**(938), 59–67 (2003).
83. L.E. Ivarsson, L.E., O.J. Karlsson, D.C. Sundberg, Influence of glass transition temperature on latex particle morphology. *Macromolecular Symposia*, **151**(1), 407–412 (2000).
84. T. Yamaoka, Tabata, Y. Ikada, *Comparison of Body Distribution of Poly(Vinyl Alcohol) with Other Water-Soluble Polymers after Intravenous Administration*. Vol. 47. Chichester: Wiley (1995).
85. T. Yamaoka, Y. Tabata, Y. Ikada, Distribution and tissue uptake of poly(ethylene glycol) with different molecular weights after intravenous administration to mice. *Journal of Pharmaceutical Sciences*, **83**(4), 601–606 (1994).
86. T. Yamaoka, Y. Tabata, Y. Ikada, Fate of water-soluble polymers administered via different routes. *Journal of Pharmaceutical Sciences*, **84**(3), 349–354 (1995).
87. D. Fischer, *et al.*, In vitro cytotoxicity testing of polycations: Influence of polymer structure on cell viability and hemolysis. *Biomaterials*, **24**(7), 1121–1131 (2003).
88. L. Parhamifar, *et al.*, Polycation cytotoxicity: A delicate matter for nucleic acid therapy-focus on polyethylenimine. *Soft Matter* (2010).
89. S. Choksakulnimitr, *et al.*, In vitro cytotoxicity of macromolecules in different cell culture systems. *Journal of Controlled Release*, **34**(3), 233–241 (1995).
90. R.K. Kainthan, *et al.*, Blood compatibility of novel water soluble hyperbranched polyglycerol-based multivalent cationic polymers and their interaction with DNA. *Biomaterials*, **27**(31), 5377–90 (2006).
91. G. Yordanov, C. Dushkin, Recent advances in the preparation of drug-loaded poly(alkyl cyanoacrylate) colloidal particles for cancer treatment: Nanoprecipitation vs. polymerization. http://costd43.lcpe.uni-sofia.bg/files/Articles/Yordanov_NW2011_rev2.pdf, 2012.
92. J. Pohn, *et al.*, Computational analysis of mixing and scale-up in emulsion polymerization reactors. *Macromolecular Symposia*, **302**(1), 133–141 (2011).
93. J.R. Richards, Schnelle, P.D. Jr., Perspectives on industrial reactor control. *Chem. Eng. Prog.*, **84**(10), 32–36 (1988).
94. J.R. Richards, Congalidis, measurement and control of polymerization reactors. *Chemical Engineering and Processing*, **30**(10–12), 1447–1463 (2006).
95. F.J. Schork, P.B. Deshpande, K.W. Leffew, *Control of Polymerization Reactors*. New York: Marcel Dekker Inc. (1993).
96. H.Y. Parker, D.G. Westmoreland, H.R. Chang, ESR study of MMA batch emulsion polymerization in real time: Effects of particle size. *Macromolecules*, **29**(15), 5119–5127 (1996).
97. C.S. Chern, *et al.*, Semibatch seeded emulsion polymerization of acrylic monomers: Bimodal particle size distribution. *Journal of Macromolecular Science, Part A*, **34**(7), 1221–1236 (1997).

98. C.S. Chern, Y.-C. Chen, Characterization of polybutyl acrylate latex particles stabilized by a reactive surfactant, in particle size distribution III. *American Chemical Society*, 161–179 (1998).
99. S. Sajjadi, Study of different types of monomer emulsion feedings to semibatch emulsion polymerization reactors. *Journal of Applied Polymer Science*, **82**(10), 2472–2477 (2001).
100. J. Zeaiter, *et al.*, Operation of semi-batch emulsion polymerisation reactors: Modelling, validation and effect of operating conditions. *Chemical Engineering Science*, **57**(15), 2955–2969 (2002).
101. E.L. Eskin, *et al.*, Influence of parameters and drying conditions of lattices on the particle size of the PVC powder obtained. *Plasticheskie Massy*, **1**, 3–4 (1995).
102. T. Vatanatham, H. Forsyth, Scaleup factors of tubular emulsion polymerization reactors. *Polymer Engineering & Science*, **19**(7), 482–487 (1979).
103. H.M.K. Cheung, K. Gaddam, Ultrasound-assisted emulsion polymerization of methyl methacrylate and styrene. *Journal of Applied Polymer Science*, **76**(1), 101–104 (2000).
104. B.M. Teo, *et al.*, Ultrasound initiated miniemulsion polymerization of methacrylate monomers. *Ultrasonics Sonochemistry*, **15**(1), 89–94 (2008).
105. B.A. Bhanvase, *et al.*, Analysis of semibatch emulsion polymerization: Role of ultrasound and initiator. *Ultrasonics Sonochemistry*, **19**, 97–103 (2012).

7

Development of Self-Emulsifying Drug Delivery Systems (SEDDS) for Oral Bioavailability Enhancement of Poorly Soluble Drugs

Dimitrios G. Fatouros and Anette Müllertz

7.1 Introduction

The development of effective oral dosage forms for hydrophobic new chemical entities (NCE) continues to be a pressing issue for the pharmaceutical industry. Many of these poorly water-soluble NCEs readily permeate biological membranes, making their maximum solubility and/or dissolution rate in the gastro-intestinal tract (GIT) the rate-limiting steps to their absorption, in other words, they are Biopharmaceutical Classification System (BCS) Class 2 compounds. These biopharmaceutical properties usually result in low and variable bioavailability, which is often influenced by the dietary status of the patient (fed vs. fasted) [1].

Small intestinal fluid contains various surfactants, including bile salts (BS) and phospholipids (PL) which, in combination with dietary or endogenous fats, form mixed micelles with high solubilizing capacity for many BCS Class 2 compounds [2–7]. Several *in vitro* studies have demonstrated a correlation between drug lipophilicity and improved compound solubility in BS media [8–10].

The transition from the fasted to the fed state can affect the rate and mechanism by which drug substances, including BCS Class 2 compounds, are absorbed. This transition is accompanied by an increase in gastric retention time, changes in the pH values and ionic strength of the GI fluids, and an increase in the secretion of the BS and pancreatic juices [11] that catalyze the hydrolysis of dietary lipid to more water-soluble species (e.g. free fatty

Drug Delivery Strategies for Poorly Water-Soluble Drugs, First Edition. Edited by Dennis Douroumis and Alfred Fahr.

acids (FA) and monoglycerides (MG)), which are subsequently emulsified with BS to form mixed micelles. The critical role of the mixed micellar phase in the efficient absorption of dietary lipids and poorly soluble drug compounds is well accepted,

Liquid dosage forms such as Self-Emulsifying Drug Delivery Systems (SEDDS), which rely on the use of lipid and surfactant excipients to dissolve the drug in the formulation, are thought to improve the bioavailability of poorly water soluble drugs by dosing the drug compound in solution, thereby obviating the need for pre-absorptive dissolution in the GIT [13].

7.2 Lipid Processing and Drug Solubilization

In vivo lipid digestion is a complex process involving insoluble substrates, surfactants and several lipases. On the way through the gastrointestinal tract a lipid first encounters potential digestion in the stomach – through the aid of gastric lipase, which is responsible for approximately 10–20% of dietary triglyceride hydrolysis [14]. Gastric lipase primarily hydrolyses the *sn*3-ester bond in the triglyceride, thus forming *sn*1,2-diglycerides and free fatty acids [15].

Both free fatty acids and diglycerides have surfaces with active properties and facilitate the emulsification of lipids from food or from lipid-based formulations, before the food enters the duodenum. In the duodenum, bile is secreted and interacts with the emulsion entering from the stomach. The process of triglyceride digestion is completed under the action of the pancreatic lipase/colipase complexes [15, 16].

Hydrolysable tri- and diglycerides will segregate into the core of the emulsion particles, which will be covered by a complex mixture of surface active compounds, for example, monoglycerides, fatty acids, bile salts and phospholipids. The enzyme complexes will bind to the surface of the emulsion droplets to produce one molecule of 2-monoglyceride (MG) and two molecules of fatty acids (FA) for each triglyceride molecule (TG) [15, 16].

In vivo lipolysis is a dynamic process [17, 18] multilamellar liquid-crystalline lamellar phases are formed continuously at the surface of the emulsion particle and have a high ratio between lipolysis products and bile salts (LP:BS ratio). At a certain concentration of LP they will detach, leave the interface and transform first to multilamellar vesicles and, second, to unilamellar vesicles by increasing the incorporation of BS. By further incorporation of BS, the LP:BS ratio decreases to 1 or lower, whereby unilamellar vesicles are transformed to mixed micelles. At the gut wall, FA and MG are absorbed from the mixed micelles, and BS are continuously recycled back in the lumen and absorbed again by vesicles with a high LP: BS ratio [18]. In the fasted state the BS concentrations in the human small intestine have been reported to be between 2 and 6mM [19–21], while the bile secretion in the postprandial state induces very variable bile salt levels in the range of 10–20 mM, with single measurements up to almost 40 mM [19, 20, 22].

Medium chain triacylglycerides (MCT) and long chain triacylglycerides (LCT) are commonly used for lipid-based formulations. Often drug solubility determines the choice of oil. However, the choice of the oil (MCT or LCT) is also important during lipolysis. MCT substrates are hydrolyzed much faster than LCT, as has been reported in the literature [23, 24]. Several major differences exist between MCT and LCT with respect their *in vivo* fate, such as lipolytic products, differences in the modulation on gastric emptying [25] and the contraction of the gall bladder in humans [23]. Long chain lipolytic products delay gastric emptying and facilitate the contraction of the gall bladder to a greater extent than the medium chain lipolytic products [25].

The release of active compounds from the SEDDS-based formulation is thought to take place through two major pathways: interfacial transfer, and degradation of the vehicle [26]. Interfacial transfer can be described as a concentration gradient-driven process in which the compound diffuses from the formulation into the bulk or directly over the intestinal membrane. The rate and extent of interfacial transfer are thought to be governed by the partition coefficient and solubility in the donor (formulation) and the recipient phase particle size and, hence, the surface area of formulation [27]. The second pathway is degradation of the vehicle, inducing the release of the compound out of the vehicle. For lipid-based formulations, the most important degradation is the lipolysis of tryglicerides catalysed by pancreatic lipase. The release rate is thought to be dependent on the solubility of the compound in the formulation and rate and extent of the degradation of the vehicle.

7.3 Self-Emulsifying Drug Delivery Systems

Among the lipid-based formulations used for oral delivery, SEDDS and self-micro emulsifying drug delivery systems (SMEDDS) have gained considerable interest in the past few years (Figure 7.1). They are isotropic mixtures of oil (pure triglyceride oils, mixed glycerides), surfactant (hydrophilic or/and lipophilic), water-soluble co-solvents and the poorly soluble compound.

7.3.1 Excipients Used in SEDDS

A large variety of liquid or waxy excipients including oils, lipids, hydrophobic and hydrophilic surfactants and water-soluble co-solvents, in different combinations can be formulated to SEDDS. The following points should be considered in the formulation of

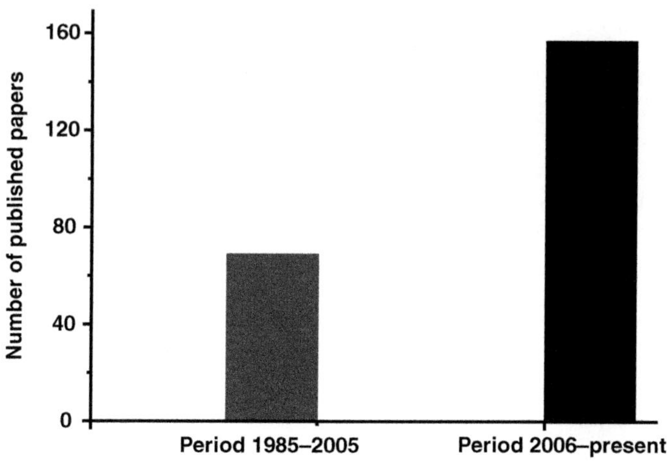

Figure 7.1 *Number of papers published between 1985 and 2005 (grey), 2006 and to the present (black) using the database Web of Science. 'Self emulsifying drug delivery systems' was the keyword.*

SEDDS: the solubility of the drug in the excipients used for the formulation, and their physical and chemical stability. Toxicity is an important issue with regard to the choice of the surfactants. Lipid-based formulations usually use non-ionic surfactants which are less toxic compared with the charged surfactants [28]. Excipient toxicity has been assumed to be of great importance to regulatory agencies, and guidelines describe the safety evaluation of pharmaceutical excipients [29].

7.3.1.1 Lipids

Triglyceride vegetable oils are widely used in food. Both long-chain triglycerides (LCT) are glyceride esters of mixed unsaturated long-chain fatty acids and medium chain tiglycerides (MCT) such as generic products of coconut oil are common building materials for SEDDS. Natural edible oils, comprising medium-chain triglycerides, are not frequently preferred in this regard, owing to their poor ability to dissolve large amounts of lipophilic drugs [30]. Partial hydrolysis of tryglicerides is used to produce semi-synthetic excipients with various proportions of monoglycerides, diglycerides and triglycerides. These derivatives exhibit enhanced solubilization capacity for the active compounds [31, 32].

7.3.1.2 Surfactants

The presence of hydrophilic surfactants enables the formulation of a nanoemulsion that can keep the drug compound in solution for a prolonged period of time at the site of absorption for effective absorption, so precipitation of the drug compound within the GI lumen is prevented [33] unless it is in the amorphous state. The two issues that govern the selection of a surfactant involve its hydrophilic–lipophilic balance (HLB) and safety. The HLB of a surfactant provides vital information on its potential utility in the formulation of SEDDS. To attain a high emulsifying performance, the emulsifier involved in the formulation of SEDDS should have high HLB and high hydrophilicity for immediate formation of droplets and rapid spreading of the formulation in aqueous media in this context. Natural co-surfactants (lecithin, peceol, akoline and medium chain monoglycerides) are safer than synthetic surfactants, however, the latter have better self-emulsification properties [34].

7.3.1.3 Co-solvents

The formulation of an effective SEDDS requires high concentrations of surfactant. Accordingly, co-solvents such as ethanol, propylene glycol and polyethylene glycol are required to enable the dissolution of large quantities of hydrophilic surfactant. However, alcohol and other volatile co-solvents have the disadvantage of evaporating into the shell of soft or hard gelatin capsules, leading to precipitation of the drug.

7.3.2 Self-Emulsification Mechanism

Self-emulsification occurs when the entropy change that favors dispersion is greater than the energy required to increase the surface area of the dispersion [35]. The free energy of the conventional emulsion is a direct function of the energy required to create a new surface between the oil and water phases and can be described by the equation:

$$\mathrm{DG} = SN_i pr_i 2S$$

where DG is the free energy associated with the process (ignoring the free energy of mixing), N is the number of droplets of radius r and S represents the interfacial energy.

The two phases of emulsion tend to separate with time to reduce the interfacial area and, subsequently, the emulsion is stabilized by emulsifying agents, which form a monolayer on the emulsion droplets, and hence reduce the interfacial energy, as well as providing a barrier to prevent coalescence [36]. The overall performance of SEDDS formulations is highly dependent on the ability of the self-emulsifying mixture to form fine particles with a uniform size and the polarity of the oil droplets to promote a fast release of the drug into the aqueous phase [33]. The polarity of the oil droplets is dictated by the hydrophilic-lipophilic balance (HLB), the molecular weight of the hydrophilic portion, the concentration of the emulsifier and the chain and degree of unsaturated fatty acids. An acceptable release rate of the drug can be achieved when small droplets are combined with the appropriate polarity [33].

7.3.3 Physicochemical Characterization of SEDDS

SEDDS formulations are typically developed by an empirical, trial and error approach; however, a carefully designed formulation will offer new insights on its performance *in vivo*. As has been outlined in the FDA guidance for process analytical technology (PAT) [37], all critical sources of any product variability due to its characteristics should be identified and explained.

7.3.3.1 Ternary Phase Diagrams

Combinations of excipients yielding SMEDDS or SNEDDS formulations are identified by the construction of ternary phase diagrams (Figure 7.2) [38]. Each point in the phase diagram represents a given combination of oil, surfactant and cosurfactant. An ideal SEDDS should exhibit the following characteristics: (i) emulsions with lipid droplet size less than several microns upon dilution with aqueous media; and (ii) the dispersed oil droplets possess

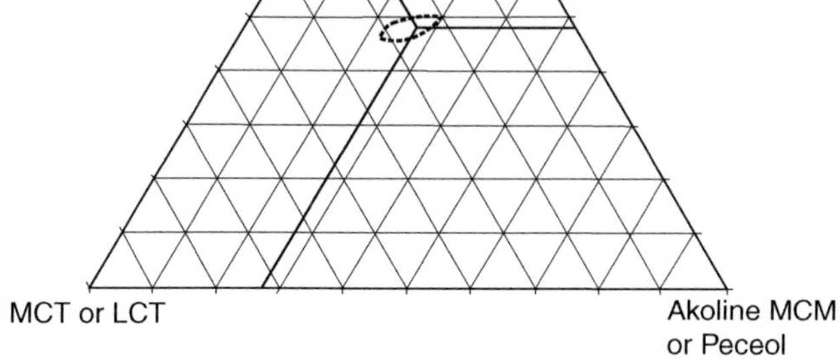

Figure 7.2 The area in the phase diagram represents microemulsions with either MCT or LCT as lipid phase, cremophor RH40 as surfactant and akoline MCM or peceol as co-surfactant obtained after addition of 250 mL water to 1 g of SMEDDS giving a droplet size of less than 100 nm. The SMEDDS in the area are mono-phasic at room temperature. Reproduced with permission from [38].

sufficient polarity and enhanced solubility to promote rapid transfer of the drug into the aqueous phase.

7.3.3.2 Visual Inspection

Mixtures corresponding to different compositions of the phase diagram are dispersed in distilled water or biorelevant media and are visually inspected [39].

7.3.3.3 Turbidity Measurements

Such an approach offers information on the efficient self-emulsification process by establishing whether the dispersion has reached equilibrium rapidly and in a reproducible time during dispersion [40,41].

7.3.3.4 Droplet Size

The entrance onto the market of a self-micro-emulsifying drug delivery system (SMEDDS) (Neoral) of cyclosporine which produces a finely dispersed micro-emulsion when diluted in an aqueous phase emphasized the impact of the droplet size – compared to formulations tested before. Scattering techniques (dynamic light scattering (DLS) [42], small angle neutron scattering (SANS) [43] and small angle X-ray scattering (SAXS) [44] can offer useful information on the structure of the micro-emulsions. Such studies can be performed both in aqueous [9] and biorelevant media [45,46].

7.3.3.5 Cryogenic Transmission Electron Microscopy Studies

Vitreous ice Cryo-Transmission electron microscopy (Cryo-TEM) has proven to be an excellent tool to provide insight on the mechanisms of the interactions between vesicles and micelles during digestion [47,48]. The advantage of Cryo-TEM is the avoidance of any fixation of the sample on the grid which can create artifacts induced by staining, and thus keeps the sample close to the original state [49,50]. Furthermore, absence of any staining eliminates the risk of extraction of lipid material during the fixation of the sample.

By way of example, Cryo-TEM work coupled with SAXS provides a complementary means of monitoring structural changes in pharmaceutically relevant formulations of the self-nano-emulsifying drug delivery system (SNEDDS) [51] (Figure 7.3). The Cryo-TEM analysis shows that spherical emulsion droplets are formed upon dilution of the SNEDDS emulsion in a Trizma maleate buffer, pH 6.5 (1% w/v) (Figure 7.3a). The SNEDDS size distribution was determined by the Image Analysis program (Figure 7.3b). The average diameter of the oil droplets obtained this way was 33.7 $\pm$ 8.4 nm. These values were in very good agreement with the average size of the oil-cores of the droplets of $\bar{D} = 37.8$ nm as determined from the SAXS data by the fitting of a structural model for polydispersed spherical oil-droplets surrounded by a lipid/surfactant layer (Figure 7.3c). The SAXS analysis furthermore showed that the surrounding lipid/surfactant layer has a thickness of on average 2.5 nm and that the obtained droplet size distribution has a standard deviation, σ, of 6.7 nm, corresponding to $\sigma/\bar{D} = 18\%$.

7.3.3.6 Determination of Self-Emulsification Time

This enables an estimation of the time taken for emulsification. Once emulsification is complete, samples are taken for particle sizing by photon correlation spectroscopy, and self-emulsified systems are compared with homogenized systems [52].

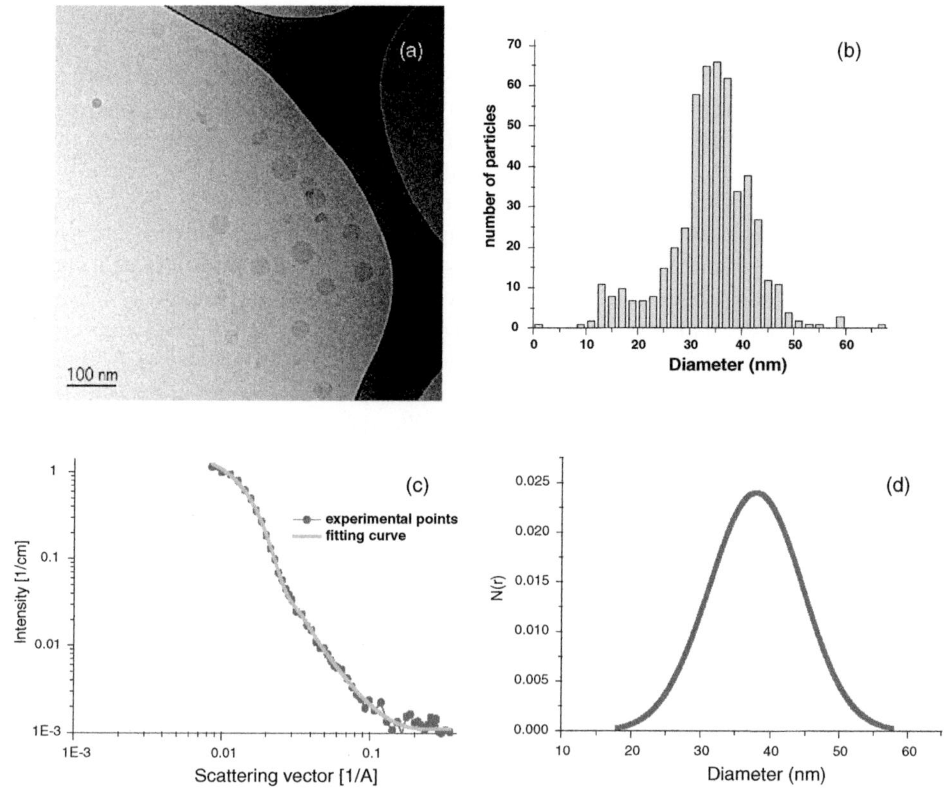

Figure 7.3 *(a) Cryo-TEM image of SNEDDS dispersed in 2 mM Trizma buffer pH 6.5 giving a final concentration of 1% w/v. (b) Size distribution of SNEDDS in a Trizma buffer pH 6.5. (c) SAXS data of SNEDDS dispersed in 2 mM Trizma buffer pH 6.5 with a final concentration of 1% w/v. Experimental points and fit results of the polydisperse core-shell model for the droplets. (d) Droplets size distribution, N(r), obtained from the SAXS model fits. Reproduced from [51] with permission from Springer Science and Business Media.*

7.3.4 Drug Incorporation in SEDDS

Drug loading can influence both the physical characteristics and the long-term physical stability of a lipid-based formulation. A plot of formulation viscosity as a function of drug concentration can be used to establish the approximate saturation solubility, which occurs in the region above the inflection point of the plotted data [53]. The potential of increasing the thermodynamic activity of drugs by using supersaturated formulations has been demonstrated as a mean of enhancing bioavailability and has been successfully employed in SEDDS [54,55].

Polyvinylpyrrolidone (PVP) [56], Hydropropylmethylcellulose (HPMC) and methylcellulose [57] are excellent crystal growth inhibitors and are effective in maintaining the supersaturated state of the drugs. Supersaturable formulations are thermodynamically stable dosage forms and pass to a supersaturated state only after administration *in vivo*.

7.4 *In Vitro* Digestion Model

At present, a rational formulation development strategy for oral lipid-based formulations is hampered not only by a lack of the mechanisms by which these formulations enhance drug bioavailability, but also by the limited number of *in vitro* methods for directing formulation development [58–60]. The dynamic lipolysis model represents one such tool for predicting formulation performance. The dynamic lipolysis model is a readily accessible *in vitro* means of determining the behavior of poorly soluble compounds under conditions simulating those of the GIT under fasted or fed conditions. This model uses bile salts, phospholipids and the generation of lipolysis products (LP) to reproduce the solubilizing environment of the GIT.

The lipolysis set-up, as it has been described previously [61, 62], is illustrated in Figure 7.4. The reaction medium consists of a mixture of BS, PL, buffer and lipid substrate (e.g. dietary lipid or lipid-based formulation incorporating the drug substance). Lipolysis is initiated by addition of the lipase solution, and the pH and free calcium concentration of the reaction mixture are maintained by the computer-controlled addition of sodium hydroxide and calcium chloride solutions, respectively.

The levels of the BS and PL used to reflect the *in vivo* intestinal condition can be varied in order to simulate the different prandial states. For simulation of the fasted state condition, an average of 5 mM and 1.25 mM PL are employed, while the levels reflecting 20 mM and 5 mM of BS and PL, respectively have been used to simulate the fed state [61–63].

The model permits sampling at different time points, offering the opportunity to follow the lipolytic process as a function of time. The samples are ultracentrifuged and three distinct phases are formed (Figure 7.5).

The *in vitro* digestion model has been shown to be a useful method for the assessment of the *in vivo* performance of lipid-based formulations [64–71].

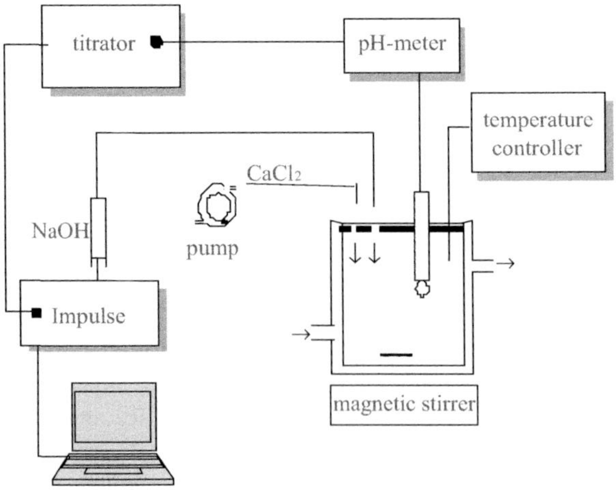

Figure 7.4 *The Dynamic Lipolysis Apparatus consists of a thermostatically controlled, double-wall reaction vessel, a computer-controlled pH-stat with an auto-burette for the addition of NaOH, and a peristaltic pump for the controlled addition of calcium chloride. Experiments are performed under continuous agitation via magnetic stirring.*

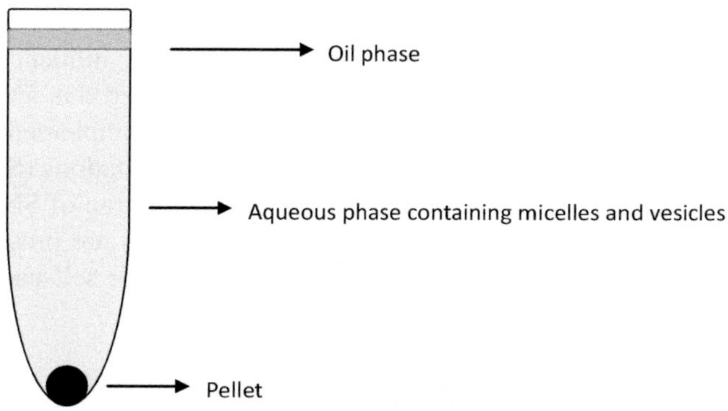

Figure 7.5 *Typical phases formed after ultracentrifugation of samples obtained after lipid digestion. The pellet mainly consists of calcium soaps of fatty acids and precipitated drug, a mixed micellar and vesicular phase and an oil upper phase. An interface containing a few layers (mesogenic phases) is located between the oil upper phase and the mixed micellar phase.*

A rank order correlation by comparing the AUC or C_{max} values with the amount of drug solubilized in the aqueous phase over time has been successfully demonstrated [67–69].

The potential of the dynamic lipolysis model to simulate the absorption of a poorly soluble model drug compound, probucol, from three lipid-based formulations containing the same excipients at different ratios has been validated [72] (Figure 7.6): an oil solution, a self-emulsifying drug delivery system generating nanoemulsions (SNEDDS), and one generating micro-emulsions (SMEDDS) have been tested in the *in vitro* lipolysis model. The release profiles were comparable with plasma profiles obtained in a previous bioavailability study conducted with mini-pigs that had received a high fat-content meal: the rank order of probucol release (SMEDDS~SNEDDS>oil formulation) was found to be similar to the rank order of bioavailability from the *in vivo* study.

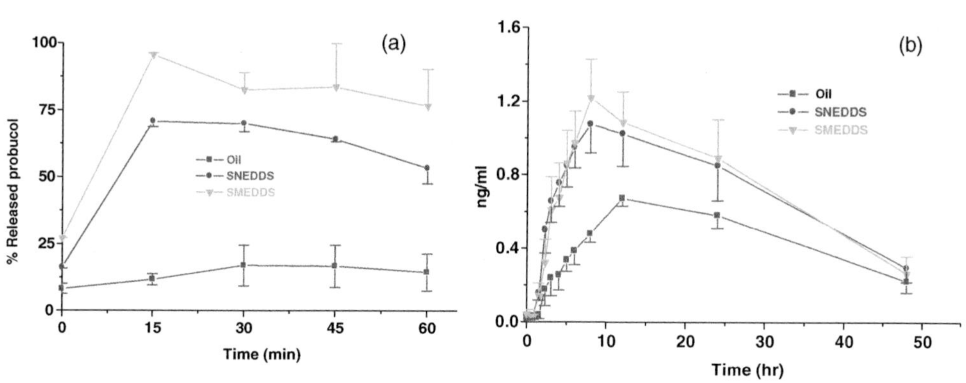

Figure 7.6 *(a) The percentage of drug released into the aqueous phase as a function of time using the in vitro dynamic lipolysis model for the oil formulation (■), SNEDDS formulation (•), SMEDDS formulation (▼), (mean ± S.E., n = 3). (b) Plasma concentrations of probucol after administration of different lipid formulations in the fed state for the oil formulation (■), the SNEDDS formulation (•), the SMEDDS formulation (▼) (mean ± S.E., n = 5). Reproduced with permission from [72].*

The intermediate phases produced during lipid digestion can play a significant role in drug solubilization and trafficking in the gastrointestinal tract, influencing the overall performance of the formulation [73, 74]. It has been demonstrated that Cryo-TEM work coupled with data from the *in vitro* lipolysis model provides a complementary means for monitoring structural changes in pharmaceutically relevant formulations (SNEDDS) [75]. The approach has been used to monitor the morphological evolution of SNEDDS formulations over time (Figure 7.7). The results revealed that micelles are present throughout the lipolysis process. As lipolysis progressed, oil droplets from the self-nano-emulsifying

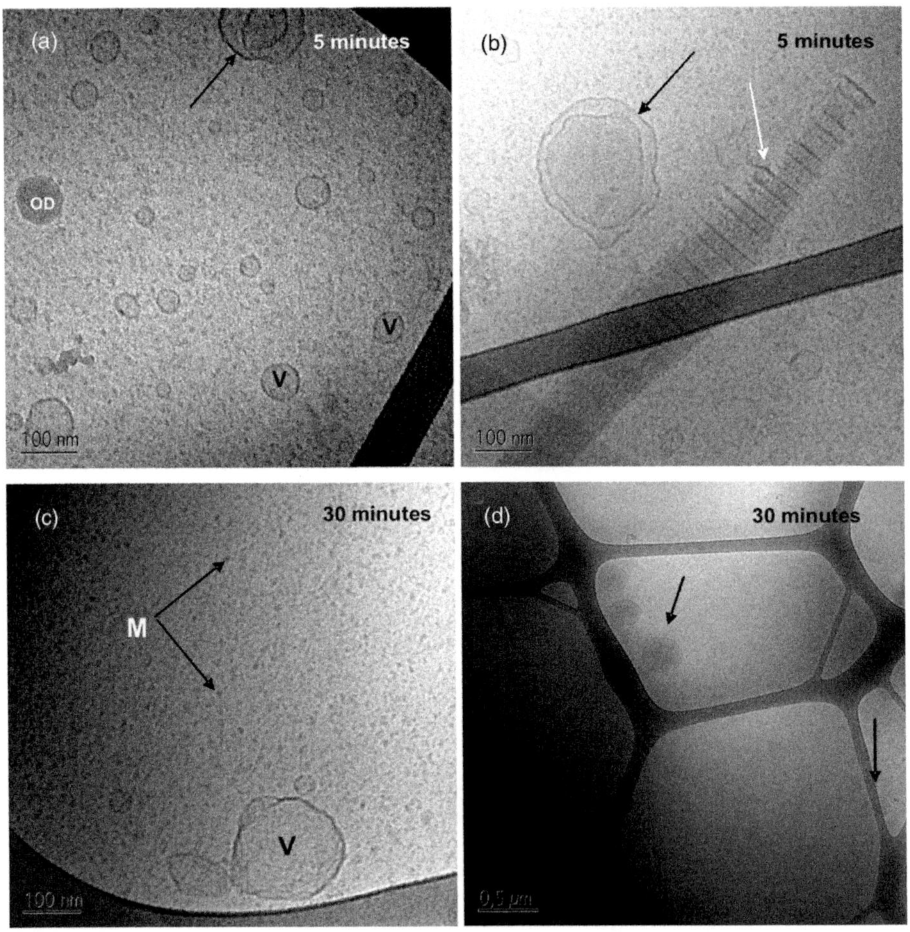

Figure 7.7 *Cryo-TEM micrographs of lipolytic products 5 minutes after the addition of lipase. (a) Unilamellar spherical vesicles and bilamellar vesicle (black arrow), and micelles (M). (b) Bilamellar vesicle (black arrow) with a rippled and angular membrane surface and ladder-shaped structures originating from the lipase (white arrow). Scale bar = 100 nm. (c) and (d): Cryo-TEM micrographs of lipolytic products 30 minutes after the addition of lipase. Unilamellar vesicles (indicated by black arrows) and micelles are visible. Scale bar represents 100 nm (c) and 500 nm (d) respectively. Reproduced with permission from [75].*

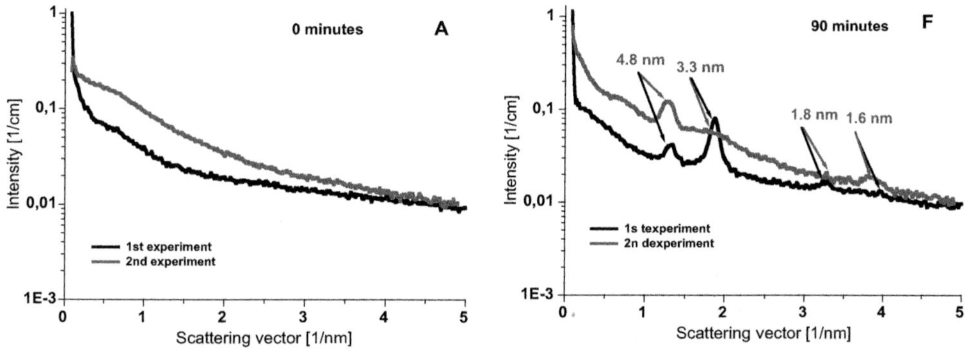

Figure 7.8 *SAXS spectra recorded for the SNEDDS formulation during digestion at 0, 5, 15, 30, 60 and 90 minutes. The intensity is plotted versus the wave vector. For the peaks* $q = 2\pi/d$*, where d is the spacing between the lattice planes. The sequence of phase development is the following: A. no phases (0 min), F. Lamellar and Hexagonal phase (90 min.) [55].*

drug delivery system were seen to transform to spherical or elongated unilamellar vesicles; bilamellar and open vesicles were detected in low numbers.

Small angle X-ray scattering studies using the same formulation revealed that a lamellar phase forms immediately after the initiation of lipolysis, with a hexagonal phase developing much later. Monitoring of the relative proportions of these phases demonstrated the dynamic nature of the lipolysis process [51]. The formation of these phases is believed to be driven by the lipase which continuously hydrolyzes triglycerides from the oil-core of the nanoemulsion droplets into mono- and di-glycerides and into fatty acids (Figure 7.8).

7.5 Enhancement of Oral Absorption by SEDDS

The Sandimmune and Sandimmune Neoral formulations of CsA are perhaps the best-known examples of a marketed lipid-based and surfactant-based systems [76]. Cyclosporine was introduced in 1981 in Europe in a self-emulsifying formulation (Sandimmune) containing Labrafil M 1944 CS (polyoxyethylated oleic glycerides), olive oil and ethanol [77]. This formulation disperses, when diluted with water, into a polydisperse oil-in-water macro-emulsion. In 1994, a new self-microemulsifying formulation (Sandimmune Neoral, referred to as Neoral in the following) was introduced, which emulsifies spontaneously into a micro-emulsion with a particle size smaller than 100nm. This formulation contains Cremophor RH40 (polyoxyl hydrogenated castor oil), corn oil glycerides, propylene glycol and ethanol [77]. The improved dispersion characteristics of Neoral have been suggested to be responsible for the increased absorption.

In the following, data from *in vivo* studies of SEDDS of poorly soluble compounds are presented. Table 7.1 summarizes the type of study and the formulations used.

The choice of surfactants and lipids is a crucial factor for the *in vivo* fate of the formulation. The lipids play a significant role since they can increase the drug solubility in lumen, can change the physical [97] and the biochemical barrier function [98] of the GI tract and they can stimulate the lymphatic transport [99]. Formulations with lipidic excipients

Table 7.1 *Type of study and the formulations used.*

Compound	Formulation	*In vivo* observation	References
Atorvastatin	Smedds formulations (Cremophor RH40, propylene glycol, Labrafil estol, Labrafac), tablet SEDDS (Trancutol, Pluronic L64, tocopherol acetetate), conventional capsule	Relative bioavailability in dogs increased significantly for all SMEDDS formulations	78
Coenzyme Q_{10}	SEDDS (Myvacet 9-45, Labrasol, Lauroglycol), powder formulation	Relative bioavailability in dogs twofold higher from SEDDS	79
Cyclosporin	Neoral®, Sandimmune®	Relative bioavailability in humans, Neoral better than Sandimmune, reduced food effect and linearity, reduced inter- and intrapatient variability	80–83
Danazol	MC-SMEDDS (Captex 355, Campul MCM, Cremophor EL) LC-SMEDDS (Soybean oil, Maisine 35-l, Cremophor EL, ethanol) LCT solutions	Relative bioavailability in dogs: LCT ~LC-SMEDDS>MC-SMEDDS	68
Halofantrine	MC-SEDDS (Captex 355, Capmul MCM Cremophor EL, ethanol) LC-SMEDDS (Soyabean oil, Maisine 35-l, Cremophor EL, ethanol) MC-SMEDDS (Captex 355, Capmul MCM, Cremophor EL, ethanol) MC-SMEDDS (Captex 355, Capmul MCM, Cremophor EL, ethanol) LC-SMEDDS (LML lipid, Maisine 35-l, Cremophor EL, ethanol) MLM-SMEDDS (MLM lipid, Maisine-35-l, Cremophor EL, ethanol) LML-SMEDDS (LML lipid, Maisine 35-l, Cremophor EL, ethanol)	Absolute bioavailability in dogs: all SEDDS formulations higher than tablet Lymphatic transport in dogs: LC-SMEDDS> MC-SMEDDS Lymphatic transport in dogs: MLM-SMEDDS>LML-SMEDDS	84 85 86

(continued)

Table 7.1 (Continued)

Compound	Formulation	*In vivo* observation	Reference
Itroconazole	SEDDS (Transcutol, Pluronic L64, tocopherole acetate), convetional capsule	Relative bioavailability in rats increased and reduced food effect from SEDDS	87
Ontazolast	SEDDS (Gelucire 44/14, Peceol) SEDDS (Gelucire 44/14, Peceol)	Absolute bioavailability in rats increased for all lipid formulations	88
PNU-9135	Supersaturable co-solvent (S-co-solvent) and supersaturable SEDDS (HPMC, Cremophor EL, Pluronic L44, PEG 400, long-chain glyceride lipid, DMA) formulations compared to co-solvent (PG) and Tween solutions	Relative bioavailability five- to sixfold enhancement for S-co-solvent, S-SEEDS and Tween 80 relative to co-solvent	89
Progesterone	SEDDS (mono-di-glycerides: Polysorbate 80), aq. suspension	Relative bioavailability in dogs ninefold higher for SEDDS	90
Seolcalcitol	MC-SMEDDS (fr. Coconut oil, Akoline MCM, Cremophor RH40), MCT, propylene glycol MC-SMEDDS (fr. Coconut oil, Akoline MCM, Cremorphor RH40) LC-SMEDDS (Sesame oil, Peceol, Cremophor RH40)	Absolute bioavailability in mini pigs comparable for all formulations Absolute bioavailability in rats comparable for both formulations	91 92
Silymarin	SMEDDS (Tween 80, ethyl alcohol, ethyl linoleate) PEG 400 solution, PEG suspension	Relative bioavailability in rabbits two- and fiftyfold higher from SMEDDS than that of PEG 400 solution and suspension	93
Simvastatin	SMEDDS (Capryol, Cremophor EL, Carbitol) and tablet	Relative bioavailability in dogs 1.5-fold higher from SEDDS	94
Tocotrienols	Two SEDDS solutions (Tween 80 and labrasol) and LCT solution	Relative bioavailability in humans two- to threefold higher for SEDDS	95
WIN 54954	SEDDS (Neobee M5 (MCT), Tagat TO) PEG 600 solution	Relative bioavailability in dogs: no changes to BA	53
Vitamin E	SEDDS (Tween, Span, Palm oil)	Relative bioavailability in humans SEDDS twofold higher than oil solution	96

exhibit higher bioavailability than commonly used PEG solutions which are water-miscible and can lead to drug precipitation upon dilution in the gastrointestinal tract [53, 93]. The emulsion droplet size can affect the drug absorption. A number of studies have reported higher bioavailability with decreased particle size when comparing self-emulsifying formulations [80, 84].

A claimed advantage of SNEDDS formulations is that the food effect on poorly soluble drugs can be diminished, as has been demonstrated by Mueller *et al.* for cyclosporine [82]. Other poorly soluble drugs for which lipid-based formulations have reduced the food effect on drug absorption have been reported for danazol [68] and seolcalcitol [91].

The degree of emulsification and dispersibility of the formulation can significantly improve its overall performance as has been illustrated for danazol where the amount found in the aqueous phase, for the (LC-SMEDDS) was much higher compared to a MC-SMEDDS [68]. Similar results were obtained for ontazolast when administered to rats with different lipid-based formulations, and among others the SEDDS formulations significantly increase drug bioavailability [88].

Deckelbaum *et al.* [100] studied the influence of lipid fatty acid chain length of self-emulsification. The better self-emulsification properties of MCT compared to LCT were attributed to the higher water solubility and the smaller molecular size of the former.

Surfactants can induce changes in the intestinal mucosal permeability since water-soluble surfactants can solubilize membrane components while the water-insoluble ones can penetrate and fluidize the membranes [101]. Attempts have been made to evaluate the toxicity of pharmaceutical excipients of SEDDS and SMEDDS with Caco-2 cell monolayeres [102, 103].

The ability of different lipid types to stimulate lymphatic drug transport has led to an increase in intestinal lymphatic drug transport [104]. The effect of MCT and LCT to SMEDDS formulations containing halofantrine were studied in conscious dogs. As expected, the amount of halofantrine transported with the LC-SMEDDS is higher compared with the MC-SMEDDS. This is due to the transport of long-chain lipid in the lymph, whereas medium-chain fatty acids are transported via the portal blood, bound to albumin. The poor performance of the MC-SMEDDS was attributed to mono- and diglycerides which are not lymphotropic [84].

7.6 Conclusion

The oral route is the preferred one for the administration of drugs. Since a lot of potential drug candidates are poorly soluble in water, resulting in poor and variable oral bioavailability, many approaches have been employed in order to produce formulations with a high and reproducible bioavailability. SEDDS and SMEDDS formulations offer the opportunity to manipulate the drug absorption profile and improve their overall performance compared with other carriers intended for oral delivery. A detailed physico-chemical characterization is a prerequisite since chemical and physical instability can limit the overall performance of the formulations. Toxicity is another important issue with regard to the choice of the surfactants. Despite the plethora of research publications dealing with orally administered, lipid-based formulations of poorly soluble compounds, the performance evaluation of such formulations relies solely on *in vivo* bioavailability data. A major challenge is the development of a universally applicable and readily accessible *in vitro* model for assessing *in vivo* performance, as this will provide a much needed screening tool for the design and optimization of lipid-based formulations. The *in vivo* studies have clearly demonstrated the superiority of SEDDS and SMEDDS formulations over the non-self-dispersing lipid-based formulations.

References

1. L.E. Schmidt, K. Dalhoff, Food-drug interactions. *Drugs*, **62**, 1481–1502 (2002).
2. T.R. Bates, M. Gibaldi, J.L. Kanig, Solubilising properties of bile salt solutions. II. Effect of inorganic electrolyte, lipids, and a mixed bile salt system on solubilisation of glutethimide, griseofulvin, and hexestrol. *J Pharm Sci*, **55**, 901–906 (1966).
3. T.R. Bates, S.L. Lin, M. Gibaldi, Solubilization and rate of dissolution of drugs in the presence of physiologic concentrations of lysolecithin. *J Pharm Sci*, **56**, 1492–1495 (1967).
4. L. Martis, N.A. Hall, A.L. Thakkar, Micelle formation and testosterone solubilisation by sodium glycocholate. *J Pharm Sci*, **61**, 1757–1761 (1972).
5. M. Rosoff, A.T.M. Serajuddin, Solubilization of diazepam in bile-salts and in sodium cholate-lecithin-water phases. *Int J Pharm*, **6**, 137–146 (1980).
6. M.A. Kassem, A.G. Mattha, A.E.M. Elnimr, S.M. Omar, Study of the influence of sodium taurocholate (STC) and sodium glycocholate (SGC) on the mass-transfer of certain drugs: Digoxin. *Int J Pharm*, **12**, 1–9 (1982).
7. A.T. Serajuddin, P.C. Sheen, D. Mufson, D.F. Bernstein, M.A. Augustine, Physicochemical basis of increased bioavailability of a poorly water-soluble drug following oral administration as organic solutions. *J Pharm Sci*, **77**, 325–329 (1988).
8. S.D. Mithani, V. Bakatselou, C.N. TenHoor, J.B. Dressman, Estimation of the increase in solubility of drugs as a function of bile salt concentration. *Pharm Res*, **13**, 163–167 (1996).
9. T.S. Wiedmann, W. Liang, L. Kamel, Solubilization of drugs by physiological mixtures of bile salts. *Pharm Res*, **19**, 1203–1208 (2002).
10. T.S. Wiedmann, L. Kamel, Examination of the solubilization of drugs by bile salt micelles. *J Pharm Sci*, **91**, 1743–1764 (2002).
11. J.B. Dressman, G.L. Amidon, C. Reppas, V.P. Shah, Dissolution testing as a prognostic tool for oral drug absorption: Immediate release dosage forms. *Pharm Res*, **15**, 11–22 (1998).
12. C.J. Porter, W.N. Charman, *In vitro* assessment of oral lipid based formulations. *Adv Drug Deliv Rev*, **50**, S127–S147 (2001).
13. C.J. Porter, N. Trevaskis, W.N. Charman, Lipids and lipid-based formulations: optimizing the oral delivery of lipophilic drugs. *Nat Rev Drug Discov*, **3**, 321–348 (2007).
14. F. Carriere, J.E. Barrowman, L. Verger, R. Laugier, Secretion and contribution to lipolysis of gastric and pancreatic lipases during a test meal in humans. *Gastroenterology*, **105**, 876–888 (1993).
15. M.E. Lowe, The triglyceride lipases of the pancreas. *J Lipid Res*, **43**, 2007–2016 (2002).

16. J.S. Patton, M.C. Carey, Inhibition of human pancreatic lipase-colipase activity by mixed bile salt-phospholipid micelles. *Am J Physiol*, **241**, G328–G336 (1981).
17. J.S. Patton, M.C. Carey, Watching fat digestion. *Science*, **204**, 145–158 (1979).
18. M.W. Rigler, R.E. Honkanen, J.S. Patton, Visualization by freeze fracture, in vitro and in vivo, of the products of fat digestion. *J Lipid Res*, **8**, 836–857 (1986).
19. A. Tangerman, E.W. van Schaik, E.W. van der Hoek, Analysis of conjugated and unconjugated bile acids in serum and jejunal fluid of normal subjects. *Clin Chim Acta*, **159**, 123–132 (1986).
20. B.L. Pedersen, A. Mullertz, H. Brondsted, H.G. Kristensen, A comparison of the solubility of danazol in human and simulated gastrointestinal fluids. *Pharm Res*, **7**, 891–944 (2000).
21. K. Kleberg, F. Jacobsen, D.G. Fatouros, A. Müllertz, Biorelevant media simulating fed state intestinal fluids: Colloid phase characterization and impact on solubilization capacity. *J Pharm Sci*, **99**, 3522–3532 (2010).
22. E.M. Persson, A.S. Gustafsson, A.S. Carlsson, R.G. Nilsson, L. Knutson, P. Forsell, G. Hanisch, H. Lennernäs, B. Abrahamsson, The effects of food on the dissolution of poorly soluble drugs in human and in model small intestinal fluids. *Pharm Res*, **22**, 2141–2151 (2005).
23. R.L. Deckelbaum, J.A. Hamilton, A. Moser, G. Bengtsson-Olivecroma, E. Butbul, Y.A. Carpentier, A. Gutman, T. Olivecroma, Medium-chain versus long-chain triacylglycerol emulsion hydrolysis by lipoprotein lipase and hepatic lipase: Implications for the mechanisms of lipase action. *Biochemistry*, **29**, 1136–1142 (1990).
24. J.O. Christensen, K. Schultz, B. Mollgaard, H.G. Kristensen, A. Mullertz, Solubilization of poorly water-soluble drugs during in vitro lipolysis of medium and long chain triaglycerols. *Eur J Pharm Sci*, **3**, 87–96 (2004).
25. J.N. Hunt, M.T. Knox, A relation between the chain length of fatty acids and the slowing of gastric emptying. *J Physiol*, **194**, 327–336 (1968).
26. P.C. de Smidt, M.A. Campanero, I.F. Troconiz, Intestinal absorption of penclomedine from lipid vehicles in the conscious rat: Contribution of emulsification versus digestibility. *Int J Pharm*, **270**, 109–118 (2004).
27. N.A. Armstrong, K.C. James, Drug release from lipid based dosage forms. II. *Inter J Pharm*, **6**, 195–204 (1980).
28. C.W. Pouton, S.J.H. Porter, Formulation of lipid based delivery systems for oral administration: Materials, methods and strategies. *Adv Drug Del Syst*, **60**, 625–637 (2008).
29. FDA Guidance for Industry Nonclinical Studies for the Safety Evaluation of Pharmaceutical Excipients. U.S. Department of Health and Human Services (2005).
30. J.M. Odeberg, P. Kaufmann, K.G. Kroon, P. Höglund, Lipid drug delivery and rational formulation design for lipophilic drugs with low oral bioavailability, applied to cyclosporine. *Eur J Pharm Sci*, **20**, 375–382 (2003).
31. P.P. Constantinides, Lipid microemulsions for improving drug dissolution and oral absorption: Physical and biopharmaceutical aspects. *Pharm Res*, **12**, 1561–1572 (1995).
32. D.J. Hauss, S.E. Fogal, J.V. Ficorilli, C.A. Price, T. Roy, A.A. Jayaraj, J.J. Keirns, Lipid-based delivery systems for improving the bioavailability and lymphatic transport of a poorly water-soluble LTB4 inhibitor. *J Pharm Sci*, **87**, 164–169 (1998).

33. H. Shah, M.T. Carvajal, C.I. Patel, M.H. Infeld, A.W. Malick, Self emulsifying drug delivery systems (SEDDS) with polyglycolyzed glycerides for improving in vitro dissolution and oral absorption of lipophilic drugs. *Int J Pharm*, **106**, 15–23 (1994).
34. R.N. Gursoy, S. Benita, Self-dispersing lipid formulations for improving oral absorption of lipophilic drugs. *Eur J Pharm Biopharm*, **50**, 179–188 (2000).
35. H. Reiss. Entropy induced dispersion of bulk liquids. *J Colloid Interface Sci*, **53**, 61–70 (1975).
36. D.Q.M. Craig, S.A. Barker, D. Banning, S.W. Booth, An investigation into the mechanism of self emulsification using particle size analysis and low frequency dielectric spectroscopy. *Int J Pharm*, **114**, 103–110 (1995).
37. Guidance for Industry. PAT—A framework for Innovative Pharmaceutical Development, Manufacturing, and Quality Assurance. U.S. Department of Health and Human Services, FDA/CDER/ CVM/ORA/CGMPs, http://www.fda.gov/cder/guidance/6419fnl.pdf (2004).
38. M. Grove, A. Müllertz, J.L. Nielsen, G.P. Pedersen, Bioavailability of seocalcitol II: Development and characterisation of self-microemulsifying drug delivery systems (SMEDDS) for oral administration containing medium and long chain triglycerides. *Eur J Pharm Sci*, **28**, 233–242 (2006).
39. L. Sek, B.J. Boyd, W.N. Charman, C.J. Porter, Examination of the impact of a range of pluronic surfactants on the in-vitro solubilisation behaviour and oral bioavailability of lipidic formulations of atovaquone. *J Pharm Pharmacol*, **58**, 809–820 (2006).
40. A.S. Zidan, O.A. Sammour, M.A. Hammad, N.A. Megrab, M.J. Habib, M.A. Khan, Quality by design: Understanding the formulation variables of a cyclosporine A self-nanoemulsified drug delivery systems by Box-Behnken design and desirability function. *Int J Pharm*, **332**, 55–63 (2007).
41. A.S. Zidan, O.A. Sammour, M.A. Hammad, N.A. Megrab, M.J. Habib, M.A. Khan, Quality by design: Understanding the product variability of a self-nanoemulsified drug delivery system of cyclosporine *A J Pharm Sci*, **96**, 2409–2423 (2007).
42. P.P. Constantinides, J.P. Scalart, Formulation and physical characterization of water-in-oil microemulsions containing long versus medium chain length glycerides. *Int J Pharm*, **158**, 57–68 (1997).
43. J. Bergenholtz, A.A. Romagnoli, N.J. Wagner, Viscosity, microstructure and interparticle potential of AOT/H2O/n-decane inverse microemulsions. *Langmuir*, **11**, 1559–1570 (1995).
44. O. Regev, S. Ezrahi, A. Aserin, N. Garti, E. Wachtel, E.W. Kaler, A. Khan, Y. Talmon, A study of the microstructure of a four-component nonionic microemulsion by cryo-TEM, NMR, SAXS and SANS. *Langmuir*, **12**, 668–674 (1996).
45. F.S. Nielsen, E. Gibault, H. Ljusberg-Wahren, L. Arleth, J.S. Pedersen, A. Müllertz, Characterization of prototype self-nanoemulsifying formulations of lipophilic compounds. *J Pharm Sci*, **96**, 876–892 (2007).
46. C. Ditner, R. Bravo, G. Imanidis, M.A. Kuentz, Systematic dilution study of self-microemulsifying drug delivery systems in artificial intestinal fluid using dynamic laser light backscattering. *Drug Dev Ind Pharm*, **35**, 199–208 (2009).
47. P. Vinson, Y. Talmon, A. Walter, Vesicle-micelle transition of phosphatidylcholine and octyl glucoside elucidated by cryo-transmission electron microscopy. *Biophys J*, **56**, 669–681 (1989).

48. A. Walter, P.K. Vinson, A. Kaplun, Y. Talmon, Intermediate structures in the cholate-phosphatidylcholine vesicle-micelle transition. *Biophys J*, **60**, 1315–1325 (1991).
49. J.R. Bellare, H.T. Davis, L.E. Scriven, Y. Talmon, Controlled environment vitrification system. An improved sample preparation technique. *J Elec Micr Tech*, **10**, 87–111 (1988).
50. J. Dubochet, M. Adrian, J. Chang, J.C. Homo, J. Lepault, A.W. McDowell, P. Schultz, Cryo-electron microscopy of vitrified specimens. *Q Rev Biophys*, **21**, 129–228 (1988).
51. D.G. Fatouros, G.R. Deen, L. Arleth, B. Bergenstahl, F.S. Nielsen, J.S. Pedersen, A. Müllertz, Structural development of self nano emulsifying drug delivery systems (SNEDDS) during in vitro lipid digestion monitored by small-angle x-ray scattering. *Pharm Res*, **24**, 1844–1853 (2007).
52. C.W. Pouton, Formulation of self emulsifying drug delivery systems. *Adv. Drug Deliv Rev*, **25**, 47–58 (1997).
53. S.A. Charman, W.N. Charman, M.C. Rogge, A.W. Boddy, B.M. Berger, Self-emulsifying drug delivery systems: Formulation and biopharmaceutical evaluation of an investigational lipophilic compound. *Pharm Res*, **9**, 87–93 (1992).
54. P. Gao, B.D. Rush, W.P. Pfund, T. Huang, J.M. Bauer, W. Morozowich, M.S. Kuo, M.J. Hageman, Development of a supersaturable SEDDS (S-SEDDS) formulation of paclitaxel with improved oral bioavailability. *J Pharm Sci*, **92**, 2386–2398 (2003).
55. P. Gao, W. Morozowich, Development of supersaturatable self-emulsifying drug delivery system formulations for improving the oral absorption of poorly soluble drugs. *Expert Opin Drug Deliv*, **3**, 97–110 (2006).
56. A.P. Simonelli, S.C. Mehta, W.I. Higuchi, Inhibition of sulfathiazole crystal growth by polyvinylpyrrolidone. *J Pharm Sci*, **59**, 633–638 (1970).
57. S.L. Raghavan, A. Trividic, A.F. Davis, J. Hadgraft, Effect of cellulose polymers on supersaturation and in vitro membrane transport of hydrocortisone acetate. *Int J Pharm*, **193**, 231–237 (2000).
58. E.S. Kostewicz, U. Brauns, R. Becker, J.B. Dressman, Forecasting the oral absorption behavior of poorly soluble weak bases using solubility and dissolution studies in biorelevant media. *Pharm Res*, **19**, 345–349 (2002).
59. J.B. Dressman, C. Reppas, *In vitro-in vivo* correlations for lipophilic, poorly water-soluble drugs. *Eur J Pharm Sci*, **11**, S73–S80 (2000).
60. V.H. Sunesen, B.L. Pedersen, H.G. Kristensen, A. Mullertz, *In vivo in vitro* correlations for a poorly soluble drug, danazol, using the flow-through dissolution method with biorelevant dissolution media. *Eur J Pharm Sci*, **24**, 305–313 (2005).
61. N.H. Zangenberg, A. Mullertz, H.G. Kristensen, L. Hovgaard, A dynamic *in vitro* lipolysis model. I. Controlling the rate of lipolysis by continuous addition of calcium. *Eur J Pharm Sci*, **14**, 115–122 (2001).
62. N.H. Zangenberg, A. Mullertz, H.G. Kristensen, L. Hovgaard, A dynamic *in vitro* lipolysis model. II: Evaluation of the model. *Eur J Pharm Sci*, **14**, 237–244 (2001).
63. E. Galia, E. Nicolaides, D. Hörter, R. Löbenberg, C. Reppas, J.B. Dressman, Evaluation of various dissolution media for predicting in vivo performance of class I and II drugs. *Pharm Res*, **15**, 698–705 (1998).
64. E.M. Persson, A.S. Gustafsson, A.S. Carlsson, R.G. Nilsson, L. Knutson, P. Forsell, G. Hanisch, H. Lennernäs, B. Abrahamsson, The effects of food on the dissolution of

poorly soluble drugs in human and in model small intestinal fluids. *Pharm Res*, **22**, 2141–2151 (2005).

65. J.P. Reymond, H. Sucker, *In vitro* model for cyclosporin intestinal absorption in lipid vehicles. *Pharm Res*, **5**, 673–676 (1988).
66. J.P. Reymond, H. Sucker. J. Vonderscher, *In vivo* model for cyclosporin intestinal absorption in lipid vehicles. *Pharm Res*, **5**, 677–679 (1988).
67. C.J. Porter, A.M. Kaukonen, A. Taillardat-Bertschinger, B.J. Boyd, J.M. O'Connor, G.A. Edwards, W.N. Charman, Use of in vitro lipid digestion data to explain the in vivo performance of triglyceride-based oral lipid formulations of poorly water-soluble drugs: Studies with halofantrine. *J Pharm Sci*, **93**, 1110–1121 (2004).
68. C.J. Porter, A.M. Kaukonen, B.J. Boyd, G.A. Edwards, W.N. Charman, Susceptibility to lipase-mediated digestion reduces the oral bioavailability of danazol after administration as a medium-chain lipid-based microemulsion formulation. *Pharm Res*, **21**, 1405–1412 (2004).
69. A. Dahan, A. Hoffman, Use of a dynamic *in vitro* lipolysis model to rationalize oral formulation development for poor water soluble drugs: Correlation with *in vivo* data and the relationship to intra-enterocyte processes in rats. *Pharm Res*, **23**, 2165–2174 (2006).
70. A. Dahan, A. Hoffman, The effect of different lipid based formulations on the oral absorption of lipophilic drugs: The ability of *in vitro* lipolysis and consecutive *ex vivo* intestinal permeability data to predict *in vivo* bioavailability in rats. *Eur J Pharm Biopharm*, **67**, 96–105 (2007).
71. A. Larsen, R. Holm, M.L. Pedersen, A. Müllertz, Lipid-based formulations for danazol containing a digestible surfactant, Labrafil M2125CS: in vivo bioavailability and dynamic in vitro lipolysis. *Pharm Res*, **25**, 2769–1777 (2008).
72. D.G. Fatouros, F.S. Nielsen, D. Douroumis, L.J. Hadjileontiadis, A. Müllertz, *In vitro-in vivo* relation in lipid based formulations using the *in vitro* dynamic lipolysis model combined with fuzzy neural networks. *Eur J Pharm Biopharm*, **69**, 887–898 (2008).
73. G.A. Kossena, W.N. Charman, B.J. Boyd, D.E. Dunstan, C.J.H. Porter, Probing drug solubilization patterns in the gastrointestinal tract after administration of lipid-based delivery systems: A phase diagram approach. *J Pharm Sci*, **93** 332–348 (2004).
74. G.A. Kossena, W.N. Charman, B.J. Boyd, C.J.H. Porter, Influence of the intermediate digestion phases of common formulation lipids on the absorption of a poorly water-soluble drug. *J Pharm Sci*, **94**, 481–492 (2005).
75. D.G. Fatouros, B. Bergenstahl A. Müllertz, Morphological observations on a lipid-based drug delivery system during in vitro digestion. *Eur J Pharm Sci*, **31**, 85–94 (2007).
76. D.W. Holt, E.A. Mueller, J.M. Kovarik, J.B. van Bree, K. Kutz, The pharmacokinetics of Sandimmun Neoral: A new oral formulation of cyclosporine. *Transplant Proc*, **26**, 2935–2939 (1994).
77. B.A. Klyashchitsky, A.J. Owen, Drug delivery systems for cyclosporine: Achievements and complications. *J Drug Target*, **5**, 443–458 (1998).
78. H. Shen, M. Zhong, Preparation and evaluation of self-microemulsifying drug delivery systems (SMEDDS) containing atorvastatin. *J Pharm Pharmacol*, **58**, 1183–1191 (2006).

79. T.R. Kommuru, B. Gurley, M.A. Khan, I.K. Reddy, Self-emulsifying drug delivery systems (SEDDS) of coenzyme Q10: Formulation development and bioavailability assessment. *Int J Pharm*, **212**, 233–246 (2001).
80. J.M. Kovarik, E.A. Mueller, J.B van Bree, W. Tetzloff, K. Kutz, Reduced inter- and intraindividual variability in cyclosporine pharmacokinetics from a microemulsion formulation. *J Pharm Sci*, **83**, 444–446 (1994).
81. E.A. Mueller, J.M Kovarik, J.B van Bree, W. Tetzloff, J. Grevel, K. Kutz, Improved dose linearity of cyclosporine pharmacokinetics from a microemulsion formulation. *Pharm Res*, **11**, 301–304 (1994).
82. E.A. Mueller, J.M. Kovarik, J.B. van Bree, J. Grevel, P.W. Lucker, K. Kutz, Influence of a fat-rich meal on the pharmacokinetics of a new oral formulation of cyclosporine in a crossover comparison with the market formulation. *Pharm Res*, **11**, 151–155 (1994).
83. J. Gonzalez-Llaven, J.A. Palma-Aguirre, R. Garcia-Arreola, V. Brizuela, J. Nava-Rangel, S. Moran-Lira, J. Vela-Ojeda, G. Castaneda-Hernandez, F.J. Flores-Murrieta, Comparative bioavailability evaluation of two cyclosporine oral formulations in healthy Mexican volunteers. *Arch Med Res*, **30**, 315–319 (1999).
84. S.M. Khoo, A.J. Humberstone, C.J. Porter, G.A. Edwards, W.N. Charman. Formulation design and bioavailability assessment of lipidic self-emulsifying formulations of halofantrine. *Int J Pharm*, **167**, 115–164 (1998).
85. S.M. Khoo, D.M. Shackleford, C.J. Porter, G.A. Edwards, W.N. Charman, Intestinal lymphatic transport of halofantrine occurs after oral administration of a unit-dose lipid-based formulation to fasted dogs. *Pharm Res*, **20**, 1460–1465 (2003).
86. R. Holm, C.J. Porter, A. Müllertz, H.G. Kristensen, W.N. Charman, Structured triglyceride vehicles for oral delivery of halofantrine: Examination of intestinal lymphatic transport and bioavailability in conscious rats. *Pharm Res*, **19**, 1354–1361 (2002).
87. J.-Y. Hong, J.-K. Kim, Y.-K. Song, J.-S. Park, C.-K. Kim, A new self-emulsifying formulation of itroconazole with improved dissolution and oral absorption. *J Cont Rel*, **110**, 332–338 (2006).
88. D.J. Hauss, S.E. Fogal, J.V. Ficorilli, C.A. Price, T. Roy, A.A. Jayaraj, J.J. Keirns, Lipid based delivery systems for improving the bioavailability and lymphatic transport of a poorly soluble water soluble LTB4 inhibitor. *J Pharm Sci*, **87**, 164–169 (1998).
89. P. Gao, M.E. Guyton, T. Huang, J.M. Bauer, K.J. Stefanski, Q. Lu, Enhanced oral bioavailability of a poorly water soluble drug PNU-91325 by supersaturable formulations. *Drug Dev Ind Pharm*, **30**, 221–229 (2004).
90. C. Tuleu, M. Newton, J. Rose, D. Euler, R. Saklatvala, A. Clarke, S. Booth, Comparative bioavailability study in dogs of a self-emulsifying formulation of progesterone presented in a pellet and liquid form compared with an aqueous suspension of progesterone. *J Pharm Sci*, **93**, 1495–1502 (2004).
91. M. Grove, A. Müllertz, G.P. Pedersen, J.L. Nielsen, Bioavailability of seocalcitol III. Administration of lipid-based formulations to mini pigs in the fasted and fed state. *Eur J Pharm Sci*, **31**, 8–15 (2007).
92. M. Grove, A. Müllertz, J.L. Nielsen, G.P. Pedersen, Bioavailability of seocalcitol II: Development and characterisation of self-microemulsifying drug delivery systems (SMEDDS) for oral administration containing medium and long chain triglycerides. *Eur J Pharm Sci*, **28**, 233–242 (2006).

93. W. Wu, Y. Wang, L. Que, Enhanced bioavailability of silymarin by self-microemulsifying drug delivery system. *Eur J Pharm Biopharm*, **63**, 288–294 (2006).
94. B.K. Kang, J.S. Lee, S.K. Chon, S.Y. Jeong, S.H. Yuk, G. Khang, H.B. Lee, S.H. Cho, Development of self-microemulsifying drug delivery systems (SMEDDS) for oral bioavailability enhancement of simvastatin in beagle dogs. *Int J Pharm*, **274**, 65–73 (2004).
95. S.P. Yap, K.H. Yuen, Influence of the lipolysis and droplet size on tocotrienol absorption from self-emulsifying formulations. *Int J Pharm*, **281**, 67–78 (2004).
96. T. Julianto, K.H. Yuen, A.M. Noor, Improved bioavailability of vitamin E with a self emulsifying formulation *Int J Pharm*, **200**, 53–57 (2000).
97. B.J. Aungst, Intestinal permeation enhancers. *J Pharm Sci*, **89**, 429–442 (2000).
98. L. Benet, The drug efflux-metabolism alliance: Biochemical aspects. *Adv Drug Deliv-Rev*, **50**, S3–S11 (2001).
99. C.J. Porter, W.N. Charman, Uptake of drugs into the intestinal lymphatics after oral administration. *Adv Drug Deliv Rev*, **25**, 71–89 (2007).
100. R.J. Deckelbaum, J.A. Hamilton, A. Moser, G. Bengtsson-Olivecrona, E. Butbul, Y.A. Carpentier, A. Gutman, T. Olivecrona, Medium-chain versus long-chain triacylglycerol emulsion hydrolysis by lipoprotein lipase and hepatic lipase: Implications for the mechanisms of lipase action. *Biochemistry*, **29**, 1136–1142 (1990).
101. R.N. Gursoy, S. Benita, Self-emulsifying drug delivery systems (SEDDS) for improved oral delivery of lipophilic drugs. *Biomed Pharmacother*, **58**, 73–182 (2004).
102. T. Gershanik, E. Haltner, C.M. Lehr, S. Benita. Charge-dependent interaction of self-emulsifying oil formulations with Caco-2 cells monolayers: Binding, effects on barrier function and cytotoxicity. *Int J Pharm*, **211**, 29–36 (2000).
103. N. Gursoy, J.S. Garrigue, A. Razafindratsita, G. Lambert, S. Benita, Excipient effects on in vitro cytotoxicity of a novel paclitaxel self-emulsifying drug delivery system. *J Pharm Sci*, **92**, 2411–2418 (2003).
104. C.J. Porter, W.N. Charman, Intestinal lymphatic drug transport: An update. *Adv Drug Deliv Rev*, **50**, 61–80 (2001).

8

Novel Top-Down Technologies: Effective Production of Ultra-Fine Drug Nanocrystals

C.M. Keck, S. Kobierski, R. Mauludin and R.H. Müller

8.1 Introduction: General Benefits of Drug Nanocrystals (First Generation)

In less than 10 years drug nanocrystals have made it to become one of the most frequently used approaches for the formulation of poorly soluble drugs. The first patent which can be regarded as the invention of the 'pharmaceutical nanocrystals for oral bioavailability enhancement' was filed by G. Liversidge and co-workers in 1991 [1], followed by R.H. Müller *et al.* [2]. The first oral nanocrystal product, Emend, appeared on the market in 2000. This is a record compared to most other formulation technologies, for example, it took more than 20 years for liposomes to appear on the pharmaceutical market as products [3]. In 2012, we can look back on 20 years of nanocrystal history [4].

Drug nanocrystals are the formulation approach of choice when simpler approaches such as, for example, solubilization, self-emulsifying drug delivery systems (SEDDS) and micro-emulsions, do not work. The formulation principle should be 'formulate it as simple as possible'. Of course it is easier to formulate a micro-emulsion pre-concentrate and fill a capsule for oral administration. However, many drugs nowadays do not work with these approaches, especially because many new chemical entities (NCEs) are simultaneously poorly soluble in water and in organic media (e.g. solvents, oils) [5]. Therefore, there is plenty of room for the nanocrystal technology.

By definition, a nanocrystal is a nanoparticle composed of 100% drug and sized of the nanometer range (a few nanometers to 1000 nm (=1 μm), the borderline of the

Drug Delivery Strategies for Poorly Water-Soluble Drugs, First Edition. Edited by Dennis Douroumis and Alfred Fahr.

microparticles). They can be produced by top-down techniques (starting from a large drug crystal and reducing its size [1,2,6,7] and by bottom-up technologies, that means aggregating molecules to form nanocrystals (classical precipitation) [8–11]. After production, the nanocrystals are typically dispersed in liquid, these suspensions are called nanosuspensions. In the same way as macrosuspensions, the nanocrystals need to be stabilized by surfactants or steric stabilizers. Most oral nanocrystal formulations are solid dosage forms (e.g. tablets, capsules with pellets), that means, the liquid needs to be removed. However, there is also an oral nanosuspension on the market, the product Megace ES by the company Bristol-Myers Squibb. Intravenous nanosuspensions can be liquid, i.e. ready-to-use or a lyophilisate to be reconstituted prior to injection. In dermal nanocrystal formulations, the nanocrystals are typically dispersed in the water phase of an o/w cream or lotion, for example, the dermal cosmetic products by Juvena Switzerland [12]. Some very interesting physico-chemical features of nanocrystals are [4, 13, 14]:

- increase in dissolution velocity dc/dt, due to the increase in surface area compared to microcrystals;
- increase in saturation solubility c_S, due to an increase in dissolution pressure achieved by the transfer of the material to the nanodimension;
- adhesion to surfaces/membranes because of an increase of surface to mass ratio (the larger surface leads to increased van der Waals interactions with any surface).

The physical background to these effects is described in detail in [15]. Based on these general features, nanocrystals can be exploited to formulate drugs of class II of the Biopharmaceutical Classification System (BCS). The class II drugs are well permeable, but their bioavailability is limited due to the low solubility and related low dissolution velocity [16]. Nanosuspensions can also be used for intravenous delivery when nanoemulsions or solubilisates would require too large an administration volume [17, 18]. Recently nanocrystals were also employed for dermal delivery [19–21]. The increased concentration gradient leads to an increased diffusion pressure into the skin. Drug penetrated into the skin will be replaced immediately in the cream by the fast dissolving nanocrystals. There are also other interesting routes of administration, that up to now have been not yet or only partially investigated, the main reason being that these areas are less financially attractive than oral, i.v. and dermal products. An example is drug delivery to the eye, exploiting the increased retention time of the adhesive nanosuspensions [22–24]. Other possibilities are various injections, for example, intra-articular, intra-peritoneal, intra-muscular and subcutaneous.

8.2 Ultra-Fine Drug Nanocrystals (≪100 Nm) and Their Special Properties

As outlined above, nanocrystals range from a few nanometer to about 1000 nanometer in size (pharmaceutical definition). At present there is a controversial discussion which size has a nanoparticle. There are also definitions that a nanoparticle is below 100 nm, and a 'real' nanoparticle is below 20 nm (e.g. as stated by some colleagues in colloid/polymer science). The definition of nanoparticles as smaller than 100 nm can be found in the legislation, such as in Europe for Cosmetics [25] and by regulatory bodies such as the FDA or the BSI [26]. In Europe, cosmetic products need to be labelled as nano products when they

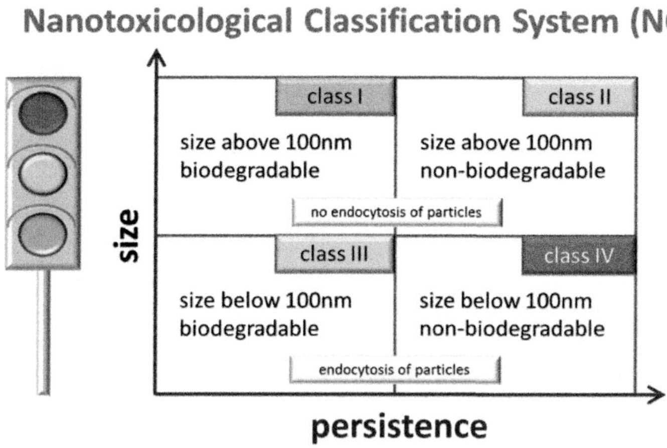

Figure 8.1 *Nanotoxicological classification system (NCS) grouping the nanoparticles into two size classes (>100 nm–1,000 nm, <100 nm) and into biodegradable and non-biodegradable nanoparticles leading to four classes of risk I to IV, simplified as a traffic light system. Each of the four classes can be biocompatible (B) or non-biocompatible (NB), yielding a total of eight classes from I-B and I-NB to IV-B and IV-NB. Modified after [4].*

contain these particles. The size limit has its origin in the increase of the potential toxicity of such nanoparticles. Larger particles can only be internalized by macrophages, hence by a very limited number of cells in the body, which are in addition often difficult to access for particles. Therefore particles >100 nm have *a priori* a low or no toxic risk. Particles below 100 nm can be taken up by all cells by endocytosis, which means their potential risk is higher. These differences were used to propose a nanotoxicological classification system (NCS), [4, 21], analogous to the BCS. It differentiates nanoparticles according to their size (>100 nm, <100 nm) and biodegradability/non-biodegradability (Figure 8.1).

Of importance is that the general benefits outlined above are size-dependent, that means, the smaller the nanoparticle/nanocrystal is, the higher is the increase in dissolution velocity and saturation solubility. This appears very logical with regard to the surface-related increase in dissolution velocity, the smaller the particle, the larger is the surface area. It is a kind of linear relationship. In contrast to this, there is an exponential increase in saturation solubility with decreasing nanocrystal size being most pronounced below approximately 100 nm (Figure 8.2).

The increase in saturation solubility c_s also affects the dissolution velocity, with the increase in dissolution velocity becoming more pronounced. According to the Noyes-Whitney law, the dissolution velocity dc/dt is proportional to the concentration gradient c_S-c_X/h (c_X – bulk concentration, h – diffusional distance) [27]. Based on this, the ultra-fine drug nanocrystals, defined as nanocrystals below 100 nm, possess superior solubility and dissolution properties, when compared to larger nanocrystals. These special properties make them very interesting for certain pharmaceutical formulations or applications, especially when they are in the size range of about 20–50 nm with very fast dissolution.

Ultra-fine drug nanocrystals are interesting for all drugs which are extremely poorly soluble and do not reach a sufficiently high dissolution velocity when formulated as a larger nanocrystal. Fast dissolution is also important when drugs exhibit an absorption

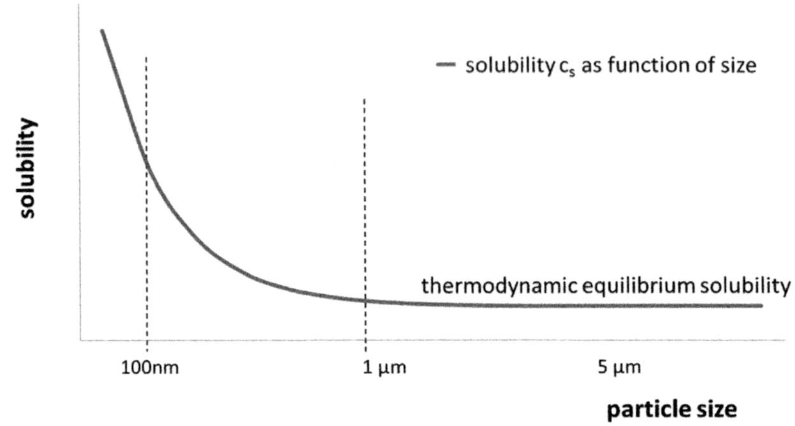

Figure 8.2 *Increase in saturation solubility as a function of decreasing particle size.*

window in the gastrointestinal tract. Dissolution should ideally be completed before the crystals have passed the absorption window. Fast dissolving nanocrystals are very important for intravenous formulations. When injecting nanocrystals of a size of a few hundred nm, they will only partially dissolve within the first 1–5 minutes after injection. Therefore, the majority will be taken up by the macrophages of the liver, typically up to 90% of the injected particle dose within 5 minutes [28]. This leads to a different pharmacokinetics compared to injecting the drug in a solubilized solution. A nice example is Taxol, as the solution contains the surfactant Cremophor EL which is responsible for anaphylactic reactions [29, 30]. Formulating and injecting the drug paclitaxel as a nanosuspension (size about 330 nm) dramatically increased the tolerability, at similar efficiency against the mamma carcinoma tumours MDA-MB 435 and 436 compared to Taxol [31], but the pharmacokinetic profile changed. In the Taxol groups of the two tumours, 2/3 resp. 1/3 of the animals (nude mice NMRI nu/nu) died on day 4 at a dose of 3 × 40 mg/kg, but all animals survived in the nanosuspension groups at doses as high as 3 × 80 mg/kg and 3 × 100 mg/kg (injection in left/right tail vein).

Based on these considerations, ultra-fine drug nanocrystals are the first choice for certain pharmaceutical formulations. However, the problem is that they are more difficult to produce. This chapter provides a short overview of the basic technologies used for the production of the first generation of drug nanocrystals (>100 nm). It describes in detail how these basic processes have been modified to enable production of ultra-fine drug nanocrystals (<100 nm). Special emphasis is given to the fact that the technologies are industrially feasible, a prerequisite for getting pharmaceutical products onto the market.

8.3 Production of First Generation Nanocrystals: A Brief Overview

8.3.1 Hydrosols

The so-called hydrosols were developed by Sucker [8] and are prepared by classical precipitation. The drug is dissolved in a solvent, and the solvent is added to a miscible non-solvent. The drug precipitates in crystals. There are various problems associated with precipitation.

The crystals tend to grow to the micrometer range, and the crystal growth needs to be terminated by appropriate precipitation conditions. The produced nanosuspensions contain solvents which normally need to be removed. For preservation, a lyophilization is recommended [9]. To our knowledge, no product based on this technology is on the pharmaceutical market.

8.3.2 NanoMorph®

The nanomorph technology comprises drug nanoparticles which are not crystalline but amorphous. Amorphous materials have in general a higher solubility than crystalline ones, in addition, the solubility increases with size in the nano range. Therefore, the amorphous drug nanoparticle is the ideal particle for increasing solubility and dissolution velocity. The particles are also produced by precipitation, a process developed by Auweter and co-workers at BASF [11]. The technology runs the risk that the particles can re-crystallize during the shelf life of the product, thus requiring a sound formulation technology. By now the technology has been applied in food and nutrition products by BASF (beta-carotene), the pharmaceutical rights belong to the drug delivery company Soliqs in Ludwigshafen (previously Knoll, now owned by Abbott Lab.). Soliqs owns amorphous drug nanoparticles (Nanomorph) but also crystalline formulation technology (smartCrystal®)

8.3.3 Nanocrystals™ by Bead Milling

In the 1990s, the company NanoSystems (nowadays: élan) employed a bead (pearl) milling process to generate drug nanocrystals below 400 nm [1]. Bead mills consist of a milling chamber containing small milling pearls available in different sizes from about 0.05 mm to about 1 mm and larger. A suspension of drug is added to the bead mill, then the milling beads are agitated and grind the drug crystals. As a rule of thumb the size of the nanocrystals obtained is about a factor 1000 smaller than the size of the milling beads used. For example, 200 μm (0.2 mm) beads will lead to crystals in the range of about 200 nm. Of course, this is only a rule of thumb; the final achieved size also depends on the hardness of the drug crystals and the amount of imperfections in these crystals. To produce 50 nm crystals, milling pearls with a size of about 50 μm are required. It is very tedious to work with such fine beads, with regard to the process itself but also to separate nanocrystals from the beads. This becomes particularly problematic when moving to the larger industrial scale. Most of the pharmaceutical products on the market are produced by élan using the bead mill process [15].

8.3.4 DissoCubes® by High Pressure Homogenization

Alternatively, nanocrystals can be produced by high pressure homogenization. A macrosuspension is prepared which is passed through a piston-gap homogenizer at pressures of up to 1500 bar (21756 psi, 150 MPa) up to 20 homogenization cycles [2,4,32,33]. The crystals are broken by the cavitation process in the homogenizer, in addition, size reduction takes place by collision of the crystals against crystals and parts of the wall in the homogenization chamber [2]. The trade name for the product is DissoCubes®. The technology was acquired by SkyePharma PLC in 1999 to become one of their core technologies for oral delivery of poorly soluble drugs. Alternatively, jet stream homogenizers can be used, preferentially the Microfluidizer by Microfluidics/USA. However, the microfluidizer process needs up to about 70 homogenization cycles [34] which is not production-friendly. The Microfluidizer process belonged to the Canadian company RTP which was also acquired by SkyePharma.

8.3.5 NANOEDGE by Baxter

The Baxter company developed a combination process, classical precipitation followed by a second step of high energy input, for example, high pressure homogenization [35, 36]. The drug is precipitated with a size in the nanometer range or a size of a few micrometer. This suspension is then subjected to a high energy process, typically high pressure homogenization. The second high energy step avoids further crystal growth which occurs in the precipitation suspension if this second step is not applied. A disadvantage of this process is that it involves solvents, identical to the production of hydrosols. It is costly to remove them, especially on a large industrial scale. Considering this, it is definitely easier to start the production process with a finely jet-milled material, coming close in size to the precipitated material. The Baxter company is focusing on intravenous nanosuspensions, an example being itraconazole [37, 38].

8.3.6 Summary of First Generation Production Technologies

All the technologies above allow – within reasonable efforts – nanocrystals to be produced in the size range of about 200 nm to 1000 nm. In general, the processes become more tedious when moving towards 100 nm, especially when going below 100 nm. Limitation in the bead milling process depends in the final stage on the size of the milling material. Even running up to 20 passages changes very little in the size of the bulk population after about 10 passages. Meanwhile small milling beads of 50 μm are available, but there is a lack of large-scale production lines using such small beads, so developing and building them would be relatively costly. Therefore larger quantities of ultra-fine drug nanoparticles cannot be produced with reasonable efforts. High pressure homogenization is limited by the pressure applicable in production lines under industrial conditions. For many drugs it was observed that the smallest achievable size was in the range 150–200 nm when applying high pressure homogenization. The 'standard' production conditions that are often applied are 1500 bar and 20 homogenization cycles, sometimes even 40 homogenization cycles. Increasing the homogenization pressure from 1500 to 3000 bar and above had very little effect on further reducing the size [39]. The same can be assumed for the NANOEDGE process by Baxter. The patent emphasizes that by the second high pressure homogenization step, the crystal size from the precipitation is maintained, and particle growth is avoided [35]. Therefore, there is a clear need for process modifications available making ultra-fine drug nanoparticles more accessible.

8.4 Production of Ultra-Fine Drug Nanocrystals: Smartcrystals

8.4.1 Fine-Tuned Precipitation

Preparation of small nanoparticles below 20 nm is relatively straight forward for a number of inorganic materials. For example, gold nanoparticles are prepared by simply reducing chloroauric acid with sodium citrate [40]. The precipitated gold particles are in the range of about 10–20 nm. Magnetites for magnet resonance tomography (MRT) can be precipitated with different sizes, ranging, for example, between about 5–15 nm [41,42]. Commercial products on the market are Ferridex® (AMAG Pharmaceuticals) or Resovist from Bayer (previously Schering). In contrast to this, precipitation of organic drug nanoparticles is more difficult. In the recent review by D'Addio and Prud'homme, the prerequisites for a controlled precipitation to target-specific sizes are described [43]. However, clearly the problem remains to remove the solvents from such processes.

8.4.2 The SmartCrystal Concept

The finally achieved size in a diminution process depends very much on the hardness of the starting material, and, in addition, on the number of imperfections in a crystal to be reduced in size. The approach of the smartCrystal process is to modify the starting material in such a way that it can be better broken down by high pressure homogenization. This means that the smartCrystal technology is basically a combination technology, a first pre-treatment step is combined with a second high pressure homogenization step. Modifications can be made to the starting material in a first step, alternatively in a special variant the dispersion medium of the crystals is modified. In contrast to the classical high pressure homogenization using water, the dispersion medium can be oil or a mixture of water with a water-miscible solvent (e.g. water-ethanol = Nanopure process). Table 8.1 gives an overview of the different process combinations.

8.4.2.1 The H42 Process

The H42 process is a combination of spray-drying followed by high pressure homogenization. In the last step of the synthesis of the drug the crystallization step is replaced by spray-drying the drug solution [44]. The spray-drying leads to a drug material which is still crystalline, but is much easier to break into particles. There is no detectable difference in crystallinity by X-ray diffraction, but the diminution in high pressure homogenization is much faster. An identical size reduction to HPH can be achieved in 1–5 cycles compared to 20–40 cycles applying homogenization of precipitated/jet-milled material (Figure 8.3).

Of course, applying higher cycle numbers in H42 leads to smaller crystals compared to HPH. The degree of size reduction achieved depends on the spray-drying conditions. It is assumed that the spray-drying leads to imperfections in the crystal not being detectable by X-ray diffraction. At least an amorphous fraction larger than 5% in the spray-dried material

Table 8.1 *Overview of the different smartCrystal processes.*

Process	Pre-treatment	Main treatment
Nanopure	–	HPH
H 42	spray-drying	HPH
H 69	precipitation	HPH
H 96	lyophylization	HPH
CT	bead milling	HPH

Note: HPH – high pressure homogenization.

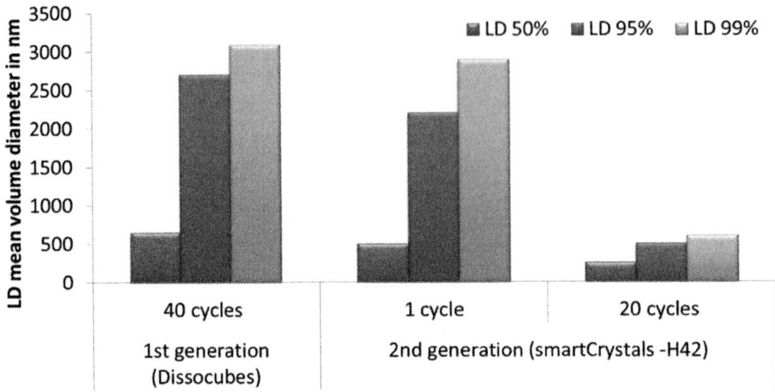

Figure 8.3 *Laser diffractometry (LD) diameters 50% to 99% of a model drug applying 40 cycles of high pressure homogenization (left) versus just 1 and 20 cycles in the H42 process. Modified after [45].* (See colour plate section.)

can be excluded, because 5% is the detection limit for X-ray diffraction. The material can be made even more fragile by adding additives like surfactants or steric stabilizers to the drug solution before spray-drying. These molecules are included in the formed crystals, making them obviously even more easily broken down [45, 46].

8.4.2.2 The H69 Process: 'CaviPrecipitation'

The basic idea behind this process is that the crystals are broken down at a time when the crystal structure is relatively weak. This is the case when a crystal is just forming in the precipitation process. The molecules are in the state of orientating to form a perfect crystalline structure. To apply breaking forces instantly, the precipitation should take place in the cavitation zone of a homogenizer. The crystals in the formation process are directly hit by the cavitation and collision forces. Alternatively precipitation can be initiated just prior to the suspension reaching the cavitation zone. The time between initiation of the precipitation and reaching the cavitation zone can be a few μs up to seconds or more [47]. For the precipitation, the solvent containing the drug and the non-solvent are brought together in a parallel flow system. Both liquids flow in the same direction, having a joint interface at which the precipitation starts. Depending on the flow velocity and the Reynold's number, the two liquid flows stay separated more or less or a more or less fast mixing occurs (Figure 8.4).

Prednisolone nanocrystals were produced using the H69 process [47]. Prednisolone was dissolved in ethanol, then the ethanol solution was admixed to an aqueous surfactant solution circulating through a microfluidizer. The velocity of admixing the ethanol solution was 1.5 ml/min. Immediately after admixing, the circulating surfactant solution turned turbid due to the formed nanocrystals. The size of the formed nanocrystals during circulation was monitored by PCS as a function of the circulation time (Figure 8.5).

The particle size was very fast below 100 nm and dropped continuously until it reached 22 nm. Nanocrystals of such a small size have a very high dissolution pressure, consequently a strong increase in saturation solubility. Further homogenization of the particles led to their

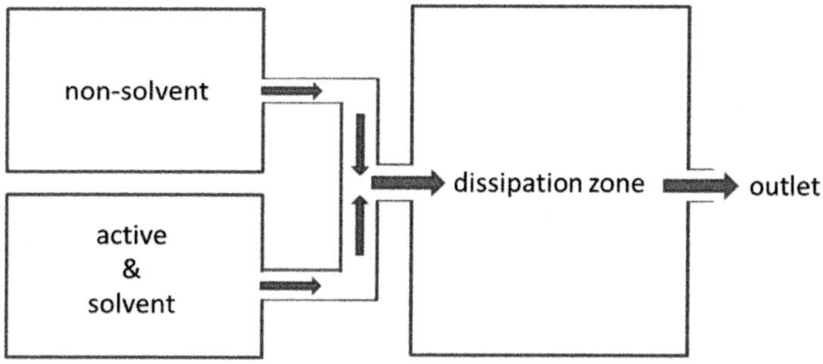

Figure 8.4 Basic set-up of a CaviPrecipitation system. The drug is dissolved in solvent, a non-solvent is provided in a separate vessel. The non-solvent is admixed to the drug solution directly in the zone and cavitation (e.g. by homogenization, ultrasound, static blenders, etc.) occurs or just before reaching this zone. Reproduced after [47].

complete dissolution. A clear solution was obtained because the saturation solubility was so strongly increased that the drug completely dissolved in the water-ethanol mixture.

This clear solution was stored at room temperature, after approximately 4 hours after the first large prednisolone microcrystals appeared. This is well in agreement with the theory. The drug nanocrystals of about 20 nm could dissolve in the water-ethanol mixture because their kinetic solubility was much higher than the thermodynamic equilibrium solubility. Such kinetic 'supersaturated' solutions are meta-stable. After a few hours re-crystallization was initiated and very large crystalline needles started forming. The crystal growth took place relatively slowly, large crystals formed because the system passed the Ostwald-Mier range slowly [48]. Figure 8.6 shows the principal situation for kinetic and thermodynamic solubility and crystal formation process. The Microfluidics company has realized this principle of precipitation followed immediately by cavitation in their new microfluidizer for the generation of small crystals (Figure 8.7).

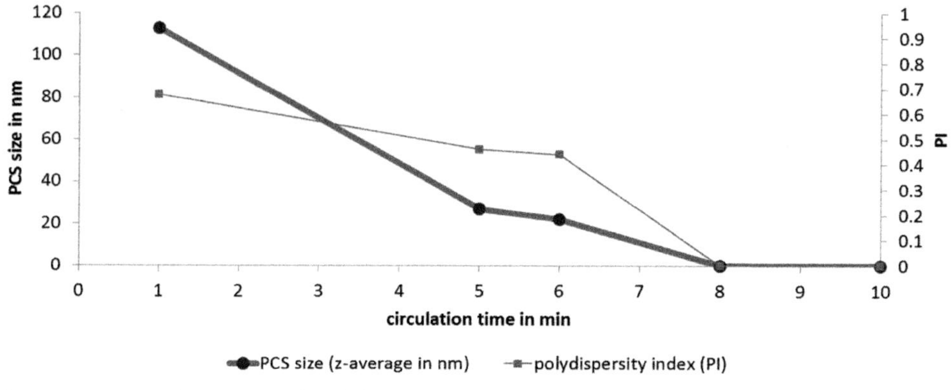

Figure 8.5 PCS diameter and polydispersity index (PI) of prednisolone nanocrystals produced by H69. The size is plotted as a function of circulation time, during the first 10 minutes. Ethanolic drug solution was continuously admixed prior to the cavitation zone of the microfluidizer. Reproduced after [47].

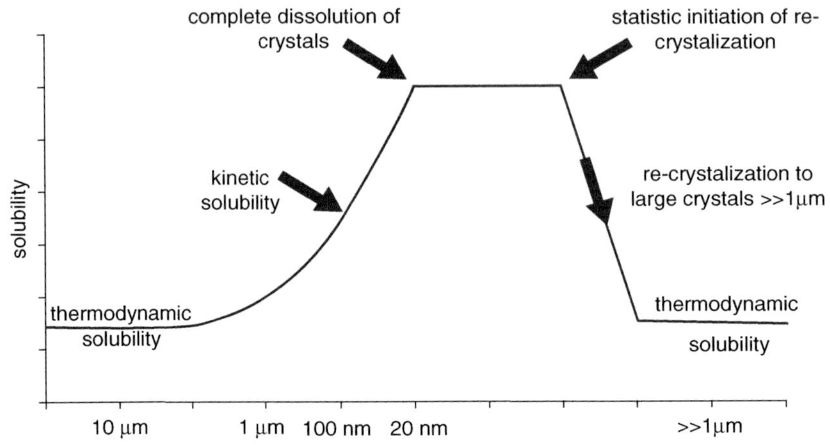

Figure 8.6 *Change in kinetic solubility as a function of size reduction, dissolution of the very small nanocrystals (20 nm) and subsequent re-crystallization, and return of the system to the thermodynamic solubility.*

8.4.2.3 *The H96 Process*

Another approach to modify the starting material is the use of lyophilization. Identical to H42 in the last step of the synthesis, the drug is not crystallized. In case of H96, the drug solution is lyophilized. Again, additives can be given to the drug solution before lyophilization. The ability of this process to generate ultra-fine drug nanoparticles was shown for amphotericin B [49]. Amphotericin B was dissolved in the DMSO, the solution was lyophilized and then the lyophilisate was dispersed in an aqueous surfactant solution

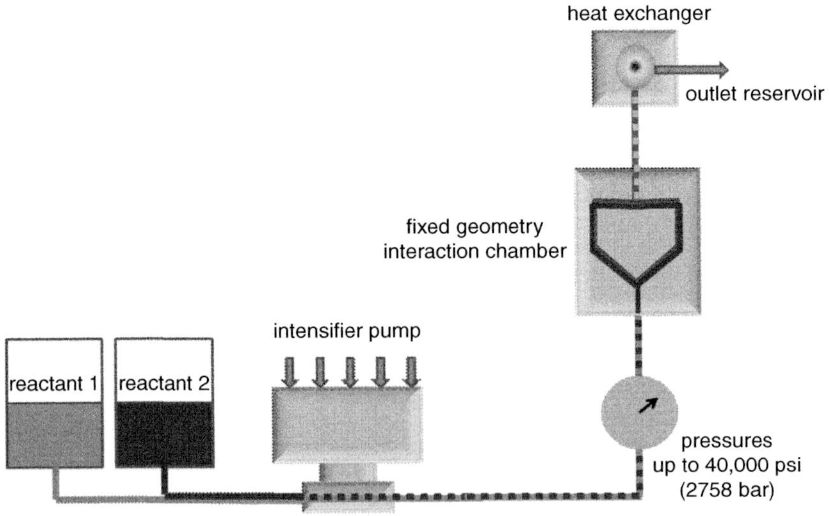

Figure 8.7 *Design of the new microfluidizer functioning according to the H69 process. There are two inlet reservoirs which mix solvent and non-solvent prior to the intensifier pump, which feeds the precipitate to the interaction chamber for grinding. Modified after [57].*

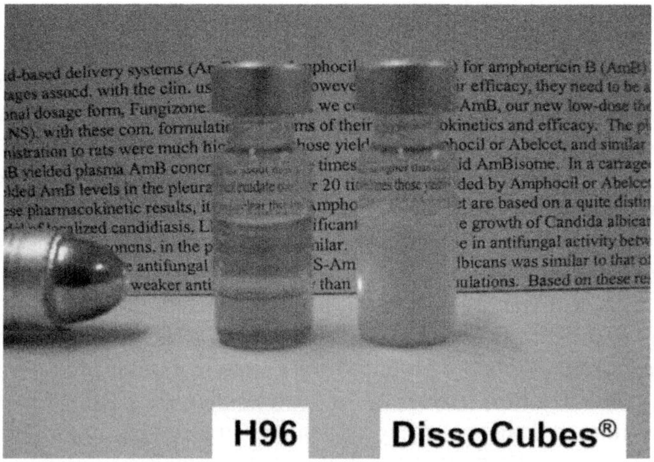

Figure 8.8 *Amphotericin B nanosuspensions: the left nanosuspension is translucent due to its small nanocrystal size – produced by H96 technology. The scattered laser beam proves the existence of the nanocrystals. The right nanosuspension is about 300 nm and turbid produced with high pressure homogenization only (= DissoCubes technology.) Reproduced with permission after [58].* (See colour plate section.)

containing 1.1% (w/w) sodium cholate. This dispersion was performed using a high speed stirrer, followed by high pressure homogenization. After 5 cycles, a suspension with a mean photon correlation spectroscopy diameter of about 60 nm was achieved [49]. If particles are much smaller than the wavelength of the visible light, the suspensions become transparent. Figure 8.8 shows a conventional amphotericin B suspension produced with normal high pressure homogenization, in comparison to a translucent 60 nm amphotericin B suspension produced with the H96 combination process.

The efficiency of this process depends very much on the lyophilization conditions, but also on the solvent used for lyophilization and potential additives such as surfactants and stabilizers. The ideal parameters cannot yet be predicted, they need to be found empirically.

8.4.2.4 Bead Mill and HPH Combination Technology (CT)

The last combination is bead milling followed by high pressure homogenization. This has basically two advantages:

- increased physical stability of the produced nanosuspension;
- potential further size reduction.

It was found empirically that nanosuspensions produced with a pearl mill often possessed a higher physical stability when they were treated in a subsequent process by high pressure homogenization. The physical stability was increased despite no size reduction being achieved when applying a low homogenization pressure of, for example, 100 bar or 300 bar [50]. The stability against electrolytes was investigated by using an electrolyte challenge test. Calcium chloride solutions were added to a bead-milled nanosuspension and to the same nanosuspension after an additional homogenization step. The increase in size was measured by laser diffractometry (LD). For example, the LD diameters of 99% for the drug Apigenin nanosuspension remained unchanged practically for 30 minutes with the

CT nanosuspension, but increased from 2.447 μm to 4.373 μm when only bead milling was applied.

In addition, it could also be shown that the physical long-term stability as a nanosuspension was improved. The improved physical stability of this second generation of nanocrystals is very important for oral administration. The electrolytes of the gut reduce the zeta potential of the drug nanocrystals, leading to aggregation. Aggregated nanocrystals lose their special 'nanoproperties', and the oral bioavailability decreases. Therefore nanocrystals ideally should not, or by only very little, aggregate in the gastric and intestinal fluids.

Depending on the hardness of the drug nanocrystals, the subsequent second homogenization step can further distinctly reduce the nanocrystal size in the case of softer drugs. The smart feature of the CT process is that it is currently running on a large industrial scale in industry. Nanocrystals are being produced for cosmetic dermal products under licence from PharmaSol GmbH Berlin. Examples of such nanocrystals are rutin and hesperidin (INCI: submicron rutin crystals, smartCrystal-lemon extract). They are found, for example, in products of the line JUVEDICAL from Juvena Switzerland, and platinum rare from la prairie, respectively [21].

8.4.2.5 Capsulution's LBL Plus Nanocrystals

The layer-by-layer technology is a process in which surfaces are coated by alternate adsorption of oppositely charged polyelectrolytes (PE). For example, a negatively charged surface is coated first with a positively charged PE (polycation), then in a second layer with a negatively charged one (polyanion), followed again by a positive one, and so on (Figure 8.9). Such a single PE layer can easily have a thickness of about 2–4 nm, by adding 5 layers a

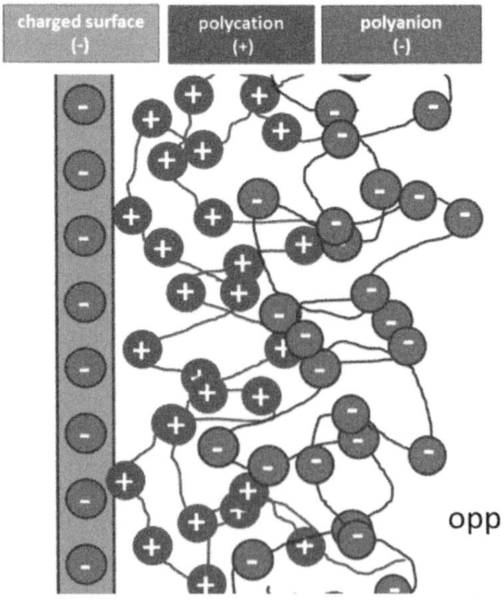

Figure 8.9 LBL coating process on a negative surface, adsorbing oppositely charged polyelectrolytes. Reproduced after Capsulution Pharma AG Berlin, product information. (See colour plate section.)

thickness of >10 nm of the LBL assembly can be achieved. Recently this LBL technology was transferred to nanocrystals by the company Capsulution Pharma AG in Berlin [51–55].

The advantage of multiple layers is that they provide an excellent steric barrier against aggregation in the body fluids. According to theoretical considerations the minimum thickness of a steric barrier for good stabilization is 10 nm [28, 56]. Adsorbing just a single polymer layer on a nanocrystal, this thickness normally cannot be achieved with a single polymer. Therefore the single polymer stabilized nanocrystals still show some instability. The solution is multiple LBL coating which has the potential to generate perfect stable crystals. In the Capsulution process the LBL coating is performed during the bead milling process without the interim removal of excess PE, that means it is a continuous industrially friendly process [51]. The LBL technology can also be applied to ultra-fine nanocrystals to make them more physically stable in the body.

8.5 Conclusion

Nanocrystals have established themselves as an effective formulation approach for poorly soluble drugs in the markets for oral administration (pharma) and dermal application (cosmetics), and quite a number of products are now in the clinical phases. Ultra-fine drug nanocrystals with a size below 100 nm are clearly superior to the larger nanocrystals of the first generation regarding saturation solubility and dissolution velocity. This is of great importance for special pharmaceutical delivery challenges, for example, fully exploiting an absorption window in oral administration or mimicking the injection of a solution with intravenous nanosuspension.

In particular, intravenous nanosuspensions might be the key to success, because accumulation of drug nanocrystals in the liver changes the pharmacokinetics, and might even target toxicity to the liver, which would exclude the use of nanocrystals. With H42, H69 and H96 production, processes are available to provide ultra-fine nanocrystals on the lab scale. However, these processes still need to be established on a larger scale. From the basic pharmaceutical processes involved – spray-drying, precipitation, lyophilization – a scale-up is possible.

Attention needs to be paid to the costs for these technologies. However, the number of new chemical entities reaching the market is unchanged or has even been decreasing in the past few years. If these technologies enable an interesting, highly effective new chemical entity to be put as a product on the market, covering at the same time a real therapeutic need, the costs will not play a role. These technologies might prevent a company from losing a valuable drug candidate.

The combination technology of bead milling and high pressure homogenization was established on the large scale easily, because each of the single processes previously existed as industrially viable. The nanocrystals of the first generation can be considered the 'basic model', now we need to work on the 'smarter versions' with improved properties and performance to tackle more challenging pharmaceutical delivery problems.

References

1. G.G. Liversidge, K.C. Cundy, J.F. Bishop, D.A. Czekai, Surface modified drug nanoparticles. United States Patent 5,145,684 (1992).

2. R.H. Müller, R. Becker, B. Kruss, K. Peters, Pharmaceutical nanosuspensions for medicament administration as systems with increased saturation solubility and rate of solution. United States Patent 5,858,410 (1999).
3. J.E. Diederichs, R.H. Müller, Liposome in Kosmetika und Arzneimitteln. *Pharm Ind*, **56**, 267–275 (1994).
4. R.H. Müller, S. Gohla, C.M. Keck, State of the art of nanocrystals: Special features, production, nanotoxicology aspects and intracellular delivery. *Eur J Pharm Biopharm*, **78**, 1–9 (2011a).
5. S. Stegemann, F. Leveiller, D. Franchi, H. de Jong, H. Linden, When poor solubility becomes an issue: From early stage to proof of concept. *Eur J Pharm Sci*, **31**, 249–261 (2007).
6. G.G. Liversidge, P. Conzentino, Drug particle size reduction for decreasing gastric irritancy and enhancing absorption of naproxen in rats. *Int J Pharm*, **125**, 309–313 (1995).
7. R.H. Müller, K. Mäder, K. Krause, Verfahren zur schönenden Herstellung von hochfeinen Micro-/Nanopartikeln. PCT Application PCT/EP00/06535 (2000).
8. M. List, H. Sucker Pharmaceutical colloidal hydrosols for injection. GB Patent 2,200,048 (1988).
9. P. Gassmann, M. List, A. Schweitzer, H. Sucker, Hydrosols: Alternatives for the parenteral application of poorly water soluble drugs. *Eur J Pharm Biopharm*, **40**, 64–72 (1994).
10. E. Liliemark, B. Sjostrom, J. Liliemark, C. Peterson, N. Kallberg, B.S. Larsson, Targeting of teniposide to the mononuclear phagocytic system (MPS) by incorporation in liposomes and submicron lipid particles: An autoradiographic study in mice. *Leukem Lymphom*, **18**, 113–118 (1995).
11. H. Auweter, H. Bohn, R. Heger, D. Horn, B. Siegel, K. Siemensmeyer, Precipitated water-insoluble colorants in colloid disperse form. United States Patent 6, 494–924 (2002).
12. C.M. Keck, S. Kobierski, R. Mauludin, R.H. Müller, Second generation of drug nanocrystals for delivery of poorly soluble drugs: Smartcrystals technology. *Dosis*, **24**, 124–128 (2008).
13. G. Buckton, A.E. Beezer, The relationship between particle size and solubility. *Int J Pharm*, **82**, R7–R10 (1992).
14. M. Mosharraf, C. Nyström, The effect of particle size and shape on the surface specific dissolution rate of microsized practically insoluble drugs. *Int J Pharm*, **122**, 35–47 (1995).
15. R.H. Müller, J.U. Junghanns, Drug nanocrystals/nanosuspensions for the delivery of poorly soluble drugs. In *Nanoparticulates as Drug Carriers*, V.P. Torchilin, (ed.). London: Imperial College Press, pp. 307–328 (2006).
16. G.L. Amidon, H. Lennernas, V.P. Shah, J.R. Crison, A theoretical basis for a biopharmaceutic drug classification: The correlation of in vitro drug product dissolution and in vivo bioavailability. *Pharm Res*, **12**, 413–420 (1995).
17. J. Möschwitzer, G. Achleitner, H. Pomper, R.H. Müller, Development of an intravenously injectable chemically stable aqueous omeprazole formulation using nanosuspension technology. *Eur J Pharm Biopharm*, **58**, 615–619 (2004).

18. C.M. Keck, R.H. Müller, SmartCrystals: Review of the second generation of drug nanocrystals. In *Handbook of Materials for Nanomedicine*. V. Torchilin, M. M. Amiji (eds). London: Pan Stanford Publishing, pp. 555–580 (2010).
19. S. Kobierski, K. Ofori-Kwakye, R.H. Muller, C.M. Keck, Resveratrol nanosuspensions for dermal application: Production, characterization, and physical stability. *Pharmazie*, **64**, 741–747 (2009).
20. P.R. Mishra, L. Al Shaal, R.H. Müller, C.M. Keck, Production and characterization of Hesperetin nanosuspensions for dermal delivery. *Int J Pharm*, **371**, 182–189 (2009).
21. R.H. Müller, R. Shegokar, S. Gohla, C.M. Keck, Nanocrystals: Production, cellular drug delivery, current and future products. In *Intracellular Delivery: Fundamentals and Applications, Fundamental Biomedical Technologies*. A. Prokop (ed.). New York: Springer, pp. 411–432 (2011).
22. C. Bucolo, A. Maltese, G. Puglisi, R. Pignatello, Enhanced ocular anti-inflammatory activity of ibuprofen carried by an Eudragit RS100 nanoparticle suspension. *Ophthalmic Res*, **34**, 319–323 (2002).
23. K. Adibkia, M.R. Siahi Shadbad, A. Nokhodchi, A. Javadzedeh, M. Barzegar-Jalali, J. Barar, G. Mohammadi, Y. Omidi. Piroxicam nanoparticles for ocular delivery: Physicochemical characterization and implementation in endotoxin-induced uveitis, *J Drug Target*, **15**, 407–416 (2007).
24. M.A. Kassem, A.A. Abdel Rahman, M.M. Ghorab, M.B. Ahmed, R.M. Khalil, Nanosuspension as an ophthalmic delivery system for certain glucocorticoid drugs. *Int J Pharm*, **340**, 126–133 (2007).
25. *Amtsblatt der Europäischen Union,* Verordnung (EG) Nr. 1223/2009 des Europäischen Parlaments und des Rates vom 30. November 2009 über kosmetische Mittel/Neufassung zur Richtlinie 76/768/EWG des Rates vom 27. Juli 1976. Artikel 2k (2009).
26. British Standards (BSI), PAS 71: Vocabulary - Nanoparticles. *Publicly Available Specification (PAS)*, 2 (2005).
27. A.A. Noyes, W.R. Whitney, The rate of solution of solid substances in their own solutions. *J. Am Chem Soc*, **19**, 930–934 (1897).
28. R.H. Müller, *Colloidal Carriers for Controlled Drug Delivery and Targeting*, Stuttgart: Wissenschaftliche Verlagsgesellschaft mbH/Boston: CRC Press (1991).
29. R. Panchagnula, Pharmaceutical aspects of paclitaxel. *Int J Pharm*, **172**, 1–15 (1998).
30. L.B. Michaud, V. Valero, G. Hortobagyi, Risks and benefits of taxanes in breast and ovarian cancer. *Drug Safety*, **23**, 401–428 (2000).
31. B. Böhm, Herstellung und Charakterisierung von Nanosuspensionen als neue Arzneiform für Arzneistoffe mit geringer Biofervügbarkeit. PhD thesis, Freie Universität Berlin (1999).
32. C.M. Keck, R.H. Müller, Drug nanocrystals of poorly soluble drugs produced by high pressure homogenisation. *Eur J Pharm Biopharm*, **62**, 3–16 (2006).
33. R.H. Müller, J. Möschwitzer, F.N. Bushrab, Manufacturing of nanoparticles by milling and homogenisation techniques. In *Nanoparticle Technology for Drug Delivery*, R. B. Gupta, U. B. Kompella (eds). New York: Taylor & Francis Group, pp. 21–52 (2006b).
34. S. Khan, G.W. Pace, Composition and method of preparing microparticles of water-insoluble substances. United States Patent 6,337,092 (2002).

35. J.E. Kipp, J.C.T. Wong, M.J. Doty, C.L. Rebbeck, Microprecipitation method for preparing submicron suspensions. United States Patent 6,607,784 (2003).
36. J.E. Kipp, J.C.T. Wong, M.J., Doty, J. Werling, C.L. Rebbeck, S. Brynjelsen, Method for preparing submicron particle suspensions. United States Patent 6,884,436 (2005).
37. B. Rabinow, J. Kipp, P. Papadopoulos, J. Wong, J. Glosson, J. Gass, C.S. Sun, T. Wielgos, R. White, C. Cook, K. Barker, Wood, Itraconazole IV nanosuspension enhances efficacy through altered pharmacokinetics in the rat. *Int J Pharm*, **339**, 251–260 (2007).
38. B.E. Rabinow, Nanosuspensions in drug delivery. *Nat Rev Drug Discov*, **3**, 785–796 (2004).
39. M. Fichera, S. Wissing, R.H. Müller, Effect of 4000 bar homogenisation pressure on particle diminution in drug suspensions. Paper presented at International Meeting on Pharmaceutics, Biopharmaceutics and Pharmaceutical Technology, Nürnberg, pp. 679–680 (2004).
40. J. Turkevich, P.C. Stevenson, J. Hillier, A study of the nucleation and growth processes in the synthesis of colloidal gold. *Discuss Faraday Soc*, **11**, 55–75 (1951).
41. A. Bee, R. Massart, S. Neveu, Synthesis of very fine maghemite particles. *Journal of Magnetism and Magnetic Materials*, **149**, 6–9 (1995).
42. F. Thünemann, D. Schütt, L. Kaufner, U. Pison, H. Möhwald, Maghemite nanoparticles protectively coated with poly(ethylene imine) and poly(ethylene oxide)-block-poly(glutamic acid). *Langmuir*, **22**, 2351–2357 (2006).
43. S.M. D'Addio, R.K. Prud'homme, Controlling drug nanoparticle formation by rapid precipitation. *Adv Drug Deliv Rev*, **63**, 417–426 (2011).
44. J.P. Möschwitzer, Method for the production of ultrafine submicronic suspensions. EP2006/002176, WO/2006/094808 (2006).
45. J.P. Möschwitzer, Method for the production of ultra-fine submicron suspensions. *DE*, **10** 2005 011 786.4 (2005).
46. J. Möschwitzer, R.H. Müller, New method for the effective production of ultrafine drug nanocrystals. *J Nanosci Nanotech* **6**, 6, 3145–3153 (2006).
47. R.H. Müller, J. Möschwitzer, Method and device for producing very fine particles and coating such particles. PCT/EP2006/009930, WO/2007/051520 (2007).
48. C.M. Lommer, Lösungen. In: *Hagers Handbuch der Pharmazeutischen Praxis: Methoden*, E. Nürnberg, P. Surmann (eds). Heidelberg: Springer (1991).
49. J. Möschwitzer, A. Lemke, Method for carefully producing ultrafine particle suspensions and ultrafine particles and use thereof. PCT/EP2006/003377, WO/2006/108637 (2007).
50. R.D. Petersen, Nanocrystals for use in topical cosmetic formulations and method of production thereof. PCT/EP2007/009943 (2007).
51. Capsulution Pharma AG, *Improved Nanoparticulate Compositions of Poorly Soluble Compounds.* EP 2127193 A1 (2008).
52. C. Dunmann, M. Grabo, R.H. Müller, LBL$^{plus\ TM}$: Polymer multilayer coated nanocrystals for the delivery of poorly soluble drugs. Paper presented at Annual Conference of the European Association of Pharma Biotechnology (EAPB), Berlin, 62 (2011).
53. C. Dunmann, M. Grabo, R.H. Müller, Large scale production of LBL^{plus} nanocrystals. Paper presented at AAPS Annual Meeting, Washington, DC, W4159 (2011).
54. C. Dunmann, M. Grabo, R.H.,Müller, Scale up of LBL^{plus} nanocrystal production. Paper presented at Int. Symp. Control. Rel. Bioact. Mater. 38, National Harbor, Maryland, 791 (2011).

55. C. Dunmann, M. Grabo, R.H. Müller, LBL^{plus}: Capsulution's multilayer nanocrystals for the delivery of poorly soluble drugs. Paper presented at Int. Symp. Control. Rel. Bioact. Mater. **38**, National Harbor, Maryland, 801 (2011).
56. G.R. Harper, M.C. Davies, S.S. Davis, T.F. Tadros, D.C. Taylor, M.P. Irving, J.A. Waters, Steric stabilization of microspheres with grafted polyethylene oxide reduces phagocytosis by rat Kupffer cells in vitro. *Biomat*, **12**, 695–700 (1991).
57. T. Panagiotou, S.V. Mesite, R.J. Fisher, Production of norfloxacin nanosuspensions using microfluidics reaction technology through solvent/antisolvent crystallization. *Ind Eng Chem Res*, **48**, 1761–1771 (2009).
58. J. Möschwitzer, Drug nanocrystals prepared by high pressure homogenisation: The universal formulation approach for poorly soluble drugs. PhD thesis, Freie Universität Berlin (2006).

9

Nanosuspensions with Enhanced Drug Dissolution Rates of Poorly Water-Soluble Drugs

Dennis Douroumis

9.1 Introduction

An increased number of active pharmaceutical ingredients (APIs) developed by the pharmaceutical industry possess low aqueous solubility. More than 40% of the newly synthesized drugs are poorly water-soluble, which demonstrate low bioavailability, according to the Biopharmaceutics Classification System (BCS) [1–3]. According to the BCS system for classes II and IV, the dissolution rate is the limiting factor for the drug absorption rate. The increase in the solubility or dissolution rate of poorly water-soluble drugs is often challenging. Quite often in drug discovery, high-throughput screening approaches overlook a significant number of compounds because of their low solubility or rate of dissolution. For poorly water-soluble compounds, simple approaches can lead to a thermodynamically stable increase in solubility by: (i) pH adjustment when the active compound is a weak electrolyte; (ii) a change in the solvent polarity by using co-solvents (PEG-300/400, Cremophor RH-40, etc.); (iii) micelle formation (e.g. surfactants); or (iv) complexation of the active compound (e.g. cyclodextrins). However, these approaches are not always successful due to the increased additive concentrations required to achieve the desired solubility increase, or the extent of the adverse effects, especially for parenteral administration. Therefore, there is an enormous need to develop technologies that can improve the dissolution rates for oral, parenteral, pulmonary and other administration routes.

An alternative approach to overcome the pitfalls of the above methods is the engineering of nanoparticulate formulations of the active substance, which can result in an increase in

Drug Delivery Strategies for Poorly Water-Soluble Drugs, First Edition. Edited by Dennis Douroumis and Alfred Fahr.

the apparent solubility or the dissolution rate. According to the Kelvin equation [4, 5], the apparent solubility S_{app} of spherical particles increases exponentially in comparison to the equilibrium solubility S_0 with decreasing particle radius.

$$\ln\left({}^{S_{app}}/_{S_0}\right) = (RT)^{-1}\, 2\gamma V/r \tag{9.1}$$

where γ represents the interfacial energy, V the molar volume and r the particle radius at a given temperature T. From the above equation it can be clearly seen that with sizes over a few microns the increase in solubility is negligible, but for nanometer-sized particles (r < 1 μm), the solubility enhancement can be substantial. Hence, the engineering of stable nanoparticulate formulations of poorly water-soluble drugs results in a high concentration S_{app} of the active substance in the dynamic equilibrium.

By implementing a different approach, the Noyes–Whitney [4] equation gives the rate of dissolution of a solid compound and the important factors of the drug dissolution can be identified.

$$\frac{dM}{dt} = \frac{DA}{h}\left(S_{app} - C_t\right) \tag{9.2}$$

where *dM/dt* is the mass transfer from the solid to the solute state, *D* the diffusion coefficient, *A* the specific surface, *h* the thickness of a 'stagnant layer' of liquid in the immediate contact with the dissolving solid and *C* the bulk concentration at any time during the dissolution. The particle size of an active compound is an important factor that determines the dissolution rate. The dissolution rate is directly proportional to the specific surface area of the drug, which in turn increases with decreasing particle size, especially for nanoparticulate systems.

It is evident, from Equations 9.1 and 9.2, that engineering of formulations containing nanosized particles can be implemented to overcome the delivery barriers of poorly water-soluble drugs. A pharmaceutical nanosuspension is defined as a submicron colloidal dispersion of drug particles, which is stabilized by suitable stabilizers (surfactants and/or polymers) [6]. In general, the nanosuspension techniques are categorized as 'bottom-up' processes and 'top-down' processes related to the production principle [7]. The 'top down' processes involve breaking down larger particles by using mechanical means such as milling (using jet mills or pear-ball mills) and high-pressure homogenization. Although 'top-down' processes are a relatively simple way to produce nanoparticles, several drawbacks are related to time consumption, intensive energy use, introduction of impurities, inadequate control of particle sizes and electrostatic effects [8]. The 'bottom-up' technologies involve assembling and controlling precipitation [9] at the nanometer range where the active compound is first dissolved in an organic solvent and then precipitated through a non-solvent addition in the presence of stabilizers. Such processes include antisolvent precipitation using static mixers or ultrasonication, microfluidic reactors, spray-freezing into a liquid process (SFL) and rapid expansion from supercritical to aqueous solution (RESAS or RESOLV). This chapter focuses on 'bottom-up' processes employed to engineer stable drug nanoparticulate formulations via supersaturation and precipitation.

9.2 Crystal Growth and Nucleation Theory

In the classical nucleation theory, phase separation is energetically favourable and occurs at the onset of supersaturation by modifying the boundary conditions (temperature or pressure) or by homogeneous mixing with another component [9] to reduce the system-free energy. The driving force for nucleation and crystal growth to occur is supersaturation, which is defined as the difference in chemical potential between a molecule in solution and that in the bulk of the crystal phase (Equation 9.3).

$$\Delta\mu = \mu_\varsigma - \mu_c \tag{9.3}$$

where μ_ς is the chemical potential of a molecule in solution and μ_c the chemical potential of a molecule in the bulk crystal. When $\Delta\mu > 0$, the solution is said to be supersaturated and nucleation is possible, while when $\Delta\mu < 0$, the solution is unsaturated and dissolution will take place.

The supersaturation ratio, S_r, is defined as the ratio of the particle solubility at the interface, C_s, to the bulk solubility, C_∞,

$$S_r = \frac{C_s}{C_\infty} \tag{9.4}$$

The *nucleation* is defined as a series of processes by which the atoms or molecules of a reactant phase are rearranged into a nuclei of the product phase large enough to have the ability to grow irreversibly to a macroscopically larger size [10–12]. Nucleation can be homogeneous or heterogeneous in the presence or the absence of foreign particles in the solution. In homogeneous nucleation, local concentration fluctuations lead to the formation of primary nuclei (primary nucleation) by a series of bimolecular collisions that form aggregates until a critical cluster size is reached and the nuclei become stable.

The nucleation theory states that the change in the Gibbs free energy (ΔG) required to form a nucleus of radius r is given by the following equation [13]:

$$\Delta G_r = \Delta G_V + \Delta G_S = -\left(\frac{4\pi r^3}{3V}\right) k_B T ln S_r + 4\pi r^2 \gamma \tag{9.5}$$

Where: r is the size of the cluster, V the specific volume of a growth unit, γ the solid-liquid interfacial energy, k_B the Boltzmann constant, and S_r the relative supersaturation.

The first term in this equation is a bulk term (ΔG_V), which is negative and decreases as r^3, and the second term is a surface term (ΔG_S), which is positive and increases as r^2 (Figure 9.1).

As can be seen in Figure 9.1, as a result of the surface and volume term, the ΔG passes through a maximum, denoted as ΔG*. Taking the derivative with respect to r_c and setting this equal to zero allows the calculation of r_c which is called the critical nuclei size.

$$r_c = \frac{2V\gamma}{k_B ln S_r} \tag{9.6}$$

The solubility at the nanoparticles interface is given by the Kelvin equation [14, 15] written in terms of the local supersaturation at the particle surface:

$$ln S_r \equiv ln \frac{C_s}{C_\infty} = \frac{2\gamma M}{\rho R T r} \tag{9.7}$$

where M is the molecular mass of the solute, ρ is the density, R is the gas constant and T is the absolute temperature.

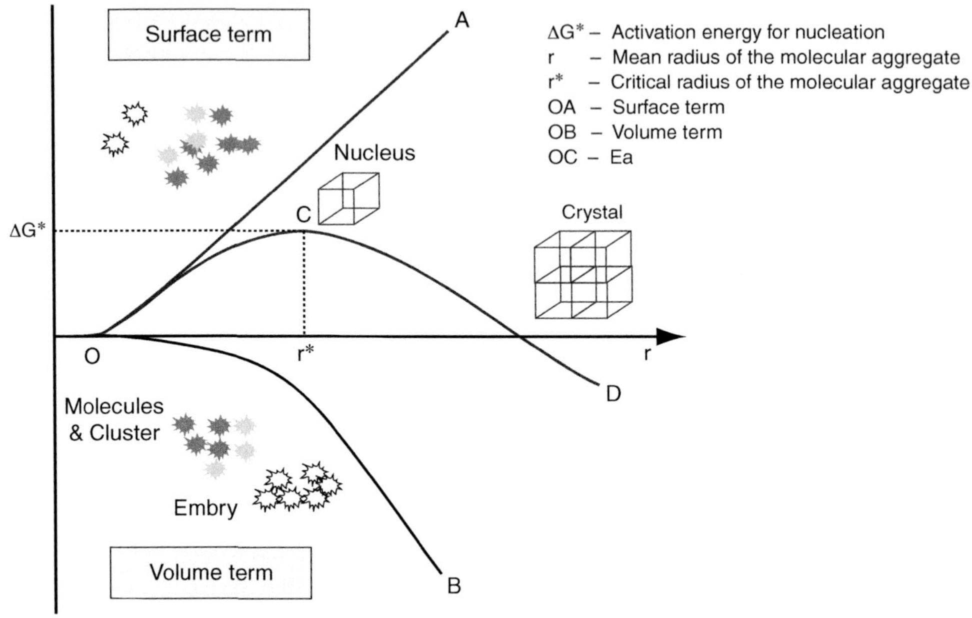

Figure 9.1 *Gibbs free energy change of nucleation as a function of the cluster size. r* is the critical radius of molecular aggregate at which nuclei spontaneously grow. Formation of nuclei is a compromise between the volume and surface term. The volume term favours aggregation whereas the surface term allows dissolution. Nuclei result when there is high volume–surface ratio. Below r*, aggregates dissolve whereas above r, nuclei form macroscopic crystals. Reproduced with permission from [12].*

The rate of nuclei formation J follows an Arrhenius relationship [13] and is defined as the number of nuclei that grow larger than the critical size and so become crystals, and is given by the equation

$$J = J_0 exp\left(-\frac{\Delta G_{cr}}{kT}\right) \tag{9.8}$$

by substituting Equations 9.6 and 9.7 into Equation 9.5, the ΔG_{cr} and the nucleation rate B are then given by:

$$J = J_0 exp\left(-\frac{16\pi\gamma^3 v^2}{3k^3T^3\,[ln\,(S_r)]^2}\right) \tag{9.9}$$

where v is the mole volume. It is apparent that nucleation kinetics are governed by both the temperature and local supersaturation.

Once an ordered structure is formed by nucleation, nuclei grow into macroscopic crystals where the growth units (atoms, ions or molecules) diffuse from the surrounding supersaturated solution to the surface of the nuclei and are incorporated into the lattice. This process is known as crystal growth. The crystal growth from supersaturated solutions can be either volume-diffusion controlled, when the diffusion of the solute from the bulk

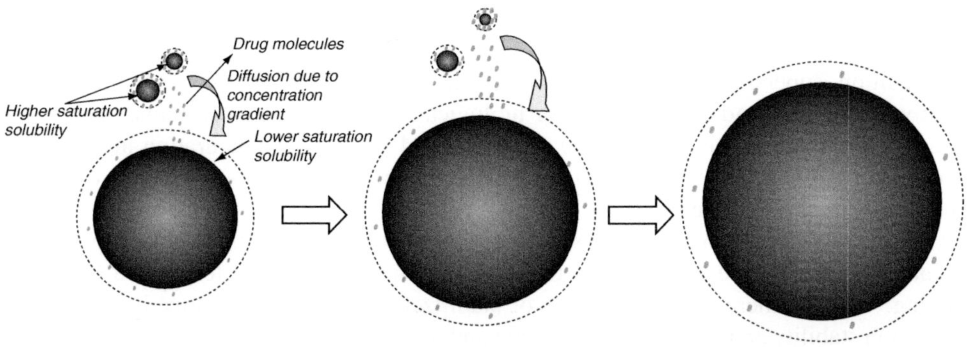

Figure 9.2 *Ostwald ripening. Reproduced with permission from [17].* (See colour plate section.)

solution to the bulk surface is rate limiting, or surface-integration controlled [16] if the incorporation into a crystal is the slowest process due to diffusion-resistance. The process of crystal growth is ruled by: (i) internal factors such as the three-dimensional crystal structure, and crystal defects; and (ii) external factors such as temperature, supersaturation, solvent and impurities.

In colloidal suspensions the crystal growth is known as Ostwald ripening (Figure 9.2), a process where large particles grow at the expense of smaller particles leading to increased particle size distribution and polydispersity. Small particles have enhanced saturation solubility arising from their high curvature and create a drug concentration gradient, whereby molecules diffuse from the higher concentration surrounding small particles to areas around larger particles with lower drug concentration [17]. Consequently, a supersaturate solution is formed around large particles facilitating drug crystallization onto large particles as shown in Figure 9.2. The dependence of solubility, c, on the particle radius, r, is described by the modified Kelvin equation known as the Ostwald–Freundlich equation [18]:

$$c(r) \equiv c_{\infty} exp\left[{}^{2\gamma M}/_{\rho RTr}\right] \tag{9.10}$$

where c_{∞} is the bulk solubility of the dispersed phase, γ is the interfacial energy, M is the molecular weight of the solute with density ρ, R is the gas constant, and T is the absolute temperature. As can be seen, for nanometer-sized particles, the effect is substantial whereby the saturation solubility is dramatically increased. The occurrence of Ostwald ripening can be inhibited to produce stable nanosuspensions by narrowing the initial size distribution, i.e., having a smaller difference between the higher-solubility small particles and the lower-solubility large particles.

9.3 Creating Supersaturation and Stable Nanosuspensions

In the pharmaceutical industry supersaturation and crystallization of active compounds can be achieved by temperature quenching through cooling saturated drug solutions to stimulate supersaturation. However, approaches such as using cooling vessels cannot be employed in the engineering of drug nanoparticles due to the resultant non-uniform and low supersaturation.

In addition, supersaturation facilitated by quenching is limited by the thermal stability of active substances and especially so for biological compounds (e.g. proteins, peptides). Other methods that regulate the solute solubility and lead to supersaturation are the pH change and addition of a solvent that lowers the solubility of the solute. However, such approaches have had limited success.

A common approach to produce a stable supersaturated suspension is the incorporation of stabilizers, which are absorbed on the nanoparticles' surface and lessen the Ostwald ripening. To date, there is a wide range of stabilizers that can be employed to enhance suspension stability and are classified as ionic and non-ionic stabilizers. Wu *et al.* [17] provided an excellent literature review on selected stabilizers used to generate stable aqueous nanosuspensions manufactured by various techniques. Typical ionic stabilizers include sodium lauryl sulphate (SLS), sodium dodecyl sulphate (SDS) and lecithin, while non-ionic surfactants are pluronics®, Tweens, Cremophors (RH40, RH60, EL), polyethylene glycols (PEG300, 400), polysorbates (60, 80), d-α tocopherol polyethylene glycol 1000 succinate, polyvinyl alcohol (PVA), polyvinylpyrrolidone (PVP) and celluloses (HPC, HPMC).

The main role of the stabilizers is to provide short- and long-term stability and to contribute to the nanoparticle formation during the manufacturing process by preventing particle agglomeration. Stabilizers can be used to stabilize suspensions either by providing electrostatic or steric repulsion. The mechanism of stability from electrostatic forces is explained by the DLVO theory formulated by Derjaguin, Landau, Verwey and Overbeek [19, 20]. The main principle of the DLVO theory is that the stability of a suspension is determined by the sum (V_T) of attractive (V_A) and repulsive forces (V_R) between individual particles. The mutual attraction of particles is caused by London–van der Waals forces and the mutual repulsion is caused by the interaction of the electrical double layers (EDL) surrounding each particle (Figure 9.3). The attraction forces dominate at both large and small distances and increase markedly at small distances leading to a deep attraction (primary minimum) while the repulsive forces decrease exponentially with the increase in the distance between the particles. The total energy of interaction between two spherical particles of radius (α) with thin double layers at distance (H) is given by the Derjaguin approximation:

$$\Delta G_R = \frac{64\eta_o \pi \alpha k T \gamma^2}{\kappa^2} \exp(-\kappa H) - \frac{A_{121}\alpha}{12H} \tag{9.11}$$

Where η_O is the concentration of counterions, k is the Boltzmann constant, T is the absolute temperature, γ is the surface potential, κ^{-1} is the double-layer thickness and A_{121} is the Hamaker constant.

On the contrary, steric stabilization differs from electrostatic stabilization in that it is not a matter of the balance of the attraction and repulsion forces between the particles. The absorbed polymer layers should be thick enough to hold the particles far enough apart and thus make the dispersion-force attraction negligible. The source of steric repulsion among suspended particles is the increase of the free energy due to the overlap of the adsorbed layers. The coiled length of the stabilizer chain (e.g. polymer) provides a long-range repulsion whose thickness determines the suspension stability. In order to stabilize

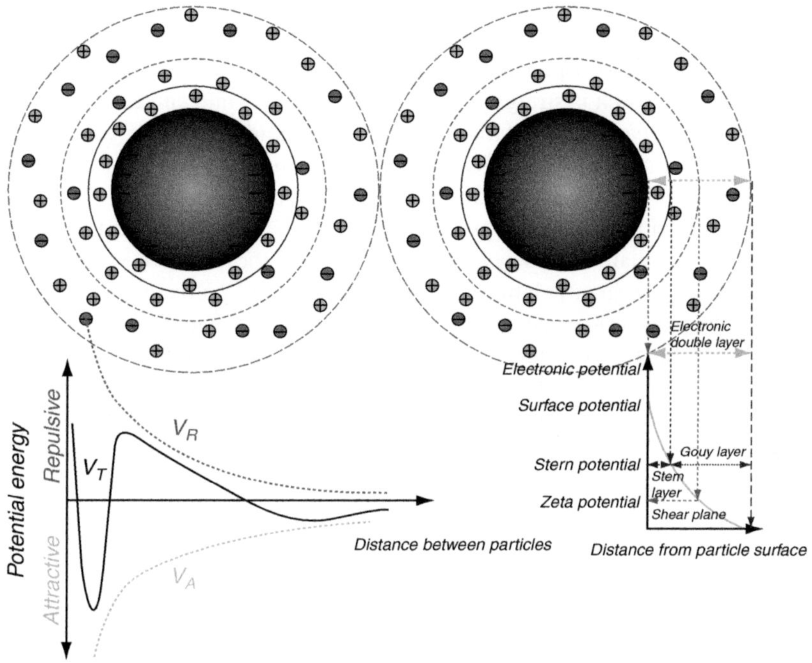

Figure 9.3 *Classical DLVO theory. Attractive forces are dominant at very small and large distances, leading to primary and secondary minimum, while repulsive forces are prevailing at intermediate distances and create net repulsion between the dispersed particles, thus preventing particle agglomeration. Reproduced with permission from [17].* (See colour plate section.)

a suspension, the efficient thickness of the absorbed layer (L) should be such that the total free energy of attraction at the closest approach (H = 2L), is less than the average kinetic energy of the per particle (kT) and thus no sticking takes place on particle collision. The free energy is given by the following equation:

$$\Delta G_{121}^{d} = -\frac{A_{121}\alpha}{12H} \quad (9.12)$$

In some cases the two types of stabilization can be combined, which is called electrosteric. It is known that steric repulsion has a shallow minimum at some distance from the particle and no minimum, while electrostatic stabilization has only a primary minimum at close distance. The combination of the two types results in a repulsion at all distances, apart from the secondary minimum and the suspension is more stable. In general, the most effective steric stabilizers are, block or graft copolymers as both contain anchoring groups and stabilizing chains. The reason is that anchoring groups have a strong affinity for the particle surface and usually are insoluble in aqueous media, while stabilizing chains have a strong affinity for the medium. Several examples of stabilizers will be discussed later in the chapter, including their nature and effect on particle stability and crystal growth.

9.4 Antisolvent Precipitation Via Mixer Processing

The manufacturing of nanosuspension through antisolvent precipitation is considered a 'bottom-up' approach which is straightforward, cost-effective and can easily be scaled up. In principle, the antisolvent precipitation engages the rapid mixing of two streams, the organic phase containing the water-insoluble active substance and the antisolvent phase which eventually leads to the precipitation of the drug nanoparticles [21]. The antisolvent is usually water with added stabilizers and is miscible to the organic solvent. As mentioned previously, stabilizers maintain suspension stability and they can be salts (e.g. citrate, tartrate), polyelectrolytes (e.g. gelatine), and surfactants (e.g. poloxamers). The additives are absorbed on the particle surface to provide electrostatic repulsion or steric stabilization. However, it is advisable to avoid surfactants with a high tendency to form micelles because drug solubility is increased and the Ostwald ripening is accelerated. The selection of the appropriate solvent to achieve the highest supersaturation is an important parameter one should consider. The organic solvent must present an excellent solubilizing capacity for the active substance, complete miscibility with the antisolvent aqueous phase and a low boiling point for the removal from the nanosuspension. In addition, the solvents should be toxicologically safe and thus it is preferable to use small chain alcohols, acetone or tetrahydrofuran.

The antisolvent approach was employed by List [22] in the production of nanosized hydrosols. Pharmaceutical hydrosols are defined as colloidal systems of active drug substances in water with a particle size between 1 and 1000 nm. The hydrosol suspensions were further studied by Gassmann [23], who achieved rapid mixing and supersaturation by injecting an ethanolic drug solution into a surfactant aqueous solution (e.g. poloxamer) under vigorous stirring. The rapid mixing leads to fast diffusion of the solvent into the antisolvent, thus to high supersaturation which results in fast nucleation rates and subsequently produces submicron particles. According to the authors, hydrosols can be produced with drug yields up to 5 mg/ml while the hydrosol principle applies to drugs with an aqueous solubility of 10^{-3}–10^{-4} ml/L. Indeed the particle size of the formed hydrosols is inversely proportional to either the solubility or supersaturation in the antisolvent water as depicted in Figure 9.4 [24].

Hydrosols can be prepared at batch scale by using mixing chambers [26, 27], static mixers [28] and Y- or T-mixers [26, 29] where the precipitation and particle size reduction take place in a turbulent flow pipe. The advantages of such mixing devices involve controllable and reproducible particle sizes, narrow size distribution, temperature control for thermal labile drugs, continuous and easy-to-scale-up processing. Several configurations have been reported in the literature and are discussed below.

A typical example is the method developed by Horn and co-workers to produce nanodispersed carotenoid hydrosols using a mixing chamber process [28]. Two mixing chambers are combined where in the first one the drug microdispersion undergoes thermal treatment (ca. 200°C) for a short residence time ($\ll$0.5 s) and turbulent mixing with a hydrophilic solvent to produce a molecularly dispersed solution. In the second chamber the main precipitation process takes place under increased pressure (50 bars) and amorphous spherical hydrosols are obtained. The hydrosol particles were uniform with a particle size range between 50–100 nm. By using the mixing chambers the authors managed to prepare diverse carotenoid preparations with different colour intensities and increased

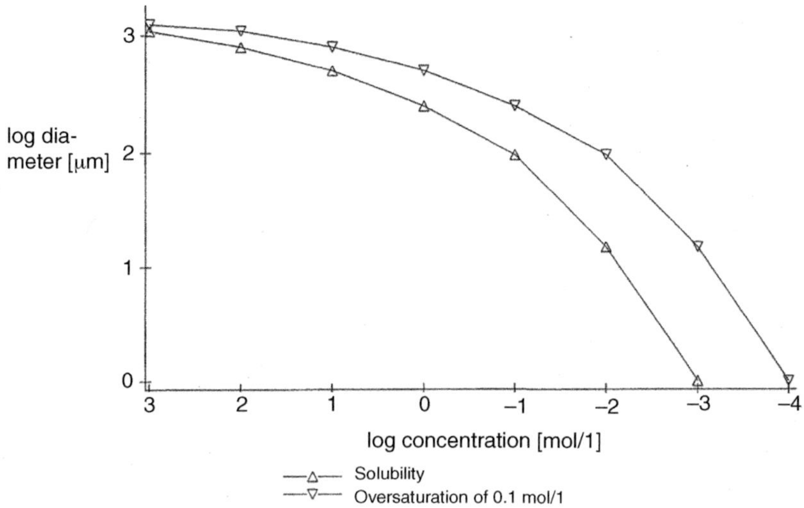

Figure 9.4 *Dependence of particle size after precipitation on the solubility and supersaturation of the substance. Reproduced with permission from [24].*

bioavailability. Cyclosporine (CyA) spray-dried hydrosols were also prepared by Gassmann *et al.* [21] with a drug content of 3 mg/ml and a particle size of 80 nm (120 after reconstitution). Animal trials conducted for 48 hrs and hydrosol formulations were labelled with tritium for comparison with the marketed product Sandimmune® I.V (infusion). The results showed no substantial difference between the two formulations and drug distribution was almost identical after 5 min. administration. The effectiveness of the CyA formulations was attributed to their amorphous character and increased surface which facilitate increased dissolution and rapid distribution into the various organs. The CyA hydrosols are believed to act as a micellar solution in the body.

More recently, Douroumis and Fahr [30] employed a static mixer (SMX DN3, 3.2 mm × 3.2 mm) apparatus with 30 loose elements welded into a stainless steel tube to produce hydrosols of various water-insoluble active substances. The tube was designed in such a way that the liquid confluence should take place in the middle of the first element while an offset of the mixer elements at 90° endured homogeneous mixing over the entire pipe cross-section. The drugs used in this study included progesterone (PRG), b-methasone valerate-17 (BMZ), carbamazepine (CBZ) and oxcarbazepine (OXC) formulated with surfactants (stabilizers) and lipids. The authors examined the effect of various processing parameters such as the element number, the temperature of the organic phase, the $V_{org/aq}$ ratio, and the $R_{drug/excipient}$ ratio on the obtained particle size and the final drug loading. Similar to List and Zucker, the authors observed that the drug aqueous solubility is crucial for the formation and stability of hydrosols. The influence of the mixer elements was evaluated by conducting experiments with 10, 20 and 30 mixers showing a significant effect with decreased particle sizes when the number of mixers was increased (Figure 9.5). Higher temperatures of the organic phase resulted in lower particle sizes, while more hydrophilic actives required increased solvent amounts. By using 30 mixers the required

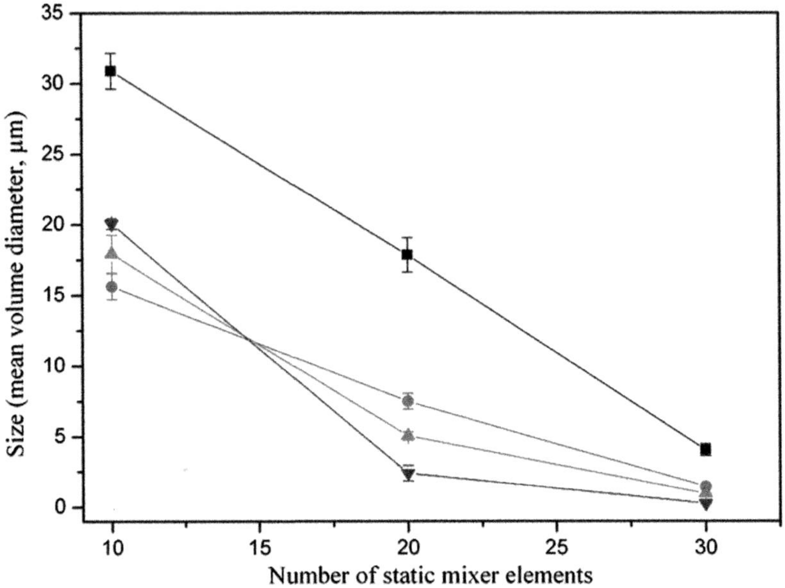

Figure 9.5 *Influence of mixer elements on the obtained particle size (■ CBZ, •PRG, ▲OXC and ▼BMZ). Reproduced with permission from [30].*

excipient amounts to produce stable hydrosols were reduced and achieved drug loading was maximized. However, nanosuspension could not be produced for every active substance and, for instance, CBZ (logP 1.24) hydrosols showed the lowest particle size at 4.09 μm. In contrast, BMZ (logP 3.98) hydrosols were stabilized with Lipoid S75 with a particle size of 250 nm.

In a more thorough study the same authors [31] used the modified Shepard's method (MSM), an advance data analysis technique to optimize the CyA hydrosol nanosuspensions. A three-factor, central composite orthogonal 2^3 + star, experimental design (DOE) was applied to investigate the process-independent variables, including the gelatin amounts, the lipoid S75 amounts and the ratio of the aqueous/organic phase. The hydrosol particle size was selected as the dependent process variable. The DOE interpretation showed a significant influence ($p < 0.05$) for both gelatin and lipoid S75 independent variables and the optimum excipient amount ranges were defined. The suitable organic/aqueous ratio was identified at 10–16 without, however, having a significant influence on the particle size. Further hydrosol optimization was achieved by employing the MSM [32] trivariate quadratic function which successfully interpolates bivariate datasets. The MSM approach was able to predict hydrosol compositions with 290 nm particle size while the obtained experimental size was 300 nm. This powerful technique can also predict the particle size not only at a single location in the diagram but also for an array of coordinate positions. As a result, two-dimensional graphs are obtained through trivariate interpolation that predicts more than one minima or maxima of the investigated system. CyA was found to be in the amorphous state in the hydrosol suspensions and consequently increased dissolution rates were observed compared to the pure crystalline CyA. The maximum drug content was 2.31 mg/mL with a size of 300 nm compared to the 3 mg/mL and 80 nm size of those formulated by List and Sucker. These differences are attributed to the effect of the selected stabilizers on the hydrosol particle size and stabilization.

By using another type of static mixer (SMV DB25) that facilitates turbulent flow, Dong *et al.* [33] developed a continuous antisolvent precipitation process for the production of spirinolactone (Class II) hydrosols. The authors investigated the effects of total flow rates (the sum of solvent and antisolvent rate), drug concentration and number of mixing elements on the size of the precipitated drug particles. The total flow rates of the two mixing streams were 1.0, 2.0 and 3.0 L/min but a plateau was reached at 1.0 L/min without any further reduction of the particle size beyond this rate. The size of the precipitated particles was also affected by the drug concentration, and by increasing the drug concentration from 5 to 10 and 20 mg/ml, the particle sizes increased respectively from 482 ± 56, 733 ± 98 and 1233 ± 42 nm respectively. Although high drug concentrations create high supersaturation that leads to faster nucleation rates and smaller particles, it is possible to induce agglomeration leading eventually to bigger particles. By using the aforementioned configuration, the authors were able to successfully produce submicron hydrosols of 500nm with drug concentrations as high as 10 mg/ml. Similar results were extracted with those of Douroumis and Fahr [30] regarding the number of mixing elements, and the size of the hydrosols was significantly affected by the structure and number of the elements. The authors claimed that the main reason for the production of nanohydrosols was the increased number of mixing elements and the type of the SMV elements that are designed for turbulent mixing. However, the discrimination between mixer designs for laminar and turbulent flow requires further investigation. The reason is that Dong *et al.* used six SMV elements – using 12 and 18 could not further reduce the particle size – with a total length of 150 mm compared to the SMX elements (96 mm total length) used by Douroumis and Fahr. Although the spironolactone hydrosols were found to be amorphous, the drug dissolution rate did not exceed 37% after 60 min. Nevertheless the dissolution rates of the lyophilized hydrosols were higher than those of the bulk drug (≈15%).

In a comparative study, the same group [34] used static mixers and batch antisolvent processing to develop fenofibrate hydrosols for oral administration. For the purposes of the study, hydroxylpropyl methyl cellulose (HPMC) and sodium dodecyl sulfate (SDS) were used as stabilizers and lactose (or mannitol) as redispersant after the spray-drying of the nanosuspensions. According to the experimental findings, both approaches produced similar particle sizes, 318 ± 19 nm for the batch processing and 328 ± 22 nm for the static mixing. The reconstituted particles showed a substantial increase in the particle size which was in the submicron range (553 nm). Interestingly, fenofibrate in the form of hydrosols remained partially crystalline when the dried batches were analysed by X-ray diffraction. However, the spray-dried nanohydrosols exhibited faster dissolution rate (84.2% dissolved within 5 min) than the physical mixture with (31.7%) and without spray-drying (9.7%). A major drawback of both approaches was the rapid crystal growth of the fenofibrate nanoparticles which were stable for only 2 min. and subsequently drug crystals of 2.5 μm were developed within 6 min.

In a recent study, a new mixer designed for impingement mixing of the solvent were then fed and antisolvent solution was introduced by Dong *et al.* [35]. The produced nanoparticles are then fed to a spray-drier to produce nanoparticulate powders. The main advantage of this mixer type is the reduced mixing time of the miscible fluid streams and the controllable processing parameters. For the purposes of the study the organic phase consisted of spironolactone dissolved in acetone and the aqueous phase of HPMC and SDS was used as the antisolvent. A careful observation of the particle morphology revealed that the impingement mixing facilitated a 'two-step crystallization model', where nanoparticles of 300 nm with a spherical shape became cubic within 30 min. Although the produced nanoparticles were in a crystalline state after freeze-drying, the dissolution profiles showed high rates (100%) within the first 10 min.

The antisolvent precipitation approach has also been used to produce microparticulate hydrosols [36–38] with increased dissolution rates. By using static mixers Douroumis and Fahr [36] prepared Oxcarbazepine (OXC) microcrystals through controlled precipitation of an OXC methanolic solution and an aqueous methycellulose (MC) solution. The authors employed a two-factor, central composite orthogonal 2^2 + star, experimental design to explore the effect of the MC amount and the aqueous/organic ratio (dependent variables). The analysis of the experimental findings revealed a significant effect of both dependent variables. The use of the static mixers formation facilitated the creation of a hydrophobic drug surface with MC absorbed on the surface through a protective layer. The crystal growth was inhibited and the particle size remained stable by steric stabilization as the MC chain length prevents the suspended particles from coming close and sticking. Dissolution studies of micronized OXC hydrosol batches of different particle sizes (2.1, 4.5 and 10.5 μm) presented increased dissolution rates with 55–75% of the active dissolved in the first 10 min. and 80–95% in 120 min. Equal amounts of bulk OXC showed a 45% dissolution rate within 2 hrs.

A controlled antisolvent precipitation approach used by Li *et al.* [39] produced hexagonal and tetragonal prednisolone microcrystals stabilized by hydroxypropyl-methylcellulose (HPMC). In this study, operation parameters, such as the solvent and anti-solvent types, the temperature and the stirring speed were explored. A range of organic solvents such as N-Methyl-2-pyrrolidinone (NMP), ethanol, methanol and acetone were used as solvents but only NMP was adopted as the suitable one. Uniform particles with controlled size were obtained when 0.2% HPMC aqueous solution was used as the antisolvent. Interestingly, the particle morphology could be controlled by adjusting the temperature of the antisolvent. Thus, at 14°C it was possible to produce microcrystals with a size of 1.6 μm and a hexagonal structure while at 4°C particles showed tetragonal morphology. The increase of the stirring rates from 2500 to 10000 rpm facilitated further particle size reduction for both hexagonal and tetragonal microparticles. Eventually, the manufactured microcrystals displayed higher dissolution rates than the commercial prednisolone. The authors attributed the increased dissolution rates to particle size reduction, uniformity and enhanced specific surface area (5.41 and 4.89 m^2/g for the hexagonal and tetragonal microcrystals compared to 3.73 m^2/g for the commercial micronized drug).

The same approach introduced by Gassmann *et al.* [23] was employed by Cunha-Filho *et al.* [37] to develop antitumour b-Lapachone [bLAP] microcrystals with HPMC as stabilizer. The precipitation occurred with HPMC at 0.05% as the stabilizing agent and a solvent ethanol/water ratio 1:4. The drug load varied from 85–95% depending on the selection of the drying process. The micronized powders demonstrated a significant increase in dissolution rates (100%) within the first 5 min. for the 50–100 μm fractions.

9.5 Antisolvent Precipitation by Using Ultrasonication

Antisolvent precipitation has recently been implemented to generate nanosuspensions through a T-mixer combined with an ultrasonic probe [40]. In this configuration both streams of the solvent are mixed at different flow rates and ratios in the T-mixer (90° angle) to achieve variation in mixing intensities. Once the streams are mixed, ultrasound is applied through a probe. A wide range of actives with different aqueous solubility such as suchasitraconazole (ITZ), ascorbylpalmitate (ASC), fenofibrate (FNB), griseofulvin (GF), and sulfamethoxazole (SFMZ) were processed to produce ultra-fine particles by using different organic solutions in acetone, dimethylsulfoxide (DMSO), tetrahydrofuran (THF) and ethanol (EtOH). The authors observed that although supersaturation increases with decreasing aqueous solubility, the observed nucleation rates do not follow the same trend, because drug molecules diffuse differently from the bulk phase to the interface of the ternary mixture (drug–solvent–water mixture) for each solvent. The different diffusion rates also affect the efficiency of the stabilizers (e.g. PEG) and might increase or reduce their ability to act as growth inhibitors. Thus solvent polarity affects the nucleation process and also precipitation.

In another study, Xia *et al.* [41] combined a precipitation ultrasonication method to develop nitrendipine nanosuspensions. Different process parameters were systematically investigated including drug concentration (20–100 mg/ml), stabilizer concentration (0.1–1.5% w/v), ultrasonic power input (200–580W) and time length (5–20 min.). The statistical analysis of the drug and polyvinyl alcohol (PVA) concentrations revealed the optimum range at 20–30mg/ml and 0.15–0.4% respectively. At low drug concentrations the drug nanocrystals showed narrow size distribution but a rapid size increase was observed within 15 min. At high drug concentrations a high diffusion-controlled growth and agglomeration rate was observed due to the increased supersaturation leading to the creation of microparticles (8 μm). Similarly low stabilizer amounts provided insufficient surface coverage, resulting in rapid crystal growth while high concentrations increased the solution viscosity and prevented the transmission of ultrasonic vibration and the diffusion during precipitation. The optimum PVA amounts led to sufficient crystal coverage and provided steric repulsion. The particle stabilization was attributed to the hydrodynamic boundary layer surrounding the nanocrystals that was absorbed by hydrogen bonding between the drug and the polymer.

The variation of the ultrasonic power showed reduced particle sizes with the increase in the power input. No further particle decrease was observed when ultrasonication exceeded 400W. Increased ultrasonication times generated smaller particle sizes when the dispersions were sonicated for up to 15 min. The temperature of the PVA solution showed the precipitation of smaller drug crystals when the temperature was changed from 35 to 3°C. At higher temperatures, supersaturation was reduced upon mixing due to the increased nitrendipine solubility, which leads to lower nucleation rates [42]. In addition, the temperature affected the crystal morphology where needle-shaped crystals of 60 μm appeared at 35°C and spherical ones of 9 μm at 3°C. The produced nanoparticles showed enhanced dissolution rates with 90% drug dissolved within 2 min., in contrast to the raw drug particles, of which only 23.4% dissolved after 2 hrs. The superiority of the nitrendipine nanosuspensions over the reference preparation was confirmed by bioavailability studies conducted on Wister rats. The nanosuspension drug plasma concentrations showed C_{max} and $AUC_{0\rightarrow 12}$

values of 6.1 and 5.0-fold greater than those of the control preparation. The enhanced bioavailability was attributed to the rapid drug absorption from the gastrointestinal track as a result of the increased dissolution rate and high surface area [43]. Another reason was the increased adhesion surface area between the drug nanoparticles and the gut-absorbing membranes [44].

9.6 Nanoprecipitation Using Microfluidic Reactors

The possibility of using an interesting bottom-up technology for the preparation of stable aqueous nanosuspensions through microfluidic reactors has been explored by several research groups. It was introduced by van der Woerd *et al.* [45] and involves the flow of fluids in micro-channel devices, which were also used by other researchers for enzyme assays [46,47]. In such systems the fluid behavior differs from the conventional flow theory mainly because of the small volumes inserted in the channels (100–500 nL) and their laminar flow as the miscible fluids move in parallel without turbulence in the microchannels [48]. The dimensionless Reynolds number is usually far below 2000 where the flow changes from laminar to turbulent and thus the mixing of the fluids' components is mainly because of diffusion. Microfluidic reactors achieve rapid mixing (e.g. 0.7 s) and therefore rapid diffusion rates including increased surface-to-volume ratios that facilitate drug–excipient interactions.

A microfluidic reactor was used by Ali *et al.* [49] for the creation of hydrocortisone (HC) nanosuspensions through controlled precipitation. The group explored experimental parameters including solvent–antisolvent flow rates, internal diameters of the microreactors, drug concentrations and inlet angles. For the purposes of the study, a 'Y–shape' microreactor, as shown in Figure 9.6, was used as it provides lower micromixing times (τ_m) compared to the crystallization times (τ_f), which is critical for the formation of uniform nanoparticles [50]. From the investigated processing parameters, only the flow rate was proved to have a profound effect. High solvent/antisolvent flow rates formed smaller particle sizes due to the high supersaturation achieved by the enhanced mixing and short residence time. By using sharper inlet angles (10 and 25°), it was also possible to slightly reduce the obtained particle sizes since this configuration allows minimal disruption of the confluent flows. The use of a mixture of stabilizers (0.2% HPMC, 0.2% PVP, 0.05% SDS) facilitated the precipitation of stable nanoparticles with sizes of 260 ± 32 nm, which was increased to 500 ± 64 nm upon further dilution. The developed nanosuspensions were evaluated in terms of stability under storage for three months and the particle size remained relatively stable without any particle aggregates. Only a slight increase of the size to 687 nm with a polydispersity index of 0.235 was observed.

The same researchers further developed the microfluidic reactor with a slight modification to improve HC nanosuspensions for ophthalmic delivery [51]. On this occasion, the reactor was combined with a probe sonicator immersed in the aqueous stabilizing solution (0.2%, w/v, PVP + 0.5%, w/v, HPMC, 0.1%, w/v Tween 80). The produced nanoparticles (295 ± 32 nm) were administrated to male normotensive albino rabbits and the intra-ocular pressure (IOP) was determined. The injected HC formulations caused a remarkable increase of IOP in normotensive rabbits reaching peak levels (max. 57.60 $\pm$ 13.1%), which then slowly returned to the pretreatment values. The HC nanosuspension showed a sustained

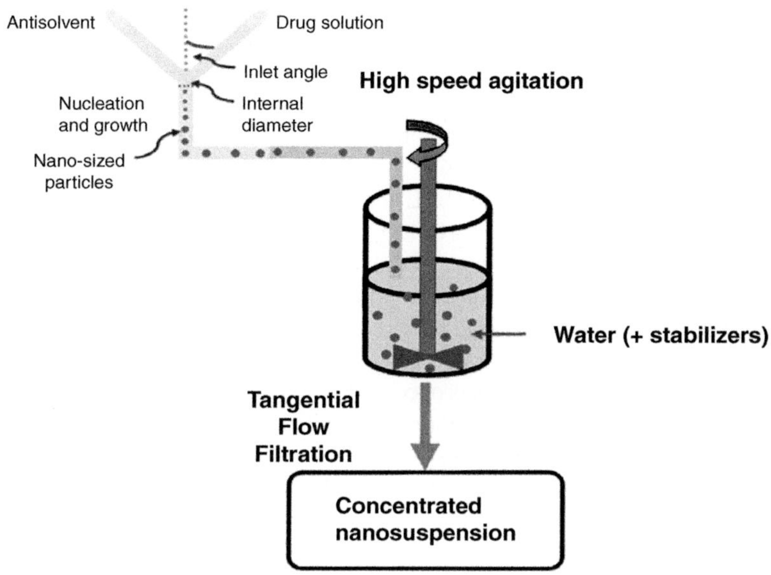

Figure 9.6 *Preparation of hydrocortisone nanosuspension by precipitation process. Reproduced with permission from [49].* (See colour plate section.)

action for 8–9 h and they actually proved more efficient to formulations prepared by ball milling (4–5 h). In addition, the nanosuspensions showed higher AUC values compared to the milled nanoparticles. The use of a microfluidic reactor combined with a probe sonicator provided stable nanosuspensions with improved bioavailability and prolonged drug release to the corneal membrane. The microfluidic precipitation appears to be a promising method for the development of nanosuspension but there are more options still to be explored.

9.7 Particle Engineering by Spray: Freezing into Liquid

A novel cryogenic spray-freezing into liquid (SFL) process was developed by Rogers *et al.* [52, 53] to enhance the dissolution rates of poorly soluble drugs by producing nanostructured drug particles. The SFL technology involves the atomization of a feed liquid phase which contains the water-insoluble active into a cryogenic liquid to produce frozen nanoparticles (or microparticles) that are subsequently frozen. By spraying the solution into the cryogenic liquid, crystal growth is inhibited while the liquid–liquid impingement results in intense micronization. The cryogen is usually nitrogen because it is inexpensive, environmentally-friendly, inert, and can be used at atmospheric pressure. The SLF powders are then separated from the liquid nitrogen by using fine sieves following lyophilization.

The SLF process is quite advantageous compared to conventional technologies as it creates high-surface area droplets and ultra-rapid freezing rates are achieved. The ultra-rapid freezing rates minimize the time for phase separation of the solutes and the drug molecules are dispersed homogenously through the solidified stabilizers. The solvents used for the active substances are not required to be miscible with the cryogen. The developed drug nanoparticles also present a high surface area and porosity resulting in increased drug dissolution rates.

Hu *et al.* [54] used the SFL technology to engineer danazol and cabramazepine nanoparticles. Both drugs were dissolved in THF and mixed with an aqueous solution of Poloxamer 407, PVP K-15, and/or SLS prior to spraying through a nozzle into the liquid N_2. The produced nanoparticles were partially crystalline and presented high porosity with increased dissolution rates. Both drugs processed by SFL demonstrated dissolution rates of 98–100% within 10 min compared to the slow rates (10–15%) of the bulk micronized drugs. However, the aggregates obtained from the SFL process did not form nanoparticles.

The process was further improved for the engineering of danazol with nanostructured aggregates and porous morphology [55]. The SFL was able to produce nanoparticles by adjusting the amounts of danazole, PVP K-15 (stabilizer) and amounts of acetonitrile/methylene chloride (CAN/MCL). It was proved that a mixture of CAN/MCL is preferable for the SFL processing due to the increased drug solubility. The obtained danazol particles were amorphous even at high drug loadings because the rapid freezing rates accomplished by SFL trapped the danazol in an amorphous state and prevented crystal growth. However, with drug loadings above 75%, the engineered particles were in the range of 3 μm and less porous. In addition, the danazol nanoparticles presented high surface area and contact angle in a decreasing order of 50 > 66 > 75 > 91% drug loadings. As a result, the dissolution rates of all SFL powders were significantly enhanced compared to the bulk micronized drug substance. Only 30% of the micronized danazole was dissolved in 2 min. while 95% was dissolved for the SFL danazol/PVP K-15 powders (33–75% loading).

The same group optimized a similar ultra-rapid freezing process (URF) [56] where the feed solution is applied to a cryogenic solid substrate cooled to a temperature range. The URF process offers fast heat transfer rates as a result of the intimate contact between the solution and the cold solid surface, but without the complexity of cryogen evaporation (the Leidenfrost effect). The main advantage of URF over the SFL technology is that the droplets are applied directly onto the cryogenic substrate and premature freezing is not a concern, allowing the use of high freezing point solvents. In this study, danazol/PVP K-15 formulations were again processed by URF to produce nanoparticles with 100nm diameters and increased dissolution rates.

9.8 Precipitation by Rapid Expansion from Supercritical to Aqueous Solution

Supercritical fluids have been extensively used for nanoparticle engineering not only in the pharmaceutical industry but also in the nutraceutical, cosmetic and specialty chemistry industries. It is beyond the scope of this chapter to discuss the various concepts and configurations of the supercritical fluid technologies [57], thus we will concentrate on those approaches related to the antisolvent precipitation. The precipitation by rapid expansion from supercritical to aqueous solution (called RESAS or RESOLV) is a novel approach to produce drug nanoparticles [58–64]. In most cases, CO_2 is chosen as the supercritical fluid because it is environmentally-friendly, nonflammable, inexpensive, and in essence nontoxic. Another advantage of CO_2 is the fact that it has relatively mild critical conditions with a critical temperature (Tc) = 31°C and critical pressure (Pc) = 73.8 bar, which allows processing at moderate temperatures of thermal labile substances. In RESAS, the active compound is placed in one or more tubes (extraction columns) and it comes in contact with a known volume of CO_2 to form a solution. In all cases, the gaseous CO_2 is condensed, subcooled, and pressurized to the desired pressure. Usually the devices are combined with a preheater to facilitate uniform heating and prevent hot spots that eventually could lead to drug crystallization within the coil. Prior to the RESAS processing, the preheater and the nozzle are pressurized with CO_2 to prevent plugging of the nozzle. The equilibrated CO_2 flow is then diverted into the tubes to push the solution through the nozzle which is sprayed in a preheated stabilizing aqueous solution (e.g. surfactant).

The RESAS technology was used by Young *et al.* [58, 59] to produce stable phospholipid nanoparticles of CyA with high drug payloads. By manipulating the processing parameters, such as temperature and suspension concentration, they achieved sizes with a mean diameter of 500 nm. Phospholipid compositions of Lipoid E80 (from chicken egg white) and Phospholipon 100H (hydro-genated soybean), which both consist primarily of dipalmi-toylphosphatidylcholine (DPPC) and surfactants such as Tween 80, Pluronic F127, and Myrj 52 were used for the purposes of the study. The obtained particle sizes of the phospholipid-stabilized suspensions were noticeably lower than those produced with micellar-forming surfactants for the same surfactant concentrations and drug/surfactant ratios. The source of the lipid (egg vs soybean) affected the performance of Lipoid E80 and Phospholipon 100H with the former producing the smallest particles. The effect of drug concentrations processed at 25°C, 45°C, and 80°C was studied thoroughly and the particle size increased significantly with drug concentration. This behavior was attributed to the surfactant depletion and particle collision rates which increase with particle concentration.

The ability of the phospholipid-surfactant compositions to stabilize the particle size was affected by the temperature of the aqueous medium. It was found that at temperatures above 30°C no substantial change of the particle size occurred but when the stabilizing solution was at 14°C, the particle size increased significantly. The main reason for this behavior is the rearrangement of the phospholipid vesicles from the rigid gel-like state to the fluid liquid-crystalline state. Lipoid E80 and Phospholipon 100H have melting transitions below 30°C and when the temperature increases, the fatty chains became more flexible, allowing membrane diffusion. The temperature of the preheater led to smaller particle sizes when the temperature increased from 30°C to 80°C by creating higher supersaturation which eventually led to faster nucleation rates. Finally, the particle size of the produced nanosuspensions decreased with smaller nozzle sizes but increased with low nozzle flow rates.

The same process was employed [60] to form stable submicron phytosterol nanoparticles and the drug/CO_2 solution was expanded into aqueous solutions of four different surfactants. Initial experimentation showed negligible solubility of phytosterol in Lutrol F68 but for sodium lauryl sulfate (SLS), Tween 80 and Solutol HS15, solubility increased with the surfactant concentration. As expected, the surfactant type altered the nanosuspension particle size with Tween 80 and Solutol HG15 stabilized particles at high drug concentrations (10.0 g/L). SLS and Lutrol F68 were able to stabilize nanoparticles at lower drug concentrations of 5.6 g/L and 2.4 g/L respectively. The use of nozzles with smaller diameters led to nanosuspensions with higher drug concentrations. The particle size was further reduced by increasing the surfactant concentrations with the exception of Lutrol F68 where no influence was observed. Interestingly, the increased surfactant amounts formed bimodal particle size distribution. Bimodal distributions with smaller particle sizes were

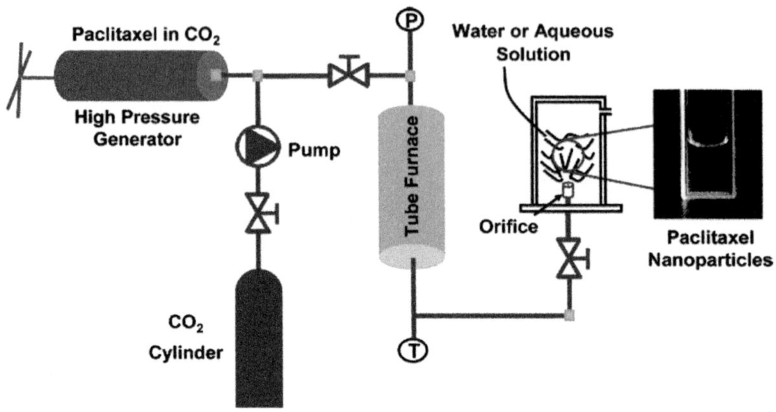

Figure 9.7 *Experimental RESOLV set-up. Shown in the figure is an aqueous suspension of stabilized paclitaxel nanoparticles. Reproduced from [64] with permission of the American Chemical Society.* (See colour plate section.)

also observed at high flow rates due to the high turbulence. The long-term stability studies (4–12 months) showed good nanosuspension stability with a small particle size increase for all surfactants.

The RESOLV technology was applied to produce paclitaxel nanosuspensions [64] as shown in Figure 9.7. The produced nanoparticles were stabilized with different PVP (Mw~40000, or 360000) grades which had an effect on the particle size. The PVP_{40K} showed a particle size of 38 ± 9 nm and 530 ± 85 nm while the PVP_{40K} was of 200 ± 30 nm. The antiproliferative effect of paclitaxel nanosuspensions was evaluated on human malignant breast cell line MDA-MB-231 [65]. The IC50 values after 48h incubation were 100 nm of equivalent paclitaxel concentrations in the nanoparticle samples in comparison to those reported in the literature for paclitaxel dissolved in DMSO. The cytotoxicity of small or large nanoparticles did not show any substantial differences but in comparison with ethanolic paclitaxel solutions, the nanosuspensions appeared to be more effective.

The rapid expansion from supercritical to aqueous solution technology has been also used successfully to develop nanosuspensions [62, 63] or modified [66] to form drug nanoparticles (e.g. indomethacin) with enhanced dissolution rates, demonstrating the potential as a bottom-up approach.

9.9 Conclusion

The antisolvent precipitation technologies can be successfully applied for the formation of nanosuspensions of various active compounds. In this chapter we discussed a wide range of these bottom – up technologies analyzing their advantages and disadvantages. The successful formulation of nano-suspensions requires the presence of extremely water insoluble drugs but it also depends on various processing parameters such as the selection of the antisolvent, the type of the stabilizer(s) and the design of the mixing devices. In conclusion, the development of novel antisolvent platforms has recently attracted industrial interest for the formulation of water insoluble drugs.

References

1. G.L. Amidon, H. Lennernas, V.P. Shah, J.R. Crison, A theoretical basis for a biopharmaceutic drug classification: the correlation of in vitro drug product dissolution and in vivo bioavailability. *Pharm Res*, **12**, 413–420 (1995).
2. R. Löbenberg, G.L. Amidon, Modern bioavailability, bioequivalence and biopharmaceutics classification system: New scientific approaches to international regulatory standards. *Eur J Pharm Biopharm*, **50**, 3–12 (2000).
3. M. Martinez, L. Augsburger, T. Johnston, W.W. Jones, Applying the biopharmaceutics classification system to veterinary pharmaceutical products. Part I. Biopharmaceutics and formulation considerations. *Adv Drug Deliv Rev*, **54**, 805–824 (2002).
4. S.H. Yalkowsky, *Solubility and Solubilization in Aqueous Media*. New York: Oxford University Press (1999).
5. M. Mosharraf, C. Nystrom, Solubility characterization of practically insoluble drugs using the Coulter counter principle. *Int J Pharm*, **122**, 35–47 (1995).
6. V.B. Patravale, A.A. Date, R.M. Kulkarni, Nanosuspensions: a promising drug delivery strategy. *J Pharm Pharmacol*. **56**, 827–840 (2004).
7. C.M. Keck, R.H. Müller, Drug nanocrystals of poorly soluble drugs produced by high pressure homogenization. *Eur Pharm Biopharm*, **62**, 3–16 (2006).
8. H.S. Ali, P. York, N. Blagden. Preparation of hydrocortisone nanosuspension through a bottom-up nanoprecipitation technique using microfluidic reactors. *Int J Pharm*, **375**(1–2), 107–113 (2009).
9. D. Horn, J. Rieger. Organic nanoparticles in the aqueous phase – theory, experiment and use. *Angew Chem Int Ed*, **40**, 4330–4361 (2001).
10. D. Kashchiev, *Nucleation: Basic Theory with Application*. Oxford: Butterworth-Heinemann (2000).
11. J.K. Guillory Generation of polymorphs, hydrates, solvates and amorphous solids. In *Polymorphism in Pharmaceutical Solids*. Brittain, H.G. (ed.). New York: Marcel Dekker Inc (1999).
12. V.K. Nanjwad, F.V. Manvi, M. Shamrez Ali, K. Basavaraj, B.K. Nanjwade, M.M. Maste. New Trends in the Co-crystallization of Active Pharmaceutical Ingredients *JAPS* (08; 2011: 01-05)
13. J.W. Mullin, *Mechanism of Crystallisation*. London: Butterworth & Co. Ltd (1972).
14. L.F. Knapp, The solubility of small particles and the stability of colloids, *Trans Faraday Soc*, **17**, 457–465 (1922).
15. L.M. Skinner, J.R. Sambles, The Kelvin equation – a review, *Aerosol Sci*, **3**, 199–210 (1972).
16. N. Rodríguez-Hornedo, D. Murphy, Significance of controlling crystallization mechanisms and kinetics in pharmaceutical systems. *J Pharm Sci*, **88**(7), 651–660 (1999).
17. L. Wu, J. Zhang, W. Watanabe, Physical and chemical stability of drug nanoparticles. *Adv Drug Deliv Rev*, **63**(6), 456–469 (2011).
18. T. Carstensen, *Solubility, Advanced Pharmaceutical Solids*. New York: Marcel Dekker, pp. 27–50 (2001).
19. B.V. Derjaguin, L. Landau, Theory of the stability of strongly charged lyophobic sols and of the adhesion of strongly charged particles in solutions of electrolytes. *Acta Phys Chim*, **14**, 633–662 (1941).

20. E.J.W. Verwey, Theory of the stability of lyophobic colloids, *J Phys Chem*, **51**(3), 631–636 (1947).
21. P. Gassmann, M. List, A. Schweitzer, H. Sucker, Hydrosols – alternatives for the parenteral application of poorly water soluble drugs, *Eur J Pharm Biopharm*, **40**, 64–72 (1994).
22. M. List, Hydrosole, eine intravenöse Arzneiform zur Herstellung von Injektionen und Infusionen in Wasser schwer löslicher Wirkstoffe, thesis, Basel (1987).
23. P. Gaßmann, Herstellen und Stabilisieren von Hydrosolen zur intravenösen Applikation, thesis, Basel (1990).
24. A. Mersmann, Gezielte Kristallisation, in D. Essig, H. Stumpf (eds), *Flüssige Arzneiformen schwerlöslicher Arzneistoffe*. Stuttgart: Wissenschaftliche Verlagsgesellschaft m.b.H (1990).
25. D. Horn, E. Lüddecke, Preparation and characterization of nano-sized carotenoid hydrosols. In *Fine Particles Science and Technology: From Micro to Nanoparticles*, Pelizzetti, E. (ed.). Dordrecht: Kluwer, pp. 761–775 (1996).
26. A.J. Mahajan, D.J. Kirwan, Rapid precipitation of biochemicals. *J Phys D: Appl. Phys*, **26**, B176–B180 (1993).
27. A.J. Mahajan, D.J. Kirwan, Nucleation and growth-kinetics of biochemicals measured at high supersaturations. *J Crystal Growth*, **144**, 281–290 (1994).
28. H. Sucker, Hydrosole, eine Alternative für die parenterale Anwendung von schwer löslichen Wirkstoffen. In *Pharmazeutische Technologie: Moderne Arzneiformen*. R.H. Müller, G.E. Hildebrand (eds). Stuttgart: WVG, pp. 383–391 (1998).
29. D.J. Kirwan, C.J. Orella, Crystallization in the pharmaceutical and bioprocessing industries. In *Handbook of Industrial Crystallization*. A.S. Myerson (ed.). Boston: Butterworth-Heinemann, pp. 219–235 (1993).
30. D. Douroumis, A. Fahr, Nano- and micro-particulate formulations of poorly water-soluble drugs by using a novel optimized technique. *Eur J Pharm Biopharm*, **63**, 173–175 (2006).
31. D. Douroumis, S. Scheler, A. Fahr, Using a modified shepards method for optimization of a nanoparticulate cyclosporin A formulation prepared by a static mixer technique. *J Pharm Sci*, **97**, 919–930 (2008).
32. R.J. Renka, Multivariate interpolation of large sets of scattered data. *ACM Trans Math Softw*, **14**, 139–148 (1988).
33. Y. Dong, W.K. Ng, J. Hu, S. Shen, R.B. Tan, A continuous and highly effective static mixing process for antisolvent precipitation of nanoparticles of poorly water-soluble drugs. *Int J Pharm*, **386**, 256–261 (2010).
34. J. Hu, W.K. Ng, Y. Dong, S. Shen, R.B. Tan, Continuous and scalable process for water-redispersible nanoformulation of poorly aqueous soluble APIs by antisolvent precipitation and spray-drying. *Int J Pharm*, **404**(1–2), 198–204 (2011).
35. Y. Dong, W.K. Ng, S. Shen, S. Kim, R.B. Tan, Controlled antisolvent precipitation of spironolactone nanoparticles by impingement mixing. *Int J Pharm*, **410**(1–2), 175–179 (2011).
36. D. Douroumis, A. Fahr, Enhanced dissolution of oxcarbazepine microcrystals using a static mixer process. *Colloids Surf B Biointerfaces*, **59**(2), 208–214 (2007).
37. M.S. Cunha-Filho, R. Martínez-Pacheco, M. Landín, Dissolution rate enhancement of the novel antitumoral beta-lapachone by solvent change precipitation of microparticles. *Eur J Pharm Biopharm*, **69**(3), 871–877 (2008).

38. F.P. Maximiano, L.M. de Paula, V.P. Figueiredo, I.M. de Andrade, A. Talvani, L.C. Sá-Barreto, M.T. Bahia, M.S. Cunha-Filho, Benznidazole microcrystal preparation by solvent change precipitation and in vivo evaluation in the treatment of Chagas disease. *Eur J Pharm Biopharm*, **78**(3), 377–384 (2011).
39. X.S. Li, J.X. Wang, Z.G. Shen, P.Y. Zhang, J.F. Chen, J. Yun, Preparation of uniform prednisolone microcrystals by a controlled microprecipitation method. *Int J Pharm*, **342**(1–2), 26–32 (2007).
40. C. Beck, S.V. Dalvi, R.N. Dave, Controlled liquid antisolvent precipitation using a rapid mixing device. *Chem Eng Sci*, **65**, 5669–5675 (2010).
41. D. Xia, P. Quan, H. Piao, H. Piao, S. Sun, Y. Yin, F. Cui, Preparation of stable nitrendipine nanosuspensions using the precipitation-ultrasonication method for enhancement of dissolution and oral bioavailability. *Eur J Pharm Sci*, **40**(4), 325–334 (2010).
42. M.E. Matteucci, M.A. Hotze, K.P. Johnston, R.O. Williams, Drug nanoparticles by antisolvent precipitation: mixing energy versus surfactant stabilization. *Langmuir*, **22**, 8951–8959 (2006).
43. R.J. Hintz, K.C. Johnson, The effect of particle size distribution on dissolution rate and oral absorption. *Int J Pharm*, **51**, 9–17 (1989).
44. L. Jia, H. Wong, C. Cerna, S.D. Weitman, Effect of nanonization on absorption of 301029: ex vivo and in vivo pharmacokinetic correlations determined by liquid chromatography/mass spectrometry. *Pharm Res*, **19**, 1091–1096 (2002).
45. M. van der Woerd, D. Ferree, M. Pusey, The promise of macromolecular crystallization in microfluidic chips. *J Struct Biol*, **142**, 180–187 (2003).
46. A.G. Hadd, D.E. Raymond, J.W. Halliwell, S.C. Jacobson, J.M. Ramsey, Microchip device for performing enzyme assays. *Anal Chem*, **69**, 3407–3412 (1997).
47. C.B. Cohen, E. Chin-Dixon, S. Jeong, T.T. Nikiforov, A microchip-based enzyme assay for protein kinase A. *Anal Biochem*, **273**, 89–97 (1999).
48. D.B. Weibel, G.M. Whitesides, Applications of microfluidics in chemical biology. *Curr Opin Chem Biol*, **10**, 584–591 (2006).
49. H.S. Ali, P. York, N. Blagden, Preparation of hydrocortisone nanosuspension through a bottom-up nanoprecipitation technique using microfluidic reactors. *Int J Pharm*, **375**(1–2), 107–113 (2009).
50. H. Zhao, J.X. Wang, Q.A. Wang, J.F. Chen, J. Yun, Controlled liquid antisolvent precipitation of hydrophobic pharmaceutical nanoparticles in a microchannel reactor. *Ind Eng Chem Res*, **46**, 8229–8235 (2007).
51. H.S. Ali, P. York, A.M. Ali, N. Blagden, Hydrocortisone nanosuspensions for ophthalmic delivery: A comparative study between microfluidic nanoprecipitation and wet milling. *J Control Release*, **149**(2), 175–181 (2011).
52. T.L. Rogers, J. Hu, Z. Yu, K.P. Johnston, R.O. Williams 3rd, A novel particle engineering technology: spray-freezing into liquid. *Int J Pharm*, **242**(1–2), 93–100 (2002).
53. T.L. Rogers, A.C. Nelsen, J. Hu, J.N. Brown, M. Sarkari, T.J. Young, K.P. Johnston, R.O. Williams 3rd, A novel particle engineering technology to enhance dissolution of poorly water soluble drugs: spray-freezing into liquid. *Eur J Pharm Biopharm*, **54**(3), 271–280 (2002).
54. J. Hu, T.L. Rogers, J. Brown, T. Young, K.P. Johnston, R.O. Williams 3rd. Improvement of dissolution rates of poorly water soluble APIs using novel spray freezing into liquid technology. *Pharm Res*, **19**(9), 1278–1284 (2002).

55. J. Hu, K.P. Johnston, R.O. Williams 3rd, Rapid dissolving high potency danazol powders produced by spray freezing into liquid process. *Int J Pharm*, **271**(1–2), 145–154 (2004).
56. K.A. Overhoff, J.D. Engstrom, B. Chen, B.D. Scherzer, T.E. Milner, K.P. Johnston, R.O. Williams 3rd. Novel ultra-rapid freezing particle engineering process for enhancement of dissolution rates of poorly water-soluble drugs *Eur J Pharm Biopharm*, **65**(1), 57–67 (2007).
57. J. Jung, M. Perrut, Particle design using supercritical fluids: Literature and patent survey. *J Super Fluid*, **20**, 179–219 (2001).
58. T.J. Young, S. Mawson, K.P. Johnston, I.B. Henriksen, G.W. Pace, A.K. Mishra, Rapid expansion from supercritical to aqueous solution to produce submicron suspensions of water-insoluble drugs. *Biotechnol Prog*, **16**(3), 402–407 (2000).
59. T.J. Young, K.P. Johnson, G.W. Pace, A.K. Mishra, Phospholipid-stabilized nanoparticles of cyclosporine A by rapid expansion from supercritical to aqueous solution. *AAPS PharmSciTech*, **1**, E115 (2004).
60. M. Türk, R. Lietzow, Stabilized nanoparticles of phytosterol by rapid expansion from supercritical solution into aqueous solution. *AAPS PharmSciTech*, **5**(4), e56 (2004).
61. P. Pathak, M.J. Meziani, T. Desai, Y.P. Sun Nanosizing drug particles in supercritical fluid processing. *J Am Chem Soc*, **126**(35), 10842–10843 (2004).
62. A. Sane, M.C. Thies, The formation of fluorinated tetraphenylporphyrin nanoparticles via rapid expansion processes: RESS vs RESOLV. *J Phys Chem B*, **109**(42), 19688–19695 (2005).
63. S. He, B. Zhou, S. Zhang, Z. Lei, Z. Zhang, Preparation of nanoparticles of Magnolia bark extract by rapid expansion from supercritical solution into aqueous solutions. *J Microencapsul*, **28**(3), 183–189 (2011).
64. P. Pathak, G.L. Prasad, M.J. Meziani, A.A. Joudeh, Y.P. Sun, Nanosized paclitaxel particles from supercritical carbon dioxide processing and their biological evaluation. *Langmuir*, **23**(5), 2674–2679 (2007).
65. N.M. Patel, S. Nozaki, N.H. Shortle, P. Bhat-Nakshatri, T.R. Newton, S. Rice, V. Gelfanov, S.H. Boswell, R.J. Goulet Jr, G.W. Sledge Jr, H. Nakshatri, Paclitaxel sensitivity of breast cancer cells with constitutively active NF-kappaB is enhanced by IkappaBalpha super-repressor and parthenolide. *Oncogene*, **19**(36), 4159–4169 (2000).
66. Y. Tozuka, Y. Miyazaki, H. Takeuchi, A combinational supercritical CO_2 system for nanoparticle preparation of indomethacin. *Int J Pharm*, **86**(1–2), 243–248 (2010).

10

Microemulsions for Drug Solubilization and Delivery

X.Q. Wang and Q. Zhang

10.1 Introduction

With the development of combinatorial chemistry, high throughput screening and cell-based activity assays, more and more new chemical entities (NCEs) are currently in development. However, up to 40% of the NCEs have been suggested to be 'poorly water-soluble' [1]. To improve the bioavailability of these poorly water-soluble NCEs, many pharmaceutical strategies have been applied [2], such as micronization, solid dispersion technology, lipid-based formulations, nanocrystal technology, nanoparticulate technology, microemulsifying technology and self-microemulsifying drug delivery system (SEDDS), etc.

The concept of microemulsions was first introduced by Hoar and Schulman [3], who noticed that coarse macroemulsions stabilized by an ionic surfactant became transparent after addition of a medium chain length alcohol. It was in 1959 that Schulman and coworkers first proposed the word 'microemulsion' [4]. And now, the application areas of microemulsions have increased dramatically because of their important merits.

Naturally, the main advantanges of microemulsions include [5]: (1) microemulsions are thermodynamically stable systems. They are easy to prepare and possess excellent long-term stability; (2) microemulsions act as supersolvents of drug. They can solubilize large amounts of both water-soluble and oil-soluble drugs due to the existence of microdomains of different polarity within the same single phase solution; (3) the mean diameter of microemulsion droplets is below 0.22 μm and hence such systems can be sterilized by filtration. Microemulsions also have low viscosity as compared to other emulsions, so they can also be used for the parenteral route; (4) microemulsions have been found to improve the bioavailability of sparingly soluble lipophilic drugs, peptide and protein drugs, and

Drug Delivery Strategies for Poorly Water-Soluble Drugs, First Edition. Edited by Dennis Douroumis and Alfred Fahr.

drugs labile in the conditions such as the GI tract after oral administration; and (5) the use of microemulsions as delivery systems can improve the efficacy of the drug as the total dose is reduced, thereby minimizing the side-effects.

Of course, there are certain limitations which should be considered before the use of microemulsions in the field of pharmaceuticals and as drug delivery systems. The components, mainly the surfactants and cosurfactants should be pharmaceutically acceptable and biocompatible. Toxicity may occur at high surfactant concentration. For high molecular weight drugs, the solubilization capacity may be limited. The evaporation or removal of the cosurfactant may influence the long-term stability of the microemulsions.

When we talk about microemulsions, the definition of microemulsions should be made clear. Microemulsions are systems consisting of water, oil and amphiphile, which are single optically isotropic and thermodynamically stable liquid solutions [6].

Microemulsions are distinctly different from emulsions. The definition of emulsions by the International Union of Pure and Applied Chemistry (IUPAC) states: 'In an emulsion, liquid droplets and/or liquid crystals are dispersed in a liquid' [7]. In general, this includes conventional emulsions (macroemulsions), multiple emulsions, nanoemulsions, etc. Nanoemulsions are sometimes confused with microemulsions. Nanoemulsions are emulsions with the droplet in the nanoscale (typically in the range 20–200 nm). They are also called mini-emulsions, ultrafine emulsions or submicron emulsions in some of the literature. Due to their characteristic size, nanoemulsions appear transparent or translucent to the naked eye like microemulsions. However, there is a fundamental difference between microemulsions and nanoemulsions: microemulsions are equilibrium systems (i.e. thermodynamically stable), while nanoemulsions are non-equilibrium systems with a spontaneous tendency to separate into the constituent phases [8]. Since the confusion is already here in the literature [9], we will not distinguish the definitions and applications of microemulsions and nanoemulsions in this chapter.

Microemulsions are also different from micelles. In recent years, polymeric micelles have been widely investigated in pharmaceutical science. They are characterized by a core-shell structure, and insoluble drugs can be incorporated into micelles by chemical conjugation or by physical entrapment [10]. Microemulsions sometimes are considered as swollen micelles with oil/water droplets inside. Both microemulsions and micelles are thermodynamically stable systems. In general, microemulsions are superior to simple micellar solutions in terms of their solubilization potential [11]. Figure 10.1 is the schematic representation of o/w

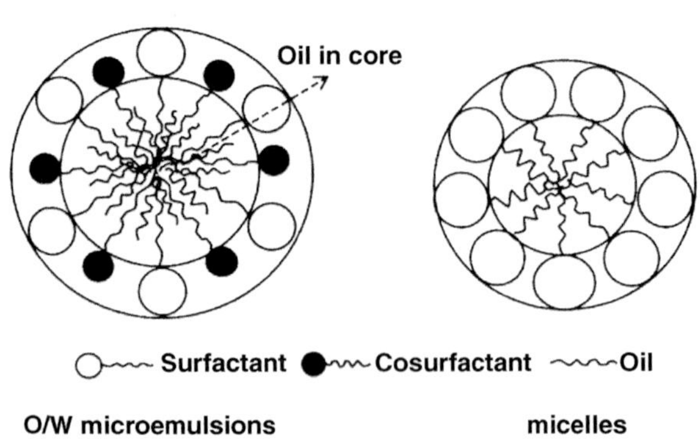

Figure 10.1 O/w microemulsions and micelles.

(oil-in-water) microemulsions and micelles. The characteristic comparisons of microemulsions, macroemulsions, multiple emulsions, nanoemulsions and polymeric micelles are listed in Table 10.1.

10.2 Microemulsion Formation and Phase Behavior

10.2.1 Theories of Microemulsion Formation

The mechanisms for microemulsion formation are varied. Up to now, there have been mainly three theories to explain the formation of microemulsions: interfacial or mixed film theories, solubilization theories and thermodynamic treatments. Interfacial or mixed film theory and solubilization theory can illustrate the formation of microemulsions qualitatively. They have been discussed in detail in some other publications [12]. Thermodynamic treatment theory can describe the formation of microemulsions quantitatively. Various thermodynamic factors contributing to the formulation of microemulsions have been suggested [5]. Ruckenstein and Chi in 1977 [13] expressed the change of the Gibbs free energy of microemulsion formulation in terms of three individual contributions as:

$$\Delta G = \Delta G_1 + \Delta G_2 + \Delta G_3$$

where ΔG_1 is the interfacial free energy including a positive term due to the creation of charged interface and a negative term due to the formation of electrical double layer; ΔG_2 is the free energy of interdroplet interaction composed of a negative term from the van der Waals attraction and a positive term from repulsive double layer interactions; and ΔG_3 is the entropy term accounting for the dispersion of microemulsion droplets in the continuous medium [5]. The resulting 'ΔG' is plotted as a function of droplet radius 'r' as shown in Figure 10.2.

Curves A and B correspond to unstable emulsion systems because the minimal free energy of the systems occurs at infinite radius (phase separation). Curve B may represent a kinetically stable system since a potential barrier must be overcome prior to final phase separation. Curve C shows that microemulsions with the most stable droplet size (r*) can be formed at negative minimum ΔG. When interfacial tension is less than 2×10^{-2} mN/m, stable microemulsions can form. It is seen that the dispersion entropy predominantly contributes to the thermodynamic stability of microemulsions.

10.2.2 Structure of Microemulsions

The representation of the three most commonly encountered microemulsion microstructures are: (i) oil-in-water (o/w); (ii) bicontinuous; and (iii) water-in-oil (w/o) microemulsions (Figure 10.3). To illustrate why these microstructures formed, double film theory, R-ratio theory and geometry theory are usually adopted. Geometry theory was introduced by Robbins *et al.* to illustrate the structure of microemulsions [14]. In his opinion, the relative sizes of the hydrophilic and hydrophobic groups of the surfactant molecules must be correctly balanced for the given oil and aqueous solution to produce microemulsions [14]. They will determine the type of microemulsions, o/w, w/o or bicontinuous, through their influence on molecular packing and film curvature. The packing ratio (P) can be calculated using the following equation:

$$P = (V/a_0 l c).$$

Table 10.1 *Comparison of microemulsions, macroemulsions, multiple emulsions, nanoemulsions and polymeric micelles.*

Property	Microemulsions	Macroemulsions	Multiple emulsions	Nanoemulsions	Polymeric micelles
Appearance	Transparent (or translucent)	Cloudy	Cloudy	Transparent (or translucent)	Transparent (or translucent)
Optical Isotropy	Isotropic	Anisotropic	Anisotropic	Anisotropic	Isotropic
Droplet size	20–200 nm	>500 nm	~μm	20–200 nm	<100 nm
Stability	Thermodynamically stable, long shelf-life	Thermodynamically unstable, will eventually phase separate	Thermodynamically unstable, will eventually phase separate	Thermodynamically unstable, will eventually phase separate	Thermodynamically stable, long shelf-life
Phase	Monophasic	Biphasic	Bi/Triphasic	Biphasic	Monophasic
preparation	Mild conditions	Require a larger input of energy	Require a larger input of energy	Require a very large input of energy	Mild conditions
Viscosity	Low	High	High	Low	Low

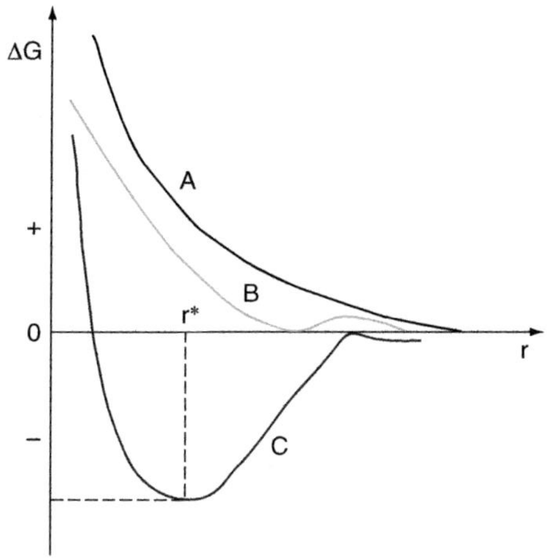

Figure 10.2 *The change of the Gibbs free energy of microemulsion formation (ΔG) relating to the particle size (r). Adapted from [5].* (See colour plate section.)

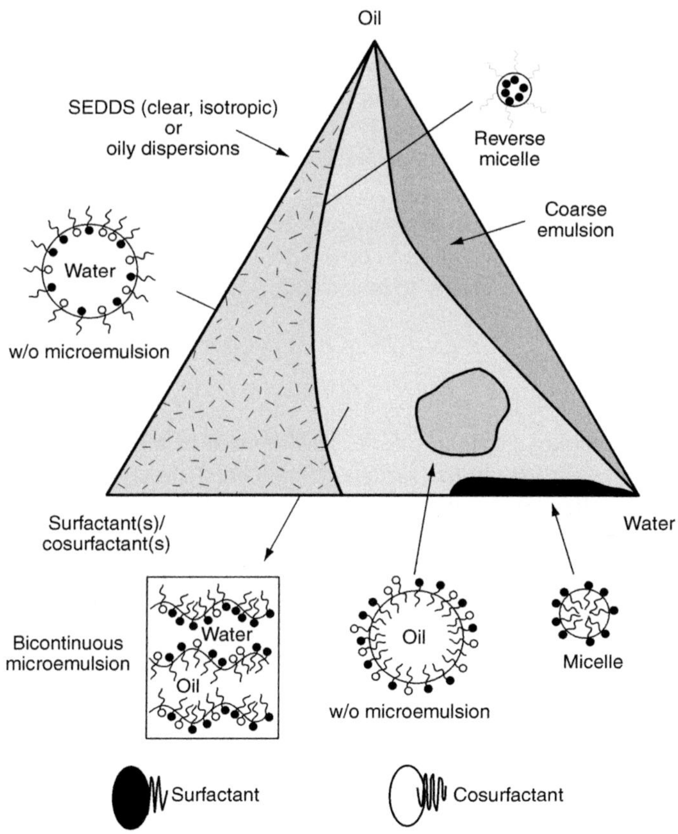

Figure 10.3 *A hypothetical pseudo-ternary phase diagram of an oil-surfactant(s)/ cosurfactant(s)-water system with emphasis on the microemulsion, micelle and emulsion phases. Within the phase diagram, different fields where w/o microemulsions, o/w microemulsions and bicontinuous microemulsion have formed are shown, along with the micelles, reverse micelles, SEDDS and coarse emulsion phases. Adapted from [11].*

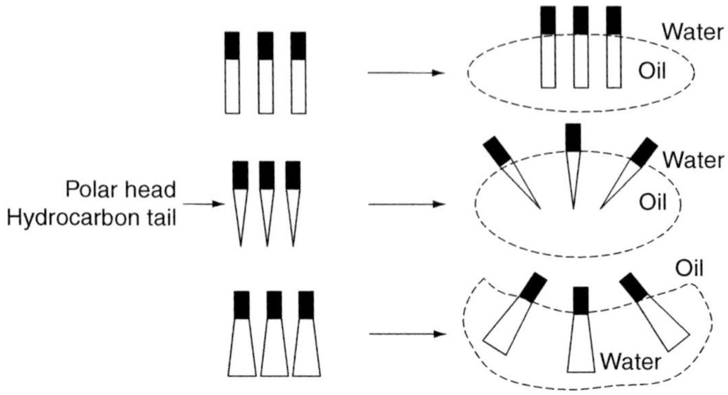

Figure 10.4 *A steric model correlating the shape of the surfactant molecules to the spontaneous curvature of the interface. Adapted from [15].*

where V is the partial molar volume of the hydrophobic portion of the surfactant, a_0 is the effective optimal head group area and lc is the critical length of the hydrophobic chain. If the effective polar part is bulkier than the hydrophobic part ($P < 1$), o/w structures are favored to form. The interface curves spontaneously towards water. The w/o structures will form when the interface curves are in the opposite direction ($P > 1$). When $P = 1$, the curvature equals zero, and either biocontinuous or lamellar structures may form, according to the rigidity of the film [15]. The relationship between the molecular packing and film curvature is illustrated in Figure 10.4. The oil and cosurfactant component in the formulation can influence the curvature by their ability to penetrate, which may swell the tail group region of the surfactant monolayer.

10.2.3 Phase Behavior

Phase behavior studies play a crucial role in development of microemulsions. Phase diagrams are normally used to study the various components and conditions influencing the kind of assembly that they ultimately adopt. Mainly, four important key components have been identified to determine the type of microemulsion. These are oil, water, surfactant (or mixed surfactants) and cosurfactant. Accordingly, phase diagrams to illustrate their interaction usually include quaternary, ternary, pseudoternary and pseudobinary phase diagrams [5]. The pseudo-ternary phase diagram is most widely used in the research. In general, to construct a pseudo-ternary phase diagram, a fixed (weight, volume or mole) ratio must be chosen for any two of the components and one of the triangle vertices represents 100% of this binary mixture. Phase diagrams of such multicomponent systems are generally based on constant ratios of surfactant-to-water or cosurfactant-to-surfactant. However, in investigations of solubilization, the use of the cosurfactant-to-oil constant ratio is also quite common. Such diagrams enable us to directly and conveniently follow the variation of the surfactant amount needed to solubilize a given amount of water [16].

A schematic representation of pseudo-ternary phase diagram of an oil/surfactant and cosurfactant/water system is presented in Figure 10.3. Microemulsions are one of a number of association structures (including emulsions, micelles, reverse micelles and coarse emulsions, etc.) that can form, depending on the chemical compositions and concentration of each component [11].

10.3 HLB, PIT and Microemulsion Stability

A surfactant is a key component in the formation of microemulsions. The tendency toward w/o or an o/w microemulsions is dependent on the properties of the oil and surfactant, as well as the ratio of oil to water. The hydrophile–lipophile balance (HLB) is an empirical formula that can be used to select surfactants for microemulsions. Non-ionic and zwitterionic surfactants are often considered for microemulsion formulation since they are less toxic. Water-in-oil microemulsions are formed using emulsifiers within the HLB range of 3–8 while oil-in-water microemulsions are formed within the range of 8–18 [11]. The choice of emulsifiers is determined by the average HLB requirement of the proposed microemulsions. In some cases, mixed surfactants are used to prepare stable microemulsions.

The phase inversion temperature (PIT), also known as the cloud point (T_{Cloud}), or the hydrophilic lipophilic balance (HLB) temperature, is another important concept in the microemulsion study [8,17]. Thermal stability or temperature sensitivity of microemulsions can be monitored through measuring the PIT. For o/w and w/o microemulsions, PIT has been defined as the temperature at which the solubility of the surfactant molecule changes from water soluble to oil soluble or vice versa. At this temperature, the net interfacial curvature of the surfactant molecule is changed from positive (o/w microemulsions) to negative (w/o microemulsions) by passing through a zero curvature (bicontinuous microemulsions). The phenomenon is based on the changes in the solubility of polyoxyethylene-type nonionic surfactants with temperature. These types of surfactants become lipophilic with increasing temperature because of the dehydration of the polyoxyethylene chains. At low temperature, o/w microemulsions are prepared with the ethoxylate type surfactant. When it is heated, the microemulsions invert to w/o microemulsions. At the PIT, the interfacial tension is extremely low, about 10^{-2}–10^{-5} mN/m. According to this character, the PIT method is used to prepare microemulsions. However, the phenomenon of phase inversion with temperature can also cause problems for the exploitation of microemulsions stabilized by nonionic surfactants as drug delivery systems, especially for intravenously administered microemulsions, where autoclaving is the preferred method of sterilization. In order to avoid destabilizing the microemulsions due to the phenomenon of phase transition resulting from temperature, the intended temperature of the microemulsion application should be 30 K below the PIT. Sterilization by filtration remains an option for microemulsions because of their low viscosity.

10.4 Microemulsion Physico-Chemical Characterization

The characters of microemulsions can be monitored through visually inspecting and measuring electrical conductivity, viscosity, droplet size, density, surface tension and diffusion coefficient, etc.

Visual observation is one of the first and foremost techniques used to establish the phase boundaries and distinguish microemulsions from coarse emulsions, liquid crystals and gels. To further differentiate microemulsions from lamellar liquid crystals, one can take advantage of other experimental techniques such as polarized light microscopy and viscosity measurements. In general, microemulsions display Newtonian flow with low viscosity while liquid crystals exhibit non-Newtonian flow with relatively high viscosity. Furthermore, viscosity can also be used to study the microstructure of microemulsions. Normally o/w microemulsions have a lower viscosity compared to w/o microemulsions in which the more viscous oil forms the external phase. Measurement of electrical conductivity is another valuable technique for understanding the structural changes taking place within a microemulsion system. Microemulsions with water as a continuous phase show high electrical conductivity as compared to microemulsions with the insulating oil as a continuous phase [18].

To further investigate the microstructures of microemulsions, electron microscopy is often used. Transmission electron microscopy (TEM) is the most important technique for the study of microstructures, which directly produces images at high resolution and can capture any coexistent structures and microstructure transitions [19]. However, conventional TEM is prone to creating different artefacts in the case of surfactant solutions. Both drying and staining processes in TEM can affect the structure and morphology of the sample. Therefore, great care should be taken during interpretation of the obtained images [20]. As a comparison, freeze-fracture transmission electron microscopy (FF-TEM) has emerged as a method for observing diverse systems such as surfactant aggregates, polymer and polymer–surfactant solutions, and microemulsions. The technique does not require a drying process and yields additional information about the internal structure [20]. The technique consists of physically breaking apart (fracturing) a frozen biological sample; the structural detail exposed by the fracture plane is then visualized by the vacuum deposition of platinum–carbon to make a replica for examination under TEM. However, the freezing of the microemulsions must be achieved sufficiently rapidly to avoid phase separation or crystallization. In other words, artefacts may easily occur due to insufficient freezing rate or re-deposition of solvent molecules onto the sample plane after fracturing. In contrast to the above-mentioned methods, cryofixation transmission electron microscopy (Cryo-TEM) allows the direct investigation of colloids in the vitrified frozen-hydrated state, and overcomes many of the problems associated with artefact formation. As with FF-TEM, information about the internal structure of the microemulsions can be obtained. This technique uses direct imaging, in which thin portions of the specimen are directly investigated in the frozen hydrated state by using a cryostage in the TEM [21]. The development of glass-forming microemulsions that do not break down during cooling, and in which neither the disperse nor the matrix phase crystallizes during the cooling process, has provided a way for direct studies of microemulsion structure. Klang *et al.* [22] used Cryo-TEM to visualize the differences in formulation properties caused by incorporation of cyclodextrins, used as stabilizing agents, in microemulsions. The images showed that cyclodextrins seemed to induce fundamental changes such as additional multilamellar structures in microstructure.

Another technique used to study the microstructure of microemulsions is the scattering method, such as small angle neutron scattering (SANS), small angle X-ray scattering (SAXS), static light scattering (SLS) and dynamic light scattering (DLS). Over the whole range of colloidal dimensions of 1–100 nm, SANS is unique in determining the particle size and the interaction of different components in the system. For simplified analysis, monodispersity and spherical shape are good approximations for the understanding of

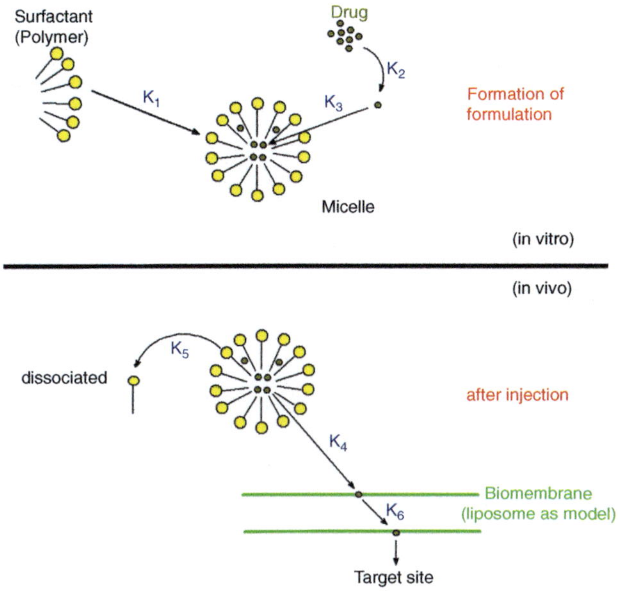

Figure 1.16 Upper part: *Illustration of the* in vitro *formation of a carrier system from surfactants and drug molecules. The drug molecules partition into the micelles or, alternatively, form an aggregate on their own. Different rate constants k_1, k_2 and k_3 are involved in the kinetics of this process.* Lower part: *Illustration of the* in vivo *transfer of drug molecules from a drug-loaded carrier to a biomembrane target. Again, different rate constants k_4, k_5 and k_6 are involved in the kinetics of this process.*

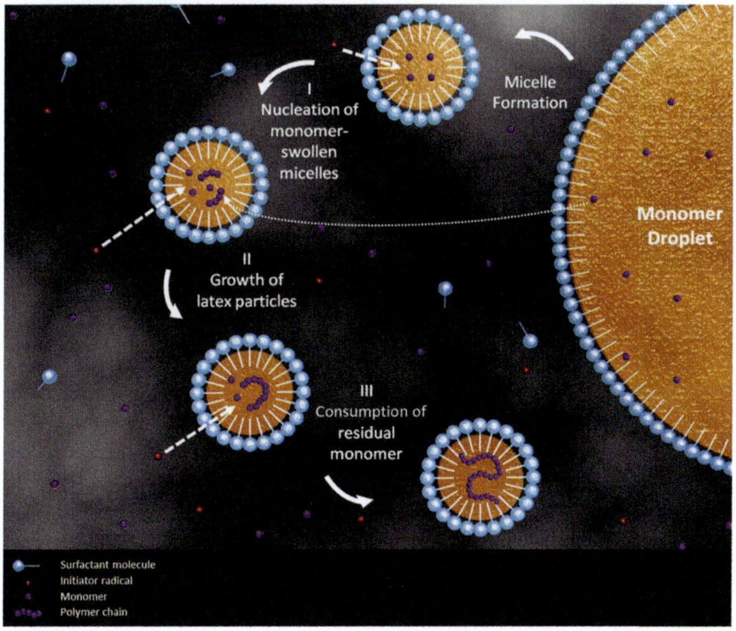

Figure 6.2 *The emulsion polymerization process [27].*

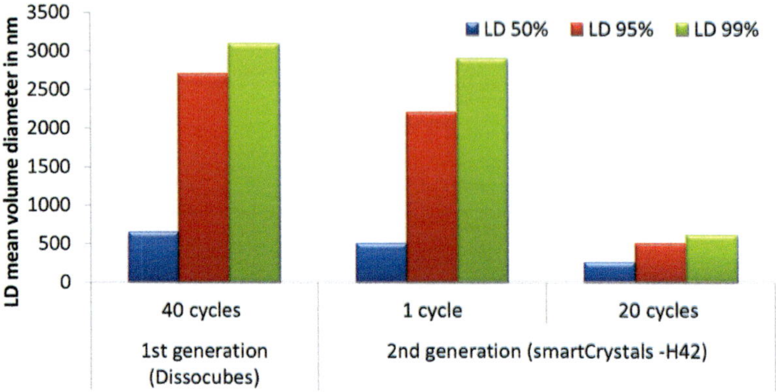

Figure 8.3 *Laser diffractometry (LD) diameters 50% to 99% of a model drug applying 40 cycles of high pressure homogenization (left) versus just 1 and 20 cycles in the H42 process. Modified after Jan Möschwitzer, reproduced with permission.*

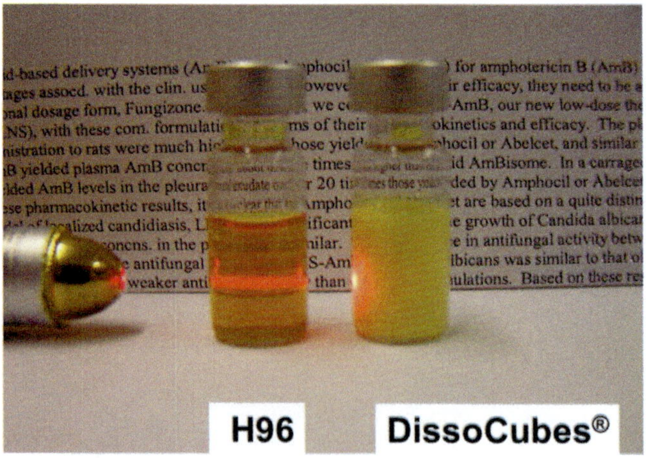

Figure 8.8 *Amphotericin B nanosuspensions: the left nanosuspension is translucent due to its small nanocrystal size – produced by H96 technology. The scattered laser beam proves the existence of the nanocrystals. The right nanosuspension is about 300 nm and turbid produced with high pressure homogenization only (= DissoCubes technology.) Reproduced with permission after [58].*

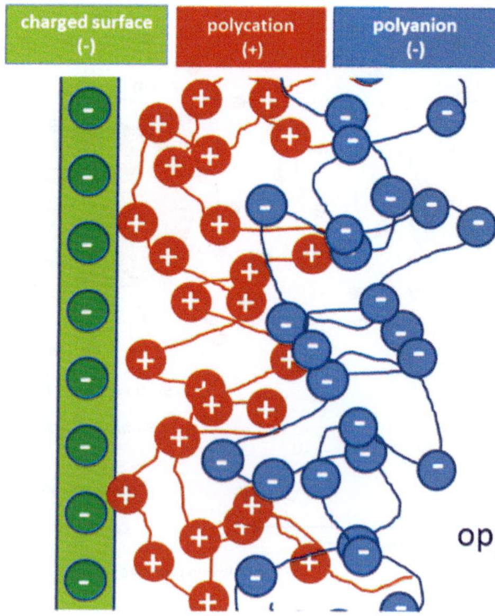

Figure 8.9 *LBL coating process on a negative surface, adsorbing oppositely charged polyelectrolytes. Reproduced after Capsulution Pharma AG Berlin, product information.*

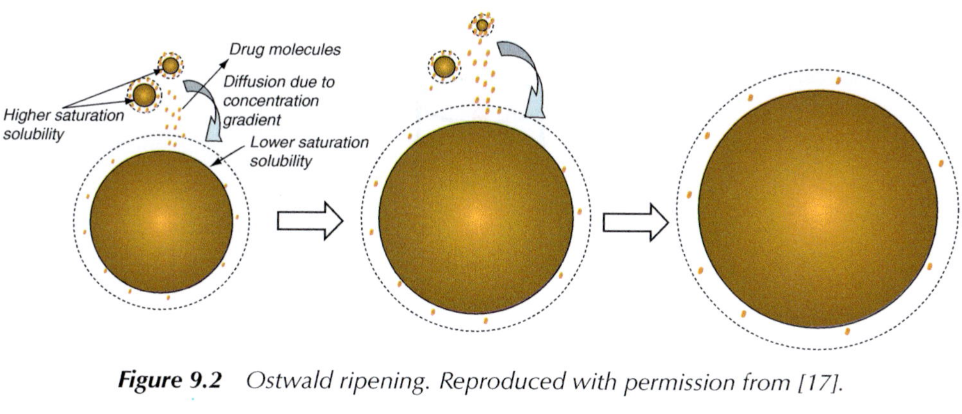

Figure 9.2 *Ostwald ripening. Reproduced with permission from [17].*

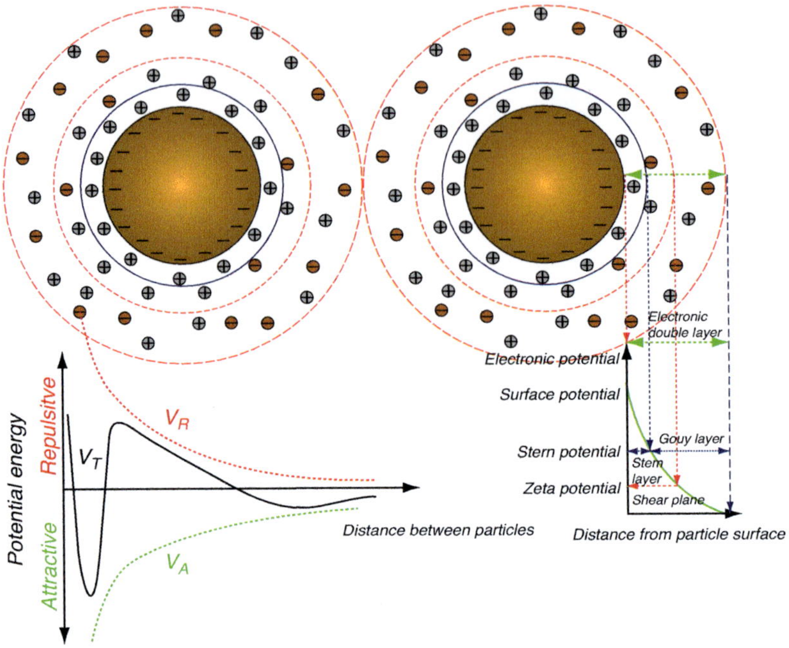

Figure 9.3 *Classical DLVO theory. Attractive forces are dominant at very small and large distances, leading to primary and secondary minimum, while repulsive forces are prevailing at intermediate distances and create net repulsion between the dispersed particles, thus preventing particle agglomeration. Reproduced with permission from [17].*

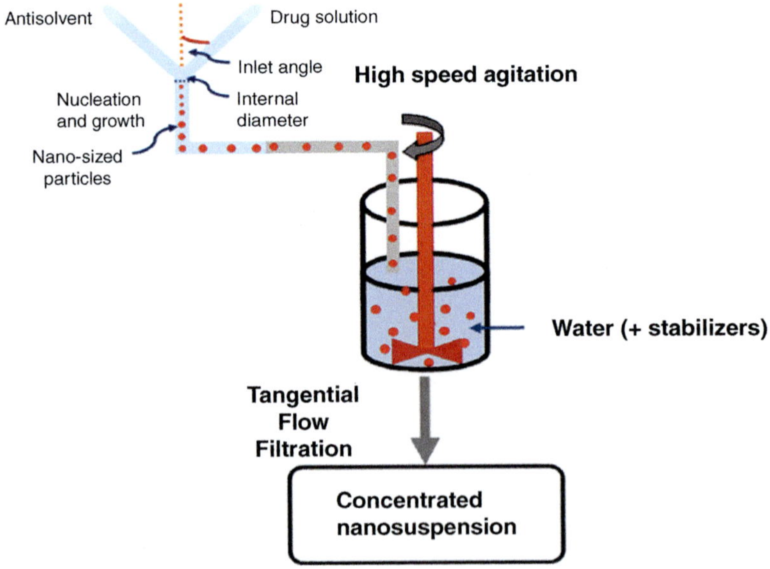

Figure 9.6 *Preparation of hydrocortisone nanosuspension by precipitation process. Reproduced with permission from [49].*

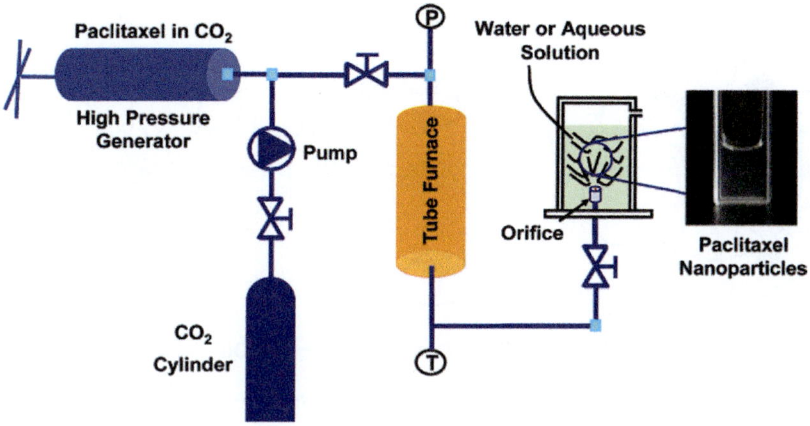

Figure 9.7 *Experimental RESOLV set-up. Shown in the figure is an aqueous suspension of stabilized paclitaxel nanoparticles. Reproduced with permission from [64].*

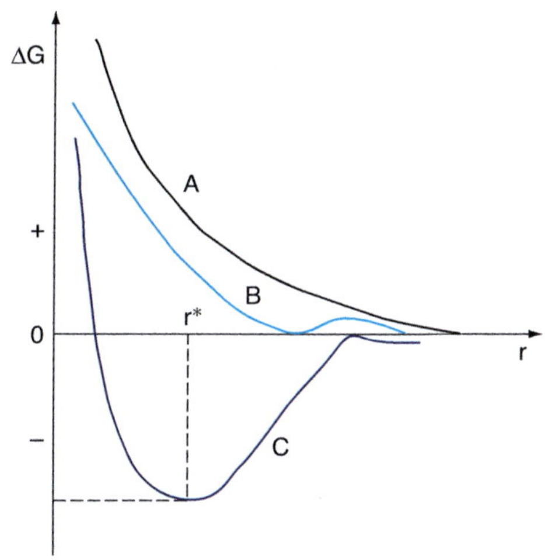

Figure 10.2 *The change of the Gibbs free energy of microemulsion formation (ΔG) relating to the particle size (r). Adapted from [5].*

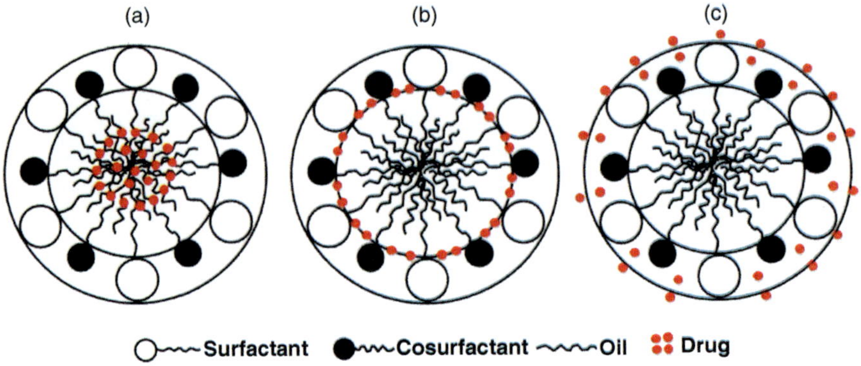

Figure 10.9 *Three possible solubilization sites in o/w microemulsions. (a) microemulsion interior; (b) palisade layer; (c) microemulsion surface.*

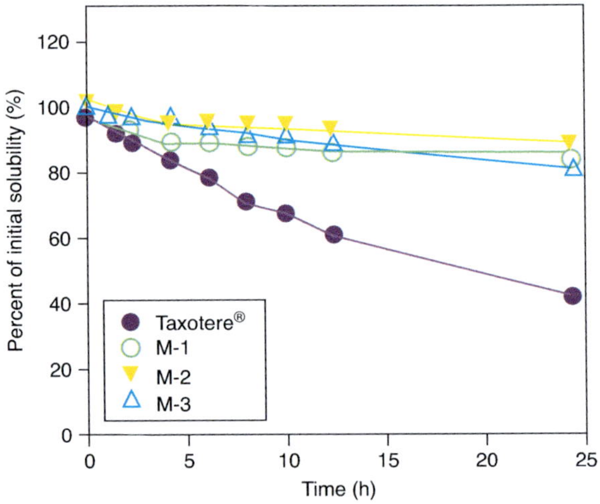

Figure 10.12 *Change of docetaxel solubility in Taxotere® and three microemulsion formulations after 20 times dilution by saline solution at 37°C. M-1: microemulsion formulation composed of oil (Capryol 90) 30.3%, surfactant and cosurfactant (Cremophor EL/propylene glycol, 2:1) 38.6% and water 31.1%. M-2: microemulsion formulation composed of oil (Capryol 90) 28.2%, surfactant and cosurfactant (Cremophor RH40/Transcutol, 2:1) 35.7% and water 36.1%. M-3: microemulsion formulation composed of oil (Capryol 90) 29.4%, surfactant and cosurfactant (Cremophor EL/Transcutol, 2:1) 37.3% and water 33.3%. Adapted from [37].*

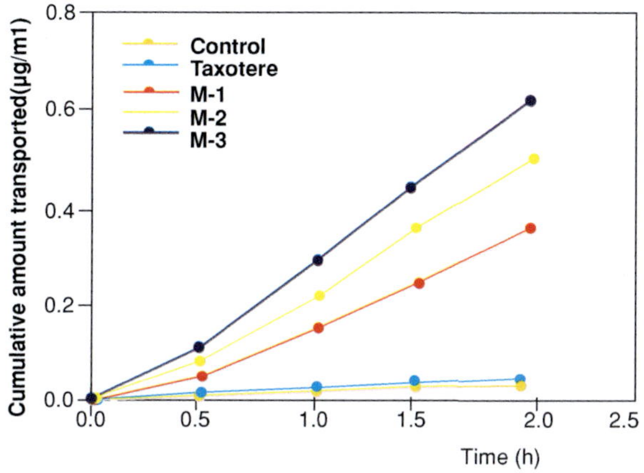

Figure 10.15 *Apical to basolateral transport profiles of docetaxel from Taxotere® or microemulsions across the Caco-2 cell monolayers at 37°C. The control for the transport study was docetaxel (10 μg/ml) in transport medium containing 4% (v/v) DMSO. ME-1: microemulsion formulation composed of oil (Capryol 90) 30.3%, surfactant and cosurfactant (Cremophor EL/ propylene glycol, 2:1) 38.6% and water 31.1%. ME-2: microemulsion formulation composed of oil (Capryol 90) 28.2%, surfactant and cosurfactant (Cremophor RH40/Transcutol, 2:1) 35.7% and water 36.1%. ME-3: microemulsion formulation composed of oil (Capryol 90) 29.4%, surfactant and cosurfactant (Cremophor EL/Transcutol, 2:1) 37.3% and water 33.3%. Adapted from [37].*

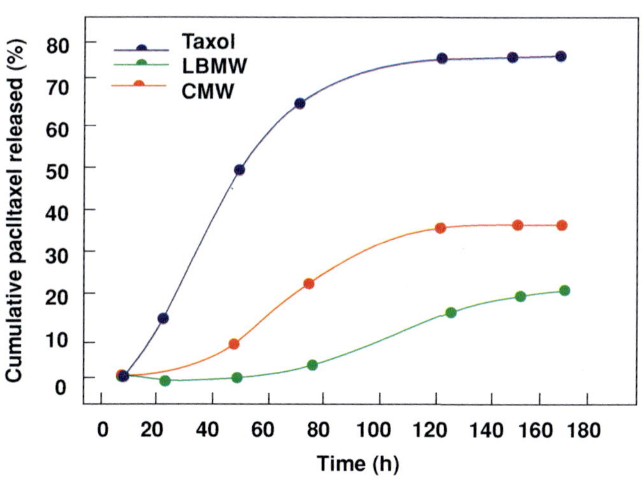

Figure 10.17 *Release profiles of paclitaxel from three formulations. Dialysis bag containing 12 mg/g of paclitaxel which was released into 0.5% sodium lauryl sulphate in water over 168 h. The formulations were Taxol®, LBMW (lecithin:butanol:myvacet:water/20:10:68:2) and CMW (capmul:myvacet:water/24:74:2). Adapted from [50].*

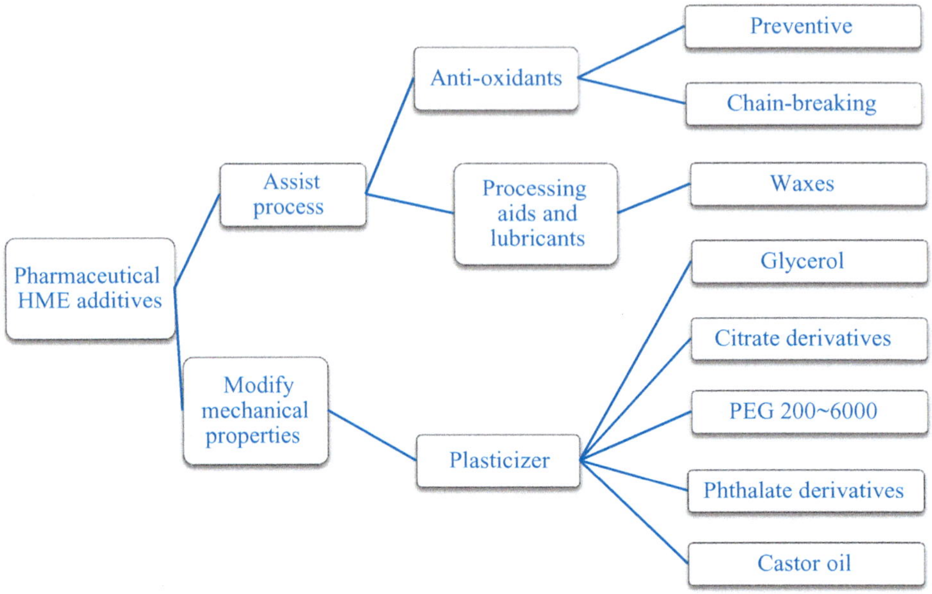

Figure 11.3 *Categorization of commonly used additives in pharmaceutical extrusion processing.*

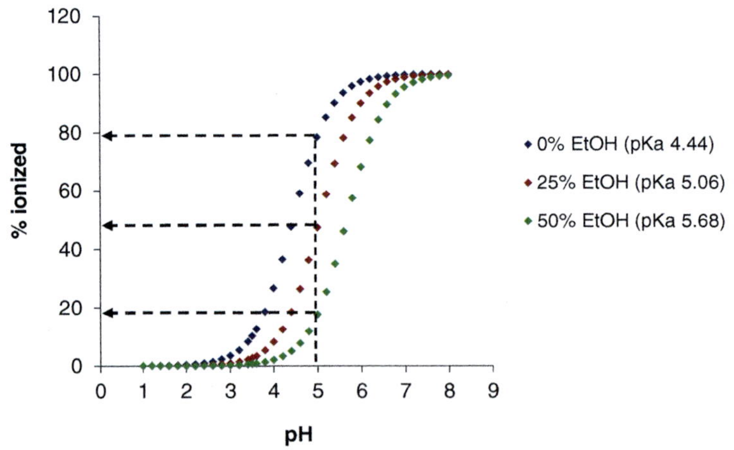

Figure 12.2 *Change of pK_a with increasing % of ethanol for ibuprofen.*

Figure 13.1 *Different types of natural dendritic patterns represented by circled areas as observed on various creatures. (a) ice flick; (b) tree moss agate; (c) tree branching; (d) drainage system; (e) dendritic cell; and (f) lizard's claw. Reprinted with permission from John Wiley & Sons, Ltd.*

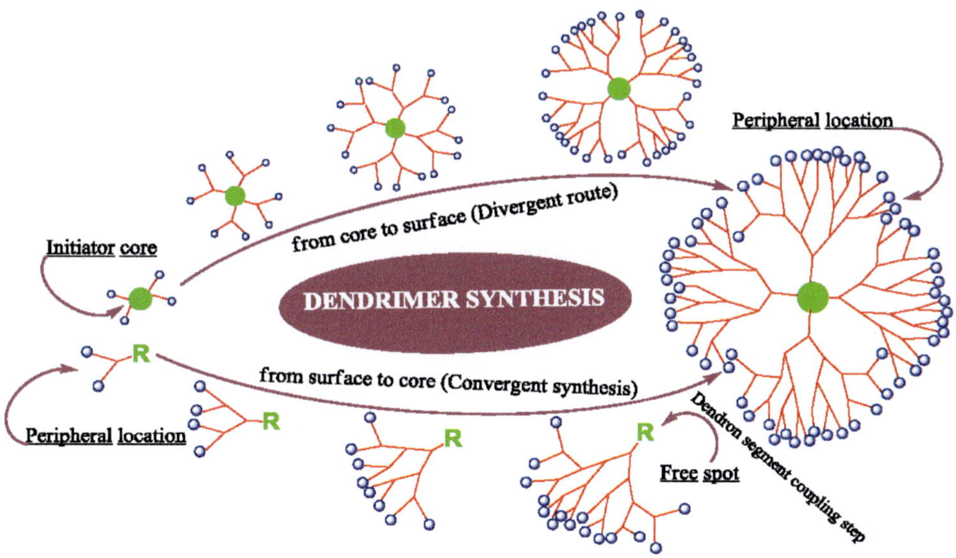

Figure 13.2 *Methodologies for synthesis of dendritic architectures (i) Divergent route, which involves progress of reaction scheme from a polyvalent core and continues racially outwards by successive repetition of reaction steps, (ii) Convergent route, in which reaction synthesis proceeds inwards from the periphery to the focal point.*

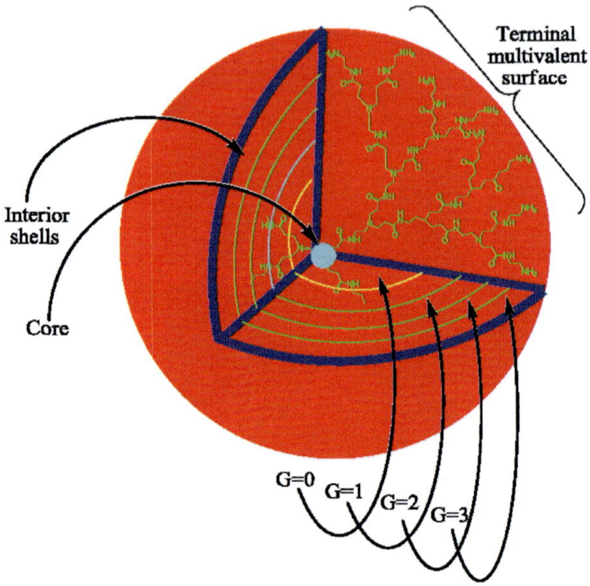

Figure 13.3 Three-dimensional architecture of dendrimer with chief architectural domains, namely, core, interior and surface.

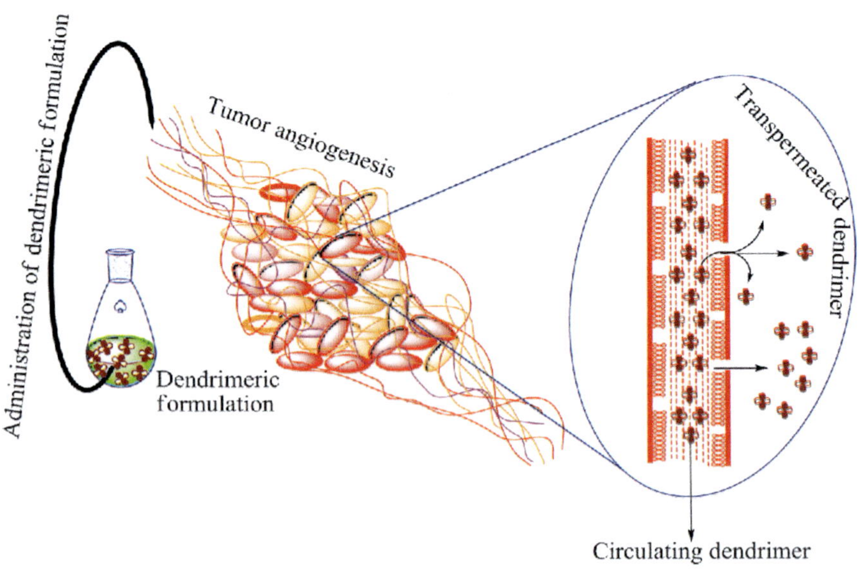

Figure 13.4 Entry of drug-dendrimer complex through pores in the tumor vasculature.

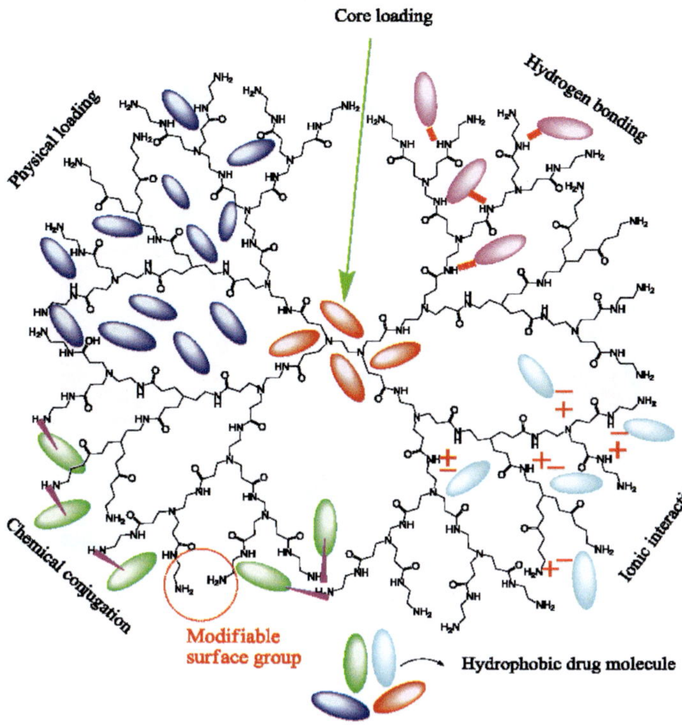

Figure 13.5 Modes of drug encapsulation inside a dendrimer to increase the water solubility of hydrophobes.

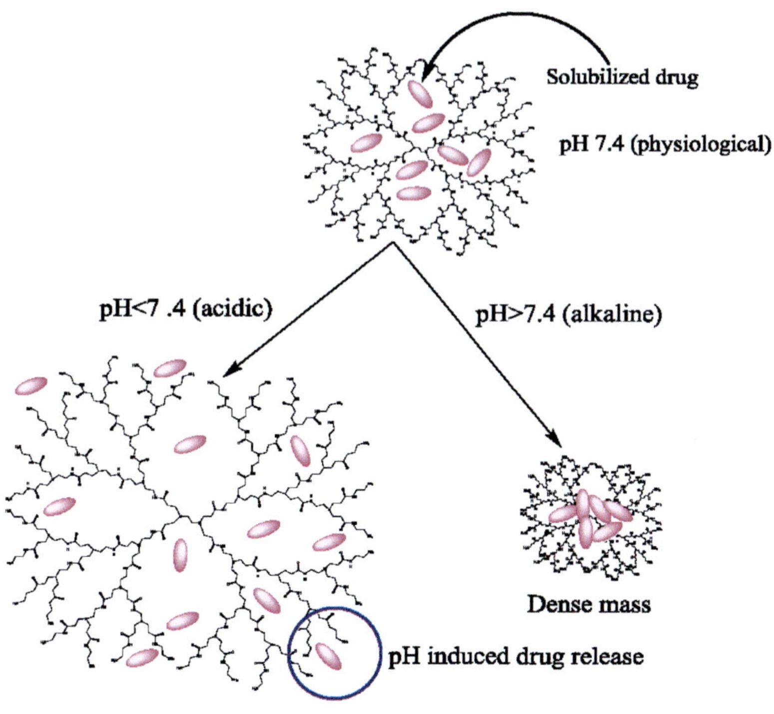

Figure 13.6 Mechanism of drug release from drug-dendrimer complex as a function of the pH of the environment.

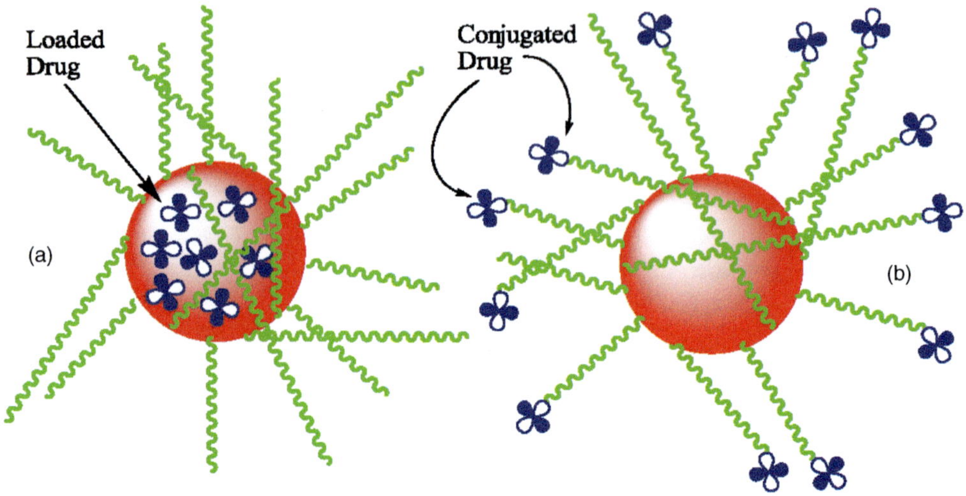

Figure 13.12 PEGylated dendrimer with (a) physical drug loading; and (b) drug loading by covalent conjugation.

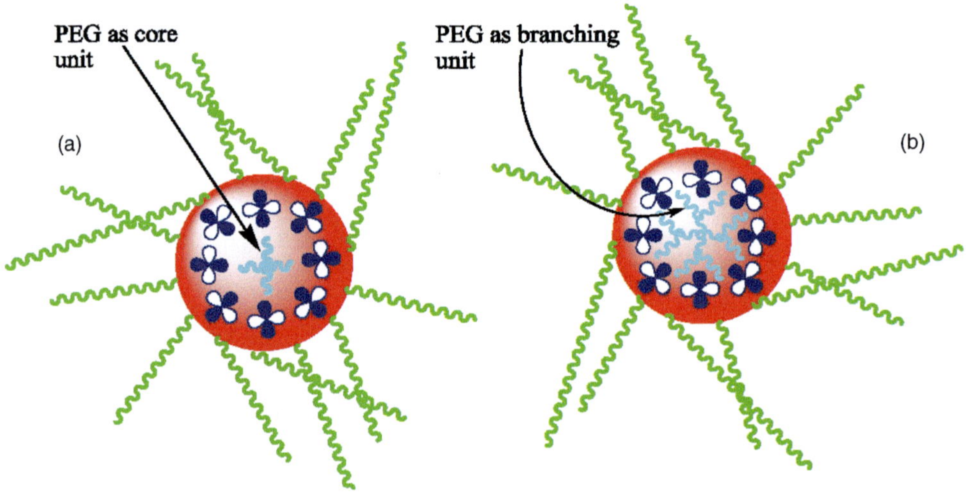

Figure 13.14 PEGylated dendrimer with (a) PEG as central core; and (b) PEG as branching units.

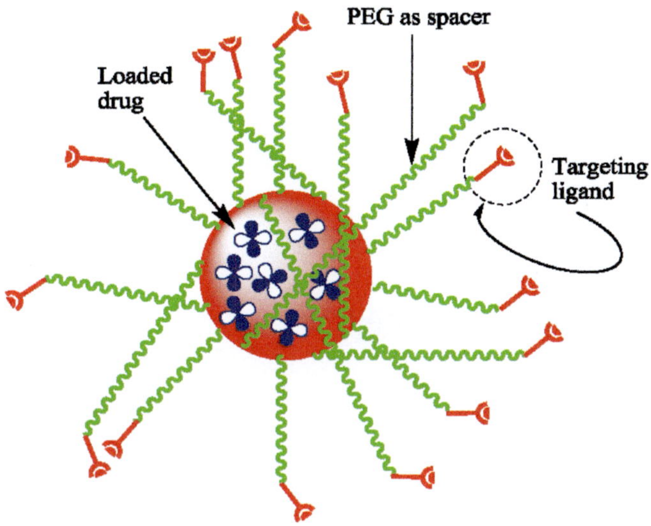

Figure 13.15 Hybrid PEGylated dendrimer with simultaneous solubilization and site-specific drug targeting.

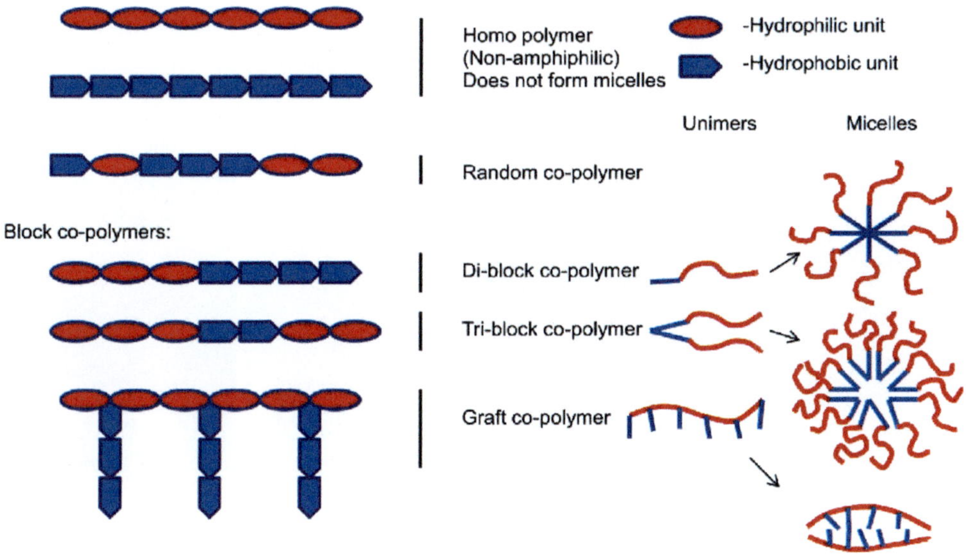

Figure 14.2 Main structural types of co-polymers and micelles formed from amphiphilic co-polymers.

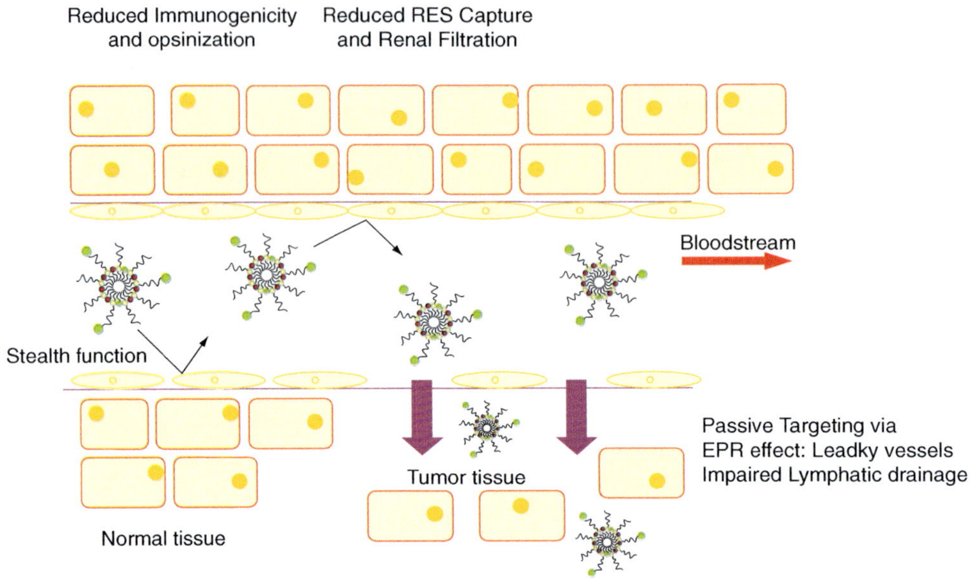

Figure 14.6 *The Enhanced Permeability and Retention (EPR) effect or 'passive' targeting of the nanocarriers. Nano-sized carriers including micelles cannot extravasate through the normal endothelium, but can penetrate through a compromised, 'leaky' vasculature characteristic of target pathological sites such as tumors, resulting in the accumulation of carrier in tumors relative to normal tissues.*

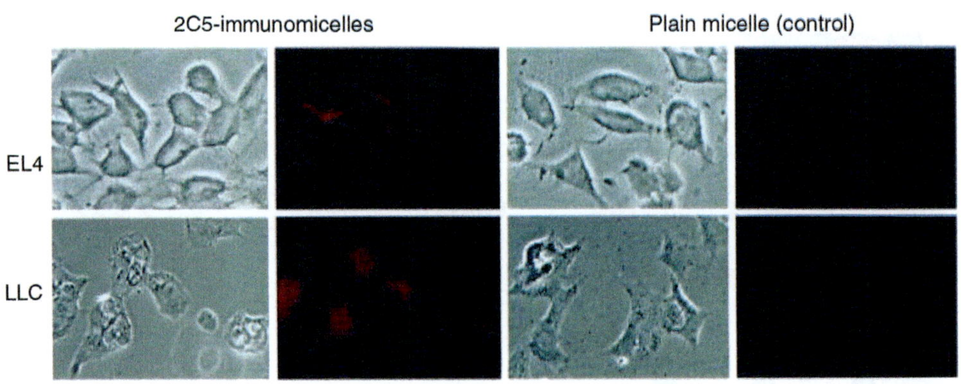

Figure 14.8 *Microscopy of the binding of Rh-labeled 2C5-grafted immunomicelles to EL4 T lymphoma cells (Top), LLC cells (bottom), (Left) Bright-field light microscopy. (Right) Fluorescent microscopy indicating the targeting potential of 2C5-immunoliposomes towards cancer cell-specific antigens. Modified from Torchilin et al., PNAS U.S.A., 2003.*

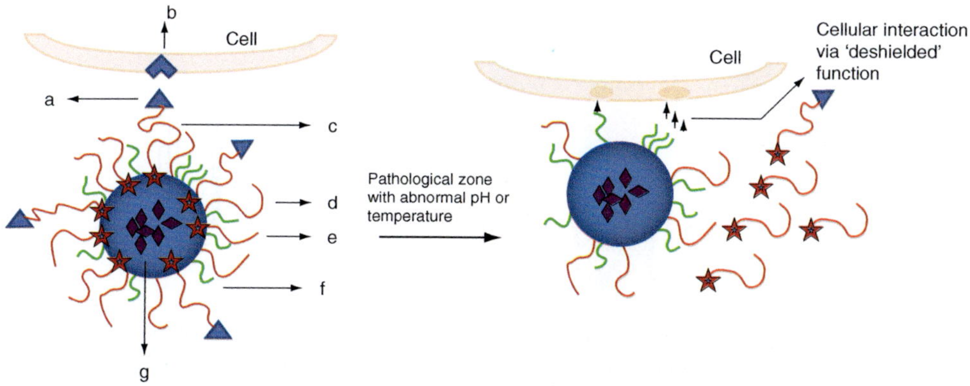

Figure 14.10 Concept of the interaction of a multifunctional stimulus-sensitive micelle with the target cell. a. Target-specific ligand; b. Binding site of the target-specific ligand on the cell surface; c. Long PEG spacer attached to micellar core by stimuli-sensitive bond; d. Hydrophilic segment such as PEG of the micelle-forming block co-polymer; e. Protective stimulus-sensitive hydrophilic segment of the block co-polymer which could be detached under the action of local stimuli in the pathological site to expose a second function; f. Cell-penetrating function; g. Drug-loaded core of the polymeric micelle.

PEG-coated micelles, after successful accumulation in the target site by active and passive targeting, degrade the stimulus-sensitive bond exposing the second cell-penetrating function for the direct interaction with the cell membrane. Reprinted with permission from (Salonen et al., 2005).

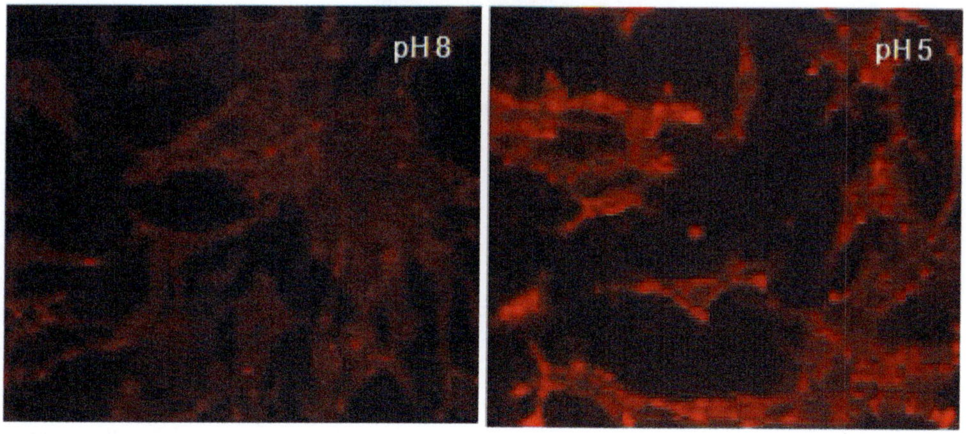

Figure 14.11 Internalization of rhodamine-labeled TAT-p-bearing micelles by NIH-3T3 mouse fibroblast cells after pre-incubation of micelles at pH 8 and pH 5 for 30 min. Pre-incubation at low pH cleaves the protective PEG-chains, exposing the hidden cell-penetrating TAT-p function Modified from [278].

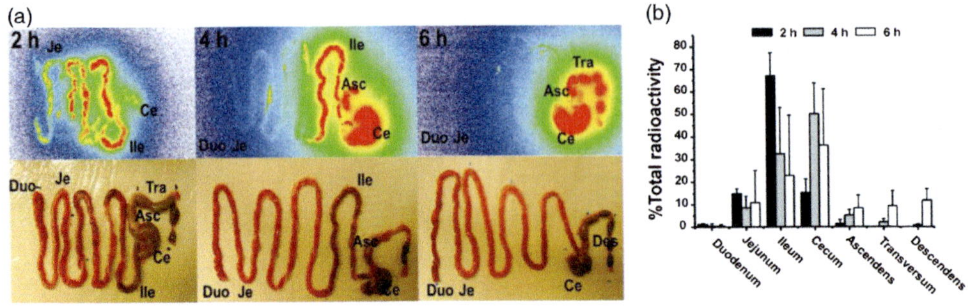

Figure 15.7 *(a) Ex vivo macroautoradiographs and respective photographs of the rat GI tracts at 2, 4, and 6 h after oral administration of* ^{18}*F-hydrocarbonized PSi nanoparticles (abbreviations: Duo, duodenum; Je, jejunum; Ile, ileum; Ce, cecum; Asc, ascendens; Tra, transversum; Des, descendens). (b) Radioactive distribution of* ^{18}*F-hydrocarbonized PSi nanoparticles after oral administration in the lower GI tract as quantified from autoradiography. Reprinted with permission from [12].*

Figure 17.1 *Colorized SEM images of spray-dried cyclosporin-loaded PLGA particles [43].*

structural changes in the microdispersions [23]. Leaver *et al.* studied the structural features of microemulsions with non-ionic surfactants by SANS [24]. The results showed a structure change sequence of normal spheres to planar bilayers via cylinders with decreasing water at a fixed temperature. It was realized that the temperature change at a constant surfactant-oil composition increased the reversal of the mean curvature of the polar-apolar interface [23]. SAXS is a widely used technique for direct structural investigations of the systems with the inner structuration falling in the colloidal domain. Tomšič *et al.* [25] used the SAXS method to study the structural properties of the quaternary microemulsion system Tween 40®/Imwitor 308®/isopropylmyristate/water and the fifth component systems obtained by adding the drug ketoprofen. The SAXS analysis indicated that the addition of smaller amounts of the ketoprofen, which was sensitive to radiation, did not change the inner structuration of these microemulsions significantly. The quaternary system containing 46.2 wt% of water seems to be very near the composition at which the transition from the bicontinuous to the lamellar structure of the microemulsions occurred.

Except for the size determination, DLS is also a good method for measuring the translational diffusion coefficient of microemulsion droplets, and is useful in the investigation of concentrated dispersions. Spernath *et al.* [26] used DLS to investigate the effect of water content and surfactant/oil phase ratio on the average droplet size of microemulsion samples. It was found that the droplet size increased upon both the water dilution and the increase of oil/surfactant ratio in the course of the transformation from bicontinuous to o/w microemulsions. The DLS also showed that the presence of phosphatidylcholine at the interface did not affect the size of the o/w droplets, while the presence of diclofenac at the interface decreased the o/w droplet size by an average of 50%.

Nuclear magnetic resonance (NMR) has become a standard technique for establishing phase diagrams of surfactant systems including both the solution phase and liquid-crystalline phases. A significant advantage of the NMR method is that it is not necessary to achieve a macroscopic phase separation, and two- or three-phase characters can be detected with single-phase domains on the micrometer scale. With the development of the pulsed-gradient spin-echo NMR (PGSE-NMR) technique, it is feasible to measure rapidly and simultaneously the self-diffusion coefficients [27]. Graf *et al.* [28] developed two microemulsion systems comprising isopropyl myristate (IPM), soybean lecithin, water, ethanol and decyl glucoside (DG) or capryl-caprylyl glucoside (CCG) as surfactant. Viscosity and conductivity measurements exhibited a continuously increasing trend with increasing water fraction. They also carried out self-diffusion NMR measurements on a system containing DG. By comparing the self-diffusion coefficients of the individual components with those measured for the pure components, identification of the microemulsion type was achieved. From the results, a change in ethanol diffusion was found only in the samples containing liquid crystals, diffusion of IPM decreased slightly with decreasing oil concentration. All these findings indicate the formation of solution-type microemulsions.

10.5 Components of Microemulsion Formulations

The formulation of microemulsions usually involves a combination of three to five components: oil, water, surfactant, cosurfactant and drug. The choice of ingredients is challenging, as most of the oils, surfactants and cosurfactants used in basic research are not suitable due to biocompatibility issues.

10.5.1 Oils

Selection of an appropriate oily phase is very important because it not only influences the selection of the other ingredients of microemulsions but also has an effect on the amount of lipophilic drug finally solublized. It can also affect the fraction of lipophilic drug transported via the intestinal lymphatic system, thereby increasing absorption from the GI tract. In general, the oils which have maximum solubilizing potential for the selected drug candidate should be selected as the oily phase for the formulation of microemulsions. This helps to achieve the maximal drug loading in the microemulsions. At the same time, the ability of the selected oil to yield systems with a larger microemulsion existence region is also important. Usually, oils with long hydrocarbon chains (or high molecular volume) such as soybean oil, have a large capacity for the solubilization for lipophilic moieties. However, they are difficult to microemulsify. As for the oils with shorter chain (or low molecular volume) such as medium chain triglycerides (MCT), or fatty acid esters (like ethyl oleate), they are easy to microemulsify. The choice of the oily phase is often a compromise between its ability to solubilize the drug and its ability to facilitate formation of microemulsions with the desired characteristics. In certain cases, a mixture of oils is used to meet both requirements [29].

The commonly used oils are listed in Table 10.2. Medium-chain triglycerides (MCT) with different degrees of saturation used for the formulation of microemulsions are particularly attractive for formulating microemulsions for i.v. administration. This is because: (i) they are stable food grade products and generally regarded as safe by the Food and Drug Administration; and (ii) microemulsions incorporating these low viscosity oils can be formulated over a wide range of compositions at ambient temperatures [30]. Novel semi-synthetic medium-chain triglyceride oils with surfactant properties are widely replacing the regular medium-chain triglyceride.

Fanum [31] reported how the type of oils affected the diclofenac solubilization. The solubilization capacity of diclofenac depended on the oil type and its ability to penetrate the interface. Two microemulsion systems were prepared based on different oils: R(+)-limonene (LIM) and isopropylmyristate (IPM). The oil and surfactant phase of the formulations of the two microemulsion systems was composed of sucrose laurate/ethoxylated mono-di-glyceride/LIM/ethanol (3/3/2/2, weight ratios) or sucrose laurate/ethoxylated mono-di-glyceride/IPM/ethanol (3/3/2/2, weight ratios). To study the solubilization capacity (SC) difference between the two microemulsion formulations, the mixtures of oil and surfactant phases were diluted by water along the N60 (N60 is the water dilution line where the weight ratios of sucrose laurate/ethoxylated mono-diglyceride/oil/ethanol are maintained 3/3/2/2 at 25°C). Dilution with the aqueous phase caused progressive deformation of the water-in-oil swollen micelles and the droplets became bicontinuous structured domains. Upon further dilution, full inversion occurred and oil-in-water nanodroplets were formed. The solubilization capacity of sodium diclofenac underwent a different change with the increase of the water volume fraction (Ø) and is shown in Figure 10.5. In a micellar solution containing mixed surfactants and oil phase, the solubilization capacity of sodium diclofenac was 15,000 and 16,000 ppm (15 and 16 wt%) for LIM- and IPM-based systems,

Table 10.2 *Examples of oils, surfactants, cosurfactants used in the preparation of microemulsions.*

Compounds (commercial name)	HLB
Oils	
Olive oil	
Corn oil	
Sesame oil	
Soyabean oil	
Peanut oil	
Oleic acid	
Ethyl oleate	
Isopropyl Myristate (IPM)	
Medium chain triglyceride (Labrafac CC®, Labrafac Lipophile WL 1349®)	
Polyglycerol ester of fatty acids (Plurol Oleique CC 497®)	
Polyoxyethylated glycerides (Labrafil M 2125 CS®)	
Polyoxyethylated oleic glyceride (Labrafil M 1944CS®)	
Propylene glycol monocaprylic ester (Sefsol-218®)	
Propylene glycol dicaprylic ester (Sefsol-228®)	
Mixed diesters of caprylic/capric acids in propylene glycol (Captex 200®)	
Triglycerides of caprylic/capric acids (Captex 355®)	
Glycerine plus mixed triester with capric/caprylic acids (Miglyol 812®)	
Distilled acetylated monoglycerides (Myvacet 9-45®)	
Surfactants	
Polysorbate 20 (Tween 20)	6.7
Polysorbate 80 (Tween 80)	15.0
Sorbitan monooleate (Span 80)	4.3
Polyoxyethylene(40) castor oil ether (Cremophor EL 40)	13.0
Polyoxyethylene(40) hydrogenated castor oil (Cremophor RH 40)	12.5
Polyoxyethylene(60) hydrogenated castor oil (Cremophor RH 60)	14.0
Polyoxyethylene(77)-polyoxypropylene(29)-polyoxyethylene(77) copolymer (Pluronic F68)	ca. 29
Polyoxyethylene(19)-polyoxypropylene(43)-polyoxyethylene(19) copolymer (Pluronic P84)	ca. 14
Decaglyceryl monolauric ester	15.5
Hexaglyceryl monolauric ester	14.5
Mono/diglycerides of caprylic/capric acid in glycerol (Capmul MCM®)	5.5–6
Polyglycolized glycerides (Labrafac CM10®)	10
Polyoxyethylated glycerides (Labrafil M 2125 CS®)	3–4
Polyoxyethylated oleic glyceride (Labrafil M 1944CS®)	3–4
Cosurfactants	
Ethanol	
Propanol	
Glycerin	
1,2-Propylene glycol	
Polyethylene glycol (300 or 400)	
Diethyleneglycol monoethyl ether (Transcutol)	

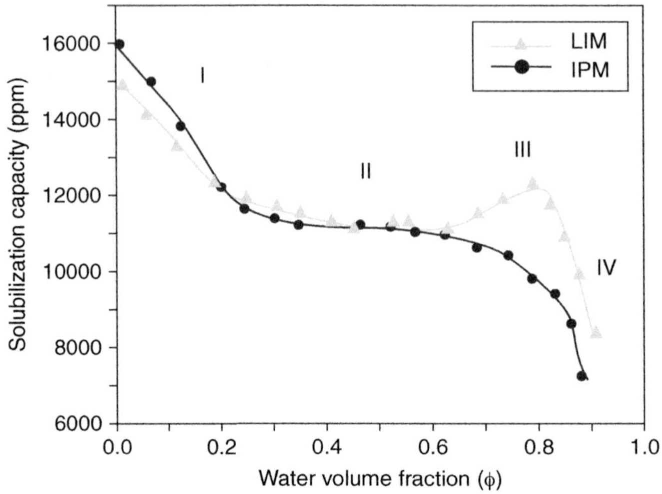

Figure 10.5 *Solubilization capacity (SC) of diclofenac sodium as a function of the water volume fraction (Ø) along the dilution line N60 in the sucrose laurate/ethoxylated mono-di-glyceride/oil/ethanol systems (3/3/2/2) at 25°C. The oils were R(+)-limonene (LIM) and isopropylmyristate (IPM). Adapted from [31].*

respectively. Upon dilution with water, up to 0.2 water volume fraction, the solubilization capacity dropped dramatically from 15,000 to 11,800 in the LIM-based system and from 16,000 to 11,600 ppm in the IPM-based microemulsions, respectively. However, upon further dilution to 0.60 to 0.70 water volume fraction, the solubilization capacity continued to decrease in the system based on IPM while in the system based on LIM, it increased to a maximum, then decreased for water volume fractions above 0.8. All this illustrated that the type of oils influenced the solubilization capacity.

10.5.2 Surfactants

Choice of surfactants is critical for the formulation of microemulsions [30]. The surfactants should favor the microemulsification of the oily phase and should also possess a good solubilizing potential for the drugs. At the same time, because there is a high content of surfactant in the formulation (it requires a high surfactant concentration, usually in the region of 20% and higher), attention should be paid to the toxicity of the surfactants. Therefore, the surfactant chosen must be of the appropriate hydrophile–lipophile character to provide the correct curvature at the interfacial region for the desired microemulsions type, o/w (HLB 8–18), w/o (HLB 3–8) or bicontinuous, and it should lower the interfacial tension to a very small value to aid the dispersion processes during the preparation of the microemulsions.

Pharmaceutically acceptable surfactants are mainly nonionic surfactants. They are insensitive to the changes in pH and electrolytic concentration, also with relatively low toxicity. The surfactants normally used are listed in Table 10.2. Among the nonionic surfactants, sucrose esters and polyoxyethylene alkyl ethers are newly developed surfactants because of their biocompatibility and relative safety [31]. Cho *et al.* [32] developed a new formulation of propofol microemulsions using both polyethylene glycol 660 hydroxystearate (Solutol HS 15) and poloxamers as nonionic surfactants. In preclinical studies, this microemulsion formulation was more tolerable in animals (higher LD_{50}) than Diprivan®. Unlike the Diprivan®, all the blood levels of lipids, triglyceride, cholesterol, low density lipoprotein and high density lipoprotein were consistently in the normal ranges after administration of the propofol microemulsions prepared.

Most commercially available surfactants are composed of a mixture of surfactants. Understanding the behavior of the mixed surfactants in the presence of water and oil is an important issue for the optimal design and development of mixed surfactants-based microemulsions for industrial applications. Fanun [16] studied the diclofenac solubilization capacity of mixed nonionic surfactants microemulsions. Sucrose laurate (L1695, HLB = 16) and ethoxylated mono-di-glyceride (EMDG, HLB = 13.5) were chosen as the surfactants and isopropylmyristate (IPM) was selected as the oil phase. Two systems were prepared to study the effects of mixed surfactants, which were: A, W/L1695/EMDG/IPM; B, W/L1695/EMDG/(IPM + ethanol). Figure 10.6 represents the variation of the total monophasic region (A_T) as a function of EMDG content in the mixed surfactants for the two systems, A and B, investigated. The results showed that the addition of an ethoxylated surfactant to sucrose esters and a short-chain alcohol (ethanol) to the oil phase increases the microemulsion area significantly. For system A, the monomeric solubilities of sucrose

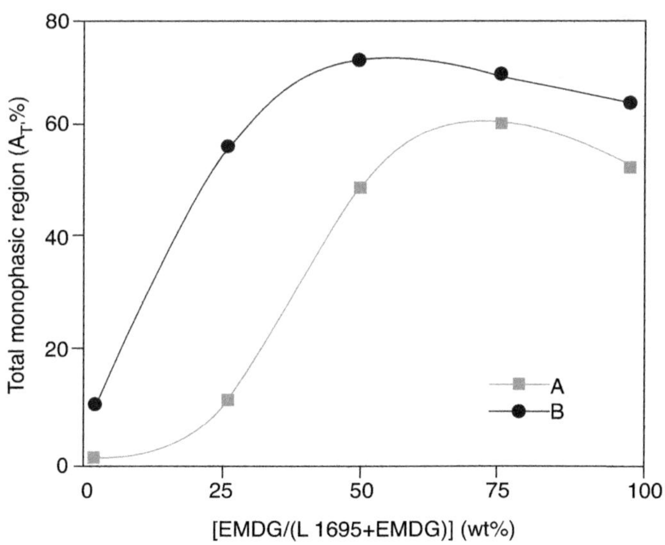

Figure 10.6 *Variation of the total monophasic region (A_T) as a function of the EMDG content in the mixed surfactants for the systems A and B studied. System A: W/L1695/EMDG/IPM; System B: W /L1695/EMDG/(IPM + ethanol). W: water; EMDG: ethoxylated mono-di- glyceride; L1696: sucrose laurate; IPM: isopropylmyristate. Adapted from [16].*

laurate were too small in both water (W) and isopropylmyristate (IPM, oil phase). In the ternary system of W/L1695/IPM, one phase microemulsion was observed at high temperatures ($T > 73°C$) and the no microemulsion phase region was observed in the ternary system W/L1695/IPM at 25°C. When EMDG was added to the ternary system W/L1695/IPM, the phase behavior changed. For EMDG content in the mixture of (EMDG + L1695) equal to 25 wt%, one phase microemulsion region was formed. The microemulsion region appeared after the addition of about 30 wt% of water at a content of mixed surfactants above 60 wt%. The A_T equaled 10%. At equal amounts of EMDG and L1695 in the surfactants mixture (i.e. EMDG content = 50 wt%), the one phase microemulsion region appeared from the first drop of water added. This one phase region extended to a maximum water solubilization (W_m) equal to 35 wt% at the dilution line N60 and the A_T equaled 49%. By increasing the EMDG weight ratio in the mixed surfactants to 75 wt%, the monophasic region extended to a maximum W_m equal to 60 wt% along the dilution line N60. The A_T continued to increase to the value of 60%. In the ternary system W/EMDG/IPM, the total monophasic area (AT) of the microemulsion phase region reached the value of 52%. In system B, similar results were obtained. The A_T was higher than that in system A at each point resulting from the introduction of ethanol (cosurfactant). The change in phase behavior caused by the addition of mixed surfactants and ethanol indicated a synergistic effect for solubilization. The synergism phenomenon might be a result of better interface organization (orientation) of the mixed surfactants around the oil droplet that allowed better interfacial solubilization (enhanced partitioning of the surfactant at the interface).

10.5.3 Cosurfactants

Most of the time, a surfactant alone cannot lower the oil–water interfacial tension sufficiently to yield microemulsions. The addition of an amphiphilic short chain molecule or cosurfactant to bring about the surface tension close to zero is necessary [29]. The most important role of cosurfactants in the formation of microemulsions is to interact with the surfactant monolayers at the interface, thereby affecting their geometric packing, increasing the interfacial fluidity by penetrating the surfactant film and consequently creating a void space among the surfactant molecules.

Short-chain alcohols (C_2 to C_{10}) are often added as cosurfactants to form microemulsions, for example, ethanol, propylene glycol, glycerol, polyethylene glycol 400, polyethylene glycol 600, isopropyl alcohol, isobutyl alcohol and Transcutol P [33]. The number of carbons in the alcohol chain can influence its solubility capacity. S.K. Mehta *et al.* [34] studied the surfactant efficiency (Qm) of the microemulsions composed of Brij 96/ethyl oleate (EO) with different cosurfactants (Figure 10.7). The value of Qm was the highest for butanol (C_4, 0.80) and the lowest for ethanol (C_2, 0.5). EO, when replaced with isopropyl palmitate (IPP) or isopropyl myristate (IPM), showed a decrease in the value of Qm. Low values for pentanol and hexanol were due to the presence of more hydrophobic cosurfactant limiting the flexibility of the surfactant film. The ability of the cosurfactant to penetrate the amphiphilic film decreased as it became more hydrophobic. However, ethanol and propanol had lower Qm values because the shorter-chain alcohols were too soluble in the aqueous phase and therefore ineffective as cosurfactants [34].

Figure 10.8 depicts the variation of water solubility (W_s) with the number of carbons in the alcohol chain (n_c) in the preconcentrate of the microemulsions composed of Brij 96,

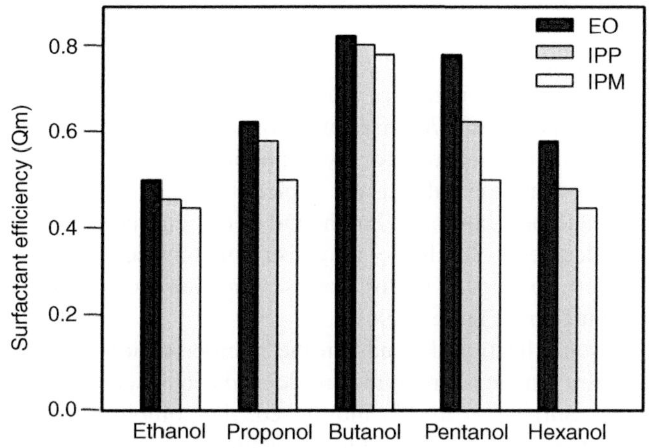

Figure 10.7 *Surfactant efficiency (Qm) as a function of the cosurfactant chain length along three microemulsion formulations with different oils. EO: ethyl oleate; IPP: isopropyl palmitate; IPM: isopropyl myristate. Adapted from [34].*

alcohol and EO/IPP/IPM [34]. As n_c increased, W_s went through maxima. The maximum water solubility was observed for butanol in all three different oils. Since the solubility of n-butanol in water was intermediate, it was expected to partition preferentially toward the interfacial layer. As a result, the increased number of molecules of butanol at the interface increased the flexibility of the interface, and thus helped to maintain the isotropic nature of the microemulsions on dilution without phase separation.

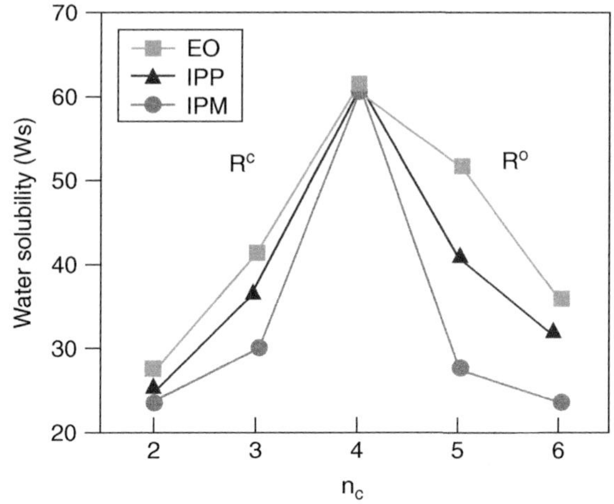

Figure 10.8 *Variation of water solubility (W_s) with alcohol chain length (n_c), for a preconcentrate of Brij 96, alcohol, and three selected oils EO, IPP and IPM. EO: ethyl oleate; IPP: isopropyl palmitate; IPM: isopropyl myristate. Adapted from [34].*

10.5.4 Drugs

Microemulsions are usually considered to be efficient solubilizers of a wide range of lipophilic substances. The solubilizing power of microemulsions depends on their structure. The solubilizing capacity of w/o microemulsions for water-soluble drugs is generally less than that of o/w microemulsions for hydrophobic drugs [35]. Three solubilization sites available in the o/w microemulsions are depicted in Figure 10.9. Hydrophobic compounds are considered to be incorporated in the interior of microemulsions (Figure 10.9a). Some solubilized molecules (solubilizate) may distribute themselves among the surfactant molecules near the surface (Figure 10.9b); poor solubilizate molecules may adsorb at the interface of microemulsions (Figure 10.9c).

Furthermore, the solubilization depends on the microemulsion composition. First, the solubilization of the drug in the microemulsions depends on the concentration of surfactant and cosurfactant. As Surabhi *et al.* [5] introduced, the solubilization volume (V) of a spherical microemulsion droplet is related to the radius of the swollen core of the droplet by $V = AR/3$, where A = total interfacial area; R = radius of the swollen core. The interfacial area A depends on the emulsifier concentration. At constant total interfacial area (or emulsifier concentration), solubilization is directly related to the droplet size and hence the curvature of the interface. As previously noted, the curvature is related to the concentration of cosurfactant.

On the other hand, the character of the drugs cannot be ignored. The effect of the drugs on the microemulsion stability and structure depends on the properties of the drug. For example, for high-melting substances and high molecular mass drugs, solubilization is usually limited. Solubilization is also strongly affected by the surface activity of the drugs. For example, sodium salicylate has been found to significantly alter the stability region of microemulsions made of lecithin (Figure 10.10). Furthermore, Carlfors *et al.* studied the microemulsions prepared by water, isopropyl myristate, and nonionic surfactant mixtures as well as their solubilization for lidocaine. By using NMR self-diffusion measurements, it was found that lidocaine lowered the phase inversion temperature. This

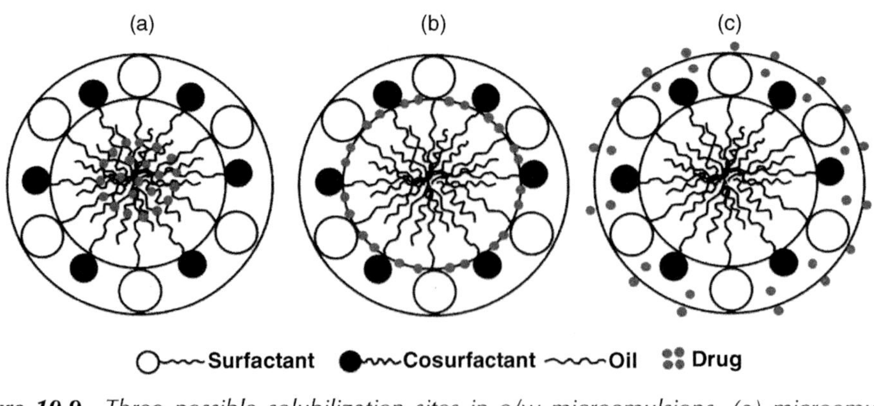

Figure 10.9 *Three possible solubilization sites in o/w microemulsions. (a) microemulsion interior; (b) palisade layer; (c) microemulsion surface.* (See colour plate section.)

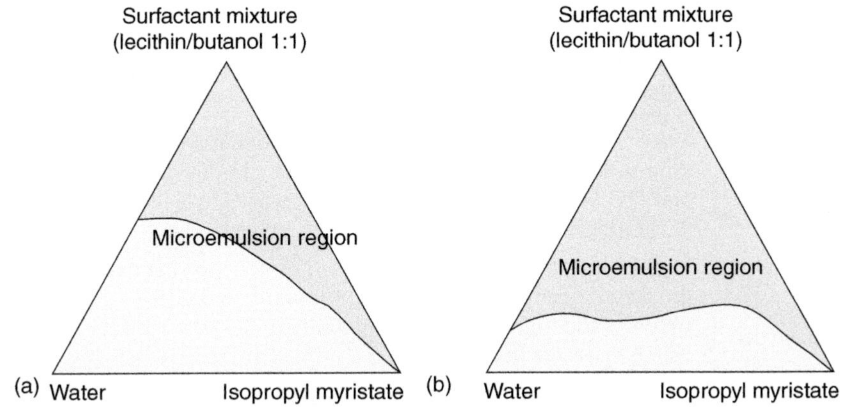

Figure 10.10 *Pseudo-ternary phase diagrams at room temperature of quaternary systems containing (a) lecithin, butanol, isopropyl myristate and water, or (b) lecithin, butanol, isopropyl myristate and a 10 wt% aqueous solution of sodium salicylate. The ratio of lecithin/butanol was 1:1. Adapted from [35].*

might be resulting from the increasing effective oil volume which decreased in the curvature toward the oil as well as the formation of the reversed structure [35].

10.6 Preparation Methods

Because there are so many factors influencing the formation of microemulsions, it is difficult to predict a formulation for a specific drug. Generally, the formulation should be optimized by phase diagrams before preparation of the microemulsions. Since microemulsions are thermodynamically stable, they can be prepared simply by blending oil, water, surfactant and cosurfactant with mild agitation. The usual method of preparing microemulsions is to dissolve the surfactant in oil and add the solution to water with gentle shaking. The solution first becomes translucent and then optically clear after a few seconds. With nonionics, the surfactants may be dissolved in water first. The order of mixing the components is generally not considered to be critical since microemulsions form spontaneously [5]. Microemulsions preparation requires only the availability of the most basic mixing equipment. It is easy to scale up into a pharmaceutical product [15].

10.7 *In Vitro* and *In Vivo* Biological Studies

10.7.1 Microemulsions Used as an Oral Delivery System for Poorly Water-Soluble Compounds

Among the various drug delivery systems, the microemulsion system is considered an ideal solution for the oral delivery of poorly water-soluble compounds. Table 10.3 summarizes some drugs employing the microemulsion formulations to improve their oral bioavailability.

Table 10.3 *Examples of bioavailability enhancement of poorly water-soluble drugs after oral administration of microemulsions.*

No.	Compound	Enhancement character	References
1	Paclitaxel	The *in situ* permeability of paclitaxel through male CD-IGS rat intestine was three- and eleven-fold higher from LBM and CM, respectively, than that from Taxol®; AUC were 1.2-fold, 2.9-fold higher than Taxol®, respectively	36
2	Docetaxel	Higher apical to basolateral transport of docetaxel across the Caco-2 cell monolayers (0.624 μg/cm^2 vs 0.025 μg/cm^2) and higher oral bioavailability in SD rats (34.42% vs 6.63%) than Taxotere®	37
3	Docetaxel	Studies in SD rats showed that the docetaxel-loaded microemulsions (Capryol 90/Cremophor RH 40/Transcutol/water) produced a significantly higher AUC (1689.4 vs 905.23 ng/ml·h) and a longer $t_{1/2\beta}$ (3.49 vs 1.64 h) compared with Taxotere®	38
4	Nitrendipine	The absorption of nitrendipine was improved 7.70-fold and 6.43-fold by employing polyoxyethylene hydrogenated castor oil 60 (HCO60) microemulsions compared with 0.5 wt% methylcellulose suspension in normal and fasted rats, respectively	39
5	Fenofibrate	The optimal formulation consists of 25% Capryol 90, 27.75% Cremophore EL, 9.25% Transcutol P and 38% water (w/w), with a maximum solubility of fenofibrate up to 40.96 mg/ml. The pharmacokinetic study performed in dogs showed that the oral relative bioavailability of fenofibrate microemulsions in ME-3 and ME-4 was 1.63 and 1.30-fold higher than that of the Lipanthy® capsule	40
6	Ibuprofen	The AUC ratio following oral administration of ibuprofen as an O/W microemulsions (MCT/DGMO-C/HCO-40/PBS: 5/1/9/5/80, w/w) to that of suspension (0.5% CMC) was 8.7-fold in fasted SD rats	41
7	Ibuprofen	The optimal microemulsion formulation consists of 17% Labrafil M 1944CS, 28% Cremophor RH40/Transcutol P (3:1, w/w), and 55% water, with a maximum solubility of ibuprofen up to 60.3 mg/ml. The relative bioavailability of ibuprofen in microemulsions was 1.9-fold higher than that of the granule formulation in rats	42
8	Cyclosporin A	The microemulsion was formed by pre-microemulsion concentrate (Cremophor EL®: Transcutol®: Captex 355® (10:5:4)) 0.2 ml dispersing into 1 ml of saline. The C_{max} and AUC of cyclosporin A after oral administration of this cyclosporin A loaded microemulsions were 3.5- and 3.3-fold increased compared with Sandimmun®, respectively. The absolute bioavailability was increased about 3.3- and 1.25-fold compared with Sandimmun® and Sandimmun Neoral®	43

Paclitaxel is an antineoplastic agent with clinical activity against ovarian, breast, non-small cell lung carcinomas and AIDS-related Kaposi's sarcoma, and it acts by stabilizing abnormal microtubule structures [30]. Due to its poor oral systemic availability as a result of its limited aqueous solubility (10.8 μg/ml) and high lipophilicity ($K_{o/w} = 311$), it is administered as an intravenous infusion to patients as Taxol®. Taxol® is a clear to colorless formulation in 5 ml vials containing paclitaxel (6 mg/ml) in Cremophor EL (polyoxyethylated castor oil) and ethanol (1:1, v/v). There are some drawbacks for Taxol® because of the use of surfactant Cremophor EL. Nornoo *et al.* [36] developed a few Cremophor-free oral microemulsions of paclitaxel to enhance its permeability and oral absorption. The *in situ* permeability of paclitaxel through rat intestine was three- and eleven-fold higher from LBM (lecithin: butanol: myvacet oil, 48.5%) microemulsions and CM (capmul: myvacet oil, 20%) microemulsions than that from the control clinical formulation, Taxol® (CE, Cremophor: ethanol), respectively. When paclitaxel was orally administrated to rats in a single oral dose 20 mg/kg, the AUC of paclitaxel in CM was 2.9-fold higher than CE.

Docetaxel is an antitumor drug belonging to the second generation of the taxoid family. Because of its practically insoluble property (4.93 μg/ml in water), high affinity to the multidrug efflux pump P-glycoprotein (P-gp) and hepatic first pass metabolism, the oral bioavailability of docetaxel is limited (<5% in mice). Yin *et al.* [37] prepared docetaxel microemulsions and evaluated the enhancement of oral bioavailability. One of the microemulsion formulations was composed of 29.4% Capryol 90 as oil, 37.3% mixture of surfactant and cosurfactant (Cremophor EL/Transcutol, 2:1) and 33.3% water. Figure 10.11 shows the plasma concentration profiles of docetaxel after oral administration of Taxotere® and the microemulsion formulation prepared, as well as that of intravenous administration of Taxotere® to rats. The AUC of docetaxel in the microemulsion formulation increased 5.2-fold compared with that of docetaxel in the orally administrated Taxotere® (389.43 ng·h/ml

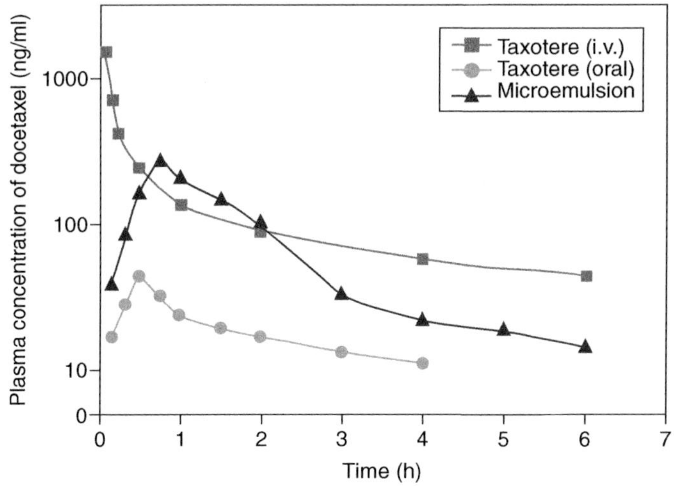

Figure 10.11 *Plasma concentration-time profiles of docetaxel after bolus intravenous injection of Taxotere® (8 mg/kg), oral administration of Taxotere® or microemulsions (equivalent to 10 mg/kg as docetaxel) to rats. Adapted from [37].*

vs. 74.98 ng·h/ml). Moreover, it showed a significantly higher absolute oral bioavailability of docetaxel (34.42%), compared to that of orally administered Taxotere® (6.63%).

Mandal *et al.* [44] developed an oral microemulsions-based drug delivery system for enhancing the bioavailability of Saquinavir, a poorly soluble drug. The optimized microemulsion formulation was Labrafac CM10 (4.0%), Tween 80 (36.0%), polyethylene glycol 400 (9.0%) and distilled water (51%). Results of *in vivo* studies showed significantly higher AUC in microemulsions than that in the suspension group. The relative bioavailability of the drug from microemulsions was found to be 57.7%, which revealed an increase of 10.68 times as compared with the commercially available tablets.

Hu *et al.* [40] reported a microemulsion system for oral administration to improve the solubility and bioavailability of fenofibrate. The optimal formulation consisted of 25% Capryol 90, 27.75% Cremophore EL, 9.25% Transcutol P and 38% water (w/w), with a maximum solubility of fenofibrate up to 40.96 mg/ml. The pharmacokinetic study was performed in dogs and compared with the Lipanthy® capsule. The results showed that microemulsions significantly increased the C_{max} and AUC compared to that of the Lipanthy® capsule. The oral bioavailability of fenofibrate microemulsions (FEN-MEs) in ME-3 (Capryol 90/Cremophore EL and Transcutol P/water: 25:37:38) and ME-4 (Labrafil M 1944 CS/Cremophore EL and Transcutol P/water: 26:39:35) were 1.63 and 1.30-fold higher than that of the capsule.

The reasons why microemulsions increase the oral bioavailability of poorly water-soluble drugs may be illustrated from their *in vitro* and *in vivo* biological characters.

10.7.1.1 Solubilization Capacity Improvement

The combination of surfactants with oils offers an advantage for microemulsions in terms of drug solubilization capacities for lipophilic compounds because of the extra locus for solubilization provided by the oil phase. For example, the solubility of paclitaxel in a myvacet oil containing microemulsions was increased 1389-fold compared to its aqueous solubility [30]. The optimal formulation for fenofibrate microemulsions had a maximum solubility of fenofibrate up to 40.96 mg/ml [40].

When microemulsions are administrated orally, they will be diluted by water in the gastrointestinal tract, which could lead to drug precipitation. Thus, the stability of the solubilization capacity in the diluting process is very important. Yin *et al.* [37] reported that the microemulsions of docetaxel exhibited satisfactory solubilization capacity for at least 24 h, while its solubility in Taxotere® decreased dramatically after 2 h despite going through the same dilution process. On the other hand, microemulsions will be subject to the enzymes in the gastrointestinal tract, resulting in reduced drug solubilization capacity. The *in vitro* digestion study conducted on microemulsions showed that 60–80% of docetaxel remained solubilized in the oil phase following the 30 min. lipolysis process (Figure 10.12). The high solubilization capacity of the microemulsions during both the diluting process and enzymatic hydrolysis process is helpful in the high dispersion and high absorptionof the drug.

10.7.1.2 Improvement of the Dissolving Rate of Low Water Solubility Compounds

The *in vitro* drug release from microemulsions is conducted usually by ultrafiltration, dialysis or the Franz diffusion cell methods [37, 45, 46]. In general, microemulsions can

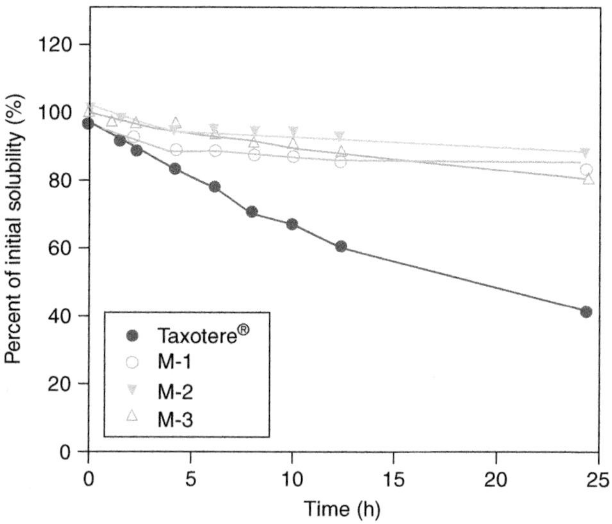

Figure 10.12 *Change of docetaxel solubility in Taxotere® and three microemulsion formulations after 20 times dilution by saline solution at 37°C. M-1: microemulsion formulation composed of oil (Capryol 90) 30.3%, surfactant and cosurfactant (Cremophor EL/propylene glycol, 2:1) 38.6% and water 31.1%. M-2: microemulsion formulation composed of oil (Capryol 90) 28.2%, surfactant and cosurfactant (Cremophor RH40/Transcutol, 2:1) 35.7% and water 36.1%. M-3: microemulsion formulation composed of oil (Capryol 90) 29.4%, surfactant and cosurfactant (Cremophor EL/Transcutol, 2:1) 37.3% and water 33.3%. Adapted from [37].* (See colour plate section.)

increase the drug release better than that of conventional formulations. Mandal [33] developed a microemulsion formulation for the solubility enhancement of flunarizine (Capmul microemulsion). After 8 h, flunarizine released from plain solution, Flunarizine* (Marketed Formulation) and Capmul microemulsions is 56.76%, 51 (19% and 78.53% respectively (Figure 10.13). Capmul microemulsions with faster release of flunarizine as compared to plain solution and Flunarizine* might be due to the smaller droplet size of microemulsions (12.3 nm).

10.7.1.3 Improve the In Vitro *Intestinal Permeability*

The *in vitro* transport study of microemulsions can be evaluated by Caco-2 cell monolayers [37] or intestinal mucosa [33]. Mandal *et al.* [44] reported on the *in vitro* intestinal permeability data of Saquinavir. The drug transported across the intestinal mucosa at a faster rate from the microemulsion system than from the tablets. The total percentage transported was much higher for the microemulsion system than for the tablet dosage form. After 5 hours of transportation, 87.42% of the drug was diffused from the microemulsion system, as compared with 67.13% diffused from the commercially available tablets. Nornoo *et al.* [36] studied a few paclitaxel microemulsion formulations and evaluated their *in situ* permeability of paclitaxel through male CD-IGS rat intestine. They all showed permeation enhancement than that from the control clinical formulation, Taxol® (Figure 10.14). For LBM (lecithin:

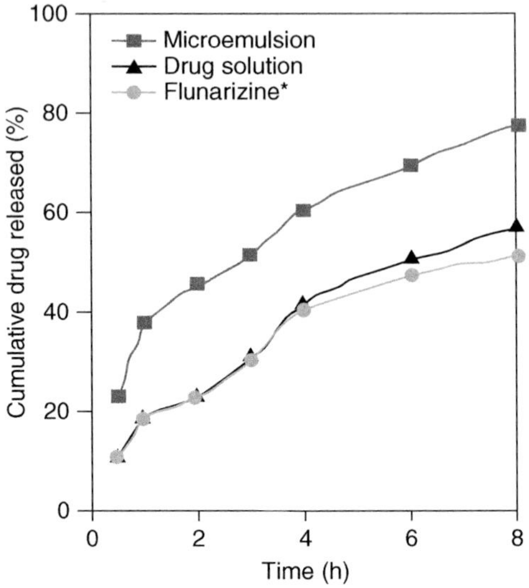

Figure 10.13 In vitro *drug release profile of flunarizine from microemulsions, flunarizine solution and marketed formulation Flunarizine*. Adapted from [33].*

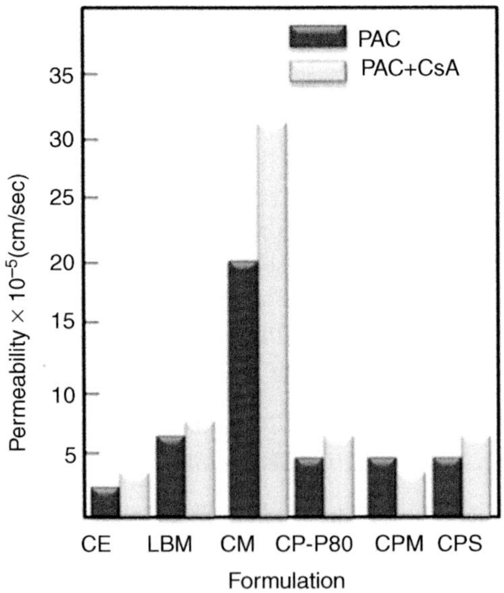

Figure 10.14 *Paclitaxel permeability from five microemulsions (LBM, CM, CP-P80, CPM and CPS) compared to CE with and without CsA. Surfactant or surfactant blends combined with oil for the microemulsions are as follows: LBM, lecithin: butanol (67:33): myvacet oil; CM, capmul MCM: myvacet oil; CP-P80, capryol 90: polysorbate 80 (30:70): myvacet oil; CPM, capmul MCM: polysorbate 80 (30:70): myvacet oil; CPS, centromix CPS: butanol (67:33): myvacet oil. CE: Taxol®. Cyclosporine A (CsA) is the P-gp/CYP3A4 inhibitor. Adapted from [36].*

butanol (67:33)/ myvacet oil/water) and CM (capmul MCM/myvacet oil/water), three- and eleven-fold higher enhancements were obtained, respectively. One of the proposed mechanisms for permeation enhancement is based on the structural and fluidity changes in the mucosal membrane, induced by the surfactants and oils [15]. Lindmark *et al.* [47] reported that medium chain fatty acids (MCFAs) were used to enhance the permeability of mucosal tissues to hydrophilic drugs. The study showed that capric acid induced redistribution of the tight junction proteins ZO-1 and occludin. The other reason may be attributed to the P-gp inhibitory effect of the surfactants or oils.

10.7.1.4 Efflux Inhibitory

A few poorly water-soluble drugs such as paclitaxel and docetaxel are P-gp substances. Microemulsions are not only useful drug delivery systems in terms of drug solubility and stability, but also can act as P-gp/CYP450 inhibitors because of the components used, such as the surfactant polysorbate 20, polysorbate 80, acacia, solutol HL 15, vitamin E TPGS, Cremophor EL, etc. [48,49].

Yin *et al.* [37] reported that the docetaxel permeability coefficient of the microemulsions was significantly higher than that of the control or Taxotere® in the A to B transport (Figure 10.15). In particular, the P_{app} value of docetaxel from one microemulsion was around 43 times higher than that from the control group (0.5 ml of docetaxel (10 μg/ml) in the transport medium containing 4% (v/v) DMSO added as the control) (7.68×10^{-5} cm/s vs

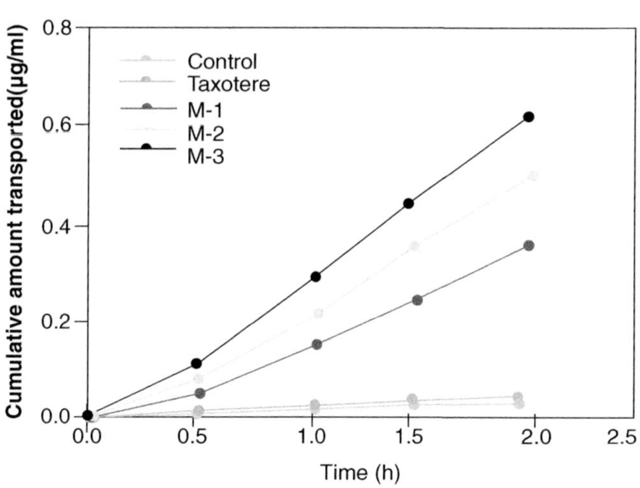

Figure 10.15 *Apical to basolateral transport profiles of docetaxel from Taxotere® or microemulsions across the Caco-2 cell monolayers at 37°C. The control for the transport study was docetaxel (10 μg/ml) in transport medium containing 4% (v/v) DMSO. ME-1: microemulsion formulation composed of oil (Capryol 90) 30.3%, surfactant and cosurfactant (Cremophor EL/ propylene glycol, 2:1) 38.6% and water 31.1%. ME-2: microemulsion formulation composed of oil (Capryol 90) 28.2%, surfactant and cosurfactant (Cremophor RH40/Transcutol, 2:1) 35.7% and water 36.1%. ME-3: microemulsion formulation composed of oil (Capryol 90) 29.4%, surfactant and cosurfactant (Cremophor EL/Transcutol, 2:1) 37.3% and water 33.3%. Adapted from [37].* (See colour plate section.)

0.18×10^{-5} cm/s). In the B to A transport, the permeability coefficient of the microemulsions significantly decreased compared to the control or Taxotere®. However, the efflux of docetaxel from the microemulsions represented by the P_{app} value of B to A was significantly lower than that from the control group (7.43×10^{-5} cm/s vs 10.44×10^{-5} cm/s). On top of that, no significant difference between the A to B and B to A P_{app} values was observed for the microemulsion formulation, implying the suppression of P-gp by the microemulsions. Therefore, in addition to the enhancement of drug solubility by the microemulsion formulation, inhibition of the P-gp efflux system could cause the increase in A to B permeability. The inhibition of P-gp might be the result of the surfactants. As reported, Cremophor EL may inhibit the function of P-gp by affecting membrane fluidity. And, Cremophor could specifically bind to the hydrophobic domain of the P-gp that may change its secondary and/or tertiary structure and reduce its function. Besides Cremophor, the enhanced permeation of docetaxel possibly also comes from the other components of microemulsions. As reported in the literature, the transport P_{app} values of several poorly water-soluble model compounds were greatly improved by Transcutol, and its permeation-enhancing effect on Caco-2 cell monolayers was much stronger than that of same amount of propylene glycol.

10.7.1.5 Lymphatic Transport

The intestinal lymph is a specialized absorption site where highly lipophilic xenobiotics and drugs can gain access to the systemic circulation and avoid potential hepatic first-pass metabolism. Puerarin is a poorly water-soluble drug with low bioavailability in animals and humans upon oral administration. Wu *et al.* [50] prepared puerarin microemulsions and evaluated the contribution of the lymphatic route to the intestinal absorption and transport of puerarin into the systemic circulation. The total amount of the administered dose collected in the thoracic lymph after 8 h for puerarin microemulsion systems was 83.07 ± 7.53 μg, equivalent to 0.06% of the administered dose. And most of the puerarin present in lymph ($85.6 \pm 2.4\%$) was measured in the cumulated floating fractions containing triglyceride-rich particles. Figure 10.16 shows the mean plasma concentration–time profile over a period of 8 h in control and lymph duct-cannulated rats after oral puerarin microemulsions and i.v. injection of puerarin. The puerarin concentrations in plasma were lower after lymph duct cannulation when compared with the non-lymph-cannulated rats. The $AUC_{0\text{–}8\,h}$ for puerarin was 2.5-fold higher in the non-lymph-cannulated rats when compared to lymph duct cannulated rats after oral puerarin microemulsion administration. The $AUC_{0\text{–}8\,h}$ ratio estimated based on the plasma AUC in lymph duct-cannulated rats to an intravenous control was 16% and the $AUC_{0\text{–}8\,h}$ ratio for non-lymph-cannulated rats to an intravenous control was 40%. The result indicates that absorption into the intestinal lymph contributed to around 25% of the systemic bioavailable puerarin. The absorption of drug is mainly intestinal. The components of microemulsions promote the intestinal lymphatic transport of drugs. It is generally unanimously ageed that the gut associated lymphoid tissue (GALT), including the isolated lymphatic follicles or Peyer's patches, is important in the particulate absorption process. Lymphatic transport and association with chylomicrons and other triglyceride-rich lipoproteins may lead to many pharmacokinetic and pharmacodynamic consequences.

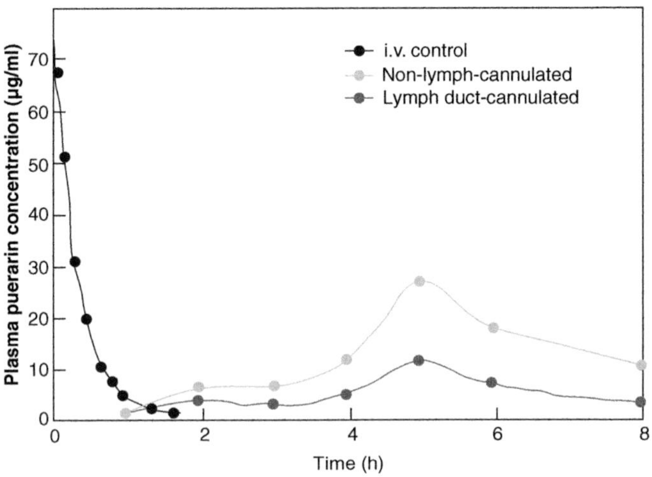

Figure 10.16 *Puerarin plasma concentration–time profile in lymph duct-cannulated and control rats. Puerarin was measured in the plasma of thoracic lymph duct-cannulated rats and of control non-lymph-cannulated rats after oral administration of puerarin microemulsions or in control rats after i.v. injection of puerarin. Adapted from [50].*

10.7.2 Microemulsions Used as a Parenteral Delivery System for Poorly Water-Soluble Compounds

Microemulsions not only deliver poorly water-soluble drugs via oral administration, they have also been developed as a novel vehicle for parenteral delivery of the hydrophobic drugs. Table 10.4 summarizes some examples for the parenteral delivery of the hydrophobic drugs after intravenous administration of microemulsions.

A serious of parenteral paclitaxel microemulsions have been introduced recently. He *et al.* [51] reported parenteral paclitaxel microemulsions containing lecithin, ethanol, poloxamer 188 and a very small amount of Cremophore® EL as compared to that used in Taxol®. The microemulsion formulation showed less hypersensitivity reactions and 1.59-fold AUC increased as compared to Taxol®.

Nornoo *et al.* [38] reported two Cremophor-free microemulsion systems LBMW (lecithin:butanol:myvacet:water/20:10:68:2) and CMW (capmul:myvacet:water/24:74:2). The developed microemulsions were found to be safer as compared to Taxol® with respect to erythrocyte toxicity. Paclitaxel in CMW and LBMW microemulsion systems exhibited sustained release characteristics when compared to that in Taxol®. As Figure 10.17 shows, the initial release rates of paclitaxel from Taxol® (1.22%/h) was faster than that from CMW (0.29%/h) and LBMW (0.05%/h). The cumulative paclitaxel released from Taxol® (78%) was 1.9 and 3.7 times of that from CMW (40.3%) and LBMW (21.2%), respectively. This may be attributed to the fact that paclitaxel is solubilized in the oil component of the microemulsions. This sustained release characteristic would be advantageous in cancer treatment since prolonged paclitaxel levels in the systemic circulation and longer exposure of paclitaxel to cancer cells are desirable. The pharmacokinetic studies indicated that

Table 10.4 Examples of the parenteral delivery of the hydrophobic drugs after intravenous administration of microemulsions.

No.	Compound	Advantages	References
1	Paclitaxel	Less Cremophor EL, less hypersensitivity reactions; Higher AUC (34.98 vs 21.98 μg/ml·h) and longer circulation time for the microemulsions compared with that in the Taxol® group in rats.	51
2	Paclitaxel	Two Cremophor-free microemulsion systems LBMW (lecithin:butanol:myvacet:water) and CMW (capmul:myvacet:water) were developed. In SD rats paclitaxel in LBMW and CMW remained in the systemic circulation five and two times longer and was eight- and three-fold larger volume of distribution than paclitaxel from Taxol®.	30, 38
3	Docetaxel	Animal studies in SD rats showed that the docetaxel-loaded microemulsions (Capryol 90/Cremophor RH 40/Transcutol/water) produced a significantly higher area under the curve (AUC) (1689.4 vs 905.23 ng/ml·h) and a longer $t_{1/2\beta}$ (3.49 vs 1.64 h) compared with Taxotere®.	52
4	Flurbiprofen	Oil-in-water microemulsions prepared with ethyl oleate and Tween 20 solubilized flurbiprofen up to 10 mg/ml, which was an eight-fold increase compared with the flurbiprofen solubility in PBS solution, which implied the reduction of injection volume. The pharmacokinetics of flurbiprofen after intravenous administration of flurbiprofen-loaded microemulsions to rats was not significantly different from that of flurbiprofen in PBS.	53
5	Flurbiprofen	Microemulsions composed of ethyl oleate, lecithin, DSPE-PEG, Flurbiprofen (8:3:1:1.2) and ethanol as a cosolvent. The $t_{1/2}$, AUC and MRT of flurbiprofen incorporated in the microemulsions increased 1.5-, 1.9- and 1.6-fold compared with those of flurbiprofen solution (4.19 vs 2.78; 102.39 vs 54.08; 5.54 vs 3.40); 1.4-, 1.3-, and 1.4-fold compared with those of Lipfen® (4.19 vs 3.04; 102.39 vs 79.12; 5.54 vs 3.92). At the same time reduced uptake into RES-rich organs.	54
6	Amphotericin B	Microemulsion formulation (M-AmB) reduced toxicity compared to the conventional micellar dispersion formulation Fungizone (D-AmB). At a dose of 1 mg/kg D-AmB had a greater AUC (mg/l · h) and a smaller CL (l/h) (32.28 and 0.082, respectively) than M-AmB (21.89 and 0.11, respectively).	55, 56
7	Vincristine	Higher plasma AUC (3.3 vs 0.2 μg/ml per h), tumor AUC (76.5 vs 12.9 μg/ml per h), lower acute toxicity and greater potential antitumor effects than free vincristine in mice.	57
8	All-trans-retinoic acid (ATRA)	Improved the solubility and stability of ATRA while maintaining its pharmacokinetic profile and anti-cancer efficacy.	58
9	Itraconazol	The hemolysis test showed the microemulsion formulation to be nontoxic to red blood cells. The pharmacokinetic profiles showed higher AUC values as compared to the cyclodextrin-based formulation.	59
10	Norcantharidin (NCTD)	The optimal microemulsion formulation consisted of 7% water, 45% soybean lecithin/ethanol (2:1), and ethyl oleate. Pharmacokinetics study showed that NCTD microemulsions had relatively longer circulating time and higher liver targeting efficiency (6.10% vs 3.66%) than NCTD injection in mice.	50

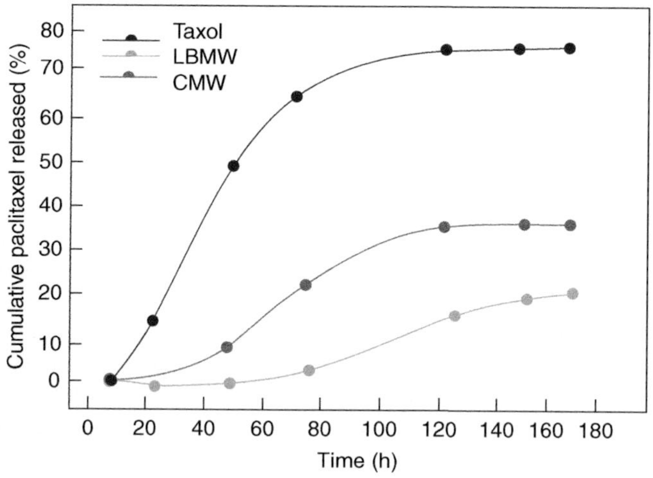

Figure 10.17 *Release profiles of paclitaxel from three formulations. Dialysis bag containing 12 mg/g of paclitaxel which was released into 0.5% sodium lauryl sulphate in water over 168 h. The formulations were Taxol®, LBMW (lecithin:butanol:myvacet:water/20:10:68:2) and CMW (capmul:myvacet:water/24:74:2). Adapted from [50].* (See colour plate section.)

paclitaxel microemulsions exhibited two- to five-fold longer circulation time, and three- to eight-fold wider tissue distribution as compared to that of Taxol®.

Yin *et al.* [52] optimized microemulsion formulations of docetaxel and systematically evaluated them *in vitro* and *in vivo*. In the microemulsion systems selected from the pseudo-ternary phase diagrams, the maximal solubility of docetaxel was as high as 30 mg/ml, which was a significant increase compared with its aqueous solubility. The droplet size of the microemulsions was around 30 nm with a narrow distribution. The *in vitro* cytotoxicity of the three microemulsions against three cancer cell-lines was not significantly different from that of Taxotere®. Additionally, animal studies showed that the docetaxel-loaded microemulsions (Capryol 90/Cremophor RH 40/Transcutol/water) produced a significantly higher area under the curve (AUC) (1689.4 vs 905.23 h·ng/ml) and a longer $t_{1/2\beta}$ (3.49 vs 1.64 h) compared with Taxotere®.

Propofol (2,6-diisopropylphenol) is a water-insoluble intravenous anesthetic that is available as an oil-in-water lipid emulsion with 10% soybean oil (Diprivan®). The oil-in-water macroemulsions are unstable, which may produce potentially dangerous increases in fat particle sizes and hence may cause fatal pulmonary fat embolism [60]. Aquafol is a colorless liquid containing 1% propofol, 8% polyethylene glycol 660 hydroxystearate (Solutol HS 15) as a nonionic surfactant, and 5% tetrahydrofurfuryl alcohol polyethylene glycol ether (Glycofurol) as a cosurfactant. Polyethylene glycol 660 hydroxystearate consists of polyglycol monoesters and diesters of 12-hydroxystearic acid (hydrophobic group) with approximately 30% free polyethylene glycol (hydrophilic group). Tetrahydrofurfuryl alcohol polyethylene glycol is an approximately equimolar mixture of the polymers tetrahydrofurfuryl alcohol ethylene glycol ether (n1-glycofurol) and tetrahydrofurfuryl alcohol diethylene glycol ether (n2-glycofurol). Kim *et al.* [61] evaluated the pharmacokinetics, pharmacodynamics, and safety of Aquafol compared with lipid emulsion propofol (Diprivan®). The results illustrated that both formulations were bioequivalent. The observed time to peak effect after a bolus dose of both formulations was 1.5 min. There was no significant difference between these two treatments in plasma concentration of propofol at loss of consciousness, time to loss of consciousness after a bolus, and time to recovery of consciousness after discontinuation of infusion. The population pharmacokinetics and pharmacodynamics revealed a variety of differences between the two formulations. Aquafol showed similar safety profile with Diprivan®.

Park *et al.* [62] prepared oil-in-water (o/w) microemulsions that were suitable for parenteral administration to improve the solubility of flurbiprofen, a poorly water-soluble drug. The microemulsions were composed of ethyl oleate, Tween 20 and isotonic solution. The maximum solubility of flurbiprofen in the microemulsion system was found to be 10 mg/ml, which was an eight-fold increase compared with the flurbiprofen solubility in PBS solution. Increased solubility of flurbiprofen in this microemulsions implies the reduction of injection volume. The pharmacokinetics of flurbiprofen after intravenous administration of flurbiprofen-loaded microemulsions to rats was not significantly different from that of flurbiprofen in PBS.

10.8 Recent Developments and Future Directions

10.8.1 Develop Cremophor-Free Microemulsions

Cremophor is a series of important surfactants widely adopted in microemulsions resulting in the easy formation of microemulsions. However, Cremophor has been implicated in serious anaphylactic reactions in certain individuals (such as Taxol®), making it necessary to pre-medicate patients with antihistamines and corticosteroids to reduce the frequency of the reactions to an acceptable level. Also, Cremophor may affect red blood cells because they penetrate biological membranes to cause an increase in permeability and cell damage. These effects depend on the quantitative ratio of the surfactant to the membrane. At low surfactant:cell ratios, the erythrocytes are protected against osmotic shock. At higher ratios, hemolysis is induced and a further increase causes solubilization of the membrane. Rouleaux formation of red blood cells and a change in the shape of white blood cells were observed in blood smears in a study of Taxol® on human red blood cells [30].

To diminish the toxic side-effects of Cremophor, biocompatible, Cremophor-free microemulsions have been investigated. As introduced above, with medium chain triglycerides as the oil phase and the biocompatible phospholipids like lecithin as a surfactant, Nornoo *et al.* [30] developed the promising Cremophor-free microemulsions LBMW (lecithin: butanol: myvacet: water/20: 10: 68: 2) and CMW (capmul: myvacet: water/24: 74: 2). The microemulsion system prepared had a droplet size in the 110 nm range and could incorporate 12 mg/g of paclitaxel. The solubility of paclitaxel was increased at least 1389-fold in the microemulsions over its aqueous solubility. These microemulsions utilized biocompatible surfactants such as lecithin and capmul that are used in parenteral emulsions. The presence of the surfactants in the formulation will aid in the even mixing of the formulation with the blood, and facilitate the controlled and extended release of the drug without precipitation. The cytotoxicity of paclitaxel was evidenced by the decrease in cell numbers when treated with these formulations. Insignificant hemolysis was observed at formulation to blood ratios less than 0.025 and the hemolytic potential of the LBMW formulation was lower than the commercially available Taxol®. At the volume ratio of 0.025, 87.6%, 72% and 61% of cells will remain unlysed with LBMW, CMW and Taxol®, respectively. *In vivo* pharmacokinetic studies in male Sprague–Dawley rats after intravenous administration revealed that paclitaxel in LBMW and CMW remained in the systemic circulation 5 and 2 times longer and was 8 and 3 times larger the volume of distribution than paclitaxel from Taxol® [38].

10.8.2 Dried O/W Emulsions for Oral Delivery of Poorly Soluble Drugs

As the most widely used lipid-based formulations, emulsion is used to improve the dissolution rate and increase the bioavailability of poorly water-soluble drugs by eliminating the dissolution step and amplifying the specific surface area. In addition, emulsion can extend the gastric emptying to provide longer dissolution time. Unfortunately, creaming, flocculation, coalescence, and phase separation are often observed in emulsion, which gives physicochemical stability problems during storage. Also, relatively large volume and compliance problems of emulsion also limit its application as the delivery vehicles for oral drugs [63].

Recently, dry emulsion, a formulation containing none or small amounts of nonionic surfactants, has been suggested as one way to circumvent such disadvantages of conventional emulsions. Dry emulsion formulations are prepared by drying o/w emulsions containing soluble solid carriers (e.g. dextrin, hydroxypropyl methylcellulose), or a combination of a surface-active protein and one or more water-soluble components (e.g. sodium caseinate and maltodextrin), or insoluble solid carriers (e.g. magnesium aluminometasilate, colloidal silica) in the aqueous phase. After drying, the solid carriers encapsulate the dispersed lipid phase and form a matrix [64]. The o/w emulsion can be reconstituted from the powder before dosing or on contact with the intestinal fluids, thereby maintaining an improved bioavailability relative to a simple lipid solution or dispersion of the drug. It has been successfully applied as a potential oral drug delivery system for poorly soluble drugs to improve bioavailability, as well as to improve photostability and enzymatic stability [65–67].

The process of solidification of emulsion into dry emulsion can be performed by spray drying, lyophilization, dry absorption or rotary evaporation, etc.

10.8.2.1 Spray-Dried Emulsions

Spray drying is defined as a process by which a liquid solution is sprayed into a hot air chamber to evaporate the volatile fraction. Prior to spray drying, an oil-in-water emulsion can be formulated and spray dried to remove the aqueous phase [67]. The droplet size distribution is affected by the carrier composition, the lipid content, the lipid type and the drying process variables. This technology solves the stability problems associated with classic emulsions (phase separation, contamination by microorganism, etc.) during storage and also helps avoid using harmful or toxic organic solvents. Dry emulsions may be redispersed in water before use [67]. What is more, the solid powder obtained by spray drying can be dry filled into hard shell capsules or alternatively compressed into tablets.

Christensen *et al.* [68] developed a dried emulsion system applying three HPMC types as solid carrier and emulsifier, and fractionated coconut oil as lipid phase. They first prepared the o/w emulsion by homogenizing it in a high speed colloid mill. Then the liquid o/w emulsions were spray dried in a laboratory spray dryer. In order to obtain dry emulsions suitable for tablet processing, they investigated various density increasing agents including dextrin, sucrose, colloidal silicon dioxide, lactose and calcium hydrogen phosphate. By adding sucrose, the reconstitution properties were preserved and a monodisperse droplet size distribution was obtained. By wet granulation using ethanol as a binder, free-flowing and compactable dried emulsions were obtained and simultaneously the reconstitution properties were preserved [69].

Using 5-phenyl-1,2-dithiole-3-thione (5-PDTT) as a model drug, Dollo *et al.* [65] developed a physically stabilized dry emulsion dosage form, in which maltodextrin, sodium caseinate and Miglyol 812 were used as the matrix material, emulsifying agent and oil phase, respectively. They found that the incorporation of 5-PDTT in the oil phase did not affect the surface morphology of the powder, the reconstitution, the droplet size, as well as the drug-releasing properties. Furthermore, such a system allowed a three-fold improvement of 5-PDTT relative bioavailability in rabbits after oral administration, compared with cyclodextrin complex.

Hansen *et al.* [70] prepared directly compressible powders to improve the oral bioavailability of a poorly water-soluble drug (Lu 28-179) by the spray drying method. They first prepared o/w emulsions. The excipients for the power contained a water-soluble carrier (sugar), a water-insoluble carrier (magnesium alumino-metasilicate) and a combined emulsifier and film-forming agent (gelatine). Lu 28-179 was dissolved in medium-chain triglycerides (MCT) to obtain the lipid phase. Trehalose dihydrate and gelatine were dissolved in water containing the NaOH added during pH adjustment. Following mixing of the two phases and homogenization, Neusilin US2 was then added. The powder was obtained by spray drying the o/w emulsions in a pilot plant spray dryer. They also compared the oral bioavailability of the directly compressible dry emulsion (DC-DE) as a tablet, the HPMC-DE powder and a simple lipid solution. The T_{max} of MCT solution, HPMC-DE, DC-DE were 2.0, 3.3, 3.5 h, respectively. C_{max} were 9.61, 18.2, 12.7 nmol/l, and $AUC_{0-\infty}$ were 104, 187, 160 nmol h/l, respectively. The absolute bioavailability decreased in the order of HPMC-DE (0.11), DC-DE (0.10) and MCT solution (0.06). Thus, tablets prepared using the directly compressible dry emulsion may be an alternative to the cohesive HPMC dry emulsion powder or a liquid solution for oral delivery of a poorly water-soluble drug.

Amlodipine, with poor photostability and aqueous solubility, was chosen by Jang *et al.* [66] to develop spray-dried dry emulsions, employing Labrafil M 1944 CS and dextrin as oil phase and matrix material, respectively. Dispersing the dry emulsion in distilled water formed an emulsion with a mean droplet size 1.4-fold larger than that of the homogenized emulsion before spray-drying (0.24 ± 0.30 μm vs 0.17 ± 0.02 μm). Comparing the dry emulsion formulation and plain powder, 94.4% vs 33.1% of amlodipine remained intact after 24 h UV irradiation. After oral administration to rats, dry emulsion formulation resulted in 2.6- and 2.9-fold higher C_{max} and $AUC_{0-24\,h}$ of amlodipine compared with amlodipine powder.

10.8.2.2 Lyophilized Dry Emulsions

Lyophilized dry emulsion tablets share the advantages of freeze-dried dosage forms, such as rapid reconstitution, good preservation and high stability. The tablets also disintegrate rapidly in the mouth upon contact with saliva and therefore do not need to be swallowed, which usually results in improved patients' compliance and acceptability [71].

Ahmed *et al.* [71] developed a fast-disintegrating lyophilized dry emulsion tablet that enhanced the *in vitro* dissolution and *in vivo* absorption of griseofulvin. They used a gelatin solution (2%, w/v) as the water phase, medium chain triglycerides (Miglyol) or sesame oil as the oil phase, HPMC or a blend of Tween 80/Span 80 as emulsifers. Griseofulvin was added to the oil phase to obtain a suspension of griseofulvin in oil. The emulsifiers (HPMC or a blend of Tween 80/Span 80 in the ratio 4:1) were added to the aqueous phase. The oil phase was then added to the water phase and homogenized. The resulting emulsions were poured into a tablet blister pack, transferred to a freezer at −22°C. The frozen tablets were placed in a lyophilizer for 24 h using a Novalyphe-NL 500 freeze dryer with a condenser temperature of −45°C and a pressure of 7×10^{-2} mbar. Results obtained from dissolution studies showed that the lyophilized dry emulsion tablets of griseofulvin improved the dissolution rate of griseofulvin compared to the plain drug. After a single oral dose administration in four healthy subjects using a randomized cross-over design, it showed that the absorption rate of griseofulvin from the lyophilized dry emulsion tablet was faster than that from the reference tablet and had significantly higher C_{max} (more than three times) and earlier t_{max} by 4 h. The extent of absorption expressed by AUC was 85% larger compared to the commercial tablet.

Corveleyn *et al.* [72] concluded that maltodextrin was a useful excipient in the formulation of lyophilized dry emulsion. Using hydrochlorthiazide (HCT) as a model drug, lyophilize-dried emulsion tablets made of maltodextrin and medium chain triglycerides were found to release 35.1% HCT after 10 min. compared to 24.1% from a conventional tablet.

Margulis-Goshen *et al.* [73] developed simvastatin nanoparticles from microemulsions and prepared the nanoparticles into tablets. The method was based on the following steps:

1. The active pharmaceutical molecule was dissolved in a water-immiscible solvent that had a high evaporation rate.
2. The microemulsions were formed by the addition of proper surfactants, cosurfactants, and water to this organic solution.
3. All the solvents (water and organic) were evaporated by lyophilizing. The resultant particles of the poorly soluble drug had nanometric dimensions and high wettability, which makes them easily re-dispersible in water.

The results showed that after freeze-drying more than 95% of the drug was present in amorphous particles, smaller than 100 nm. The tablets containing the flakes of simvastatin nanoparticles showed tremendous enhancement in dissolution profile compared with conventional tablets. X-ray diffraction revealed that the resulting flakes of simvastatin nanoparticles were initially amorphous, but a slow crystallization process took place when the product was stored at room temperature.

10.8.2.3 Dry Absorbed Emulsions

A dry absorbed emulsion is made up of a w/o emulsion whose two phases are stabilized by two adsorbent powders with suitable polarity. The first powder adsorbs the aqueous phase and the second absorbs the oily phase. The finished form, without removing water, is a free-flowing powder whose particle size ranges from 100 to 1000 nm [74]. It can be of interest, because of its manufacturing process (without organic solvent), its stability and its sustained-release effect.

Chambin *et al.* [74, 75] developed a dry absorbed emulsions system, including theophylline as model drug, castor oil as the lipid phase, calcium stearate as emulsifier, hydrophilic silica (Aerosil 380®) and hydrophobic silica (Aerosil R974®) as adsorbent powders. They obtained a free-flowing powder with nonporous particles of size from 125 to 710 μm, with a small specific surface area and a spherical shape. They also analyzed their kinetic release and found that discrete dry absorbed emulsions particles showed an extended release, in comparison with the native drug powder, which depended on both drug solubility in the medium and particle size. Furthermore, the Higuchi model fitted well in theophylline release, confirming that dry absorbed emulsions could be classified as a matrix with a delivery mechanism controlled by drug diffusion. Yan *et al.* [76] prepared dry absorbed emulsions loaded with methotrexate. The releasing assay showed an obviously sustained release of methotrexate from the dry absorbed emulsions.

10.8.2.4 Rotary Evaporation (Lipid-Based Emulsifiable Glass)

Shively *et al.* [77] described a solid-state lipid-based emulsifiable glass system in which no emulsifying agent was utilized and high temperature was avoided. Without the emulsifying agent, common toxicities and environmental concerns were not problematic. However, it also meant that conventional theories of emulsification do not completely describe this system. Contrary to conventional emulsion theory, the addition of water to the glass spontaneously results in the self-emulsification of the oil phase with the formation of either an o/w or a w/o/w emulsion.

The system was prepared by adding a matrix material (sucrose or maltodextrin) and an oil phase to a round-bottom vacuum flask. Double-distilled deionized water was then added. The flask was fitted to a rotary evaporator and partially immersed in a water bath (60°C), while a vacuum was applied (5 mm Hg). The mixture was slowly rotated until a dry film or foam was produced. The particle size characteristics indicated a unimodal distribution of particles with an average droplet diameter of approximately 1.5 μm [77]. Porter *et al.* [78] extended the emulsifiable glass concept to surfactant-enhanced formulations. With cyclosporin as the model drug, the surfactant-enhanced emulsifiable glass formulations exhibited marked improvements in drug-loading capacity, redispersion characteristics and ease of processing. The preparing method included a pre-emulsification step where the oil was initially emulsified in the sucrose solution prior to rotary evaporation and glass formation. Compared with the emulsifiable glass system, the surfactant-enhanced emulsifiable glass system offered great advantages: (i) rapid, efficient and complete emulsification; (ii) a four-fold increase in drug-loading capacity; and (iii) a two-fold decrease in processing time [78].

10.8.3 Self-Microemulsifying Drug Delivery System (SMEDDS)

A self-microemulsifying drug delivery system (SMEDDS) is an anhydrous system of microemulsions. It has also been referred to as microemulsion preconcentrate by some researchers. It is composed of oil, surfactant and cosurfactant and has the ability to form o/w microemulsions when dispersed in aqueous phase under gentle agitation. The agitation required for the self-emulsification comes from the stomach and intestinal motility. The nanosized droplets have very high surface to volume ratios which are able to efficiently solubilize the drug. The drug is released in a more reproducible manner which will become less dependent on the gastrointestinal physiology and the fed/fasted state of the patient. SMEDDS will be discussed in Chapter 7 of this volume.

References

1. C.J.H. Porter, C.W. Pouton, J.F. Cuine, W.N. Charman, Enhancing intestinal drug solubilisation using lipid-based delivery systems. *Adv Drug Deliv Rev*, **60**, 673–691 (2008).
2. S. Talegaonkar, A. Azeem, F.J. Ahmad, R.K. Khar, S.A. Pathan, Z.I. Khan, Microemulsions: A novel approach to enhanced drug delivery. *Recent Pat Drug Deliv Formul*, **2**, 238–257 (2008).
3. T.P. Hoar, J.H. Schulman, Transparent water-in-oil dispersions: the oleopathic hydromicelle. *Nature*, **152**, 102–103 (1943).
4. J.H. Schulman, W. Stoeckenius, L.M. Prince, Mechanism of formation and structure of microemulsions by electron microscopy. *J Phys Chem*, **63**, 1677–1680 (1959).
5. S. Khanna, O.P. Katare, A. Nasa, A. Garg, Microemulsions: Developmental aspects. *Research Journal of Pharmaceutical, Biological and Chemical Sciences*, **1**, 683–706 (2010).
6. I. Danielsson, B. Lindman, The definition of microemulsion. *Colloids and Surfaces*, **3**, 391–392 (1981).
7. J.M. Gutiérrez, C. González, A. Maestro, I. Solè, C.M. Pey, J. Nolla, Nano-emulsions: New applications and optimization of their preparation. *Curr Opin Colloid Interface Sci*, **13**, 245–251 (2008).
8. C. Solans, P. Izquierdo, J. Nolla, N. Azemar, M.J. Garcia-Celma, Nano-emulsions. *Curr Opin Colloid Interface Sci*, **10**, 102–110 (2005).
9. N. Anton, T.F. Vandamme, Nano-emulsions and Micro-emulsions: Clarifications of the critical differences. *Pharm Res*, **28**, 978–985 (2011).
10. M.C. Jones, J.C. Leroux, Polymeric micelles: A new generation of colloidal drug carriers. *Eur J Pharm Biopharm*, **48**, 101–111 (1999).
11. P.P. Constantinides, Lipid microemulsions for improving drug dissolution and oral absorption: Physical and biopharmaceutical aspects. *Pharm Res*, **12**, 1561–1572 (1995).
12. T.F. Tadros, Microemulsions: An overview, in: *Proc. 4th Int. Symp. Surfactants in Solution*, K.L. Mittal, B. Lindman (eds), New York: Plenum, pp. 1501–1532 (1984).
13. E. Ruckenstein, J.C. Chi, Stability of microemulsions. *J Chem Soc Faraday Trans 2*, **71**, 1690–1707 (1975).
14. M.L. Robbins, Theory of microemulsion. *Journal of AIChE Symp Ser*, **40A**, 1–17 (1974).
15. P.K. Ghosh, R.S.R. Murthy, Microemulsions: A potential drug delivery system. *Current Drug Delivery*, **3**, 167–180 (2006).
16. M. Fanun, Conductivity, viscosity, NMR and diclofenac solubilization capacity studies of mixed nonionic surfactants microemulsions. *J Mol Liq*, **135**, 513 (2007).

17. T. Tadros, P. Izquierdo, J. Esquena, C. Solans, Formation and stability of nanoemulsions. *Adv Colloid Interface Sci*, **108–109**, 303–318 (2004).
18. S. Zargar-Shoshtari, J. Wen, R.G. Alany, Formulation and physicochemical characterization of Imwitor 308 based self microemulsifying drug delivery systems. *Chem Pharm Bull*, **58**, 1332–1338 (2010).
19. V. Bali, M. Ali, J. Ali, Study of surfactant combinations and development of a novel nanoemulsion for minimising variations in bioavailability of ezetimibe. *Colloids and Surfaces B: Biointerfaces*, **76**, 410–420 (2010).
20. V. Klang, N.B. Matskob, C. Valenta, F. Hofer, Electron microscopy of nanoemulsions: An essential tool for characterisation and stability assessment. *Micron*, article in press (2011).
21. J. Kuntsche, J.C. Horst, H. Bunjes, Cryogenic transmission electron microscopy (cryo-TEM) for studying the morphology of colloidal drug delivery systems. *Int J Pharm*, **417**, 120–137 (2011).
22. V. Klang, N. Matsko, A.M. Zimmermann, E. Vojnikovic, C. Valenta, Enhancement of stability and skin permeation by sucrose stearate and cyclodextrins in progesterone nanoemulsions. *Int J Pharm*, **393**, 152–160 (2010).
23. S.P. Moulik, B.K. Paul, Structure, dynamics and transport properties of microemulsions. *Adv Colloid Interface Sci*, **78**, 99–195 (1998).
24. M.S. Leaver, U. Olsson, H. Wennerström, R. Strey, U. Würz, Phase behaviour and structure in a non-ionic surfactant-oil-water mixture. *J Chem Soc Faraday Trans*, **91**, 4269–4274 (1995).
25. M. Tomšič, F. Podlogar, M. Gašperlin, M. Bešter-Rogač, A. Jamnik, Water-Tween 40®/Imwitor 308®-isopropyl myristate microemulsions as delivery systems for ketoprofen: Small-angle X-ray scattering study. *Int J Pharm*, **327**, 170–177 (2006).
26. A. Spernath, A. Aserin, L. Ziserman, D. Danino, N. Garti, Phosphatidylcholine embedded microemulsions: Physical properties and improved Caco-2 cell permeability. *J Control Release*, **119**, 279–290 (2007).
27. H.L. Rosano and M. Clausse (eds), *Microemulsion Systems*. New York: Marcel Dekker, pp. 129–144 (1987).
28. A. Graf, E. Ablinger, S. Peters, A. Zimmer, S. Hook, T. Rades, Microemulsions containing lecithin and sugar-based surfactants: Nanoparticle templates for delivery of proteins and peptides. *Int J Pharm*, **350**, 351–360 (2008).
29. A.A. Date, M.S. Nagarsenker, Parenteral microemulsions: An overview. *Int J Pharm*, **355**, 19–30 (2008).
30. A.O. Nornoo, D.W. Osborne, D.S.L. Chow, Cremophor-free intravenous microemulsions for paclitaxel I: Formulation, cytotoxicity and hemolysis. *Int J Pharm*, **349**, 108–116 (2008).
31. M. Fanun, Oil type effect on diclofenac solubilization in mixed nonionic surfactants microemulsions. *Colloids and Surfaces A: Physicochem Eng. Aspects*, **343**, 75–82 (2009).
32. J. Cho, J.C. Cho, P. Lee, M. Lee, E. Oh, Formulation and evaluation of an alternative triglyceride-free propofol microemulsion. *Arch Pharm Res*, **33**, 1375–1387 (2010).
33. S. Mandal, S.D. Mandal, N. Surti, V.B. Patel, Development of microemulsion formulation for the solubility enhancement of flunarizine. *Der Pharmacia Lettre*, **2**, 227–236 (2010).

34. S.K. Mehta, G. Kaur, R. Mutneja, K.K. Bhasin, Solubilization, microstructure, and thermodynamics of fully dilutable U-type Brij Microemulsion. *J Colloid Interface Sci*, **338**, 542–549 (2009).
35. P. Kumar (ed.), *Handbook of Microemulsion Science and Technology*. New York: Marcel Dekker, Inc., pp. 755–771 (1999).
36. A.O. Nornoo, H. Zheng, L.B. Lopes, B. Johnson-Restrepo, K. Kannan, R. Reed, Oral microemulsions of paclitaxel: In situ and pharmacokinetic studies. *Eur J Pharm Biopharm*, **71**, 310–317 (2009).
37. Y.M. Yin, F.D. Cui, C.F. Mu, M.K. Choi, J.S. Kim, S.J. Chung, C.K. Shim, D.D. Kim, Docetaxel microemulsion for enhanced oral bioavailability: Preparation and in vitro and in vivo evaluation. *J Control Release*, **140**, 86–94 (2009).
38. A.O. Nornoo, D.S.L. Chow, Cremophor-free intravenous microemulsions for paclitaxel II. Stability, in vitro release and pharmacokinetics. *Int J Pharm*, **349**, 117–123 (2008).
39. K. Kawakami, T. Yoshikawa, T. Hayashi, Y. Nishihara, K. Masuda, Microemulsion formulation for enhanced absorption of poorly soluble drugs II. In vivo study. *J Control Release*, **81**, 75–82 (2002).
40. L. Hu, H. Wu, F. Niu, C. Yan, X. Yang, Y. Jia, Design of fenofibrate microemulsion for improved bioavailability. *Int J Pharm*, **420**, 251–255 (2011).
41. H. Araya, M. Tomita, M. Hayashi, The novel formulation design of O/W microemulsion for improving the gastrointestinal absorption of poorly water soluble compounds. *Int J Pharm*, **305**, 61–74 (2005).
42. L. Hu, J. Yang, W. Liu, L. Li, Preparation and evaluation of ibuprofen-loaded microemulsion for improvement of oral bioavailability. *Drug Delivery*, **18**, 90–95 (2011).
43. Z. Gao, H. Choi, H. Shin, K. Park, S.J. Lim, K.J. Hwang, C.K. Kim, Physicochemical characterization and evaluation of a microemulsion system for oral delivery of cyclosporin. *A Int J Pharm*, **161**, 75–86 (1998).
44. S. Mandal, S.S. Mandal, N. Surti, V.B. Patel, Design and development of saquinavir microemulsion for the oral bioavailability enhancement. *International Journal of PharmTech Research*, **1**(4), 1442–1448 (2009).
45. Z. Zeng, G. Zhou, X. Wang, E. Z. Huang, X. Zhan, J. Liu, S. Wang, A. Wang, H. Li, X. Pei, T. Xie, Preparation, characterization and relative bioavailability of oral elemene o/w microemulsion. *Int J Nanomed*, **5**, 567–572 (2010).
46. A.R. Dixit, S.J. Rajput, S.G. Patel, Preparation and bioavailability assessment of SMEDDS containing valsartan. *AAPS PharmSciTech*, **11**, 314–321 (2010).
47. T. Lindmark, Y. Kimura, P. Artursson, Absorption enhancement through intracellular regulation of tight junction permeability by medium chain fatty acids in Caco-2 cells. *J Pharmacol Exp The*, **284**, 362–369 (1998).
48. Y.L. Lo, C.Y. Hsu, J.D. Huang, Comparison of effects of surfactants with other MDR reversing agents on intracellular uptake of epirubicin in Caco-2 cell line. *Anticancer Res*, **18**, 3005–3009 (1998).
49. L.E. Buckingham, M. Balasubramanian, R.M. Emanuele, K.E. Clodfelter, J.S. Coon, Comparison of solutol HS 15, Cremophor EL and novel ethoxylated fatty acid surfactants as multidrug resistance modification agents. *Int J Cancer*, **62**, 436–442 (1995).

50. H. Wu, A. Zhou, C. Lu, L. Wang, Examination of lymphatic transport of puerarin in unconscious lymph duct-cannulated rats after administration in microemulsion drug delivery systems. *Eur J Pharm Sci*, **42**, 348–353 (2011).
51. L. He, G. Wang, Q. Zhang, An alternative paclitaxel microemulsion formulation: Hypersensitivity evaluation and pharmacokinetic profile. *Int J Pharm*, **250**, 45–50 (2003).
52. Y. Yin, F. Cui, C. Mu, S.J. Chung, C.K. Shim, D.D. Kim, Improved solubility of docetaxel using a microemulsion delivery system: Formulation optimization and evaluation. *Asian J Pharm Sci*, **4**, 331–339 (2009).
53. K.M. Park, C.K. Kim, Preparation and evaluation of flurbiprofen-loaded microemulsion for parenteral delivery. *Int J Pharm*, **181**, 173–179 (1999).
54. K.M. Park, M.K. Lee, K.J. Hwang, C.K. Kim, Phospholipid-based microemulsions of flurbiprofen by the spontaneous emulsification process. *Int J Pharm*, **183**, 145–154 (1999).
55. B. Brime, P. Frutos, P. Bringas, A. Nieto, M.P. Ballesteros, G. Frutos, Comparative pharmacokinetics and safety of a novel lyophilized amphotericin B lecithin-based oil-water microemulsion and amphotericin B deoxycholate in animal models. *J Antimicrob Chemoth*, **52**, 103–109 (2003).
56. B. Brime, G. Molero, P. Frutos, G. Frutos, Comparative therapeutic efficacy of a novel lyophilized amphotericin B lecithin-based oil-water microemulsion and deoxycholate-amphotericin B in immunocompetent and neutropenic mice infected with Candida albicans. *Eur J Pharm Sci*, **22**, 451–458 (2004).
57. J. Wang, K. Takayama, T. Nagai, Y. Maitani, Pharmacokinetics and antitumor effects of vincristine carried by microemulsions composed of PEG-lipid, oleic acid, vitamin E and cholesterol. *Int J Pharma*, **251**, 13–21 (2003).
58. S.R Hwang, S.J. Lim, J.S. Park, C.K. Kim, Phospholipid-based microemulsion formulation of all-trans-retinoic acid for parenteral administration. *Int J Pharm*, **276**, 175–183 (2004).
59. Y.S. Rhee, C.W. Park, T.Y. Nam, Y.S. Shin, S.C. Chi, E.S. Park, Formulation of parenteral microemulsion containing itraconazole. *Arch Pharm Res*, **30**, 114–123 (2007).
60. L. Zhang, X. Sun, Z.R. Zhang, An investigation on liver-targeting microemulsions of norcantharidin. *Drug Delivery*, **12**, 289–295 (2005).
61. K.M. Kim, B.M. Choi, S.W. Park, S.H. Lee, L.V. Christensen, J. Zhou, B.H. Yoo, H.W. Shin, K.S. Bae, S.E. Kern, S.H. Kang, G.J. Noh, Pharmacokinetics and pharmacodynamics of propofol microemulsion and lipid emulsion after an intravenous bolus and variable rate infusion. *Anesthesiology*, **106**, 924–934 (2007).
62. K.M. Park, C.K. Kim, Preparation and evaluation of flurbiprofen-loaded microemulsion for parenteral delivery. *Int J Pharm*, **181**, 173–179 (1999).
63. J. Zhang, Y. Gao, S. Qian, X. Liu, H. Zu, Physicochemical and pharmacokinetic characterization of a spray-dried malotilate emulsion. *Int J Pharm*, **414**, 186–192 (2011).
64. T. Hansen, P. Holm, M. Rohde, K. Schultz, In vivo evaluation of tablets and capsules containing spray-dried o/w-emulsions for oral delivery of poorly soluble drugs. *Int J Pharm*, **293**, 203–211 (2005).
65. G. Dollo, P.L. Corre, A. Guerin, F. Chevanne, J.L. Burgot, R. Leverge, Spray-dried redispersible oil-in-water emulsion to improve oral bioavailability of poorly soluble drugs. *Eur J Pharm Sci*, **19**, 273–280 (2003).

66. D.J. Jang, E.J. Jeong, H.M. Lee, B.C. Kim, S.J. Lim, C.K. Kim, Improvement of bioavailability and photostability of amlodipine using redispersible dry emulsion. *Eur J Pharm Sci*, **28**, 405–411 (2006).
67. Z. Ge, X. Zhang, L. Gan, Y. Gan, Redispersible, dry emulsion of lovastatin protects against intestinal metabolism and improves bioavailability. *Acta Pharmacol Sin*, **29**, 990–997 (2008).
68. K.L. Christensen, G.P. Pedersen, H.G. Kristensen, Technical optimisation of redispersible dry emulsions. *Int J Pharm*, **212**, 195–202 (2001).
69. K.L. Christensen, G.P. Pedersen, H.G. Kristensen, Preparation of redispersible dry emulsions by spray drying. *Int J Pharm*, **212**, 187–194 (2001).
70. T. Hansen, P. Holm, K. Schultz, Process characteristics and compaction of spray-dried emulsions containing a drug dissolved in lipid. *Int J Pharm*, **287**, 55–66 (2004).
71. I.S. Ahmed, M.H. Aboul-Einien, In vitro and in vivo evaluation of a fast-disintegrating lyophilized dry emulsion tablet containing griseofulvin. *Eur J Pharm Sci*, **32**, 58–68 (2007).
72. S. Corveleyn, J.P. Remon, Formulation of a lyophilized dry emulsion tablet for the delivery of poorly soluble drugs. *Int J Pharm*, **166**, 65–74 (1998).
73. K. Margulis-Goshen, S. Magdassi, Formation of simvastatin nanoparticles from microemulsion. *Nanomedicine: Nanotechnology, Biology, and Medicine*, **5**, 274–281 (2009).
74. O. Chambin, C. Bellone, D. Champion, M.H. Rochat-Gonthier, Y. Pourcelot, Dry adsorbed emulsion: 1. Characterization of an intricate physicochemical structure. *J Pharm Sci*, **89**, 991–999 (2000).
75. O. Chambin, V. Bérard, M.H. Rochat-Gonthier, Y. Pourcelot, Dry adsorbed emulsion: 2. Dissolution behaviour of an intricate formulation. *Int J Pharm*, **235**, 169–178 (2002).
76. S. Yan, D. Su, C. Pi, A study on the preparation and properties of the methotrexate dry adsorbed emulsion. *Journal of Shenyang Pharmaceutical University*, **16**(4), 258–261 (1999).
77. M.L. Shively. Characterization of oil-in-water emulsions prepared from solid-state emulsions: Effect of matrix and oil phase. *Pharm Res*, **10**, 1153–1156 (1993).
78. C.J.H. Porter, S.A. Charman, R.D. Williams, M.V. Bakalova, W.N. Charman, Evaluation of emulsifiable glasses for the oral administration of cyclosporin in beagle dogs. *Int J Pharm*, **141**, 227–237 (1996).

11

Hot Melt Extrusion: A Process Overview and Use in Manufacturing Solid Dispersions of Poorly Water-Soluble Drugs

Shu Li, David S. Jones and Gavin P. Andrews

11.1 Introduction: Present Challenges to Oral Drug Delivery

Over the past twenty years the way in which new drug entities are developed and selected has changed dramatically. Principally, the emergence of high throughput screening (HTS) and combinatorial chemistry has resulted in a significant increase in the volume of new candidate compounds having acceptable therapeutic activity [1, 2]. Interestingly, during this period, the number of drug compounds with poor aqueous solubility has risen considerably and consequently there now exists a large number of potentially useful active agents in pharmaceutical pipelines that have poor bioavailability.

The Biopharmaceutical Classification System (BCS) categorizes a drug candidate into one of four categories depending upon solubility and permeability. Drugs are regarded as possessing 'low solubility' (BCS Classes II and IV) if the highest dose is insoluble in 250 mL of aqueous media over the pH range of 1∼7.5. Additionally, if such compounds exhibit high absorption following dissolution, the drug candidate may be regarded as having a bioavailability limited by aqueous dissolution. Such compounds are classified as BCS Class II [3]. Currently it is estimated that more than 70% of new chemical entities are BCS Class II and have a bioavailability limited by poor aqueous dissolution.

In general, solubility issues and associated bioavailability problems can be overcome through formulation design. In principle, dissolution of a crystalline compound

Drug Delivery Strategies for Poorly Water-Soluble Drugs, First Edition. Edited by Dennis Douroumis and Alfred Fahr.

is the result of the disruption of the crystal lattice, solvation or hydration of the solute molecules, and finally the accommodation of each solute molecule in to solvent. Thus, an amorphous counterpart or any other metastable form (polymorph) that possesses a less-ordered crystal arrangement serves as a method to promote apparent solubility and dissolution rate, due to the lowered energy barrier. Additionally, as clearly identified in the Noyes–Whitney equation [4], one of the most logical and simplistic methods to enhance dissolution rate would be to increase contact surface area by reducing particle size [5].

Currently there is a wide range of known formulation strategies to enhance poor drug solubility, including crystal habit modification (salt formation, metastable polymorphism and co-crystal formation); co-solvent/solvent evaporation; increase of specific area by particle size reduction (nanoparticle dispersions, micronization); pH modification and adjustment; self-emulsification (SEDDS); Cyclodextrin complexation; surfactant solubilization; amorphization, solid–solid drug dispersion, and most recently, the use of ordered mesoporous silica [6–8].

Of the above techniques, particle size reduction, salt formation, and surfactants/co-solvent solubilization are commonly used to increase the dissolution rate [9]. However, practical limitations surrounding these methods remain a significant drawback. The reduction of drug particle size is limited by the availability of conventional processes, and also produces a product that is difficult to handle. Moreover, the formation of salts is restricted to weakly acidic or weakly basic drugs, and is not feasible for neutral drug compounds. Furthermore, reconversion of the salt back into its respective acid or base form often prevents realization of the solubility enhancement [10]. Lipid-based SEDDS lead to enhanced drug absorption accompanied by inhibiting P-glycoprotein-mediated drug efflux and pre-absorptive metabolism, hence preventing hepatic first-pass metabolism [11]. Therefore, there has only been a limited amount of SEDD formulations commercialized because of the high dependency of SEDDs on *in situ* environment [8, 12]. Despite the extensive use of solvent methods (co-solvent evaporation, spray drying, ordered mesoporous silica), the ecological and subsequent economic issues associated with the employment and the disposal of a large amount of organic solvent have not been truly resolved [5]. Furthermore, as a result of the increased complexity of the drug compounds in regard to their molecular structure, reduced solubility within suitable processing solvents has become an issue [8].

Amorphization through which poorly water-soluble drugs are formulated as amorphous high-energy forms can provide improved solubility and dissolution rate and a consequent improvement in bioavailability [10, 13, 14]. However, the approach is fraught with difficulties from a stability perspective as the high-energy amorphous drug tends to relax and recrystallize to a low-energy crystalline form.

Solid drug dispersions have gained increased attention over the past ten years and been widely investigated throughout the past half a century for two major reasons: solid dispersions provide more significant enhancement in solubility or dissolution rate compared with traditional crystal habit modification strategies; and solid dispersions also retard recrystallization/agglomeration of drug molecules or clusters due to molecular interactions and steric hindrance within the carrier matrices. The major challenges in formulating a solid dispersion are defining the miscibility behavior between the selected drug(s) and carrier polymers, the maintenance of the immediate post-treatment status of the formulated matrix, and the lack of suitable models to predict the miscibility and stability.

11.2 Solid Drug Dispersions for Enhanced Drug Solubility

Undoubtedy, solid dispersions are an attractive means of enhancing drug solubility. The potential of solid dispersions has long been known, however, limited stability has been a key drawback [5, 15]. The very first reported solid dispersion exhibiting enhanced *in vivo* absorption was a eutectic mixture of sulfathiazole and urea [16]. A eutectic mixture, generally speaking, is a system comprising a single composition of two compounds that solidify at a lower temperature relative to any other composition of the same two compounds. Upon cooling, the two compounds comprising a eutectic will recrystallize simultaneously. Rapid/quench cooling from a molten mixture is the most commonly used method to prepare a eutectic mixture. In such prepared solid eutectic mixtures, poorly water-soluble drugs are usually present as very fine crystalline particulates suspended in hydrophilic carriers [16]. The dissolution behavior of drugs formulated in this way is modified by both the reduction of the drug particle size and through incorporation within a hydrophilic polymeric carrier. The use of eutectic mixtures has been, since conception, considered the keystone of using solid dispersions to improve drug dissolution.

Following the initial concept of embedding drugs within polymeric carriers, other matrix platforms were investigated as potential carriers. Solid dispersions using mannitol, in which the drug was presented as a molecular mixture, have been reported [17, 18]. In a series of articles published by Goldberg, it was shown that the use of molecular dispersions could offer the greatest potential to enhance drug solubility and dissolution rate [19–22]. Such molecular dispersions were classified as solid solutions. Eutectic mixtures and solid solutions published during this period were later regarded as first-generation solid dispersions [15]. Matrix carriers used in the preparation of first-generation solid dispersions were typically low molecular weight crystalline compounds – sugars and urea. The rigid crystalline structure of the carriers post-processing was beneficial in kinetically retaining the dispersed drug throughout the matrix. Conversely, however, such structure limited dissolution due to the significant energy barrier required to dissolve the carrier itself. Moreover, in these crystalline systems, drugs were mostly present in the crystalline state [23]. Interestingly amorphous carriers were beginning to show distinct advantages [24].

In these second-generation solid dispersions, carriers were amorphous materials exhibiting irregular short-range order. Specifically, polymeric carriers possessed an irregular distribution of their polymeric chains and were capable of dispersing drug molecules on a molecular level, a formulation commonly referred to as a glass solution, wherein a glassy compound functions as a solid solvent. A glass solution has the molecular conformation of a 'frozen' high temperature solution [25] possessing a high viscosity due to the presence of an amorphous high Tg polymer. The formation of a glass solution is advantageous in terms of solubility and dissolution. From a kinetic perspective, the high viscosity of glass solutions hinders molecular motion due to the slow relaxation of polymer chains. Moreover, homogeneity on a molecular level offers maximum size reduction, hence increasing the surface area to a maximum [26].

Amorphous solid suspensions are amorphous dispersions in which phase separation occurs between amorphous drug clusters and amorphous carriers. An amorphous solid suspension usually occurs when the drug and polymer have limited miscibility, or when the drug is oversaturated within the matrix [15, 27]. In either case, immiscibility may be a result of limited molecular interaction between the drug and the polymer preventing molecular level dispersion. Consequently domains of amorphous phase-separated drug clusters result. In both glass solutions and amorphous suspensions, drugs are typically incorporated into matrix polymers at a highly concentrated level as a result of forced solubilization (i.e. hot-melt extrusion [28], lyophilization [29]). Such oversaturation results in a high propensity to recrystallize upon cooling post processing or during storage. To overcome these challenges, a polymeric carrier that is highly miscible with the selected drug and inhibits recrystallization during drug release or storage should be selected [30]. The most common and widely accepted technique to assess drug–polymer miscibility is thermal analysis [26,29]. While there has been a significant volume of work done predicting drug–polymer miscibility and solubility, there still remains a number of limitations associated with methods currently being employed [31–36].

A more recent development in this area has been the formulation of third generation solid dispersions incorporating surfactants [37,38]. These solid dispersions are designed to obtain the highest degree of solubility enhancement and to prevent morphological changes during administration and aging [39]. Figure 11.1 illustrates the different categories of solid dispersion developed over the past 50 years.

To date, the most common technologies applied to prepare solid dispersions are fusion methods, co-evaporation (co-precipitation), freeze drying, co-milling, and embedding of drug by means of spray drying. In this chapter, the core technology discussed is pharmaceutical hot melt extrusion (HME).

In the past decade there has been an increasing interest in HME as a technology capable of producing melt extrudates with enhanced solubility. HME is a non-ambient process wherein a raw material blend can be converted into a product with uniform appearance and density by continuously forcing it through a (pair of) rotating screw(s) and a discharge die

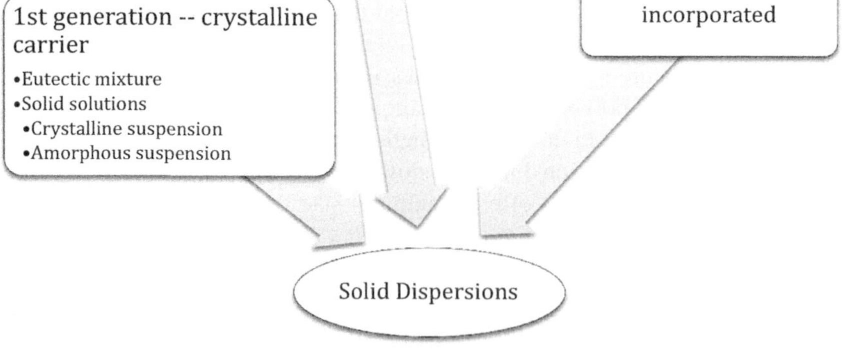

Figure 11.1 Categorization of solid dispersions developed over the past 50 years.

under controlled conditions (temperature, screw speed, throughput) [40,41]. The combination of heat and mechanical force facilitates contact between the drug and the matrix carrier, thus helping to form solid dispersions. More recently, lipophilic materials [42], cyclodextrin [43,44], and ethylene vinyl acetate [45] have been investigated as viable HME excipients for the preparation of novel solid dispersions. For drugs that are difficult to stabilize in their amorphous form, the formation of solid crystal suspensions [46] or melt-extruded cocrystals [47,48] with improved dissolution properties have been demonstrated to be feasible using HME technology.

11.3 Hot Melt Extrusion (HME) as a Drug Delivery Technology

11.3.1 Historical Review of HME

The origin of extrusion can be dated back to 1795, when a hand-operated piston-die combination was built to manufacture a seamless lead pipe. The machinery was soon adapted to produce soap, macaroni and building materials. During the mid-late nineteenth century, screw extrusions were constructed for the cable industry (earliest schematic drawing of a single-screw extruder) and sausage manufacture (first known twin-screw extruder) [49]. The first documented screw-extruder designed specifically for thermoplastic materials was built in 1935, in Germany, by Paul Troester [40,41]. Roberto Colombo (co-rotating) and Carlo Pasquetti (counter-rotating) built the first twin screw extruders for polymer processing [49]. Also around this time, electrical heating elements were designed and attached to the extruder to replace steam-heating units. Since then, the extruder has undoubtedly become one of the most important pieces of equipment used in the polymer industry. This is reflected in its comprehensive use in a wide range of industries includes plastics, metal, ceramic, foods, and more recently, pharmaceuticals.

The adaption of HME for the pharmaceutical industry was first documented in 1971 with hot extrusion moulding being used in the preparation of sustained release medications [50]. However, it was not until almost twenty years later that HME received significant interest as a viable technique for the manufacture of pharmaceutical products [51–55].

11.3.2 Equipment

11.3.2.1 Types of Extruder

Hot melt extruders used in the pharmaceutical industry are typically screw extruders with a rotating screw assembled in the centre of a heated barrel to deliver formulation matrices in a continuous fashion, or ram extruders with a ram (or a piston) pressurizing softened material through a shaping die [50,56]. The regulatory requirements for a pharmaceutical-class extruder include inertness of the metallurgy of all contact surfaces, as well as instrument configuration for cleaning and process validation under GMP. A pharmaceutical instrument will often require modification relative to its plastic processing counterpart, in order to achieve acceptable levels of cleanliness and reproducibility.

11.3.2.2 Ram Extruders

Ram or plunger extruders are simple in design, rugged and discontinuous in their mode of operation. During ram extrusion, material is fed from a hopper and packed into a cylinder in repeated increments by a reciprocating plunger. The material is transported through the barrel where it is softened and fused into the desired profile. The output of such a machine is proportional to the length and frequency of the ram strokes. The die length, electrical heater capacity, hydraulic system power and maximum force, and the strength of the construction materials determine the equipment capacity. One of the main disadvantages of ram extrusion is their intermittent operation and very poor melting and mixing characteristics, making it extremely difficult to obtain a reproducible pharmaceutical product. Consequently there has been more interest in screw extrusion that offers high mixing capabilities and continuous operation.

11.3.2.3 Screw Extruders

In contrast, a screw extruder offers continuous conveying, intense mixing and higher homogeneity of the melt (extrudates). Moreover, in screw extrusion, mechanical energy is introduced through the rotation of the shaft (screws) and converted to thermal energy in the form of viscous dissipation and hence a uniform melt temperature. Screw extruders can be categorized as single-screw extruders, or twin (multi)-screw extruders. This classification is simply based on the number of rotating shafts in the barrel.

11.3.2.3.1 Single-Screw Extruders From an operational perspective, a single-screw extruder is primarily a drag-induced pump with a relatively wide distribution of material residence time, frictional drag in the solid conveying zone and viscous drag in the melt conveying zone [57]. It is well suited for processing highly viscous and coherent materials. For mixed blends with dissimilar viscosities, which is normally the case in pharmaceutical extrusion processes, the dragging phenomenon might cause discontinuous distribution and insufficient dispersion of one component in another, as the differing viscosities of components may lead to slip. The limited mixing capability of single screw extrusion has been shown through incorporation of a coloured pigment to the melt during processing [58, 59]. This resulted in extruded strands that were striped and unevenly coloured. Conversely, in the same experiment only with a twin-screw extruder, a smooth and uniformly coloured product was produced, indicating the improved homogeneity achievable using this process.

To overcome insufficiency in mixing, single-screw extruders are generally designed with extended barrel length so that processing times are longer. However, longer residence times are often undesirable when extruding pharmaceuticals, due to the increased risk of chemical and physical degradation.

Within a single screw extruder, material conveying is largely dependent upon the frictional properties of the fed solid and the viscous properties of the melt. The raw materials in the pharmaceutical industry are mostly in a pulverized form, which can be unfavourable. Moreover, feed problems can be observed due to undesirable flow and frictional properties. In a worst-case scenario the friction between the processing material and the barrel wall could be lower than the friction between the material and the screw surface. In such cases a stagnant layer may remain on the screw surface and potentially rotate with the screw shaft. The lack of forward transport, together with excessive residence time in the barrel, could also cause significant product degradation. Although there are concerns with regard to product homogeneity and potential flow problems within a single-screw set-up, undoubtedly single-screw extruders are advantageous in terms of financial considerations and instrument simplicity. Moreover, the melt flow patterns in a single-screw extrusion process are fairly easy to describe [41].

11.3.2.3.2 Twin-Screw Extruders Twin-screw extruders (TSE) have a pair of identical or mirrored screws placed in a figure of eight section barrel. The original objective of this design was to negate the disadvantages caused by slip page in single-screw extruders. After the establishment of special thrust-bearing designs in the late 1960s, the mechanical reliability of twin-screw extruders had become comparable to their single-screw competitors. Unlike single-screw extruders, TSEs exhibit more complex melt flow patterns. The complexity is mostly considered beneficial since it provides intense material displacement, fast heat transfer, large melting and devolatilization capacity, and good control over stock temperature. Apart from all these, the screw geometry and relative motion during processing deliver positive displacement and thus forward transport of materials. The degree of positive conveying is to some extent dependent on the degree of intermeshing geometry and the direction of the screw rotation. By altering the mode of rotation, twin-screw extruders can be categorized as co-rotating (in the same direction), or counter-rotating (in the opposite direction). Meanwhile, by adjusting the relative distance between two screw shafts, twin-screw extruders may also be classified as intermeshing or non-intermeshing.

In general, a close intermeshing, counter-rotating (CICT) screw geometry offers the most positive mass conveying [41, 49, 57, 59]. This positive forward conveying allows processing of 'hard to feed' materials, i.e., drug powders, slippery additives, etc., and effectively avoids the formation of stagnant layers, thus helping deliver mixed materials in a more uniform fashion, and significantly narrows the residence time distribution. It is, however, worth knowing that improved conveying characteristics are usually accompanied by poor axial mixing. Therefore, if material blending, chemical reaction or devolatilization are required, non-intermeshing counter-rotating TSEs are more often used [60]. The flow and conveying mechanisms in these extruders are similar to that in single-screw extruders, except that there is potential material exchange from one screw to another.

To combine the advantages of the afore-mentioned screw types, closely intermeshing co-rotating (CICO) extruders are often used. The intermeshing nature permits sufficient mass displacement although this forward conveying is not as positive as that of a CICT extruder. Nonetheless, CICO extruders are non-conjugated with relatively large openings between the channels of the two screws, wherein materials subjected into the intermeshing region tend to flow into the channel of the adjacent screw. This material exchange between the intermeshing channel regions of the two screws moves the processed mass in an open figure of eight and simultaneous axial mixing [60, 61].

11.3.3 Screw Geometry

The rotating motion of the screw inside the barrel is the source of forward transport and simultaneous homogenization of a processed product. A standard screw and its features are shown in Figure 11.2. The dimension of a screw is most commonly given as a length to diameter (L/D) ratio. To overcome the backpressure, almost all plasticating extruder screws are designed with a tapered section where the channel depth gradually decreases. Along the length of the barrel, the screw geometry is divided into three sections: (i) the feeding section which possesses the widest channel depth to ease the fed mass flow; (ii) the compression section with decreased channel depth to generate higher pressure and facilitate melting; and

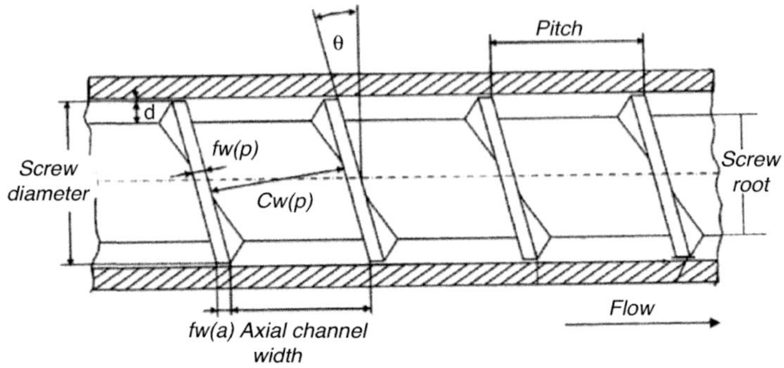

Figure 11.2 *Schematic extrusion screw geometry, where theta is the helix angle; d is the channel depth; fw(p) is the perpendicular flight width; fw(a) is the axial flight width; Cw(p) is the perpendicular channel width; and the distance between the flight tip and the inner barrel surface is the flight clearance.*

(iii) the metering section that soothes the pulsating flow and assures the greatest extent of uniformity [62]. Although, in certain applications in the pharmaceutical industry, conical screws and barrel designs in which the screw channel remains constant while the barrel diameter decreases along its length, are also popular [63].

Melting and mixing are the two most significant elements in a hot melt extrusion process, especially in pharmaceutical applications wherein all ingredients exhibit discrete thermal and rheological properties. Features like helix angle, pitch, number of flights, flight clearance, and flight width may all be optimized for specific requirements. For example, the most common helix angle is 17.66° (square pitch), whereas optimizing for plastification necessitates a higher helix angle [41].

Twin-screw extruders have been proven versatile for their higher degree of dispersive mixing (to be discussed in later sections) and self-cleansing properties relative to single-screw extruders. The motion of the twin screws is considered self-wiping or cleansing and the strength of the wiping mechanism is dependent upon specific design. There are three broadly classified screw element types in TSEs, namely a conveying/forwarding element, a mixing element and a zoning element, which can be changed and self-assembled for modular extruders. Conveying elements function to transport materials forward. Mixing elements perform in different ways: dispersive mixing elements encourage morphological breakdown whereas distributive mixing elements accommodate the original morphology evenly without alteration. Zoning elements are used to isolate one function from another: They can be righted, non-righted, restrictive, or neutral and can even incorporate more or less mixing functionality [64].

11.3.4 HME Processing

HME processing involves processing a blend of active compounds and pharmaceutical-class matrix carriers at elevated temperatures, with intense mixing (preferably at molecular level when formulating solid dispersions), and with precise shaping at the discharge end, in a single continuous step. Twin-screw extrusion, in particular, has been increasingly utilized to perform such processes as a result of the recognition of its superior capability of forceful incorporation in a timely manner. Notwithstanding the advantages that twin-screw extruders afford over single screw extruders, both of the two designs possess similar fundamental elements in terms of functional processing. From a process engineering point of view, a standard extruder ought to be divided into six main function zones: solid conveying zone, melting/plastification zone, melt conveying zone, devolatilization, mixing zone, and discharge die forming. However, one should bear in mind these functional zones are, to some extent, overlapping rather than discrete. The boundary between each of the two zones is somewhat dependent upon the properties of the materials being extruded, as well as the operating conditions [41]. Therefore, these functional zones are to be differentiated from screw geometric sections/zones, which are simply fixed as feed section, compression section, and metering section for a typical screw design.

11.3.4.1 Feeding/Solid Conveying

A melt extrusion process commences by introducing materials into the machine, i.e. feeding. At this point, the materials are still a physical mixture of solids. Solid conveying begins at this point with melting typically not occurring until about 3~4 flights after material entry [41]. The solid flow in a feed hopper is either gravity induced or through auxiliary forcing screws. With regard to gravity-induced feeding auxiliaries, particle properties are critical as 'mass/hopper flow' is favourable, whereas 'arching/bridging flow' or 'funnel/tube flow' are to be avoided. Such gravity-induced feed systems are usually employed in single-screw extruders. Thus all raw materials in a feed blend for a single-screw extrusion process are required to be similar in shape, size and density. However, high pressures generated during flood feeding have been shown to cause agglomeration of ingredients and hinder dispersion and distribution at later stages in the barrel [41].

Conversely, force feeding has the capability of working with a wider range of materials. Materials are packed tightly to form a solid bed and slide down the helical channel as a result of screw rotation. Twin-screw extruders are usually equipped with metering feeders in order to maintain the desired starve feeding rate, as well as to avoid accumulation of material. The throughput of the process is then controlled primarily by the volume flow rate of the force feeder, but not by the screw speed. There is also less risk of high-pressure generation in the solid conveying and melting zones compared with gravity-fed single-screw extruders.

Once the solid flow has reached the feed port (discharge end of hopper channel), the particles will fall down and situate themselves in the screw channels. At this point, the outer-barrel feeding ends and in-barrel solid transport starts. The solids will then be conveyed forward either as a result of drag effects (in single-screw extruders), or as a consequence of positive onward displacement (in twin-screw extruders).

11.3.4.2 Melting/Plastification

The melting/plastification zone starts once melting occurs. Early studies in the polymer industry revealed that the first trace of melt was always present at the barrel-solid boundary due to the extra frictional heat generated here, relative to the screw-solid boundary [41]. It is therefore, to be noted that melting may occur even if the melt temperature is lower than the melting temperature of the material. It is then ideal to have both barrel temperature sensors and melt temperature sensors assembled on the extruder for more accurate control of the process settings. The melting function of a screw extruder is essential in determining the global performance of the machine, as incomplete melting may cause inconsistency in pumping capacity and thus poor product characteristics. Melting within a screw extruder has been reported to be a combination of conduction melting and dissipative mix melting [65]. The former is a direct result of the thermal energy input, while the latter is directly influenced by the mechanical energy input. However, both mechanisms are mostly present simultaneously to achieve sufficient melt within an extruder barrel. In addition, the melting mechanism has been reported to be a function of screw style [41]. In fact, contiguous solids melting (CSM) is most often observed in extruders that feature a continuous, uninterrupted channel for material flow, like single-screw extruders, whereas dispersed solids melting (DSM) is more likely to occur in twin-screw extruders.

11.3.4.3 Melt Conveying

At the point when melting/plastification is completed, the matrix melt then enters the melt-conveying zone. In this section, the melt is typically pumped towards the discharge die against considerable pressure. The study of melt-conveying mechanisms in a pharmaceutical extruder is restricted due to the fact that formulations being extruded are non-Newtonian and non-isothermal. This complexity prevents the adaption of simplified mathematical models used for the melt conveying of Newtonian fluids.

11.3.4.4 Devolatilization

Devolatilization/venting is a function that is not necessarily present in all extruders. The volatiles to be removed in the venting zone are usually entrapped air, remaining monomers, residual moisture, and decomposed products that may cause undesirable bubbles, holes, or a foamy structure in the extrudates. Evaporation from the melt on the interior barrel wall has been shown to be of minor importance relative to that from the melt in the screw pitch [42, 66–68]. Thus, to maximize devolatilization efficiency, the venting zone typically possesses a coarse pitch screw flight with low degree of fill, providing an appropriate pressure-drop and maximum surface/volume ratio.

11.3.4.5 Mixing

Generally speaking, the mixing zone is where the non-uniformity of a composition is reduced and homogenized. Mixing is an essential and must-have function especially for screw extruders, as opposed to devolatilization. The mixing zone extends from the start of the melting/plastification zone to the end of the die. This wide range and overlap with the melting and melt conveying zones may be problematic as the melt properties of different ingredients can result in considerably non-uniform melting histories. Thus, even if a perfectly mixed fluid enters a die, non-uniformities might still be expected as it exits [41]. Therefore, careful control of the extrusion process parameters (e.g. barrel temperature settings, screw speed, screw element design, and feeding rate) is required in order to promote sufficient melting and improve uniformity. Due to the highly viscous nature of polymers, the mixing motion within polymer melts is always convective mixing by laminar flow, or is generally referred to as laminar mixing [40, 41, 57, 69]. This mixing involves adjacent layers sliding relative to each other without interpenetration. The action of mixing in a screw extruder is generally caused by the shear and elongational flow. The reduction in non-uniformity by this relative motion is known as *distributive* mixing.

During the melt screw extrusion process, solid particles will elastically deform under a finite shear stress, and tend to recover their initial dimensions on removal of this stress. To obtain the reduced domain size of the minor phase (e.g. drug) with permanent deformation or fracture, it is necessary to process beyond the finite stress before distributive mixing occurs. The type of mixing that results in reduced minor phase sizes is also referred to as *dispersive* mixing [57]. Practically in order to exceed the finite shear stress, the screw design is of the greatest importance as this minimum critical stress is usually not achieved in a single-screw extruder. In contrast, the structure of a twin-screw extruder provides a larger area for shearing in the clearance between two screws. Moreover, twin-screw extruders are often starve-fed, as described previously. An increased screw speed will produce a decreased screw volume at a constant feed rate, which reduces the overall shear stress. When the screw speed is kept constant, an increased feed rate means increased screw volume that subjects a greater proportion of the melt mass to shearing, hence promoting higher shear stress. In addition, the shear stress is also dependent upon the barrel and the melt temperatures since higher temperatures result in lower melt viscosity, thus decreased shear stress [57, 65].

11.3.4.6 Die Shaping

In many respects, the die shaping/forming zone plays an invaluable role in the whole process as the final shape of melt extrudates are determined in this section. At present, extrusion die geometry has reached a high level of sophistication and efficiency. Whichever geometry is chosen, this functional zone is essentially and always a pressure-consuming section. The flow channel geometry of the die itself, the rheology of the polymer melt, the temperature of the melt, and the flow rate through the die can all affect the die-head pressure [41]. A well-designed extruder must be capable of engineering a homogeneous polymer melt with the required melt flow rate and pressure in order to overcome the resistance of the die and push the melt through this discharge point. The pressure drop as the melt exits the dic is then dependent upon the melt flow rate and temperature. It is preferable for the die-head pressure to be minimized in order to prevent dramatic temperature increases due to viscous dissipation [65]. Nevertheless, die flow instabilities, i.e., sharkskin and melt fracture, may occur if the melt velocity in this region is very high [70].

11.3.4.7 Downstream Auxiliary Processes

A wide range of downstream processing is available following HME to tailor the appearance and the properties of the final extrudates. Such auxiliary processes include external cooling systems, downstream feeding ports, pelletizing, and film/lamination systems. Of these, downstream feeding is of great importance especially for thermally labile components. By feeding at this later stage, residence times may be reduced and hence barrel degradation avoided.

11.3.5 Product Characteristics

The characterization of extrudate properties includes the inspection of the chemical stability of the active ingredients and other excipients during the extrusion processes; morphological identification and physical/chemical stability of extrudates upon aging; the drug release (*in vitro* and *in vivo*) performance; and mechanical properties analyses. The common methods routinely used to characterize hot melt extrudates are listed in Table 11.1.

Table 11.1 *Characterization methods commonly used for pharmaceutical hot melt extrudates.*

Chemical stability	Extrudate morphology	Drug release performance	Miscibility of drug/Polymer	Storage stability
TGA	DSC	DVS	DSC	DSC
HPLC	PXRD	HPLC	FTIR	PXRD
FTIR	FTIR	UV	Raman	FTIR
	Raman	*In vitro/In vivo* dissolution	Chemical imaging	Raman
			DMTA	
	PLM			Chemical imaging
	SEM/TEM			PLM
	AFM			AFM
	NMR			SEM/TEM
				NMR

The chemical stability of all ingredients, especially the active pharmaceutical ingredient (API), during the hot melt extrusion process must be pre-examined and closely monitored. Since the HME process is a solvent-free treatment where elevated temperature is involved, the potential component degradation would be mostly thermal degradation. The use of thermogravimetric analysis (TGA) offers an indication of moisture content as well as the thermal degradation temperature of each ingredient through simple sample weight loss. This information is beneficial in narrowing the HME process temperature window. The determination of drug chemical stability after melt extrusion could be conducted using high-pressure liquid chromatography (HPLC) through evaluation of the purity of the drug peak.

To understand drug release mechanisms, it is essential to first understand the morphology or crystallography of the extruded system. The combination of differential scanning calorimetry (DSC) and powder X-ray diffraction (PXRD) is typically used to identify whether there is residual API crystallinity. A melting event indicative of melting of the drug in the DSC thermograms or diffraction bands that match the drug's characteristic fingerprint region in the XRD pattern can indicate the presence of crystalline content in the extrudate. Generally, melt extrudates are examined using such techniques and compared to physical mixtures of the formulated systems. The DSC thermograms can even be used to quantitatively calculate the energy required or liberated during a thermal transition event, which is advantageous in determining the miscibility of a binary system at elevated temperatures. However, the sensitivity of both DSC [35] and PXRD [62, 71] techniques are limited by crystalline particle size and concentration. Therefore, other techniques may be needed to complement DSC and PXRD.

Microscopic analysis is typically considered the most straightforward way to examine the surface morphology of melt extrudates. Polarized light microscopy (PLM) and atomic force microscopy (AFM) offer optical observations of the presence of crystalline particles, while electronic microscopy (Scanning EM and Transmission EM) probe crystalline and amorphous domains. Solid-state nuclear magnetic resonance (ssNMR) has also shown potential in probing API crystallinity within extruded matrices.

In addition to the afore-mentioned techniques, spectroscopic analysis may be used to detect API crystallinity and address interactions between molecules. Morphological alterations (the lack of crystal structure) or the presence of intermolecular interactions both may cause changes in the bonding between functional groups. Variation in peak position and shape may be identified in Infrared (IR) or Raman spectra. Moreover, spectral differences can be used to construct chemical images (i.e. Raman mapping), indicating the distribution of a reference compound. These techniques may also be used to study the storage stability as a function of both temperature and humidity.

11.3.6 Materials Commonly Used in HME for Solubility Enhancement

11.3.6.1 Regulatory Considerations

The materials used to produce hot melt extruded solid dosage forms are the same pharmaceutical compounds used for the preparation of other traditional formulations. The most important consideration in selecting materials for pharmaceutical HME is that all components must be GRAS (Generally Recognized as Safe) listed and exempted from the usual FFDCA (Federal Food, Drug, and Cosmetic Act) food additive tolerance requirements.

11.3.6.2 Drug Compounds

Active ingredients for this purpose are those exhibiting poor bioavailability limited by aqueous solubility, namely BCS Class II compounds. The chemical stability of the drug candidates at elevated temperatures and during forceful mechanical deformation is critical to the pharmaceutical efficacy of the extruded dosage form. Thus thermal analysis prior to the extrusion process is essential. Through careful selection of formulation excipients and the incorporation of proper processing additives, many thermo-sensitive drug compounds have also been demonstrated to be 'melt extrudable'. The formation of a molecular solution, or other solid dispersion type, is, to some extent, dependent upon the unique properties of the drug. The chemical structure, stability and physical properties of the drug substances may impact the selection of other functional excipients and process conditions.

11.3.6.3 Formulation Excipients

Thermal stability is the prerequisite for any material that is to be melt extruded. A hot melt extruded dosage form typically consists of a drug embedded in a carrier matrix incorporating one or more meltable materials along with other functional excipients. The meltable materials are normally low melting point waxes or thermoplastic polymers. These substances melt within the extruder barrel to mediate the formation of molten solid dispersions, and then solidify at the exit of the extruder. Thus the selection and use of various 'meltable' excipients can be decisive in dictating the HME process conditions and in shaping the specific product properties of the melt-extruded dosage forms. The drug release kinetics of melt extrudates is greatly dependent upon the use of formulation excipients. For drug substances embedded in waxy materials, such as microcrystalline wax, corn, and potato starch, or even waxy-like semi-crystalline polyethylene glycol/oxide (PEG/O), the drug release mechanism is typically erosion-based, or a complex mixture of diffusion and relaxation phenomena [72]. The presence of physically separate drug entities at the dissolution site prevents agglomeration of the drug particles, thus enhancing dissolution performance. Despite the risk of forming a eutectic mixture, the use of wax-based materials in pharmaceutical extrusion is advantageous in terms of non-toxicity, low cost, low melting viscosity, fast solidification upon cooling, and relative inertness [62].

For formulations extruded with thermoplastic polymers, on the other hand, the formation of water-soluble drug–polymer complexes in aqueous solution is usually known to improve the aqueous solubility of poorly water-soluble drugs [73]. During the HME process, the drugs can be solubilized in the polymeric matrix. The saturation limit (solubility) of the drug in a chosen polymer is to a great extent dependent on drug–polymer interactions. Water-soluble polymers such as polyvinyl pyrrolidone (PVP) and copolymers (PVP-VA) [74–77], methyl cellulose (MC) and derivatives (HPMC, HPMC-AS) [78–82], cellulose derivatives (HPC) [83], polymethacrylate copolymers (Eudragit®) [28, 84–88], various grades of polyethylene glycols (PEG, PEO) [89, 90], and poloxamers have been extensively investigated as matrix excipients for pharmaceutical extrusion. Among various classes of biodegradable polymers, the thermoplastic aliphatic poly (esters) such as poly (lactide) (PLA), poly (glycolide) (PGA) and copolymer of lactide and glycolide, poly (lactide-co-glycolide) (PLGA) have also appeared in the pharmaceutical extrusion literature [91]. Investigations of sustained release platforms for highly water-soluble drug candidates have employed water-insoluble [92] and non-biodegradable polymers to retard the formulation burst at the initial stages of dissolution [45]. Other functional excipients such as drug release modifying agents (disintegrants, thickening agents), bulk agents, and miscellaneous additives may also be incorporated in the extrusion process to impart specific designed properties of the extrudates in a way similar to that in other traditional formulations, as long as the chemical stability of such excipients is guaranteed during the melt process.

11.3.6.4 Processing Additives

Additives used in the polymer industry are usually categorized according to their functionality rather than on a chemical basis [93]. In order to guarantee the chemical stability of the ingredients used during extrusion, processing additives, that, for example, can either lower the required temperature or assist melt flow within the barrel. Figure 11.3 exemplifies the typical aspects of processing additives often used in pharmaceutical extrusion processes [94].

An anti-oxidant is a substance capable of inhibiting or preventing the initiation of free radical chain reactions in the presence of oxygen. Based on the mechanism, anti-oxidants are usually classified as preventive or chain breaking. Both mechanisms take advantage of anti-oxidants by their propensity to undergo rapid oxidation themselves, thus consuming the oxygen present in the surrounding environment. Some of the preventive anti-oxidants, such as ascorbic acid (vitamin C), are reducing agents that can attack and thereby neutralize reactive oxygen species. These agents are sometimes referred to as oxygen scavengers. Another type of preventive anti-oxidants – chelating agents, such as citric acid – contribute to anti-oxidant defence by chelating transition metals and preventing them from catalyzing the production of free radicals. Chain-breaking anti-oxidants may be subcategorized as hindered phenols (comprising a benzene ring and an alcoholic hydroxyl group, i.e., vitamin E, butylated hydroxytoluene), and aromatic amines. They both remove the free radical intermediates by sacrificing their weak O-H or N-H bonds to self-oxidize and prevent further propagation [95].

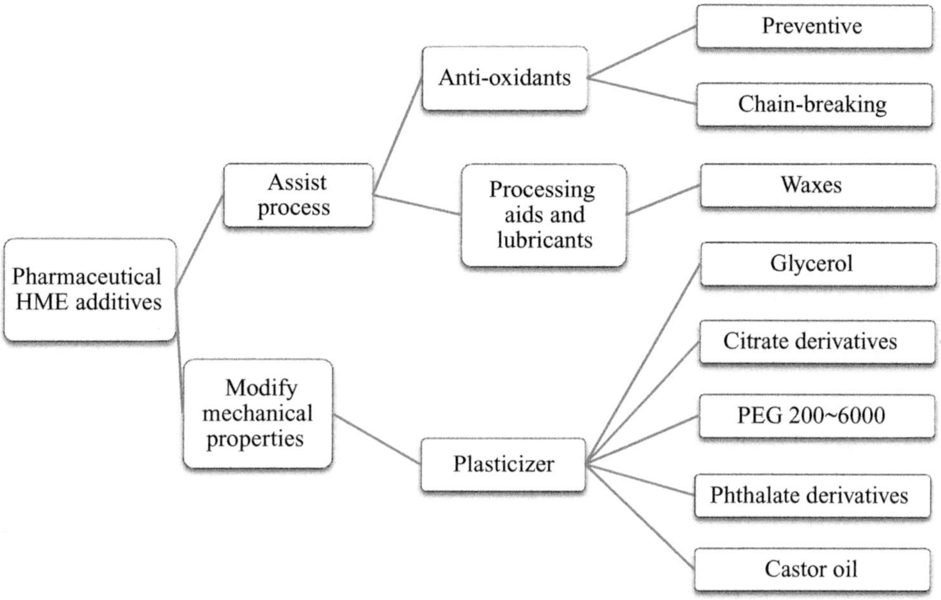

Figure 11.3 *Categorization of commonly used additives in pharmaceutical extrusion processing.* (See colour plate section.)

If one considers anti-oxidants as paramount in the melt extrusion process against oxygen, then plasticizers assure the processing stability from an interior perspective. Commonly used plasticizers are listed in Figure 11.3. They are usually small molecular weight substances with relatively low melting points. In the polymer industry, plasticization is where the plasticizer interacts with the polymer chains to uniformly distribute itself within the polymer matrix, hence creating additional free volume. In doing so, the glass transition temperatures (T_g) and the melt viscosity of a polymer can be decreased [94, 96]. Therefore, with the incorporation of plasticizers, the HME process may be conducted at lower temperatures or at the same temperature but with lower mechanical energy, thus maintaining the stability of all components during the HME process. Previous work in the polymer industry has shown that the most effective plasticizer typically closely resembles in structure the polymer it most efficiently plasticizes. Thus, polymers possessing large amounts of hydroxyl groups (i.e. polar cellulose ethers, cellulose esters, PEO, etc.) are best plasticized by materials that also contain hydroxyl groups, such as polyethylene/polypropylene glycol and glycerol; whereas polymers that are less polar (i.e. Eudragit®, ethylcellulose, cellulose acetate phthalate as examples) are better plasticized by organic esters such as citrate and phthalate derivatives [80, 85, 94, 97, 98].

The selection of a proper plasticizer for a chosen polymer may also be rationalized and predicted through the comparison of the solubility parameters of both the polymer and plasticizer. In some cases, the drug may impart a plasticizing effect to the matrix carrier during processing [99–101]. The advent of third-generation solid dispersions has inspired the use of various surfactants as plasticizers during the HME process. In doing so, in the formulation design of poorly water-soluble drugs, the surfactants may facilitate and stabilize the process in addition to being a solubilizer [38, 102, 103].

In spite of the benefit that plasticizers offer processing-wise, one should also be aware of their potential drawbacks. The presence of a plasticizer within a polymeric matrix not only depresses the T_g. This may be useful during processing but may reduce stability during drug dissolution and aging. The enhanced global mobility, as a result of the increased free volume and decreased chain entanglement, may undesirably alter the drug release profile or affect the aging stability of the formulation. The performance of plasticizers during high-temperature processing and long-term storage has also gained much attention as diminished drug dissolution profiles have been reported with the evaporation of highly volatile plasticizers. Equivalent plasticizing effects of pressurized carbon dioxide (supercritical CO_2) in polymer processing have been well established and utilized [104–111]. The temporary residing nature of the supercritical gas during the hot melt extrusion process eliminates issues related to volatility during storage. In addition, supercritical CO_2, as a novel physical blowing agent (PBA), evaporates rapidly from the melt extrudates during expansion through the die. Controlling the process parameters can modify the strength of the foaming effect and the pore size left within the remaining formulation.

11.4 Solubility Enhancement Using HME

The use of HME as a pharmaceutical formulation process was originally aimed at sustained release platforms for highly water-soluble drugs. When HME appeared in the pharmaceutical literature for the first time, various acid or alkaline soluble thermoplastic resins were investigated with regard to their dissolution behavior in artificial gastric-intestinal (GI) fluids [50]. It was demonstrated that the satisfactory dissolution rates of acid-soluble epoxy resins in simulated gastric fluid were beneficial to release the initial dose of a drug in the stomach. Further to initial investigations, HME was considered a potentially useful process to embed a drug in a hydrophilic matrix at a particle size that was significantly reduced, thus enhancing the water solubility, and subsequently the bioavailability of poorly water-soluble drugs [112, 113].

Melt-extruded formulations may be generally regarded as solid dispersions, regardless of drug morphology within the extrudates. The mechanism of embedding the drug in a polymeric matrix during extrusion may be regarded as either interactive and/or physical embedding. In the former concept, intermolecular interactions between the drug and the matrix polymer are the driving force to disperse and stabilize an active ingredient in the matrix carrier(s). Physical embedding involves mixing within the extruder and polymeric viscosity preventing re-agglomeration in the congealed melt.

11.4.1 Product Structure

11.4.1.1 Extrudate Morphology

In the HME process, the drug and excipient are physically blended and subjected to a non-ambient process where all components are intensely mixed in the molten state. A melt extruded dosage form may be a modified crystalline particulate dispersion (crystalline precipitation), a molecular solid dispersion (true solid solution/glass solution), or an amorphous dispersion (amorphous suspension) [26]. The morphology of the drug compound in the final extrudate is dependent on the formulation factors, i.e. drug loading, the interactive properties of the materials being processed; and the selection of the process parameters, i.e. temperature settings, screw element configuration and screw speed.

11.4.2 HME Matrix Carriers

Previously we have discussed different kinds of matrix carriers commonly used in HME. We further discuss the role of these carriers within HME.

11.4.2.1 Water-Soluble Carriers

Particle aggregation, morphological instability, and poor wettability are the three main considerations in formulating poorly water-soluble drug compounds using HME. Manipulation of particle size is a simple and useful way of altering the properties of a poorly soluble compound. However, particles with a reduced size are known to aggregate. The formation of metastable polymorphs or even amorphous counterparts is also well recognized to radically enhance the rate of dissolution and apparent solubility. However, these drug forms are highly susceptible to recrystallization. The incorporation of pharmaceutical excipients in the above case may help to exert a stabilizing barrier. In this regard, hydrophilic excipients are preferred in order to improve the wettability of the drug particles. While hydrophilic excipients may improve wettability, significantly they may draw moisture into the formulation, thus plasticizing the matrix structure, hence reducing physical stability.

11.4.2.2 Physical Embedding Using HME Diluents

For excipients that are non-interactive/immiscible with the dispersed drug, molecular dispersions may occur during the HME processing due to the high thermal and mechanical energy input. Prevention of molecular mixing during processing is a challenge, and problematic from a storage perspective since it is known that immiscibility between both components will lead to relaxation and drug precipitation. Therefore, using crystalline drug suspensions in such carriers might offer an advantage over molecular dispersions or amorphous embeddings.

During HME processing, one can extrude the fed blend at a temperature that is high enough to soften the matrix excipients, but still low enough to maintain the solid state of the drug compound. By doing so, only the friction between the drug, the excipient, and the particles induced by the mechanical energy input of the rotating screws, is responsible for breaking the drug particles and the subsequent size reduction. The resultant fine drug particulates would be expected to be evenly distributed throughout the entire matrix as a consequence of the mass transport in the HME conveying zone. The preferred carrier candidates to 'dilute' drug particulates are hydrophilic small molecular crystalline materials such as sugars [114, 115], sugar alcohols [46], starch and microcrystalline cellulose. The final extrudates would be compressed physical mixtures of fine crystalline drug particulates and crystalline excipients. Nonetheless, the skilful choice of hydrophilic semi-crystalline or amorphous polymer combinations with additives can also contribute to 'diluted' fine particulates for a more desirable dissolution profile [74]. Data already in the literature suggests the presence of deaggregated and dispersed engineered drug particles in amorphous polymer [78,79,116–119]. When the extrudates congeal post processing, the viscous nature and the rigid chain structure of the polymer may hinder the global mobility of the dispersed fine drug particulates. Dissolution rate enhancement is the main improvement for poorly water-soluble drugs engineered as 'diluted' fine particulates due to size reduction. Moreover, in an ideal case, the drug particulate would be intimately covered with hydrophilic polymer, thus improving wettability [120, 121].

11.4.2.3 Inter-Molecular Interaction with Drug Molecules

The presence of the drug in a thermodynamically unstable state is beneficial from a dissolution perspective wherein the crystal lattice energy does not need to be broken prior to dissolution. Therefore the energy barrier associated with crystal lattice disruption during solubilization is removed. Interestingly, the rapid dissolution of drug under such conditions may produce a supersaturated concentration of the drug within the dissolution media and thus precipitation [122]. The maintenance of this high-energy status both during aging and release, therefore, is practically challenging. Such challenges have been overcome through engineered interactive drug–polymer matrix systems. For many years there has been debate over the exact stabilization mechanism of polymers in such dispersions [123–128]. One thing that is considered paramount is drug–polymer miscibility. An ideal miscible system is that in which only a single phase is present, and requires intermolecular forces between chemical moieties of both the drug and the carrier molecules [123, 127, 129–131]. In more recent data, vitrified homogeneous systems are also referred to as glassy molecular alloys [132, 133]. The improvement of the dissolution rates of poorly water-soluble drugs from miscible molecular dispersions is achieved due to the significant size change of the drug compound (molecular level dispersion), accompanied by improved wettability attributable to the surrounding hydrophilic matrix. Furthermore, the intermolecular interactions between the drug and polymer molecules may tailor the dissolution profile through the control of the drug dissolution rate [88]. The intensity of interactions modifies the dissolution mechanism from 'carrier-controlled' to 'drug-controlled' for low drug loadings [121, 134]. Therefore, in molecular solid dispersions, favourable drug–polymer interactions are critical to the formation and stabilization of solid–glass solutions. Extensive work has been published with regard to the prediction of drug–excipient interactions prior to formulation, or the estimation of such interactions post-manufacture.

11.4.2.3.1 Prediction of Drug–Polymer Miscibility The most commonly used predictive method to date to examine drug–polymer miscibility is solubility parameters [135–139], or the Flory-Huggins interaction parameter, derived from thermal methods [140–144]. The former is useful in large volume screening through the comparison of the solubility parameter difference ($\Delta\delta$) between the drug and the polymer. Forster *et al.* demonstrated experimentally that drug–polymer systems with a $\Delta\delta < 7.0$ are highly likely to be miscible and form molecular solid dispersions during HME; whereas a $\Delta\delta > 10.0$ denoted considerable immiscibility and such systems would not be expected to form molecular solid dispersions [139]. The use of the Flory-Huggins interaction parameter is considerably more labour-intensive, being experimentally derived. An intimately but physically mixed binary system consisting of drug and excipient is usually subjected to a heat ramp using a differential scanning calorimeter (DSC) at various drug fractions. In such an analysis, allowing the physically mixed system to attain equilibrium is crucial. A miscible system would exhibit additional depressed melting of the drug, whereas an immiscible system shows only linear melting depression.

11.4.2.3.2 Evaluation of System Miscibility in Hot Melt Extrudates Thermal analysis is often the most straightforward approach to examine miscibility. As a general rule, an intimately mixed binary system would exhibit a single T_g, indicative of a single amorphous phase. In contrast, a binary matrix with more than one amorphous phase or at least a combination of a true molecular dispersion and amorphous phase separation would exhibit two T_gs. Such phase differences may be simply reflected in a DSC thermogram by the number of glass transition temperatures (T_g) observed. A distinctive single T_g that is intermediate between the T_g of the amorphous drug and that of the amorphous polymer, according to the Gordon-Taylor equation, has been commonly accepted as general evidence of a homogeneous single phase at the molecular level [76,145–147]. Although the detection limit of DSC [148, 149] has prevented investigation of separate phase regions smaller than 30 nm, other techniques such as chemical imaging/Raman mapping [35], or transmission electronic microscopy (TEM) may also be used for comprehensive studies.

With regard to molecular drug–polymer interactions, H-bonding is most often observed within molecular dispersions. In the solid state, hydrogen bonding may be identified using Raman and Infrared (IR) spectroscopy, as the peak arrangement of an H donor group is very sensitive to the extent of association. The formation of hydrogen bonding is typically accompanied by the lengthening of the X-H bond in the H donor group. This lengthening would be represented by a downward band shift or broadening. The concomitant decrease in wave number of an electronegative group identifies itself as the corresponding H acceptor [150, 151]. Moreover, Karavas *et al.* have demonstrated the use of ^1H NMR to identify the nature of the interactions between felodipine and PVP or PEG in binary glassy/solid solutions [118, 134].

11.4.2.4 Formation of Complexes

Before the use of Cyclodextrin (CD) in pharmaceutical extrusion applications, CD had been well understood to enhance the aqueous solubility of poorly water-soluble compounds through the formation of host–guest complexes. CD is a toroidal compound made up of bonded sugar molecules with their hydroxyl groups facing out, while leaving a cavity in the centre [152]. The considerably less hydrophilic interior of the toroid may thus host hydrophobic molecules, as the hydrophilic exterior imparts solubilization of the whole complex to the adjacent aqueous environment.

More recently HME has been implemented as a potential technology to enhance drug solubility using CD complexes. Owing to the high temperature and artificial agitation associated with the HME process, the global mobility of the drug molecules would be intensified. The increased contact between the drug molecules and CD during extrusion favours complex formation. Recently, another complexation mechanism – a counter-charged drug–excipient blend has been used to form a polyelectrolyte complex during HME [153].

11.4.3 HME for the Manufacture of Pharmaceutical Co-Crystals

Due to the extreme conditions inside the HME barrel (elevated temperature, lack of humidity, high pressure, and global movement of the melt mass), melt extrusion may be considered an engineering reactor. The modern application of twin-screw melt extrusion technology has been demonstrated for the manufacture of pharmaceutical co-crystals, a novel crystal engineering approach to improved drug properties [48, 154].

11.5 Representative Case Studies with Enhanced Solubility

11.5.1 Increased Dissolution Rate Due to Size Reduction or De-Aggregation

More than ten years ago, HME was tabled as an idea for enhancing drug solubility by way of particle size reduction. In this regard, Hülsmann *et al.* extruded 17β-Estradiol hemihydrate (17β-E2) solid dispersions in polymeric matrices [155]. PEG 6000, PVP (Kollidon® 30:vinylpyrrolidone), and PVA 64 (Kollidon® VA64, vinylpyrrolidone/vinylacetate-copolymer 6:4) were used as matrix polymers; additives with low melting properties such as Sucroester® WE15 and Gelucire® 44/14 were incorporated as thermal lubricants and release modifying agents. *In vitro* dissolution tests revealed a 10-32-fold improvement in dissolution after only 60 min. release for extruded solid dispersions, relative to the pure drug. The tablets prepared from PVP solid dispersions all met USP dissolution rate requirements for immediate release dosage forms – over 75% dissolved in 60 min. Interestingly, crystalline 17β-E2 was present in the melt extrudates, confirmed by X-ray diffraction analysis. The authors thus concluded the enhancement of dissolution rate was a result of the improved wettability due to size reduction and incorporation in hydrophilic matrices.

A more advanced study by Miller *et al.* further investigated the deaggregation and dispersion capabilities of HME using a single-screw Randcastle Microtruder [116]. Itraconazole, the chosen poorly water-soluble model drug, was pre-engineered as micronized particles embedded in PVP and HPMC respectively. The micronized particle formulations were then 1:1 (w/w) extruded with a 7:3 (w/w) mixture of Poloxamer 407:PEO 200M. SEM images provided visual confirmation that the micronized particle formulations were deaggregated and homogeneously dispersed within the extrudates. The micronized particles existed as highly aggregated and tightly adhered fine primary particles with aggregate diameters exceeding 50 μm. The *in vitro* dissolution profiles illustrated a further improvement in dissolution rate.

Most recently, Thommes *et al.* demonstrated a novel melt extrusion-based approach to improve dissolution properties by solid crystal suspensions [46]. With the adaption and modification of the concept of first-generation solid dispersions, crystalline drug suspensions in crystalline matrices were formulated. Three poorly water-soluble drugs, namely griseofulvin, phenytoin and spironolactone, were co-processed respectively with a highly hydrophilic sugar alcohol, mannitol, in a hot melt twin-screw extrusion process. DSC thermograms of physical mixtures indicated the immiscibility of griseofulvin and mannitol in the molten state. PXRD patterns and DSC thermograms of the melt-extruded solid dispersions signalled unchanged drug crystallinity following extrusion. The drug dissolution rates from extruded crystalline suspensions of all three drugs were considerably faster than the pure drug. The drug release attributes of these crystalline suspensions, according to the solid-state charateristics, were the result of the physical configuration of the extrudates rather than that of drug–excipient interactions. The authors also suggested that, by governing the shear strength and mechanism, it would be possible to control the crystalline drug particle size in the suspension and hence the dissolution rate.

11.5.2 Increased Dissolution Rate Due to Drug Morphology Change

Although long-term instability remains a major issue, the formation of high-energy amorphous or molecular solid dispersions is usually thought to have greater order of magnitude in terms of improving the gastrointestinal solubility compared to the downsized crystalline particles. An informative piece of work conducted by Albers *et al.* [86] elucidated the drug release mechanisms from extruded solid dispersions. In their study, Celecoxib (CEL) was extruded with the cationic butylated methacrylate copolymer Eudragit® EPO at a 1:1 w/w ratio. The authors demonstrated in their work that the formation of a glass solution or a crystalline drug suspension during HME process was highly reliant on the temperature settings of the barrel segments. Insufficient melt (low temperature settings) led to crystalline CEL particles suspended within glassy Eudragit® EPO, which exhibited a drug-controlled release mechanism. A glass solution formed at the higher temperature settings revealed a carrier-controlled release mechanism, which was to some extent independent of the initial particle size and the physical form of the drug. The *in vitro* dissolution study under non-sink conditions illustrated considerable initial supersaturation with both the glass solution and the crystalline suspension, although the extent of supersaturation with the former was nearly seven-fold higher than the latter.

Verreck *et al.* prepared a drug–polymer ratio of 40/60 w/w Itraconazole solid dispersion in HPMC 2910 5 mPas using the HME process [78]. The resultant extrudate was an amorphous solid dispersion wherein complete suppression of Itraconazole crystallinity was indicated in both the DSC thermogram and the PXRD pattern. A significantly increased dissolution rate was observed with the extrudates physically mixed HPMC and crystalline Itraconazole. A subsequent stability test indicated that the extruded amorphous dispersion was stable for up to 6 months when packed in aluminium bags. Six *et al.* conducted a further development of the study in the following year [79]. The latter study confirmed the morphology change of Itraconazole from a stable crystalline form to high-energy amorphous form after the HME process. A significantly improved dissolution performance was observed for the solid dispersion, with more than 80% of the drug released within 30 min. However, it was revealed through the use of hermetically sealed DSC pans, that phase separation had occurred in this system and an HPMC-rich phase was observed.

Tang *et al.* conducted a series of studies on the solubility enhancement of a calcium channel blocker, Nimodipine, through the formation of solid dispersions using HME [84, 87, 156, 157]. HPMC, PVP/VA, and Eudragit® EPO were used individually or together as the matrix polymers. Increased Nimodipine water-solubility was observed in solid dispersions with all three polymers. A subsequent study incorporated pulverized Nimodipine-Eudragit® EPO extrudates within semi-solid capsules (SSC) and thus yielded a rapid, pH-independent dissolution of Nimodipine with prevented recrystallization during release. More recently, the same group have demonstrated a combination of Eudragit® EPO and PVP/VA that was melt-extruded at various weight fractions with Nimodipine. DSC and PXRD indicated the melt-extruded solid dispersions to be amorphous, which markedly increased the solubility and dissolution rate of Nimodipine. The presence of PVP/VA also prevented the recrystallization of Nimodipine in acidic dissolution media to some extent compared to pure Eudragit® EPO solid dispersions.

Another calcium channel blocker, felodipine, was also extruded with Eudragit® EPO and showed dramatically increased dissolution rate and extent (360-fold enhancement in the case of 10% drug loading) in simulated gastric fluids within the first 20 min. of the drug release [101]. Complete conversion from crystalline felodipine to amorphous dispersion was confirmed using DSC and PXRD. However, a fast recrystallization of felodipine was observed immediately following rapid dissolution. The use of a water-insoluble polymer Eudragit® NE to inhibit rapid recrystallization of supersaturated felodipine concentration was evaluated and 5% of Eudragit® NE was determined to be most sufficient in inhibiting recrystallization without decreasing the extent of felodipine dissolution.

Andrews *et al.* [76] melt extruded Bicalutamide (BL) with PVP at various drug-polymer ratios 1:10, 2:10, 3:10 (w/w) using a twin-screw extruder. Extensive DSC studies on both physical mixtures and extrudates offered a significant understanding of both the system and the process: first, a significant decrease in the T_g of PVP in the first heat run of all extrudates upon increasing the BL concentration revealed the solid-state plasticizing effect of BL on PVP; then, the presence of a single T_g in the second heat run was indicative of the miscibility between BL and PVP and the formation of molecular solid dispersions during HME; finally, the absence of recrystallization of amorphous BL neither during melting nor cooling showed the inhibiting effect of PVP against BL recrystallization. The existence of amorphous BL in the extrudates was also confirmed by PXRD. Moreover, FTIR and Raman spectra strongly suggested intermolecular interactions between BL and PVP within the extrudates. Such highly miscible melt-extruded molecular dispersion presented significantly enhanced dissolution properties in comparison to unprocessed crystalline drug: 8.93-fold increase for 1:10 extrudates, 8.05- and 7.53-fold respectively for 2:10 and 3:10 extrudates. The authors also stated that the rate and extent of BL dissolution were highly dependent upon the PVP weight fraction. These molecular dispersions were experimentally proved to be stable for up to 12 months at ambient conditions (20°C, 40%RH). Although higher humidity was found to destabilize these extrudates, interestingly, the extrudates' dissolution performance remained unchanged.

11.5.3 Controlled or Prolonged Release with Enhanced Release Extent

Zhu *et al.* [158] successfully extruded enteric tablets containing indomethacin and acrylic polymers. Indomethacin was found to have a solid-state plasticizing effect on Eudragit® RL PO by decreasing the T_g of the co-melt powder blend. The strong bonding between indomethacin and Eudragit® RD 100, as attributed to the presence of quaternary ammonium groups in Eudragit® RD 100, was shown to be responsible for the prolonged release profile and incomplete drug release. The addition of acrylic polymers that contained anionic groups (Eudragit® L 100 or Eudragit® S 100) weakened the interaction between the drug and Eudragit® RD 100, thus enhancing dissolution performance. The authors proved that HME was a viable method to facilitate the formation of molecular dispersion of the drug molecules within a continuous polymer matrix, and that this could be used to tailor the drug diffusivity from the extrudates and hence improve drug release.

11.5.4 Complexation to Enhance Dissolution Performance

Hot melt extrusion for the formation of Cyclodextrin-Class II drug complexes was patented by Baert *et al.* over ten years ago. The authors demonstrated successful dissolution rate improvement of Itraconazole in gastric acid by melt extruding with 2-Hyroxypropyl-β-Cyclodextrin (HP-β-CD) [159]. Subsequently, a ternary mixture consisting of drug, HP-β-CD and HPMC was designed and optimized to improve the dissolution performance of Itraconazole in 0.1N HCl [160]. It was concluded by the authors that the addition of

HPMC to the Itraconazole/ HP-β-CD complex significantly improved the melt properties during process. Fukuda *et al.* compared the solubility enhancing capabilities between the chemical modified Sulfobutyl ether β-Cyclodextrin (SBE_7-β-CD, Captisol®) and its parent β-Cyclodextrin. It was noted by the authors that the dissolution rate of Ketoprofen from melt extruded SBE_7-β-CD matrix was significantly higher than that from the parent β-CD extrudates due to greater electronic interactions between SBE_7-β-CD and Ketoprofen [43]. More recently, Yano and Kleinebudde [44] achieved an improvement of the dissolution behavior of Indomethacin with the complexation with HP-β-CD in both melt and wet extrusion processes. The main mechanism of significant dissolution enhancement was found to be attributable to the complete amorphization of Indomethacin through the melt extrusion process.

In more recent studies, researchers have innovatively investigated the formation and characteristics of polyelectrolyte complexes of the poorly soluble acid drug Naproxen and the basic polymer EUDRAGIT® E PO prepared via HME [153]. The same complexation strategy was transferred to a second model drug Furosemide. The formation of a furosemide-EUDRAGIT® E PO polyelectrolyte complex was also established by FTIR and Raman spectroscopy. Drug release studies for both polyelectrolyte complexes showed instantaneous release with the addition of pH neutral alkali-halogen electrolytes. The dissolution performances of the two formulated poorly water-soluble model drugs were comparable to that of their soluble ionic salts.

11.5.5 Co-Crystal Formation

As a solvent-free, continuous and scalable novel technology, twin-screw extrusion (TSE) has been considered a viable technique to prepare pharmaceutical co-crystals. This idea was first demonstrated by Medina *et al.* [154] with Caffeine and AMG 517 as model drugs. The formation of the two co-crystals and the loss of the original drugs were confirmed using PXRD and DSC. The co-crystallization mechanism was noted as the result of highly efficient mixing, close material packing and the subsequently improved surface contact between the drug and the co-former particles within the extruder barrel during processing. A more comprehensive piece of work was accomplished with the formation of Ibuprofen-nicotinamide co-crystals in the twin-screw extrusion process [48]. The crystallography was verified using both PXRD and DSC. The contour and size of the extruded co-crystal agglomerates were examined using scanning electronic microscopy (SEM). The resultant co-crystals exhibited favourable mechanical properties and significant improvements in dissolution when compared to unprocessed ibuprofen particles. The authors also noted that by increasing the mixing and shear intensity of the extrusion, the co-crystal yield was facilitated. It was thus evident that mixing was crucial in determining the crystallography of the final product. Nonetheless, a process temperature that was higher than the melting point of ibuprofen, and a relatively low screw speed were also realized to be preferable for the co-crystallization in the melt extrusion process.

11.6 Conclusion

Hot melt extrusion is a developing yet challenging method to enhance aqueous drug solubility. The formation of solid dispersions of poorly water-soluble drugs, specifically of molecular dispersions, wherein the drug is dispersed at a molecular level in a matrix carrier, results in high-energy systems that are beneficial in dissolution but detrimental for physical stability considerations. To date, a significant volume of work has been published demonstrating the potential of HME to enhance drug solubility. Moreover, there are many articles describing the extruded solid dispersion structure; the appreciation of the mechanism by which solubility improvement occurs; the evaluation of the product stability upon aging; and more recently, the prediction of the drug/polymer miscibility as well as the solid dispersion performance during dissolution and storage.

However, it is also clear that pharmaceutical hot melt extrusion technology today still remains more like an art rather than a science. Although it has long been proven as a suitable way of manufacturing solid dispersions with improved aqueous solubility for poorly water-soluble drugs, the reason for the solubility enhancement is somehow more qualitatively than quantitatively described within the literature. The absence of systematic models for the selection of the most appropriate excipients for a chosen drug is also noteworthy within current literature. If indeed possible the construction of such a model requires a thorough understanding of the nature of the interaction between the embedded drug and its surrounding excipients. The expansion of current characterization techniques or the development of novel techniques to accurately and quantitatively determine the properties of extrudates is also of great importance.

References

1. P. Gribbon, S. Andreas, High-throughput drug discovery: What can we expect from HTS? *Drug Discov Today*, **10**(1), 17–22 (2005).
2. C.A. Lipinski, F. Lombardo, B.W. Dominy, P.J. Feeney, Experimental and computational approaches to estimate solubility and permeability in drug discovery and development settings. *Adv Drug Deliv Rev*, **46**(1–3), 3–26 (2001).
3. G.L. Amidon, H. Lennernäs, V.P. Shah, J.R. Crison, A theoretical basis for a biopharmaceutic drug classification: The correlation of *in vitro* drug product dissolution and *in vivo* bioavailability. *Pharmaceutical Research*, **12**(3), 413–420 (1995).
4. A.A. Noyes, W.R. Whitney, The rate of solution of solid substances in their own solutions. *Journal of the American Chemical Society*, **19**, 930–934 (1897).
5. C. Leuner, J. Dressman, Improving drug solubility for oral delivery using solid dispersions. *European Journal of Pharmaceutics and Biopharmaceutics*, **50**(1), 47–60 (2000).
6. Y. Kawabata, K. Wada, M. Nakatani, S. Yamada, S. Onoue, Formulation design for poorly water-soluble drugs based on biopharmaceutics classification system: Basic approaches and practical applications. *International Journal of Pharmaceutics*, **420**, 1–10 (2011).
7. V.R. Vemula, L. Venkateshwarlu, L. Srikanth, Solubility enhancement techniques. *International Journal of Pharmaceutical Sciences Review and Research*, **5**, 41–45 (2010).
8. A. Singh, A.Z. Worku, G. Van den Mooter, Oral formulation strategies to improve solubility of poorly water-soluble drugs. *Expert Opinion on Drug Delivery*, **8**(10), 1361–1378 (2011).

9. D.A. Wadke, A.T.M. Serajuddin, H. Jacobson, Preformulation testing. In: *Pharmaceutical Dosage Forms: Tablets*. H.A. Lieberman, L. Lachman, J.B. Schwartz (eds). Volume 1, New York: Marcel Dekker, pp. 1–73 (1989).
10. A.T.M. Serajuddin, Solid dispersion of poorly water-soluble drugs: Early promises, subsequent problems, and recent breakthroughs. *Journal of Pharmaceutical Sciences*, **88**(10), 1058–1066 (1999).
11. D.J. Hauss, Oral lipid-based formulations. *Advanced Drug Delivery Reviews*, **59**(7), 667–676 (2007).
12. C.J.H. Porter, C.W. Pouton, J.F. Cuine, W.N. Charman, Enhancing intestinal drug solubilisation using lipid-based delivery systems. *Advanced Drug Delivery Reviews*, **60**(6), 673–691 (2008).
13. B.C. Hancock, G. Zografi, Characteristics and significance of the amorphous state in pharmaceutical systems. *J Pharm Sci*. **86**(1), 1–12 (1997).
14. B.C. Hancock, M. Parks, What is the true solubility advantage for amorphous pharmaceuticals? *Pharmaceutical Research*, **17**(4), 397–404 (2000).
15. T. Vasconcelos, B. Sarmento, P. Costa, Solid dispersions as strategy to improve oral bioavailability of poor water soluble drugs. *Drug Discov Today*, **12**(23–24), 1068–1075 (2007).
16. K. Sekiguchi, N. Obi, Studies on absorption of eutectic mixture. I. A comparison of the behaviour of eutectic mixture of sulfathiazole and that of ordinary sulfathiazole in man, *Chemical & Pharmaceutical Bulletin*, **9**, 866–872 (1961).
17. G. Levy, Effect of particle size on dissolution and gastrointestinal absorption rates of pharmaceuticals. *American Journal of Pharmacy and the Sciences Supporting Public Health*, **135**, 78–92 (1963).
18. J.L. Kanig, Properties of fused mannitol in compressed tablets. *J Pharm Sci*, **53**(2), 188–192 (1964).
19. A.H. Goldberg, M. Gibaldi, J.L. Kanig, Increasing dissolution rates and gastrointestinal absorption of drugs via solid solutions and eutectic mixtures. I. Theoretical considerations and discussion of the literature. *Journal of Pharmaceutical Sciences*, **54**, 1145–1148 (1965).
20. A.H. Goldberg, M. Gibaldi, J.L. Kanig, Increasing dissolution rates and gastrointestinal absorption of drugs via solid solutions and eutectic mixtures. II. Experimental evaluation of a eutectic mixture: Urea-acetaminophen system. *Journal of Pharmaceutical Sciences*, **55**, 482–487 (1966).
21. A.H. Goldberg, M. Gibaldi, J.L. Kanig, M. Mayersohn, Increasing dissolution rates and gastrointestinal absorption of drugs via solid solutions and eutectic mixtures. III. chloramphenicol-urea system. *Journal of Pharmaceutical Sciences*, **55**, 487–492 (1966).
22. A.H. Goldberg, M. Gibaldi, J.L. Kanig, M. Mayersohn, Increasing dissolution rates and gastrointestinal absorption of drugs via solid solutions and eutectic mixtures. IV. chloramphenicol-urea system. *Journal of Pharmaceutical Sciences*, **55**(6), 581–583 (1966).
23. W.L. Chiou, S. Riegelman, Preparation and dissolution characteristics of several fast-release solid dispersions of griseofulvin. *Journal of Pharmaceutical Sciences*, **58**, 1505–1510 (1969).

24. A.P. Simonelli, S.C. Mehta, W.I. Higuchi, Dissolution rates of high energy polyvinylpyrrolidone (PVP)-sulfathiazole coprecipitates. *Journal of Pharmaceutical Sciences*, **58**, 538–549 (1969).
25. G. Van den Mooter, The use of amorphous solid dispersions: A formulation strategy to overcome poor solubility and dissolution rate. *Drug Discovery Today: Technologies*, (2012).
26. D.J. van Drooge, W.L.J. Hinrichs, M.R. Visser, H.W. Frijlink, Characterization of the molecular distribution of drugs in glassy solid dispersions at the nano-meter scale, using differential scanning calorimetry and gravimetric water vapour sorption techniques. *Int J Pharm*, **310**(1–2), 220–229 (2006).
27. W.L. Chiou, S. Riegelman, Pharmaceutical applications of solid dispersion systems. *Journal of Pharmaceutical Sciences*, **60**(9), 1281–1302 (1971).
28. S. Qi, P. Belton, K. Nollenberger, N. Clayden, M. Reading, D. Craig, Characterisation and prediction of phase separation in hot-melt extruded solid dispersions: A thermal, microscopic and NMR relaxometry study. *Pharmaceutical Research*, **27**(9), 1869–1883 (2010).
29. M. Vasanthavada, W. Tong, Y. Joshi, M.S. Kislalioglu, Phase behavior of amorphous molecular dispersions I: Determination of the degree and mechanism of solid solubility. *Pharmaceutical Research*, **21**(9), 1598–1606 (2004).
30. H. Konno, T. Handa, D.E. Alonzo, L.S. Taylor, Effect of polymer type on the dissolution profile of amorphous solid dispersions containing felodipine. *European Journal of Pharmaceutics and Biopharmaceutics*, **70**, 493–499 (2008).
31. P.J. Marsac, T. Li, L.S. Taylor, Estimation of drug-polymer miscibility and solubility in amorphous solid dispersion using experimentally determined interaction parameters. *Pharmaceutical Research*, **26**(1), 139–151 (2008).
32. A.C.F. Rumondor, I. Ivanisevic, S. Bates, D.E. Alonzo, L.S. Taylor, Evaluation of drug-polymer miscibility in amorphous solid dispersion systems. *Pharmaceutical Research*, **26**(11), 2523–2534 (2009).
33. Y. Sun, J. Tao, G.G.Z. Zhang, L. Yu, Solubilities of crystalline drugs in polymers: An improved analytical method and comparison of solubilities of indomethacin and nifedipine in PVP, PVP/VA, and PVAc. *Journal of Pharmaceutical Sciences*, **99**(9), 4023–4031 (2010).
34. D. Lin, Y. Huang, A thermal analysis method to predict the complete phase diagram of drug-polymer solid dispersions. *International Journal of Pharmaceutics*, **399**, 109–115 (2010).
35. F. Qian, J. Huang, Q. Zhu, R. Haddadin, J. Gawel, R. Garmise, *et al.* Is a distinctive single T_g a reliable indicator for the homogeneity of amorphous solid dispersion? *Int J Pharm*. **395**(1–2), 232–235 (2010).
36. Y.Y. Zhao, P. Inbar, H. Chokshi, A.W. Malick, D.S. Choi, Prediction of the thermal phase diagram of amorphous solid dispersions by Flory-Huggins theory. *Journal of Pharmaceutical Sciences*, **100**(8), 3196–3207 (2011).
37. R.M. Dannenfelser, H. He, Y. Joshi, S. Bateman, A.T. Serajuddin, Development of clinical dosage forms for a poorly water soluble drug I: Application of polyethylene glycol-polysorbate 80 solid dispersion carrier system. *Journal of Pharmaceutical Sciences*, **93**, 1165–1167 (2004).

38. A.N. Ghebremeskel, C. Vemavarapu, M. Lodaya, Use of surfactants as plasticizers in preparing solid dispersions of poorly soluble API: Stability testing of selected solid dispersions. *Pharmaceutical Research*, **23**(8), 1928–1936 (2006).
39. C.W. Pouton, Formulation of poorly water-soluble drugs for oral administration: Physicochemical and physiological issues and the lipid formulation classification system. *European Journal of Pharmaceutical Sciences*, **29**, 278–287 (2006).
40. Z. Tadmor, I. Klein, *Engineering Principles of Plasticating Extrusion*. New York: Van Nostrand Reinhold (1970).
41. C. Rauwendaal, *Polymer Extrusion*. Revised 4th edn. Gramann, P.J., Davis, B.A., Osswald, T.A., (eds). Munich: Hanser Publishers (2001).
42. J. Liu, F. Zhang, J.W. McGinity, Properties of lipophilic matrix tablets containing phenylpropanolamine hydrochloride prepared by hot-melt extrusion. *European Journal of Pharmaceutics and Biopharmaceutics*, **52**(2), 181–190 (2001).
43. M. Fukuda, D.A. Miller, N.A. Peppas, J.W. McGinity, Influence of sulfobutyl ether β-cyclodextrin (captisol®) on the dissolution properties of a poorly soluble drug from extrudates prepared by hot-melt extrusion. *Int J Pharm*, **350**(1–2), 188–196 (2008).
44. H. Yano, P. Kleinebudde, Improvement of dissolution behavior for poorly water-soluble drug by application of cyclodextrin in extrusion process: Comparison between melt extrusion and wet extrusion. *AAPS PharmSciTech*, **11**(2), 885–893 (2010).
45. A. Almeida, S. Possemiers, M.N. Boone, T.D. Beer, T. Quinten, L.V. Hoorebeke, *et al.* Ethylene vinyl acetate as matrix for oral sustained release dosage forms produced via hot-melt extrusion. *European Journal of Pharmaceutics and Biopharmaceutics*, **77**, 297–305 (2011).
46. M. Thommes, D.R. Ely, M.T. Carvajal, R. Pinal, Improvement of the dissolution rate of poorly soluble drugs by solid crystal suspensions. *Molecular Pharmaceutics*, **8**, 727–735 (2011).
47. C. Medina, D. Daurio, K. Nagapudi, F. Alvarez-Nunez, Manufacture of pharmaceutical co-crystals using twin screw extrusion: A solvent-less and scalable process. *Journal of Pharmaceutical Sciences*, **99**(4), 1693–1696 (2009).
48. R. Dhumal, A. Kelly, P. York, P. Coates, A. Paradkar, Cocrystalization and simultaneous agglomeration using hot melt extrusion. *Pharmaceutical Research*, **27**(12), 2725–2733 (2010).
49. L.P.B.M. Janssen, *Twin Screw Extrusion*. Churchill, S.W. (ed.). New York: Elsevier Scientific (1978).
50. M.A. El-Egakey, M. Soliva, P. Speiser, Hot extruded dosage forms I: Technology and dissolution kinetics of polymeric matrices. *Pharmaceutica Acta Helvetiae*, **46**(1), 31–52 (1971).
51. W. Prapaitrakul, D.L. Sprockel, P. Shivanand, M. Sen, Development of a drug delivery system through melt extrusion. 4th AAPS, Atlanta, pp. S–98 (1989).
52. R. Mank, H. Kala, M. Richter, Preparation of drug containing extrusion pellets with a thermoplastic base. 2. The optimization of drug release. *Pharmazie*, **45**, 592–593 (1990).
53. P. Shivanand, A.S. Hussain, D.L. Sprockel, Factors affecting release of KCl from melt extruded polyethylene disks. *Pharmaceutical Research*, **8**, 185–192 (1991).

54. N. Follonier, E. Doelker, E.T. Cole, Evaluation of hot-melt extrusion as a new technique loading for the production of polymer based pellets for sustained release capsules containing high loading of freely soluble drugs. *Drug Development and Industrial Pharmacy*, **20**, 1323–1339 (1994).
55. O. Sprockel, M. Sen, P. Shivanand, W. Prapaitrakul, A melt extrusion process for manufacturing matrix drug delivery systems. *International Journal of Pharmaceutics*, **155**, 191–199 (1997).
56. J. Breitenbach, Melt extrusion: From process to drug delivery technology. *European Journal of Pharmaceutics and Biopharmaceutics*, **54**(2), 107–110 (2002).
57. M.J. Stevens, J.A. Covas, *Extruder Principles and Operation*. 2nd edn. London: Chapman & Hall (1995).
58. A.W.D. Ferns, Twin-screw machines for polymer compounding operations. *Plastics Polymer*, **42**(8), 149–157 (1974).
59. M. Mollan, Historical overview. In: *Pharmaceutical Extrusion Technology*. I. Ghebre Sellassie, C. Martin (eds). New York: Marcel Dekker, Inc, pp. 1–18 (2003).
60. C. Rauwendaal, 10 twin screw extruders. In *Polymer Extrusion*. Revised 4th edn. Munich: Hanser, pp. 576–633 (2001).
61. M.J. Stevens, J.A. Covas, 10 twin-screw extruders. In *Extruder Principles and Operation*. 2nd edn. London: Chapman & Hall, pp. 316–351 (1995).
62. M.M. Crowley, F. Zhang, M.A. Repka, S. Thumma, S.B. Upadhye, S.K. Battu, *et al.* Pharmaceutical applications of hot-melt extrusion: Part I. *Drug Development and Industrial Pharmacy*, **33**(9), 909–926 (2007).
63. Thermo scientific HAAKE MiniLab II: A micro compounder for small sample amount [Internet]. Available from: http://www.thermo.com/eThermo/CMA/PDFs/Product/productPDF_6857.pdf.
64. W. Thiele, Twin-screw extrusion and screw design. In: *Pharmaceutical Extrusion Technology*. I. Ghebre Sellassie, C. Martin (eds). New York: Marcel Dekker, Inc, pp. 69–98 (2003).
65. A. Dreiblatt, Process design. In: *Pharmaceutica Extrusion Technology*. I. Ghebre Sellassie, C. Martin (eds). New York: Marcel Dekker, Inc, pp. 153–169 (2003).
66. J.A. Biesenberger, S. Lee, A fundamental study of polymer melt devolatilization. part I: Some experiments on foam-enhanced devolatilization. *Polymer Engineering & Science*, **26**(14), 982–988 (1986).
67. J.A. Biesenberger, S. Lee, A fundamental study of polymer melt devolatilization: III more experiments on foam-enhanced DV. *Polymer Engineering & Science*, **27**(7), 510–517 (1987).
68. J.A. Biesenberger, S.K. Dey, J. Brizzolara, Devolatilization of polymer melts: Machine geometry and scale factors. *Polymer Engineering & Science*, **30**(23), 1493–1499 (1990).
69. M. Kaufman, *Plastics and Polymers*. New York (1969).
70. F.N. Cogswell, Stretching flow instabilities at the exits of extrusion dies. *Journal of Non-Newtonian Fluid Mechanics*, **2**(1), 37–44 (1977).
71. H.G. Brittain (ed.), *Physical Characterization of Pharmaceutical Solids*. New York: Marcel Dekker Inc. (1995).
72. S. Bialleck, H. Rein, Preparation of starch-based pellets by hot-melt extrusion. *European Journal of Pharmaceutics and Biopharmaceutics*, **79**(2), 440–448 (2011).

73. T. Loftsson, H. Friðriksdóttir, T.K. Guðmundsdóttir, The effect of water-soluble polymers on aqueous solubility of drugs. *Int J Pharm*, **127**(2), 293–296 (1996).
74. S. Hülsmann, T. Backensfeld, S. Keitel, R. Bodmeier, Melt extrusion – an alternative method for enhancing the dissolution rate of 17β-estradiol hemihydrate. *European Journal of Pharmaceutics and Biopharmaceutics*, **49**(3), 237–242 (2000).
75. R.J. Chokshi, H.K. Sandhu, R.M. Iyer, N.H. Shah, Characterization of physico-mechanical properties of indomethacin and polymers to assess their suitability for hot-melt extrusion process as a means to manufacture solid dispersion/solution. *Journal of Pharmaceutical Sciences*, **94**(11), 2463–2474 (2005).
76. G.P. Andrews, O.A. AbuDiak, D.S. Jones, Physicochemical characterization of hot melt extruded bicalutamide-polyvinylpyrrolidone solid dispersions. *J Pharm Sci*, **99**(3), 1322–1335 (2009).
77. I. Tho, B. Liepold, J. Rosenberg, M. Maegerlein, M. Brandl, G. Fricker, Formation of nano/micro-dispersions with improved dissolution properties upon dispersion of ritonavir melt extrudate in aqueous media. *European Journal of Pharmaceutical Sciences*, **40**(1), 25–32 (2010).
78. G. Verreck, K. Six, G. Van den Mooter, L. Baert, J. Peeters, M.E. Brewster, Characterization of solid dispersions of itraconazole and hydroxypropylmethylcellulose prepared by melt extrusion – part I. *International Journal of Pharmaceutics*, **251**, 165–174 (2002).
79. K. Six, H. Berghmans, C. Leuner, J. Dressman, K. Van Werde, J. Mullens, *et al.* Characterization of solid dispersions of itraconazole and hydroxypropylmethylcellulose prepared by melt extrusion – part II. *Pharmaceutical Research*, **20**(7), 1047–1054 (2003).
80. E. Mehuy, J.P. Remon, C. Vervaet, Production of enteric capsules by means of hot-melt extrusion. *European Journal of Pharmaceutical Sciences*, **24**, 207–212 (2005).
81. Z. Dong, A. Chatterji, H. Sandhu, D.S. Choi, H. Chokshi, N. Shah, Evaluation of solid state properties of solid dispersions prepared by hot-melt extrusion and solvent co-precipitation. *Int J Pharm*, **355**(1–2), 141–149 (2008).
82. I. Ghosh, J. Snyder, R. Vippagunta, M. Alvine, R. Vakil, W. Tong, *et al.* Comparison of HPMC based polymers performance as carriers for manufacture of solid dispersions using the melt extruder. *Int J Pharm*, **419**(1–2), 12–19 (2011).
83. S. Prodduturi, K.L. Urman, J.U. Otaigbe, M.A. Repka, Stabilization of hot-melt extrusion formulations containing solid solutions using polymer blends. *AAPS PharmSciTech*, **8**(2), E1–E10 (2007).
84. X. Zheng, R. Yang, X. Tang, L. Zheng, Part I: Characterization of solid dispersions of nimodipine prepared by hot-melt extrusion. *Drug Development and Industrial Pharmacy*, **33**(7), 791–802 (2007).
85. G.P. Andrews, D.S. Jones, O. Abu-Diak, C.P. McCoy, A.B. Watts, J.W. McGinity, The manufacture and characterisation of hot-melt extruded enteric tablets. *European Journal of Pharmaceutics and Biopharmaceutics*, **69**, 264–273 (2008).
86. J. Albers, R. Alles, K. Matthée, K. Knop, J.S. Nahrup, P. Kleinebudde, Mechanism of drug release from polymethacrylate-based extrudates and milled strands prepared by hot-melt extrusion. *European Journal of Pharmaceutics and Biopharmaceutics*, **71**(2), 387–394 (2009).

87. X. Fu Jijun, S. Li, X. Wang, S. Zhang, X. Tao, X. Zhao, *et al.* Nimodipine (NM) tablets with high dissolution containing NM solid dispersions prepared by hot-melt extrusion. *Drug Development and Industrial Pharmacy*, **37**(8), 934–944 (2011).
88. O. Abu-Diak, D.S. Jones, G.P. Andrews, An investigation into the dissolution properties of celecoxib melt extrudates: Understanding the role of polymer type and concentration in stabilizing supersaturated drug concentrations. *Mol Pharmaceutics*, **8**(4), 1362–1371 (2011).
89. M.M. Crowley, A. Fredersdorf, B. Schroeder, S. Ucera, The influence of guaifenesin and ketoprofen on the properties of hot-melt extruded polyethylene oxide films. *European Journal of Pharmaceutical Sciences*, **22**(5), 409–418 (2004).
90. L. Li, O. AbuBaker, Z.J. Shao, Characterization of poly(ethylene oxide) as a drug carrier in hot-melt extrusion. *Drug Development and Industrial Pharmacy*, **32**(8), 991–1002 (2006).
91. R.J. Chokshi, H. Zia, Hot-melt extrusion technique: A review. *Iranian Journal of Pharmaceutical Research*, **3**, 3–16 (2004).
92. M.M. Crowley, B. Schroeder, A. Fredersdorf, S. Obara, M. Talarico, S. Kucera, Physicochemical properties and mechanism of drug release from ethyl cellulose matrix tablets prepared by direct compression and hot-melt extrusion. *Int J Pharm*, **271**(1–2), 77–84 (2004).
93. L. Mascia, *The Role of Additives in Plastics*. London: Edward Arnold (1974).
94. P. Sakellariou, R.C. Rowe, Interactions in cellulose derivative films for oral drug delivery. *Progress in Polymer Science*, **20**, 889–942 (1995).
95. M.A. Repka, J.W. McGinity, Influence of vitamin E TPGS on the properties of hydrophilic films produced by hot-melt extrusion. *Int J Pharm*, **202**(1–2), 63–70 (2000).
96. S.M. Aharoni, Increased glass transition temperature in motionally constrained semicrystalline polymers. *Polymers for Advanced Technologies*, **9**(3), 169–201 (1998).
97. E. Mehuys, C. Vervaet, J.P. Remon, Hot-melt extruded ethylcellulose cylinders containing HPMC-gelucire core for sustained drug delivery. *Journal of Controlled Release*, **94**, 273–280 (2004).
98. L.D. Bruce, N.H. Shah, A.W. Malick, M.H. Infeld, J.W. McGinity, Properties of hot-melt extruded tablet formulations for the colonic delivery of 5-aminosalicylic acid. *European J Pharm Biopharm*, **59**, 85–97 (2005).
99. C. De Brabander, G. Van den Mooter, C. Vervaet, J.P. Remon, Characterization of ibuprofen as a nontraditional plasticizer of ethyl cellulose. *J Pharm Sci*, **91**(7), 1678–1685 (2002).
100. C. Wu, J.W. McGinity, Influence of methylparaben as a solid-state plasticizer on the physicochemical properties of eudragit RS PO hot-melt extrudates. *European Journal of Pharmaceutics and Biopharmaceutics*, **56**(1), 95–100 (2003).
101. K. Nollenberger, A. Gryczke, C. Meier, J. Dressman, M.U. Schmidt, S. Brühne, Pair distribution function X-ray analysis explains dissolution characteristics of felodipine melt extrusion products. *J Pharm Sci*, **98**(4), 1476–1478 (2009).
102. G. Van den Mooter, I. Weuts, T. De Ridder, N. Blaton, Evaluation of inutec SP1 as a new carrier in the formulation of solid dispersions for poorly soluble drugs. *Int J Pharm*, **316**(1–2), 1–6 (2006).

103. A.N. Ghebremeskel, C. Vemavarapu, M. Lodaya, Use of surfactants as plasticizers in preparing solid dispersions of poorly soluble API: Selection of polymer–surfactant combinations using solubility parameters and testing the processability. *Int J Pharm*, **328**(2), 119–126 (2007).
104. J.S. Chiou, J.W. Barlow, D.R. Paul, Plasticization of glassy polymers by CO_2. *Journal of Applied Polymer Science*, **30**(6), 2633–2642 (1985).
105. M.D. Elkovitch, D.L. Tomasko, J.L. Lee, Supercritical carbon dioxide assisted blending of polystyrene and poly(methyl methacrylate). *Polymer Engineering & Science*, **39**(10), 2075–2084 (1999).
106. D.L. Tomasko, H. Li, D. Liu, X. Han, M.J. Wingert, J.L. Lee, *et al.* A review of CO2 applications in the processing of polymers. *Industrial & Engineering Chemistry Research*, **42**(35), 6431–6456 (2003).
107. M.J. Bortner, D.G. Baird, Absorption of CO_2 and subsequent viscosity reduction of an acrylonitrile copolymer. *Polymer*, **45**(10), 3399–3412 (2004).
108. G. Verreck, A. Decorte, K. Heymans, J. Adriaensen, D. Cleeren, A. Jacobs, *et al.* The effect of pressurized carbon dioxide as a temporary plasticizer and foaming agent on the hot stage extrusion process and extrudate properties of solid dispersions of itraconazole with PVP-VA 64. *European Journal of Pharmaceutical Sciences.* **26**(3–4), 349–358 (2005).
109. G. Verreck, A. Decorte, H. Li, D. Tomasko, A. Arien, J. Peeters, *et al.* The effect of pressurized carbon dioxide as a plasticizer and foaming agent on the hot melt extrusion process and extrudate properties of pharmaceutical polymers. *The Journal of Supercritical Fluids*, **38**(3), 383–391 (2006).
110. M. Sauceau, C. Nikitine, E. Rodier, J. Fages, Effect of supercritical carbon dioxide on polystyrene extrusion. *The Journal of Supercritical Fluids*, **43**(2), 367–73 (2007).
111. M. Sauceau, J. Fages, A. Common, C. Nikitine, E. Rodier, New challenges in polymer foaming: A review of extrusion processes assisted by supercritical carbon dioxide. *Progress in Polymer Science*, **36**(6), 749–766 (2011).
112. L.E.C. Baert, D. Thone, G. Verreck, inventors; Antifungal compositions with improved bioavailability (1997).
113. L.E.C. Baert, G. Verreck, D. Thone, inventors; Janssen Pharmaceutical NV, assignee. Antifungal compositions with improved bioavailability. USA patent US7081255 (2006).
114. L.V.J. Allen, V.A. Yanchick, D.D. Maness, Dissolution rates of corticosteroids utilizing sugar glass dispersions. *Journal of Pharmaceutical Sciences*, **66**(4), 494–497 (1977).
115. M. Meshali, A. Ghanem, Y. Ibraheem, Enhanced absorption and dissolution of trimethoprim from sugar glass dispersions. *Pharmaceutica Acta Helvetiae*, **58**(2), 62–64 (1983).
116. D.A. Miller, J.T. McConville, W. Yang, R.O. Williams 3rd, J.W. McGinity, Hot-melt extrusion for enhanced delivery of drug particles. *Journal of Pharmaceutical Sciences*, **96**(2), 361–376 (2006).
117. G. Van den Mooter, I. Weuts, T. De Ridder, N. Blaton, Evaluation of inutec SP1 as a new carrier in the formulation of solid dispersions for poorly soluble drugs. *Int J Pharm.* **316**(1–2), 1–6 (2006).

118. E. Karavas, E. Georgarakis, M.P. Sigalas, K. Avgoustakis, D. Bikiaris, Investigation of the release mechanism of a sparingly water-soluble drug from solid dispersions in hydrophilic carriers based on physical state of drug, particle size distribution and drug-polymer interactions. *European Journal of Pharmaceutics and Biopharmaceutics*, **66**, 334–347 (2007).
119. J.P. Lakshman, Y. Cao, J. Kowalski, A.T.M. Serajuddin, Application of melt extrusion in the development of a physically and chemically stable high-energy amorphous solid dispersion of a poorly water-soluble drug. *Mol Pharmaceutics*, **5**(6), 994–1002 (2008).
120. S.C. Dyar, M. Mollan, Melt-extruded particulate dispersions. In: *Pharmaceutical Extrusion Technology*, I. Ghebre Sellassie, C. Martin (eds). New York: Marcel Dekker, Inc, pp. 261–276 (2003).
121. Q.M.C. Duncan, The mechanisms of drug release from solid dispersions in water-soluble polymers. *Int J Pharm*, **231**(2), 131–144 (2002).
122. M.E. Matteucci, J.C. Paguio, M.A. Miller, Williams, K.P. Johnston, Highly supersaturated solutions from dissolution of amorphous itraconazole microparticles at pH 6.8. *Molecular Pharmaceutics*, **6**(2), 375–378 (2008).
123. L.S. Taylor, G. Zografi, Spectroscopic characterization of interactions between PVP and indomethacin in amorphous molecular dispersions. *Pharmaceutical Research*, **14**(12), 1691–1698 (1997).
124. D. Law, S.L. Krill, E.A. Schmitt, J.J. Fort, Y. Qiu, W. Wang, *et al.* Physicochemical considerations in the preparation of amorphous ritonavir–poly(ethylene glycol) 8000 solid dispersions. *J Pharm Sci*, **90**(8), 1015–1022 (2001).
125. D. Law, E.A. Schmitt, K.C. Marsh, E.A. Everitt, W. Wang, J.J. Fort, *et al.* Ritonavir–PEG 8000 amorphous solid dispersions: In vitro and in vivo evaluations. *J Pharm Sci*, **93**(3), 563–570 (2004).
126. T. Miyazaki, S. Yoshioka, Y. Aso, S. Kojima, Ability of polyvinylpyrrolidone and polyacrylic acid to inhibit the crystallization of amorphous acetaminophen. *J Pharm Sci*, **93**(11), 2710–2717 (2004).
127. F. Qian, J. Huang, M.A. Hussain, Drug–polymer solubility and miscibility: Stability consideration and practical challenges in amorphous solid dispersion development. *J Pharm Sci*, **99**(7), 2941–2947 (2010).
128. F. Qian, J. Huang, Q. Zhu, R. Haddadin, J. Gawel, R. Garmise, *et al.* Is a distinctive single T_g a reliable indicator for the homogeneity of amorphous solid dispersions? *International Journal of Pharmaceutics*, **395**, 232–235 (2010).
129. J.L. White, P. Mirau, Probing miscibility and intermolecular interactions in solid polymer blends using the nuclear overhauser effect. *Macromolecules*, **26**, 3049–3054 (1993).
130. C. Lau, S. Zheng, Z. Zhong, Y. Mi, Miscibility, intermolecular interactions, and thermal behavior of poly(hydroxy ether of bisphenol A)/poly(ethyl oxazoline) blends macromolecules. *Int J Pharm*, **31**, 7291–7297 (1998).
131. C.W. Kim, J.K. Park, T.S. Hwang, Effect of intramolecular interactions on the miscibility windows of PVC blends with modified SAN copolymers. *Polymer Engineering & Science*, **36**(4), 535–540 (2004).
132. J.F. Willart, E. Dudognon, N. Descamps, T. Larsson, M. Descamps, Lactose-based molecular alloys obtained by solid state vitrification. *J Non Cryst Solids*, **352**(42–49), 4475–4479 (2006).

133. J.F. Willart, N. Descamps, V. Caron, F. Capet, F. Danède, M. Descamps, Formation of lactose-mannitol molecular alloys by solid state vitrification. *Solid State Commun*, **138**(4), 194–199 (2006).
134. E. Karavas, G. Ktistis, A. Xenakis, E. Georgarakis, Effect of hydrogen bonding interactions on the release mechanism of felodipine from nanodispersions with polyvinylpyrrolidone. *European Journal of Pharmaceutics and Biopharmaceutics*, **63**, 103–114 (2006).
135. B.C. Hancock, P. York, R.C. Rowe, The use of solubility parameters in pharmaceutical dosage form design. *International Journal of Pharmaceutics*, **148**, 1–21 (1996).
136. B.C. Hancock, G. Zografi, Characteristics and significance of the amorphous state in pharmaceutical systems. *Journal of Pharmaceutical Sciences*, **86**, 1–12 (1997).
137. H. Suzuki, H. Sunada, Influence of water-soluble polymers on the dissolution of nefedipine solid dispersions with combined carriers. *Chemical & Pharmaceutical Bulletin*, **45**, 1688–1693 (1997).
138. D. Greenhalgh, A. Williams, P. Timmins, P. York, Solubility parameters as predictors of miscibility in solid dispersions. *Journal of Pharmaceutical Sciences*, **88**, 1182–1190 (1999).
139. A. Forster, J. Hempenstall, I. Tucker, T. Rades, Selection of excipients for melt extrusion with two poorly water-soluble drugs by solubility parameter calculation and thermal analysis. *Int J Pharm*, **226**(1–2), 147–161 (2001).
140. P.J. Marsac, T. Li, L.S. Taylor, Estimation of drug-polymer miscibility and solubility in amorphous solid dispersions using experimentally determined interaction parameters. *Pharmaceutical Research*, **26**(1), 139–151 (2008).
141. J. Tao, Y. Sun, G.G.Z. Zhang, L. Yu, Solubility of small molecular crystals in polymers: D-mannitol in PVP, indomethacin in PVP/VA, and nifedipine in PVP/VA. *Pharmaceutical Research*, **26**(4), 855–864 (2008).
142. D. Lin, Y. Huang, A thermal analysis method to predict the complete phase diagram of drug-polymer solid dispersions. *International Journal of Pharmaceutics*, **399**, 109–115 (2010).
143. Y. Sun, J. Tao, G.G.Z. Zhang, L. Yu, Solubilities of crystalline drugs in polymers: An improved analytical method and comparison of solubilities of indomethacin and nifedipine in PVP, PVP/VA, and PVAc. *Journal of Pharmaceutical Sciences*, **99**(9), 4023–4031 (2010).
144. Y.Y. Zhao, P. Inbar, H.P. Chokshi, A.W. Malick, D.S. Choi, Prediction of the thermal phase diagram of amorphous solid dispersions by Flory-Huggins theory. *Journal of Pharmaceutical Sciences*, **100**(8), 3196–3207 (2011).
145. O. Olabisi, L.M. Robeson, M.T. Shaw, *Polymer-Polymer Miscibility*. San Diego: Academic Press, Inc. (1979).
146. E. Karavas, G. Ktistis, A. Xenakis, E. Georgarakis, Miscibility behaviour and formation mechanism of stabilized felodipine polivinylpyrrolidone amorphous solid dispersions. *Drug Development and Industrial Pharmacy*, **31**, 473–489 (2005).
147. A.C.F. Rumondor, I. Ivanisevic, S. Bates, D.E. Alonzo, L.S. Taylor, Evaluation of drug-polymer miscibility in amorphous solid dispersion systems. *Pharmaceutical Research*, **26**(11), 2523–2534 (2009).
148. S. Krause, I. Magdy, Phase separation in styrene-alpha-methyl styrene block copolymers. *Polymer Science and Technology*, **10**, 231–243 (1977).

149. S. Wies, A. Geyer, W. Eysel, The limit of detection in differential scanning calorimetry. *Journal of Thermal Analysis and Calorimetry*, **38**(3), 277–278 (1992).
150. D. Lin-Vien, N.B. Colthup, W.G. Fateley, J.G. Grasselli, *The Handbook of Infrared and Raman Characteristic Frequencies of Organic Molecules*. San Diego: Academic Press, Inc (1991).
151. X.C. Tang, M.J. Pikal, L.S. Taylor, A spectroscopic investigation of hydrogen bond patterns in crystalline and amorphous phases in dihydropyridine calcium channel blockers. *Pharmaceutical Research*, **19**(4), 477–483 (2002).
152. T. Loftsson, inventor; Cyclodextrin complexation. US patent 5472954 (1995).
153. C. Kindermann, K. Matthée, J. Strohmeyer, F. Sievert, J. Breitkreutz, Tailor-made release triggering from hot-melt extruded complexes of basic polyelectrolyte and poorly water-soluble drugs. *European Journal of Pharmaceutics and Biopharmaceutics*, **79**(2), 372–381 (2011).
154. C. Medina, D. Daurio, K. Nagapudi, F. Alvarez-Nunez, Manufacture of pharmaceutical co-crystals using twin screw extrusion: A solvent-less and scalable process. *Journal of Pharmaceutical Sciences*, **99**(4), 1693–1696 (2010).
155. S. Hülsmann, T. Backensfeld, S. Keitel, R. Bodmeier, Melt extrusion – an alternative method for enhancing the dissolution rate of 17β-estradiol hemihydrate. *European Journal of Pharmaceutics and Biopharmaceutics*, **49**(3), 237–242 (2000).
156. J. Fu, L. Zhang, T. Guan, X. Tang, H. He, Stable nimodipine tablets with high bioavailability containing NM-SD prepared by hot-melt extrusion. *Powder Technol*, **204**(2–3), 214–221 (2010).
157. J. Fu, L. Xu, X. Tao, M. Su, M. Zhao, H. He, *et al.* The inhibition effect of high storage temperature on the recrystallization rate during dissolution of nimodipine–kollidon VA64 solid dispersions (NM-SD) prepared by hot-melt extrusion. *J Pharm Sci*, **100**(5), 1643–1647 (2011).
158. Y. Zhu, N.H. Shah, M.A. Waseem, M.H. Infeld, J.W. McGinity, Controlled release of a poorly water-soluble drug from hot-melt extrudates containing acrylic polymers. *Drug Development and Industrial Pharmacy*, **32**, 569–583 (2006).
159. L.E.C. Baert, J. Peeters, G. Verreck, inventors; Process for preparing a solid mixture comprising one or more cyclodextrins and an insoluble active ingredient characterized in that said process comprises a melt-extrusion step, wherein the active ingredient is embedded into the cyclodextrin carrier. US patent 6365188 (2002).
160. B. Rambali, G. Verreck, L. Baert, D.L. Massart, Itraconazole formulation studies of the melt-extrusion process with mixture design. *Drug Development and Industrial Pharmacy*, **29**(6), 641–652 (2003).

12

Penetration Enhancers, Solvents and the Skin

Jonathan Hadgraft and Majella E. Lane

12.1 Introduction

The skin is the largest and most accessible organ of the body and it is surprising how difficult it is to treat common skin complaints such as eczema and psoriasis. This is becoming an increasingly significant problem as the incidence of these conditions becomes more prevalent. This is a result of changes in lifestyle and living conditions. The major problem in treating the skin is its unique barrier function which has evolved to prevent the body from losing excessive water. In order to understand the role of solubility in formulation approaches to topical and transdermal delivery, it is important to have a basic knowledge of the structure of the skin and how this impacts on permeation. In general terms, the skin is a multi-layered membrane overlying sub-cutaneous fat. The lowest layer is the dermis and above this is the epidermis. The epidermis is composed of two major layers, the lower of the two being the viable tissue. It is this layer and the underlying dermis that contain the most common target regions for topical delivery. The blood vessels also traverse the junction of the epidermis and dermis and it is these that must be reached in order to achieve transdermal delivery. Thus the formulation factors that need to be considered for both topical and transdermal delivery are very similar.

The top outermost layer of the skin is the stratum corneum. This thin layer of ~15 μm [1] provides the major barrier to the absorption of topical and transdermal drugs. It consists of corneocytes 'riveted' together by corneodesmosomes. The corneocytes are dense keratinized polygonal cells (usually pentagonal or hexagonal in shape) and they overlap to resemble a brick wall [2]. The 'mortar' between the bricks is a complex mixture of lipids such as ceramides, cholesterol, cholesterol sulphate, free fatty acids and esters,

Drug Delivery Strategies for Poorly Water-Soluble Drugs, First Edition. Edited by Dennis Douroumis and Alfred Fahr.

both of which usually possess long alkyl chains [3]. In terms of the barrier properties, the important feature is that they are structured into bi-layer arrays [4]. It is also pertinent to note that the route of permeation through the stratum corneum is thought to be through the intercellular spaces, the 'mortar' [5]. Therefore, a permeant has to pass along a tortuous route and has to cross, sequentially, a number of bi-layers. It is this combination of features that gives the stratum corneum its unique barrier properties. There are very few xenobiotics that possess properties which will allow fast and significant permeation through the stratum corneum. It is clear that one of the important features of a permeant or solvent is its solubility in the skin lipids. Over 50 years ago, Hadgraft and Somers [6] wrote 'maximal percutaneous absorption occurs when the medicament combines lipoid solubility with a moderate solubility in water'. It should be evident to a formulation scientist that penetration of a solvent into the skin lipids will alter their properties and affect the solubility of the permeant in this region. It is important, therefore, to understand the role of solubility as well as diffusion within the skin.

12.2 Interactions of Solvents and Enhancers with the Skin

In order to understand how solvents/enhancers can interact with the skin it is useful to understand Fick's laws of diffusion. The simplest form can be written:

$$J = KDc/h \tag{12.1}$$

where J is the rate of diffusion across the skin, K is the partition coefficient between the skin and the applied solution, c is the concentration of the diffusant in the applied solution and h is the diffusional path length. In this form of the equation the concentration on the receptor (blood) side of the skin is assumed to be very small and therefore tends to zero.

In most cases the driving force for diffusion is the concentration. More accurately, it is the chemical potential which is related to the activity state of the diffusant in the solution. At saturation the activity is 1, below saturation it is less than 1. It is possible to create metastable supersaturated states, in which case the activity is above 1. In these latter cases, penetration enhancement is caused by the high activity state. However, it must be remembered that these states are thermodynamically unstable.

Penetration enhancement can be achieved by increasing the partition coefficient (K) or by increasing the diffusion coefficient (D). It may be difficult to determine how exactly a penetration enhancer is operating and it is possible that it has an effect both on K and D. Experimentally these can be very difficult to separate. A final and less common means of enhancing permeation is extraction of the skin lipids, for example, a mixture of methanol: chloroform extracts non-bonded lipids from the intercellular channels.

In the first instance, it is easier to consider the case where the formulation is considered not to interact with the skin, i.e. the components remain on the skin surface. Such cases were examined by Katz and Poulsen in the late 1960s and early 1970s [7]. If a saturated solution of the diffusant is considered, its concentration will be c_{sat}. Katz and Poulsen examined the solubility and partition behavior of two steroids, fluocinolone acetonide and its acetate ester, in different water propylene glycol mixtures [7, 8]. In water the steroids are very sparingly soluble, but in mixtures of water propylene glycol (PG) there is an exponential increase in solubility as the PG concentration increases. This is mirrored by a decrease in the partition behavior. For optimum delivery the product Kc_{sat} has to be maximum. This will occur at different mixtures, depending on the steroid. For example, for fluocinolone acetonide, it is about 25% PG, whereas for the acetate ester the optimum occurs around 70% PG. This was confirmed by release studies, *in vitro* penetration across skin and vasoconstriction studies *in vivo*.

Another two component solvent mixture can be envisaged in which there is a volatile component, e.g. isopropanol and a non-volatile component, e.g. isopropyl myristate. If the steroid is more soluble in the volatile component than the non-volatile one, then as it evaporates it can create a situation in which the steroid is supersaturated in the residual phase and therefore the driving force for penetration across the skin will be greater and enhanced delivery will occur [9]. However, the statements made above apply for the simple case in which the solvents do not interact with the skin. Clearly they can and a number of different cases are possible.

12.2.1 Small Solvents

Ethanol is a small volatile solvent which will evaporate quickly from the skin surface, however, if evaporation is prevented, as in the case of the original reservoir transdermal patch containing fentanyl (Durogesic®), the ethanol will permeate into and across the skin and it acts as an enhancer for the fentanyl. The main reason that Durogesic® patches stop delivering the active is that the reservoir of ethanol depletes over the application time of 3 days. The ethanol appears to penetrate into and across the skin and appears to act in a genuine 'solvent' type capacity keeping the fentanyl in solution.

Similar solvency is created by propylene glycol and the presence of this in the skin also aids partition into the skin lipids for certain compounds, thus it has been used extensively as a penetration enhancer, especially in combination with other components. It should also be noted that it has a certain amount of volatility and will disappear from the skin surface, both as a result of penetration and evaporation. This is particularly significant for the application of topical products where the dosing is of the order of 2 mg/cm^2. This corresponds to a layer about 20 μm thick, i.e. similar to the thickness of the stratum corneum. Depletion of the solvent as a result of evaporation and penetration can restrict its efficiency. For example, in permeation studies on oxybutynin from a finite dose, oxybutynin was left on and in the skin because the PG in which it was dissolved disappeared from the skin, leaving about 80% of the active 'stranded' and therefore bio-unavailable [10]. Similar solvents include diethylene glycol, dipropylene glycol and the ether glycols such as diethylene glycol monomethyl ether, Transcutol® and butylene glycol.

12.2.2 Solvents with Longer Carbon Chains

Because of the nature of the skin lipids, it is likely that longer chain materials will penetrate into the lipids and intercalate with them, particularly if they have straight chains and a polar head group. Azone® has been extensively studied though not commercialized as a penetration enhancer. Its structure has a seven-membered ring with a 12 carbon chain.

Structure activity studies on this compound by Boddé *et al.* [11] showed that a carbon chain of 12 or 14 was optimum for penetration enhancement activity. Looking at the structure of Azone®, it is probable that the ring structure perturbs the packing of the polar head groups of the skin lipids and the chain length anchors it in place. Azone® disrupts the lipid chains of the skin lipids, making them more fluid, and hence diffusion of the permeant is enhanced. From studies on perdeuterated Azone, it appears that Azone intercalates homogeneously into the skin lipids rather than phase separating [12]. It is possible that the solvent power of Azone® also promotes penetration. It is a lipophilic molecule and, unlike propylene glycol, its residence time in the stratum corneum is quite prolonged since it will not partition well into the underlying aqueous viable epidermis [13].

It is interesting to note that combinations of propylene glycol and Azone have synergistic effects on the promotion of penetration. This has been attributed to PG increasing K and Azone increasing D; therefore the effect is multiplicative [14]. Multiplicative effects have also been seen with permeation enhancers that increase D and the application of supersaturated solutions [15, 16].

Azone has been extensively studied whereas other long straight alkyl chain enhancers such as isopropyl myristate (IPM), isopropyl palmitate (IPP), isopropyl stearate (IPS), isostearyl isostearate (ISS), lauryl alcohol (LA), or methyl laurate (MA) have not. It is likely that they act by increasing the solubility of the permeant in the skin and maybe by making the lipid environment more fluid. They are lipophilic in nature and therefore they are likely to have prolonged residence time in the skin. The effects on permeation enhancement are likely to be more similar, the more similar the structures are related, e.g. IPM and IPP are very similar but different from LA, MA, IPS and ISS.

Another, but related, series of permeation enhancers are those related to oleic acid (OA). This structure has a *cis* double bond about halfway along the carbon chain. Infra-red studies using perdeuterated oleic acid imply that the presence of the double bond induces phase separation and that pools of oleic acid form in the skin lipids and this promotes permeation of the diffusant either through the liquid oleic acid pools or through the defects between the oleic acid and the endogenous skin lipids. Examples of penetration enhancers currently used in commercial products include both types of molecule (Table 12.1).

Vaughan [17] characterized a set of cosmetic ingredients in terms of their simple solubility parameter. The scale was felt to be useful in the selection of solvents in topical formulations. Subsequently, a more complete description of the solubilizing properties of solvents has been proposed by Hansen [18] and this describes the properties in a three-dimensional (3D) space. In this analysis, solvents are represented in a three-dimensional array called the Hansen Solubility Parameter (HSP) space. Three parameters (D, P, and H) are required to describe whether the particular solvents are miscible and whether alone or in combination they will dissolve a solute (API). D, P and H describe the nonpolar, atomic (dispersion) interactions, E_D, permanent dipole-permanent dipole molecular interactions, E_P, and the hydrogen bonding (electron interchange) molecular interactions, E_H. The solvents/enhancers can be classified in terms of their 3D Hansen solubility parameter. Some of these are shown in the 3D plot (Figure 12.1).

It is interesting to note that where combinations of enhancers are used, they tend to be from the top right of the 3D plot with those from the bottom left. The utility of this approach has been described in more detail by Wiechers [19] where he discusses formulation for efficacy.

For a series of infinite dose experiments, Guy and co-workers [20] showed that the partition coefficient of ibuprofen was close to unity, irrespective of the formulation used. This suggests that formulation components enter the stratum corneum lipids and alter their solubility characteristics such that they are dominated by the presence of the solvents

Table 12.1 *Penetration enhancers currently used in commercial formulations.*

Active	Patch trade name	Enhancer
Fentanyl	Fentalis Matrifen	Ethanol Dipropylene glycol
Nitroglycerin	Minitran	Ethyl oleate Glyceryl monolaurate
Estradiol	Estraderm Estradot	Isopropyl palmitate Dipropylene glycol Oleyl alcohol
	Fematrix	Diethyl-m-toluamide
	Oestrogel	Ethanol
	Progynova	Ethyl oleate Glycerol monolaurate Isopropyl myristate
Ethinyl estradiol, norelgestromin	Evra	Lauryl lactate
Oxybutynin	Kentera	Triacetin
Testosterone	Andropatch	Ethanol Glycerol mono-oleate Methyl oleate
	Axiron	Octyl salicylate
	Intrinsa	Sorbitan oleate
	Testim	Ethanol Pentadecalactone Propylene glycol
	Testogel	Ethanol Isopropyl myristate
	Tostran	Ethanol Isopropyl alcohol Oleic acid Propylene glycol

entering the skin. This again exemplifies the importance of solubility in the formulation of topical formulations. The next section considers the role of solvents in ibuprofen delivery across skin for a set of commonly used penetration enhancers/co-solvents.

12.3 Skin Permeation Enhancement of Ibuprofen

Ibuprofen is a weakly acidic drug with a pK_a of approximately 4.4 [21]. Because of its non-steroidal anti-inflammatory activity, it is formulated for topical delivery in a range of creams, gels and sprays for the relief of soft tissue injury. As it is a relatively insoluble active, suitable solvents are required to engineer formulations which contain adequate ibuprofen

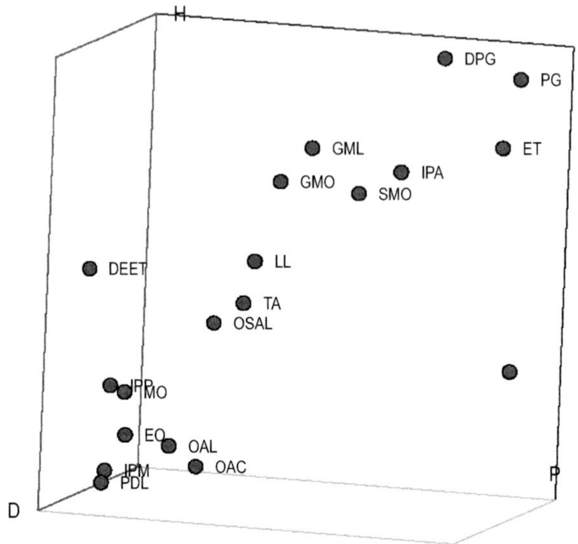

Figure 12.1 *Penetration enhancers plotted in Hansen 3D space.*

for therapeutic efficacy. The following sections summarize studies on various solvents selected to enhance skin delivery of ibuprofen. Both infinite and finite dose conditions are considered.

12.3.1 Infinite Dose Conditions

12.3.1.1 Ethanol

The skin partitioning of the ionized and neutral form of a weak acid can differ by several orders of magnitude and the addition of a co-solvent such as ethanol can modify the dissociation constant of the weak acid at a given pH. The precise outcome will depend on the nature of the co-solvent, the pH of the formulation and the pK_a of the drug. The influence of increasing amounts of ethanol on the solubility, ionization, partition and permeability characteristics of ibuprofen in skin was investigated by Watkinson *et al.* [22].

Flux values were calculated by monitoring the cumulative amount (Q) of drug diffused as a function of time (t). The permeability coefficient (k_p) was derived from the linear portion of this profile when steady state transport had been achieved. Partition (Kh) and diffusion parameters (D/h^2) were derived by fitting the entire Q versus t profile to Equation 12.2 [23]

$$Q = c_v \alpha \left[\beta t - \frac{1}{6} - \frac{2}{\pi^2} \sum_{n=1}^{\infty} \frac{(-1)}{h^2} \exp(n^2 \pi^2 \beta t) \right] \qquad (12.2)$$

where K is the membrane-vehicle partition coefficient of the drug, D is its diffusivity in the membrane of thickness h, α represents Kh, and β represents D/h^2. Addition of just 25% ethanol increased the solubility of ibuprofen to 342 mg/mL, almost 4000-fold higher than its aqueous solubility. In pure ethanol the solubility was 5500-fold higher than in

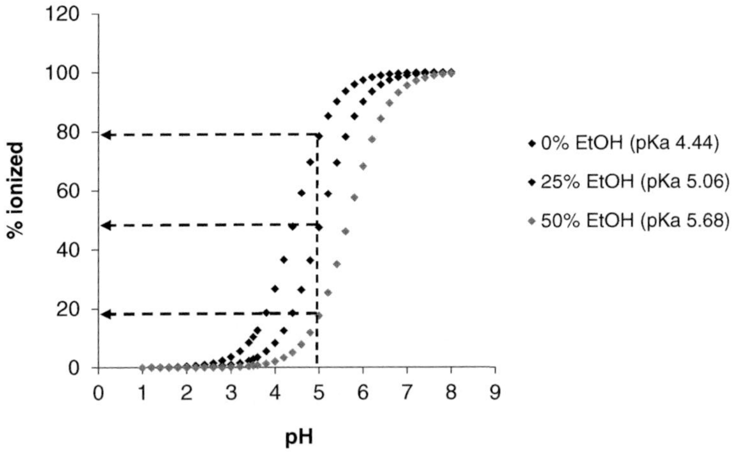

Figure 12.2 *Change of* pK_a *with increasing % of ethanol for ibuprofen.* (See colour plate section.)

water, demonstrating the effective solvent power of ethanol. At pH 5, a plot of fraction of ibuprofen ionized as a function of pH (Figure 12.2) shows that increasing amounts of ethanol shift the pK_a from 4.4 to 5.7. In a 50:50 ethanol water formulation, the percentage of the ionized form of ibuprofen decreased to ~15%.

The flux through skin increased in a stepwise manner as the amount of ethanol in the formulation increased (Table 12.2). The low flux value for the pure ethanol solution might be attributed to two possible events: evaporation of ethanol during the experiment, or

Table 12.2 *Steady-state fluxes, thermodynamic and kinetic parameters, permeability coefficients for the permeation of ibuprofen from saturated formulations containing ethanol and water through human epidermis at 32°C*

Vehicle Ethanol/ water (v/v)	J (μg/cm²/hr) ± SD	K*h (cm) ± SD	D/h^2 (s) ± SD	k_p (cm/s) ± SD
0/100	24.4 ± 2.69	1.32 ± 0.65	$7.49 \times 10^{-5} \pm 2.83 \times 10^{-5}$	$6.38 \times 10^{-5} \pm 2.95 \times 10^{-5}$
25/75	48 ± 19.34	$3.89 \times 10^{-4} \pm 2.51 \times 10^{-4}$	$3.02 \times 10^{-3} \pm 6.59 \times 10^{-3}$	$3.94 \times 10^{-8} \pm 1.55 \times 10^{-8}$
50/50	273.8 ± 32.71	$9.67 \times 10^{-4} \pm 2.15 \times 10^{-3}$	0.147 ± 0.22	$1.46 \times 10^{-7} \pm 7.28 \times 10^{-8}$
75/25	293.3 ± 23.63	$1.76 \times 10^{-3} \pm 7.82 \times 10^{-4}$	$1.22 \times 10^{-4} \pm 8.40 \times 10^{-5}$	$1.67 \times 10^{-7} \pm 1.32 \times 10^{-8}$
100/0	59.8 ± 23.93	$6.83 \times 10^{-3} \pm 5.85 \times 10^{-3}$	$1.88 \times 10^{-5} \pm 7.19 \times 10^{-6}$	$1.04 \times 10^{-7} \pm 7.43 \times 10^{-8}$

Note: Mean ± SD, $n = 5$.

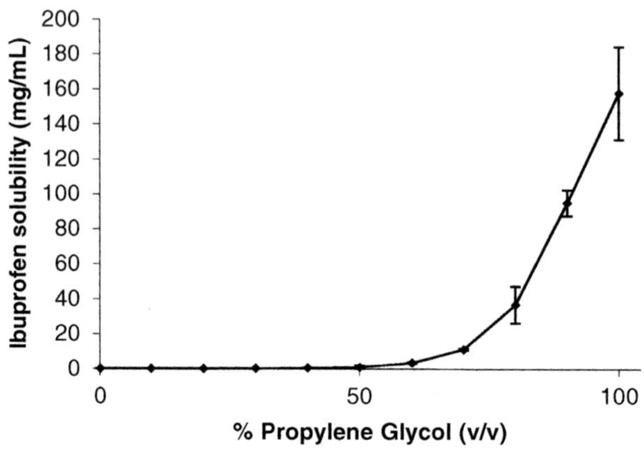

Figure 12.3 *Solubility of ibuprofen in PG and water mixtures at 32°C.*
Note: Mean ± SD, n = 3.

dehydration of the stratum corneum. There was no difference in the partition parameters for vehicles containing 25% to 100% ethanol. There was a significant difference between the pure water formulation and the other formulations, suggesting that ethanol alters the manner in which ibuprofen partitions into skin, increasing the flux. Diffusion parameters for ethanol and water formulations did not vary significantly, suggesting that the diffusional parameter does not play a significant part in the enhancement of permeation for these systems.

12.3.1.2 Propylene Glycol

The influence of propylene glycol (PG) on ibuprofen permeation on partition and diffusion parameters in skin was also investigated [24]. Figure 12.3 shows the exponential increase in ibuprofen solubility with increasing PG content. Above 60% v/v PG there is a marked increase in the solubility of ibuprofen; the drug's solubility in pure PG is ~1750 times greater than its aqueous solubility. This exponential increase has also been reported by Manrique *et al.* [25] who investigated the solubility of ibuprofen at several temperatures in PG–water co-solvent mixtures.

The observed solubilities were lower than those previously reported for ethanol/water systems [22] since PG is a more polar solvent than ethanol [26, 27].

For the binary PG:water systems, as the amount of PG increased, so did the permeation rate (Table 12.3). The flux increase is particularly noticeable between 25% and 50% PG. There was no significant difference in the ibuprofen fluxes from the PG/water 50:50 mixture and from pure PG. The thermodynamic parameters derived for the formulations containing 0 and 25% PG were not significantly different but there is a significant difference in these parameters compared with the 50% PG mixture and pure PG. The kinetic parameter does not alter significantly across the range of formulations used in the study.

As the amount of PG in the formulation increased, so did the permeation rate in skin for binary PG/water systems. The flux increase is particularly noticeable between 25–50% PG. Overall, flux values from PG/water were lower than from corresponding ethanol:water

Table 12.3 Steady-state fluxes, thermodynamic and kinetic parameters and permeability coefficients of ibuprofen through human epidermis from saturated formulations containing PG and water at 32°C.

PG:Water (v/v)	J ($\mu g/cm^2/hr$) ± SD	K*h (μm)	D/h^2 (hr^{-1})	k_p (cm/hr)
0/100	24.4 ± 2.69	1183 ± 503	0.23 ± 0.02	0.27 ± 0.03
25/75	35.5 ± 9.77	460 ± 260	0.7 ± 0.3**	0.26 ± 0.08
50/50	99.1 ± 22.77*	2.5 ± 0.2*	0.4 ± 0.1**	$9.58 \times 10^{-4} \pm 4.86 \times 10^{-5}$***
100/0	133.8 ± 36.1*	1.36 ± 0.15*	0.6 ± 0.4**	$8.24 \times 10^{-4} \pm 1.75 \times 10^{-4}$***

Notes: Mean ± SD, $n = 5$.
*$p < 0.05$ (comparison to 0/100 and 25/75 systems).
**$p < 0.05$ (comparison to 0/100 system).
***$p < 0.05$ (comparison to 0/100 and 25/75 systems).

mixtures [22, 27]. The derived kinetic parameters are independent of the vehicle composition. In contrast, the thermodynamic parameters derived for the formulations containing 90% and 25% PG are significantly different from those with 50% and 100% PG, and the values of k_p reflect these differences too. The results were also consistent with *in vivo* data [28] which showed that ibuprofen delivery into the stratum corneum increased with increasing percentage of PG in PG/water formulations and that the differences observed were because of changes in the partition coefficient, not the drug's diffusivity across the stratum corneum. In addition, these workers also observed a clear decrease in K with increasing amounts of PG in the vehicle. The primary influence of PG in the simple PG/water systems therefore, appears to be on the solubility and partitioning behavior of ibuprofen.

12.3.1.3 Ethanol: Propylene Glycol: Water

Ternary systems comprising of ethanol, propylene glycol and water have recently been investigated [24]. The solubilities of ibuprofen in ethanol/PG/water systems studied are given in Table 12.4. These values are approximately 1500-fold and 3000-fold greater, for the 25:25:50 system and the 50:25:25 systems respectively, than the aqueous solubility of ibuprofen (0.09 mg/ml, [22]).

Table 12.4 Solubility of ibuprofen in ethanol/PG/water mixtures at 32°C.

Ethanol:PG:Water	Solubility (mg/ml)
25:25:50	133 ± 25
50:25:25	292 ± 5

Note: Mean ± SD, $n = 3$.

Table 12.5 *Steady-state fluxes, thermodynamic and kinetic parameters and permeability coefficients of ibuprofen through human epidermis from saturated formulations containing ethanol, PG and water through human epidermis at 32°C.*

EtOH:PG: Water (v/v)	J (μg/cm²/hr) ± SD	K*h (μm) ± SD	D/h² (hr⁻¹) ± SD	k_p (cm/hr) ± SD
25/25/50	77.6 ± 9.04	1.07 ± 0.94	0.63 ± 0.25	$6.70 \times 10^{-4} \pm 9.82 \times 10^{-5}$
50/25/25	203.5 ± 36.24*	2.83 ± 0.30*	0.25 ± 0.02*	$7.09 \times 10^{-4} \pm 1.24 \times 10^{-4}$

Notes: Mean ± SD, $n = 5$.
*$p < 0.05$.

The results of permeation studies conducted using ternary solvent formulations applied to human skin are given in Table 12.5.

Ibuprofen flux was almost threefold higher from the ethanol-rich formulation relative to that from the vehicle with less of the volatile solvent. The thermodynamic and kinetic parameters were significantly different for the two vehicles ($p < 0.05$).

12.3.2 Finite Dose Conditions

When very small topical doses are applied, there is considerable depletion of both the active and the solvent, the latter both by diffusion into, and across the skin, and also by evaporation if the solvent is volatile. Because of this depletion it is likely that the amount of active that gets into the stratum corneum is more important than the role of conventional partition. This is exemplified in the case of a study of finite doses of ibuprofen applied to the skin *in vitro* in a variety of solvents.

The ibuprofen was applied in a co-solvent mixture of isopropanol (95%) and a non-volatile solvent (5%) at a total dose of 3.6 μl/cm². The experimental design was such that when the isopropanol evaporated, it left a residual phase in which the ibuprofen was at saturation. The non-volatile solvents selected were: dipropylene glycol (DIPG), glycerol (GLY), isopropyl myristate (IPM), polyethylene glycol 200 and polyethylene glycol 400 (PEG 200; PEG 400), propylene glycol (PG), propylene glycol monolaurate (PGML), octyl salicylate (OSAL), Transcutol® (TC®) and triacetin (Tri). The amounts of ibuprofen permeating the skin over a 24-hour period (Q_{24H}) were monitored and the results are shown in Figure 12.4.

Figure 12.4 shows that at these low dose rates it appears that the amount of drug absorbed is related to the solubility; although the drug will be at saturation initially and the driving force for diffusion should be the same, there will be rapid depletion as it enters the outer layer of the stratum corneum and the thermodynamic activity will drop rapidly from 1. This exemplifies the differences between finite and infinite doses and the need to conduct experiments that are relevant to the actual use of the formulation by patients or in a clinical setting.

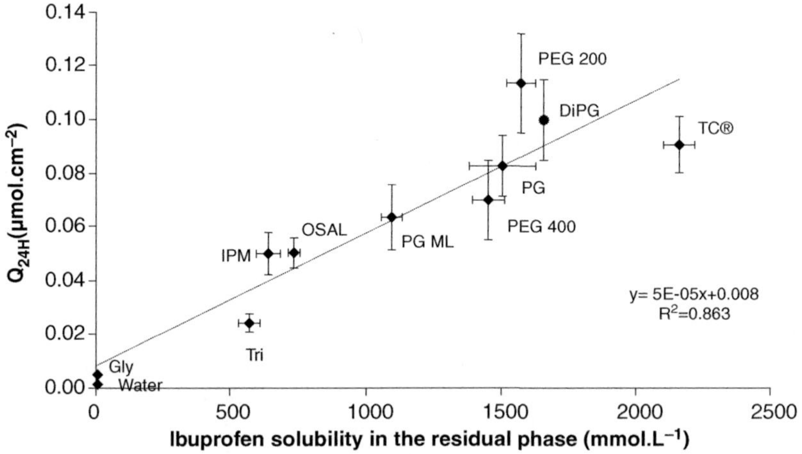

Figure 12.4 *The relationship between the amounts of ibuprofen permeated in vitro across human skin after 24 hours and the solubility of ibuprofen in the non-volatile solvent [29].*

12.4 Conclusion

One the most important characteristics of a drug molecule with respect to effective skin delivery is its solubility in the skin lipids. The interaction of a solvent or vehicle with the skin may alter the solubility of a drug in the lipids and effects on drug diffusion will also occur, depending on the solvent(s). This chapter outlines the important role of solubility for penetration enhancement in skin. Examples of penetration enhancers in commercial transdermal formulations are provided. The effects of solvents on drug permeation are considered for both infinite and finite dose conditions for a weakly acidic drug, ibuprofen. Attempts to delineate whether solvents influence partition and diffusion, or both, are discussed for infinite dose conditions. For finite dose conditions, solubility of the active in the vehicle appears to be the dominant factor influencing skin permeation.

References

1. P.J. Caspers, G.W. Lucassen, E.A. Carter, H.A. Bruining, G.J. Puppels, In vivo confocal Raman microspectroscopy of the skin: Noninvasive determination of molecular concentration profile, *J Invest Dermatol*, **116**(3): 434–442 (2001).
2. A.S. Michaels, S.K. Chandrasekaran, J.E. Shaw, Drug permeation through human skin: Theory and in vitro experimental measurement, *AIChE*, **21**: 985–996 (1975).
3. D.T. Downing, M.E. Stewart, P.W. Wertz, S.W. Colton 6th, J.S. Strauss, Skin lipids, *Comp Biochem Physiol B*, **76**(4): 673–678 (1983).
4. J.A. Bouwstra, G.S. Gooris, J.A. van der Spek, W. Bras, Structural investigations of human stratum corneum by small-angle X-ray scattering, *J Invest Dermatol*, **97**(6): 1005–1012 (1991).
5. W.J. Albery, J. Hadgraft, Percutaneous absorption: In vivo experiments, *J Pharm Pharmacol*, **31**(3): 140–147 (1979).

6. J.W. Hadgraft, G.F. Somers, Percutaneous absorption, *J Pharm Pharmacol*, **8**: 625–634 (1956).
7. M.A. Katz, B.J. Poulsen, Absorption of drugs through the skin. In *Handbuch der Experimentellen Pharmakologie*, Volume 28, Concepts in Biochemical Pharmacology, B. Brodie, J. Gillette (eds), Berlin: Springer-Verlag, pp. 103–74 (1971).
8. B.J. Poulsen, E. Young, V. Coquilla, M. Katz, Effect of topical vehicle composition on the in vitro release of fluocinolone acetonide and its acetate ester, *J Pharm Sci*, **57**(6): 928–933 (1968).
9. M.F. Coldman, B.J. Poulsen, T. Higuchi, Enhancement of percutaneous absorption by the use of volatile: Nonvolatile systems as vehicles, *J Pharm Sci*, **58**(9):1098–1102 (1969).
10. P. Santos, A.C. Watkinson, J. Hadgraft, M.E. Lane, Oxybutynin permeation in skin: The influence of drug and solvent activity, *Int J Pharm*, **384**(1–2): 67–72 (2010).
11. H.E. Boddé, M. Ponec, A.P. Ijzerman, A.J. Hoogstraate, M.A. Salomons, J.A. Bouwstra, In vitro analysis of QSAR in wanted and unwanted effects of azacycloheptanones as transdermal penetration enhancers. In *Pharmaceutical Skin Penetration Enhancement*, K.A. Walters, J. Hadgraft (eds), New York: Marcel Dekker, Inc. pp. 199–214 (1993).
12. J.E. Harrison, A.C. Watkinson, D.M. Green, J. Hadgraft, K. Brain, The relative effect of Azone and Transcutol on permeant diffusivity and solubility in human stratum corneum, *Pharm Res*, **13**(4): 542–546 (1996).
13. J.W. Wiechers, B.F. Drenth, J.H. Jonkman, R.A. de Zeeuw, Percutaneous absorption and elimination of the penetration enhancer Azone in humans, *Pharm Res*, **4**(6): 519–523 (1987).
14. P.K. Wotton, B. Mollgaard, J. Hadgraft, A. Hoelgaard, Vehicle effect on topical drug delivery. III. Effect of Azone on the cutaneous permeation of metronidazole and propylene glycol, *Int J Pharm*, **24**: 19–26 (1985).
15. M.L. Francoeur, G.M. Golden, R.O. Potts, Oleic acid: Its effects on stratum corneum in relation to (trans)dermal drug delivery, *Pharm Res*, **7**(6): 621–627 (1990).
16. M.A. Pellett, M.S. Roberts, J. Hadgraft, Supersaturated solutions evaluated with an in vitro stratum corneum tape stripping technique, *Int J Pharm*, **151**: 91–98 (1997).
17. C.D. Vaughan, Using solubility parameters in cosmetic formulations, *J Soc Cosmet Chem*, **36**: 319–333 (1985).
18. C.M. Hansen (ed.), *Hansen Solubility Parameters: A User's Handbook*, 2nd edn, Boca Raton, FL: CRC Press, pp. 1–24 (2007).
19. J.W. Wiechers, C.L. Kelly, T.G. Blease, J.C. Dederen, Formulating for efficacy, *Int J Cosmet Sci*, **26**(4): 173–182 (2004).
20. R.H. Guy *et al.* (1997).
21. A. Avdeef, K.J. Box, J.E. Comer, C. Hibbert, K.Y. Tam, pH-metric logP 10: Determination of liposomal membrane-water partition coefficients of ionizable drugs, *Pharm Res*, **15**(2): 209–215 (1998).
22. R.M. Watkinson, C. Herkenne, R.H. Guy, J. Hadgraft, G. Oliveira, M.E. Lane, Influence of ethanol on the solubility, ionization and permeation characteristics of ibuprofen in silicone and human skin. *Skin Pharmacol Physiol*, **22**(1): 15–21 (2009).
23. O. Diez-Sales, A.C. Watkinson, M. Herraez-Dominguez, C. Javaloyes, J. Hadgraft, A mechanistic investigation of the in vitro human skin permeation enhancing effect of Azone®, *Int J Pharm*, **129**: 33–40 (1996).

24. R.M. Watkinson, R.H. Guy, J. Hadgraft, M.E. Lane, Optimisation of cosolvent concentration for topical drug delivery – II: Influence of propylene glycol on ibuprofen permeation, *Skin Pharmacol Physiol,* **22**(4): 225–230. (2009).
25. Y.J. Manrique, D.P. Pacheco, F. Martinez, Thermodynamics of mixing and solvation of ibuprofen and naproxen in proplylene glycol + water cosolvent mixtures, *Journal of Solution Chemistry*, **37**: 165–181 (2008).
26. S.H. Yalkowsky, *Solubility and Solubilization in Aqueous Media*, New York: American Chemical Society and Oxford University Press (1999).
27. D. van der Merwe, J.E. Riviere, Cluster analysis of the dermal permeability and stratum corneum/solvent partitioning of ten chemicals in twenty-four chemical mixtures in porcine skin, *Skin Pharmacol Physiol*, **19**: 198–206 (2006).
28. C. Herkenne, A. Naik, Y.N. Kalia, J. Hadgraft, R.H. Guy, Effect of PG on ibuprofen absorption into human skin in vivo, *J Pharm Sci*, **97**: 185–197 (2007).
29. M.E. Lane, *et al.* unpublished data (2012).

13

Dendrimers for Enhanced Drug Solubilization

Narendra K. Jain and Rakesh K. Tekade

13.1 Introduction

The therapeutic effectiveness of a drug molecule chiefly relies upon its bioavailability and ultimately upon its solubility [1]. Any drug to be absorbed must be present in the form of an aqueous solution at the absorption site [2]. The solubility of active pharmaceutical ingredients (APIs) plays a crucial role in drug disposition as the chief conduit for drug absorption is a function of permeability and solubility. According to a recent report, most of the existing as well as upcoming APIs identified by high-throughput methods of screening have pronounced solubility problems [3]. Approximately 30% of marketed APIs are classified as BCS Class II compounds (i.e. high permeability and low solubility), and 10% of marketed APIs are classified as BCS Class IV compounds (i.e. low permeability, low solubility) [4]. Currently only 8% of new drug candidates have both high solubility and permeability [5]. Of the rest, approximately 70% of New Chemical Entities (NCEs) under development may be classified as BCS Class II compounds, 20% of which are classified as BCS Class IV compounds [5]. Although many of these entities are highly potent compounds, because of their very low aqueous solubility, they are not taken forward to formulation development. Resolving these issues demands the development of technologies for enhancing drug solubility. A great number of solubilization strategies are currently under constant exploration to solubilize insoluble drugs in aqueous or desired solvents.

Drug Delivery Strategies for Poorly Water-Soluble Drugs, First Edition. Edited by Dennis Douroumis and Alfred Fahr.

13.2 Current Solubilization Strategies

Several formulation strategies to improve the solubilization of poorly water-soluble drugs and to further improve their bioavailability have been reported in the literature. These methodologies can be broadly classified as: (i) chemical methods, which include developing more soluble prodrugs or improving solubility through salt formation; and (ii) physical methods, which include micronization or nanosizing, polymorph production, crystal habit alteration, complexation, solid dispersions, micellar solubilization, pH adjustment, complexation, co-solvency, hydrotropy, etc. [6]. However, it should be noted that these techniques are of limited utility in solubility enhancement because of allied limitations and associated side effects. Co-solvency was anticipated to be the universal answer; however, the application was restricted due to safety concerns as well as precipitation of the solubilized drug upon *in vivo* administration. The uses of cyclodextrins are limited after being known to be nephrotoxic [7]. Other principal solubilization strategies also bring their own constraints such as restricted options in pH modification; stringent obligations of preserving critical micellar concentration in surfactants-based solubilization, and so on [8]. All these dilemmas have forced the development of some novel approaches for drug solubilization.

A wide variety of colloidal and vesicular systems such as liposomes, microspheres, nanoparticles and microemulsions have been investigated for their potential as drug carriers in attempts to overcome the problem of solubility. But the major problems associated with these techniques lie in the lack of universal applicability to all drugs [9]. During the past few decades in particular, pharmaceutical science has witnessed the exploration of a resourceful nanosystem, which successfully has resolved the problems associated with various approaches described earlier. These architectures were claimed to be 'Dendrimer' in the literature.

13.3 Origin of Dendrimers

The origin of a three-dimensional, dendritic branching concept can be traced back to the 1940s in the preliminary introduction of network theory by Flory wherein he introduced the term network cell, which he defined as the most elementary component in a molecular network arrangement [10–12]. The idea was carried forward one step further by Graessley, who described assembly of these networks as micro networks [13]. Extending the concept of Flory's statistical treatment of Gaussian-coil networks, analogous species that are part of an open, branched/dendritic organization, are known as branch cells and dendritic assemblies [14, 15]. It was the year 1978 when Vogtle *et al.* synthesized extremely branched cascades such as Oligoamines by repetitive synthesis technique [16]. Initially, only a small number of laboratories showed their curiosity in research and development of these aesthetically pleasing macromolecules. Fortunately in recent years many leading research laboratories around the world have realized that dendritic polymers have something very special to offer [1, 17–23]. Because of hard work and persistent investigation in this fascinating arena, a vista, developed into a superfamily with more than hundreds of compositionally diverse dendrimers with enormous pharmaceutical prospective, has opened up [24–27]. Later on, development of these molecular designs together with advanced synthetic techniques gave rise to larger dendritic structures, and these classes of molecules are now commonly known as dendrimers.

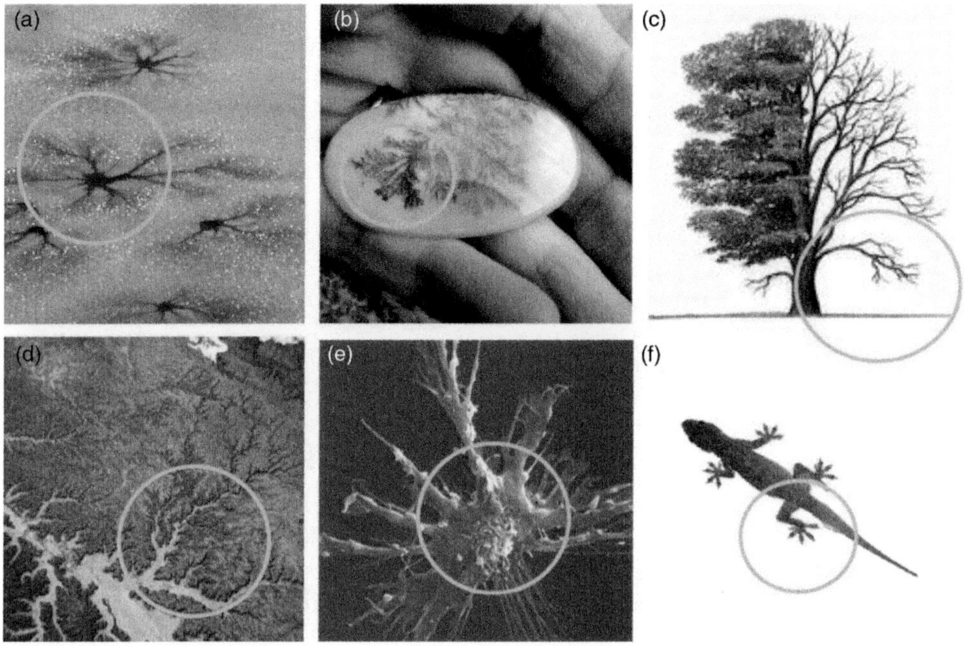

Figure 13.1 Different types of natural dendritic patterns represented by circled areas as observed on various creatures. (a) ice flick; (b) tree moss agate; (c) tree branching; (d) drainage system; (e) dendritic cell; and (f) lizard's claw. Reprinted with permission from John Wiley & Sons, Ltd. (See colour plate section.)

13.4 What Are Dendrimers?

The phrase dendrimer is derived from the Greek word *dendron*, meaning 'tree' and *meros* meaning 'part.' It can be defined as synthetic, highly branched, spherical and monodisperse macromolecules with three-dimensional nanometric configurations. In the literature, they are also called pet names such as, 'Arborals,' 'Cascade molecules,' 'Dendritic structures' [17,28]. Also because of their nanoscopic size and monodispersity they are often referred to as 'nanoscopic compounds' [29]. The dendritic architecture is conceivably one of the most omnipresent topologies observed. A number of examples explain the dendritic prototype in both living (tree branching, roots, vasculatory systems, dendritic cells, etc.) as well as in non-living systems (lightning patterns, snow crystals, erosion, etc.) (Figure 13.1). All these architectural domains of dendrimer skillfully bestow universal peculiarity of providing the best possible energy [30].

13.5 Synthesis of Dendritic Architecture

Synthetic methodologies to craft dendritic molecule in the laboratory have been the subject of enormous attention for many decades. A lack of a simplistic synthetic approach to build these molecules has hampered progress in this arena until roughly 30 years ago. In 1978,

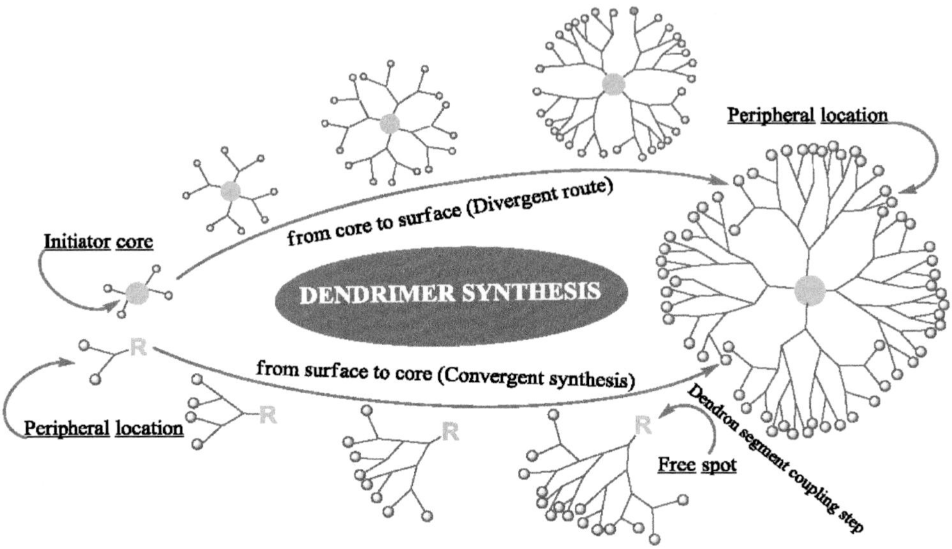

Figure 13.2 *Methodologies for synthesis of dendritic architectures (i) Divergent route, which involves progress of reaction scheme from a polyvalent core and continues racially outwards by successive repetition of reaction steps, (ii) Convergent route, in which reaction synthesis proceeds inwards from the periphery to the focal point.* (See colour plate section.)

Buhleier *et al.* developed an iterative cascade method for the synthesis of low molecular weight branched architectures [31]. Using chemistry and conditions, which are less prone to cyclization and more suitable for repetitive growth, Tomalia *et al.* reported the synthesis and characterization of the first family of dendrimers in 1984–1985 [17]. Since then, the influential hard work underway to build these macromolecules has followed two major strategies [32, 33] (Figure 13.2).

The first, introduced by Tomalia, was the divergent method in which growth of a dendron originates from a core position. This strategy involves assembling monomeric units in a radial, branch-upon-branch motif according to certain dendritic rules and principles [34, 35]. Advantages of this method are the ability to modify the surface of dendrimer molecules by changing the end groups at the outermost layer. This procedure is repeated until the dendrimer of the desired dimension is produced [36, 37]. The second method, established by Hawker and Fréchet, follows the convergent growth process [38]. This proceeds from the dendrimer surface to a reactive focal point, leading to the design of a reactive dendron. To affect a dendritic architecture, several dendrons are then reacted with a polyvalent core to produce the product. Using these two chief synthetic strategies, over one hundred compositionally different dendrimer families have been synthesized and over 1000 differentiated chemical surface modifications have been reported [39, 40].

Divergent dendrimer synthesis necessitates surplus monomer loading and lengthy chromatographic separations, predominantly at advanced generations. In contrast, convergent strategies are typically limited to the construction of only lower generation dendrimers due to the nanoscale steric concerns encountered when attaching the dendrons to the core.

Simplifying the synthetic preparation of dendrimers thus has been a foremost concern and a problem for the commercial exploitation of these unique structures in industrial premises that require large quantities of low-cost materials [41].

One of the inadequacies associated with divergent growth method is that the resulting dendrimer generation has only a single type of functional group. Convergent growth would overcome such limitations. The convergent method involves two stages: first, a reiterative coupling of protected/deprotected branch to produce a focal point functionalized dendron; and second, a divergent core anchoring stage to produce various multidendron dendrimers. Advantages of this method include precise control over molecular weight and synthesis of dendrimers having functionalities in precise positions and number.

Using these two synthetic strategies, more than 100 compositionally different dendrimer families have been synthesized and these are reviewed in the literature [23]. Commercial quantities (kg) of controlled macromolecular structures with low polydispersities are now routinely synthesized using traditional organic reagents and monomers, such as ethylenediamine and alkyl acrylates.

13.6 Structure and Intrinsic Properties of Dendrimeric Compartments

As the dendritic structure grows, several fundamental compartments arise. The structure of dendrimers can be considered to be a nano-container consisting of three critical architectural domains: (i) the central core, which in higher generation dendrimers is protected from the surroundings, creating a microenvironment; (ii) the interior shells, just beneath the surface having a well-defined microenvironment; and (iii) the multivalent surface (nano-scaffoldings) with a high number of potential reactive sites [42]. The first two domains represent the nano-environments, which are protected from outside by the dendrimers' exterior. The high density of exo-presented surface functionalities makes the dendritic surface well suited as a nano-scaffold where the close proximity of functional groups is important (polyvalency) or for receptor-mediated targeting purposes.

The branching elements are illustrated by the term 'generation,' starting with the central branched core molecule as generation 0 (G_0) and with each consecutive addition of branching points successive architectures are termed G_1, G_2, G_3 and so on. With each succeeding generation, the number of end groups increases exponentially. Dendritic macromolecules tend to linearly augment in diameter and adopt a more globular shape with increasing dendrimer generation [43]. The addition of successive layers (generations) gradually increases the molecular size and amplifies the number of surface groups present. As PAMAM dendrimers grow through generations 1–10, their size increases from 1.1 to 12.4 nm [39]. These dimensions have been compared to those of proteins (3–8 nm), linear polymers (5–20 nm) and viruses (25–240 nm) (Figure 13.3). The nanoscale branching architecture of the dendrimers provides several advantages over linear polymers such as rapid cellular entry, reduced macrophage uptake and targetability (Table 13.1).

The dendrimers are big enough to cross through the normal blood vessels but are of optimal size to enter the more porous tumor. It is very likely that the majority of the other delivery systems engaged for the identical purpose would not voluntarily gain entrance to tumors from the vasculature, due to their larger diameters (between 40–100 nm), which is too large to cross vascular pores [40, 44]. Dendrimer is of equivalent dimensions to

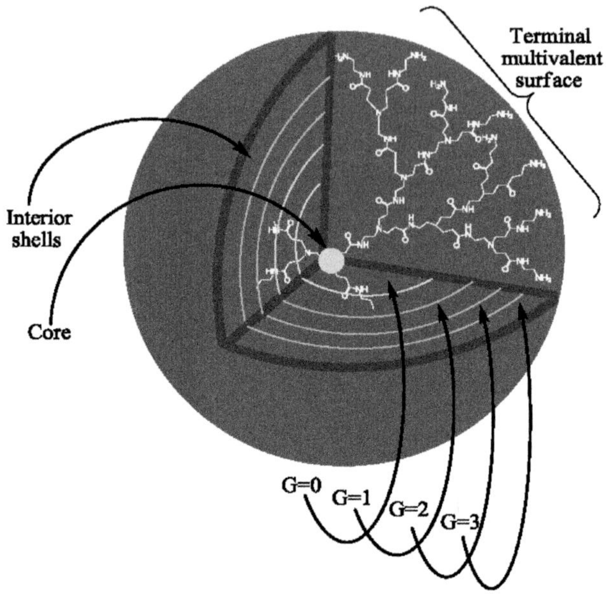

Figure 13.3 *Three-dimensional architecture of dendrimer with chief architectural domains, namely, core, interior and surface.* (See colour plate section.)

miniature serum protein and so can gain entry through pores in the vasculature and infuse tumor cells directly (Figure 13.4).

Modification of these domains can be done by the available synthetic strategies to impart a specific function to the architecture. On the other hand, the interior is well suited to host–guest interaction and encapsulation of guest molecules. Our group recently presented a comprehensive view of various dendritic scaffolds explored for varying purposes, ranging from solubilization to hybrid dendrimer-mediated targeting, in *Chemical Reviews* [23].

The properties of dendrimers are dependent predominantly on the functional groups present on its periphery [30, 36, 45, 46] and, for the case in hand, a dendrimer can be more water-soluble when its end-group is hydrophilic, like the -COOH group. It is possible to design a water-soluble dendrimer with internal hydrophobicity, which would allow it to carry a hydrophobic drug in its interior [47]. Careful examination of the available literature suggests that ionic interaction, hydrogen bonding and hydrophobic interactions are the chief possible mechanisms by which dendrimers exert their solubilizing effect.

13.7 Dendrimers in Solubilization

Most of the drug molecules employed in the pharmaceutical field are hydrophobic in nature and this property, in particular, poses a major formulation challenge. This drawback of hydrophobic drugs can be ameliorated by dendrimeric scaffolding, which can be used to encapsulate as well as to solubilize the drugs because of the pability of such scaffolds to participate in extensive hydrogen bonding with water. In 1985, Newkome first reported

Table 13.1 Comparison of properties of linear and dendritic architectures.

Properties	Linear polymer	Dendritic polymer
Structure		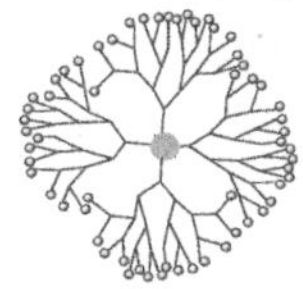
Shape	They resemble random coil configurations. They are non-compact	The structure of dendritic architectures changes as a function of core, generation and molecular weight. They are globular
Structural control	Low	Very high
Polydispersity	Monodisperse	Polydisperse
Crystalline nature	Linear polymers are semicrystalline/ crystalline materials, which exhibit higher glass temperatures	They are non-crystalline, amorphous materials with lower glass temperatures
Solubility	They bear lower solubility that in general decreases with molecular weight	Its solubility increases with increase in molecular weight
Rheological properties	Linear polymers have entanglement directed rheological properties with high shear sensitivity	They exhibit Newtonian type of rheology with no shear sensitivity
Intrinsic viscosity	Intrinsic viscosity follows logarithmic-increase with molecular weight	Exhibits viscosity maxima and minima plateaus
Electronic conductivity	Shows anisotropic electronic conductivity	Shows isotropic electronic conductivity
Mobility	Exhibits segmental and molecular mobility	Dendrimer mobility involves the kinetic flow unit

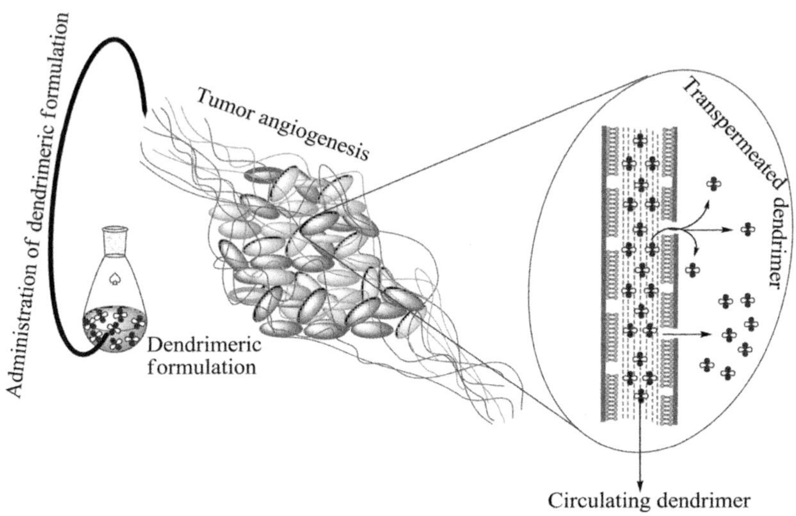

Figure 13.4 Entry of drug-dendrimer complex through pores in the tumor vasculature. (See colour plate section.)

the unimolecular micellar nature of dendrimers and soon after this report the solubilizing properties of dendrimers came into existence [36, 48].

The micellar and container properties of dendrimers were additionally demonstrated by dissolving some organic dyes such as pyrene, as revealed in a number of reports [46, 49]. Inverted micellar behavior of the dendrimers was also reported by Stevelmans *et al.* using PPI dendrimers and conjugating its surface with hydrophobic alkyl chains [50, 51]. These explorations opened up an innovative area of dendrimer applications, specifically, the solubilization of hydrophobes. This solubilization performance of dendrimers was found to be superior compared to some leading solubilizing agents, as evidenced by a study on solubilization of Niclosamide [51]. Dendrimers enhance the solubility of hydrophobic moieties possibly due to the hydrophobic interactions, hydrogen bonding and electrostatic interaction between the terminal functional groups of the dendrimers and the hydrophobes (Figure 13.5).

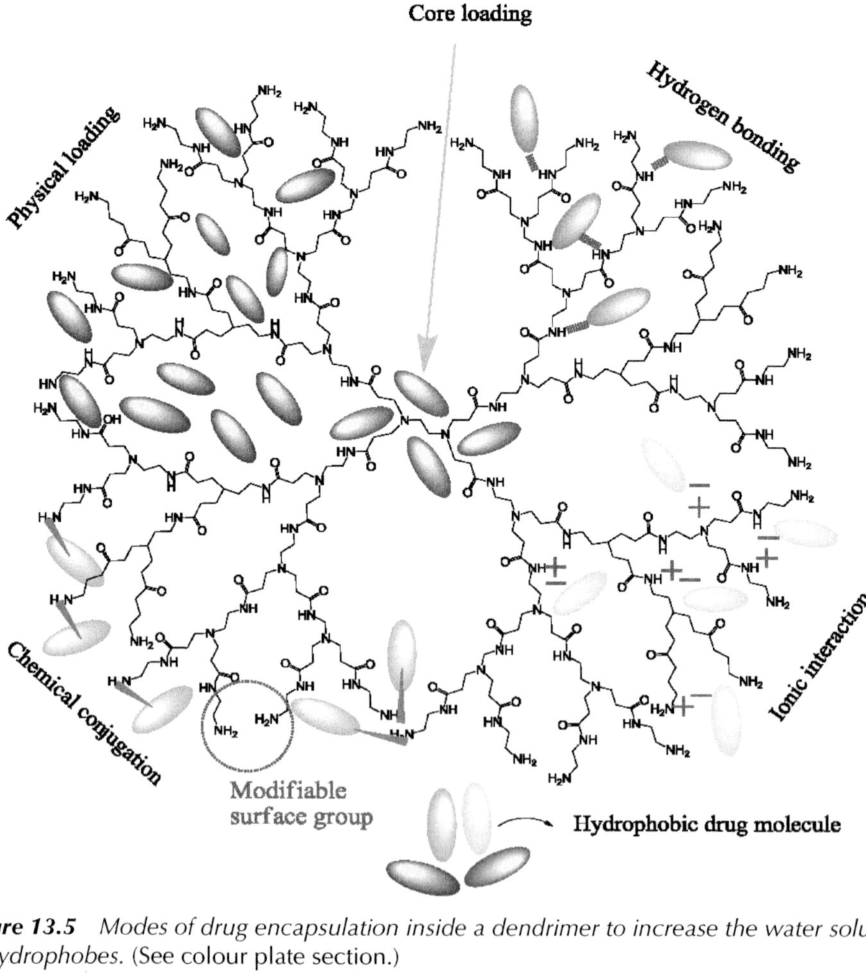

Figure 13.5 *Modes of drug encapsulation inside a dendrimer to increase the water solubility of hydrophobes.* (See colour plate section.)

The entry of dendrimer therapeutics into human clinical (Phase I) trials as exemplified by dendrimer-based nanopharmaceuticals (VivaGel), DNA delivery vectors (Superfect), nanodiagnostics (Stratus) and other model therapeutics elegantly narrates its emerging role in pharmacotherapeutics [43,52]. This enormous recognition of dendritic architectures in a short span of time may be attributed to its numerous properties, namely, its nanoscopic size, low polydispersity, stable unimolecular micellar architecture, high drug-loading capacity, associated solubilizing attributes, transfection efficiencies, modifiable terminal functionality, protein mimicry, and so on [46,53,54]. These architectures are capable of admirably solubilizing numerous hydrophobic drug molecules such as Aceclofenac, paclitaxel, ibuprofen, Nifedipine, Piroxicam, and Indometacin [55,56].

Devarakonda *et al.* compared the solubility potential of PAMAM dendrimers and cyclodextrins in the solubility enhancement of Niclosamide, and observed that the amine-terminated dendrimers have a superior ability to enhance the solubility of Niclosamide as compared to cyclodextrins [51,57]. Loading of drug molecules inside dendrimers may be used to provide a means of drug solubilization as well as controlling its release fate in the circulation.

13.8 Factors Affecting Dendrimer-Mediated Solubilization and Drug Delivery

The open structural design of the dendritic architecture offers the potential for dendrimers to interact with labile or poorly soluble drugs. This has led numerous investigators to explore the conditions required to optimally encapsulate drug molecules within the branches of a dendrimer. The outcomes of these tactically designed explorations inferred that the dendrimer-mediated solubilization depends on several factors, including the nature of the dendritic core, the dendrimer generation, the pH of the solution, the concentration of the dendrimer in the solution, etc.

13.8.1 Nature of the Dendritic Core

The actual internal geometry of a dendrimer is likely to be determined by the type of core used, the branching molecule, the nature of the bonds and the nature of the surface groups. The excellence of dendrimer chemistry is that it offers immense flexibility in the selection of the core for synthesis. A more extended and elongated core ensures larger internal cavities and imparts greater flexibility to the unimolecular dendritic structure [58]. The influence of available larger internal cavities due to the substitution of a larger core can be obvious from the enhancement in the aqueous solubility of pyrene in closely related work reported by Hawker *et al.* [42] and Liu *et al.* [46]. Hawker and coworkers synthesized a dendritic architecture using 3,5-dihydroxy benzyl alcohol as the building block and these architectures were used to augment the solubility of pyrene. The aqueous solubility of pyrene was found to increase by about 120-fold [42].

In the same direction, Liu *et al.* also synthesized analogous unimolecular micelles. The difference lay in the use of a building block where, instead of using 3,5-dihydroxybenzyl alcohol, these workers used 4,4-bis(4-hydroxyphenyl) pentanol as the building block. This time the solubility of pyrene increased approximately 356-fold [46] as against 120-fold in the previous study by Hawker and coworkers [42].

13.8.2 Dendrimer Generation

Dendrimer size is also related to the three-dimensional shape; lower generation dendrimers tend to be open and amorphous structures whereas higher generations adopt a spherical conformation capable of incorporating drug molecules. Researchers have investigated and reported the effect of dendrimer generation on solubility enhancement. It was observed that the solubility of a hydrophobe increased with increasing dendrimer generation. Generation-dependent enhancement in solubility could be attributed to this conformational change. With the rise in dendrimer generation, the void spaces within the core are amplified and hence the pability to solubilize the drug molecules is enhanced.

Kaanumalle *et al.* reported that an aqueous solubility of pyrene increased by 5-, 8-, and 24-fold with generations 1, 2, and 3, respectively with poly(alkyl aryl ether) dendrimers [59]. A similar trend was reported by several workers wherein dendrimer-mediated solubilization was found to be proportional to the concentration of dendrimers [57, 60, 61]. There is a tendency, however, to select dendrimers of lower generation as they are less immunogenic, less cytotoxic and more biocompatible compared to higher generations.

13.8.3 Nature of the Dendrimer Surface

It is now obvious that the cytotoxicity of the dendrimer relies on the chemistry of its core, but is also most sturdily influenced by the nature of the dendrimer surface. For instance, the cytotoxicity of cationic dendrimers on Clone-9 cells on a melamine-based dendrimer library including amine, guanidine, carboxylate, sulfonate, or phosphonate modification, inferred that cationic dendrimers were more cytotoxic than anionic dendrimers [62].

In the same way, quaternized PAMAM–OH derivatives exhibited a lower intensity of cytotoxicity than PAMAM–NH_2 because of shielding of the internal cationic charges by surface hydroxyl groups [63]. Modification of the surface of PAMAM 4.0G with arginine or lysine accounted for the increased toxicity compared to native PAMAM when incubated with HEP G2 cells or 293 human embryonic kidney cells [64]. The authors attributed this effect to increased charge density and molecular weight. The authors also observed that anionic PAMAM dendrimers and DAB dendrimers, with –COOH surface were not cytotoxic towards B16F10, CCRF or HepG2 cells up to concentrations of 5 mg/ml [65].

A potentially toxic dendrimer core is more accessible when presented to cells as a low generation species due to the more open architecture. In addition, low generation species have more accessible surface terminal groups, and these result in steric hindrance due to crowding at higher generations. Increased branching (generation) and a larger surface coverage with biocompatible terminal groups like PEG are being widely used to create less toxic dendrimers [65–67].

13.8.4 Dendrimer Concentration

The dendrimer-mediated solubilization increases linearly with the increase in the concentration of the dendrimer in the solution, similar to other solubilizing agents [46, 68, 69]. However, the concentration of the dendrimer employed as a solubilizing agent must be judiciously justified considering the allied toxicity and biocompatibility of the resulting drug–dendrimer complex. For instance, in an investigation Tekade *et al.* observed insignificant cytotoxicity for PPI dendrimer after 48 hrs, which was considered to be safer even at higher concentration of 10 μM [70]. However, it should be noted that this amount of dendrimer is sufficient to solubilize and deliver a pharmacologically effective concentration of a drug [71].

13.8.5 pH of Solution

Interaction between hydrophobes and amines (internal, peripheral, or tertiary) of dendrimers is a key mechanism accountable for solubility augmentation and subsequent drug delivery. Protonation behavior and the pH of the dendrimer affect the solubilizing power as well as the drug-carrying capacity of dendrimers. The protonation of nitrogen, whether at the periphery or at dendrimer interiors, is influenced by pH [72]. In an investigation, Asthana *et al.* found that the augmentation in the aqueous solubility of Flurbiprofen employing G4 PAMAM dendrimer was maximum at pH 7, less at pH 10, and least at pH 2 [73].

Milhem *et al.* also studied the solubility enhancement of weakly acidic Ibuprofen using G4 PAMAM dendrimers as a function of pH. The amplification in solubility was highest at pH 10.5, whereas at pH 2, no noteworthy rise in solubility was recorded. The authors concluded that at extremely basic pH, weakly acidic Ibuprofen was ionized and interacted electrostatically with surface amine groups of dendrimers, whereas at pH 2, this interaction was not possible because Ibuprofen largely remained un-ionized [74].

Along the same lines, Yiyun *et al.* also explored the solubility enhancement propensity of Nicotinic acid and Ketoprofen using PAMAM dendrimers. A concentration- and generation-dependent solubility enhancement was observed. The solubility was highest at pH 6.0 and least at pH 3.0. The authors proposed an electrostatic interaction between the surface amine groups of the dendrimer and the carboxylic group of guests, and also hydrogen bonding between the atoms of the Nicotinic acid and the tertiary amine groups of the dendrimers, as possible mechanisms for the solubility enhancement [60,75]. These studies clearly highlight pH as an important factor to be considered during dendrimer-mediated solubilization.

13.8.5.1 pH-Sensitive Dendrimers

By virtue of the inherent properties of dendrimers, they can even be formulated to behave as pH-sensitive. The alternation in hydrodynamic radii and architectural arrangement of a dendrimer with pH can be an imperative perception to attain sustained and controlled drug release. Under typical physiological pH (pH 7.4), the peripheral amine groups in PPI and PAMAM dendrimers are not protonated, and the branches meet the core unit. Nevertheless, once they enter the tumor vessels, the dendrimers are subjected to lower pH, as a result of which the groups at the periphery are protonated and repel each other. This leads to a change in three-dimensional conformations, which moves branches outward and thereby releases the drug at the tumor site. This avoids the release of a potent drug at non-desired locations, thereby preventing toxic and unwanted side effects [43,76,78].

Along the same lines, ester-terminated dendrimers are non-toxic, biocompatible and have shown tumor-selective affinity. Peptides such as L-lysine are also pH-sensitive and charged at lower pH. Thus peptide-based dendrimers have the potential for pH-sensitive delivery.

Modification of the surface with acetic-anhydride avoids non-specific interaction and can be attached to the ligand to direct dendrimers to neoplastic cells. Also a pH-sensitive spacer can be attached between the dendrimer and the drug. For instance, hydrazone as well as aconityl linkage can be used, that are acid labile. The amide and ester bonds can be developed that undergo hydrolysis at lower pH [79].

Baars and Meijer suggested that the acid-base reaction between the dendrimer and the guest (Coulomb attraction) drags the guest molecule into the dendrimer, while the hydrogen bonding is responsible for retaining the guest with the host [80]. However, it should be noted that the guest molecules were retained within the dendritic branching clefts by fragile ionic interaction with interior protonated amide groups. This inclusion complex was remarkably witnessed to distort at pH less than neutrality (pH < 7) [57, 81]. This is because dendrimers adopt conformations of diverse shape and density depending on the prevailing circumstances in the bulk solution (pH, ionic strength, polarity). For instance, polyamido amine (PAMAM) and polypropylene imine (PPI) dendrimers with primary amines as surface functionality display extended conformations upon lowering the pH due to the electrostatic repulsion between the protonated tertiary amines in the interior as well as between surface primary amines, thus forcing the dendritic branches apart. At pH 9, back-folding occurs as a consequence of hydrogen bonding between interior protonated tertiary amines and surface primary amines, resulting in a denser interior (Figure 13.6). Evidently, these pH-related conformational changes are dependent on the charge of the respective functional groups [82, 83]. This key principle, in particular, projects the inherent property of dendrimers that in other ways guards the release of loaded drugs in relatively higher proportion in the acidic tumorous environment where pH below neutrality persists.

13.8.6 Temperature

Temperature is one of the most imperative aspects, which directly influence the solubility of drug molecule. Using G4 PAMAM dendrimer as a model construct, Milhem *et al.* explored the influence of temperature on the solubility intensification of Ibuprofen. The investigators evaluated the solubility enhancement proficiency of Ibuprofen at different temperatures (27°, 35°, 40°, 45° and 50°C). Surprisingly, it was observed that the solubility enhancement was inversely proportional to an increase in temperature (Figure 13.7). However, no convincing justification was offered for the unusual solubility pattern [74]. Unfortunately, this aspect of dendrimer-mediated solubilization has, by some means, remained incompletely explored and demands more exhaustive investigation to assess the effect of temperature on dendrimer-mediated solubilization and the mechanisms involved therein.

13.8.7 Solvents

The capability of the solvent to solvate the dendrimer is an important aspect when examining the conformation of the dendrimer. In general, all dendrimers sense a large extent of back-folding with the decreasing solvation property of solvent. However, being more flexible, the low generation dendrimers illustrate the highest tendency of back-folding as a consequence of poor solvation compared to the higher generation dendrimers. NMR studies performed on PPI dendrimers concluded that nonpolar solvents, like benzene, poorly solvate the dendrons favoring intramolecular interactions. However, a feebly acidic solvent, like chloroform, can

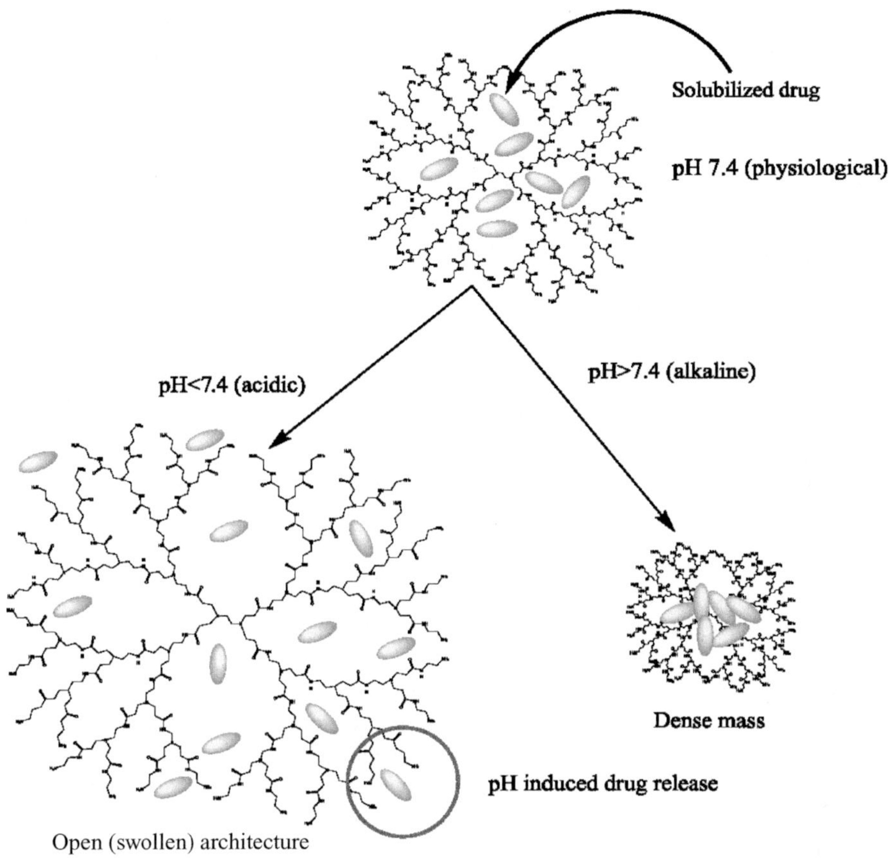

Figure 13.6 *Mechanism of drug release from drug-dendrimer complex as a function of the pH of the environment.* (See colour plate section.)

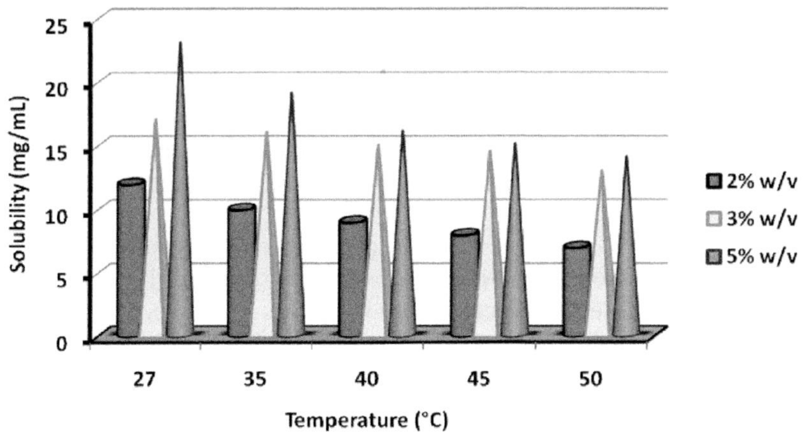

Figure 13.7 *Effect of temperature on dendrimer-mediated solubilization of Ibuprofen solubilized by employing varying concentrations of dendrimer (% w/v). Reproduced with permission from [74].*

operate as a hydrogen donor for the interior amines in a basic dendrimer like PPI, leading to an extended conformation of the dendrimer because of extensive hydrogen bonding between the solvent and the dendrimer amines [84]. Both theoretical as well as experimental studies on polar dendrimers (amine-terminated PPI and PAMAM) show the tendency that non-polar aprotic solvents provoke higher molecular densities in the core area as a result of back-folding while polar solvents solvate the arms of dendrimer and persuade a higher molecular density on the dendrimer's periphery. Back-folding of the polar surface groups may render the hydrophobic dendrimer components not suitable to the surroundings leading to a reduced surface polarity of the dendrimer. These aspects must be judiciously considered in achieving dendrimer-mediated solubilization.

13.9 Drug–Dendrimer Conjugation Approaches

13.9.1 Physical Loading: Complexation of Water-Insoluble Drugs

The most fascinating architectural applications of the dendrimers relate to its unique structural features and its tailor-made framework. This excellent quality of architectural domains of dendrimers is ascribed to its polyvalent surface, nano-scaffolding and inner clefts. Dendrimer-based formulation development requires simple drug loading as well as conjugation strategy [85, 86].

The architectural arrangement of dendrimers permits a huge loading of a variety of cargo in appreciable amounts because of the hydrophobic sectors present inside its branching clefts [23, 87]. The method of drug loading is based on the theory of ionic interaction, hydrophobic interaction, hydrogen bonding as well as the physical entrapment of the bioactive agent inside the carrier upon extensive dialysis [41, 88]. Insertion of hydrophobic drug molecules inside dendrimers is typically accomplished by a simple admixture of the polymer and drug solutions where the hydrophobic drug associates with the non-polar core through hydrophobic interactions [89, 90].

To capitalize on the loading capability of drug molecules within the dendrimer, one has to vigilantly consider the factors affecting solubilization, as discussed in the previous section. Such a move leads to higher loading of drugs in dendrimers as exemplified by a number of reports. Encapsulation of the drugs within dendrimers has been reported to generate conjugates that exhibited slower release, higher accumulation in solid tumors, high biocompatibility and lower toxicity compared to free drug counterparts [91, 92]. The drug–dendrimer complex can be characterized by various available analytical techniques, namely, HPLC, UV/V, etc. [55, 93].

According to Vogtle and co-workers, the encapsulation of a drug inside branched polymers symbolizes an earlier form of physical encapsulation to augment the aqueous solubility of poorly soluble drugs in the dendrimers' voids [31], and as a consequence, a range of non-covalent interactions including hydrophobic forces, hydrogen bonding, steric hindrance, and electrostatic interactions may be established between the guest molecules and the dendrimer [41]. These findings pave the way for the advancement of inclusion complexes where dendrimers can encapsulate hydrophobic drug molecules to improve aqueous solubility.

In another investigation, Milhem *et al.* reported encapsulation of approximately 40 Ibuprofen molecules per molecule of G4 PAMAM dendrimers, resulting in a substantial enhancement of drug solubility [74]. Kohle *et al.* studied the dendrimeric scaffold by taking the same drug as a model compound. Since this similar perception of drug encapsulation applies to all hydrophobic drugs, there is an imperative need to understand this work and its outcome. The NH_2-terminated 3.0 and 4.0G PAMAMs predominantly form a complex with the –COOH ion of Ibuprofen because of ionic interactions, while the OH-terminated Polyol appears to encapsulate Ibuprofen. Up to 78 molecules of the model drug were complexed by the PAMAM dendrimers through electrostatic interactions between the dendrimer amines and the carboxyl group of the drug. In contrast, up to 24 drug molecules were encapsulated into the hyperbranched polyol [85]. Further, Kannan *et al.* also employed a similar drug as a model construct wherein the drug payload for the complex was found to be approximately 50% by weight with PAMAM-NH_2 dendrimers and was about 30% with Polyol-OH and PAMAM-OH dendrimers [94].

In a follow-up study, the same group covalently attached 58 molecules of Ibuprofen to one molecule of fluoroisothiocynate (FITC)-labeled 4.0G PAMAM-OH dendrimer and investigated its cellular entry in human lung epithelial carcinoma A549 cells. Unlike the simple drug-loaded dendrimers, the covalently linked drug–dendrimer conjugates would be more stable *in vivo*, thus prolonging drug circulation and tissue delivery [71]. Along the same lines, Ke *et al.* also recently reported 95% encapsulation of doxorubicin in 3G PAMAM dendrimers [95].

Another area of current dendrimer research efforts is controlled drug delivery [30]. Topological encapsulation within dendrimers has been suggested as a means to transport drugs. Meijer and coworkers explored this concept of physical loading by encapsulation of polar guest molecules into a dendritic core-shell architecture called the dendritic box [27].

13.9.1.1 Dendritic Box

Meijer and co-workers [96, 98] capped the end groups of the dendrimer, thereby permanently encapsulating the guest molecule. The physical encapsulation was achieved in the final step of the dendrimer synthesis by reacting the terminal amino groups of a PPI dendrimer small molecule complexed with various activated Di-*tert*-butyl dicarbonate (*t*-Boc) or fluorenylmethyloxycarbonyl (*f*-moc) protected amino acids [96, 99]. Dendritic boxes reported by Meijer's group were also among the earliest attempts at the dendrimer-mediated non-covalent host–guest interactions. Thus, non-covalent interactions are chiefly accountable for solubilization. The size of 5.0G dendritic boxes was reported to be 5 nm. A number of dye molecules were encapsulated in the dendritic box, for example, up to 4 molecules of Bengal Rose and 8–10 molecules of 4-nitrobenzoic acid could be encapsulated per dendrimer. The shape of the selective discharge of guests from dendritic boxes was also described. Hydrolysis of the surface blocking group (*t*-BOC, *f*-MOC) of a dendritic box (with formic acid) resulted in perforation of the dendrimer, allowing the release of the entrapped moiety (Figure 13.8). It was inferred that depending on the molecular weight, guest molecules can be encapsulated and selectively released from this dendritic host [96, 98]. The larger Bengal Rose molecule remained entrapped and could only be liberated following hydrolysis of the outer shell by 12 mol·dm^{-3} HCl under reflux for 2 hr. A comparable mechanism of release was described by several workers [100]. The possible applications of this method include its utilization as a drug delivery container, a fluorescence marker, as well as IN examining the chemical and physical behavior of isolated molecules [97].

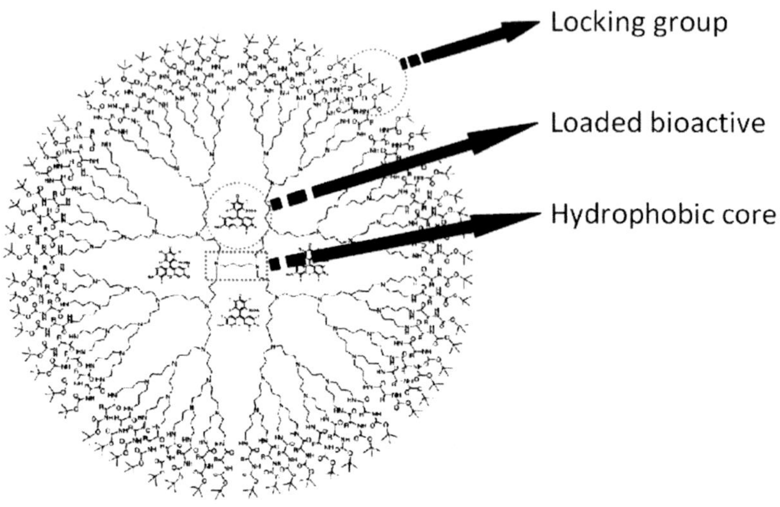

Figure 13.8 *Dendritic box with entrapped molecules: A progression of physical loading concept.*

13.9.1.2 Saponified Dendrimer: Unimolecular Micellar Architecture

Several investigators have reported the application of peripheral groups to impart water solubility to the corresponding dendrimers [101, 102]. Hawker *et al.* first established the synthesis of dendritic unimolecular micelles, employing molecules with a globular amphiphilic structural design [38, 42] analogous to that reported by Newkome in divergent schemes [28]. Saponification of ester-terminated dendrimer (Figure 13.9A), gave a dendrimer with

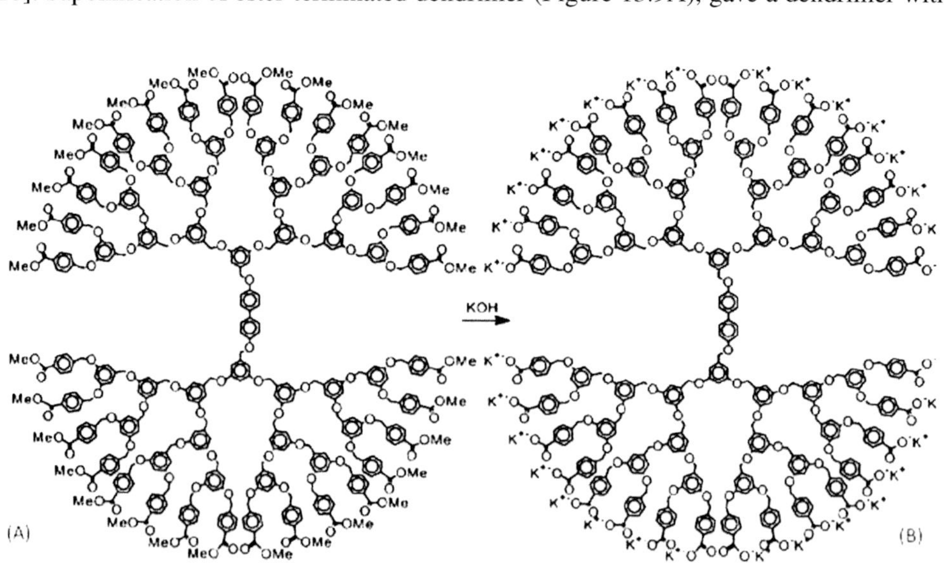

Figure 13.9 *Saponification of ester-terminated dendrimer. Reproduced from [104] with permission from the American Chemical Society.*

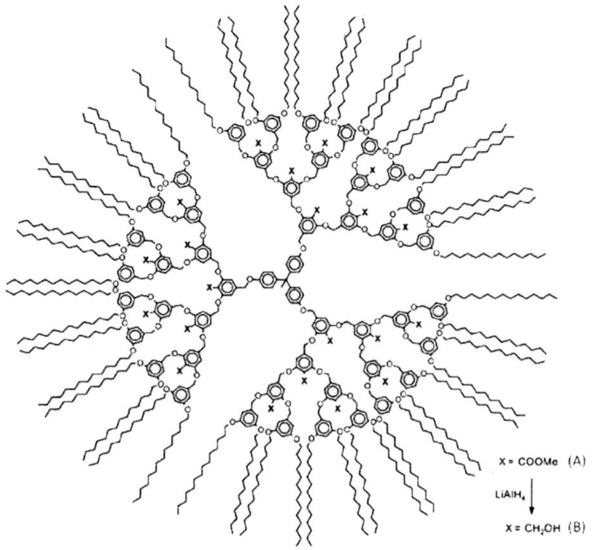

Figure 13.10 *Dendrimer-based unimolecular micelle. Reproduced from [104] with permission from the American Chemical Society.*

a polycarboxylate edge and pH reliant solubility [42]. The polycarboxylate (Figure 13.9B) also demonstrated micelle-like solubilization of polycyclic aromatic hosts. These configurations have been given the name 'unimolecular micelles' for the reason that, unlike conventional micelles, their covalently bound organization is not subject to disaggregation below a certain threshold of micellar concentration. Workers later employed these micellar belongings to devise more proficient macromolecular nanoreactors (Figure 13.10A, 13.10B), that used the diversity in polarity between the interior and exterior of dendrimer to compel the reactants toward the internal catalytic sites while simultaneously expelling the products to prevent catalyst inhibition [103]. Fréchet's group have also comprehensively explored the inclusion of drug moieties by both encapsulation and covalent attachment [30, 104].

To date, a spate of reports are available on dendrimers, which highlight its use in loading of diverse bioactive agents [66, 105]. Initially these molecules were focused on their usage for non-covalent encapsulation of drug molecules [106, 107]. But in the last decade there were numerous reports of chemically conjugated guest molecules, or targeting moieties, to the dendrimers as well [22, 108].

13.9.2 Covalent Loading: Synthesis of Drug–Dendrimer Conjugate

An optional approach for utilizing dendrimers as a drug carrier is to exploit their well-defined polyvalent organization in a covalent attachment of drug molecules to their periphery. In addition, the release of the bioactive can be adjusted by applying the concept of a location-specific degradable spacer between the drug and the dendrimer's peripheral groups. Moreover, the drug loading can be tuned by altering the number of groups on the dendrimer's periphery. Various strategies have been devised to modify dendrimers with drug

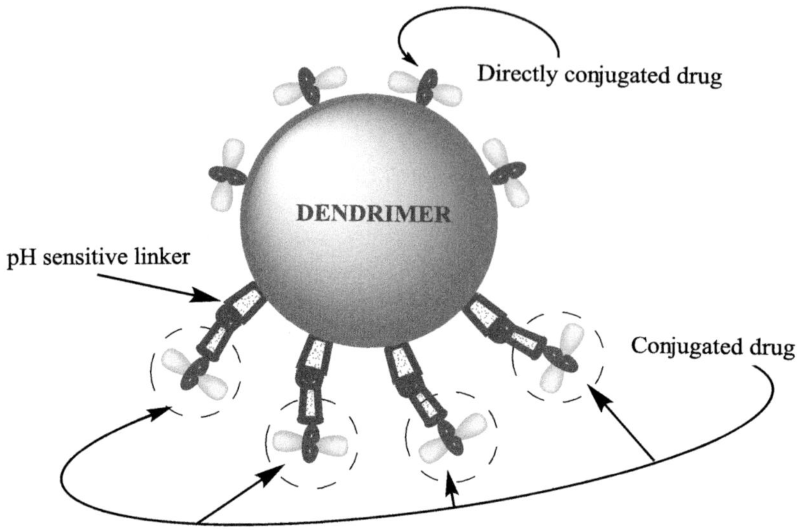

Figure 13.11 *Schematic showing a dendrimer-drug conjugate where the drug molecules (bullets) are either directly coupled to the dendrimer's surface groups (bullets without lines) or via a pH-sensitive linkage (bullets with lines).*

molecules, genetic materials, targeting ligands, dyes and imaging agents by chemical conjugation. 'Smart' drug-delivery nanodevices can be developed by conjugating appropriate targeting ligands, drugs and imaging agents to dendritic polymers, which can target, deliver and monitor the progression of therapy. Drugs can be conjugated to dendritic nanodevices through either ester or amide, or some other linkage depending on dendritic surface, which can be hydrolyzed inside the cell by endosomal or lysosomal enzymes (Figure 13.11).

5-Fluorouracil (5-FU) is a potent anticancer drug, but it has very high toxicity and relatively low solubility. Partly successful attempts have been made by investigators to address this issue [109–111]. In this respect, Zhou *et al.* reported the synthesis of a succession of dendritic polymers (0.5 to 5.5G) starting from 1,4,7,10-tetraazacyclododecane, a cyclic tetra amine core. The dendrimers were first acetylated and then reacted with 1-bromoacetyl-5-FU to form a dendrimer-FU conjugate. These conjugates were conclusively found to be greatly soluble in water and released free 5-FU at a slower rate with a concomitant reduction of its toxicity [112].

Propranolol is a poorly water-soluble drug. In a unique study D'Emanuele *et al.* reported solubility enhancement of Propranolol by chemically conjugating it with PAMAM dendrimers. The aqueous solubility of Propranolol increased by about 106-fold by conjugating it to G3 PAMAM dendrimers [113].

Duncan and co-workers have studied dendrimers for their potential in solubilization and delivery of hydrophobic agents. They conjugated 3.5G PAMAM dendrimers with Cisplatin to yield a dendrimer-platinate (dendrimer-Pt; 20-25 wt % platinum), which was highly water-soluble and released platinum slowly *in vitro*. In this way Cisplatin, a potent anticancer drug with significant toxicity and poor water-solubility, was optimized to have increased solubility, decreased systemic toxicity, and selective accumulation in solid tumors [114, 115]. More recently Bellis *et al.* reported a method for obtaining dendrimers with peripheral Cisplatin moieties by the reaction of the free amine dendrimers and potassium tetrachloroplatinate (II). For this they used the three generations of PPI dendrimers modified with α,γ-diaminobutyric acid moieties [116].

Another group of researchers synthesized polyester dendritic architectures using 2,2-bishydroxymethyl propionic acid as the monomer unit. The potent anticancer drug doxorubicin was covalently attached through an acid–labile hydrazone linkage to these dendrimers. Thus, hydrazone linkage provides a suitable system for pH-dependent release [66, 117]. In another study, Patri *et al.* compared the efficacy of covalently attached Methotrexate (MTX) with a non-covalent complex utilizing G5 PAMAM dendrimer. MTX–dendrimer conjugates were found to be more stable compared to non-covalent MTX–dendrimer complexes [89].

Gurdag *et al.* compared the anticancer activity of MTX when coupled through its amine group to the carboxylic acid surface groups of G2.5-COOH dendrimers forming stable amide linkages versus MTX coupling through its carboxylic group to the primary amine groups of G3-NH_2 dendrimers [118]. Results showed that G2.5 MTX conjugates were three-fold more cytotoxic compared to free MTX toward lymphoblastic leukemia cells, whereas G3-MTX conjugates were ten-fold less toxic than the free MTX. Similarly, Baker and co-workers compared the *in vitro* anticancer activity of G5-OH and G5-NH_2 conjugates with MTX attached via ester and amide linkages, respectively. Incubation of ester-linked G5-MTX conjugates with KB cells at a concentration of 1-100 nM MTX equivalent resulted in a ten-fold lower IC_{50} value compared to amide-linked G5-MTX conjugates, which was a result of faster hydrolysis of the ester linkages and release of the incorporated MTX drug molecules [22]. This outcome was additionally supported by an allied report from Minko's group [119].

The number of drug molecules per conjugate can be varied by using different generations of dendrimers or by changing the coupling conditions. The attachment of three model drugs, each with a different functional group (–OH, –NH_2 and –COOH) was demonstrated using a variety of potentially cleavable bonds (carbonate, carbamate and ester). Even after attachment of five cholesterol molecules, the resulting conjugates were still water-soluble. Obviously, these experiments were meant only to test design features and the dendritic structure alone was not optimized. Considering the requirement for water solubility, the poly(aryl ether) dendrimers used in this work might best be replaced by less hydrophobic structures [45].

Dhanikula and Hildgen non-covalently encapsulated 24.5% *w/w* Methotrexate in polyester-co-polyether dendrimers [120]. Recently, the same group of investigators further conjugated polyester-co-polyether dendrimers with D-glucosamine and loaded MTX inside dendritic cavities. This drug–dendrimer assembly was proposed to be suitable for the treatment of gliomas [121]. During non-covalent interactions, the hydrophilic periphery of dendrimers completely participates in solubilization, which is not feasible in the case of covalent drug conjugation on peripheral groups of dendrimer. In covalent conjugation, most of hydrophilic functional groups present on the surface of dendrimers are engaged in bonding, and such incidence may lessen the overall functionalities existing on the surface for solubilization. Consequently, in general, it is assumed that the non-covalent complexation is primarily accountable for drug solubilization [105].

The investigators have reported an enhancement of the transport of Propranolol across monolayers of the human colon adenocarcinoma cell line (*Caco-2*) following conjugation to G3 PAMAM dendrimers and to G3 PAMAM dendrimers with attached lauroyl chains (which lower cytotoxicity and enhance permeation) [113, 122]. In addition to enhancing its aqueous solubility, conjugation of Propranolol to these dendrimer carriers increased its epithelial permeation by circumventing the P-glycoprotein efflux transporter.

In one study, the dendron was based on aspartic acid units and Ara-C was conjugated via its amine group by various linkers, including amides and carbamates [123]. In another study, the bicarboxylic amino acid L-2-aminoadipic acid provided the branching unit of the dendron, and the amine of Ara-C was conjugated directly to the peripheral carboxylic acid groups of the dendron [124]. This prodrug strategy was found to improve the blood residence time of the drug and to increase its stability towards degradation. Starlike carriers of the anticancer drug, doxorubicin, have been prepared by the conjugation of poly [*N*-(2-hydroxypropyl) methacrylamide] macromonomers to PAMAM dendrimers, followed by attachment of the drug to the polymer arms by a biodegradable peptide spacer. The compact structure of the star polymer results in slower rates of enzyme-mediated drug release, thereby decreasing the cytotoxicity of the conjugate [125]. The introduction of pH-responsive linkage into dendrimer-drug conjugates appeared to further expand the application potential of this area.

13.9.2.1 pH-Responsive Drug–Dendrimer Conjugate

The utility of acid-sensitive linkages had been demonstrated by Greenfield *et al.* for Adriamycin immuno-conjugates [126]. The conjugate was reported to be stable at the physiological pH of 7.4, but found to undergo hydrolysis upon uptake of the polymers by endocytosis and subsequent trafficking to mildly acidic sub-cellular organelles such as the endosomes and lysosomes [23, 127, 128]. Keeping these reports in mind, Ihre *et al.* reported the design and synthesis of dendritic polyester systems based on the monomer unit 2,2-bis (hydroxymethyl) propanoic acid as a possible versatile drug carrier. The potent anticancer drug doxorubicin was attached via a pH-sensitive linkage, thus demonstrating the feasibility of using these polyester dendritic structures to produce polymer-drug conjugates capable of delivering the drug to a chosen low pH cancerous site [66]. Other classes of therapeutic molecules including DOX and other anticancer drugs [117, 128, 129], natural curcumin derivatives [130], pro-apoptotic peptide, and photosensitizing agents [131, 132] were coupled to a dendritic carrier, which significantly increased the solubility of the loaded drug.

In 2006, Padilla *et al.* reported the synthesis of bow-tie-shaped polyester G3-G4 dendrimers. DOX was linked to the G4 region by means of either a pH-sensitive hydrazone (HYD) or a carbamate linkage to yield G4-HYD-DOX and G4-DOX conjugates, respectively. G4-HYD-DOX conjugates were stable at pH 7.4 as designated by the release of less than 10% of the incorporated DOX as against 100% release of attached DOX upon incubation at pH 5.0 [117]. G4-HYD-DOX conjugates were found to be more cytotoxic towards colon carcinoma cells with an IC_{50} of 1.4 μg of DOX/mL in contrast to carbamate-linked G4-DOX conjugates with an IC_{50} of 2.0 μg of DOX/mL [128]. Subsequent reports confirmed the higher *in vitro* and *in vivo* anticancer activity of G4-HYD-DOX conjugates compared to amide-linked conjugates and the free drug. In addition, G4-HYD-DOX conjugates proved to be equally effective against DOX-sensitive and DOX-resistant cells, whereas free DOX was 58 times less effective in inducing apoptosis in resistant cancer cells [133]. Fluorescence microscopy studies of *Ca9-22* cells separately treated with G4.5-HYD-DOX and G4.5-amide-DOX conjugates revealed that the hydrolysis of the hydrazone linkage allows the liberated DOX molecules to enter the nucleus, whereas

G4.5-amide-DOX conjugates fail to release the incorporated drug, thus limiting its access to the nucleus and diminishing its therapeutic activity [132].

It is appealing to note that an additional study illustrated that smaller G0-DOX conjugates were competent to go through the nucleus despite the linkage chemistry [133]. While these pH-sensitive linkages signify a considerable enhancement over noncleavable conjugates for intracellular drug delivery of drugs, they only sense the acidity of the endosomal compartment but fail to differentiate between cancer cells and normal healthy ones [134]. Thus, additional selectivity of drug release from dendrimer conjugates can be accomplished by the development of novel chemical linkages that are sensitive to cancer-specific markers such as intracellular enzymes. This will permit the release of the incorporated drug only in response to these enzymes, which are exclusively articulated by cancer cells.

During the developmental stage of dendrimer synthesis several less toxic dendrimers have been developed, which have poor water solubility, namely, Polyaryl ether dendrimer reported by Hawker and Fréchet [38]. Such incidence necessitates the conjugation of solubilizing groups at their terminal groups as suggested by a few workers [45, 46]. Also, in spite of having huge biomedical applications, the use of dendritic architectures to a great extent is constrained due to RES uptake, drug leakage, immunogenicity, stability, hemolytic toxicity, hydrophobicity, etc. Recently, PEGylation has been used to make dendrimers less susceptible to drug seepage as well as to decrease their hemolytic toxicity [86]. In the past few decades in particular, scientists have mainly focused their investigation on development of more practically feasible, safe and biocompatible dendrimers.

13.10 Dendrimers' Biocompatibility and Toxicity

As a universal rule, for any polymeric carrier to be suitable for bioapplication, it is necessary that the carrier is non-toxic and non-immunogenic, and it should preferably be biodegradable. It must exhibit an innate body distribution that will permit suitable tissue targeting to the desired location, but away from sites of toxicity. In order to utilize dendrimeric architectures as Generally Regarded As Safe (GRAS) tool for drug delivery, they have to fulfill all these biological stipulations of vital significance.

To achieve this goal of making dendrimers more biocompatible, several attempts have been made, however, a notable success was observed with PEGylation technology. Many of the undesirable affects exhibited by dendrimers can be attenuated through the linkage of PEG chains on to the dendrimer periphery. PEGylated materials show greatly reduced cyto- and hemotoxicity [62, 86, 135], as well as increased circulation time in the blood and decreased accumulation in the liver and kidneys. These effects are attributed to the reduction or shielding of the positive charge on the dendrimer surface by the PEG chains [122]. This shielding benefit elicited by PEG chains was considered to be a significant move towards the development of biocompatible dendrimer.

13.10.1 PEGylation Technology: A Way to Enhance Dendrimer Solubility and Biocompatibility

The practice of organized conjugation of polyethylene glycol (PEG) chains on the surface of dendritic architecture is referred to as PEGylation. The terminal groups of dendrimer easily allow PEGylation, which improves water solubility and non-statistical attachment of drug molecules. The PEG chain bears two equivalent hydroxyl groups to which a variety of bioactive species can be attached by covalent coupling. These hydroxyl groups can still be substituted to form simple PEG analogues such as carboxymethyl, succinate, succinimido, mesylate, tosylate, bromo, amino, aldehyde, monopalmitate, octadesylamine, and stearyloxy derivatives of PEG [136, 137]. The covalent scaffold forms a stable molecular structure for the internal encapsulation of a guest, which is independent of thermodynamics and physical factors [30, 45, 86].

The main rationale behind the PEGylation of dendrimers is altering the biodistribution and pharmacokinetics of dendrimers by increasing blood circulation due to decreased RES, liver, spleen and macrophageal uptake, altering the solubility profile of dendrimers making them more soluble [137], decreasing the toxicity by shielding the peripheral –NH_2 groups [135], improved utilization as drug delivery system by increasing drug loading [138], sustained and controlled drug delivery [86], for improved transfection capability [139], and for increased drug uptake. In recent years quite many studies have been carried out with the varieties of dendrimers including its use as a drug delivery device [140–143].

The PEGylation technique excellently raises the level of dendrimer-mediated solubilization owing to enhanced entrapment and the solubilization of hydrophobic drugs molecules in additional crevices generated by PEG networks [144]. To realize these objectives PEGylation of dendrimer has been achieved in various ways and because of continual research and development, this emerging area has been organized into a dendrimer superspecialty.

13.11 Classification of PEGylated Dendrimers

The PEGylated dendrimers can be broadly classified on the basis of location as well as the orientation of PEG chains in the dendritic architectures.

13.11.1 PEGylated Dendrimer

This class of dendrimer bears PEG chains conjugated to their periphery while the drug molecules are entrapped in the inner crevices of the architecture (Figure 13.12a). Several investigators have PEGylated simple dendrimers but with different objectives [50, 86, 135]. With the aim of accomplishing both solubilization and sustained release profits from dendritic formulations, Kojima *et al.* synthesized a dendrimeric assembly by uniting the chain ends of 3.0 and 4.0G PAMAM dendrimers with PEG-monomethyl ether by means of an urethane bond. The hydrophobic anticancer drugs, Adriamycin and MTX were encapsulated inside the hydrophobic regions of the PEGylated PAMAM dendrimer. The investigators noticed that the capability of the nano-system to encapsulate hydrophobic entities was enhanced with increasing dendrimer generation as well as the extent of chains of PEG grafts. Among the methoxy PEG (*m*PEG) attached dendrimers, the highest capability was achieved with the 4.0G-mPEG$_{2000}$, which could retain 6.5 Adriamycin or 26 MTX molecules per dendrimer molecule. It was felt that Adriamycin was solubilized and complexed on the chain surface of *m*PEG, while for MTX, the encapsulation efficiency of an acidic drug was increased by a greater electrostatic interaction arising from the acid base feedback between MTX and dendrimer. The MTX-loaded PEG dendrimers released the

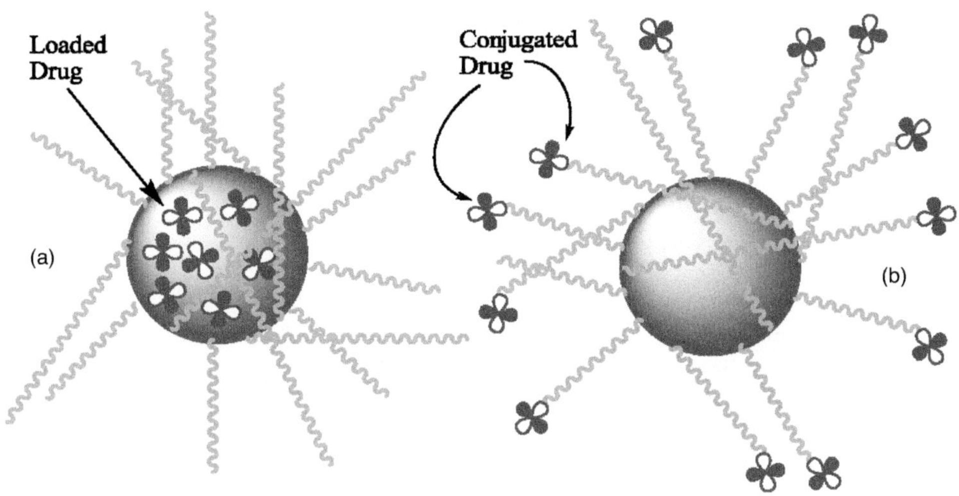

Figure 13.12 *PEGylated dendrimer with (a) physical drug loading; and (b) drug loading by covalent conjugation.* (See colour plate section.)

drug slowly in an aqueous solution of low ionic strength. Moreover, as the encapsulation via the dendrimer varied appreciably depending on the drug and the dendrimer configuration, this method would be quite tricky to apply universally to all drugs [138].

The synthesis of PEGylated dendritic systems as nanoparticulate depots for drug delivery was described by Bhadra *et al.* G4 PAMAM dendrimers were synthesized and PEGylated using *m*PEG-5000. A comparison of the properties of G4 PAMAM dendrimer and PEGylated dendrimer revealed that the PEGylated systems had a higher drug-loading capacity (12-fold for 5-fluorouracil), a slower drug release rate (1/6th), and decreased toxicity as compared to the non-PEGylated dendrimer [86].

Ooya *et al.* studied the effects of polymeric architecture on the solubilization and release of Paclitaxel, a poorly water-soluble drug. In their work they investigated the effect of the density of ethylene glycol chains on the solubility enhancement of Paclitaxel. For this, they synthesized the poly(oligo ethylene glycol) methacrylate (OEGMA), star-shaped poly(OEGMA), and poly glycerol 3.0, 4.0, and 5.0G dendrimers. The synthesized Poly(OEGMA) increased the Paclitaxel solubility, but a much more noteworthy outcome was noticed with the five-arm star poly(OEGMA). The aqueous solubility produced by 10% star-shaped poly OEGMA, and by 3.0, 4.0 and 5.0G polyglycerol dendrimers respectively, was 130-, 270-, 370- and 434-fold greater than the naked Paclitaxel solubility in water. These statistics are adequate to conclude that polyglycerol dendrimers are much more efficient in increasing the Paclitaxel solubility than the others under study. This is likely due to the increased local density of the ethylene glycol unit. Even with relatively similar molecular weight (Mw: 1690) and concentration (50 wt.%) of the 3.0G dendrimers to PEG-2000 (Mw: 2000), the Paclitaxel solubility was raised eleven-fold over that of PEG-2000. These polyglycerol dendrimers dissolved in water at high concentrations without significantly increasing the viscosity and, at 80 wt.%, were found to increase the solubility of Paclitaxel 10,000-fold [145].

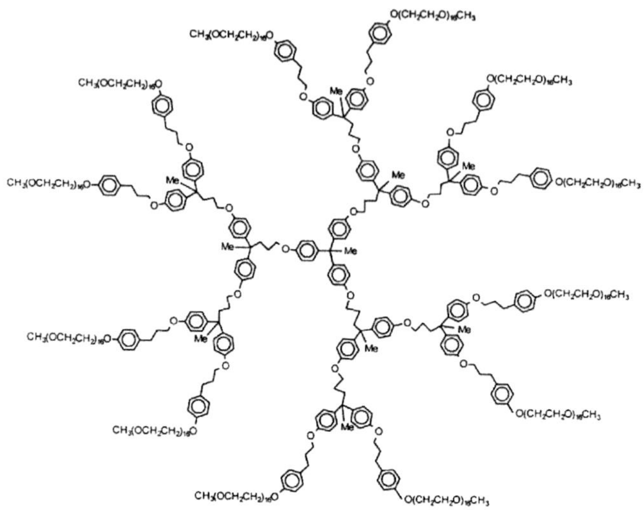

Figure 13.13 Structure of G2 water-soluble dendritic unimolecular micelles based on 4,4-bis(4V-hydroxyphenyl) pentanol building blocks and a surface shell of polyethylene glycol (PEG) chains. Reproduced with permission from [46].

In a follow-up assessment, Ooya *et al.* synthesized the same architecture and used this to examine the effect of dendritic architecture and generation on the aqueous solubilization of Paclitaxel. The Paclitaxel solubility in all the solutions of polyglycerol dendrimers, even below 10-wt %, was much higher than that in PEG-400, which is frequently engaged as a co-solvent or a hydrotropic agent. The increase in the solubility of Paclitaxel was found to be a function of dendrimer generation. ^{1}H NMR spectra of Paclitaxel before and after mixing with polyglycerol dendrimers (PGDs) in D_2O suggested that the aromatic rings and some methyne groups of Paclitaxel were surrounded by PGDs, hence providing an excellent alternative mode for the hydrotropic solubilization of a poorly soluble drug [146].

Liu *et al.* synthesized water-soluble dendritic unimolecular micelles based on 4,4-bis(4V-hydroxyphenyl) pentanol building blocks and a surface shell of PEG chains [145, 146] (Figure 13.13). The pentanol-based monomer was used to amplify the flexibility and void size of the dendritic architecture. A single dendritic molecule of G1, G2 and G3 was found to dissolve respectively, 0.39, 0.97 and 2.9 molecules of pyrene in water. A model drug (Indomethacin) was loaded to a level of 11% *w/w* in a G3 dendrimer, a value that matches roughly nine drug molecules per dendrimer. The drug-loaded dendrimer provided sustained release of Indomethacin over a period of approximately 30 hr.

Betamethasone valerate and Betamethasone dipropionate are basically water-insoluble anti-inflammatory corticosteroids, and because of this, it is difficult to entrap these compounds in hydrophilic carriers. The results of Sideratou and coworkers inferred that the concentration of encapsulated Betamethasone derivatives can be significantly enhanced in PEGylated dendrimer [137].

Although the introduction of stabilizing PEO chains on the dendrimer periphery has expanded the scope of dendritic unimolecular micelles to incorporate anticancer drugs such as 5-fluorouracil [86], Methotrexate and Doxorubicin, and can slow the drug release rates in these systems to some extent, this method has yet to be demonstrated as a general strategy [138]. A promising new approach to controlling the release of drugs from the encapsulating micellar compartment involves the use of hybrids of PEO and dendrimers with pH-sensitive hydrophobic acetal groups on the dendrimer periphery [147]. Loss of the hydrophobic groups on acetal hydrolysis at mildly acidic pH triggers disruption of the micelle and release of the payload [148].

13.11.2 Drug-Conjugated PEGylated Dendrimer

If a drug is conjugated directly to the dendrimer surface, there is a huge likelihood of the drug being folded inside the dendritic core due to hydrophobic interaction. This act shields the drug from being exposed to physiological fluids, which results in the failure of the drug release from the carrier. Drug-conjugated PEGylated dendrimers constitute the second major category of PEGylated dendrimer, which is to have PEG chains as a spacer between the drug and dendrimers (Figure 13.12b). Yang *et al.* conjugated one end of the PEG chain to the dendrimer while the distal end used penicillin-V. This move avoids steric hindrance between drug and dendrimer [149].

An early model demonstrating the advantage of the stepwise synthesis and the controlled multivalency of PEGylated dendrimers for drug delivery was put forward by Liu *et al.*, wherein it was possible by using a careful synthetic approach to attach both hydrophobic drugs and polyethylene oxide (PEO) moieties to the dendrimer periphery in a controlled manner [45]. Hence this approach can be considered a hybrid of drug-conjugated [150] and PEGylated dendrimers [86]. In addition, Liu *et al.* explored the potential of dendritic unimolecular micelles for drug delivery using both the container and sustained drug-releasing properties of such systems [46].

The covalent conjugation of drugs onto the surface of dendrimers through hydrolysable or biodegradable linkages offers the prospect for superior control over drug release that cannot be achieved by electrostatic complexation of drugs to the dendrimers. The utilization of an amide linkage provided bond stability, whereas the ester linkage of the drug to the dendrimer provided a means of controlling drug release via hydrolysis. The antimicrobial performance of penicillin released by the ester hydrolysis of the PEG-PAMAM conjugate was approximately similar to that of unmodified penicillin. In another investigation by these authors, an extended release formulation is described in which the antidepressant drug, Venlafaxine, was covalently linked to a G2.5 PAMAM dendrimer via a hydrolysable ester bond, with the aim of overcoming the problem of poor patient compliance [149].

13.11.3 PEG Cored Dendrimer

This category of PEGylated dendrimers has PEG as a core material (Figure 13.14a). The use of PEG as central core increases the distance between the branching units of the dendrimer and prevents the drug from being dumped into the dendritic core [47, 151]. An ideal PEGylated dendritic drug-carrier must be non-toxic [152], non-immunogenic, preferably biodegradable [47, 153] and present an adequate biodistribution and allow tissue targeting. Namazi *et al.* synthesized citric acid dendrimers having PEG as a core and increased the solubility of many hydrophobic drugs such as 5-Amino salicylic acid, Pyridine, Mefenamic

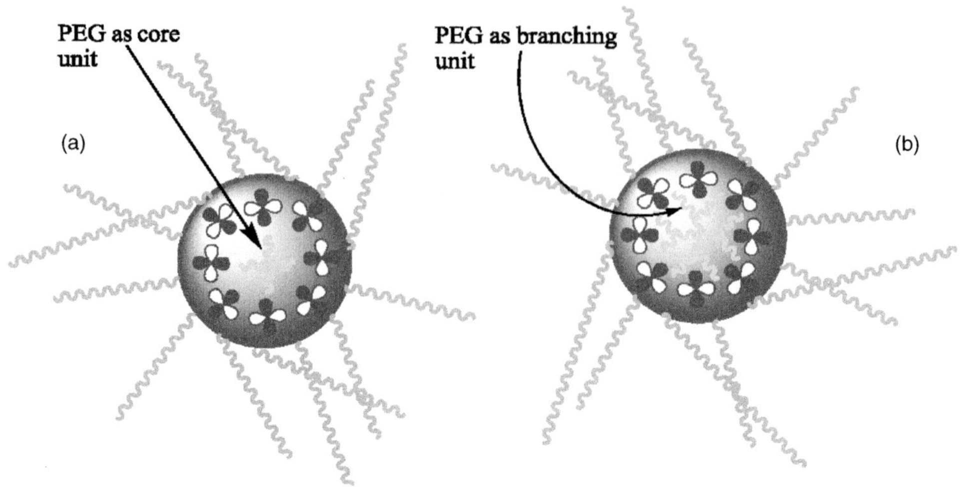

Figure 13.14 *PEGylated dendrimer with (a) PEG as central core; and (b) PEG as branching units.* (See colour plate section.)

acid, and Diclofenac [47]. Similarly, increased solubility of pyrene and Indomethacin was observed with dendritic unimolecular micelles prepared with PEGmesylate [46].

13.11.4 PEG Branched Dendrimer

Dendrimer with PEG as branching monomeric units confers a third category of PEGylated dendrimers (Figure 13.14b). Several investigators have designed dendrimers with PEG as a branching unit [154]. Berna *et al.* established that these dendrimers circulate for a long time in the bloodstream with reduced cytotoxicity [155]. Ooya *et al.* used this category of PEGylated dendrimers for enhanced solubilization and controlled release of bioactives [145].

13.11.5 PEG-Conjugated Targeted Dendrimer

The prospect of pharmaceutical applications of PEGylated dendrimers is expected to increase in future, to include a multifunctional drug delivery system for the intended diagnosis and targeting purposes, while keeping the exceptional solubility of the system in mind [57, 149]. This category involves hybrid dendrimers bearing PEG as well as targeting moiety attached to their periphery (Figure 13.15). The drug can be either in loaded as well as in conjugated form. Tansey *et al.* [153], Qualmann *et al.* [156] and Heldt *et al.* [157] synthesized such dendrimers bearing both targeting ligand as well as PEG chains.

A similar approach was designed by Singh and co-workers wherein PAMAM dendrimers up to 4.0G were synthesized and conjugated to folic acid directly or indirectly through PEG-4000 as spacer. 5-FU was encapsulated in plain PAMAM as well as dendrimer-folic acid conjugates. The system was hypothesized to have the drug (i.e. 5-FU) within its internal cavities, which are shielded by exterior groups linked to folic acid and folate-PEG ligands

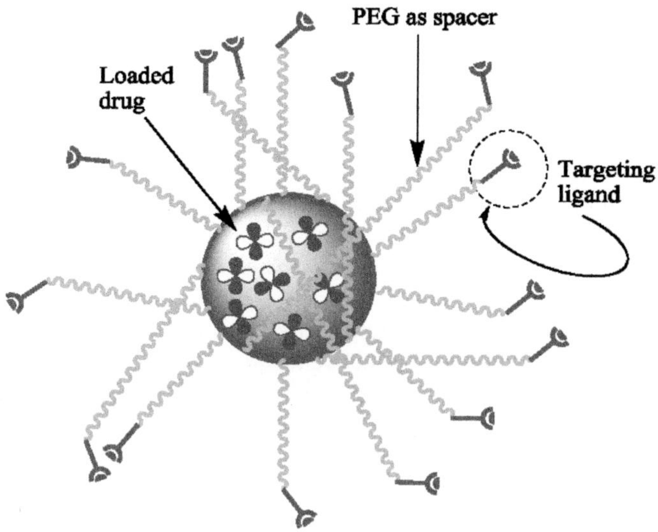

Figure 13.15 *Hybrid PEGylated dendrimer with simultaneous solubilization and site-specific drug targeting.* (See colour plate section.)

[37]. Previously also this strategy was found to be efficient in anticancer drug targeting [22, 158]. The PEGylation of dendrimers was found to extend the circulation time while folate-conjugated systems may support the site-specific delivery of the bioactives. The developed formulation might be acting through the EPR effect as well as receptor-mediated endocytosis (RME), providing synergistic action, thereby enhancing the therapeutic potential. Conclusively, engineering of the dendritic surface with a targeting ligand such as folic acid can improve the site-specific drug delivery. PEGylation of dendrimers can improve the circulation time of dendrimers. The formulation was found to bear excellent water solubility owing to the attached PEG as a spacer and the authors proposed the formulation as suitable for administering through the intravenous route [37].

13.12 Conclusion

Hydrophobicity of a drug molecule presents the chief formulation faced during product development. Consequently, solubility augmentation always had been a headache to scientists. Over the past three decades, dendrimers have grown from a concept to become a new class of polymers with versatile chemical structures. Dendrimers are well-defined, globular, synthetic polymers with numerous characteristics that make them valuable in the bio-pharmaceutical arena. In this line, the traits offered by dendrimers have tremendously focused the concentration of investigators to explore them as solubilizing agents owing to their stable micellar nature, which differentiates them from other traditional micellar architectures. Studies comparing the potential of dendrimers in solubility enhancement with micelles and cyclodextrins are available and suggest that dendrimers might prove significantly superior to both these systems. The high loading capacity of dendrimers renders them highly attractive as carriers for drug delivery. Advances in synthetic strategies have led to the development of well-controlled dendrimers structures with significantly huge number of surface groups that can be utilized to exhibit a variety of bio-physical motifs while carrying a large therapeutic payload.

Today, dendritic nanostructures are a commercially viable move towards solving the problems of hydrophobic drugs, such as poor solubility and deprived bioavailability. However, before proceeding with more applicable next-generation designs, the biocompatibility and toxicity of the dendrimers must be explored exhaustively and understood completely. More detailed experimentations correlating *in vitro* and *in vivo* performance of dendrimers can yield substantial information in these directions. We anticipate that prospective investigations will focus on the design and synthesis of novel linkers that will be selectively recognized and cleaved by enzymes in selective locations (e.g. cancer, arthritis) to achieve an additional degree of selective control over the site-specific release from dendrimer-drug conjugates. It is expected that new dendrimer structures will continue to be developed, and more types of dendrimers will be manufactured to a high scale of perfection and will become commercially available. There is an urgent need to establish the mechanisms of solubilization using dendrimers. An equally important and rewarding strategy would be to compare the solubilization propensity, the safety as well as the efficacy of different solubilization techniques in a systematic manner to arrive at a meaningful conclusion (possibly!).

References

1. R.J. Boghra, P.C. Kothawade, V.S. Belgamwar, *et al.*, Solubility, dissolution rate and bioavailability enhancement of irbesartan by solid dispersion technique, *Chem Pharm Bull*, **59**, 438–441 (2011).
2. M. Vogt, K. Kunath, J.B. Dressman, Dissolution enhancement of fenofibrate by micronization, cogrinding and spray drying: Comparison with commercial preparations, *Eur J Pharm Biopharm*, **68**, 283–288 (2008).
3. F. Guo, H. Zhong, J. He, *et al.*, Self–microemulsifying drug delivery system for improved oral bioavailability of dipyridamole: Preparation and evaluation, *Arch Pharm Res*, **34**, 1113–23 (2011).
4. P. Sharma, W.A. Denny, S. Garg, Effect of wet milling process on the solid state of indomethacin and simvastatin, *Int J Pharm*, **380**, 40–48 (2009).
5. C.Y. Wu, L.Z. Benet, Predicting drug disposition via application of BCS: transport/absorption/elimination interplay and development of a biopharmaceutics drug disposition classification system, *Pharm Res*, **22**, 11–23 (2005).
6. J.S. Valentino, W.N. Kwame, Prodrug strategies to overcome poor water solubility. *Adv Drug Del Rev.*, **59**, 677–694 (2007).
7. Y.G. Bachhav, V.B. Patravale, Formulation of meloxicam gel for topical application: *In vitro* and *in vivo* evaluation, *Acta Pharm*, **60**, 153–163 (2010).
8. S.N. Pace, G.W. Pace, L. Parikh, *et al.*, Novel injectable formulations of insoluble drugs, *Pharm Tech*, **23**, 116–134 (1999).
9. P. Lakshmi, K.G. Ashwini, Nanosuspension technology: A review. *Int J Pharm Sci*, **2**, 35–40 (2010).

10. P.J. Flory, Molecular size distribution in three dimensional polymers. II. Trifunctional Branching Units, *J Am Chem Soc*, **63**, 3091–3096 (1941).
11. P.J. Flory, Crystallinity and dimensional changes in fibrous proteins, *J Cell Physiol Suppl*, **49**, 175–183 (1957).
12. P.J. Flory, Conformations of macromolecules, *Brookhaven Symp Biol*, **13**, 89–103 (1960).
13. W.W. Graessley, Statistical mechanics of random coil networks, *Macromolecules*, **8**, 186–190 (1975).
14. P.J. Flory, Fundamental principles of condensation polymerization, *Chem Rev*, **39**, 137–197 (1946).
15. P.J. Flory, Molecular size distribution in three dimensional polymers. VI. Branched polymers containing A—R—B_{f-1} Type Units, *J Am Chem Soc*, **74**, 2718–2723 (1952).
16. F. Vogtle, S. Gestermann, R. Hesse, *et al.*, Functional dendrimers, *Prog Polym Sci*, **25**, 987–1041 (2000).
17. D.A. Tomalia, H. Baker, J. Dewald, *et al.*, A new class of polymers: starburst–dendritic macromolecules, *Polym J*, **17**, 117–132 (1985).
18. J.M.J. Fréchet. Functional polymers and dendrimers: Reactivity, molecular architecture, and interfacial energy, *Science*, **263**, 1710–1715 (1994).
19. A.K. Zimmerman, P. Chetoni, M.F. Saettone, Evaluation of pilocarpine-loaded albumin particles as controlled drug delivery systems for the eye. II. Co-administration with bioadhesive and viscous polymers. *J Cont Rel*, **33**, 31–46 (1995).
20. E.C. Wiener, S. Konda, A. Shadron, *et al.*, Targeting dendrimer–chelates to tumors and tumor cells expressing the high-affinity folate receptor, *Invest. Radiol*, **32**, 748–754 (1997).
21. J.R. Baker, A. Quintana, L.T. Riehler, *et al.*, The synthesis and testing of anti-cancer therapeutic nanodevices: Biomed, *Microdevices*, **3**, 61–69 (2001).
22. A. Quintana, E. Raczka, L. Piehler, *et al.*, Design and function of a dendrimer-based therapeutic nanodevice targeted to tumor cells through the folate receptor, *Pharm Res*, **19**, 1310–1316 (2002).
23. R.K. Tekade, P.V. Kumar, N.K. Jain, Dendrimers in oncology: An expanding horizon, *Chem Rev*, **109**, 49–87 (2009).
24. G.R. Newkome, C.N. Moorefield, J.N. Keith, *et al.*, Escamilla, Chemistry within a unimolecular micelle precursor: Boron superclusters by site- and depth-specific transformations of dendrimers, *Angew Chem, Int Ed*, **33**, 666–668 (1994).
25. J. Issberner, R. Moors, Dendrimers: From generations and functional groups to functions, *Angew Chem, Int Ed Engl*, **33**, 2413–2420 (1994).
26. R.F. Service, Dendrimers: Dream molecules approach real applications, *Science*, **267**, 458–459 (1995).
27. R. Esfand, D.A. Tomalia, Poly (amidoamine) (PAMAM) dendrimers: From biomimicry to drug delivery and biomedical applications, *Drug Disc Today*, **6**, 427–436 (2001).
28. G.R. Newkome, C.N. Moorefield, G.R. Baker, *et al.*, Unimolecular micelles, *Angew Chem, Int Ed Engl*, **30**, 1176–1178 (1991).
29. G.M. Dykes, Dendrimers: A review of their appeal and applications, *J Chem Technol Biotechnol*, **76**, 903–918 (2001).

30. M. Liu, J.M.J. Fréchet, Designing dendrimers for drug delivery. *Pharm Sci Technol, Today*, **2**, 393–401 (1999).
31. E. Buhleier, W. Wehner, F. Vogtle, "Cascade"– and "Nonskid-Chain-like" synthesis of molecular cavity topologies, *Synthesis*, **3**, 155–158 (1978).
32. K. Autumn, Y.A. Liyang, S.T. Hseih, *et al.*, Adhesive force of a single gecko foot-hair, *Nature*, **405**, 681–685 (2000).
33. U. Boas, P.M. Heegaard, Dendrimers in drug research, *Chem Soc. Rev*, **33**, 43–63 (2004).
34. D.A. Tomalia, H. Baker, J. Dewald, *et al.*, Dendritic macromolecules: Synthesis of starburst dendrimers, *Macromolecules*, **19**, 2466–2468 (1986).
35. D.A. Tomalia, Starburst dendrimers–nanoscopic supermolecules according to dendritic rules and principles. *Macromol Symp*, **101**, 243–255 (1996).
36. G.R. Newkome, Z. Yao, G.R. Baker, *et al.*, Cascade molecules: A new approach to micelles, *J Org Chem*, **50**, 2003 (1985).
37. P. Singh, U. Gupta, A. Asthana, *et al.*, Folate and Folate–PEG–PAMAM dendrimers: Synthesis, characterization, and targeted anticancer drug delivery potential in tumor bearing mice. *BioconJ Chem*, **19**, 2239–2252 (2008).
38. C.J. Hawker, J.M.J. Fréchet, Preparation of polymers with controlled molecular architecture. A new convergent approach to dendritic macromolecules, *J Am Chem Soc*, **112**, 7638–7647 (1990).
39. D.A. Tomalia, A.M. Naylor, W.A. Goddard III, Starburst dendrimers: Molecular level control of size, shapes, surface chemistry, topology and flexibility from atoms to macroscopic matter, *Angew Chem, Int Ed*, **29**, 138–175 (1990).
40. D. Fischer, Y. Li, B. Ahlemeyer, *et al.*, *In vitro* cytotoxicity testing of polycations: Influence of polymer structure on cell viability and hemolysis, *Biomaterials*, **24**, 1121 (2003).
41. A.W. Bosman, H.M. Janssen, E.W. Meijer, About dendrimers: Structure, physical properties, and applications, *Chem Rev*, **99**, 1665–1688 (1999).
42. C.J. Hawker, K.L. Wooley, J.M.J. Fréchet, Unimolecular micelles and globular amphiphiles: Dendritic macromolecules as novel recyclable solubilization agents, *J Chem Soc*, **21**, 1287–1297 (1993).
43. S. Svenson, D.A. Tomalia, Dendrimers in biomedical applications: reflections on the field. *Adv Drug Deliv Rev*, **57**, 2106–2129 (2005).
44. G. Kong, R.D. Braun, M.W. Dewhirst, Hyperthermia enables tumor-specific nanoparticle delivery: Effect of particle size, *Can Res*, **60**, 4440–4445 (2001).
45. M. Liu, K. Kono, J.M.J. Fréchet, Water-soluble dendrimer–polyethylene glycol starlike conjugates as potential drug carriers, *J Polym Sci*, **37**, 3492–3503 (1999).
46. M. Liu, K. Kono, J.M.J. Fréchet, Water-soluble dendritic unimolecular micelles: Their potential as drug delivery agents, *J Cont Rel*, **65**, 121–131 (2000).
47. H. Namazi, M. Adeli, Dendrimers of citric acid and poly (ethylene glycol) as the new drug-delivery agents, *Biomaterials*, **26**, 1175–1183 (2005).
48. G. Caminati, K. Gopidas, A.R. Leheny, *et al.*, Photochemical probes of Starburst® dendrimers and their utilization as restricted reaction spaces for electron transfer processes, *Polym Prep*, **32**, 602 (1991).

49. G. Caminati, N.J. Turro, D.A. Tomalia, Photophysical investigation of Starburst® dendrimers and their interactions with anionic and cationic surfactants, *J Am Chem Soc*, **112**, 8515–8522 (1990).
50. S. Stevelmans, J.C.M. Hest van, J.F.G.A. Jansen, *et al.*, Synthesis, characterization and guest–host properties of inverted unimolecular dendritic micelles, *J Am Chem Soc*, **118**, 7398–7399 (1996).
51. B. Devarakonda, R.A. Hill, W. Liebenberg, *et al.*, Comparison of aqueous solubilization of practically insoluble niclosamide by polyamidoamone (PAMAM) dendrimers and cyclodextrins, *Int J Pharm*, **300**, 193–209 (2005).
52. E. Gong, B. Matthews, T. McCarthy, *et al.*, Evaluation of dendrimer SPL7013, a lead microbicide candidate against herpes simplex viruses, *Antiviral Res*, **68**, 139–146 (2005).
53. M. Enomoto, A. Aida, Iron complexes capped with dendrimer-appended triazacyclononanes as the novel spatially encumbered models of non-heme iron proteins, *J Am Chem Soc*, **124**, 6099–6108 (2002).
54. A.S. Chauhan, S. Sridevi, K.B. Chalasani, *et al.*, Dendrimer–mediated transdermal delivery: Enhanced bioavailability of indomethacin, *J Cont Rel*, **90**, 335–343 (2003).
55. U. Gupta, H.B. Agashe, N.K. Jain, Polypropylene imine dendrimer mediated solubility enhancement: Effect of pH and functional groups of hydrophobes, *J Pharm Pharm Sci*, **10**, 358–367 (2007).
56. R.N. Prajapati, R.K. Tekade, U. Gupta, *et al.*, Dendimer-mediated solubilization, formulation development and *in vitro–in vivo* assessment of Piroxicam, *Mol Pharm*, **6**, 940–950 (2009).
57. B. Devarakonda, R.A. Hill, M.M. DeVilliers, The effect of PAMAM dendrimer generation size and surface functional group on the aqueous solubility of nifedipine, *Int J Pharm*, **284**, 133–140 (2004).
58. S. Uppuluri, D.R. Swanson, L.T. Piehler, *et al.*, Core-shell tecto(dendrimers), synthesis and characterization of saturated shell models, *Adv Mater*, **12**, 796–800 (2000).
59. L.S. Kaanumalle, J. Nithyanandhan, M. Pattabiraman, *et al.*, Water-Soluble Dendrimers as Photochemical Reaction Media: Chemical Behavior of Singlet and Triplet Radical Pairs Inside Dendritic Reaction Cavities, *J Am Chem Soc*, **126**, 8999–9006 (2004).
60. C. Yiyun, X. Tongwen, Polyamidoamine dendrimers used as solubility enhancers of ketoprofen. *Eur J MEd Chem*, **40**, 1390–1393 (2005b).
61. G. Pistolis, A. Malliaris, Study of poly(propylene imine) dendrimers in water by exciplex formation, *Langmuir*, **18**, 246–251 (2002).
62. H.T. Chen, M.F. Neerman, A.R. Parrish, *et al.*, Cytotoxicity, haemolysis, and acute *in vivo* toxicity of dendrimer based on melamine, candidate vehicles for drug delivery, *J Am Chem Soc*, **126**, 10044–10048 (2004).
63. H. Lee, Y.B. Lim, J.S. Choi, *et al.*, Polyplexes assembled with internally quaternized PAMAM–OH dendrimer and plasmid DNA have a neutral surface and gene delivery potency, *BioconJ Chem*, **14**, 1214–1221 (2003).
64. J.S. Choi, K. Nam, J. Park, *et al.*, Enhanced transfection efficiency of PAMAM dendrimer by surface modification with l–arginine, *J Cont Rel*, **99**, 445–456 (2004).

65. N. Malik, R. Wiwattanapatapee, R. Klopsch, *et al.*, Dendrimers: Relationship between structure and biocompatibility *in vitro*, and preliminary studies on the biodistribution of 125I-labelled polyamidoamine dendrimers *in vivo*, *J Cont Rel*, **65**, 133–148 (2000).
66. H.R. Ihre, O.L. Padilla DeJesus, J.L. Peters, *et al.*, Polyester dendritic systems for drug delivery applications: Design, synthesis, and characterization, *Bioconj Chem*, **13**, 443–452 (2002).
67. M. Xyloyiannis, O.L. Padilla DeJesus, J.M.J. Fréchet, *et al.*, PEG–dendron architecture influences endocytic capture and intracellular trafficking, *Proc. Int Symp Cont Rel Bioact Mat*, **30**, 149 (2003).
68. J.L. Lach, J. Cohen, Interaction of pharmaceuticals with Schardinger dextrins. II. Interaction with selected compounds, *J Pharm Sci*, **52**, 137–142 (1963).
69. S.B. La, T. Okano, K. Kataoka, Preparation and characterization of the micelle-forming polymeric drug indomethacin–incorporated poly(ethylene oxide)–poly(beta–benzyl L–aspartate) block copolymer micelles, *J Pharm Sci*, **85**, 85–90 (1996).
70. R.K. Tekade, T. Dutta, A. Tyagi, *et al.*, Surface-engineered dendrimers for dual drug delivery: A receptor up-regulation and enhanced cancer targeting strategy, *J Drug Target*, **16**, 758–772 (2008).
71. P. Kolhe, J. Khandare, O. Pillai, *et al.*, Preparation, cellular transport, and activity of polyamidoamine-based dendritic nanodevices with a high drug payload, *Biomat*, **27**, 660–669 (2006).
72. P.K. Maiti, T. Cagin, S.T. Lin, *et al.*, Effect of solvent and pH on the structure of PAMAM dendrimers, *Macromolecules*, **38**, 979–991 (2005).
73. A. Asthana, A.S. Chauhan, P.V. Diwan, *et al.*, Poly(amidoamine) (PAMAM) dendritic nanostructures for controlled site specific delivery of acidic anti-inflammatory active ingredient, *AAPS Pharm Sci Tech*, **6**, E536–E542 (2005).
74. O.M. Milhem, C. Myles, N.B. McKeown, *et al.*, PAMAM starburst dendrimers as solubility enhancers, *Int J Pharm*, **197**, 239–241 (2000).
75. C. Yiyun, X. Tongwen, Solubility of nicotinic acid in polyamidoamine dendrimer solutions, *Eur J MEd Chem*, **40**, 1384–1389 (2005a).
76. A.K. Patri, I.J. Majoros, J.R. Baker Jr., Dendritic polymer macromolecular carriers for drug delivery, *Curr Opin Chem Biol*, **6**, 466–471 (2002a).
77. A.K. Patri, T. Thomas, J.R. Baker, *et al.*, Antibody-dendrimer conjugate for targeted prostate cancer therapy, *Polym. Mater. Sci Eng*, **86**, 130 (2002b).
78. E.R. Gillies, J.M.J. Fréchet, Development of acid-sensitive copolymer micelles for drug delivery, *Pure Appl Chem*, **76**, 1295–1307 (2004).
79. J.C.M. VanHest, M.W.P.L. Baars, R.C. Elissen, *et al.*, Acid functionalized ampiphiles derived from polystyrene–PPI dendrimers with pH dependent aggregation, *Macromolecules*, **28**, 6689–6691(1995).
80. M.W.P.L. Baars, E.W. Meijer, Host-guest chemistry of dendritic molecules, *Top Curr Chem*, **2**, 131–182 (2000).
81. A.E. Beezer, A.S.H. King, I.K. Martin, *et al.*, Dendrimers as potential drug carriers: Encapsulation of acidic hydrophobes within water soluble PAMAM derivatives, *Tetrahedron*, **59**, 3873–3880 (2003).
82. W. Chen, D.A. Tomalia, J.L. Thomas, Unusual pH-dependent polarity changes in PAMAM dendrimers: Evidence for pH responsive conformational changes, *Macromolecules*, **33**, 9169–9172 (2000).

83. I. Lee, B.D. Athey, A.W. Wetzel, *et al.*, Structural molecular dynamics studies on polyamidoamine dendrimers for a therapeutic application: Effects of pH and generation, *Macromol*, **35**, 4510–4520 (2002).
84. M. Chai, Y. Niu, W.J. Youngs, *et al.*, Structure and conformation of DAB dendrimers in solution via multidimensional NMR techniques, *J Am Chem Soc*, **123**, 4670–4678 (2001).
85. P. Kolhe, E. Misra, R.M. Kannan, *et al.*, Drug complexation, *in vitro* release and cellular entry of dendrimers and hyperbranched polymers, *Int J Pharm* **259**, 143–160 (2003).
86. D. Bhadra, S. Bhadra, S. Jain, *et al.*, A PEGylated dendritic nanaparticulate carrier of fluorouracil, *Int J Pharm*, **257**, 111–124 (2003).
87. J. Khandare, P. Kohle, S. Kannan, Synthesis, cellular transport, and activity of PAMAM–methylprednisolone conjugates, *BioconJ Chem*, **16**, 1049–1054 (2005).
88. T. Dutta, M. Garg, V. Dubey, *et al.*, Investigation on toxicological profile of surface engineered fifth generation poly (propyleneimine) dendrimers *in vivo*, *Nanotoxicol*, **2**, 62–70 (2008).
89. A.K. Patri, J.F. Kukowska–Latallo, J.R. Baker Jr., Targeted drug delivery with dendrimers: Comparison of the release kinetics of covalently conjugated drug and non-covalent drug inclusion complex, *Adv Drug Del Rev*, **57**, 2203–2214 (2005).
90. M.T. Morgan, Y. Nakanishi, D.J. Kroll, *et al.*, Dendrimer-encapsulated camptothecins: Increased solubility, cellular uptake, and cellular retention affords enhanced anticancer activity *in vitro*, *Can Res*, **66**, 11913–11921 (2006).
91. R.K. Tekade, T. Dutta, V. Gajbhiye, *et al.*, Exploring dendrimers towards dual-drug delivery: pH responsive simultaneous kinetics, *J Microencap*, **26**, 287–296 (2009).
92. N. Malik, E.G. Evagorou, R. Duncan, Dendrimer–platinate: A novel approach to cancer chemotherapy, *Antican Drugs*, **10**, 767–776 (1999).
93. S.H. Medina, M.E.H. El-Sayed, Dendrimers as carriers for delivery of chemotherapeutic agents, *Chem Rev*, **109**, 3141–3157 (2009).
94. S. Kannan, P. Kolhe, V. Raykova, *et al.*, Dynamics of cellular entry and drug delivery by dendritic polymers into human lung epithelial carcinoma cells, *J Biomat Sci*, 15311–15330 (2004).
95. W. Ke, Y. Zhao, R. Huang, *et al.* Enhanced oral bioavailability of doxorubicin in a dendrimer drug delivery system, *J Pharm Sci*, **97**, 2208–2216 (2008).
96. J.F.G.A. Jansen, E.M.M. De Brabander-van den Berg, E.W. Meijer, Induced chirality of guest molecules encapsulated into a dendritic box, *Rec Trav Chim*, **114**, 225–230 (1995a).
97. J.F.G.A. Jansen, E.W. Meijer, E.M.M. De Brabander-van den Berg, The dendritic box: Shape–selective liberation of encapsulated guests, *J Am Chem Soc*, **117**, 4417–4418 (1995b).
98. J.F.G.A. Jansen, E.W. Meijer, E.M.M. De Brabander-van den Berg, Bengal rose–at–dendritic box, *Macromol Symp*, **102**, 27–33 (1996).
99. J.F.G.A. Jansen, E.M.M. De Brabander-van den Berg, E.W. Meijer, Encapsulation of guest molecules into a dendritic box, *Science*, **266**, 1226–1229 (1994).
100. R. Esfand, D.A. Tomalia, A.E. Beezer, *et al.*, Orford, dendripore and dendrilock concepts new controlled delivery strategies, *Polym Prep*, **41**, 1324–1325 (2000).

101. A.C.H. Ng, X. Li, D.P.K. Ng, Synthesis and photophysical properties of nonaggregated phthalocyanines bearing dendritic substituents, *Macromolecules*, **32**, 5292–5298 (1999).
102. A. Bahr, B. Felber, K. Schneider, *et al.*, Dendritic, 1,1′–binaphthalene–derived cleft–type receptors (dendroclefts) for the molecular recognition of pyranosides, *Helv Chim Acta*, **83**, 1346–1376 (2000).
103. A. Adronov, S.L. Gilat, J.M.J. Fréchet, *et al.*, Light harvesting and energy transfer in laser-dye labeled poly(aryl ether) dendrimers, *J Am Chem Soc*, **122**, 1175–1185 (2000).
104. S.M. Grayson, J.M.J. Fréchet, Convergent dendrons and dendrimers: From synthesis to applications, *Chem Rev*, **101**, 3819–3867 (2001).
105. A. D'Emanuele, D. Attwood, Dendrimer–drug interactions, *Adv Drug Del Rev*, **57**, 2147–2162 (2005).
106. J. Haensler, F.C. Szoka, Polyamidoamine cascade polymers mediate efficient transfection of cells in culture, *BioconJ Chem*, **4**, 372–379 (1993).
107. J. Satija, U. Gupta, N.K. Jain, Pharmaceutical and biomedical potential of surface engineered dendrimers, *Crit Rev Ther Drug Carrier Syst*, **24**, 261–315 (2007).
108. K. Kono, H. Akiyama, T. Takahashi, *et al.*, Transfection activity of polyamidoamine dendrimers having hydrophobic amino acid residues in the periphery, *Bioconj Chem*, **16**, 208–214 (2005).
109. S. Ahmad, S. Ozaki, T. Nagase, *et al.*, A facile method for synthesis of N–acyloxymethyl–5–Fluorouracils, as a class of antitumor agents, *Chem Pharm, Bull*, **35**, 4137–4143 (1987).
110. Y. Ohya, H. Kobayashi, T. Ouchi, Design of poly (a-malic acid)–5FU–accharide conjugate exhibiting antitumour activity, *React Polym*, **15**, 153–163 (1991).
111. J. Liao, R.X. Zhuo, Synthesis, hydrolysis, and antitumour activity of conjugates of 5-fluorouracil with poly (L–lysine), *Polym J*, **25**, 401–405 (1993).
112. R.X. Zhou, B. Du, Z.R. Lu, *In vitro* release of 5-fluorouracil with cyclic core dendritic polymer, *J Cont Rel*, **57**, 249–257 (1999).
113. A. D'Emanuele, R. Jevprasesphant, J. Penny, *et al.*, The use of a dendrimer–propranolol prodrug to bypass efflux transporters and enhance oral bioavailability, *J Cont Rel*, **95**, 447–453 (2004).
114. L. Lachman, H.A. Lieberman, J.L. Kanig, *The Theory and Practice of Industrial Pharmacy*, IIIrd edition, Bombay: Varghese Publishing House, pp. 457–468 (1991).
115. R. Duncan, N. Malik, Dendrimers: Biocompatibility and potential for delivery of anticancer agents, *Proc. Int Symp Cont Rel Bioact Mat*, **23**, 105–106 (1996).
116. E. Bellis, L. Hajba, B. Kovacs, *et al.*, Three generations of α,γ-diaminobutyric acid modified PPI dendrimers and their cisplatin-type platinum complexes, *J BioChem Biophys Meth*, **69**, 151–161 (2006).
117. O.L. Padilla, H. Ihre, L. Gagne, *et al.*, Polyester dendritic system for drug delivery application: *in vitro* and *in vivo* evaluation, *Bioconj Chem*, **13**, 453–461 (2002).
118. S. Gurdag, J. Khandare, S. Stapels, *et al.*, Activity of dendrimer–methotrexate conjugates on methotrexate-sensitive and -resistant cell lines, *Bioconj Chem*, **17**, 275–283 (2006).

119. J.J. Khandare, S. Jayant, A. Singh, *et al.*, Dendrimer versus linear conjugate: Influence of polymeric architecture on the delivery and anticancer effect of paclitaxel, *Bioconj Chem*, **17**, 1464–1472 (2006).
120. R.S. Dhanikula, P. Hildgen, Influence of molecular architecture of polyether–co–polyester dendrimers on the encapsulation and release of Methotrexate, *Biomaterials*, **28**, 3140–352 (2007).
121. R.S. Dhanikula, A. Argawal, J.F. Bouchard, *et al.*, Methotrexate loaded polyether–copolyester dendrimers for the treatment of gliomas: Enhanced efficacy and intratumoral transport capability, *Mol Pharm*, **5**, 105–116 (2008).
122. R. Jevprasesphant, J. Penny, D. Attwood, *et al.*, Transport of dendrimer nanocarriers through epithelial cells via the trans-cellular route, *J Cont Rel*, **97**, 259– 267 (2004).
123. Y.H. Choe, C.D. Conover, D. Wu, *et al.*, Anticancer drug delivery systems: multi–loaded *N*4–acyl poly(ethylene glycol) prodrugs of ara–C. II. Efficacy in ascites and solid tumors, *J Cont Rel*, **79**, 55–70 (2002).
124. O. Schiavon, G. Pasut, S. Moro, *et al.* PEG–Ara–C conjugates for controlled release, *Eur J MEd Chem*, **39**, 123–133 (2004).
125. D. Wang, Synthesis of starlike *N*–(2–hydroxypropyl)methacrylamide copolymers: potential drug carriers, *Biomacromolecules*, **1**, 313–319 (2000).
126. R.S. Greenfield, T. Kaneko, A. Daues, *et al.*, *In vitro* evaluation of adriamycin immunoconjugates synthesized using an acid-sensitive hydrazone linker, *Can Res*, **50**, 6600–6607 (1990).
127. B. Rihova, T. Etrych, M. Pechar, *et al.*, Doxorubicin bound to a HPMA copolymer carrier through hydrazone bond is effective also in a cancer cell line with a limited content of lysosomes, *J Cont Rel*, **74**, 225–232 (2001).
128. C.C. Lee, E.R. Gillies, M.E. Ox, *et al.*, A single dose of doxorubicin-functionalized bow-tie dendrimer cures mice bearing C-26 colon carcinomas, *Proc Natl Acad Sci*, **103**, 16649–16654 (2006).
129. C. Falciani, M. Fabbrini, A. Pini, *et al.*, Synthesis and biological activity of stable branched neurotensin peptides for tumor targeting, *Mol Can Ther*, **6**, 2441–2448 (2007).
130. W. Shi, S. Dolai, S. Rizk, *et al.*, Synthesis of monofunctional curcumin derivatives, clicked curcumin dimer, and a PAMAM dendrimer curcumin conjugate for therapeutic applications, *Org Lett*, **9**, 5461 (2007).
131. S. Battah, S. Balaratnam, A. Casas, *et al.*, Macromolecular delivery of 5-aminolaevulinic acid for photodynamic therapy using dendrimer conjugates, *Mol Pharm*, **6**, 876–885 (2007).
132. P.S. Lai, P.J. Lou, C.L. Peng, *et al.*, Doxorubicin delivery by polyamidoamine dendrimer conjugation and photochemical internalization for cancer therapy, *J Cont Rel*, **122**, 39–46 (2007).
133. K. Kono, C. Kojima, N. Hayashi, *et al.*, Preparation and cytotoxic activity of poly(ethylene glycol)-modified poly(amidoamine) dendrimers bearing adriamycin, *Polymer*, **29**, 1664–1675 (2008).
134. T. Kapp, P. Francke, R. Gust, Investigations on surface modified dendrimers: enzymatic hydrolysis and uptake into MCF-7 breast cancer cells, *Chem MEd Chem*, **3**, 635–641 (2008).

135. R. Jevprasesphant, J. Penny, R. Jalal, *et al.*, The influence of surface modification on the cytotoxicity of PAMAM dendrimers, *Int J Pharm*, **252**, 263–266 (2002).
136. L.D. Margerum, B.K. Campion, M. Koo, *et al.*, Gadolinium(III) D03A macrocycles and PEG coupled to dendrimers: Effect of molecular weight on physical and biological properties of macromolecular magnetic resonance imaging contrast agents, *J Alloys Comp*, **249**, 185–190 (1997).
137. Z. Sideratou, D. Tsiourvas, C.M. Paleos, Solubilization and release properties of PEGylated diaminobutane poly(propylene imine) dendrimers, *J Colloid Int Sci*, **242**, 272–276 (2001).
138. C. Kojima, K. Kono, K. Maruyama, *et al.*, Synthesis of polyamidoamine dendrimers having poly(ethylene glycol) grafts and their ability to encapsulate anticancer drugs, *BioconJ Chem*, **11**, 910–917 (2000).
139. M. Mannisto, S. Vanderkerken, V. Toncheve, *et al.*, Structure-activity relationships of Poly(L–lysine): Effects of PEGylation and molecular shape on physiochemical and biological properties in gene delivery, *J Cont Rel*, **83**, 169–182 (2002).
140. F. Aulenta, W. Hayes, S. Rannard, Dendrimers: A new class of nanoscopic containers and delivery devices, *Eur Polym J*, **39**, 1741–1771 (2003).
141. S.E. Stiriba, H. Frey, R. Haag, Dendritic polymers in biomedical applications: From potential to clinical use in diagnostics and therapy, *Angew Chem Int Ed Engl*, **41**, 1329–1334 (2002).
142. H. Frey, R. Haag, Dendritic polyglycerol: A new versatile biocompatible material, *Rev Mol Biotech*, **90**, 257–267 (2002).
143. A.R. Borges, C.L. Schengrund, Dendrimers and antivirals: A review. *Curr Drug Targets*, **5**, 247–54 (2005).
144. N.A. Pepas, K.B. Keys, M. Torres-Lugo, *et al.*, PEG-containing hydrogels in drug delivery. *J Cont Rel*, **62**: 81–87 (1999).
145. T. Ooya, J. Lee, K. Park, Effects of ethylene glycol-based graft, star-shaped, and dendritic polymers on solubilization and controlled release of paclitaxel, *J Cont Rel*, **5**, 121–127 (2003).
146. T. Ooya, J. Lee, K. Park, Hydrotropic dendrimers of generations 4 and 5: Synthesis, characterization, and hydrotropic solubilization of paclitaxel, *Bioconj Chem*, 151221–151229 (2004).
147. E.R. Gillies, T.B. Jonsson, J.M.K. Fréchet, Stimuli-responsive supramolecular assemblies of linear-dendritic copolymers, *J Am Chem Soc*, **126**, 11936–11943 (2004).
148. E.R. Gillies, J.M.J. Fréchet, Dendrimers and dendritic polymers in drug delivery, *DDT*, **10**, 35–43 (2005).
149. H. Yang, S.T. Lopina, Penicillin V-conjugated PEGPAMAM star polymers, *J Biomater Sci, Polym Ed*, **14**, 1043–1056 (2003).
150. M. Thanou, R. Duncan, Polymer-protein and polymer-drug conjugates in cancer therapy, *Cur Opin Invest Drugs*, **4**, 701–709 (2003).
151. D. Bhadra, S. Bhadra, N.K. Jain, PEGylated peptide-based dendritic nanoparticulate systems for delivery of artemether, *J Drug Del Sci Tech*, **15**, 65–73 (2005).
152. H. Petersen, P.M. Fechner, D. Fischer, *et al.*, Synthesis, characterization, and biocompatibility of polyethylenimine–graftpoly(ethylene glycol) block copolymers, *Macromolecules*, **35**, 6867–6874 (2002).

153. W. Tansey, S. Ke, X.Y. Cao, *et al.*, Synthesis and characterization of branched poly(L–glutamic acid) as a biodegradable drug carrier, *J Cont Rel*, **94**, 39–51 (2004).
154. I. Gitsov, J.M.J. Fréchet, Stimuli-responsive hybrid macromolecules: Novel amphiphilic star copolymers with dendritic groups at the periphery, *J Am Chem Soc*, **118**, 3785–3786 (1996).
155. M. Berna, D. Dalzoppo, G. Pasut, *et al.*, Novel monodisperse PEG–dendrons as new tools for targeted drug delivery: Synthesis, characterization and cellular uptake. *Biomacromolecules*, **7**, 146–153 (2006).
156. B. Qualmann, M.M. Kessels, H.J. Musiol, *et al.*, Synthesis of boron rich lysine dendrimer as protein labels in electron microscopy, *Angew Chem Int Ed Engl*, **35**, 909–911 (1996).
157. J.M. Heldt, N.F. Durand, M. Salmain, *et al.*, Preparation and characterization of poly(amidoamine) dendrimers functionalized with a rhenium carbonyl complex and PEG as new IR probes for carbonyl metallo immunoassay, *J Organomet Chem*, **689**, 4775–482 (2004).
158. Y. Choi, T. Thomas, A. Kotlyar, *et al.*, Synthesis and functional evaluation of DNA-assembled polyamidoamine dendrimer clusters for cancer cell-specific targeting, *Chem Biol*, **12**, 35–43 (2005).

14

Polymeric Micelles for the Delivery of Poorly Soluble Drugs

Swati Biswas, Onkar S. Vaze, Sara Movassaghian and Vladimir P. Torchilin

14.1 Micelles and Micellization

Micelles are spherical, supramolecular, self-assembled, nano-sized, colloidal particles with a hydrophobic core and a hydrophilic shell formed by the aggregation of self-assembling amphiphilic molecules or surfactants in solutions. Amphiphilic macromolecules or surfactants, the structural unit of polymeric micelles have distinct hydrophobic (non-polar) and hydrophilic (polar, water-loving) block domains. Often, the hydrophilic part of the molecule is called the *head* and the hydrophobic part, usually a long alkyl substituent is called the *tail*. The simultaneous existence of chemical entities having an affinity and antipathy for water within the same molecule is the key phenomenon of amphipathicity [1–4]. The presence of a hydrophilic group makes the surfactants slightly soluble in water. Aqueous exposure induces the hydrophobic and hydrophilic segments to phase-separate and form nanoscopic, supramolecular core/shell structures, at a minimum particular polymeric concentration, termed micelles [5] The size and shape of micellar aggregates vary as spheres, rods, vesicles, tubules and lamellae, depending on the balance between the hydrophilicity and lipophilicity of a given amphiphilic molecule. Surfactants are classified on the basis of the charge carried by the polar head group as anionic, cationic, non-ionic, or amphoteric. Table 14.1 shows the typical structures of the few surfactants of these individual classes.

The drastic changes in many physicochemical properties seen in an aqueous solution of amphiphilic molecules or surfactants with a hydrophobic domain take place when a specific concentration is exceeded because of the formation of oriented colloidal aggregates. The

Drug Delivery Strategies for Poorly Water-Soluble Drugs, First Edition. Edited by Dennis Douroumis and Alfred Fahr.

Table 14.1 *Chemical structure of hydrophilic groups of some common surfactants.*

Name	Type of surfactant	Structure
Carboxylate	Anionic	$R-(COO^-)_nM^{n+}$
Sulfonate	Anionic	$R-SO_3^-M^+$
Polyoxyethylene sulfate	Anionic	$R-(OCH_2CH_2)n-OSO_3^-M^+$
Sulfocarboxylate	Anionic	$^-O_3S-R-COO-M^{2+}$
Ammonium	Cationic	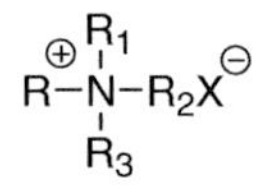
Sulfonium	Cationic	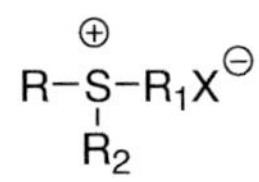
Phosphonium	Cationic	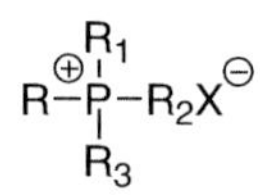
Pyridinium	Cationic	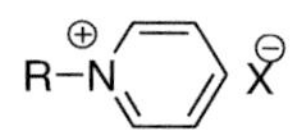
Polyoxyethylene alcohol	Non-ionic	$R-(OCH_2CH_2)_n-OH$
Polyoxyethylene ester	Non-ionic	$R-COO-(CH_2CH_2O)_n-H$
Polyoxyethylene thioether	Non-ionic	$R-S-(CH_2CH_2O)_n-H$
Triglycine	Amphoteric	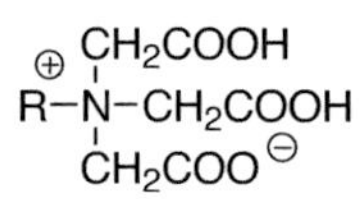
Phosphatidylcholine	Amphoteric	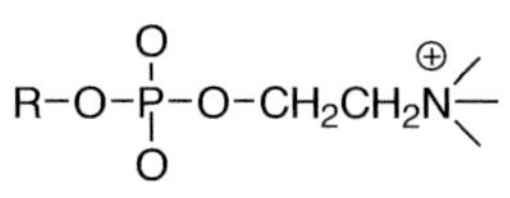

narrow concentration range over which these sudden physico-chemical changes occur is termed the critical micelle concentration (CMC) [6]. The molecular aggregates that are formed above the CMC are known as micelles (Figure 14.1). Another important term in this regard is the aggregation number or association number which is the number of amphiphilic molecules engaged to form one micellar unit. The number of unimers per micelles is decided thermodynamically and can range anywhere from dozens to approximately 200 unimers depending on the amphiphilic molecules [7].

Below the CMC, the amphiphilic molecules have a strong tendency to be absorbed at the air/water interface [8]. With the increase of the concentration of an amphiphile in the system, a point is reached when both the interface and the bulk of the solvent (water) become saturated with monomeric amphiphiles. At this point, any further increase in amphiphile concentration leads to the formation of micelles within the bulk phase and a subsequent

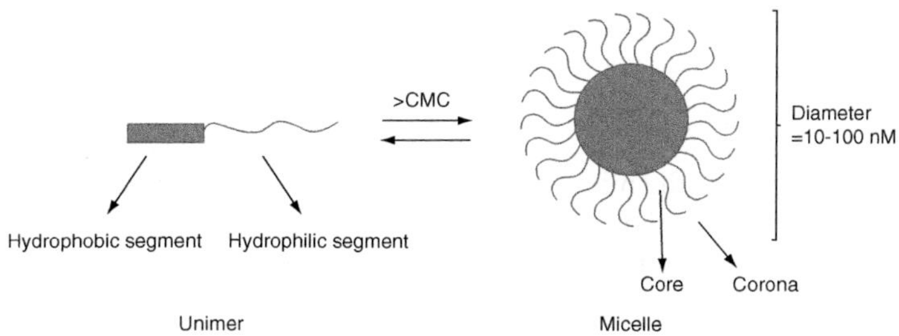

Figure 14.1 *Micelle formation from an amphiphilic unimer.*

decrease in the free energy of the system. One of the important physical properties of amphiphiles is the reduction of surface tension at the air/water interface by orienting them in the air/water monolayer. Surface tension decreases with increasing concentrations of amphiphile, while above the CMC, the surface tension remains constant. This observation indicates the complete saturation of air/water interface by polymer.

Micellization is easier for non-ionic surfactants because the aggregation takes place mainly due to the hydrophobic attraction among non-polar chains whereas hydrophilic chains are easily separated in an aqueous environment. Ionic surfactants are more difficult with regard to micellization in aqueous solutions than are nonionic surfactants of identical alkyl chain length, because higher concentrations are necessary to overcome the electrostatic repulsion between the ionic head groups of ionic surfactants during aggregation. Polymeric micelles are composed of non-ionic surfactants which form micelles at a very low CMC.

14.1.1 Factors Affecting Micellization

Micellization in an aqueous solution is governed by two opposing forces. The hydrophobic force favors expulsion of the hydrophobic tail of the amphiphile from the aqueous environment with the formation of a fluid organic droplet that constitutes the micelle core. The polar head groups extend to the aqueous medium from this hydrophobic compartment and form the corona. The force of repulsion between hydrophilic chains limits self-association to relatively smaller aggregates [9]. Therefore, the main driving force for micellization in an aqueous environment is the effective interaction between the hydrophobic parts of the surfactants' molecules, whereas electrostatic repulsive interactions are a force opposing micellization. Other forces opposing micellization include repulsive osmotic interactions between chain-like polar head groups and steric interactions between bulky head groups. Thus, the driving forces affecting micellization involve a delicate balance between the opposing forces. In addition, micellization by self-assembly of surfactants is affected by a range of additional factors, including the size of the hydrophobic moiety, the nature of the polar head group, co-solutes, pH and temperature.

The most important factor for micellization is the size of the hydrophobic moiety of the surfactant. Increasing the hydrophobic portion's size by increasing the chain length, for example, increases the hydrophobic interactions which promote micellization. As a general rule, the CMC of ionic surfactants decreases by a factor of 2, and with non-ionic surfactants, a factor of 3, on addition of one methylene group to the alkyl chain of the surfactant molecule [10]. A decrease in the CMC value is indicative of favorable micellization.

The nature of the polar head group also influences micellization. The CMC of non-ionic surfactants is generally much lower than that of ionic surfactants due to the repulsive electrostatic interaction among charged head groups opposing micellization of ionic surfactants [10]. The packing of the polar head groups in non-ionic surfactants is also important for micellization. An increase in the number of polar units in the head group results in increasing osmotic repulsion between the neighboring polar chains and tends to disturb the packing of the micellar aggregates, resulting in smaller and more spherical micelles.

Co-solutes such as hydrophobic compounds, salt, alcohols, etc. tend to affect micellization to some extent. Salt plays an important role in micellization for ionic surfactants. The counterions strongly interact with the charged head groups at the interface. The electric double layer that exists around each micelle, known as a stern layer, helps stabilize the micellar surface charge. As a result, the CMC decreases on the addition of salt. However, for nonionic surfactants, the addition of salt in low or moderate amounts has little influence on micellization.

Temperature also has an effect on micellization. Micellization takes place when the temperature of the system is greater than the critical micelle temperature (CMT), or Krafft temperature. Below this temperature, amphiphilic polymers exist as unimers, while above it, both the unimers and micelles are present in the system in dynamic equilibrium (Figure 14.1). The method for the determination of CMC can be applied to determine the CMT. Some polar head groups such as oligo(ethylene oxide) and its derivatives show a decreased water solubility as well as decreased repulsion between polar head groups resulting in favorable micellization with increasing temperature. This behavior is opposite to what is observed for most other types of surfactants and polymers, which show increasing solubility with increasing temperature. CMC and CMT provide a simple method of determining the thermodynamic parameters such as free energy and enthalpy of micelle formation. For ionic surfactant, increasing temperatures oppose micellization with the increase in entropy.

14.1.2 Thermodynamics of Micellization

The thermodynamics of micelle formation have been studied and discussed in detail elsewhere [9, 11–14]. Micellization reduces the total free energy of the solution. The removal of hydrophobic tails of the amphiphilic molecules from the water environment decreases energy. The aggregated hydrocarbons experience a loss of freedom, i.e. a decrease in entropy. There is a loss of translational and rotational entropy of the micellar aggregate relative to the monomeric amphiphile units. On the other hand, the interaction of the hydrophilic groups of the surfactant molecules with water decreases the free energy of the system. Partial removal of the hydrophobic group from the solution through micellization can result in an increase in free energy. Moreover, the hydrophilic group may also possess an electrostatic charge which introduces electrostatic repulsion by micellization. This acts as a force against micellization. Therefore, micellization is a tug-of-war between opposing free energy effects [15]. Micelles formation occurs spontaneously because of the balance between entropy and enthalpy of the system.

One can say that the micellization is entropically driven by the hydrophobic effect. The hydrophobic effect increases the entropy due to the increased disorder of water molecules that minimizes exposure of the hydrophobic segments in water. In other words, the water molecules are expelled from the hydrophobic compartment to the bulk aqueous phase.

Micellization can be explained in terms of free energy and the CMC via the following equation:

$$\Delta G^\circ = RT \ln(\mathrm{CMC}).$$

where ΔG° is the standard free energy change of micellization, R is the gas constant and T is the absolute temperature of the system. CMC can also be expressed in terms of enthalpy:

$$\ln(\mathrm{CMC}) = \frac{\Delta H^\circ}{RT} + \mathrm{Constant}$$

assuming the ΔH°, the standard change in enthalpy of micellization, and the association numbers are independent of temperature.

From a thermodynamic point of view, the formation of micelles of aggregation number, m, from monomeric amphiphile (Z) is governed by a set of equilibria:

$$\mathrm{mZ} \leftrightarrow \mathrm{Zm}$$
$$\mathrm{Km} = [\mathrm{Zm}]\,/\,[\mathrm{Z}]^{\mathrm{m},}$$

that is, expressing the equation in terms of equilibrium constant.

The free energy $\Delta G^\circ m$ for the transfer of a single amphiphile from an aqueous medium to a micelle,

$$\mathrm{m}\Delta\mathrm{Gm}^\circ = -\mathrm{RT}\ln \mathrm{K},$$

The fundamental equation for micelle formation can be obtained by replacing RT ln Km by -mΔGm°, where concentrations are expressed in mole fraction units [9].

$$\ln \mathrm{Xm} = -\mathrm{m}\Delta\mathrm{Gm}^\circ/\mathrm{RT} + \mathrm{m}\ln \mathrm{X}_1 + \ln \mathrm{m}$$

where Xm = the mole fraction of amphiphile incorporated in the micelles of aggregation number m, ΔGm°. The free energy of the transfer of a single amphiphile from an aqueous solution to a micelle of aggregation number m. X_1 = the mole fraction of monomer and Xm is the function of m at any given value of X_1.

The equation can be considered a micelle size distribution function. The co-operativity of micelle formation is reflected in the term mlnX_1 in the equation. The value of X_1 is limited to a narrow range close to the CMC if a reasonable value of Xm is to be obtained.

$$\Delta\mathrm{Gm}^\circ \text{ is the sum of two components,}$$
$$\Delta\mathrm{Gm}^\circ = \Delta\mathrm{Um}^\circ + \mathrm{Wm}.$$

ΔUm° is the hydrophobic component, representing the free energy of transfer of the hydrocarbon tail of the amphiphilic molecule from water to the hydrophobic core of the micelles. Wm is the head group compartment, representing the free energy arising from the repulsion between polar head groups. The occurrence of micellization and the concentration at which micellization occurs (the CMC) for a given surfactant system are determined by the relative balance of the forces favoring and retarding the molecular aggregation and also affect the size, shape and stability of the micelles.

14.2 Chemical Nature and Formation Mechanism of Polymeric Micelles

The surfactants, those considered for drug delivery, are characterized by at least two segments or blocks of a distinct chemical nature, where one of the segments is relatively hydrophobic or non-polar and the other segment is relatively hydrophilic or polar. These multiple segments are linked together chemically to form an individual block copolymer molecule termed a 'unimer.' Polymeric micelles represent a class of micelles that are formed from the self-assembly of the block co-polymeric unimers in an aqueous media and are especially useful as delivery systems for hydrophobic, poorly water-soluble drugs [1, 16–18]. These spherical particles are nanosized (typically in the range of 10–100 nm), supramolecular structures. In aqueous environments, the hydrophobic portion of the block co-polymer self-assembles into a semi-solid core, with the hydrophilic segment of the co-polymer forming a coronal layer. For drug delivery purposes, the hydrophobic core can act as a reservoir for poorly water-soluble drugs, while the outer shell protects the drug from rapid clearance of the nanocarrier [1]. The most important criteria for choosing a polymeric micelle for drug delivery applications are biocompatibility and/or biodegradability. Currently, the most commonly used corona-forming polymer is polyethylene glycol (PEG), with a molecular weight range from 2–15 of KDa. The core forming blocks typically consist of hydrophobic poly(propylene oxide) (PPO), poly(D,L-lactic acid) (PDLLA), poly(ε-caprolactone) (PCL), or poly(L-aspartic acid) etc.

The use of micelles for solubilization of poorly soluble drugs has attracted considerable attention in the past few decades [1, 19]. Low solubility in water appears to be an intrinsic property of many drugs especially anticancer agents, since many of them are bulky polycyclic compounds. Hydrophobicity is required to some extent in a drug molecule to cross the cell membrane [20] and to interact with a cell receptor [21, 22]. Many drug candidates with high potency identified by high throughput screening (HTS) technologies have a poor solubility in water [21,22]. Poor aqueous solubility poses such a serious problem that some leading pharmaceutical companies exclude poorly soluble drugs early in their screening process regardless of how active these compounds are in preliminary HTS [21]. Because of their lipophilic nature, most anticancer drugs are inherently water-insoluble. As an example, one of the most efficient anticancer drugs, paclitaxel (PTX), has a water solubility of 0.0015 mg/mL. While this degree of hydrophobicity is favorable for drug permeation through cell membranes, intravenous (i.v.) administration would result in drug aggregation and formation of capillary embolisms [23].

Conventional ways to address the poor solubility of drugs in order to increase their bioavailability include the excipients such as ethanol and other organic solvents, Cremophor EL (polyethoxylated castor oil), and certain surfactants [24]. Salt formation or pH adjustment, in some cases, facilitates the dissolution of poorly soluble drugs, if they contain ionizable groups [25]. More recent approaches include the use of liposomes, microemulsions and cyclodextrins [26]. One of the shortcomings of co-solvents and surfactants is the danger of drug precipitation upon dilution of a solubilized drug in the systemic circulation upon parenteral administration. Surfactants cannot retain solubilized material at concentrations lower than their critical micelle concentration (CMC) value, which is typically rather high for conventional low molecular weight surfactants [27]. The administration of many co-solvents or surfactants causes toxic and other undesirable side effects [1]. The use of liposomes, microemulsions and cyclodextrins has demonstrated some promising results with certain poorly soluble drugs, but solubilization capacity varies widely among different drugs. Thus, the solubilization of sparingly soluble pharmaceuticals represents a challenge.

Polymeric micelles, often, are more stable compared to micelles prepared from conventional detergents. Amphiphilic co-polymers have CMC values as low as 10^{-6}–10^{-7} M, resulting in stable systems that are not easily dissociable *in vivo* [28, 29]. In addition to high thermodynamic stability (signified by low CMC), micelles from certain co-polymers have high kinetic stability (retarded disintegration at concentrations lower than CMC) [30], apparently because of the presence of multiple sites capable of hydrophobic interaction within each polymer molecule. Similar to micelles formed by conventional detergents, polymeric micelles solubilize poorly water-soluble drugs by incorporating them into their hydrophobic core, thus, allowing for an increased bioavailability. High stability of polymeric micelles ensures retention of the encapsulated drugs in the solubilized form upon parenteral administration. Additionally, encapsulating the drug within the polymer core affords drug stability by hindering enzymatic degradation and inactivation of these bioactive molecules.

14.2.1 Core and Corona of the Polymeric Micelles

The two compartments, a hydrophobic core and the hydrophilic corona of the polymeric micelles, have distinct functions. The hydrophobic core solubilizes the poorly water soluble drug, protects it from degradation during circulation, improves its bioavailability, and reduces toxicity and other adverse effects. The core compartment of a pharmaceutical polymeric micelle ideally should have a high loading capacity, a controlled release profile of the incorporated drug, and good compatibility between the core forming block and the incorporated drug.

The micelle corona provides effective steric protection for the micelle and determines the micelles' hydrophilicity, the charge, length and surface density of the hydrophilic blocks, and the presence of the reactive chemical groups suitable for micellar surface modification especially for targeting purposes [19]. The corona is also responsible for micelle stabilization and interaction with plasma proteins and cell membranes. Note that an uncharged hydrophilic corona is desirable to lessen the possibility of undesired electrostatic interactions with plasma proteins. Poly(ethylene oxide) (PEO), often termed poly(ethylene glycol) (PEG) is a versatile corona-forming polymer which range from 1–15 KDa. The advantage of the PEG chain is its high aqueous solubility and low immunogenicity. The PEG cloud on the surface of polymeric micelles is highly hydrated and serves as an efficient steric protector for long systemic circulation times [31–33]. The long circulation times and small size of polymeric micelles also help in the preferential accumulation of micelles in tumor tissues through the enhanced permeability and retention (EPR) effect discussed later in this chapter. The EPR effect is based on the spontaneous penetration of long circulating macromolecules, particulate drug carriers, and molecular aggregates into the interstitium through the leaky vasculature characteristic of certain pathological sites in the body. The EPR effect is typical of solid tumors and infarcts which allow the accumulation of micelles by increased passive targeting [34–36]. The micelle corona can be functionalized for active targeting purposes to further increase the efficiency of the micelle-encapsulated drugs' uptake by organs or tissues of interest. The micelles can be targeted by attachment to their surface molecules, such as antibodies, peptides, lectins, saccharides, hormones and some low-molecular-weight compounds with specificity toward ligands characteristic of the tissues of interest [37].

The properties of the core-forming block of the polymeric micelles have the most profound impact on the CMC and the aggregation/association number of polymeric micelles. Increasing hydrophobicity and size of the hydrophobic block serves to lower the CMC [38–41]. When the hydrophobic to hydrophilic block ratio is kept constant, an increase in the molecular weight leads to a decrease in the CMC [42]. Various core-forming hydrophobic polymers have been explored to engineer polymeric micelles. Depending on the properties of the core-forming polymers, the physico-chemical properties such as micellar size, CMC, drug-loading capacity and stability will differ. Additives in the solution may affect the CMC of polymeric micelles. Some highly hydrophobic drugs incorporated in the micelle core increase the hydrophobicity of the core, drive the CMC lower and increase the association number [42–44].

14.2.2 Block Co-Polymers as Building Block of Polymeric Micelles

Various block co-polymeric unimers self-assemble in aqueous media above their CMC to form polymeric micelles. There are several types of block co-polymers, such as homo and hetero, represented in Figure 14.2, which are classified depending on the arrangements of hydrophilic and hydrophobic units in a single unimer. Homo-polymers are made up of identical monomeric units that do not have the ability to form micelles. The hydrophobic and hydrophilic monomeric units can be oriented in many different ways to provide random, block and graft co-polymers. In random co-polymers, the hydrophobic and hydrophilic units are distributed throughout the polymeric chain in a random manner. It may or may not form micellar aggregates depending on the thermodynamics of micellization. In the

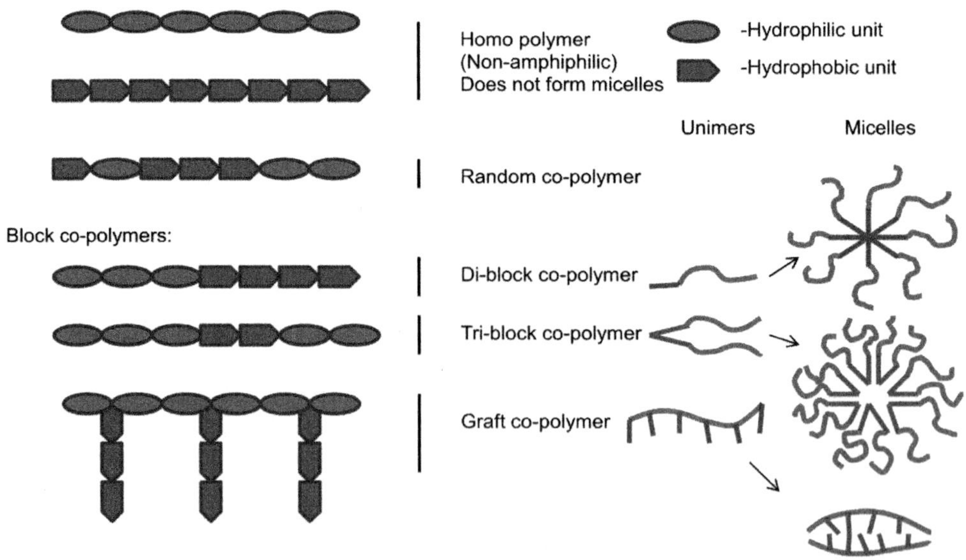

Figure 14.2 *Main structural types of co-polymers and micelles formed from amphiphilic co-polymers.* (See colour plate section.)

case of block co-polymers, the hydrophobic and hydrophilic segments are linked together in an organized fashion. The orientation of hydrophobic and hydrophilic segments in di-block, tri-block and graft co-polymers is shown in Figure 14.2. Micelles made up of di- and tri-block-co-polymers are of particular interest for the delivery of poorly water soluble drugs [32, 33, 45, 46]. To make the the hydrophilic corona in the micelles, both di-block and tri-block-co-polymers frequently contain poly(ethylene oxide) or PEO chains, commonly referred to as polyethylene glycol or PEG [47]. In di-block co-polymers, the PEG chains are simply conjugated with various hydrophobic blocks, whereas in tri-block co-polymers, both termini of a hydrophilic or hydrophobic block may be coupled with the second segment. Triple co-polymers of hydrophilic ethylene oxide units and hydrophobic propylene oxide (PPO) units are the most common examples of tri-block co-polymers used for pharmaceutical applications. Polymers of this scaffold are known as Pluronics (poly(ethylene oxide)-poly(propylene oxide)-poly(ethylene oxide) tri-block co-polymers). Table 14.2 represents the structures of the core- and corona-forming segments of the co-polymers. In grafted co-polymers, multiple hydrophobic chains are anchored along the main chain composed of hydrophilic units [48]. Within the structure of an amphiphilic polymer, monomer units with different hydrophobicities can be combined randomly, represented by two conjugated blocks (A-B type co-polymer), or be made from alternating blocks with different hydrophobicity (A-B-A type co-polymers).

14.3 Polymeric Micelles: Unique Nanomedicine Platforms

Polymeric micelles have distinct properties which differentiate them from other drug delivery vehicles. Polymer-drug conjugates, dendrimers and liposomes represent other major polymer-based nanocarrier systems for drug delivery applications, each with their particular chemical structures and biological properties [49–51]. Of these systems, polymer–drug conjugates and liposomes have a longer history of development and, to this date, have found most success in the clinic. Dendrimers are relatively new compared to micelles, liposomes and polymer–drug conjugates. Polymeric micelles are smaller in size compared to liposomes and micro particles. The lack of an interior water compartment as in liposomes, the existence of an interior hydrophobic compartment, and the protective effect of an exterior polymer are some of the unique features of these polymers. In contrast to liposomes that can incorporate a water-soluble drug in their aqueous interior compartment, micelles can accommodate only water-insoluble or poorly soluble pharmaceuticals in their hydrophobic core [1]. Both polymer–drug conjugates and dendrimers require the chemical conjugation of drug molecules to the carriers, whereas the drug molecules can be solubilized in micelle or liposomal nanocarriers [52]. The covalent conjugation of the drug molecule to the dendrimer or drug–polymer conjugate requires the presence of functionalizable chemical groups on the drug molecules as well as on the nanocarrier, limiting the application to a wider variety of drugs. In addition, specific chemical reactions such as enzymatic degradation, and acid catalyzed hydrolysis are necessary to release the drug [53]. Moreover, due to the small size of these systems (typically less than 10 nm), they can cross the basement membrane of the glomeruli of kidneys and quickly be cleared from the body. On the other hand, liposomes and micellar nanocarriers solubilize the drugs physically. As a result of their inner hydrophilic compartment, liposomes are more suitable for the delivery of water-soluble agents such as

Table 14.2 *Commonly used co-polymeric blocks used to make polymeric micelles.*

Co-polymers	Chemical structure of repeating units	Abbreviations
Hydrophilic corona:		
Poly(ethylene glycol/oxide)	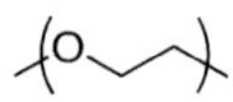	PEG, PEO
Poly(N-isopropyl acrylamide)	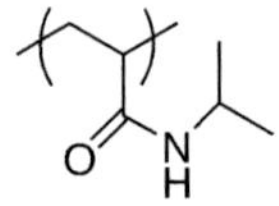	pNIPAM, NIPAM
Poly(N-Vinyl pyrrolidone)	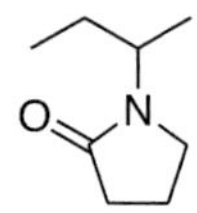	PVP
Hydrophobic core:		
Polyethers Poly(propylene oxide)	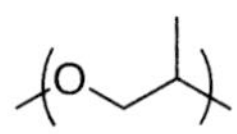	PPO
Poly(esters) Poly(L-lactide) Poly(D,L-lactide)	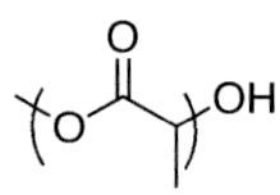	PLA
Poly(lactide-co-glycolide)	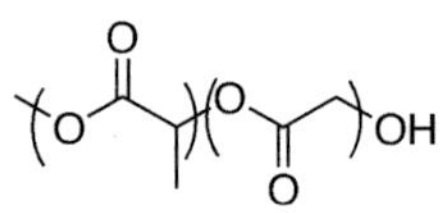	PDLLA
Poly(ε-caprolactone)	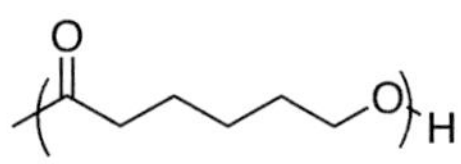	PLGA
Poly(β-amino ester)	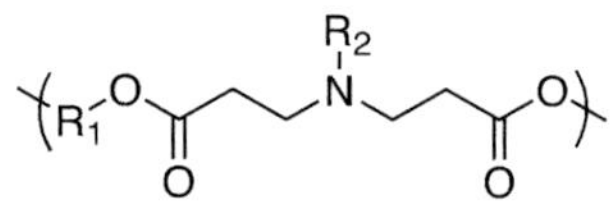	PCL
Polyamides. Poly(L-Histidine)	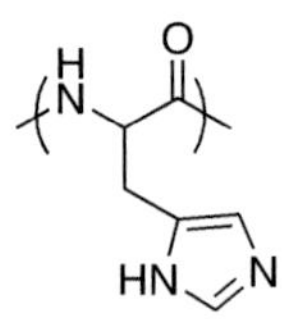	pHis
Poly(L-aspartic acid)	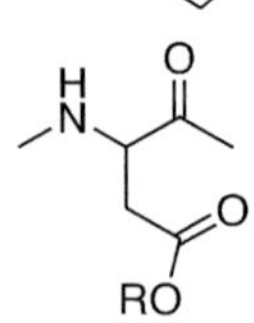	pAsp
Poly(L-glutamice acid)	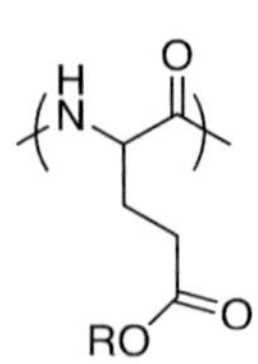	pGlu

therapeutic proteins or DNAs. Poorly soluble drugs can be solubilized in the lipid bilayer of liposomes. However, the loading capacity is limited due to membrane destabilization effects [54]. On the other hand, polymeric micelles have a structural resemblance to lipoproteins. Polymeric micelles loaded with drugs could be an alternative to the reconstituted low density lipoprotein approach. However, the polymeric micelles do not compete with any small molecules or naturally occurring particulates for receptor binding and cellular internalization. For the hydrophilic segment, the most commonly used polymer is polyethylene glycol (PEG) with molecular weight ranging from 1–15 kDa. Other hydrophilic polymers such as poly(*N*-vinyl pyrrolidone) (PVP) [55] or poly(*N*-isopropylacrylamide) (pNIPAM) [56], poly(alkylacrylic acid) [57] have also been used to form the micelle corona [56, 58, 59]. The latter two impart temperature and pH-sensitivity to the micelles. Three polymeric micelles SP1049C, NK911, and Genexol-PM have stabilizing PEG coronas that minimize opsonization of the micelles and maximize blood circulation times. Attempts have been made to deliver the potent chemotherapeutic drug paclitaxel (PTX) and doxorubicin (DOX) by loading in a micellar formulation. NK911 is the DOX-encapsulated micelle from the copolymer of PEG and DOX-conjugated poly(aspartic acid), SP1049C. Various polymers made up of the above-mentioned hydrophobic and hydrophilic segments have found wide applicability in the delivery of poorly soluble drugs. Examples of polymeric micelles of a few distinct classes used for the delivery of poorly water-soluble drugs are presented in Table 14.3.

14.3.1 Polymeric Micelles for the Delivery of Poorly Soluble Drugs

Many types of copolymers have been used to prepare polymeric micelles. However, the requirements for biocompatibility as well as biodegradability limit the choice of many synthesized copolymers for clinical applications [60]. The three most widely studied block copolymers are characterized by their hydrophobic blocks as poly(ethers), mainly poly(propylene oxide), poly(amides) and poly(esters).

14.3.1.1 Pluronics

The most extensively studied A-B-A-tri-block copolymers pluronics fall in the class of polyether block copolymers. These surfactants were introduced in the early 1950s by the Wyandotte Chemicals Corporation under the widely known name of Pluronics. These block copolymers consist of hydrophilic poly(ethylene oxide) (PEO) and hydrophobic poly(propylene oxide) (PPO) blocks arranged in an A-B-A tri-block structure: (PEO-PPO-PEO). Table 14.4 shows the structure of the A-B-A co-polymer Pluronics with varied A and B-chain lengths. Pluronics are chemically stable in a biological system, and like low molecular weight surfactants, Pluronic micelles efficiently solubilize the poorly water-soluble drugs in the hydrophobic core compartment.

Pluronics are available in a wide range of molecular weights and ratios of PEO to PPO with varying hydrophobicity. The block co-polymers with different numbers of hydrophilic ethylene oxide and hydrophobic propylene oxide units are characterized by the hydrophilic-lipophilic balance (HLB). The HLB numerically quantifies the relative hydrophobicity of a surfactant, where a higher HLB signifies greater hydrophilicity [61]. There are over 20 poloxamers available for study in a variety of molecular weights, relative block lengths, and hydrophilic-lipophilic balances. The diameter of pluronic micelles usually varies from

Table 14.3 *Examples of various micelles loaded with poorly water-soluble drugs.*

Block co-polymers	Micelle-loaded drugs	References
Pluronics	Doxorubicin, Cisplastin, Carboplatin, Epirubicin, Haloperidol	71, 73, 115
Polycaprolactone-*b*-PEG	FK506, L-685, 818, Cyclosporin-A	9, 122
Polycaprolactone-*b*-methoxy-PEG	Indomethacin, Cisplstin, paclitaxel	207, 313
Poly(caprolactone/trimethylene carbonte)-PEG	Risperidone, Ellipticin	314, 315
Poly(aspartic acid)-*b*-PEG	DOX, Cisplastin, Lysozyme, Adriamycin	198, 316, 317
Poly(β-benzyl-L-aspartate)-*b*-PEG	DOX, indomethacin, KRN, amphotericin B	28, 123, 215, 311
Poly(L-lysine)-*b*-PEG	DNA	161, 201
PEG-PE	Dequalinium, soya bean trypsin inhibitor, Camptothecin, PTX, porphyrine	143, 146, 147, 153
Poly(D,L-lactide)-*b*-methoxy-PEG	PTX, testosterone, griseofulvin.	307, 310
PEG-*b*-poly(lactic-co-glycolic acid)	PTX	312
Poly(L-histidine)-*b*-PEG	DOX	240
Chitosan conjugated to Palmitoyl	Ibuprofen	154
Dextran-*b*-poly(ε-caprolactone)	DOX	309
Poly(N-vinyl2-pyrrolidone)-*b*-poly(D,L-lactide)	Indomethacin	55
PEG-PE/EGG phosphatidyl choline (Mixed micelles)	PTX	77

Table 14.4 *Structure of A-B-A co-polymer Pluronics.*

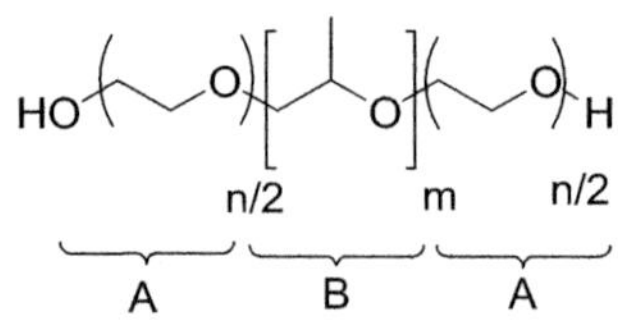

A. Ethylene glycol (EG)
B. propylene glycol (PG)

Pluronic L61	EG2-PG30-EG2	MW 1950
Pluronic P85	EG25-PG40-EG25	MW 4600
Pluronic F127	EG100-PG65-EG100	MW 12600
Pluronic F68	EG76-PG30-EG76	MW 8400

ca. 10 nm to 100 nm [23]. The core of the micelles consists of hydrophobic PPO blocks whereas the PEO chains serve as the hydrophilic corona. The unique nature of the tri-block co-polymer is that the core compartment represents a 'cargo hold' for the incorporation of various therapeutic or diagnostic agents [62]. For some of the poloxamers, the PPO core can incorporate a considerable amount (up to 20–30 weight %) of a poorly water-soluble drug [63]. The PEO shell ensures that the micelles remain in a dispersed state and decreases undesirable drug interactions with the cells and proteins.

The major advantages of using Pluronics are its safe toxicity profile and high solubility in water. Poloxamers are relatively non-toxic ($LD_{50} > 5$ g/Kg, for more hydrophilic poloxamers), highly water-soluble and form micelles easily in water, which allows easy preparation. However, these surfactants do not biodegrade upon administration, typical of a poly(ether), where elimination occurs mainly through renal clearance of the unimers [64]. The toxicity of Pluronics correlates with its hydrophobicity as seen after intravenous administration [65]. The most effective Pluronics are the most hydrophobic [65]. Therefore, what is considered a safe level of hydrophobicity depends on the severity of the disease. The half-life of Pluronic P85 depends on the aggregation state of the block copolymer and varies from 60 h for unimers to 90 h for micelles [66]. Formation of micelles decreased the uptake of the block copolymer in the liver. Pluronic F68 was administered i.v. over 48 h to dogs and resulted in an elimination half-life of approximately 9 h. More than 90% of the polymer was eliminated by the kidneys.

Pluronics are considered to be among the least toxic of surfactants and are widely used in the food, drug and cosmetic industries [62, 64, 67–69]. Pluronic micelles are very efficient in the delivery of poorly water-soluble drugs since incorporation of hydrophobic drugs into the Pluronic micelles increases drug solubility, drug stability and can improve drug pharmacokinetics and biodistribution. Many Pluronics used for drug solubilization have high ratios of PEO to PPO and are quite non-toxic, relative to many low molecular weight surfactants such as Tween-80, especially in terms of cell membrane lysis or hemolysis. The hydrophilic Pluronic F-68 (average molecular weight 8400 Da) is approved for intravenous use, and, many other Pluronics are approved by the US Food and Drug Administration (FDA) for ingestion and topical use. Pluronics have been used to solubilize haloperidol [70, 71], indomethacin [72], doxorubicin (DOX) [73], epirubicin [73], and amphotericin B (AmB) [74]. The formulation, named SP1049C, developed by Supratek Pharma, Inc. is a Pluronic mixed-micelle formulation loaded with DOX. The mixed-micelle-injectable formulation of DOX is composed of a combination of the hydrophilic Pluronic F127 and the more hydrophobic Pluronic L61 at a weight ratio of 1:8, L61 to F127. Pluronic L61, used at very low concentration (0.25% w/v of L61 in SP1049C), inhibits drug efflux protein P-gp expressed in multidrug-resistant cancer cells and therefore extends retention of DOX in cells. Hydrophilic F127 forms mixed micelles of uniform size for drug delivery with L61, and solubilizes L61 and DOX in water. SP1049C has shown greater efficacy than DOX alone against a variety of drug-resistant cancer cells *in vitro* as well as *in vivo* in DOX-resistant solid tumors. The elimination half-life of DOX in the Pluronic formulation was considerably longer (48.8 h) than free DOX (30 h) [75]. This formulation is in a phase II clinical trial for the treatment of metastatic esophageal carcinoma and soft-tissue sarcoma.

Relatively hydrophobic Pluronics have also found applications as immunological adjuvants for vaccines to produce an immune response [76–78]. Pluronics bind antigens as well as other proteins on the surface. All Pluronics with strong adjuvant activity have HLB values of less than 2. The strong lipophilicity promotes retention of the adjuvant, immunogen and host proteins on hydrophobic surfaces, where they are more effectively presented to the cells of the immune system. Pluronic L121 was a powerful adjuvant for increasing antibody formation to bovine serum albumin in mice when injected as an oil-in-water emulsion [78].

Contrary to other block copolymers which are biologically inert, some of the Pluronic co-polymers possess therapeutic potential especially useful for the treatment of cancer. Generally, the polymeric nanocarriers used for drug delivery that protects drugs from degradation, prolong exposure of the drugs to the tissue by long systemic circulation, and enhance the transport of the drugs into the cells considered to be biologically inert. However, this is not exactly true for all Pluronic co-polymers as some of them can alter specific biological responses [79–81]. Pluronic block co-polymers were shown to be potent biological response modifiers and capable of sensitizing multi-drug resistant (MDR) cancer cells [82–87] which could serve to enhance the efficacy of chemotherapeutic agents. Kabanov and coworkers demonstrated that Pluronic unimers can sensitize the cells by inhibiting drug efflux transporters, such as Pgp, the multidrug-resistance proteins [29, 88–92], and breast cancer-resistant protein [93, 94], decrease the anti-apoptotic defense in MDR cells [95], induce the release of cytochrome C and increase reactive oxygen species, inhibit the glutathione/glutathione S-transferase detoxification system [88, 89] and prevent the sequestration of drugs in acidic vehicles [38, 83, 96]. Pluronics also reduce the ATP level in MDR cells which is important for anticancer therapy since the mechanisms of drug resistance are energy-dependent. However, these effects were observed at polymer concentrations below the CMC [38, 97] suggesting that the unimers are responsible for these biological properties and not the micellar forms. An intravenous formulation of Pluronic F68 used at the submicellar concentration of 0.4% w/w, under the trade name Flocor, is in clinical trials for the treatment of diseases including sickle-cell disease and muscular dystrophy [98].

One of the characteristic physical properties of PEO-PPO-PEO block copolymers is their inverse temperature dependence in solubility and micellization. Contrary to most common co-polymers, an increase in temperature decreases the solubility of poloxamers [99]. Decreased solubility at increasing temperature results in a high concentration of polymer at the air/water interface [99]. An increase in temperature causes a sharp decrease in the CMC of the poloxamers. Therefore, this class of polymers is also described by its critical micelle temperature (CMT). The CMT is the temperature at which micelles begin to form when the concentration of the polymer is kept constant. As temperature increases, the solubility of the hydrophobic block decreases, which drives the decrease in the CMC. The typical CMT of the poloxamers ranges from 20–40°C, which is at or below body temperature [100].

Pluronic micelles loaded with the potent anticancer drug, doxorubicin, have been very extensively studied. SP1049C is the doxorubicin-loaded Pluronic micellar formulation, developed by Supratek Pharma Inc. It is a mixed micellar formulation composed of more hydrophobic Pluronic L61 (with an HLB value of 3) and hydrophilic Pluronic F127 (HLB = 29). Hydrophobic Pluronic L61, when present at concentration 0.25% w/v in saline, promotes cellular uptake and retention of DOX by inhibiting the drug efflux protein P-gp [101]. Hydrophilic F127 used at a concentration 2% w/v in saline improves the solubilization of both DOX and L61 and forms micellar aggregates of uniform particle size for drug delivery [96, 101].

Another Pluronic mixed-micellar formulation involves the physical encapsulation of the injectable general anesthetic propofol [102]. Propofol is a safe anesthetic with a short half-life, which is considered an advantage for adjusting the time of sedation. However, propofol has very poor water solubility (1×10^{-3} M). As a consequence, propofol was initially formulated with Chremofor EL (CrEL) as the solubilizing agent. However, the CrEL formulation is associated with anaphylactoid reactions. The oil-in-water emulsion of Propofol also has stability issues. Oil-in-water emulsions are physically unstable with respect to flocculation and coalescence. They are quite sensitive to temperature fluctuations and additives, including other drugs. Intravenous administration of propofol is painful, which necessitates the use of a local anesthetic such as lidocaine hydrochloride. Co-administration leads to instability of the emulsion formulation. These disadvantages have led to the preparation of a micellar formulation which is thermodynamically stable, easy to manufacture, easy to sterilize, amenable to drug co-administration and lacks any excipients that could encourage the growth of microorganisms. The Pluronics used for mixed micelle preparation to load propofol are Pluronic F68 and Pluronic F127. In this formulation, both surfactants are relatively hydrophilic with HLB values of 22 and 29 respectively. Mixed micellar formulations of Pluronics solubilize more drugs inside than Pluronic micelles made up of one composition. Thus, the mixed F68 and F127 micelles in a ratio of 3:7 solubilize propofol at 3.5–3.8% whereas Pluronics F-68 micelles at 10% w/v solubilize propofol at 0.8% and Pluronics F127 micelles at 10% w/v solubilize at 1.7%.

Antibiotics such as Amphotericin B (AmB) and Nystatin used to combat systemic fungal diseases have been administered using Pluronics-based micellar formulations. Systemic fungal diseases are a major concern owing to the rising populations of immunocompromised patients [103]. Systemic fungal disease has become one of the major factors of morbidity and mortality in those patients. For many years AmB has been the drug of choice to treat systemic fungal infection [104]. However, the major disadvantages of using AmB are its toxicity and poor water solubility. AmB exhibits both acute and chronic toxicities upon systemic administration. Acute toxicities include fever, chills, anorexia, nausea, vomiting, myalgia, arthralgias and headache. Major chronic toxicity includes renal dysfunction [104]. Efforts have been made to prepare a less toxic formulation of AmB. In its current formulation (Trade name: Fungizone), solubility of AmB is improved with the bile salt sodium deoxycholate. Lipid-based formulations such as Abelcet and Amphotec have progressed into clinical practice and have produced less renal toxicity compared to Fungizone [105]. However, all lipid-based formulations require higher doses to reach the efficacy obtained by Fungizone. This increases the cost per dose because of the expense of such lipid-based formulations. AmB has also been formulated with Pluronics F68 and F127. Amphiphilic AmB self-aggregated in the presence of Pluronic F68. However, AmB in the presence of Pluronic 127 micelles minimizes the self-aggregation. Pluronic 127 is more hydrophobic than Pluronic F68 because of the presence of a larger poly(propylene glycol) block than Pluronic F68. The hydrophobic interaction between Pluronics and AmB resulted in disaggregation of the drug. AmB-loaded F68-Pluronic micelles showed minimal inhibitory concentrations ten-fold lower than AmB in Fungizone against clinical isolates of *Candida albicans* [106]. In mice, the LD_{50} was 9.8 mg/kg, whereas LD_{50} of AmB in Fungizone was 4.0 mg/Kg. Apart from AmB, Nystatin, another antibiotic which closely resembles the structure of AmB, was also formulated using Pluronic micelles F68 and F127 for the treatment of systemic fungal diseases.

14.3.1.2 PEG-b-Poly(L-Amino Acid)

Block copolymer micelles based on PEG-*b*-poly(L-amino acid) utilize polymerized amino acids to generate hydrophobic core-forming blocks. The amino acids polymerized to prepare the micellar core for drug delivery applications are lysine, aspartic acid, glutamic acid and histidine [107–110]. The general structure of the block co-polymers of this class is shown in Table 14.2. An advantage of using amino acids as the hydrophobic core is that they are biocompatible and biodegradable by hydrolysis or enzymatic degradation into naturally occurring substances in the body. Another advantage is the easy tailoring of the polymer via conjugation between the free amine and the acid functional group of the amino acids that improves compatibility with the loaded solubilizate and the stability of the micelles [111, 112]. Generally, PEG is used as a hydrophilic corona for this class of polymers. All α-amino acids can be incorporated to make the polymer depending on the need for incorporation of the wide variety of bioactive molecules, including poorly water-soluble anticancer drugs such as paclitaxel, DOX, peptides (insulin) and proteins as well as various nucleic acid materials [113]. Sometimes, drugs can be linked to the poly(L-amino acid) part of the PEG-*b*-poly(L-amino acid) copolymer to prepare micellar polymer/drug conjugates [114]. Two example of drugs chemically encapsulated into PEG-*b*-poly(L-amino acid) micelles are DOX [108, 116–118] and methotrexate [119]. The copolymers utilized for this purposes are poly(L-aspartate) for DOX and poly(2-hydroxyalkyl-L-aspartamide) for methotrexate [119].

One of the unique properties of polymeric micelles based on a PEG-*b*-poly(aspartic acid)/DOX conjugate is that the micellar form is extremely stable kinetically with respect to dissociation after incubation for several days in a PBS buffer, as shown by HPLC studies that indicate strong cohesive interaction in their cores [30]. This unique stability was also confirmed by *in vivo* studies in mice [120, 121]. Yokoyama *et al.* were the first to show that PEG-*b*-poly(aspartic acid)/DOX conjugate polymeric micelles could stay intact in the blood and circulate for prolonged periods by evading uptake by the reticulo endothelial system (RES). In accordance with the prolonged systemic circulation, the polymeric micelle formulation accumulated preferentially at solid tumors relative to plain DOX via the enhanced permeability and retention (EPR) effect and had much less accumulation, indicating less cardiotoxicity [108]. Kwon *et al.* showed stability of the blood as well as biodistribution depend on the chemical composition of the PEG-*b*-poly(aspartic acid)/DOX polymer [121]. Longer PEG and shorter poly(aspartic acid)s favored prolonged circulation in blood. The structure of the core is also important for delivery. The micellar system made up of a PEG-*b*-poly(aspartic acid)/DOX conjugate is a better delivery system than PEG-*b*-poly(β-benzyl-L-aspartate) micelles, perhaps due to less compatibility of the core with the DOX which influences the systemic circulation time of the drug [123].

DOX molecules have also been conjugated to the PEG-*b*-poly(aspartate) via a pH sensitive hydrazone bond [117]. The hydrazone bond is stable in the systemic circulation at pH 7.4, but is cleaved at in a low pH environment of the tumor and in endosomes and lysosomes (pH 5–6). The pH-triggered drug release can improve the delivery and therapeutic efficacy of the drug. Poly(β-benzyl-L-aspartate) and poly(β-benzyl-L-glutamate) have also been utilized for physical encapsulation of AmB [124], camptothecin [125] and clonazepam [126].

A number of hydrophobic units such as acyl spacer groups or acyl side chains have been introduced into the poly(L-Lysine) to vary the hydrophobicity of the copolymer needed for drug solubilization and release [114]. A polymeric micellar system made up of the polymer PEG-*b*-poly(N-hexyl-L-aspartamide) was used to solubilize poorly water-soluble drugs [111, 127, 128]. In this case, poly(L-aspartate) was derivatized by incorporating the acyl chains. This modification produced a highly hydrophobic core with a very low CMC [111]. This micellar system has been used for the solubilization and delivery of AmB in its disaggregated form. AmB is a polyene macrolide antifungal drug which exerts its antifungal activity by creating ion channels in fungal cell membranes through selective interaction with ergosterol that is present exclusively in fungal cell membranes. This formation of ion channels results in the loss of cytosolic contents, resulting in cell death. Amphiphilic AmB has the tendency to self-aggregate. Aggregated AmB can form ion channels non-specifically in mammalian and fungal cell membranes, resulting in host–cell toxicity [129]. The toxicity of AmB could be minimized by preventing the aggregation by solubilizing AmB in the hydrophobic compartment of the micelles. The formulation of AmB by PEG-*b*-poly(N-hexyl-L-aspartamide) block co-polymer micelles has the ability to prevent aggregation of AmB. In addition, an increased level of acylation in the core resulted in increased hydrophobicity of the micelles with decreased aggregation and decreased AmB-induced hemolysis. Adams *et al.* found that the potential of this AmB-block co-polymeric micellar formulation to reduce systemic candidiasis was equivalent to a commercially available AmB formulation, but with significantly decreased toxicity in a murine model [130]. Based on the preclinical success of PEG-*b*-poly(aspartic acid)/DOX conjugate micelles, the DOX formulation NK 911 has entered a clinical trial, with DOX used incorporated by physical encapsulation as well as by chemical conjugation [131].

*14.3.1.3 PEG-*b*-Poly(esters)*

This class of block copolymer micelles has been studied extensively for drug delivery purposes. The most commonly used polymers for drug delivery applications are poly(D, or L-lactic acid), poly(ε-caprolactone) and poly(lactic-*co*-glycolic acid) which are non-toxic, biocompatible and biodegradable. Poly(ester)s are very prone to hydrolysis in biological systems and are degraded to non-toxic species [132, 133]. Poly(esters) are not amenable to chemical modifications as are poly(L-amino acids). Therefore, this class of polymers is most commonly used for physical encapsulation of poorly water-soluble drugs. The poorly soluble anticancer drug PTX has been loaded in various poly(ester) micelles to improve solubility and efficacy. PTX is a highly potent anticancer drug with the disadvantage of its poor solubility in water. The solubility of PTX in water is about 1 μg/mL. PTX, marketed as Taxol, is dissolved in a solution of cremophore EL and ethanol (1:1 v/v) and causes adverse biological effects. Methoxy-poly(ethylene glycol)-*b*-poly(D, L-lactide) has been shown to improve the solubilization of PTX to levels greater than 5000 times its solubility in water [134]. Varied compositions of polymers were readily prepared by a ring opening polymerization of D,L-lactide with stannous octoate as the catalyst. The solubilization of PTX increased with the increase in the molecular weight of PEG-*b*-poly(D,L-lactic acid). This PTX in PEG-*b*-poly(D,L-lactide) showed much less toxicity. A fivefold increase in the maximum tolerated dose was observed from (20 to 100 mg/Kg), after intraperitoneal administration in mice [135]. In another study with intravenous administration in nude mice, the tolerated PTX doses when incorporated in PEG-*b*-poly(D,L-lactic acid) micelles and Taxol were 60 mg/Kg and 20 mg/Kg respectively [136]. The levels of PTX given at 50 mg/Kg in tissues such as the liver, kidneys, spleen, lungs, heart and tumor were two- to three-fold higher compared to the standard commercial formulation Taxol administered at a dose of 20 mg/Kg in a B16 melanoma murine tumor model [136]. As a result, PTX incorporated in PEG-*b*-poly(D,L-lactic acid) micelles had significantly greater antitumor activity than Taxol in mice bearing human ovarian SKOV-3 and human breast cancer MX-1 tumors. Therefore, a higher dose of PTX can be delivered when PTX is encapsulated in PEG-*b*-poly(D,L-Lactic acid) micelles than when administered as the Taxol formulation.

PEG-*b*-PLGA micelles have been studied extensively for drug delivery purposes. This is non-toxic, biodegradable and non-immunogenic. The degradation rate and kinetics of micellar drug release can be controlled *in vitro* by altering the PEG to PLGA ratio [137]. Doxorubicin has been loaded physically and by chemical modification into PEG-*b*-PLGA micelles [138]. Physical encapsulation of DOX in these micelles was poor compared to chemical encapsulation. With physical encapsulation, loading of DOX was 0.51% w/w, and the encapsulation efficiency was 23%, compared to the chemical encapsulation of DOX, where the loading increased more than four-fold and the encapsulation efficiency achieved was 100%. For chemical encapsulation, DOX was conjugated with the PLGA-core. In addition, more sustained drug release was observed with PEG-*b*-PLGA-DOX-conjugated micelles compared to DOX-encapsulated PEG-*b*-PLGA micelles. While the DOX-PLGA-PEG micelles showed sustained release of up to 50% of the drug load over 2 weeks, release from PLGA-PEG micelles containing non-conjugated DOX lasted only for 3 days. The sustained release of DOX occurred due to the gradual hydrolysis of the carbamate bond between the PLGA backbone and DOX as well as for water-soluble DOX-PLGA oligomer conjugates [138]. DOX-PLGA-PEG-micelles showed 10 times the cytotoxicity to the human hepatoblastoma cell line (HepG2) than the free DOX due to the enhanced cellular uptake of the nanoparticles over free drugs.

14.3.1.4 Lipid-Core Micelles

Conjugates of water-soluble polymers with lipids, for example, diacyllipid-PEG conjugates, such as polyethylene glycol-phosphatidyl ethanolamine (PEG-PE) were introduced into the area of controlled drug delivery as polymeric surface modifiers for liposomes [139] that found form as very stable low CMC micelles in an aqueous environment. The structure of this type of compound is similar to that of amphiphilic copolymers of the A-B type. Instead of a hydrophobic polymer block, the micelle core-forming part is made up of phospholipid residues that are short, but strongly hydrophobic, due to the presence of two long-chain fatty acyls that provide high micellar stability and prolong circulation in the blood [140]. Conjugates of lipids with various water-soluble polymers are now commercially available, or can be synthesized [139, 141, 142]. The chemical structure of one of the most common lipid core micelles, PEG-DSPE is shown in Figure 14.3. PEG-PEs with a PEG chain molecular weight from 750 to 5000 Da form micelles spontaneously in an aqueous environment with very low CMC and a particle size of 7–35 nm. The CMC values and the particle size of micelles formed from PEG-PE with various lipids and polymeric components are shown in Table 14.5. Micelles formed from conjugates with PEG blocks of higher molecular weights have a slightly larger diameter. The highly hydrated outer PEG-corona of these micelles serves as an efficient steric protector in biological media. The extremely hydrophobic lipid component of the micelles solubilizes poorly soluble drugs, including paclitaxel, camptothecin [143], porphyrin [144], vitamin K_3 [145], soya bean trypsin inhibitor [146], and dequalinium [147]. Micelles prepared

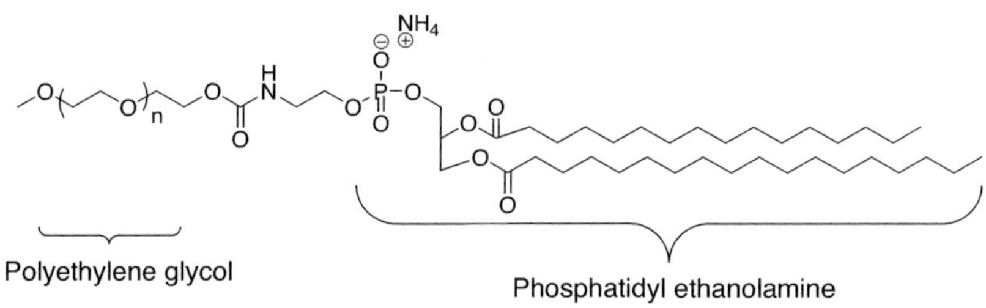

Figure 14.3 *Chemical structure of 1,2-Distearoyl-sn-Glycero-3-phosphoethanolamine-N-[Methoxy(Polyethylene Glycol)] (Ammonium Salt) (PEG-DSPE).*

from PEG-PE conjugates with shorter versions of PEG are more efficient carriers of poorly soluble drugs because of their greater hydrophobic-to-hydrophilic-phase ratio and can be loaded with drug more efficiently on a weight-to-weight basis [47]. The exposure of PEG residues on the surface prevents rapid uptake of the particles by the reticulo-endothelial system and makes them long-circulating. PEG-PE was first introduced to the drug delivery field while preparing long-circulating liposomes by the authors [139] followed by others [148, 149]. The ability of PEG-PE molecules to form micelles in an aqueous environment was observed as early as 1994 [150]. During preparation of long-circulating liposomes, it was noted that phospholipid/PEG-PE mixtures of a certain composition form micelles instead of liposomes if the PEG-PE content exceeds a certain critical limit [151, 152]. Trubetskoy and Torchilin [140] were the first to demonstrate the prolonged circulation and stability of PEG-PE micelles in the blood and the potential of these polymeric micelles for the solubilization of poorly soluble drugs. Various PEG-PE molecules were synthesized by conjugation of egg-PE to *N*-hydroxysuccinimide esters of PEG at 2000, 5000 and 12,000 da polymers. The presence of PEG protects liposomes from interaction with opsonins in the blood plasma and prevents rapid uptake by the reticulo-endothelial system. This results in a dramatic increase in liposomal circulation time [139, 148, 149]. PEG-PE is currently in use as a component of the liposomal formulation. Doxil®, which is a long-circulating liposome loaded with doxorubicin [153]. Among the micelle-forming compounds, PEG-PE have gained more attention due to some attractive properties such as good stability, longevity, an ability to accumulate in areas with a leaky vasculature such as tumors' site, via the enhanced permeability and retention effect [1–3, 23].

Table 14.5 *Particle size and CMC of commonly used micelle-forming PEG-PE polymers.*

Micelle-forming block co-polymer	Micelle size (nm)	CMC (M)
PEG-750-DSPE	7–15	1×10^{-5}
PEG-2000-DSPE	7–20	1×10^{-5}
PEG-5000-DSPE	10–40	6×10^{-6}
PEG-2000-DOPE	7–20	9×10^{-6}
PEG-5000-DOPE	10–35	7×10^{-6}

Retention of hydrophobic drugs within the core of stable PEG-PE micelles is crucial for their consideration as efficient drug delivery vehicles. To demonstrate that the incorporated drug is firmly associated with micelles, PEG-PE micelles loaded with several drugs were dialyzed against an aqueous buffer in a volume sufficient to provide sink conditions. All tested preparations retained more than 90% of the encapsulated drug during the first 7 h of incubation. The micelles retained 95%, 75% and 87% of initially incorporated chlorine e6 trimethyl ester, tamoxifen and paclitaxel respectively even after 48 h of incubation [144]. These results confirm the high stability of PEG-PE micelles and demonstrate the strong association of the drug with these micelles.

Various attempts have been made to further increase the solubilization efficiency of micelles by forming mixed micelles by the addition of another surfactant or hydrophobic material. The mixed micelles would allow better solubilization of poorly soluble drugs by a loosening of the micelle core (e.g. egg phosphatidyl choline) [155] or by increasing the volume of the lipophilic core of the mixed micelle with D-α-tocopheryl polyethylene glycol 1000 succinate/vitamin E/Pluronic [156–158]. With regards to incorporating poorly soluble drugs in PEG-PE micelles, Kim and co-workers [159] incorporated a poorly water-soluble cyclic peptide, Cyclosporin A, used for immunosuppressive therapy. Dimyristoylphosphatidylcholine was combined with PEG-PE to form the mixed micelles that solubilized the cargo Cyclosporin A in the micellar core. Torchilin and co-workers [23, 144] incorporated the poorly soluble anticancer drug paclitaxel (PTX) in PEG-PE micelles. The PTX loading in PEG-PE micelles was 1.5% w/w. The dimensions of the PEG-PE micelles (20–30 nm) after the incorporation of PTX were not changed. PTX-loaded PEG-PE micelles were further functionalized by modifying the surface with cancer-cell specific monoclonal antibody 2C5. These immunomicelles specifically bound target cells *in vitro* based on images obtained from fluorescence microscopy and circulated in the blood as long as plain micelles did. Accumulation of the PTX from immunomicelles in a solid tumor (Lewis lung carcinoma) was higher than plain PEG-PE micelles in a murine tumor model, which were better than PTX solubilized in Cremophor EL. The immunomicelles delivered 6% dose/g of tumor compared to 4% of the dose by plain PTX micelles at 2 h.

14.4 Determination of Physico-Chemical Characteristics of Polymeric Micelles

14.4.1 Critical Micelle Concentrations (CMC)

Amphiphilic molecules or surface active agents exist as single molecules or unimers at relatively low concentrations in a liquid medium. However, as their concentrations increased, aggregation takes place over a narrow concentration interval. These aggregates, composed of several dozens of amphiphilic molecules, are arranged in spherical or near spherical form are micelles. The concentration of the amphiphilic molecule at which spontaneous self-assembly or micellization starts is termed the critical micelle concentration (CMC). In an aqueous solution, the CMC value of a given surfactant can be lowered in the presence of electrolytes. Below the CMC, the amphiphilic molecules have a strong tendency to be adsorbed at the air–water interface [160]. As the concentration of amphiphilic molecules in the system is increased, a particular concentration is reached, above which amphiphilic monomers self-assemble to form a spherical micellar structure. However, the micelles are always in dynamic equilibrium with the monomeric form of the amphiphilic polymer above its CMC. Therefore, CMC is the total surfactant concentration at which a small and constant number of amphiphilic molecules are in aggregated form. During micelle formation, the free energy of the system is decreased by the removal of hydrophobic fragments from the aqueous surroundings with the formation of a micelle core. Because of the restricted freedom of movement of the hydrophobic chains, the entire system moves towards increased entropy. Polymeric micelles can achieve a very low CMC in the range of 10^{-6} to 10^{-7} M compared to low molecular weight or natural surfactants which have CMC values typically in the range of 10^{-3} to 10^{-4} M. This is particularly advantageous for drug delivery applications. The lower the CMC value, the more stable the micelles are, even at low concentrations. This is especially important from the pharmacology point of view, since upon dilution with a large volume of blood after i.v. administration, only polymeric micelles with low CMC exist. Others with higher CMC break down to their unimers and the micelles-loaded drugs precipitate out. The properties of the core-forming block have the most profound impact on the CMC. Increasing hydrophobicity and size of the hydrophobic block lowers the CMC [39].

There are several methods available to determine the CMC value of an amphiphilic compound, including surface tension, dye solubilization, light scattering and fluorescence [162–165]. Figure 14.4 presents the changes in various physical properties with increasing concentration of surfactants. However, because of sensitivity issues, only certain methods are applied to determine the CMC. Of those methods, particle size measurement by small angle light scattering, HPLC, and fluorescent spectroscopy are widely used. The determination of CMC using the fluorescent probe pyrene is the most sensitive and precise one [1, 39, 166]. Fluorescent probes, such as pyrene, have the tendency to associate with the micelles rather than water due to their poor solubility in water and their high solubility in the

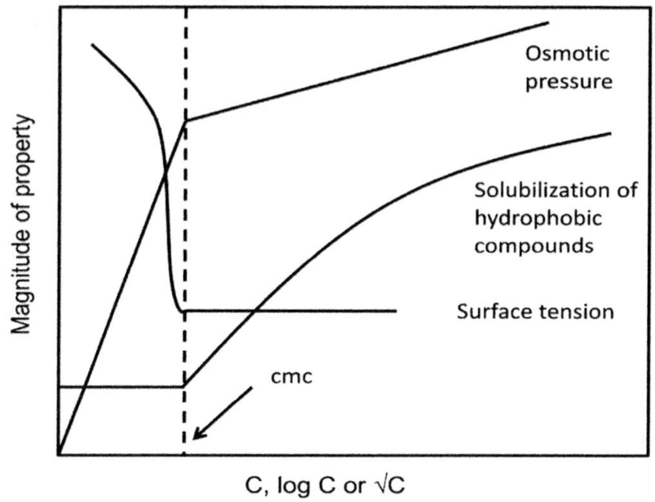

Figure 14.4 *CMC determination of amphiphile by changes in various physical properties with increasing concentrations.*

lipophilic micelle core. Pyrene, at a constant dilute concentration, is placed in surfactant solutions of varying concentrations. The high quantum yield and partitioning of pyrene in most micellar cores allows for versatility and little interference with the micellization process. The fluorescence intensity of pyrene increases dramatically when pyrene starts partitioning into the forming micelle cores, allowing for the determination of the CMC. Figure 14.4 presents a typical curve showing the fluorescence changes of a marker in the presence of varying concentrations of a micelle-forming amphiphile and demonstrates how the CMC value is calculated.

Mixed micelles are micelles made up of more than one type of amphiphile. In the ideal case, the CMC of the mixture can be calculated from the CMC values of the individual components and their molar fraction (M) in the mixture.

$$1/\mathrm{CMC} = \mathrm{M}_1/\mathrm{CMC}_1 + \mathrm{M}_2/\mathrm{CMC}_2$$

14.4.1.1 Protocol for CMC Determination

CMC determination by the pyrene method is most effective, reliable and convenient [19, 28]. In glass test tubes, a solution of pyrene (1 mg) in chloroform is added to each tube. The solvent is removed under vacuum. Each tube containing 1 mg of pyrene crystals is incubated and shaken in darkness at room temperature for 24 h with known dilutions of micelle test suspensions. A range of concentrations between 7.2×10^{-7} M and 3.6×10^{-4} M is optimum for the block co-polymers possessing low CMC numbers. The unincorporated pyrene in the micelles is filtered through 0.2 μM nylon filters. Each dilution is analyzed by fluorescence spectroscopy at ex/em 339/390 nm wavelengths to evaluate the amount of pyrene retained in the micelles and to directly correlate it with concentration of the micelles themselves.

14.4.2 Particle Size and Stability

14.4.2.1 Size

Micelles fall in the category of colloidal dispersions, where colloids or a dispersed phase are distributed uniformly in the aqueous dispersion media. The dispersed phase or colloids have a diameter of approximately 5–500 nm, whereas molecular dispersions have particle sizes under 1 nm and coarse dispersions have sizes greater than 500 nm. The particles are normally invisible by optical microscope, but are visible by electron microscope. Colloidal dispersions are normally divided into three principal groups: lyophilic, lyophobic and association colloids. The micelles belong to the group of association colloids. The morphology of the block copolymer micelles is universally spherical when in dilute solution. In general, the properties affecting the shape of micelles are the ratio of hydrophobic to hydrophilic segment size on the unimers [167, 168]. Most block co-polymers typically have larger hydrophilic segments than hydrophobic segments which creates a large head group. The large head groups with smaller hydrophobic segment give the polymeric unimers a cone shape which is amenable for spherical micelle formation [167]. A spherical shape minimizes the hydrophobic exposure to the bulk aqueous phase and represents the lowest energy conformation. However, temperature and a high concentration of unimers have effects on the morphology of the self-assembled aggregates. The surfactant poloxamers generate more exotic shapes such as rods or vesicles, due to the increase in temperature or polymeric concentration [100, 169]. Many different morphologies of micelle can also be created with the polymers poly(acrylic acid)-*b*-poly(styrene) and poly(ethylene oxide)-*b*-poly(styrene) when the hydrophobic block lengths are significantly greater than the hydrophilic block lengths [169].

There are a variety of methods available to characterize micelle morphology. Microscopy and the small angle scattering techniques are the most useful and reliable. Transmission electron microscopy (TEM) has been used since 1980 to characterize the shapes of nanoparticles [170]. It involves analyzing dry samples of dehydrated micelles which may not represent the exact micellar morphology in an aqueous environment. The cryo-TEM is the more recent advancement which allows viewing the micellar size and shape as they appear in solution [171]. Atomic force microscopy [AFM) has also been used where sample preparation is much simpler than other microscopy techniques [172]. The size of polymeric micelles varies between 10–100 nm in diameter depending on various properties of the surfactant. The overall size of the micelle is primarily dependent on the lengths of the hydrophilic and hydrophobic blocks, the total surfactant molecular weight and the relative proportion of hydrophobic and hydrophilic segment. These properties influence the packing of the unimers to form micelles and also determine the association number and therefore, the overall size of the micelles. Usually, the core size is primarily dependent on the length of the core-forming block. However, studies show when the core forming block length is fixed and the shell-forming block length is increased, the micelle association number is decreased, leading to a decrease in core size [64, 69]. Other factors, such as the presence of additives, temperature and even the surfactant concentration can affect the overall micelle size [64, 69].

Two parameters which are measured for micellar size determination purpose are micellar hydrodynamic diameter and size poly-dispersity. Dynamic light scattering technique (DLS) is a very widely used method for determination of hydrodynamic diameter as is poly-dispersity, which has sensitivity down to approximately 3 nm. Micellar size can be obtained directly in water or in an isotonic buffer without affecting the micelles physically or chemically. The parameter measured by DLS is the equivalent sphere translational diffusion coefficient (D_o). This coefficient is used with the Stokes-Einstein equation to obtain hydrodynamic radius (R_h):

$$R_h = \frac{k\mathrm{T}}{6\pi\eta D_o}$$

where k is the Boltzmann constant, T is the absolute temperature and η is the viscosity of the solvent.

Micelle size can also be determined by microscopy techniques. However, the resulting sizes obtained may not be the actual hydrated diameters of micelles in solution, depending on the sample preparation method required for microscopic visualization.

14.4.2.2 Stability

The ability of micelles to prevent disassembly upon dilution is described in terms of their thermodynamic stability. The stability of micelles plays an important role in various technological processes such as foaming, wettability, emulsification and detergency [173]. The thermodynamic stability of polymeric micelles is a major issue when the micelles are used for drug delivery purposes, especially for intravenous administration as micelles are diluted in the blood volume. The administration of micelles by the intravenous route exposes the micelles to sink conditions. The CMC is the major parameter for stability determination. CMC describes the concentration of unimers in solution above which it starts forming micelles or is the concentration of unimers below which micelles disassemble. The CMC describes the extent to which micelles can withstand dilution. Micelles with the lowest CMC withstand the greatest dilution and are the most thermodynamically stable. As the concentration of surfactant in solution drops below the CMC, micelles dissociate into unimers and re-establish a micelle–unimer thermodynamic equilibrium [113]. Polymeric micelles have very a low CMC. Because of this property, the polymeric micelles have found wide application for drug delivery purposes.

There are various factors that determine the CMC and thereby determine the micellar stability. Composition and molecular weight are two of the prime parameters of interest in the formation of stable block co-polymeric micellar systems. Analogous to low molecular weight non-ionic surfactants, micellization is facilitated by an increased length of the hydrophobic blocks and decreased length of the hydrophilic ones. Even though micelles are represented as static structures of spherical aggregates of oriented surfactant molecules, they are in dynamic equilibrium with individual monomers of surfactants that are constantly being exchanged between the micelles and the surrounding bulk phase. However, for polymeric micelles with very low CMC values, the unimer-micelle thermodynamic equilibrium shifts in favor of micelle formation from disintegration into unimers. As the concentration of the surfactant drops below the CMC, micelles dissociate into unimers to re-establish the unimer-micelle thermodynamic equilibrium [113]. Block copolymers form micelles at a very low concentration of unimers in the solution, indicating the micellar forms are stable. However, the ordinary surfactants that are in dynamic equilibrium between monomer and micellar system present only temporal stability [174].

Micellar stability can be compromised in the case of mixed micelles when compared to micelles made up of one type of surfactant. In general, the chain length of surfactants used in a given mixture of surfactants must be the same to maximize the lateral molecular interactions. When the chain length of both the surfactants is equal, the molecular packing is closer at the oil/water interface. By contrast, when the chain length of the hydrocarbon part is different, the portion of the molecule above the height of the adjacent molecules will exhibit thermal motion which causes an increase in the area/molecule [175]. Therefore, mixed polymeric micelles made up of polymers with different chain lengths results in a looser packing which accelerates the break-up of the micellar form. In addition, micelles formed by amphiphilic graft-copolymers are more likely to form aggregates, since hydrophobic chains in such micelles are less labile and more loosely packed than in micelles formed by block co-polymers, so that they interact with hydrophobic blocks from other micelles [28, 58, 176–178].

For drug delivery application of polymeric micelles, micelles can be used to solubilize poorly water-soluble drugs in the hydrophobic compartment. However, the CMC can be influenced by drug solubility, drug interactions with polymers, drug loading content and the interaction of micelles with the variety of biological components, including soluble proteins, blood cells, various biological compartments and colloidal particles.

Even though the CMC determines the micellar stability to a great extent, it does not dictate the stability of the micelle entirely. The time scale of micellar dissociation is also important in determining micellar stability [179].

The kinetics of micellization has been studied extensively by various techniques [180–182]. There are two relaxation processes involved in micellar solutions. The first one is the rapid relaxation process (relaxation time designated as τ_1, usually on the order of microseconds), which is associated with the fast exchange of monomers between micellar aggregates and the surrounding medium. The second relaxation time (τ_2, usually on the order of milliseconds or longer) is attributed to the micelle dissolution process, i.e. breaking up the micelles into its monomers particularly when the polymer concentration drops below CMC. These two relaxation times can be used to calculate two important parameters of a micellar solution: the time the monomer is associated with the micelles, and the stability or the lifetime of micelles [183–185]. Kinetics of micellar dissolution and the rate are important when the concentrations of micelles drop below the CMC, particularly after i.v. injection. Even though the entropy promotes dissociation of the micelles below the CMC, the rate of dissociation can be instantaneous, a few minutes, or as slow as a few minutes, hours or days. The dissolution kinetics become further complicated if the micelles are loaded with hydrophobic molecules such as drugs compared to plain empty micelles because the tertiary system (polymer–drug–water) exhibits different exchange rates among polymer components in micelles and free unimers in water compared to the binary system. Thus, the dissolution kinetics' rate upon dilution determines the stability of the micelles. In addition, the viscosity of the core, the glass-transition temperature (T_g), core size and chain entanglement play roles in defining the kinetic stability of the micelles [179]. For example, a micelle core formed with a lipophilic segment with a high T_g is described as 'frozen.' A micelle with a frozen core has very low segmental motion and dissociation, that enables the micelles to remain intact after dilution below CMC [30, 186]. On the other hand, a micelle core with a low T_g is described as having a liquid-like core and dissociates rapidly upon dilution below the CMC. Poloxamer micelles are considered 'moderately stable to relatively unstable' in solution because of their liquid-like core and high CMC whereas micelles made up of the core-forming hydrophobic blocks poly(styrene), poly(ε-caprolactone) or poly(β-benzyl-L-aspertate) are stable thermodynamically as well as kinetically. Micellar stability can be measured indirectly by measuring the CMC/CMT (by static light scattering, fluorescence, surface tension, dye solubilization), hydrodynamic radius (by dynamic light scattering, pulsed-gradient-spin-echo NMR), morphology (by transmission electron microscopy (TEM), atomic force microscopy (AFM), or small angle neutron/X-ray scattering (SANS/SAXS)). Nuclear magnetic resonance (NMR) can also be used to predict segmental motion and viscosity of the hydrophobic core which is an indicator of micellar kinetic stability. Peak broadening occurs in proton NMR spectra if core-forming segment blocks move very slowly. Micelle cores with undetectable segmental motion and high viscosity are considered kinetically stable. Another indicator of micellar stability upon dilution is the use of gel permeation chromatography (GPC). In this case, the eluent from the column is checked for intact micelles or unimers since the sheer force in the column and dilution by the mobile phase could cause micellar dissociation, depending on the nature of polymeric micelles.

14.5 Drug Loading

Conventional polymer micelles such as PEG-*b*-poly (D, L-Lactic acid) (PDLLA or PLA) served as excipients for the solubilization (or loading) of highly lipophilic drugs. Drug solubility in water is increased as a result of the hydrophobic interaction between hydrophobic polymer blocks and drugs [187]. These hydrophobic interactions are due to the London dispersive forces that prevail between all molecules. The London dispersive force between lipophilic drugs and a hydrophobic block is much stronger than that between the lipophilic drug and water, thus the hydrophobic effect is induced when hydrophobic molecules are mixed with water [188]. In addition to the hydrophobic effects, drug-loading capacity and efficiency depend on the miscibility between polymers and drugs. The mechanism of drug loading can be explained through the Hildebrand–Scatchard solubility parameter (δ) $\delta = \frac{\sqrt{\delta Evap}}{V}$, where, Δ E_{vap} is the Energy of vaporization, and V is the Volume of the solvent [189]. Drug loading capacity of polymeric micelles implies the mixing of the polymer with drugs, which has been described by the Flory–Huggins theory given by the equation $X_{drug\text{-}polymer} = (V_{drug}/RT)\ (\delta_{drug} - \delta_{\ polymer})\ 2$, where $X_{drug\text{-}polymer}$ is the Flory–Huggins interaction parameter between the drug and the polymer, V_{drug} is the volume of the drug, R is the ideal gas constant, T is the temperature, and δ_{drug} and $\delta_{polymer}$ are the Hildebrand–Scatchard solubility parameters of the drug and the polymer, respectively [190, 191]. The miscibility of PEG-*b*-PCL with five different drugs such as etopside, paclitaxel, plumbagin, curcumin and indomethacin followed an ascending order of $X_{drug\text{-}polymer}$ [192]. Indomethacin showed a lower value of <0.5 for the Flory–Huggins parameter, implying a better solubility. Miscibility between polymer and drugs is also influenced by another parameter, heat of mixing. The drug-loading efficiency was found to be highly dependent on the heat of mixing and the order was poly (benzyl-L-aspartate) (PBLA) > PCL > PDLLA > PGA [188]. Loading of a drug into a polymeric micelle is not only influenced by the hydrophobic effect, but also by other interactions between polymers and drugs that may enhance the miscibility. It should be noted that the hydrophilic-lipophilic balance (HLB) of block polymers is another parameter that governs the drug-loading capacity of polymer micelles. It was shown that block copolymers with a longer hydrophobic block have better drug-loading capacity confirmed by the partition coefficient of drugs into PEG-*b*-PCL [192]. On the other hand, the partition coefficient deteriorates with a longer PEG chain. Drug-loading capacity can be increased for most drugs if the miscibility of one polymer is significantly increased. One possible solution is hydrotopy.

Hydrotropy is a process in which the presence of large amounts of a second solute, called the hydrotope, significantly increases the aqueous solubility of poorly soluble compounds [193, 194]. The drug-loading capacity of polymer micelles can be improved by optimizing the polymer–drugs miscibility. Nicotinamide derivatives are excellent hydrotopes for paclitaxel [195]. Using 3.5 M aqueous solutions of N,N-diethylnicotinamide (DENA) and N-picolylnicotinamide (PNA) significantly enhanced the solubility of paclitaxel (0.3 μg/ml) up to 39 and 29 mg/ml respectively. Paclitaxel loading was increased up to 37% (w/w) with a PEG-containing polymer micelle based on a polymerizable derivative of DENA, 4-2(-vinylbenzyloxy)-N,N-DENA (VBODENA) with a loading efficiency of more than 90% (w/w) while having a loading capacity of a conventional polymer micelle of PEG-*b*-PDLLA was ~20% (w/w) of paclitaxel [196]. A polymer micelle consisting of PEG and poly[4-(2-vinylbenzylloxy)-N-PNA](PEG-*b*-PVBOPNA) spontaneously generates polymer micelles by simply adding paclitaxel to the aqueous solution without any solvent system. The sizes of both core- and corona-forming blocks also influence the efficacy of drug loading into the micelle [197]. Hydrophobic drugs' entrapment can be increased by a larger hydrophobic block, thereby generating a bigger core size. Also, the increase in the length of the hydrophilic block results in the increase of the CMC value, i.e., at a given concentration of the amphiphilic polymer in solution the smaller fraction of this polymer will be present in the micellar form and the quantity of the micelle-associated drug drops [4]. Various drugs like diazepam and indomethacin [198, 199], adriamycin [200–202], anthracycline antibiotics [73], polynucleotides [203, 204], and doxorubicin [46] were effectively solubilized by various polymeric micelles, including Pluronic (block copolymers of PEG and polypropelene glycol) [29]. Doxorubicin incorporated into Pluronic micelles showed superior properties as compared with a free drug in the experimental treatment of murine tumors (leukemia P388, myeloma, Lewis lung carcinoma) and human tumors (breast carcinoma MCF-7) in mice [46]. The whole set of micelle-forming co-polymers of PEG with poly (L-amino acids) was used to prepare drug-loaded micelles by direct entrapment of a drug into the micelle core [28, 205–207]. PEG-*b*-poly(caprolactone) co-polymer micelles were successfully used as delivery vehicles for dihydrotestosterone [197]. PEG-PE micelles can efficiently incorporate a variety of sparingly soluble and amphiphillic substances including paclitaxel, tamoxifen, camptothecin, prophyrine, vitamin K3, and others [145, 155, 208]. Numerous studies are also dealing with micellar forms of platinum-based anti-cancer drugs [209–211] and cyclosporine A [212, 213]. Mixed polymeric micelles made of positively charged polyethyleneimine and Pluronic were used as carriers for antisense oligonucleotides [214]. Loading efficiency values apparently correlate with the hydrophobicity of a drug, varying from 1.5 to 50% by weight. The addition of mixed micelle-forming compounds to polymeric micelles may improve the drug solubilization. For example, the addition of egg phosphatidylcholine (PC) to the PEG-PE-based micelle composition increased the paclitaxel encapsulation efficiency two-fold (from 15 to 33 mg of the drug per gram of the micelle-forming material) [155, 215]. This can be explained by the fact that ePC, unlike PEG-PE, does not have a large hydrophilic PEG domain, and its addition to the micelles composition results in particles with a higher hydrophobic content. Paclitaxel incorporates into lipid particles since it interacts with the hydrophobic part of amphiphiles [216, 217]. Paclitaxel in mixed PEG-PE/ePC micelles showed high cytotoxic activity against MCF-7 human mammary adenocarcinoma cells [155].

The initial amount of drug influences the entrapment efficiency. Drugs that precipitate result in a lower yield if they are loaded above the maximum capacity of the micelle. Drug-loading efficiency was also found to be dependent on the aggregation number of the copolymer [218]. A higher aggregation number was associated with more drug loading in the inner core.

14.5.1 Drug-Loading Procedures

Drug-loaded micelles can be prepared through several techniques depending upon the physicochemical properties of the block copolymer. Common drug-loading procedures include: (1) simple equilibrium; (2) chemical conjugation; (3) dialysis; (4) o/w emulsions; (5) solution casting; and (6) freeze-drying (Figure 14.5).

The simple equilibration of the drug and micelles in water may not result in high levels of incorporated drugs [186, 219]. Chemical conjugation involves a covalent bond such as an amide bond between specific groups on the drug and the hydrophobic polymer of the core. Such a bond offers resistance to enzymatic cleavage as a result of steric hindrance and is not susceptible to hydrolysis unless spacer groups are introduced [220]. This technique is useful for the insertion of hydrophilic compounds such as proteins which require the chemical hydrophobization of the molecule [146]. Dialysis or oil-in-water emulsion procedures are

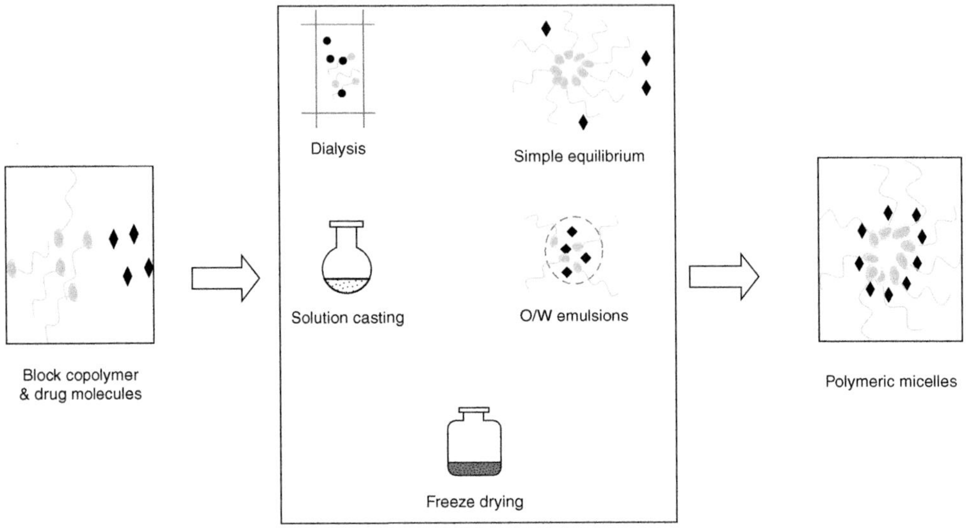

Figure 14.5 *Methods for drug loading of polymeric micelles.*

generally used for the physical entrapment of the drugs. The dialysis method uses a solvent common to both the amphiphilic copolymer and the drug (such as dimethylsulfoxide, N,N-dimethylforamide, acetonitrile, THF, acetone or dimethylacetamide) and a solvent that is exclusive to the hydrophilic part of the polymer (e.g. water). Solvent-removal induces the formation of micelles. The hydrophobic part of the polymer forms the core of the micelles incorporating the insoluble drug during the process, as the good solvent is displaced by the selective one. Complete removal of the organic solvent can be achieved by dialysis for several days. In the solution-casting method an organic phase is evaporated to form a polymeric film that favors polymer–drug interactions. This film is then rehydrated by a heated aqueous solvent to form drug-loaded micelles. In the oil-in-water emulsion method, the drug in a water-insoluble volatile solvent is added to an aqueous solution of the co-polymer to form an oil-in-water emulsion. Evaporation of solvent induces the formation of the micelle-drug conjugate. The dialysis method has an advantage over the oil-in-water emulsion method since use of toxic solvents like chlorinated solvents can be avoided. Comparison of the dialysis and oil-in-water emulsion methods for incorporation of doxorubicin in PEO-PBLA micelles showed that the emulsification method was better. The doxorubicin content of micelles was ~12% (w/w) [42] as against 8% (w/w) for the dialysis method [43]. Sant *et al.* reported that the O/W emulsion technique had a 1.5-fold increase in the drug incorporation efficiency for a variety of hydrophobic drugs into PEG-*b*-poly(alkyl acrylate-co-MAA) (PEG-*b*-P(ALA-co-MAA) micelles when compared to the dialysis method [221]. It was reasoned that, by this procedure, the drug molecules are entrapped within the droplets of organic solvent stabilized by the polymer chains and are retained accordingly upon evaporation of the organic phase. The above-discussed methods also need sterilization and freeze-drying steps in order to produce injectable formulations with a sufficient shelf-life.

Another method was recently reported in which both the polymer and the drug are dissolved in a water/tert-butanol (TBA) mixture followed by lyophilization of the solvent. Reconstitution of the resultant freeze-dried polymer drug cake with an injectable vehicle produces spontaneous formation of drug-loaded micelles. Using this technique, hydrophobic drugs such as paclitaxel and docetaxel were successfully loaded into a PVP-*b*-PDLLA copolymer, resulting in stable spherical micelles with a uniform size distribution [222]. This straightforward technique helps in producing freeze-dried, drug-loaded micelles with a sufficient shelf life that can be easily redispersed in water.

The simplest and most convenient technique for the preparation of drug-loaded PEG-PE micelles involves the simple dispersion of a dry PEG-PE-drug mixture in an aqueous buffer. A typical protocol for the preparation of drug-loaded micelles includes the following steps. Solutions of PEG-PE and a drug of interest in miscible volatile organic solvents are mixed and the organic solvents are evaporated to form a PEG-PE-drug film. The film obtained is hydrated in the presence of an aqueous buffer and the micelles are formed by intensive shaking. If the amount of a drug exceeds the solubilization capacity of micelles, the excess drug precipitates in a crystalline form and the precipitated crystals are removed by filtration.

14.6 Biodistribution and Toxicity

Nanoparticles such as micelles are employed in drug delivery to improve solubility, and reduce renal clearance for poorly soluble drugs for a prolonged pharmacological effect. However, the micellar nanocarrier may enhance the delivery of drugs to certain tissues and cause new side-effects. Therefore, it is very important to monitor the pharmacokinetics (PK) and the biodistribution of nanocarriers to understand and predict their efficacy and side-effects.

Many hydrophobic drugs injected intravenously are transformed into hydrophilic metabolites in the liver and excreted into the bile or eliminated in the urine. When the drugs are encapsulated in the micelles, they are protected from the metabolizing enzyme in the liver before release and also from renal clearance due to their increased size. The cut-off size for renal excretion is approximately 5.5 nm which is much smaller than the average micellar size, typically 10–100 nm. Therefore, a drug encapsulated in micelles experiences reduced liver metabolism and renal clearance which results in prolonged blood circulation and an increased chance of accumulation at the site of disease [223].

After the polymeric micelles are delivered to the target via the bloodstream, the extravasation process is usually slow and in a passive manner. The limited pore size of the endothelial wall in the tissue is considered the main barrier to the delivery of micelles or other nanocarriers in normal healthy tissues. However, they can accumulate significantly more in the tissues with leaky endothelial walls, including tumors, as well as in the liver, bone marrow and spleen. Micellar uptake by tumor tissues is achieved by the EPR effect owing to the leaky vasculature in the tumor tissue [224]. Table 14.6 shows the biodistribution of various micelles. The general trend observed (i.e. the enhanced uptake by liver, bone marrow and spleen) is due to the presence of macrophages present in the tissues which contributes to the significant loss of nanoparticles, from the circulation. In addition, serum proteins bind on

Table 14.6 *Biodistribution of various polymeric micelles.*

Polymer	Block ratio	Drug	Model	Biodistribution	References
PEG-*b*-PCL	6000–1000	N/A	Mouse	Blood>bone>kidney>liver>lung>brain	336
	5000–5000	N/A	Mouse	Liver>kidney>spleen,lung>heart	337
	5000–13,000	Cyclosporine	Rat	Blood>heart>plasma>liver>kidney >spleen	212
PEG-*b*-PLA	2000–1750	Paclitaxel	B-16 tumor-bearing mice	Tumor>liver>kidney,spleen>lung>,heart>plasma	136
	33,000	Paclitaxel	MDA-MB-435 tumor-bearing mice	Kidney>Liver>lung>spleen>heart>tumor >muscle>blood	338
PEG-*b*-PAsp-DOX	4300–1900	Doxorubicin	Mouse	Blood>spleen>kidney>liver>lung>heart	120
PEG-*b*-P(Asp(Bz-70))	5000–4700	Campthothecin	C-26-bearing mice	Spleen,tumor,plasma>kidney,lung>heart	339
PEG-*b*-P(Asp(Bz-75))	N/A	All trans-retinoic acid	Mice	Liver>kidney>lung>spleen	340
PVA-*b*-PLA	2480–1820	Paclitaxel	C26 tumor-bearing mice	Liver>kidney,lung>Spleen>heart>muscle	335
RGD-PEG-*b*-PLA	33,000	Paclitaxel	MDA-MB-435 tumor-bearing mice	Kidney>liver>tumor>lung,spleen>heart >muscle>blood	338
FA-PEG-*b*-P(Asp-Hyd-DOX)	N/A	Doxorubicin	KB tumor-bearing mice	Lower FA content: Blood>liver,spleen>tumor>kidney>heart Higher FA content: Liver>blood,spleen>tumor>kidney,heart	341
PEG-*b*-P(HEMAm-Lac)	N/A	N/A	14 C-bearing mice	Non-cross-linked micelle: Liver>tumor>skin>spleen>blood Cross-linked micelle: Liver>blood>skin,tumor>spleen	342
PAA-*b*-PMAA	4800–18,600 6000–7000	N/A	Mouse	Spleen>liver>kidney>blood>lung Lower PEG density: Spleen>liver>kidney>lung>blood Higher PEG density: Spleen>liver,blood>lung>kidney	343

the surface of the circulating nanoparticles, which are recognized by the scavenging receptors on the surface of the macrophages [225]. This phenomena is referred to as opsonization which is responsible for a significant loss of nanocarriers. The macrophages engulfing the nanoparticles are part of the reticuloendothelial system (RES) or mononuclear phagocyte system (MPS). Minimization of the protein binding is the key factor for the development of long-circulating liposomes [223]. A highly effective and hydrated PEG chain attached to the nanoparticle surface acts as a shield and is predicted to have an effective protein-resistant property due to its steric repulsion effect [226]. Thus, polymeric micelles with a high density of PEG shells circumvent the obstacle of opsonization and recognition by RES system and achieve prolonged blood circulation. The PK profile of nanoparticles can be modified by changing the size, charge and surface chemistry. Long systemic circulation time for polymeric micelles also depends on their CMC as well as their kinetic stability. They may not necessarily dissociate immediately after extreme dilution by an intravenous injection due to their characteristic low CMC (10^{-6}–10^{-7} M) which is 1000-fold lower than that of the surfactant micelles [28, 227]. In addition, slow dissociation kinetics of polymeric micelles also hinders micellar disassembly.

Polymeric micelles composed of PEG-*b*-poly (D, L-lactide) (PEG-*b*-PDLLA) copolymers labeled with 125I showed a remarkably prolonged blood circulation ($t_{1/2} = 18$ hours) after intravenous administration [228]. Up to 25% of the injected dose was retained in the circulation at 24 hours post-injection. The distribution volume in the central compartment (V_0) and plasma-to-blood ratio of the micelles were nearly equivalent to the blood volume and the plasma space volume respectively. This implies that polymeric micelles distribute only into the blood compartment and hardly interact with blood cells after their administration [228]. Tissue-to-blood concentration ratios for the PEG-*b*-PDLLA micelles were comparable to vascular space volume (~0.2) in normal organs (lung, kidney, liver and spleen) until 24 hours after injection. Thereafter were minimal increases in the liver and spleen corresponding to an extracellular space volume (~0.3) [228]. PEG-*b*-PDLLA polymeric micelles avoided the RES recognition as well as entrapment by hepatic sinusoidal capillaries characterized by large inter-endothelial junctions (~100 nm) despite the relatively small size (~30 nm) of the micelles. Furthermore, it was revealed that the constituent block copolymers were finally excreted into the urine since their molecular weight was lower than the threshold of glomerular filtration, suggesting the safety of polymeric micelles without a low risk of chronic accumulation in the body.

Polymeric micelle nanocarriers showed an increased accumulation in solid tumors [112, 122, 333, 334]. Various PEG-PE micellar formulations studied showed circulation half-lives in mice, rats, and rabbits of from 1.2 to 2.0 h depending on the size of the PEG block [233]. The circulation time in the blood increases as the size of the PEG block increases. This may be due to better steric protection against opsonin penetration into the hydrophilic micelle core provided by higher molecular weight PEG.

PEG-PE-based micelles prepared using different molecular weights of PEG (example: PEG750-PE, PEG2000-PE, and PEG5000-PE) accumulate efficiently in tumors via the EPR effect. It should be noted that micelles prepared with several different PEG-PE conjugates demonstrated much higher accumulation in tumors compared to non-target tissue (muscle), even in the case of an experimental Lewis lung carcinoma (LLC) in mice that are known to have a relatively small vasculature cut-off size [146, 229]. Since micelles have a small size, they may have additional advantages as tumor drug-delivery systems which utilize the EPR effect, compared to particulate carriers with larger sizes. Thus, the micelle-incorporated model protein (soybean trypsin inhibitor or STI, MW 21.5 KDa) accumulates to a higher extent in subcutaneously established murine LLC than the same protein in larger liposomes [146].

The accumulation time for all the micelles prepared from all versions of PEG-PE conjugates is characterized by peak tumor accumulation times of about 3–5 h. The largest total tumor uptake of the injected dose 5 h post injection (as AUC) was found for micelles formed by the unimers with a relatively large PEG block (PEG5000-PE). This may be explained by the fact that these micelles have the longest circulation time and less extravasation into the normal tissue than micelles prepared from smaller PEG-PE conjugates.

Biodistribution studies on solid tumor-bearing (C26) mice showed a tumor/heart concentration ratio of 12, 8.1 and 0.9 for PEO-*b*-P(Asp)-DOX with PEO chains of 12,000 and 5000 g/mol, and free DOX respectively 24 h after intravenous injection [122]. For a Pluronic formulation of doxorubicin, called as SP-1049C, biodistribution results showed an increase in AUC in the tumor and the brain [75]. The results [335] of biodistribution studies for paclitaxel encapsulated in micelles composed of PDLLA (the core-forming block) and poly(N-vinylpyrrolidone) (PVP) as the shell-forming block showed a decrease in the AUC of paclitaxel in liver, kidney, spleen and heart while the AUC in the tumor did not change significantly.

It was reported that Pluronic polymers (ABA type) could specifically inhibit the ATP production in mitochondria, resulting in the inhibition of the action of P-glycoprotein [73–75, 83, 318–320]. Multi-drug resistance was successfully circumvented both *in vitro* and *in vivo* using this co-polymer. However, it should be noted that attention must be paid to the biological activities of these carrier polymers as the mitochondrial and P-glycoprotein inhibitory activities can produce toxic side-effects in normal organs and tissues.

Another group [42, 108, 112, 120, 200, 321, 322] chemically conjugated doxorubicin to aspartic acid residues of poly (ethylene glycol)-*b*-poly(aspartic acid) block copolymer via an amide bond. This drug–block copolymer conjugate shows an amphiphilic character (PEG being hydrophilic and DOX conjugated poly(aspartic acid) as hydrophobic) can form micelles. Doxorubicin was loaded into the micellar structure by physical entrapment in the core via hydrophobic and π-π interactions with DOX conjugated poly(aspartic acid). The cytotoxicity and anti-tumoral activities were due to physically entrapped doxorubicin while the carrier polymer PEG-P(Asp(DOX)) was inactive. *In vivo* toxicities of such polymer-conjugating drug molecules should be evaluated but cannot be investigated in simple toxicity studies.

Yokoyama *et al.* evaluated the side-effects of a DOX polymer micelle system similar to the one discussed above. Hepatotoxicity was evaluated after the first drug injection. Doxorubicin polymer micelle and free doxorubicin showed the same toxicity pattern, implying the liver toxicity was no higher than the other side-effects for the DOX polymeric micelle. However, considerable attention must be paid to chronic liver toxicity that could not be evaluated, reported up to day 10.

As discussed, chronic liver toxicity was not a major problem for doxorubicin-containing polymeric micelle systems. Most investigators postulate that a considerable slowness of the metabolism of the incorporated drug compared with the free drug is a universal phenomenon of polymeric micelle drug carrier systems due to the isolated drug-containing inner core from the aqueous environment, where a liver's metabolic enzymes work. Hence, chronic liver toxicity merits careful examination not only for the polymeric micelle systems, but also for all nano-sized drug carrier systems such as PEG-coated liposomes.

In an evaluation of toxicity of a polymeric micelle carrier without any drug, a PEG-*b*-poly (amino acid) block copolymer was used as the carrier for campthothecin [125, 322–327]. Toxicities were evaluated by conducting pathological examinations of rats. No pathological abnormality was found, even for a higher dose (200 mg/kg $\times$ 5), though the group found significant activation of the mononuclear phagocytic system (MPS) in various organs like the spleen and liver. Immuno-histochemical analysis confirmed selective accumulation of the polymeric micelle in MPS. All these evaluations were done with a polymeric micelle without any drug. There was substantial damage from polymer micelles incorporating the drug. This MPS-related effect seems to be important if the polymeric micelle carrier systems are used to deliver drugs which are much less toxic than typical anticancer drugs.

Another study [328] reported the accelerated blood clearance (ABC) phenomenon for the polymeric micelle carriers. ABC is a process where the clearance rates of carrier systems from the bloodstream are raised substantially after repeated injections [329–332]. This phenomenon is well studied in the case of PEG-coated long-circulating liposomes and is due to the immunological activity that is stimulated at the first dose. For example, after the first injection, 90% of the dose may be circulating for 2 hours while less than 5% may circulate in the blood 2 hours after the second dose. Both polymeric micelles with outer PEG shells and PEG-coated liposome share the same profile, so it is important to investigate whether these polymeric micelles induce the ABC phenomenon. After injecting three kinds of polymeric micelles followed by PEG-coated liposomes, the ABC phenomenon was observed for only one polymeric micelle, indicating that some polymeric micelles can induce the ABC phenomenon. More studies are required to investigate the properties of polymeric micelles for the induction of the ABC phenomemon. To conclude, *in vivo* toxicities for the polymeric micelle drug carriers systems should be carefully studied at the basic study stage even if the toxicities are mild and do not qualify as dose-limiting toxicities.

14.7 Targeting Micellar Nanocarriers: Example: Drug Delivery to Tumors

Micellar nanocarriers are now being used for targeted delivery of therapeutics especially for the treatment of cancer. Targeting micelles to pathological organs or tissues can further increase the efficiency of the micelle-encapsulated drug. Several approaches exist to enhance the accumulation of various drug-loaded pharmaceutical nanocarriers, including micelles, in pathological zones.

14.7.1 Passive Targeting

It is now a well-established phenomenon that under certain circumstances, the endothelial lining of the blood vessel wall becomes more permeable than in normal conditions. This was clearly demonstrated for many tumors [229, 230] and in infarcted areas [35, 231]. As a result, in such areas, large molecules and even relatively large particles such as liposomes and micelles ranging from 10–500 nm in size can extravasate from the compromised leaky vasculature and accumulate inside the interstitial space. Polymeric micelles loaded with

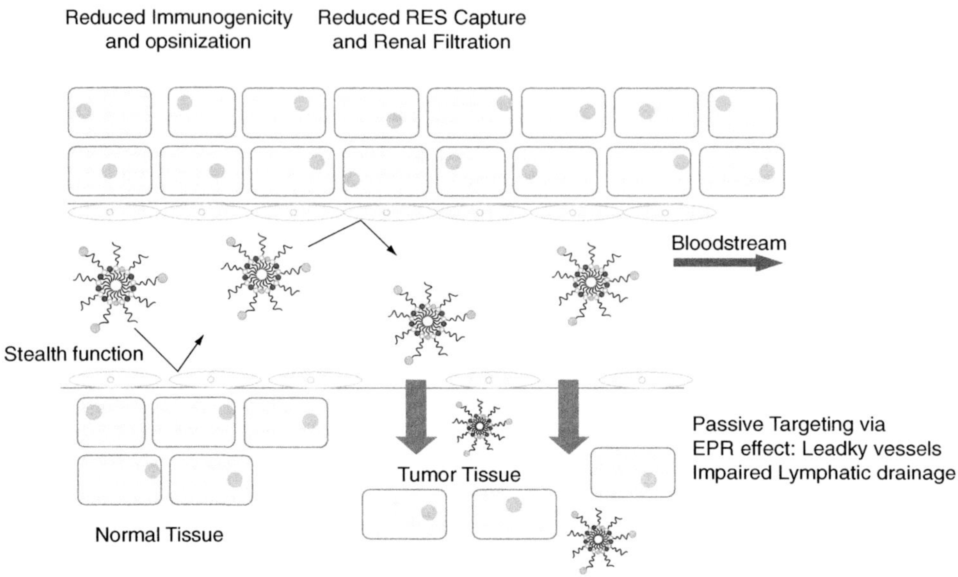

Figure 14.6 *The Enhanced Permeability and Retention (EPR) effect or 'passive' targeting of the nanocarriers. Nano-sized carriers including micelles cannot extravasate through the normal endothelium, but can penetrate through a compromised, 'leaky' vasculature characteristic of target pathological sites such as tumors, resulting in the accumulation of carrier in tumors relative to normal tissues.* (See colour plate section.)

drug can take the pay load into the area of increased vascular permeability. The phenomenon of spontaneous accumulation of nanocarriers or macromolecules within the interstitial space that takes advantage of the compromised leaky vasculature characteristic of solid tumors, infarcts, infections and inflammations or into an area of increased vascular permeability is called the enhanced permeability and retention (EPR) effect [34–36]. Figure 14.6 shows the schematic representation of the accumulation of the nanocarriers into the tumor site by the EPR effect. Passive targeting by the EPR effect requires the drug delivery system to be long-circulating in order to provide sufficient time to accumulate in the target. The usual way to keep drug carriers in the blood long enough is to wrap them with certain water-soluble polymers such as PEG [139, 140] to slow down the opsonization of the nanocarriers, including micelles, and their clearance by the reticuloendothelial system. A direct correlation between the longevity of a particulate drug carrier in the circulation and its ability to reach the target site has been observed [36, 232].

The results of the blood clearance study of various micelles clearly demonstrated their longevity: micellar formulations, such as PEG-PE-based micelles have a circulation half-life of around 2 h in mice and rats with variations dependent on the molecular weight or size of the PEG block [233]. The increase in the size of the PEG block increases the micelle circulation time in the blood probably by providing better steric protection against opsonization. Circulation times for long-circulating micelles are shorter compared to those for long-circulating PEG-coated liposomes [139] which could be due to the considerably smaller size of the micelles which facilitates its rapid extravasation. The slow dissociation of micelles into unimers and the subsequent clearance of unimers under physiological conditions could shift the micelles–unimer equilibrium towards the unimer formation [234]. Micelles are considered an efficient tumor drug-delivery system and a good candidate for passive targeting by the EPR effect because their smaller size allows easy extravasation of the tumor vascular endothelium and accumulation, compared to nanoparticles of a larger size [235, 236]. Diffusion and accumulation parameters are strongly dependent on the cut-off size of the tumor blood vessel wall which varies for different tumors [229, 235, 237]. For example, an anticancer drug adriamycin (ADR) was loaded in the polymeric micelles for delivery inside a solid tumor of murine colon adenocarcinoma [238]. Using a polymeric micelle delivery system, the tumor completely disappeared at two doses, while free ADR had only an inhibitory effect on tumor growth. Biodistribution analysis revealed that the micellar ADR accumulated at tumor sites in a highly selective manner. Since tumor vasculature permeability depends on the particular type of the tumor [229], the use of micellar drug carrier seems appropriate for tumors with vasculatures with a low cut-off size (below 200 nm). In fact, PEG-PE micelles of 15–20 nm effectively delivered a model protein to a solid tumor of Lewis lung carcinoma (LLC) which has very low cut-off size. Interestingly, even small 100 nm long-circulating liposomes did not provide an increased accumulation of liposomes-encapsulated drug in this solid tumor [239]. In case of PEG-PE-based micelles, the peak tumor accumulation time is about 5 h post-injection and the largest total tumor uptake of the injected dose within the observation period was found for micelles formed by PEG-5000-PE. This was due to the greater circulation time and low extravasation into the normal tissues compared to micelles prepared from the PEG-PE with shorter PEG chains. However, PEG-PE micelles with shorter PEG chains are better carriers for poorly soluble drugs, compared to its longer chain versions because of the greater hydrophobic-to-hydrophilic phase ratio and can be loaded with a drug more efficiently on a weight-to-weight basis.

14.7.2 Active Targeting: Functionalized Polymeric Micelles

In addition to passive targeting of micelles associated with their smaller size and long systemic circulation, the targetability of the micelles can be further enhanced by attaching targeting ligands to the micellar surface, i.e. to the water-exposed distal end of the hydrophilic blocks [4]. For example, although polymer-coated pegylated liposomes allow for passive accumulation inside a tumor by virtue of the EPR effect, the majority of these nanoparticles are cleared by the RES system, resulting in short half-lives and unwanted micelle deposition in the liver and spleen [60]. Surface-grafted nanoparticles suitable for actively targeting a disease site can accumulate more in the organ of interest. These targeting ligands fall broadly under the categories of small organic molecules, carbohydrates, antibodies, and aptamers.

Small organic molecules such as folic acid have been utilized successfully for active targeting of micelles [240–244]. Folic acid (FA) is a vital component for cell division and growth, the receptors of which are over-expressed in many malignant tissues, including ovarian, breast, and lungs due to the high rate of cell proliferation [245]. The expression levels of folate receptors in tumors have been reported to be 100–300 times higher than those observed in normal tissue [246]. Folate receptor-targeted micelles loaded with DOX showed significant accumulation of the formulation in the tumor area, resulting in marked tumor volume regression [247, 248]. DOX-loaded pH-sensitive micelles composed of poly(L-histidine)/PEG-Folate (75 wt %) and poly(L-Lactic acid)/PEG-Folate (25 wt %), the surface of which was modified with FA, were evaluated in DOX-resistant MCF-7 (MCF-7/DOX(R)) cell lines and MCF-7/DOX(R) xenografts [248]. The micelles showed more than 90% cytotoxicity in DOX-resistant MCF-7 cells. The tumor volumes in mice were significantly less than the control groups treated with free DOX or similar micelles without folate. In a study by Yoo *et al.* [247], biodegradable polymeric micelles made up of di-block copolymer of PLGA-PEG were prepared to achieve folate receptor-targeted delivery of DOX. FA was conjugated to the distal end of the PEG chain and DOX to the termini of the hydrophobic PLGA. DOX was physically entrapped to the micelles as well. The result demonstrated significant tumor volume regression in a nu/nu xenograft model and increased accumulation of DOX in tumor, compared to free DOX which suggested the proposed mixed micellar formulation could be a viable means for treating solid tumors [247].

The concept of micelle functionalization with peptide ligands has been utilized for cancer-targeted drug delivery. The most widely used peptide for cancer-cell targeting is the cRGD peptide (cyclic(Arg-Gly-Asp-D-Phe-Lys)) which targets $\alpha_v\beta_3$ integrin. Integrin, the transmembrane receptor, which plays a key role in endothelial cell survival during angiogenesis, is highly over-expressed in angiogenic vessels and therefore a potential target for anticancer therapy. The cRGD peptide binds specifically to $\alpha_v\beta_3$ integrins and is 170 times more active than the linear form (IC_{50}. 8 nM) [249]. The cRGD was conjugated to the surface of DOX-loaded poly(ε-caprolactoone)-PEG (PCL-PEG) micelles [250]. The results demonstrated that the cRGD on the surface of the micelles can selectively bind to $\alpha_v\beta_3$ integrin receptors with high affinity and it targets the tumor blood vessels studied using SLK tumor endothelial cells (derived from human Kaposi's sarcoma) that over-express the $\alpha_v\beta_3$ integrins [250, 251].

Targeting micelles to the desired organ by attachment of target tissue-specific monoclonal antibodies or proteins on the surface is another approach to active targeting of nanocarriers. These large (~150 KDa) molecules can be customized to bind specifically to the large variety of target antigens. Cancer cell-specific antinucleosome antibody 2C5 or cell-surface specific protein transferrin over-expressed on the cancer cell surface has been conjugated on the nanocarrier surface [252]. Attempts have been made to prepare micelles with attached mAbs on the surface [71, 253, 254]. Incorporation of brain-specific mAbs into haloperidol-containing micelles resulted in a drastic increase (more than by 2 orders of magnitude) in the neuroleptic action of haloperidol [71]. We have prepared PEG-PE-based immunomicelles [23] modified with the anti-nucleosome-antibody mAb, reactive towards a variety of different cancer cells [255] and antimyosin mAb 2G4, specific for infarcted areas of myocardium [256]. Antibodies were conjugated on the micelle surface by a *p*-nitrophenyl (pNP) group activated PEG-PE polymer which readily reacted with the free amine group of the antibodies. Figure 14.7 shows the synthesis scheme for preparation of pNP-PEG-PE and antibody-conjugation with pNP-PEG-PE. Both antibodies, after conjugation on micellar surface, retained their ability to interact specifically with their antigens. mAb-2C5 on the rhodamine-labeled micellar surface effectively recognized and bound effectively with various cancer cells *in vitro* including LLC (Lewis lung carcinoma), EL4 (T-lymphoma), BT20 and MCF-7 (breast adenocarcinoma) cells compared to the antibody-free micelles (Figure 14.8). *In vivo* studies with LLC tumor-bearing mice showed an increased accumulation in experimental Lewis lung

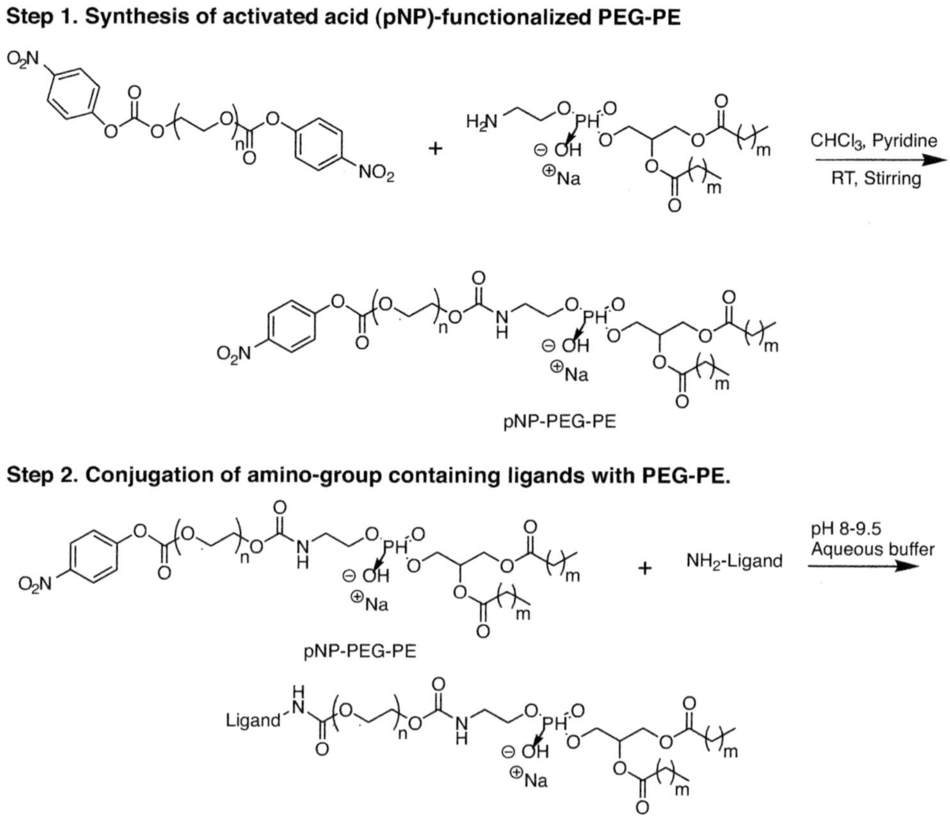

Figure 14.7 *General reaction scheme for the synthesis of ligand-conjugated PEG-PE by the reaction between an amino-group of the ligand and the pNP group of the PEG-PE.*

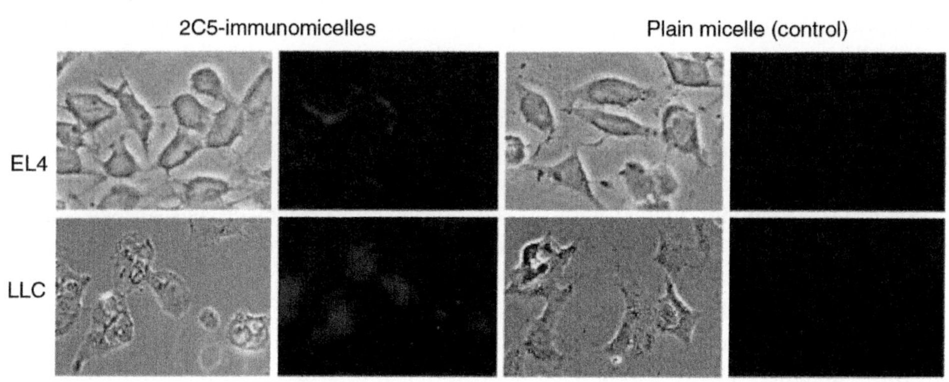

Figure 14.8 *Microscopy of the binding of Rh-labeled 2C5-grafted immunomicelles to EL4 T lymphoma cells (Top), LLC cells (bottom), (Left) Bright-field light microscopy. (Right) Fluorescent microscopy indicating the targeting potential of 2C5-immunoliposomes towards cancer cell-specific antigens. Modified from [23] with permission from National Academy of Sciences, USA.* (See colour plate section.)

carcinomas (LLCs) tumors in mice when compared with non-targeted micelles. Intravenous administration of 2C5-immunomicelles loaded with the poorly water-soluble anticancer agent Taxol into LLC-tumor-bearing mice resulted in significant tumor growth reduction compared to non-targeted micelles.

Certain carbohydrates have been utilized as targeting ligands especially in liver-specific chemotherapy [257]. The asialoglycoprotein receptor (ASGPR) is a membrane lectin receptor present in normal liver cells [258]. In addition, ASGPR is over-expressed in hepatocellular carcinoma which makes it a potential target for anticancer therapy [257]. Galactose-moiety (abbreviated as GEG) functionalized micelles made up of poly(γ-benzyl L-glutamate (PBLG)/poly(ethylene glycol) (PEG) diblock copolymer were synthesized and characterized [259]. The GEG-micelles showed specific interactions with ASGPR, over-expressed in the human hepatoma cell line (HepG2). The results demonstrated that HepG2 cells are more sensitive to PTX-loaded GEG-micelles than free PTX, whereas a murine leukemia cell line and a human hepatoma cell line (SK-Hep-01) without ASGPR are less sensitive to PTX-loaded micelles than free PTX, suggesting that the cytotoxic effect of PTX is mediated through the targeting of micelles towards ASGPR by a specific carbohydrate. Lactose molecules were also grafted onto the surface of the micelles [252]. Block copolymers with various sugar molecules on the distal end of the PEG chain in PEG/poly(D,L-lactide) (Sugar-PEG/PLA) were synthesized. Both the galactose- and lactose-installed micelles interacted specifically with ricinus communis agglutinin (RCA-1), the lectin protein. However, mannose-installed micelles showed affinity for the protein concannavalin-A.

Micelle-functionalization with nucleic acid ligands (Aptamers) for active-targeting has also been demonstrated [260, 261]. Aptamers are DNA or RNA oligonucleotides that bind specifically to proteins [262]. A high affinity aptamer for a particular protein was isolated by developing methods for the synthesis of random sequence RNA molecules and alternate cycles of ligand selection from pools of variant sequences and amplification of the bound species for the isolation of aptamers from such populations with specific ligand binding properties [262, 263]. RNA-aptamers, that bind specifically to the prostate-specific membrane antigen, a well-known prostate cancer tumor marker over-expressed in prostate acinar epithelial cells, were conjugated with the poly(Lactic acid)-*b*-PEG copolymer [261]. The micelle-aptamer bio-conjugates can efficiently target prostate LNCaP epithelial cells, which express the prostate-specific membrane antigen protein (PSMA) (with a 77-fold higher binding than controls). The targeting of the nanoparticles was reduced drastically in cell lines without the expression of the specific membrane antigens. In another study, nanoparticles made up of poly(D,L-lactic co-glycolic acid)-*b*-Poly(ethylene glycol) (PLGA-PEG) copolymer, loaded with the anticancer drug Docetaxel, and surface grafted with RNA aptamer, specific for recognizing the extracellular domain of the prostate-specific membrane antigens, were prepared and tested in *in vitro* and *in vivo* assay systems. These aptamer-functionalized nanoparticles bound to the PSMA protein on LnCaP cells resulting in significantly enhanced cellular toxicity as compared to the non-targeted ones that lacked the PSMA-aptamer. The functionalized nanoparticles, after a single intra-tumoral injection, exhibited complete tumor reduction in five of seven LNCaP xenograft nude mice with significant reduction of toxicity as measured by mean body weight loss [260]. The results hold promise for the utilization of aptamers as targeting ligands for the functionalization of drug-loaded nanocarriers in cancer chemotherapy.

14.8 Site-Specific Micellar-Drug Release Strategies

Micelles improve the pharmacokinetics of many poorly water-soluble drugs by solubilizing them in the hydrophobic corona and minimizing its degradation. However, the drug has to be released from the micelles to exhibit its function. After reaching the target site, efficient drug release from micelles becomes important for the bioavailability of the drug. Drug delivery efficiency of the micellar carrier can be improved by making micelles capable of disintegration and drug release locally at the increased temperature or at the decreased pH values typical of many pathological sites.

Many pathological processes in various tissues and organs are accompanied by some distinct physiological changes such as local temperature increase by 2–5° or a pH decrease by 1–2.5 pH units [264, 265]. Micelles made up of thermo- or pH-sensitive components possess the ability to disintegrate in target areas and release the micelle-incorporating drugs [34, 266]. Combining the EPR effect with stimuli-sensitivity could be a smart approach to the preparation of polymeric micelles. Thermo-sensitive micelles have been prepared from copolymers of poly (N-isopropylacrylamide) such as pegylated N-isopropylacrylamide/2-[mono(mono/di)lactoloxypropylmethacrylamide] [117, 267, 268]. Some related thermo-sensitive micelle compositions based on poly(N-isopropylacrylamide) have been described [267, 268]. Thermo-responsive polymeric micelles made up of poly(n-isopropylacrylamide-co-n,n-dimethylacrylamide)-*b*-poly(D,L-lactide-co-glycolide) had an increased drug (DOX) release upon temperature changes [269]. Combination of the properties of thermo-sensitivity and biodegradability in one micellar formulation has been demonstrated by using the AB block co-polymer poly(N-(2-hydroxypropyl) methacrylamide lactate) and poly(ethylene glycol) (pHPMAmDL-*b*-PEG) [270].

pH-sensitive di-block copolymers made up of PEG and *t*-butyl methacrylate, ethyl acetate or n-butyl acrylate were used for micelle formation and loaded poorly water-soluble drugs such as indomethacin, fenofibrate and progesterone [221]. The pH-sensitive micellar system demonstrated efficient loading of drugs and pH-dependent release kinetics [221]. Release of drugs from micelles increased when the pH of the release medium was increased from 1.2 to 7.2, which suggests the potential of the micelles to enhance the oral bioavailability of poorly water-soluble drugs. Micelles made up of a poly(2-ethyl-2-oxazoline)-*b*-poly(L-lactide) diblock copolymer loaded with DOX are capable of releasing the drug at the pH values typical of late endosomes of around 5.5 and secondary lysosomes at a pH around or below 5.0 [271, 272]. Block copolymer poly(ethylene oxide)-*b*-poly(allyl glycidyl ether) (PEO-PAGE) was covalently attached to DOX via a pH-sensitive hydrazone bond. The micelles formed from the conjugate released DOX faster at the low endosomal pH than at pH 7.4 [273]. Polymeric micelles made up of various pH-sensitive biodegradable block copolymers, such as mPEG-poly(D,L-Lactide)-sulfamethazine oligomer (mPEG-PDLLA-OSM) or poly(2-ethyl-2-oxazoline)-*b*-poly(L-lactide) (PEOz-PLLA) loaded with anticancer drug PTX and DOX respectively demonstrated pH-sensitive micellization and rapid release of the cargo at low pH [274, 275]. Polymeric micelles comprised of corona-forming poly(2-methacryloyloxyethyl phophorylcholine) (PMPC) and a core-forming pH-sensitive poly(2-(diisopropylamino)ethyl methacrylate) (PDPA) have been developed with demonstrated slower release kinetics for physically entrapped small hydrophobic molecules, rapid release of which could be triggered by lowering the pH to induce micellar dissociation [276]. A set of aliphatic and aromatic aldehyde-derived hydrazone-based acid-sensitive PEG-PE

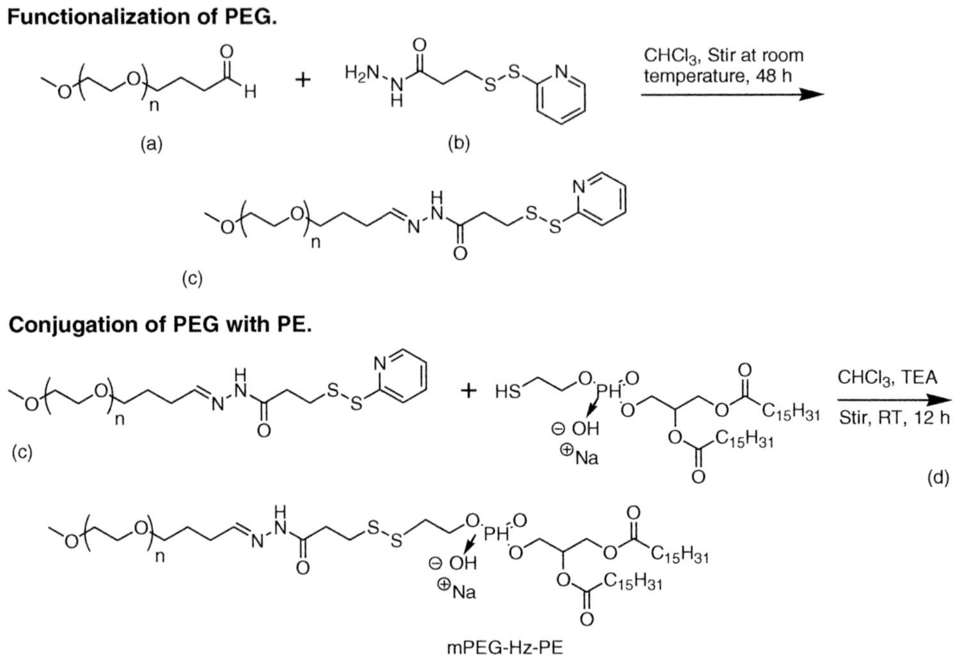

Figure 14.9 *The conjugation reaction for the formation of pH-sensitive PEG-Hz-PE. a. mPEG-butyraldehyde; b. Cross-linking agent PDPH; c. mPEG-Hz-PDP; d. PE-SH.*

conjugates was synthesized, and the micelles formed by aliphatic aldehyde-based PEG-Hz-PE conjugates were found to be highly sensitive to mildly acidic pH and reasonably stable at systemic pH (7.4) [277]. Figure 14.9 shows the representative synthesis of PEG-Hz-PE developed in our lab.

The concept of pH sensitivity was applied successfully to the preparation of a multi-functional micellar nanocarrier where a drug delivery system was built in such a way that a non-specific cell-penetrating function was hidden by a pH-sensitive pegylation using PEG-Hz-PE conjugates, which was degraded at extracellular low pH environment in a tumor, exposing cell-penetrating functionalities (Figure 14.10) [278]. A block copolymer of PEG and poly-(L-histidine) was utilized to prepare pH-sensitive micelles where histidine residues are hydrophobic at neutral pH (pKa. 6.1), but protonate at endosomal pH (5.5), resulting in a buffering effect causing micellar destabilization [110, 240, 248, 279]. The unique pH-dependent solubility of poly(β-amino ester] has been established for the preparation of pH-responsive micelles [280–282].

The penetration of drug-loaded polymeric micelles into tumor cells as well as drug release from the micelles can also be enhanced by externally applied ultrasound [56, 283]. Ultrasound is a non-invasive method which has wide applicability in diagnostic imaging. Currently, it is being adopted in various drug delivery and therapeutic applications. Ultrasound facilitates delivery of the drugs across the skin, promotes gene therapy to the targeted tissues as well as helps delivery of chemotherapeutic drugs into tumors [284]. Ultrasound triggers drug release from implanted drug depots or nanocarriers such as liposomes

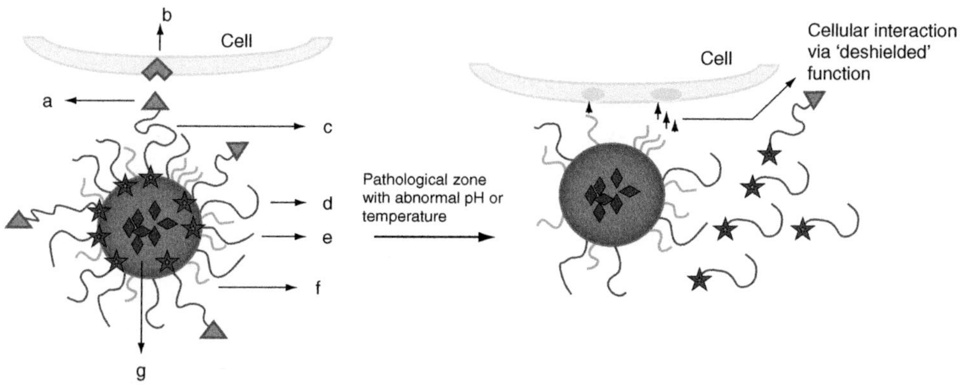

Figure 14.10 *Concept of the interaction of a multifunctional stimulus-sensitive micelle with the target cell. a. Target-specific ligand; b. Binding site of the target-specific ligand on the cell surface; c. Long PEG spacer attached to micellar core by stimuli-sensitive bond; d. Hydrophilic segment such as PEG of the micelle-forming block co-polymer; e. Protective stimulus-sensitive hydrophilic segment of the block co-polymer which could be detached under the action of local stimuli in the pathological site to expose a second function; f. Cell-penetrating function; g. Drug-loaded core of the polymeric micelle.*

PEG-coated micelles, after successful accumulation in the target site by active and passive targeting, degrade the stimulus-sensitive bond exposing the second cell-penetrating function for the direct interaction with the cell membrane. Reprinted with permission from (Salonen et al.*, 2005).* (See colour plate section.)

[283, 285]. This method is being used at different frequencies, either by itself or in combination with drugs, in the treatment of diabetes, stroke, cancer, cardiovascular diseases, infections, osteoporosis and nerve damage, etc. Ultrasound exposure causes oscillations at a frequency and amplitude determined by the ultrasound source. Ultrasound causes a local increase in the temperature of the exposed tissues due to absorption of sound waves. Tissues possessing higher ultrasound-absorption coefficients, such as bone, experience higher temperature increases compared to muscle tissues, which possess lower absorption coefficients. The increase in temperature triggers micellar degradation and subsequent drug release in appropriately designed micelles [60]. Cavitation is another phenomenon that the cells can experience involving growth and oscillation of gaseous cavities or bubbles. The cavitation exerts sheer forces and generates shock waves with high amplitude depending on the bubble size. This increases the permeability of the cell membrane and increases the uptake of gene delivery vectors.

Ultrasound-triggered drug release has been studied in several Pluronic micelles [83, 286, 287]. The effect of high-frequency ultrasound on DOX release from Pluronic micelles was observed on promyelocytic leukemia HL-60 cells, breast cancer MCF-7 cells, ovarian carcinoma, sensitive and multi-drug resistant (MDR) versions (A2780 and A2780/ADR) cells. Much higher power densities even for a short period of exposure (15–30 sec) were more effective than low frequency ultrasound (20–100 KHz) in triggering DOX release and uptake by the cells [286]. As stated before, Pluronic copolymers, in addition to forming stable micelles (the unimer-form of the polymer) inhibit the p-glycoprotein that is responsible for multi-drug resistance in many cancer cells. The presence of Pluronic unimers with DOX synergistically increased the toxicity even though uptake of DOX was lower with DOX-loaded Pluronic micelles since most of the drug was sequestered away from the cells. Ultrasound treatment destabilized the micelles and induced drug release to increase cytotoxicity [286, 287]. Treatment with ultrasound increased the DNA damage which was otherwise undetectable with DOX-Pluronic micelles. Ultrasound also increased the cell permeability [286]. *In vivo* studies with mice using DOX-loaded stable mixed micelles of Pluronics and PEG-DSPE was conducted where drug release was induced by ultrasound treatment 288]. Ultrasound improved the therapeutic efficacy of free DOX and micelle-encapsulated DOX and also increased the drug accumulation in the tumor.

Chemical-sensitive drug release is another approach in the design of intelligent polymeric micelles for drug delivery. A micellar system has been designed which has an enzyme-sensitive linkage between the hydrophilic and hydrophobic moiety. A polypeptide substrate of protein kinase A was linked to hydrophilic PEG and hydrophobic thermo-sensitive pNIPAM (poly(N-isopropylacrylamide). This polymer has a core of polypeptide and pNIPAM with a hydrophilic, PEG-corona which forms micelles at 37°C. Upon exposure to the protein kinase, the peptide becomes phosphorylated and becomes hydrophilic, which destabilizes the micelle [289]. Polymers based on hydrophobic segment poly(propylene sulfide) and hydrophilic segment PEG were found to be sensitive to oxidation [290]. The presence of hydrogen peroxide oxidizes the sulfide residues into hydrophilic sulfoxide, resulting in destabilization of the micellar system.

14.9 Intracellular Delivery of Micelles

Polymeric micelles are internalized by endocytosis rather than simple diffusion. Following cellular uptake, the micelles are trapped within the acidic endosomes and further fused with lysosomes, where degradation of micelles takes place. The efficiency of drug-loaded micelles should involve enhancement of their intracellular delivery to compensate for excessive drug degradation in lysosomes as a result of the endocytosis-mediated capture of the micelles.

PEG-PE micelles have a negative surface charge which hinders their internalization by cells. Incorporation of positively charged lipids in the PEG-PE micelles improves the uptake of drug-loaded micelles. Such positively charged micelles could more easily escape from the endosome and enter the cytoplasm. PEG-PE micelles modified with positively charged lipofectin lipid (LL), loaded with the anticancer drug PTX were prepared [291].The result demonstrated that the fluorescently labeled PEG-PE and PEG-PE/LL micelles were internalized by endocytosis. However, with PEG-PE/LL micelles, endosomes appeared to be partially disrupted and released drug-loaded micelles into the cytoplasm due to the destabilizing effect of the LL component. The conclusion of endosomal escape of the micelles was supported by the results of *in vitro* cytotoxicity studies against breast cancer cells (BT-20) and human adenocarcinoma (A2780) cells. The PTX-loaded PEG-PE/LL-micelles were significantly more cytotoxic compared to that of free PTX and PTX-loaded LL-free micelles [291].

A promising approach for the intracellular delivery of micelles is the use of cell-penetrating peptides (CPPs) [292]. Among many CPPs, in particular, the CPP derived from human immunodeficiency virus-1 transactivator protein, TAT peptide (TAT-p) has gained much attention. TAT-p-mediated cytoplasmic delivery of a wide variety of drug delivery systems including plasmid DNA [293], liposomes [294–296], micelles [297, 298] and other nanoparticles [299–301] has been reported. The mechanism of cellular internalization of TAT-p modified nanocarriers is cell-type and cargo-specific [302]. Smaller molecules, conjugated to TAT-p, seem to make their way directly into cells by energy-independent electrostatic interactions and hydrogen bonding [303]. However, large nanoparticles penetrate the cell through a macropinocytic pathway [304]. TAT-p is highly efficient in bringing nanocarriers to the cells. However, one of the major obstacles of TAT-p-assisted intracellular delivery is its lack of specificity for specific cell types which would give rise to unwanted cytotoxicity towards normal cells. To utilize the high efficiency of TAT-p-mediated intracellular delivery of drug-loaded nanocarriers without the untoward non-specific side-effects, intra-tumoral administration is the choice, so we prepared and studied PTX-loaded TAT-p-modified PEG-PE micelles [305]. TAT-p was conjugated to PEG-1000-PE and incorporated at 2.5 mole % with the micelle-forming component, PEG-750-PE. TAT-p-bearing PEG-PE micelles showed enhanced cell association compared to plain PEG-PE micelles. The strong interaction of TAT-p with the cells resulted in enhanced delivery of PTX-loaded TAT-p-modified micelles into cells and resulted in much higher cytotoxicity against MCF-7 and 4T1 cell lines. *In vivo* PTX-loaded TAT-p-functionalized micelles were also administered intra-tumoraly to avoid non-specific interaction with normal tissues and tumors were collected 48 h after drug administration. Nuclear DNA fragmentation in tumor sections undergoing apoptosis was observed using the TUNNEL assay. Significant apoptotic cell death was observed in tumors treated with PTX-loaded TAT-p-functionalized PEG-PE micelles. However, the drawback of the non-specificity of TAT-p remains to be addressed to utilize the full potential of TAT-p-assisted intracellular delivery in systemic anticancer therapy. Engineering a multifunctional nanocarrier is a step forward in the approach towards remedying the problem.

14.10 Multifunctional Micellar Nanocarriers

The newest generation of nanocarriers are being engineered so that multiple functionalities can be incorporated within a single vehicle unit for efficient delivery of therapeutics to a targeted site. Ideally, we require a nanoparticular drug delivery system to accumulate specifically in the required organ or tissue, penetrate target cells and deliver a therapeutic cargo effectively in an intracellular organelle of interest. In cancer chemotherapy, accumulation of nanocarriers including micelles can be achieved by incorporating both active and passive targeting properties. Active targeting to the tumor site may be achieved by the attachment of tumor-specific ligands such as folate, transferrin, tumor-antigen-specific mAbs, to the nanocarrier surface while passive targeting of the nanocarrier could be done by making them in a long-circulating and intracellular delivery, achieved by functionalization of the nanocarriers with CPPs. In addition, the ability to switch certain functions on and off when necessary, under the action of local stimuli, characteristic of a targeted pathological zone that has an elevated temperature or lowered pH value, is considered a potentially important feature of the multifunctional nanocarrier. However, achieving such manifold functionalities in a single unit of nanocarrier represents a challenge [278].

In our studies, we have developed a multifunctional PEG-PE-based micelle which has a tissue-specific targeting moiety such as tumor-specific anti-nucleosome mAb 2C5 or infarct-tissue specific mAb 2G4, attached to the surface [278]. These micelles were modified additionally with a secondary ligand targeting TATp moiety, attached to the surface using TATp-PEG-PE. Additionally, the main micelle component PEG-PE was made pH-sensitive by insertion of a pH-sensitive hydrazone (Hz) bond in between the PEG and PE (PEG-Hz-PE). TATp was shielded by long PEG-chains from pH-degradable PEG-2000-Hz-PE/PEG-5000-Hz-PE or by the long PEG-PE moieties used to attach mAbs on the micellar surface. It was hypothesized that micelles would be stable, can be targeted to the site of interest by specific mAbs on the surface and that TATp would remain shielded at normal systemic pH. However, at the low pH environment of a tumor, the pH-sensitive degradation of many of the PEG chains led to de-shielding of the secondary ligand, and enhanced the intracellular delivery of the payload. Figure 14.10 demonstrates the concept of interaction of multifunctional micelles with target cells. At pH 7.5 to 8, micelles demonstrated high specific binding with antibody substrates, but very limited internalization by NIH/3T3 cells. However, after brief incubation (15–20 min) at low pH (pH 5–6), these micelles lost their protective PEG shell by acidic hydrolysis of PEG-Hz-PE and were effectively internalized by cells via exposure to TAT-p moieties (Figure 14.11). Various multifunctional tumor-targeted drug delivery systems have been studied [240, 248, 279]. Bae *et al.* prepared micelles from pH-sensitive polymer PEG-pHis. Additionally, the ligand folic acid was attached to the micelle surface for target recognition [117, 118]. Enhanced cell uptake and intracellular release were observed. Gao *et al.* incorporated super-paramagnetic iron oxide nanoparticles (SPIONs) into a DOX-loaded, cRGD-targeted micelles composed of PEG-PLA copolymer. This multifunctional micelle demonstrated efficient loading of DOX, enhanced cell-recognition and uptake due to cRGD ligand and magnetic resonance imaging ultra-sensitivity [306].

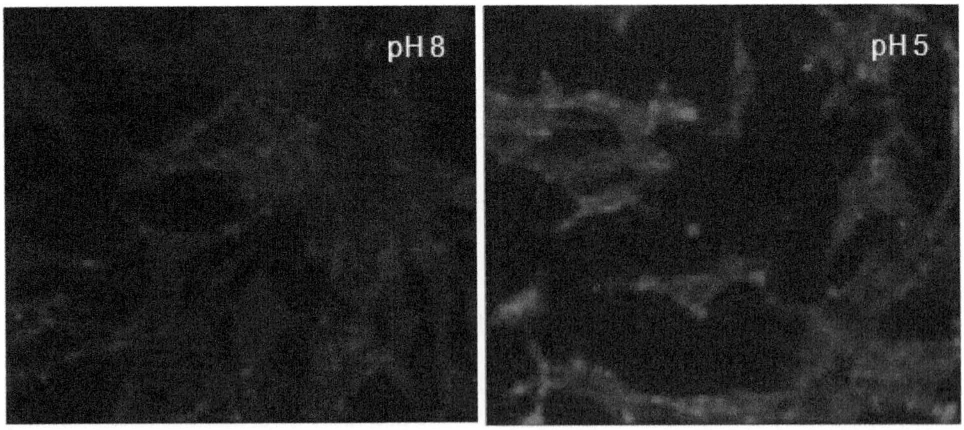

Figure 14.11 *Internalization of rhodamine-labeled TAT-p-bearing micelles by NIH-3T3 mouse fibroblast cells after pre-incubation of micelles at pH 8 and pH 5 for 30 min. Pre-incubation at low pH cleaves the protective PEG-chains, exposing the hidden cell-penetrating TAT-p function Modified from [278].* (See colour plate section.)

14.11 Conclusion

Polymeric micelles represent a promising nano-medical platform for the delivery of poorly soluble drugs, mainly for cancer chemotherapy. Their unique core-shell structure, nanosize and self-assembling property are the main advantages utilized for cancer-targeted drug delivery. The core compartment of the core–shell structure solubilizes a large variety of hydrophobic moieties including a broad variety of potent poorly water-soluble anticancer drugs, increases their bioavailability, prolongs their blood half-life, and decreases nonspecific interactions and toxicity. The shell protects the encapsulated drug from metabolic degradation and excretion hinders recognition by the RES system and makes the micelles long-circulating to allow time to accumulate at the tumor site by passive targeting via the EPR effect. The nano-size of the micelles is extremely advantageous for the treatment of tumors with a restricted vascular cut-off size. The smaller size allows preferential spontaneous accumulation of micelles at the tumor site through the compromised 'leaky' vasculature. The self-assembling nature of the nanocarrier is very amenable to changes in the micellar composition. Incorporation of the chemically active-amphiphilic polymers at a particular ratio that can attach varied target-specific ligands to the micellar surface is easy due to the self-assembling nature of the micelles. By varying the micelle composition, the size of the hydrophobic/hydrophilic blocks of the micelle-forming copolymer can be used to easily control the properties of the micelles, such as size, loading capacity, stability and longevity in the blood.

Under the paradigm of multifunctional DDS, micelles can be prepared and endowed various functionalities. The micelles can be actively targeted towards a tumor site by attachment of target-specific ligands. After successful accumulation at the target site, under the action of local stimuli such as pH and temperature, the protective shielding can be removed to expose a previously hidden secondary function such as cell-penetration functionalities. Such stimuli-responsive micelles are of great interest for the preparation of so-called smart DDS for cancer-targeted drug delivery. The biodegradability of micelles is also considered of prime importance in engineering multifunctional micelles, where controlled degradation of the hydrophilic blocks can facilitate drug liberation from the hydrophobic core. Ideally, a micellar drug delivery system selectively targets a disease site, leaving healthy tissues unaffected. However, because of the similarities of normal and cancer cells, building an efficient DDS remains challenging. Good discrimination requires a multifaceted approach to cancer therapy. In spite of tremendous advances and successes in cancer-targeted therapies, achieved with the introduction of micellar nanocarriers, many challenges still remain to achieve the anti-tumor response as selectively as Paul Ehrlich's 'magic bullet.' Deep insight into cancer biology along with polymer chemistry has brought us to the development of these multifunctional micellar systems in the hope of achieving more efficacious anticancer therapy.

References

1. V.P. Torchilin, Structure and design of polymeric surfactant-based drug delivery systems, *J Control Release*, **73**, 137–172 (2001).
2. V.P. Torchilin, PEG-based micelles as carriers of contrast agents for different imaging modalities, *Adv Drug Deliv Rev*, **54**, 235–252 (2002).

3. V.P. Torchilin, Lipid-core micelles for targeted drug delivery, *Curr Drug Deliv*, **2**, 319–327 (2005).
4. V.P. Torchilin, Micellar nanocarriers: Pharmaceutical perspectives, *Pharm Res*, **24**, 1–16 (2007).
5. P. Mukerjee, Analysis of distribution model for micellar solubilization using thermodynamics of small systems: Nonideality of solubilization of benzoic acid derivatives in nonionic surfactants, *J Pharm Sci*, **60**, 1531–1534 (1971).
6. W.C. Preston, Some correlating principles of detergent action, *J Phys Colloid Chem*, **52**, 84–97 (1948).
7. P.A. Kralchevsky, K.D. Danov, N.D. Denkov, *Handbook of Surface and Colloid Chemistry*, 2nd edn. Boca Raton, FL: CRC Press (2003).
8. A. Martin, *Physical Pharmacy*, 4th edn. Baltimore, MD: Williams and Wilkins (1993).
9. C. Tanford, Thermodynamics of micelle formation: Prediction of micelle size and size distribution, *Proc Natl Acad Sci U S A*, **71**, 1811–1815 (1974).
10. B. Lindman, H. Wennerstrom, Micelles. Amphiphile aggregation in aqueous solution, *Top Curr Chem*, **87**, 1–87 (1980).
11. C. Tanford, Ionization of cationic micelles: Effect of the detergent structure, *J Phys Chem*, **78**, 2469–2479 (1974).
12. C. Tanford, Micelle shape and size, *J Phys Chem*, **76**, 3020–3024 (1972).
13. C. Tanford, *The Hydrophobic Effect: Formation of Micelles and Biological Membranes*, New York: Wiley (1973).
14. C. Tanford, *Micellization, Solubilization, and Microemulsions*, Vol. 1, K.L. Mittal (ed.). New York: Wiley (1977).
15. D. Myers, *Surfactant Science and Technology*, 3rd edn. Hoboken, NJ: Wiley-Interscience (2006).
16. A.S. Mikhail, C. Allen, Block copolymer micelles for delivery of cancer therapy: Transport at the whole body, tissue and cellular levels, *J Control Release*, **138**, 214–223 (2009).
17. K. Kataoka, A. Harada, Y. Nagasaki, Block copolymer micelles for drug delivery: design, characterization and biological significance, *Adv Drug Deliv Rev*, **47**, 113–131 (2001).
18. R.R. Sawant, V.P. Torchilin, Multifunctionality of lipid-core micelles for drug delivery and tumour targeting, *Mol Membr Biol*, **27**, 232–246 (2010).
19. M. Jones, and J. Leroux, Polymeric micelles: A new generation of colloidal drug carriers, *Eur J Pharm Biopharm*, **48**, 101–111 (1999).
20. K. Yokogawa, E. Nakashima, J. Ishizaki, H. Maeda, T. Nagano, F. Ichimura, Relationships in the structure-tissue distribution of basic drugs in the rabbit, *Pharm Res*, **7**, 691–696 (1990).
21. C.A. Lipinski, Drug-like properties and the causes of poor solubility and poor permeability, *J Pharmacol Toxicol Methods*, **44**, 235–249 (2000).
22. C.A. Lipinski, F. Lombardo, B.W. Dominy, P.J. Feeney, Experimental and computational approaches to estimate solubility and permeability in drug discovery and development settings, *Adv Drug Deliv Rev*, **46**, 3–26 (2001).
23. V.P. Torchilin, A.N. Lukyanov, Z. Gao, B. Papahadjopoulos-Sternberg, Immunomicelles: Targeted pharmaceutical carriers for poorly soluble drugs, *Proc Natl Acad Sci U S A*, **100**, 6039–6044 (2003).

24. S.H. Yalkowsky, *Techniques of Solubilization of Drugs.* New York: Marcel Dekker (1981).
25. H.C. Ansel, L.V. Allen, N.G. Popovich, *Pharmaceutical Dosage Forms and Drug Delivery Systems*, Baltimore, MD: Lippincott, Williams, and Wilkins (1999).
26. P.P. Constantinides, K.J. Lambert, A.K. Tustian, B. Schneider, S. Lalji, W. Ma, B. Wentzel, D. Kessler, D. Worah, S.C. Quay, Formulation development and antitumor activity of a filter-sterilizable emulsion of paclitaxel, *Pharm Res*, **17**, 175–182 (2000).
27. M.J. Rosen (ed.), *Surfactants and Interfacial Phenomena*, New York: Wiley-Interscience (1989).
28. S.B. La, T. Okano, K. Kataoka, Preparation and characterization of the micelle-forming polymeric drug indomethacin-incorporated poly(ethylene oxide)-poly(beta-benzyl L-aspartate) block copolymer micelles, *J Pharm Sci*, **85**, 85–90 (1996).
29. A.V. Kabanov, E.V. Batrakova, V.Y. Alakhov, Pluronic block copolymers as novel polymer therapeutics for drug and gene delivery, *J Control Release*, **82**, 189–212 (2002).
30. M. Yokoyama, T. Sugiyama, T. Okano, Y. Sakurai, M. Naito, K. Kataoka, Analysis of micelle formation of an adriamycin-conjugated poly(ethylene glycol)-poly(aspartic acid) block copolymer by gel permeation chromatography, *Pharm Res*, **10**, 895–899 (1993).
31. V.P. Torchilin, V.S. Trubetskoy, Which polymers can make nanoparticulate drug carriers long circulating?, *Adv Drug Deliv Rev*, **16**, 141–155 (1995).
32. G.S. Kwon, Diblock copolymer nanoparticles for drug delivery, *Crit Rev Ther Drug Carrier Syst*, **15**, 481–512 (1998).
33. G.S. Kwon, K. Kataoka, Block copolymer micelles as long circulating drug vehicles, *Adv Drug Deliv Rev*, **16**, 295–309 (1995).
34. H. Maeda, J. Wu, T. Sawa, Y. Matsumura, K. Hori, Tumor vascular permeability and the EPR effect in macromolecular therapeutics: A review, *J Control Release*, **65**, 271–284 (2000).
35. T.N. Palmer, V.J. Caride, M.A. Caldecourt, J. Twickler, V. Abdullah, The mechanism of liposome accumulation in infarction, *Biochim Biophys Acta*, **797**, 363–368 (1984).
36. A.A. Gabizon, Liposome circulation timr and tumor targeting: Implications for cancer therapy, *Adv Drug Deliv Rev*, **16**, 285–294 (1995).
37. G. Gregoriadis, Targeting of drugs, *Nature*, **265**, 407–411 (1977).
38. E. Batrakova, S. Lee, S. Li, A. Venne, V. Alakhov, A. Kabanov, Fundamental relationships between the composition of pluronic block copolymers and their hypersensitization effect in MDR cancer cells, *Pharm Res*, **16**, 1373–1379 (1999).
39. I. Astafieva, A. Zhong, A. Eisenberg, Critical micellization phenomena in block polyelectrolyte solution, *Macromolecules*, **26**, 7339–7352 (1993).
40. A. Kabanov, V. Alakhov, Pluronic block copolymers in drug delivery: From micellar nanocontainers to biological response modifiers, *Crit Rev Ther Drug*, **19**, 1–73 (2002).
41. P. Alexandridis, J.F. Holzwarth, T.A. Hatton, Micellization of poly(ethylene oxide)-poly(propylene oxide)-poly(ethylene oxide) triblock copolymers in aqueous solution: Thermodynamics of copolymer association, *Macromolecules*, **27**, 2414–2425 (1994).
42. M. Yokoyama, S. Fukushima, R. Uehara, K. Okamoto, K. Kataoka, Y. Sakurai, T. Okano, Characterization of physical entrapment and chemical conjugation of

adriamycin in polymeric micelles and their design for in vivo delivery to a solid tumor, *J Control Release*, **50**, 79–92 (1998).
43. A.K. Rakshit, R.M. Palepu, Additive effects on non-ionic surfactant assemblies: A review, *Recent Dev Coll Interface Res*, **1**, 203–219 (2003).
44. R. Nagarajan, Solubilization of hydrocarbons and resulting aggregate shape transitions in aqueous solutions of Pluronic (PEO-PPO-PEO) block copolymers, *Coll Surfaces B Biointerf*, **16**, 55–72 (1999).
45. H.H. Bader, B. Ringsdorf, B. Schmidt, Water soluble polymers in medicine, *Angew Chem*, **123/124**, 457–463 (1984).
46. V.Y. Alakhov, A.V. Kabanov, Block copolymeric biotransport carriers as versatile vehicles for drug delivery, *Expert Opin Investig Drugs*, **7**, 1453–1473 (1998).
47. R.R. Sawant, V.P. Torchilin, Polymeric micelles: Polyethylene glycol-phosphatidylethanolamine (PEG-PE)-based micelles as an example, *Methods Mol Biol*, **624**, 131–149 (2010).
48. H.G. Schild, D.A.R. Tirrel, Microheterogenous solutions of amphiphilic copolymers of N-isopropylacrylamide: An investigation via fluorescence methods, *Langmuir*, **7**, 1319–1324 (1991).
49. C. Li, S. Wallace, Polymer-drug conjugates: Recent development in clinical oncology, *Adv Drug Deliv Rev*, **60**, 886–898 (2008).
50. C.C. Lee, J.A. MacKay, J.M. Fréchet, F.C. Szoka, Designing dendrimers for biological applications, *Nat Biotechnol*, **23**, 1517–1526 (2005).
51. V.P. Torchilin, Recent advances with liposomes as pharmaceutical carriers, *Nat Rev Drug Discov*, **4**, 145–160 (2005).
52. E.R. Gillies, J.M. Fréchet, Dendrimers and dendritic polymers in drug delivery, *Drug Discov Today*, **10**, 35–43 (2005).
53. A. Gopin, S. Ebner, B. Attali, D. Shabat, Enzymatic activation of second-generation dendritic prodrugs: Conjugation of self-immolative dendrimers with poly(ethylene glycol) via click chemistry, *Bioconjug Chem*, **17**, 1432–1440 (2006).
54. J. Liu, H. Lee, M. Huesca, A. Young, C. Allen, Liposome formulation of a novel hydrophobic aryl-imidazole compound for anti-cancer therapy, *Cancer Chemother Pharmacol*, **58**, 306–318 (2006).
55. A. Benahmed, M. Ranger, J.C. Leroux, Novel polymeric micelles based on the amphiphilic diblock copolymer poly(N-vinyl-2-pyrrolidone)-block-poly(D,L-lactide), *Pharm Res*, **18**, 323–328 (2001).
56. J.E. Chung, M. Yokoyama, M. Yamato, T. Aoyagi, Y. Sakurai, T. Okano, Thermo-responsive drug delivery from polymeric micelles constructed using block copolymers of poly(N-isopropylacrylamide) and poly(butylmethacrylate), *J Control Release*, **62**, 115–127 (1999).
57. W.Y. Chen, C.K. Su, C.S. Patrickios, W.R. Hertler, T.A. Hatton, Effect of block size and sequence on the micellization of ABC triblock methacrylic acid polyampholytes, *Macromolecules*, **28**, 8604–8611 (1995).
58. J.E. Chung, M. Yokoyama, T. Aoyagi, Y. Sakurai, T. Okano, Effect of molecular architecture of hydrophobically modified poly(N-isopropylacrylamide) on the formation of thermoresponsive core-shell micellar drug carriers, *J Control Release*, **53**, 119–130 (1998).

59. J.E. Chung, M. Yokoyama, T. Okano, Inner core segment design for drug delivery control of thermo-responsive polymeric micelles, *J Control Release*, **65**, 93–103 (2000).
60. D. Sutton, N. Nasongkla, E. Blanco, J. Gao, Functionalized micellar systems for cancer targeted drug delivery, *Pharm Res*, **24**, 1029–1046 (2007).
61. P. Becher, R.L. Birkmeier, The determination of hydrophile-lipophile balance (HLB) by gas-liquid chromatography (GLC), *J Am Oil, Chem Soc*, **41**, 169–172 (1964).
62. E.V. Batrakova, A.V. Kabanov, Pluronic block copolymers: Evolution of drug delivery concept from inert nanocarriers to biological response modifiers, *J Control Release*, **130**, 98–106 (2008).
63. M. Kozlov, N. Melik-Nubarov, E. Batrakova, A. Kabanov, Relationship between pluronic block copolymer structure, critical micellization concentration and partitioning coefficients of low molecular mass solutes, *Macromolecules*, **33**, 3305–3313 (2000).
64. A.V. Kabanov, V.Y. Alakhov, Pluronic block copolymers in drug delivery: From micellar nanoontainers to biological response modifiers, *Crit Rev Ther Drug*, **19**, 1–73 (2002).
65. T.P. Johnston, .C. Miller, Toxicological evaluation of poloxamer vehicles for intramuscular use, *J Parenter Sci Technol*, **39**, 83–89 (1985).
66. E.V. Batrakova, S. Li, Y. Li, V.Y. Alakhov, W.F. Elmquist, A.V. Kabanov, Distribution kinetics of a micelle-forming block copolymer Pluronic P85, *J Control Release*, **100**, 389–397 (2004).
67. M.W. Edens, *Nonionic Surfactants: Polyoxyalkylene Block Copolymers*, Vol. 60, ed. Nace V.M., New York (1996).
68. A.V. Kabanov, P. Lemieux, S. Vinogradov, V. Alakhov, Pluronic block copolymers: Novel functional molecules for gene therapy, *Adv Drug Deliv Rev*, **54**, 223–233 (2002).
69. B. Chu, Z. Zhou, *Onionic Surfactants: Polyoxy-alkylene Block Copolymers*, Vol. 60, New York: Marcel Dekker, Inc. (1996).
70. A. Kabanov, E. Batrakova, N.S. Melik-Nubarov, N.A. Fedoseev, T.Y. Dorodnich, V. Alakhov, V.P. Chekhonin, I.R. Nazarova, V.A. Kabanov, A new class of drug carriers: Micelles of poly(oxyethylene)-poly(oxypropylene) block copolymers as microcontainers for drug targeting from blood in brain, *J Control Release*, **22**, 141–158 (1992).
71. A.V. Kabanov, V.P. Chekhonin, V. Alakhov, E.V. Batrakova, A.S. Lebedev, N.S. Melik-Nubarov, S.A. Arzhakov, A.V. Levashov, G.V. Morozov, E.S. Severin, *et al.* The neuroleptic activity of haloperidol increases after its solubilization in surfactant micelles: Micelles as microcontainers for drug targeting, *FEBS Lett*, **258**, 343–345 (1989).
72. S.Y. Lin, Y. Kawashima, Kinetic studies on the stability of indomethacin in alkaline aqueous solution containing poly(oxyethylene)poly(oxypropylene) surface-active block copolymers, *Pharm Acta Helv*, **60**, 345–350 (1985).
73. E.V. Batrakova, T.Y. Dorodnych, E.Y. Klinskii, E.N. Kliushnenkova, O.B. Shemchukova, O.N. Goncharova, S.A. Arjakov, V.Y. Alakhov, A.V. Kabanov, Anthracycline antibiotics non-covalently incorporated into the block copolymer micelles: In vivo evaluation of anti-cancer activity, *Br J Cancer*, **74**, 1545–1552 (1996).

74. D. Forster, C. Washington, S.S. Davis, Toxicity of solubilized and colloidal amphotericin B formulations to human erythrocytes, *J Pharm Pharmacol*, **40**, 325–328 (1988).
75. S. Danson, D. Ferry, V. Alakhov, J. Margison, D. Kerr, D. Jowle, M. Brampton, G. Halbert, M. Ranson, Phase I dose escalation and pharmacokinetic study of pluronic polymer-bound doxorubicin (SP1049C) in patients with advanced cancer, *Br J Cancer*, **90**, 2085–2091 (2004).
76. R.L. Hunter, B. Bennett, The adjuvant activity of nonionic block polymer surfactants. III. Characterization of selected biologically active surfaces, *Scand J Immunol*, **23**, 287–300 (1986).
77. R.L. Hunter, J. McNicholl, A.A. Lal, Mechanisms of action of nonionic block copolymer adjuvants, *AIDS Res Hum Retroviruses*, **10** (Suppl 2), S95–S98 (1994).
78. R. Hunter, F. Strickland, F. Kezdy, The adjuvant activity of nonionic block polymer surfactants. I. The role of hydrophile-lipophile balance, *J Immunol*, **127**, 1244–1250 (1981).
79. A.V. Kabanov, E.V. Batrakova, S. Sriadibhatla, Z. Yang, D.L. Kelly, V.Y. Alakov, Polymer genomics: Shifting the gene and drug delivery paradigms, *J Control Release*, **101**, 259–271 (2005).
80. A. Kabanov, J. Zhu, V. Alakhov, Pluronic block copolymers for gene delivery, *Adv Genet*, **53**, 231–261 (2005).
81. A.V. Kabanov, Polymer genomics: An insight into pharmacology and toxicology of nanomedicines, *Adv Drug Deliv Rev*, **58**, 1597–1621 (2006).
82. Y. Romsicki, F.J. Sharom, The ATPase and ATP-binding functions of P-glycoprotein–modulation by interaction with defined phospholipids, *Eur J Biochem*, **256**, 170–178 (1998).
83. A.V. Kabanov, E.V. Batrakova, V.Y. Alakhov, Pluronic block copolymers for overcoming drug resistance in cancer, *Adv Drug Deliv Rev*, **54**, 759–779 (2002).
84. E.V. Batrakova, S. Li, W.F. Elmquist, D.W. Miller, V.Y. Alakhov, A.V. Kabanov, Mechanism of sensitization of MDR cancer cells by Pluronic block copolymers: Selective energy depletion, *Br J Cancer*, **85**, 1987–1997 (2001).
85. Y. Romsicki, F.J. Sharom, The membrane lipid environment modulates drug interactions with the P-glycoprotein multidrug transporter, *Biochemistry*, **38**, 6887–6896 (1999).
86. G.D. Luker, C.M. Pica, A.S. Kumar, D.F. Covey, D. Piwnica-Worms, Effects of cholesterol and enantiomeric cholesterol on P-glycoprotein localization and function in low-density membrane domains, *Biochemistry*, **39**, 7651–7661 (2000).
87. E.V. Batrakova, S. Li, V.Y. Alakhov, D.W. Miller, A.V. Kabanov, Optimal structure requirements for pluronic block copolymers in modifying P-glycoprotein drug efflux transporter activity in bovine brain microvessel endothelial cells, *J Pharmacol Exp Ther*, **304**, 845–854 (2003).
88. E.V. Batrakova, D.W. Miller, S. Li, V.Y. Alakhov, A.V. Kabanov, W.F. Elmquist, Pluronic P85 enhances the delivery of digoxin to the brain: In vitro and in vivo studies, *J Pharmacol Exp Ther*, **296**, 551–557 (2001).
89. E.V. Batrakova, S. Li, S.V. Vinogradov, V.Y. Alakhov, D.W. Miller, A.V. Kabanov, Mechanism of pluronic effect on P-glycoprotein efflux system in blood–brain barrier:

Contributions of energy depletion and membrane fluidization, *J Pharmacol Exp Ther*, **299**, 483–493 (2001).

90. G. Szakacs, J.K. Paterson, J.A. Ludwig, C. Booth-Genthe, M.M. Gottesman, Targeting multidrug resistance in cancer, *Nat Rev Drug Discov*, **5**, 219–234 (2006).
91. R. Regev, H. Katzir, D.Yeheskely-Hayon, G.D. Eytan, Modulation of P-glycoprotein-mediated multidrug resistance by acceleration of passive drug permeation across the plasma membrane, *FEBS J*, **274**, 6204–6214 (2007).
92. E.V. Batrakova, S. Li, V.Y. Alakhov, W.F. Elmquist, D.W. Miller, A.V. Kabanov, Sensitization of cells overexpressing multidrug-resistant proteins by pluronic P85, *Pharm Res*, **20**, 1581–1590 (2003).
93. T. Yamagata, H. Kusuhara, M. Morishita, K. Takayama, H. Benameur, Y. Sugiyama, Effect of excipients on breast cancer resistance protein substrate uptake activity, *J Control Release*, **124**, 1–5 (2007).
94. T. Yamagata, H. Kusuhara, M. Morishita, K. Takayama, H. Benameur, Y. Sugiyama, Improvement of the oral drug absorption of topotecan through the inhibition of intestinal xenobiotic efflux transporter, breast cancer resistance protein, by excipients, *Drug Metab Dispos*, **35**, 1142–1148 (2007).
95. T. Minko, E.V. Batrakova, S. Li, Y. Li, R.I. Pakunlu, V.Y. Alakhov, A.V. Kabanov, Pluronic block copolymers alter apoptotic signal transduction of doxorubicin in drug-resistant cancer cells, *J Control Release*, **105**, 269–278 (2005).
96. A.Venne, S. Li, R. Mandeville, A. Kabanov, V. Alakhov, Hypersensitizing effect of pluronic L61 on cytotoxic activity, transport, and subcellular distribution of doxorubicin in multiple drug-resistant cells, *Cancer Res*, **56**, 3626–3629 (1996).
97. D.W. Miller, E.V. Batrakova, T.O. Waltner, V. Alakhov, A.V. Kabanov, Interactions of pluronic block copolymers with brain microvessel endothelial cells: Evidence of two potential pathways for drug absorption, *Bioconjug Chem*, **8**, 649–657 (1997).
98. W.J. Gibbs, T.M. Hagemann, Purified poloxamer 188 for sickle cell vaso-occlusive crisis, *Ann Pharmacother*, **38**, 320–324 (2004).
99. S.L. Frey, K.Y. Lee, Temperature dependence of poloxamer insertion into and squeeze-out from lipid monolayers, *Langmuir*, **23**, 2631–2637 (2007).
100. G. Wanka, H. Hoffman, W. Ulbricht, Phase diagrams and aggregation behavior of poly(oxyethylene)-poly(oxypropylene)-poly(oxyethylene) triblock copolymers in aqueous soluions, *Macromolecules*, **27**, 4145–4159 (1994).
101. V. Alakhov, E. Klinski, S. Li, G. Pietrzynski, A.Venne, E. Batrakova, Block copolymer-based formulation of doxorubicin: From cell screen to clinical trials, *Colloids Surf A*, **16**, 113–134 (1999).
102. J. Meadows, S. Higginbottom, *Anaesthetic formulation of propofol*. U.S. Patent 20040253277. (2003).
103. S.K. Fridkin, W.R. Jarvis, Epidemiology of nosocomial fungal infections, *Clin Microbiol Rev*, **9**, 499–511 (1996).
104. W.E. Dismukes, Introduction to antifungal drugs, *Clin Infect Dis*, **30**, 653–657 (2000).
105. J. Bennett, Editorial response: Choosing amphotericin B formulations – between a rock and a hard place, *Clin Infect Dis*, **31**, 1164–1165 (2000).
106. M.S. Espuelas, P. Legrand, M.A. Campanero, M. Appel, M. Cheron, C. Gamazo, G. Barratt, J.M. Irache, Polymeric carriers for amphotericin B: In vitro activity,

toxicity and therapeutic efficacy against systemic candidiasis in neutropenic mice, *J Antimicrob Chemother*, **52**, 419–427 (2003).
107. K. Kataoka, A. Ishihara, A. Harada, H. Miyazaki, Effect of secondary structure of poly(L-Lysine) segments on the micellization of poly(ethylene glycol)-poly(L-Lysine) block copolymer partially substituted with hydrocinnamoyl group at the N^{ε}-position in aqueous milieu, *Macromolecules*, **31**, 6071–6076 (1998).
108. G. Kwon, S. Suwa, M. Yokoyama, T. Okano, Y. Sakurai, K. Kataoka, Enhanced tumor accumulation and prolong circulation times of micelle-forming poly(ethylene-oxide-aspartate)block copolymer-adriamycin conjugates, *J Control Release*, **29**, 17–23 (1994).
109. I. Oh, K. Lee, H.Y. Kwon, Y.B. Lee, S.C. Shin, C.S. Cho, C.K. Kim, Release of adriamycin from poly(gamma-benzyl-L-glutamate)/poly(ethylene oxide) nanoparticles, *Int J Pharm*, **181**, 107–115 (1999).
110. E.S. Lee, H.J. Shin, K. Na, Y.H. Bae, Poly(L-histidine)-PEG block copolymer micelles and pH-induced destabilization, *J Control Release*, **90**, 363–374 (2003).
111. M.L. Adams, G.S. Kwon, The effects of acyl chain length on the micelle properties of poly(ethylene oxide)-block-poly(N-hexyl-L-aspartamide)-acyl conjugates, *J Biomater Sci Polym Ed*, **13**, 991–1006 (2002).
112. M. Yokoyama, T. Okano, Y. Sakurai, K. Kataoka, Improved synthesis of adriamycin-conjugated poly(ethylene oxide)-poly(aspartic acid) block copolymer and formation of unimodal micellar structure with controlled amount of physically entrapped adriamycin, *J Control Release*, **32**, 269–277 (1994).
113. G.S. Kwon, Polymeric micelles for delivery of poorly water-soluble compounds, *Crit Rev Ther Drug Carrier Syst*, **20**, 357–403 (2003).
114. K. Dorn, G. Hoerpel, H. Ringsdorf (eds), *Polymeric Anti-Tumor Agents on a Molecular and Cellular Level*, New York: Marcel Dekker (1985).
115. V. Alakhov, *et al.*, Hypersensitization of multidrug resistant human ovarian carcinoma cells by pluronic P85 block copolymer, *Bioconjug Chem*, **7**(2), 209–216 (1996).
116. N. Nishiyama, K. Kataoka, Polymeric micelle drug carrier systems: PEG-PAsp(DOX) and second generation of micellar drugs, In *Advances in Experimental Medicine and Biology*, New York: Plenum, pp. 155–177 (2003).
117. Y. Bae, N. Nishiyama, S. Fukushima, H. Koyama, M. Yasuhiro, K. Kataoka, Preparation and biological characterization of polymeric micelle drug carriers with intracellular pH-triggered drug release property: Tumor permeability, controlled subcellular drug distribution, and enhanced in vivo antitumor efficacy, *Bioconjug Chem*, **16**, 122–130 (2005).
118. Y. Bae, S. Fukushima, A. Harada, K. Kataoka, Design of environment-sensitive supramolecular assemblies for intracellular drug delivery: Polymeric micelles that are responsive to intracellular pH change, *Angew Chem Int Ed Engl*, **42**, 4640–4643 (2003).
119. Y. Li, G.S. Kwon, Methotrexate esters of poly(ethylene oxide)-block-poly(2-hydroxyethyl-L-aspartamide). Part I: Effects of the level of methotrexate conjugation on the stability of micelles and on drug release, *Pharm Res*, **17**, 607–611 (2000).
120. M. Yokoyama, T. Okano, Y. Sakurai, H. Ekimoto, C. Shibazaki, K. Kataoka, Toxicity and antitumor activity against solid tumors of micelle-forming polymeric anticancer drug and its extremely long circulation in blood, *Cancer Res*, **51**, 3229–3236 (1991).

121. G.S. Kwon, M. Yokoyama, T. Okano, Y. Sakurai, K. Kataoka, Biodistribution of micelle-forming polymer-drug conjugates, *Pharm Res*, **10**, 970–974 (1993).
122. C. Allen, *et al.*, Polycaprolactone-*b*-poly(ethylene oxide) block copolymer micelles as a novel drug delivery vehicle for neurotrophic agents FK506 and L-685,818, *Bioconjug Chem*, **9**(5), 564–572 (1998).
123. K. Kataoka, T. Matsumoto, M. Yokoyama, T. Okano, Y. Sakurai, S. Fukushima, K. Okamoto, G.S. Kwon, Doxorubicin-loaded poly(ethylene glycol)-poly(beta-benzyl-L-aspartate) copolymer micelles: Their pharmaceutical characteristics and biological significance, *J Control Release*, **64**, 143–153 (2000).
124. B. Yu, T. Okano, K. Kataoka, S. Sardari, G. Kwon, In vitro dissociation and antifungal efficacy and toxicity for amphotericin B-loaded poly(ethylene oxide)-block-poly(β-benzyl-L-aspartate) micelles, *J Control Release*, **56**, 285–291 (1998).
125. P. Opanasopit, M. Yokoyama, M. Watanabe, K. Kawano, Y. Maitani, T. Okano, Block copolymer design for camptothecin incorporation into polymeric micelles for passive tumor targeting, *Pharm Res*, **21**, 2001–2008 (2004).
126. Y. Jeong, J. Cheon, S. Kim, J. Nath, Y. Lee, Y. Sung, Clonazepam release from core-shell type nanoparticles in vitro, *J Control Release*, **51**, 169–178 (1998).
127. M.L. Adams, G.S. Kwon, Relative aggregation state and hemolytic activity of amphotericin B encapsulated by poly(ethylene oxide)-block-poly(N-hexyl-L-aspartamide)-acyl conjugate micelles: Effects of acyl chain length, *J Control Release*, **87**, 23–32 (2003).
128. A. Lavasanifar, J. Samuel, G.S. Kwon, The effect of fatty acid substitution on the in vitro release of amphotericin B from micelles composed of poly(ethylene oxide)-block-poly(N-hexyl stearate-L-aspartamide), *J Control Release*, **79**, 165–172 (2002).
129. J. Bolard, P. Legrand, F. Heitz, B. Cybulska, One-sided action of amphotericin B on cholesterol-containing membranes is determined by its self-association in the medium, *Biochemistry*, **30**, 5707–5715 (1991).
130. M.L. Adams, D.R. Andes, G.S. Kwon, Amphotericin B encapsulated in micelles based on poly(ethylene oxide)-block-poly(L-amino acid) derivatives exerts reduced in vitro hemolysis but maintains potent in vivo antifungal activity, *Biomacromolecules*, **43**, 750–757 (2003).
131. T. Nakanishi, S. Fukushima, K. Okamoto, M. Suzuki, Y. Matsumura, M. Yokoyama, T. Okano, Y. Sakurai, K. Kataoka, Development of the polymer micelle carrier system for doxorubicin, *J Control Release*, **74**, 295–302 (2001).
132. K.S. Anseth, A.T. Metters, S.J. Bryant, P.J. Martens, J.H. Elisseeff, C.N. Bowman, In situ forming degradable networks and their application in tissue engineering and drug delivery, *J Control Release*, **78**, 199–209 (2002).
133. X.S. Wu, N. Wang, Synthesis, characterization, biodegradation, and drug delivery application of biodegradable lactic/glycolic acid polymers. Part II: biodegradation, *J Biomater Sci Polym Ed*, **12**, 21–34 (2001).
134. H. Burt, X. Zhang, P. Toleikin, L. Embree, W. Hunter, Development of copolymers of poly(D,L-lactide) and methoxypolyethylene glycol as micellar carriers of paclitaxel, *Coll Surfaces B Biointerf*, **16**, 161–171 (1999).
135. X. Zhang, H.M. Burt, D. Von Hoff, D. Dexter, G. Mangold, D. Degen, A.M. Oktaba, W.L. Hunter, An investigation of the antitumour activity and biodistribution of polymeric micellar paclitaxel, *Cancer Chemother Pharmacol*, **40**, 81–86 (1997).

136. S.C. Kim, D.W. Kim, Y.H. Shim, J.S. Bang, H.S. Oh, S. Wan Kim, M.H. Seo, In vivo evaluation of polymeric micellar paclitaxel formulation: toxicity and efficacy, *J Control Release*, **72**, 191–202 (2001).
137. K. Huh, Y. Cho, K. Park, PLGA-PEG block copolymers for drug formulations, *Drug Del Tech*, **3**, 44–49 (2003).
138. H.S. Yoo, T.G. Park, Biodegradable polymeric micelles composed of doxorubicin conjugated PLGA-PEG block copolymer, *J Control Release*, **70**, 63–70 (2001).
139. A.L. Klibanov, K. Maruyama, V.P. Torchilin, L. Huang, Amphipathic polyethyleneglycols effectively prolong the circulation time of liposomes, *FEBS Lett*, **268**, 235–237 (1990).
140. V.S. Trubetskoy, V. Torchlin, Use of polyoxyethylene-lipid conjugates as long-circulating carriers for delivery of therapeutic and diagnostic agents, *Adv Drug Deliv Rev*, **16**, 311–320 (1995).
141. V.P. Torchilin, M.I. Shtilman, V.S. Trubetskoy, K. Whiteman, A.M. Milstein, Amphiphilic vinyl polymers effectively prolong liposome circulation time in vivo, *Biochim Biophys Acta*, **1195**, 181–184 (1994).
142. V.P. Torchilin, T.S. Levchenko, K.R. Whiteman, A.A. Yaroslavov, A.M. Tsatsakis, A.K. Rizos, E.V. Michailova, M.I. Shtilman, Amphiphilic poly-N-vinylpyrrolidones: synthesis, properties and liposome surface modification, *Biomaterials*, **22**, 3035–3044 (2001).
143. M. Li, A. Chrastina, T. Levchenki, V.P. Torchilin, Micelles from poly(ethylene glycol)-phosphatidyl ethanolamine conjugates (PEG-PE) as pharmaceutical nanocarriers for poorly soluble drug camptothecin, *J Biomed Nanotechnol*, **1**, 190–195 (2005).
144. Z. Gao, A.N. Lukyanov, A. Singhai, V.P. Torchilin, Diacyllipid-polymer micelles as nanocarriers for poorly soluble anticancer drugs, *Nanoletters*, **2**, 979–982 (2002).
145. J. Wang, D.A. Mongayt, A.N. Lukyanov, T.S. Levchenko, V.P. Torchilin, Preparation and in vitro synergistic anticancer effect of vitamin K3 and 1,8-diazabicyclo[5,4,0]undec-7-ene in poly(ethylene glycol)-diacyllipid micelles, *Int J Pharm*, **272**, 129–135 (2004).
146. V. Weissig, K.R. Whiteman, V.P. Torchilin, Accumulation of protein-loaded long-circulating micelles and liposomes in subcutaneous Lewis lung carcinoma in mice, *Pharm Res*, **15**, 1552–1556 (1998).
147. V. Weissig, C. Lizano, V.P. Torchilin, Micellar delivery system for dequalinium-a lipophilic cationic drug with anticarcinoma activity, *J Liposome Res*, **8**, 391–400 (1998).
148. J. Senior, C. Delgado, D. Fisher, C. Tilcock, G. Gregoriadis, Influence of surface hydrophilicity of liposomes on their interaction with plasma protein and clearance from the circulation: Studies with poly(ethylene glycol)-coated vesicles, *Biochim Biophys Acta*, **1062**, 77–82 (1991).
149. G. Blume, G. Cevc, Molecular mechanism of the lipid vesicle longevity in vivo, *Biochim Biophys Acta*, **1146**, 157–168 (1993).
150. V.P. Torchilin, V.G. Omelyanenko, M.I. Papisov, A.A. Bogdanov, Jr., V.S. Trubetskoy, J.N. Herron, C.A. Gentry, Poly(ethylene glycol) on the liposome surface: on the mechanism of polymer-coated liposome longevity, *Biochim Biophys Acta*, **1195**, 11–20 (1994).

151. F.K. Bedu-Addo, P. Tang, Y. Xu, L. Huang, Interaction of polyethyleneglycol-phospholipid conjugates with cholesterol-phosphatidylcholine mixtures: Sterically stabilized liposome formulations, *Pharm Res*, **13**, 718–724 (1996).
152. K. Edwards, M. Johnsson, G. Karlsson, M. Silvander, Effect of polyethyleneglycol-phospholipids on aggregate structure in preparations of small unilamellar liposomes, *Biophys J*, **73**, 258–266 (1997).
153. A.A. Gabizon, Pegylated liposomal doxorubicin: Metamorphosis of an old drug into a new form of chemotherapy, *Cancer Invest*, **19**, 424–436 (2001).
154. G.B. Jiang, *et al.*, Novel polymer micelles prepared from chitosan grafted hydrophobic palmitoyl groups for drug delivery, *Mol Pharm*, **3**(2), 152–60 (2006).
155. Z. Gao, A.N. Lukyanov, A.R. Chakilam, V.P. Torchilin, PEG-PE/phosphatidylcholine mixed immunomicelles specifically deliver encapsulated taxol to tumor cells of different origin and promote their efficient killing, *J Drug Target*, **11**, 87–92 (2003).
156. H. Arimura, Y. Ohya, T. Ouchi, Formation of core-shell type biodegradable polymeric micelles from amphiphilic poly(aspartic acid)-block-polylactide diblock copolymer, *Biomacromolecules*, **6**, 720–725 (2005).
157. R.D. Dabholkar, R.M. Sawant, D.A. Mongayt, P.V. Devarajan, V.P. Torchilin, Polyethylene glycol-phosphatidylethanolamine conjugate (PEG-PE)-based mixed micelles: Some properties, loading with paclitaxel, and modulation of P-glycoprotein-mediated efflux, *Int J Pharm*, **315**, 148–157 (2006).
158. L. Li, Y.B. Tan, Preparation and properties of mixed micelles made of Pluronic polymer and PEG-PE, *J Colloid Interface Sci*, **317**, 326–331 (2008).
159. M.K. Lee, L. Choi, M.H. Kim, C.K. Kim, Pharmacokinetics and organ distribution of cyclosporin A incorporated in liposomes and mixed micelles, *Int J Pharm*, **191**, 87–93 (1999).
160. A. Martin, *Physical Pharmacy*, 4th edn. Baltimore, MD: Williams and Wilkins (1993).
161. S. Katayose, K. Kataoka, Water-soluble polyion complex associates of DNA and poly(ethylene glycol)-poly(L-lysine) block copolymer, *Bioconjug Chem*, **8**(5), 702–707 (1997).
162. B. Chu, Laser light scattering. In *Basic Principles and Practice*, 2nd edn. New York: Academic Press (1991).
163. A. Jada, B. Siffert, G. Reiss, Adsorption at the solid-solution interface and micelle formation in water of a PEO-PS-PEO triblock copolymer, *Colloids Surf A*, **75**, 203–209 (1993).
164. F.M. WinniK, S.T.A. Regismond, Fluorescence methods in the study of the interactions of surfactants with polymers, *Colloids Surf A*, **118**, 1–39 (1996).
165. A. Patist, S.S. Bhagwat, K.W. Penfield, P. Aikens, D.O. Shah, On the measurement of critical micelle concentrations of pure and technical-grade nonionic surfactants, *J Surfactants Deterg*, **3**, 53–58 (2000).
166. N.J. Turro, C.J. Chung, Photoluminescent probes for water soluble polymers: Pressure and temperature effect on a polyol surfactant, *Macromolecules*, **17**, 2123–2126 (1984).
167. P.C. Heimenz, R. Rajagopalan, *Principles of Colloid and Surface Chemistry*, 3rd edn. New York: Marcel Dekker (1997).
168. L. Zhang, A. Eisenberg, Thermodynamic vs kinetic aspects in the formation and morphological transitions of crew-cut aggregates produced by self-assembly

of polystyrene-*b*-poly(acrylic acid) block copolymers in dilute solution, *Macromolecules*, **32**, 2239–2249 (1999).
169. C. Allen, D. Maysinger, and A. Eisenberg, Nano-engineering block copolymer aggregates for drug delivery, *Colloids Surf B*, **16**, 3–27 (1999).
170. G. Reiss, Micellization of block copolymers, *Prog Polym Sci*, **28**, 1107–1170 (2003).
171. M. Goldraich, Y. Talmon, Direct-imaging cryo-transmission electron microscopy in the study of colloids and polymer solution, in *Amphiphilic Block Copolymers*, Alexandridis, P., and Lindman, B. (eds), pp. 253–280, Amsterdam: Elsevier (2000).
172. E. Allenmann, J. Leroux, R. Gurny, Biodegradable nanoparticles of poly(lactic acid) and poly(lactic-co-glycoliv acid) for parental administration. In *Pharmaceutical Dosage Forms: Disperse Systems*. New York: Marcel Dekker, pp. 163–193 (1998).
173. S.G. Oh, D.O. Shah, Micellar lifetime: Its relevance to various technological processes, *J Dispersion Sci Technol*, **15**, 297–316 (1994).
174. Y.H. Bae, H. Yin, Stability issues of polymeric micelles, *J Control Release*, **131**, 2–4 (2008).
175. D.O. Shah, J.H. Schulman, Influence of calcium, cholesterol, and unsaturation on lecithin monolayers, *J Lipid Res*, **8**, 215–226 (1967).
176. M. Yokoyama, Block copolymers as drug carriers, *Crit Rev Ther Drug Carrier Syst*, **9**, 213–248 (1992).
177. Z. Gao, A. Eisenberg, A model of micellization for block copolymers in solutions, *Macromolecules*, **26**, 7353–7360 (1993).
178. F.M. Winnik, A.R. Davidson, G.K. Hamer, H. Kitano, Amphiphilic poly(N-isopropylacrylamide) prepared by using a lipophilic radical initiator: Synthesis and solution properties in water, *Macromolecules*, **25**, 1876–1880 (1992).
179. S.R. Croy, G.S. Kwon, Polymeric micelles for drug delivery, *Curr Pharm Des*, **12**, 4669–4684 (2006).
180. E.A.G. Aniansson, S.N. Wall, Kinetics of step-wise micelle association, *J of Phys Chem*, **78**, 1024–1030 (1974).
181. E.A.G. Aniansson, S.N. Wall, Kinetics of step-wise micelle association. Correction and improvement, *J Phys Chem*, **79**, 857–858 (1975).
182. E.A.G. Aniansson, S.N. Wall, M. Almgren, H. Hoffman, I. Kielmann, W. Ulbricht, R. Zana, J. Lang, C. Tondre, Theory of the kinetics of micellar equilibria and quantitative interpretation of chemical relaxation studies of micellar solutions of ionic surfactants, *J Phys Chem*, **80**, 905–922 (1976).
183. D. Attwood, A.T. Florence, *In Surfactant Systems*. London: Chapman and Hall, Ltd. (1983).
184. R. Leung, D.O. Shah, *J Colloid Interface Sci*, **113** (1986).
185. A. Patist, V. Chhabra, R. Pagidipati, R. Shah, D.O. Shah, Effect of chain length compatibility on micellar stability in sodium dodecyl sulfate/alkyltrimethylammonium bromide solutions, *Langmuir*, **13**, 432–434 (1997).
186. G.S. Kwon, M. Naito, M. Yokoyama, T. Okano, Y. Sakurai, K. Kataoka, Physical entrapment of adriamycin in AB block copolymer micelles, *Pharm Res*, **12**, 192–195 (1995).
187. G. Gaucher, M.H. Dufresne, V.P. Sant, N. Kang, D. Maysinger, J.C. Leroux, Block copolymer micelles: Preparation, characterization and application in drug delivery, *J Control Release*, **109**, 169–188 (2005).

188. S. Kim, Y. Shi, J.Y. Kim, K. Park, and J.X. Cheng, Overcoming the barriers in micellar drug delivery: Loading efficiency, in vivo stability, and micelle-cell interaction, *Expert Opin Drug Deliv*, **7**, 49–62 (2010).
189. J.H. Hildebrand, A critique of the theory of solubility of non-electrolytes, *Chem Rev*, **44**, 37–45 (1949).
190. F.P. Jones, *Principles of Polymer Chemistry*, Ithaca, NY: Cornell University Press (1953).
191. R.F.M. Liu, G.S. Kwon, *Micellelization and Drug Solubility Enhancement Part 2*. Boca Raton, FL: CRS Press (2008).
192. K. Letchford, R. Liggins, H. Burt, Solubilization of hydrophobic drugs by methoxy poly(ethylene glycol)-block-polycaprolactone diblock copolymer micelles: theoretical and experimental data and correlations, *J Pharm Sci*, **97**, 1179–1190 (2008).
193. T.L.S. Ooya, K.M. Huh, K. Park, *Hydrotropic Nanocarriers for Poorly Soluble Drugs*. Amsterdam: Springer (2006).
194. R.E. Coffman, D.O. Kildsig, Hydrotropic solubilization–mechanistic studies, *Pharm Res*, **13**, 1460–1463 (1996).
195. J. Lee, S.C. Lee, G. Acharya, C.J. Chang, K. Park, Hydrotropic solubilization of paclitaxel: Analysis of chemical structures for hydrotropic property, *Pharm Res*, **20**, 1022–1030 (2003).
196. K.M. Huh, S.C. Lee, Y.W. Cho, J. Lee, J.H. Jeong, K. Park, Hydrotropic polymer micelle system for delivery of paclitaxel, *J Control Release*, **101**, 59–68 (2005).
197. C. Allen, J. Han, Y. Yu, D. Maysinger, A. Eisenberg, Polycaprolactone-*b*-poly(ethylene oxide) copolymer micelles as a delivery vehicle for dihydrotestosterone, *J Control Release*, **63**, 275–286 (2000).
198. S.Y. Lin, Y. Kawashima, The influence of three poly(oxyethylene)poly(oxypropylene) surface-active block copolymers on the solubility behavior of indomethacin, *Pharm Acta Helv*, **60**, 339–344 (1985).
199. S.Y. Lin, Y. Kawashima, Pluronic surfactants affecting diazepam solubility, compatibility, and adsorption from i.v. admixture solutions, *J Parenter Sci Technol*, **41**, 83–87 (1987).
200. M. Yokoyama, M. Miyauchi, N. Yamada, T. Okano, Y. Sakurai, K. Kataoka, and S. Inoue, Characterization and anticancer activity of the micelle-forming polymeric anticancer drug adriamycin-conjugated poly(ethylene glycol)-poly(aspartic acid) block copolymer, *Cancer Res*, **50**, 1693–1700 (1990).
201. S. Katayose, K. Kataoka, Remarkable increase in nuclease resistance of plasmid DNA through supramolecular assembly with poly(ethylene glycol)-poly(L-lysine) block copolymer, *J Pharm Sci*, **87**(2), 160–163 (1998).
202. M. Yokoyama, A. Satoh, Y. Sakurai, T. Okano, Y. Matsumura, T. Kakizoe, K. Kataoka, Incorporation of water-insoluble anticancer drug into polymeric micelles and control of their particle size, *J Control Release*, **55**, 219–229 (1998).
203. A.V. Kabanov, I.R. Nazarova, I.V. Astafieva, E.V. Batrakova, V.Y. Alakhov, A.A. Yaroslavov, V.A. Kabanov, Micelle Formation and Solubilization of Fluorescent-Probes in Poly(Oxyethylene-B-Oxypropylene-B-Oxyethylene) Solutions, *Macromolecules*, **28**, 2303–2314 (1995).

204. A.V. Kabanov, S.V. Vinogradov, Y.G. Suzdaltseva, V. Alakhov, Water-soluble block polycations as carriers for oligonucleotide delivery, *Bioconjug Chem*, **6**, 639–643 (1995).
205. K. Kataoka, G.S. Kwon, M. Yokoyama, T. Okano, and Y. Sakurai, Block-copolymer micelles as vehicles for drug delivery, *Journal of Controlled Release*, **24**, 119–132 (1993).
206. G.M.N. Kwon, M. Yokoyama, T. Okano, Y. Sakurai, K. Kataoka. Micelles based on AB block copolymers of poly(ethylene oxide) and poly(beta.-benzyl L-aspartate), *Langmuir*, **9**, 945–949 (1993).
207. G.S. Kwon, T. Okano, Polymeric micelles as new drug carriers, *Advanced Drug Delivery Reviews*, **21**, 107–116 (1996).
208. V.P. Torchilin. Polyethylene glycol based micelles as carriers of therapeutic and diagnostic agents, *STP Pharma, Sco* **6**, 79–86 (1996).
209. P. Xu, E.A. Van Kirk, S. Li, W.J. Murdoch, J. Ren, M.D. Hussain, M. Radosz, Y. Shen, Highly stable core-surface-crosslinked nanoparticles as cisplatin carriers for cancer chemotherapy, *Colloids Surf B Biointerfaces*, **48**, 50–57 (2006).
210. A.A. Exner, T.M. Krupka, K. Scherrer, J.M. Teets, Enhancement of carboplatin toxicity by Pluronic block copolymers, *J Control Release*, **106**, 188–197 (2005).
211. H. Cabral, N. Nishiyama, S. Okazaki, H. Koyama, K. Kataoka, Preparation and biological properties of dichloro(1,2-diaminocyclohexane)platinum(II) (DACHPt)-loaded polymeric micelles, *J Control Release*, **101**, 223–232 (2005).
212. H.M. Aliabadi, D.R. Brocks, A. Lavasanifar, Polymeric micelles for the solubilization and delivery of cyclosporine A: Pharmacokinetics and biodistribution, *Biomaterials*, **26**, 7251–7259 (2005).
213. H.M. Aliabadi, A. Mahmud, A.D. Sharifabadi, A. Lavasanifar, Micelles of methoxy poly(ethylene oxide)-*b*-poly(epsilon-caprolactone) as vehicles for the solubilization and controlled delivery of cyclosporine A, *J Control Release*, **104**, 301–311 (2005).
214. S.V. Vinogradov, E.V. Batrakova, S. Li, A.V. Kabanov, Mixed polymer micelles of amphiphilic and cationic copolymers for delivery of antisense oligonucleotides, *J Drug Target*, **12**, 517–526 (2004).
215. A. Krishnadas, I. Rubinstein, H. Onyuksel, Sterically stabilized phospholipid mixed micelles: In vitro evaluation as a novel carrier for water-insoluble drugs, *Pharm Res*, **20**, 297–302 (2003).
216. S.V. Balasubramanian, R.M. Straubinger, Taxol-lipid interactions: Taxol-dependent effects on the physical properties of model membranes, *Biochemistry*, **33**, 8941–8947 (1994).
217. B.G. Yu, T. Okano, K. Kataoka, G. Kwon, Polymeric micelles for drug delivery: Solubilization and haemolytic activity of amphotericin B, *J Control Release*, **53**, 131–136 (1998).
218. S.A. Hagan, A.G.A. Coombes, M.C. Garnett, S.E. Dunn, M.C. Davies, L. Illum, S.S. Davis, S.E. Harding, S. Purkiss, P.R. Gellert. Polylactide–poly(ethylene glycol) copolymers as drug delivery systems. 1. Characterization of water dispersible micelle-forming systems, *Langmuir*, **12**, 2153–2161 (1996).
219. G. Kwon, M. Naito, M. Yokoyama, T. Okano, Y. Sakurai, K. Kataoka, Block copolymer micelles for drug delivery: Loading and release of doxorubicin, *Journal of Controlled Release*, **48**, 195–201 (1997).

220. Č.K. Karel Ulbrich, Z. Tuzar, J. Kopeček. Solution properties of drug carriers based on poly[N-(2-hydroxypropyl)methacrylamide] containing biodegradable bonds, *Makromol Chem*, **188**, 1261–1272 (1987).
221. V.P. Sant, D. Smith, J.C. Leroux, Novel pH-sensitive supramolecular assemblies for oral delivery of poorly water soluble drugs: Preparation and characterization, *J Control Release*, **97**, 301–312 (2004).
222. E. Fournier, M.H. Dufresne, D.C. Smith, M. Ranger, J.C. Leroux, A novel one-step drug-loading procedure for water-soluble amphiphilic nanocarriers, *Pharm Res*, **21**, 962–968 (2004).
223. S.D. Li, L. Huang, Pharmacokinetics and biodistribution of nanoparticles, *Mol Pharm*, **5**, 496–504 (2008).
224. L. Brannon-Peppas, J.O. Blanchette, Nanoparticle and targeted systems for cancer therapy, *Adv Drug Deliv Rev*, **56**, 1649–1659 (2004).
225. P. Opanasopit, M. Nishikawa, M. Hashida, Factors affecting drug and gene delivery: Effects of interaction with blood components, *Crit Rev Ther Drug Carrier Syst*, **19**, 191–233 (2002).
226. S.I. Jeona, J.H.L.J.D. Andradea, P.G. De Gennes, Protein–surface interactions in the presence of polyethylene oxide: I. Simplified theory, *Journal of Colloid and Interface Science*, **142**, 149–158 (1991).
227. Y. Yamamoto, K. Yasugi, A. Harada, Y. Nagasaki, K. Kataoka, Temperature-related change in the properties relevant to drug delivery of poly(ethylene glycol)-poly(D,L-lactide) block copolymer micelles in aqueous milieu, *J Control Release*, **82**, 359–371 (2002).
228. Y. Yamamoto, Y. Nagasaki, Y. Kato, Y. Sugiyama, K. Kataoka, Long-circulating poly(ethylene glycol)-poly(D,L-lactide) block copolymer micelles with modulated surface charge, *J Control Release*, **77**, 27–38 (2001).
229. S.K. Hobbs, W.L. Monsky, F. Yuan, W.G. Roberts, L. Griffith, V.P. Torchilin, R.K. Jain, Regulation of transport pathways in tumor vessels: Role of tumor type and microenvironment, *Proc Natl Acad Sci U S A*, **95**, 4607–4612 (1998).
230. R.K. Jain, Transport of molecules, particles, and cells in solid tumors, *Annu Rev Biomed Eng*, **1**, 241–263 (1999).
231. V.P. Torchilin, A.L. Klibanov, L. Huang, S. O'Donnell, N.D. Nossiff, B.A. Khaw, Targeted accumulation of polyethylene glycol-coated immunoliposomes in infarcted rabbit myocardium, *FASEB J*, **6**, 2716–2719 (1992).
232. H. Maeda, T. Sawa, T. Konno, Mechanism of tumor-targeted delivery of macromolecular drugs, including the EPR effect in solid tumor and clinical overview of the prototype polymeric drug SMANCS, *J Control Release*, **74**, 47–61 (2001).
233. A.N. Lukyanov, Z. Gao, L. Mazzola, V.P. Torchilin, Polyethylene glycol-diacyllipid micelles demonstrate increased acculumation in subcutaneous tumors in mice, *Pharm Res*, **19**, 1424–1429 (2002).
234. V.S. Trubetskoy, G.S. Gazelle, G.L. Wolf, V.P. Torchilin, Block-copolymer of polyethylene glycol and polylysine as a carrier of organic iodine: Design of long-circulating particulate contrast medium for X-ray computed tomography, *J Drug Target*, **4**, 381–388 (1997).
235. F. Yuan, M. Dellian, D. Fukumura, M. Leunig, D.A. Berk, V.P. Torchilin, R.K. Jain, Vascular permeability in a human tumor xenograft: Molecular size dependence and cutoff size, *Cancer Res*, **55**, 3752–3756 (1995).

236. F. Yuan, M. Leunig, S.K. Huang, D.A. Berk, D. Papahadjopoulos, R.K. Jain, Microvascular permeability and interstitial penetration of sterically stabilized (stealth) liposomes in a human tumor xenograft, *Cancer Res*, **54**, 3352–3356 (1994).
237. W.L. Monsky, D. Fukumura, T. Gohongi, M. Ancukiewcz, H.A. Weich, V.P. Torchilin, F. Yuan, R.K. Jain, Augmentation of transvascular transport of macromolecules and nanoparticles in tumors using vascular endothelial growth factor, *Cancer Res*, **59**, 4129–4135 (1999).
238. M. Yokoyama, T. Okano, Y. Sakurai, S. Fukushima, K. Okamoto, K. Kataoka, Selective delivery of adriamycin to a solid tumor using a polymeric micelle carrier system, *J Drug Target*, **7**, 171–186 (1999).
239. M.J. Parr, D. Masin, P.R. Cullis, M.B. Bally, Accumulation of liposomal lipid and encapsulated doxorubicin in murine Lewis lung carcinoma: The lack of beneficial effects by coating liposomes with poly(ethylene glycol), *J Pharmacol Exp Ther*, **280**, 1319–1327 (1997).
240. E.S. Lee, K. Na, Y.H. Bae, Polymeric micelle for tumor pH and folate-mediated targeting, *J Control Release*, **91**, 103–113 (2003).
241. P.R. Dash, M.L. Read, K.D. Fisher, K.A. Howard, M. Wolfert, D. Oupicky, V. Subr, J. Strohalm, K. Ulbrich, L.W. Seymour, Decreased binding to proteins and cells of polymeric gene delivery vectors surface modified with a multivalent hydrophilic polymer and retargeting through attachment of transferrin, *J Biol Chem*, **275**, 3793–3802 (2000).
242. C.P. Leamon, D. Weigl, R.W. Hendren, Folate copolymer-mediated transfection of cultured cells, *Bioconjug Chem*, **10**, 947–957 (1999).
243. C.P. Leamon, P.S. Low, Folate-mediated targeting: From diagnostics to drug and gene delivery, *Drug Discov Today*, **6**, 44–51 (2001).
244. J.H. Jeong, S.H. Kim, S.W. Kim, T.G. Park, In vivo tumor targeting of ODN-PEG-folic acid/PEI polyelectrolyte complex micelles, *J Biomater Sci Polym Ed*, **16**, 1409–1419 (2005).
245. S.D. Weitman, A.G. Weinberg, L.R. Coney, V.R. Zurawski, D.S. Jennings, B.A. Kamen, Cellular localization of the folate receptor: Potential role in drug toxicity and folate homeostasis, *Cancer Res*, **52**, 6708–6711 (1992).
246. J.F. Ross, P.K. Chaudhuri, M. Ratnam, Differential regulation of folate receptor isoforms in normal and malignant tissues in vivo and in established cell lines: Physiologic and clinical implications, *Cancer*, **73**, 2432–2443 (1994).
247. H.S. Yoo, T.G. Park, Folate receptor targeted biodegradable polymeric doxorubicin micelles, *J Control Release*, **96**, 273–283 (2004).
248. E.S. Lee, K. Na, Y.H. Bae, Doxorubicin loaded pH-sensitive polymeric micelles for reversal of resistant MCF-7 tumor, *J Control Release*, **103**, 405–418 (2005).
249. S.L. Goodman, G. Holzemann, G.A. Sulyok, H. Kessler, Nanomolar small molecule inhibitors for alphav(beta)6, alphav(beta)5, and alphav(beta)3 integrins, *J Med Chem*, **45**, 1045–1051 (2002).
250. N. Nasongkla, X. Shuai, H. Ai, B.D. Weinberg, J. Pink, D.A. Boothman, J. Gao, cRGD-functionalized polymer micelles for targeted doxorubicin delivery, *Angew Chem Int Ed Engl*, **43**, 6323–6327 (2004).
251. F. Samaniego, D. Young, C. Grimes, V. Prospero, M. Christofidou-Solomidou, H.M. DeLisser, O. Prakash, A.A. Sahin, S. Wang, Vascular endothelial growth factor and Kaposi's sarcoma cells in human skin grafts, *Cell Growth Differ*, **13**, 387–395 (2002).

252. E. Jule, Y. Nagasaki, K. Kataoka, Lactose-installed poly(ethylene glycol)-poly(d,l-lactide) block copolymer micelles exhibit fast-rate binding and high affinity toward a protein bed simulating a cell surface: A surface plasmon resonance study, *Bioconjug Chem*, **14**, 177–186 (2003).
253. S. Vinogradov, E. Batrakova, S. Li, A. Kabanov, Polyion complex micelles with protein-modified corona for receptor-mediated delivery of oligonucleotides into cells, *Bioconjug Chem*, **10**, 851–860 (1999).
254. M. Yokoyama, T. Okano, Y. Sakurai, K. Kataoka, S. Inoue, Stabilization of disulfide linkage in drug-polymer-immunoglobulin conjugate by microenvironmental control, *Biochem Biophys Res Commun*, **164**, 1234–1239 (1989).
255. L.Z. Iakoubov, V.P. Torchilin, A novel class of antitumor antibodies: Nucleosome-restricted antinuclear autoantibodies (ANA) from healthy aged nonautoimmune mice, *Oncol Res*, **9**, 439–446 (1997).
256. B.A. Khaw, J.A. Mattis, G. Melincoff, H.W. Strauss, H.K. Gold, E. Haber, Monoclonal antibody to cardiac myosin: Imaging of experimental myocardial infarction, *Hybridoma*, **3**, 11–23 (1984).
257. J.R. Wands, H.E. Blum, Primary hepatocellular carcinoma, *N Engl J Med*, **325**, 729–731 (1991).
258. G. Ashwell, J. Harford, Carbohydrate-specific receptors of the liver, *Annu Rev Biochem*, **51**, 531–554 (1982).
259. Y.I. Jeong, S.J. Seo, I.K. Park, H.C. Lee, I.C. Kang, T. Akaike, C.S. Cho, Cellular recognition of paclitaxel-loaded polymeric nanoparticles composed of poly(gamma-benzyl L-glutamate) and poly(ethylene glycol) diblock copolymer endcapped with galactose moiety, *Int J Pharm*, **296**, 151–161 (2005).
260. O.C. Farokhzad, J. Cheng, B.A. Teply, I. Sherifi, S. Jon, P.W. Kantoff, J.P. Richie, R. Langer, Targeted nanoparticle-aptamer bioconjugates for cancer chemotherapy in vivo, *Proc Natl Acad Sci U S A*, **103**, 6315–6320 (2006).
261. O.C. Farokhzad, S. Jon, A. Khademhosseini, T.N. Tran, D.A. Lavan, R. Langer, Nanoparticle-aptamer bioconjugates: A new approach for targeting prostate cancer cells, *Cancer Res*, **64**, 7668–7672 (2004).
262. A.D. Ellington, J.W. Szostak, In vitro selection of RNA molecules that bind specific ligands, *Nature*, **346**, 818–822 (1990).
263. C. Tuerk, L. Gold, Systematic evolution of ligands by exponential enrichment: RNA ligands to bacteriophage T4 DNA polymerase, *Science*, **249**, 505–510 (1990).
264. G. Helmlinger, F. Yuan, M. Dellian, R.K. Jain, Interstitial pH and pO2 gradients in solid tumors in vivo: High-resolution measurements reveal a lack of correlation, *Nat Med*, **3**, 177–182 (1997).
265. I.F. Tannock, D. Rotin, Acid pH in tumors and its potential for therapeutic exploitation, *Cancer Res* **49**, 4373–4384 (1989).
266. F. Kohori, K. Sakai, T. Aoyagi, M. Yokoyama, Y. Sakurai, T. Okano, Preparation and characterization of thermally responsive block copolymer micelles comprising poly(N-isopropylacrylamide-*b*-DL-lactide), *J Control Release*, **55**, 87–98 (1998).
267. H. Yan, K. Tsujii, Potential application of poly(N-isopropylacrylamide) gel containing polymeric micelles to drug delivery systems, *Colloids Surf B Biointerfaces*, **46**, 142–146 (2005).

268. H. Wei, X.Z. Zhang, Y. Zhou, S.X. Cheng, R.X. Zhuo, Self-assembled thermoresponsive micelles of poly(N-isopropylacrylamide-*b*-methyl methacrylate), *Biomaterials*, **27**, 2028–2034 (2006).
269. S.Q. Liu, Y.W. Tong, Y.Y. Yang, Incorporation and in vitro release of doxorubicin in thermally sensitive micelles made from poly(N-isopropylacrylamide-co-N,N-dimethylacrylamide)-*b*-poly(D,L-lactide-c o-glycolide) with varying compositions, *Biomaterials*, **26**, 5064–5074 (2005).
270. O. Soga, C.F. van Nostrum, M. Fens, C.J. Rijcken, R.M. Schiffelers, G. Storm, W.E. Hennink, Thermosensitive and biodegradable polymeric micelles for paclitaxel delivery, *J Control Release*, **103**, 341–353 (2005).
271. M.C. Jones, M. Ranger, J.C. Leroux, pH-sensitive unimolecular polymeric micelles: Synthesis of a novel drug carrier, *Bioconjug Chem*, **14**, 774–781 (2003).
272. C.H. Wang, G.H. Hsiue, Polymeric micelles with a pH-responsive structure as intracellular drug carriers, *J Control Release*, **108**, 140–149 (2005).
273. M. Hruby, C. Konak, K. Ulbrich, Polymeric micellar pH-sensitive drug delivery system for doxorubicin, *J Control Release*, **103**, 137–148 (2005).
274. W.S. Shim, S.W. Kim, E.K. Choi, H.J. Park, J.S. Kim, D.S. Lee, Novel pH sensitive block copolymer micelles for solvent free drug loading, *Macromol Biosci*, **6**, 179–186 (2006).
275. G.H. Hsiue, C.H. Wang, C.L. Lo, J.P. Li, J.L. Yang, Environmental-sensitive micelles based on poly(2-ethyl-2-oxazoline)-*b*-poly(L-lactide) diblock copolymer for application in drug delivery, *Int J Pharm*, **317**, 69–75 (2006).
276. C. Giacomelli, L. Le Men, R. Borsali, J. Lai-Kee-Him, Brisson, A. S. P. Armes, A.L. Lewis, Phosphorylcholine-based pH-responsive diblock copolymer micelles as drug delivery vehicles: Light scattering, electron microscopy, and fluorescence experiments, *Biomacromolecules*, **7**, 817–828 (2006).
277. A.A. Kale, V.P. Torchilin, Design, synthesis, and characterization of pH-sensitive PEG-PE conjugates for stimuli-sensitive pharmaceutical nanocarriers: The effect of substitutes at the hydrazone linkage on the ph stability of PEG-PE conjugates, *Bioconjug Chem*, **18**, 363–370 (2007).
278. R.M. Sawant, J.P. Hurley, S. Salmaso, A. Kale, E. Tolcheva, T.S. Levchenko, V.P. Torchilin, 'SMART' drug delivery systems: Double-targeted pH-responsive pharmaceutical nanocarriers, *Bioconjug Chem*, **17**, 943–949 (2006).
279. E.S. Lee, K. Na, Y.H. Bae, Super pH-sensitive multifunctional polymeric micelle, *Nano Lett*, **5**, 325–329 (2005).
280. D.M. Lynn, M.M. Amiji, R. Langer, pH-responsive polymer microspheres: Rapid release of encapsulated material within the range of intracellular pH. *Angew Chem Int Ed Engl*, **40**, 1707–1710 (2001).
281. A. Potineni, D.M. Lynn, R. Langer, M.M. Amiji, Poly(ethylene oxide)-modified poly(beta-amino ester) nanoparticles as a pH-sensitive biodegradable system for paclitaxel delivery, *J Control Release*, **86**, 223–234 (2003).
282. D. Shenoy, S. Little, R. Langer, M. Amiji, Poly(ethylene oxide)-modified poly(beta-amino ester) nanoparticles as a pH-sensitive system for tumor-targeted delivery of hydrophobic drugs. 1. In vitro evaluations, *Mol Pharm*, **2**, 357–366 (2005).
283. N. Rapoport, W.G. Pitt, H. Sun, J.L. Nelson, Drug delivery in polymeric micelles: From in vitro to in vivo, *J Control Release*, **91**, 85–95 (2003).

284. S. Mitragotri, Healing sound: The use of ultrasound in drug delivery and other therapeutic applications, *Nat Rev Drug Discov*, **4**, 255–260 (2005).
285. S. Miyazaki, C. Yokouchi, M. Takada, External control of drug release: Controlled release of insulin from a hydrophilic polymer implant by ultrasound irradiation in diabetic rats, *J Pharm Pharmacol*, **40**, 716–717 (1988).
286. A. Marin, H. Sun, G. A. Husseini, W. G. Pitt, D. A. Christensen, N. Y. Rapoport, Drug delivery in pluronic micelles: Effect of high-frequency ultrasound on drug release from micelles and intracellular uptake, *J Control Release*, **84**, 39–47 (2002).
287. G.A. Husseini, D.A. Christensen, N.Y. Rapoport, W.G. Pitt, Ultrasonic release of doxorubicin from Pluronic P105 micelles stabilized with an interpenetrating network of N,N-diethylacrylamide, *J Control Release*, **83**, 303–305 (2002).
288. Z.G. Gao, H.D. Fain, N. Rapoport, Controlled and targeted tumor chemotherapy by micellar-encapsulated drug and ultrasound, *J Control Release*, **102**, 203–222 (2005).
289. Y. Katayama, T. Sonoda, M. Maeda, A polymeric micelle responding to the protein kinase A signal, *Macromolecules*, **34**, 8569 (2001).
290. A. Napoli, M. Valentini, N. Tirelli, M. Muller, J.A. Hubbell, Oxidation-responsive polymeric vesicles, *Nat Mater*, **3**, 183–189 (2004).
291. J. Wang, D. Mongayt, V.P. Torchilin, Polymeric micelles for delivery of poorly soluble drugs: Preparation and anticancer activity in vitro of paclitaxel incorporated into mixed micelles based on poly(ethylene glycol)-lipid conjugate and positively charged lipids, *J Drug Target*, **13**, 73–80 (2005).
292. S.R. Schwarze, A. Ho, A. Vocero-Akbani, S.F. Dowdy, In vivo protein transduction: Delivery of a biologically active protein into the mouse, *Science*, **285**, 1569–1572 (1999).
293. A. Astriab-Fisher, D. Sergueev, M. Fisher, B.R. Shaw, R.L. Juliano, Conjugates of antisense oligonucleotides with the TAT and antennapedia cell-penetrating peptides: Effects on cellular uptake, binding to target sequences, and biologic actions, *Pharm Res*, **19**, 744–754 (2002).
294. V.P. Torchilin, R. Rammohan, V. Weissig, T.S. Levchenko, TAT peptide on the surface of liposomes affords their efficient intracellular delivery even at low temperature and in the presence of metabolic inhibitors, *Proc Natl Acad Sci U S A*, **98**, 8786–8791 (2001).
295. T.S. Levchenko, R. Rammohan, N. Volodina, V.P. Torchilin, TAT peptide-mediated intracellular delivery of liposomes, *Methods Enzymol*, **372**, 339–349 (2003).
296. M.M. Fretz, G.A. Koning, E. Mastrobattista, W. Jiskoot, G. Storm, OVCAR-3 cells internalize TAT-peptide modified liposomes by endocytosis, *Biochim Biophys Acta*, **1665**, 48–56 (2004).
297. V.A. Sethuraman, Y.H. Bae, TAT peptide-based micelle system for potential active targeting of anti-cancer agents to acidic solid tumors, *J Control Release*, **118**, 216–224 (2007).
298. R.R. Sawant, R.M. Sawant, V.P. Torchilin, Mixed PEG-PE/vitamin E tumor-targeted immunomicelles as carriers for poorly soluble anti-cancer drugs: Improved drug solubilization and enhanced in vitro cytotoxicity, *Eur J Pharm Biopharm*, **70**, 51–57 (2008).
299. K.S. Rao, M.K. Reddy, J.L. Horning, V. Labhasetwar, TAT-conjugated nanoparticles for the CNS delivery of anti-HIV drugs, *Biomaterials*, **29**, 4429–4438 (2008).

300. M. Lewin, N. Carlesso, C.H. Tung, X.W. Tang, D. Cory, D.T. Scadden, R. Weissleder, TAT peptide-derivatized magnetic nanoparticles allow in vivo tracking and recovery of progenitor cells, *Nat Biotechnol*, **18**, 410–414 (2000).
301. M. Zhao, M.F. Kircher, L. Josephson, R. Weissleder, Differential conjugation of TAT peptide to superparamagnetic nanoparticles and its effect on cellular uptake, *Bioconjug Chem*, **13**, 840–844 (2002).
302. M. Zorko, U. Langel, Cell-penetrating peptides: Mechanism and kinetics of cargo delivery, *Adv Drug Deliv Rev*, **57**, 529–545 (2005).
303. E. Vives, J.P. Richard, C. Rispal, B. Lebleu, TAT peptide internalization: Seeking the mechanism of entry, *Curr Protein Pept Sci*, **4**, 125–132 (2003).
304. J.S. Wadia, R.V. Stan, S.F. Dowdy, Transducible TAT-HA fusogenic peptide enhances escape of TAT-fusion proteins after lipid raft macropinocytosis, *Nat Med*, **10**, 310–315 (2004).
305. R.R. Sawant, R.M. Sawant, A.A. Kale, and V.P. Torchilin, The architecture of ligand attachment to nanocarriers controls their specific interaction with target cells, *J Drug Target*, **16**, 596–600 (2008).
306. N. Nasongkla, E. Bey, J. Ren, H. Ai, C. Khemtong, J.S. Guthi, S.F. Chin, A.D. Sherry, D. A. Boothman, J. Gao, Multifunctional polymeric micelles as cancer-targeted, MRI-ultrasensitive drug delivery systems, *Nano Lett*, **6**, 2427–2430 (2006).
307. T.Y. Kim, *et al.*, Phase I and pharmacokinetic study of Genexol-PM, a cremophor-free, polymeric micelle-formulated paclitaxel, in patients with advanced malignancies, *Clin Cancer Res*, **10**(11), 3708–3716 (2004).
308. I.G. Shin, *et al.*, Methoxy poly(ethylene glycol)/epsilon-caprolactone amphiphilic block copolymeric micelle containing indomethacin. I. Preparation and characterization, *J Control Release*, **51**(1), 1–11 (1998).
309. H. Sun, *et al.*, Shell-sheddable micelles based on dextran-SS-poly(epsilon-caprolactone) diblock copolymer for efficient intracellular release of doxorubicin, *Biomacromolecules*, **11**(4), 848–854 (2010).
310. H. Uchino, *et al.*, Cisplatin-incorporating polymeric micelles (NC-6004) can reduce nephrotoxicity and neurotoxicity of cisplatin in rats, *Br J Cancer*, **93**(6), 678–687 (2005).
311. M. Yokoyama, *et al.*, Incorporation of water-insoluble anticancer drug into polymeric micelles and control of their particle size, *J Control Release*, **55**(2–3), 219–229 (1998).
312. C. Zhan, *et al.*, Cyclic RGD conjugated poly(ethylene glycol)-co-poly(lactic acid) micelle enhances paclitaxel anti-glioblastoma effect, *J Control Release*, **143**(1), 136–142 (2010).
313. X. Shuai, *et al.*, Core-cross-linked polymeric micelles as paclitaxel carriers, *Bioconjug Chem*, **15**(3), 441–448 (2004).
314. L. Ould-Ouali, *et al.*, Self-assembling PEG-p(CL-co-TMC) copolymers for oral delivery of poorly water-soluble drugs: A case study with risperidone, *J Control Release*, **102**(3), 657–668 (2005).
315. J. Liu, F. Zeng, C. Allen, Influence of serum protein on polycarbonate-based copolymer micelles as a delivery system for a hydrophobic anti-cancer agent. *J Control Release*, **103**(2), 481–497 (2005).

316. V.I. Slepnev, *et al.*, Micelles of poly(oxyethylene)-poly(oxypropylene) block copolymer (pluronic) as a tool for low-molecular compound delivery into a cell: Phosphorylation of intracellular proteins with micelle incorporated. (1996).
317. A. Harada *et al.*, Novel polyion complex micelles entrapping enzyme molecules in the core: Preparation of narrowly-distributed micelles from lysozyme and poly(ethylene glycol)–poly(aspartic acid) block copolymer in aqueous medium, *Macromolecules*, **31**, 288–294 (1998).
318. A.V. Kabanov, E.V. Batrakova, V.Y. Alakhov, An essential relationship between ATP depletion and chemosensitizing activity of Pluronic block copolymers, *J Control Release*, **91**(1–2), 75–83 (2003).
319. A.K. Sharma, *et al.*, Prevention of MDR development in leukemia cells by micelle forming polymeric surfactant, *J Control Release*, **131**(3), 220–227 (2008).
320. A.V. Kabanov, Micelles of amphiphilic block copolymers as vehicles for drug delivery. In *Amphiphilic Block Copolymers: Self-assembly and Applications*. P. Alexandridis, B. Lindman (eds), Amsterdam: Elsevier, pp. 1–31 (1997).
321. S.I. Masayuki Yokoyama, K. Kataoka, N. Yui, Y. Sakurai, Preparation of adriamycin-conjugated poly(ethylene glycol)-poly(aspartic acid) block copolymer: A new type of polymeric anticancer agent, *Makromol Chem Rapid Commun*, **8**(9), 431–435 (1987).
322. M. Yokoyama, *et al.*, Polymer design and incorporation methods for polymeric micelle carrier system containing water-insoluble anti-cancer agent camptothecin, *J Drug Target*, **12**(6), 373–384 (2004).
323. T. Yamamoto, *et al.*, What are determining factors for stable drug incorporation into polymeric micelle carriers? Consideration on physical and chemical characters of the micelle inner core, *J Control Release*, **123**(1), 11–18 (2007).
324. P. Opanasopit, *et al.*, Influence of serum and albumins from different species on stability of camptothecin-loaded micelles, *J Control Release*, **104**(2), 313–321 (2005).
325. M, Watanbe *et al.*, Preparation of campthothecin-loaded polymeric micelles and evaluation of their incorporation and circulation stability, *Int J Pharm*, **308**: 183–189 (2006).
326. K. Kawano, *et al.*, Enhanced antitumor effect of campthothecin loaded in long-circulating polymeric micelles, *J Control Release*, **112**(3), 329–332 (2006).
327. A. Hayama, *et al.*, Polymeric micelles modified by folate-PEG-lipid for targeted drug delivery to cancer cells in vitro, *J Nanosci Nanotechnol*, **8**(6), 3085–3090 (2008).
328. H. Koide, *et al.*, Particle size-dependent triggering of accelerated blood clearance phenomenon, *Int J Pharm*, **362**(1–2), 197–200 (2008).
329. E.T. Dams, *et al.*, Accelerated blood clearance and altered biodistribution of repeated injections of sterically stabilized liposomes, *J Pharmacol Exp Ther*, **292**(3), 1071–1079 (2000).
330. P. Laverman, *et al.*, Factors affecting the accelerated blood clearance of polyethylene glycol-liposomes upon repeated injection, *J Pharmacol Exp Ther*, **298**(2), 607–612 (2001).
331. T. Ishida, *et al.*, Spleen plays an important role in the induction of accelerated blood clearance of PEGylated liposomes, *J Control Release*, **115**(3), 243–250 (2006).
332. T. Ishida, and H. Kiwada, Accelerated blood clearance (ABC) phenomenon upon repeated injection of PEGylated liposomes, *Int J Pharm*, **354**(1–2), 56–62 (2008).

333. N. Nishiyama, *et al.*, Novel cisplatin-incorporated polymeric micelles can eradicate solid tumors in mice, *Cancer Res*, **63**(24), 8977–8983 (2003).
334. T. Hamaguchi, *et al.*, NK105, a paclitaxel-incorporating micellar nanoparticle formulation, can extend in vivo antitumour activity and reduce the neurotoxicity of paclitaxel, *Br J Cancer*, **92**(7), 1240–1246 (2005).
335. D. Le Garrec, *et al.*, Poly(N-vinylpyrrolidone)-block-poly(D,L-lactide) as a new polymeric solubilizer for hydrophobic anticancer drugs: In vitro and in vivo evaluation, *J Control Release*, **99**(1), 83–101 (2004).
336. Y.J. Park, *et al.*, Radioisotope carrying polyethylene oxide-polycaprolactone copolymer micelles for targetable bone imaging, *Biomaterials*, **23**(3), 873–879 (2002).
337. J. Liu, F. Zeng, C. Allen, In vivo fate of unimers and micelles of a poly(ethylene glycol)-block-poly(caprolactone) copolymer in mice following intravenous administration, *Eur J Pharm Biopharm*, **65**(3), 309–319 (2007).
338. Z. Hu, *et al.*, Arg-Gly-Asp (RGD) peptide conjugated poly(lactic acid)-poly(ethylene oxide) micelle for targeted drug delivery, *J Biomed Mater Res A*, **85**(3), 797–807 (2008).
339. K. Kawano, *et al.*, Enhanced antitumor effect of camptothecin loaded in long-circulating polymeric micelles, *J Control Release*, **112**(3), 329–332 (2006).
340. S. Kawakami, *et al.*, Biodistribution characteristics of all-trans retinoic acid incorporated in liposomes and polymeric micelles following intravenous administration, *J Pharm Sci*, **94**(12), 2606–2615 (2005).
341. Y. Bae, N. Nishiyama, K. Kataoka, In vivo antitumor activity of the folate-conjugated pH-sensitive polymeric micelle selectively releasing adriamycin in the intracellular acidic compartments, *Bioconjug Chem*, **18**(4), 1131–1139 (2007).
342. C.J. Rijcken, *et al.*, Hydrolysable core-crosslinked thermosensitive polymeric micelles: Synthesis, characterisation and in vivo studies, *Biomaterials*, **28**(36), 5581–5593 (2007).
343. H.Y. Tian, *et al.*, Biodegradable cationic PEG-PEI-PBLG hyperbranched block copolymer: Synthesis and micelle characterization, *Biomaterials*, **26**(20), 4209–4217 (2005).

15

Nanostructured Silicon-Based Materials as a Drug Delivery System for Water-Insoluble Drugs

Vesa-Pekka Lehto, Jarno Salonen, Hélder A. Santos and Joakim Riikonen

15.1 Introduction

Mesoporous materials have become popular in the past few years due to their unique properties that can be employed in biomedical applications. Their functionality can be varied greatly by tuning the well-defined nanostructure and surfaces. For example, the behavior of porous silicon (PSi) *in vivo* can be altered from bioinert and bioactive to resorbable by varying the porosity and surface chemistry of the silicon sample. The mesoporous materials can be fabricated to be structurally and chemically stable even under the harsh conditions of the stomach and gastrointestinal (GI) lumen. A major feature of silicon-based materials is that the surface chemistry of the pore walls and the external surfaces of the mesoporous particles can be independently modified, although this feature is not yet fully exploited. However, the feature is becoming more important as in the future more sophisticated applications are being developed; the pore walls are functionalized in order to accommodate the payload molecules and the external surfaces of the carrier are functionalized in order to respond to the surrounding environment. In addition, the materials are also biocompatible and relatively inexpensive. All these features render the inorganic mesoporous materials potential candidates as a drug carriers and biomaterial for therapeutic applications.

In this chapter, the use of mesoporous silicon-based materials to improve the drug dissolution behavior for oral drug delivery is surveyed. The focus is on PSi, but the consideration is partly valid also for silica-based (SiO_2) mesoporous materials. Basically, the principle

Drug Delivery Strategies for Poorly Water-Soluble Drugs, First Edition. Edited by Dennis Douroumis and Alfred Fahr.

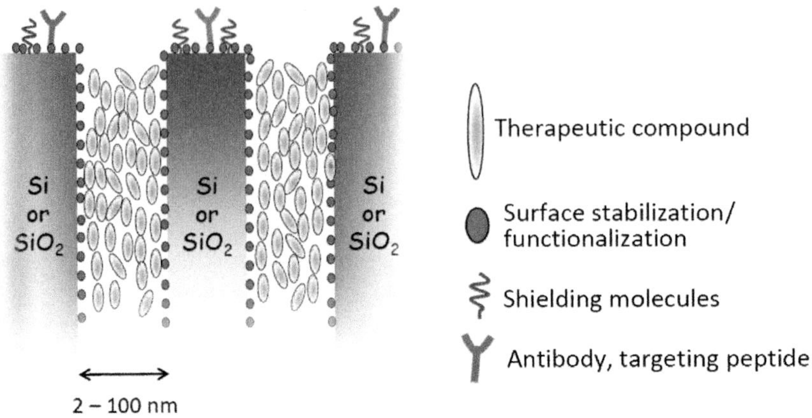

Figure 15.1 *Mesoporous materials as drug carriers. For the fast release of the payload molecules, the pores need to be small enough to prevent the formation of well-ordered crystals of the drug molecules but large enough to permit the fast diffusion of the molecules. The pore walls and the external surfaces of the particles are modified separately: shielding (e.g. with polyethylene glycol, PEG) and targeting moieties are introduced only on the external surfaces.*

to improve the dissolution behavior of the payload drugs is the same for both materials: by confining the drug in a space only few times larger than the drug molecules themselves (Figure 15.1), the physicochemical properties of the drug can be dramatically changed. Both the physical and chemical properties determine the functionality of the mesoporous carrier which makes the material versatile – the material can be designed to be suitable practically for any type of payload molecules without compromising the payload capacity and the stability of the molecules.

15.2 Control of Particle Size and Pore Morphology

The performance of drug carriers has traditionally been considered to be related to the chemical properties of the carrier. However, there is growing recognition that also the physical properties affect the behavior of the biomaterials in biological conditions. The physical characteristics of the biomaterials essential to their performance and biofate are widely recognized to be the size, shape, mechanical properties and the surface texture including microstructure and the porosity of the surface. Thus, to optimize the behavior of the drug carrier and its performance in the body for various applications, it is beneficial for the delivery system that both the physical and the chemical properties of the carrier can be freely varied with no restrictions depending on, e.g. the payload molecule.

Mesoporous (2 nm $<$ pore diameter $<$ 50 nm) inorganic materials, such as silicon-based PSi and silicon dioxide-based MCM-41 and SBA-15, are solid materials comprising a nanoporous structure with different morphologies depending on the fabrication method. PSi is a top-down nanomaterial, where the nanoporous structure is produced in the bulk silicon material, while mesoporous silicas are bottom-up nanomaterials produced via chemical synthesis. The nanopores are able to absorb/encapsulate relatively large amounts of bioactive molecules. The pore diameters can be tuned according to the size of the payload molecule together with the surface chemistry. The characteristics for both the PSi and porous silica materials are the high specific surface area (up to 1000 m^2/g) and the high pore volume (up to 2 cm^3/g), which are the critical parameters for obtaining high loading degrees of the drugs. The stability of the porous materials in biological environment depends on the porosity, surface chemistry and the material itself (silicon/silica). These properties make them potentially suitable for various drug delivery systems (DDS).

There are several different methods employed to produce PSi [1,2]. The most commonly used is electrochemical etching of Si single crystals, but there is also increasing interest in applying purely chemical etching methods such as stain etching, which is beneficial for large-scale production of PSi [3]. Unfortunately, chemical etching does not provide the same versatile possibility of adjusting the pore morphology of PSi as the electrochemical etching does. However, some new promising results for understanding the stain etching process have recently been reported [4] and, on the other hand, in certain applications exact control of the morphology is not even necessary.

There are many sophisticated etching set-ups for electrochemical etching [5], but in the simplest one, a strip of silicon wafer and a Pt wire working as electrodes are placed in a HF-based electrolyte solution and an etching current is applied between these electrodes. Although the mechanism of pore formation is not generally agreed, it is known to be a combination of electronic and chemical factors. The process needs holes at the surface to polarize the surface Si atoms. In p-type Si, holes are the major current carriers, but in n-type Si additional illumination during the etching is needed to generate the holes at the surface. In the next step, the polarized surface atoms are attacked by HF, to dissolve them and leave the generated new surface covered with hydrogen. The process produces gaseous hydrogen, which should be taken into account when considering the ventilation of the space where the etching system is located.

The dissolution of Si could be direct where H_2SiF_6 and molecular hydrogen H_2 with a dissolution valence close to 2 are produced, or indirect where the dissolution takes place through the oxidation of the Si surface producing H_2SiF_6 and H_2O [6]. The latter has the dissolution valence of about 4 and is related to the use of high current densities in or near the electropolishing range. In Figure 15.2, the different phases of the anodization

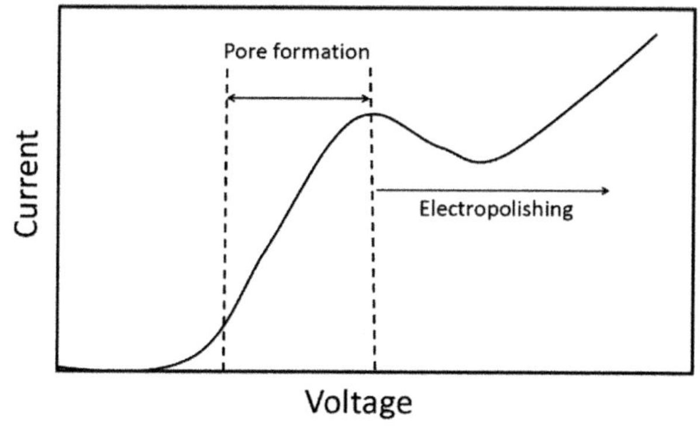

Figure 15.2 *Current-voltage profile for the Si/HF electrochemical etching system.*

are illustrated using a typical current-voltage curve. In the pore formation phase, the etching rate is proportional to both the current density and the HF concentration. After a certain voltage value, which depends on the type of Si and electrolyte, the current starts to decrease, pore formation stops and electropolishing process begins at the interface. This change from the pore formation to the electropolishing phase due to the increased current density is commonly used to detach the porous layer produced from the Si substrate at the end of the etching. Thoroughly porous samples can be used in many applications, e.g. in drug delivery, where biodegradation is an important property of the drug carrier.

A number of fabrication parameters affect the homogeneity and reproducibility of the porous layer. The most important parameters are the stability of the temperature and the current density, as well as the good and homogeneous electrical contact to the Si wafer. The other parameters affecting the product include chemical composition of the electrolyte, the doping type and the resistivity of the Si wafer, the crystallographic orientation of the Si wafer, duration of the etching, electrolyte stirring, illumination intensity and wavelength, the surface ratio of the cathodic electrode and the etched area. Due to the above-mentioned parameters and the complicated interdependency between them, complete control of the etching process is difficult. However, there are some empirical rules for PSi formation and some effects on the pore morphology are quite well characterized and documented.

The pore diameter increases with increasing current density and decreasing HF concentration. The pore density increases with increasing doping density [7]. In the p-type Si, the increase in the doping density also increases the pore diameter, but in the n-type Si the relation is not as straightforward. Usually, the increase in doping density decreases the pore diameter in the n-type Si, but large variations in the pore diameter have been reported. This could be related to the fact that in n-type Si the effect of the current density and HF concentration is more pronounced [8].

The electrolytes commonly used include ethanol in order to reduce the surface tension of the electrolytes. Without reduced surface tension, the formation of hydrogen bubbles at the Si/HF interface may lead to a heterogeneous layer and significantly affect the pore morphology. The addition of ethanol has similar effects to decreasing the HF concentration, but it usually leads to even more columnar, smoother and larger pores with better homogeneity than the simple decrease of the HF concentration.

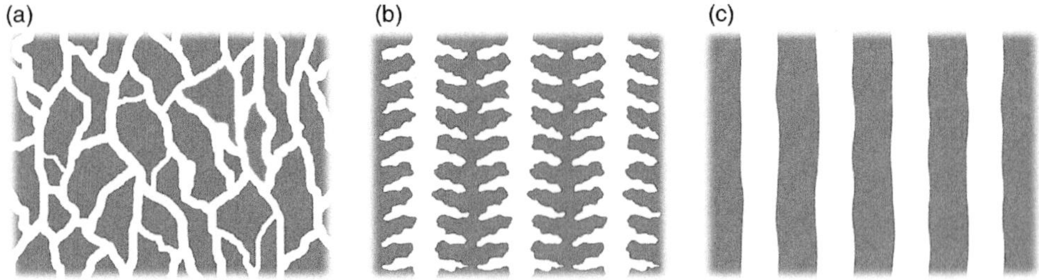

Figure 15.3 *The main type of pores commonly observed in PSi materials as a result of both the type of silicon and the fabrication parameters used: (a) sponge-like; (b) branching; and (c) perpendicular without interconnections.*

The pore types commonly observed in PSi can be classified into three groups (Figure 15.3). A sponge-like structure is usually produced from p-type Si using low current density with relatively high HF concentrations (Figure 15.3a). It has small pore diameter (sometimes even less than 2 nm) and high surface area compared to the pore volume. Branching morphology is the most common morphology of PSi (Figure 15.3b). It can be obtained from all types of Si by varying the fabrication parameters. The branching depends on the etching current density. With low current density, the formed subpores are aligned along the ⟨100⟩ axis perpendicular to the main pores, while with higher current density a fir tree type branching is obtained. In this type of morphology, the pore diameters between 5–20 nm are dominant. The third type of the pores is well orientated columnar pores perpendicular to the initial surface (if ⟨100⟩ wafers are used) with smooth pore walls without interconnections between pores (Figure 15.3c). These types of pores can be produced using p+-type Si together with high current density and low HF concentration. The pores are typically larger than 15 nm in diameter and can have various shapes [9].

Due to the several fabrication parameters affecting the pore morphology, there is a wide variation in the nanostructure of the PSi produced. The pore diameter from 1 nm up to several μm can be produced separately or the pore walls of macropores (pore diameter >50 nm) can be decorated with micropores (pore diameter < 2 nm). The pore morphology can be fluctuated during the etching process by varying the current density to produce, e.g. ink bottle type pores or optical multilayer structures with specific properties for sensing purposes. In addition, the morphology of the pores can be adjusted even after the etching process by thermal annealing [10]. All these properties together make PSi a very versatile material for many applications, including drug delivery.

After the pore etching process, an appropriate porous layer (e.g. with specified pore size) has been etched on the Si surface (substrate) which can then be detached from the substrate by increasing the etching current abruptly. In oral delivery applications, microparticles are normally used. From PSi films, microparticles of different size fractions are produced by milling and subsequent sieving to obtain the desired particle size. Typically, the values of the surface area, pore volume and porosity obtained for microparticles vary in the ranges of 200–500 m^2/g, 0.5–2.0 cm^3/g and 50–80%, respectively.

In the case of intravenous administration or when mucoadhesive properties of nanoparticles need to be employed, nanoparticles can be produced by sonication of the PSi films [11] but a better yield is obtained using a more sophisticated etching procedure [12]. This procedure consists of three different current pulses, where the first low current pulse is used to produce a similar mesoporous silicon structure as in case of the microparticles. The second pulse is shorter, with a high current density near the electropolishing region, which produces a highly porous and mechanically fragile layer. The third, zero current pulse, is used to remove the possible electrolyte concentration gradients in the pores formed during the high current pulse before starting a new etching cycle. After drying the electrolyte off, the production of nanoparticles is performed by ball milling the porous film in an appropriate liquid. Because of the layered structure of the porous film, nanoparticles of certain sizes are obtained; the fragile layers predetermine the fracture sites on the plane parallel to the surface of the porous film. Wet milling is used instead of dry milling in order to increase the grinding efficacy and to facilitate the collection of the nanoparticles. The final size separation and changes of the suspension media can be done with centrifugation or filtration.

15.3 Surface Functionalization

An essential requirement for using PSi in drug delivery applications is that the material should be sufficiently stable, i.e., its properties should not be altered uncontrollably during the drug loading, storage or usage. Pristine as-anodized PSi cannot meet this requirement, and, therefore, surface stabilization is a crucial step in the PSi fabrication process.

15.3.1 Stabilization

The weakest bonds on silicon surfaces are the Si-Si backbonds, which are very susceptible to chemical reactions [13]. The surface of PSi after pore formation in an HF-containing solution is hydrogen terminated, which provides the stabilization of the the surface to some extent. The surface is covered by hydrides Si-H_x ($x = 1$–3) and is chemically very homogeneous (>99% coverage) and hydrophobic [13]. However, the hydride-covered surface is not stable even in ambient conditions and the material undergoes slow oxidation in contact with atmospheric oxygen or water vapor. Due to the relative strength of the Si-H bond, oxygen and water react primarily with the weak Si-Si bonds producing backbond oxidation O_y-Si-H_x and Si-O-Si [14]. The silanol groups (Si-OH) are mainly produced in reaction with liquid or gaseous water. Immersion in aqueous solutions causes a rapid oxidation and dissolution of the material [15, 16].

The stabilization of the PSi surface is important as the oxidation of the hydride surface causes temporal changes in the chemical, optical and structural properties of PSi. However, the preferred extent of stabilization is application dependent. For example, the biodegradability of PSi can be tuned by surface stabilization. There are two main ways to stabilize the surface: stabilization with oxygen or carbon. Oxidation stabilizes the surface by producing Si-O bonds on the PSi surface, where the bond strength is considerably higher compared to Si-Si or Si-H bonds and forms, therefore, more stable structure. Stabilization with carbon produces Si-C bonds, which are more stable than Si-O bonds despite the lower bond strength. The stability is due to the non-polar nature of the Si-C bond [17], which reduces the probability of hydrolysis.

15.3.1.1 Stabilization with Oxygen

Oxidation is the most common way to stabilize PSi surfaces because it is relatively easy to perform. The oxidized surface is also advantageous due to its biocompatibility and biodegradability [18]. The oxidized PSi surface contains typically Si-O-Si, O_y-Si-H_x, Si-H and Si-OH species and their relative concentrations depend on the oxidation method used.

Thermal oxidation is performed at elevated temperatures in a gaseous environment containing oxygen. The degree of oxidation and the appearance of the previously mentioned surface species can be controlled by the oxidation temperature and duration. Rapid backbond oxidation of Si-H_x groups begins at 250°C and no Si-H_x groups without backbond oxide were found in a sample oxidized at 350°C [19, 20]. Hydroxyl groups (Si-OH) begin to form at 300°C [20]. Backbonded O_y-Si-H_x groups oxidize into Si-OH groups at 440–730°C [19]. The complete oxidation of the PSi structure into SiO_2 can take place at 800°C or above [21].

Another commonly used method to oxidize PSi is chemical oxidation with liquid oxidants. Chemical oxidation generates a thin oxide layer, in contrast to thermal oxidation, which can be used even to thoroughly oxidize the material. As mentioned earlier, water can also be used to oxidize the PSi surface. However, better results are achieved when stronger oxidants in aqueous solutions such as H_2O_2 and HNO_3 are used [22, 23]. Care should be taken when using alkaline oxidants such as NH_3 or pyridine due to the fact that PSi reacts strongly in highly alkaline solutions [24].

15.3.1.2 Stabilization with Carbon

The main advantage of stabilizing PSi with carbon is the extreme stability of the Si-C bond, allowing highly persistent PSi surfaces to be generated. In addition, the Si-C bond is the ideal bond to introduce organic molecules into PSi in order to take advantage of the vast possibilities of organic chemistry. Therefore, the attached carbons can be used not only to stabilize the PSi but also to further functionalize it. Stabilization of PSi with carbon is typically achieved by three methods: silanization, hydrosilylation and thermal carbonization. The reactions involved in these treatments are summarized in Figure 15.4.

In silanization (Figure 15.4a), alkoxysilanes are normally covalently bonded to PSi surface. Typically, the reaction takes place between alkoxysilanes and Si-OH groups on the oxidized PSi surface and forms a covalent -Si-O-Si- bond [16]. It has been estimated that, depending on the reaction conditions, up to 60% of the Si-OH groups on the PSi surface can react with the silanes [25].

Hydrosilylation (Figure 15.4b) involves insertion of a (terminal) unsaturated carbon bond into the Si-H bond at elevated temperatures (90–120°C) [17]. Several kinds of organic molecules can be bonded to the PSi surface through hydrosilylation including linear alkenes, dienes, esters and carboxylic acids [26, 27]. The efficiency of the treatment has been estimated to be 50%, i.e., half of the surface hydrides have been consumed by the reaction [28, 29]. It should be noted, however, that not all hydrides can react with the alkene due to steric reasons.

In thermal carbonization (Figures 15.4c and 15.4d), hydrogen-terminated PSi reacts with acetylene gas at high temperatures [30]. The resulting surface has a different nature depending on the temperature at which the treatment is performed. Treatments under continuous acetylene flush at temperatures between 400°C and 600°C produce hydrophobic hydrocarbon surfaces (hydrocarbonization). When acetylene, which is adsorbed on hydrogen terminated or thermally hydrocarbonated surfaces, decomposes at temperatures between 600°C and 950°C, a nonstoichiometric silicon carbide layer is formed on the PSi surface with a hydrophilic nature. Both treatments produce high coverage of carbonized surface.

The stability of the carbonized surfaces may differ. Silanized surfaces tend to be the most unstable of the three carbonized surfaces due to the polarized Si-O bonds that are susceptible to nucleophilic attack [31]. On the contrary, thermal carbonization produces a good surface coverage and superior stability due to the absence of Si-O and Si-H species [32]. The stability of the thermally carbonized surface increases with the increase of the treatment temperature [30].

15.3.2 Biofunctionalization

Biofunctionalization of PSi aims to create the desired interface between the inorganic material and the biological system providing additional functionality to the material. Such functionalizations often involve covalent binding of proteins or peptides on the surface of the

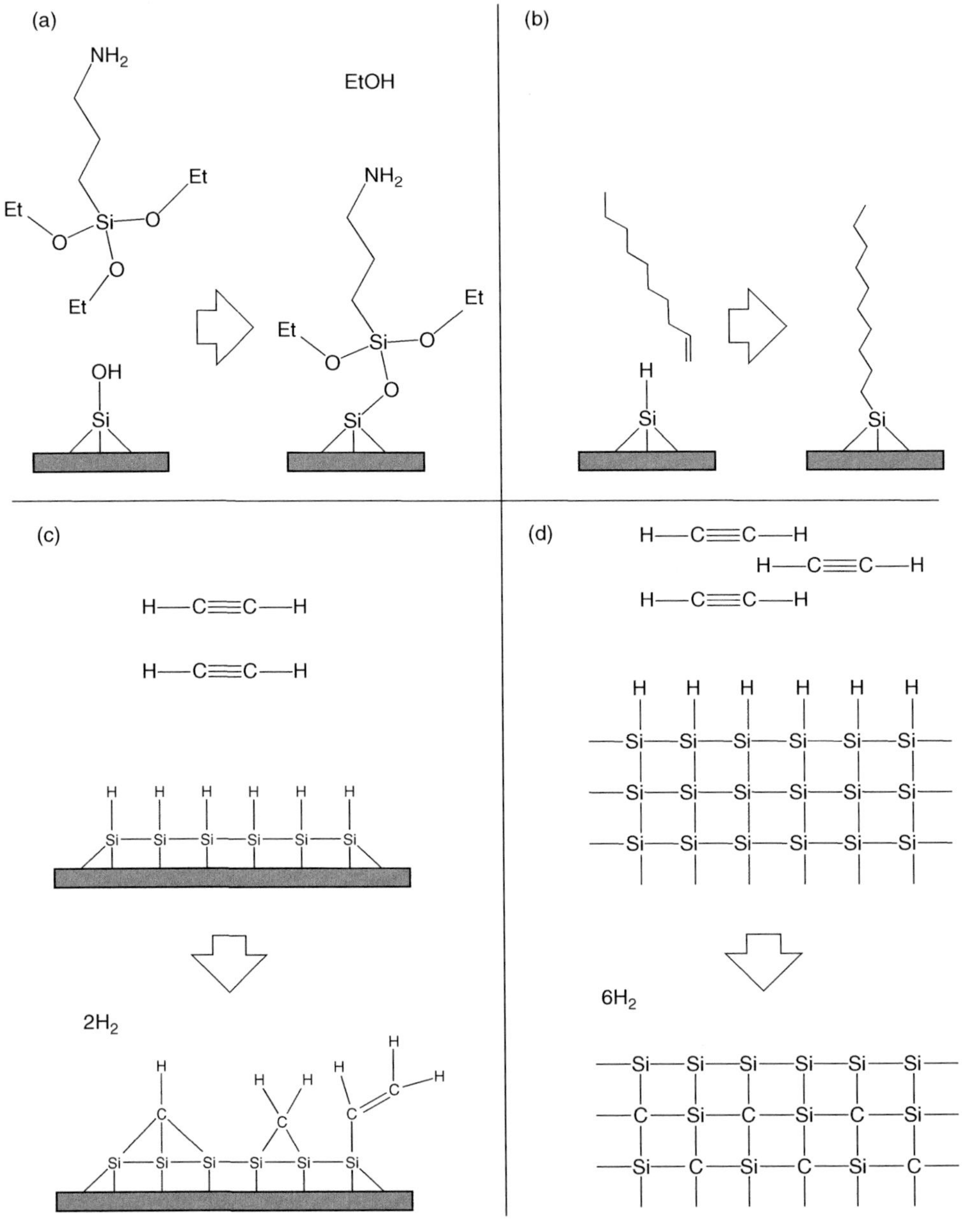

Figure 15.4 *Silanization (a); hydrosilylation; (b) and thermal carbonization at low (c) and high (d) temperature of PSi to produce stable carbon-containing Si surfaces.*

PSi. On the other hand, biofunctionalization can also aim to reduce the interactions between the PSi materials and the biological systems. For example, PEGylation has often been used to reduce protein adsorption on materials and to avoid the rapid systemic clearance by the mononuclear phagocyte system (MPS).

In order to produce a stable biofunctionalized surface, functionalization should be performed on the stabilized PSi surface. The typical route for the biofunctionalization is to modify the surface with molecules containing carboxylic acid or amine groups and conjugate the biomolecule to these groups via an amide bond or gross linker. For example, thermally oxidized PSi was silanized with aminopropyltriethoxysilane (APTES) and bovine serum albumin was subsequently linked to the amine group by cross-linking with glutaraldehyde [33]. In another study, hydrogen-terminated PSi was hydrosilylated by undecylenic acid and the acid was conjugated to a peptide via activation by N-Hydroxysuccinimide (NHS) and 1-ethyl-3-(3-dimethylaminopropyl) carbodiimide (EDC) [34].

The study also showed that the efficiently hydrosilylated surface was highly stable, but it was not able to bind high amounts of protein due to crowding of the functional groups on the hydrosilylated surface. The binding of the peptide was better on the less effectively hydrosilylated surface, but its stability was compromised due to poorer stabilization of the surface.

15.4 Biocompatibility and Cytotoxicity

In biological applications the drug delivery system itself should be safe (non-toxic) and biocompatible, and also enable the safe drug distribution in the body. The biocompatibility of a material is highly anatomically dependent, i.e., it does not simply depend on the intrinsic properties or composition of the material itself, and it can be very dependent on local phenomena or affect the entire organism [35]. Besides the functionality of the delivery system, also pre-clinical tests (e.g. biodistribution, immune response, toxicity, reactive oxygen species, clearance, etc.) have to be evaluated prior to any clinical trial in order to understand the material properties affecting the interactions with the surrounding environment, which in turn affect the fate of the drugs (systemic distribution) and its release, and thus, the overall toxicity, the biocompatibility and safety of the carrier.

In vitro toxicity studies are used to assess the behavior and survival (viability) of different cell types in the presence of biomaterials. Cell culturing can be used to screen the toxicity and the study aims mainly at the detection of biological activity of tested substances or materials in a wide range of different cell types. The parameters and end-points used to measure toxicity include, e.g. vital staining, cytosolic enzyme release and organ-specific cell functions, such as alterations in the cell membrane and metabolism integrity. The following end-points have been extensively measured in the literature as cytotoxicity parameters in the presence of biomaterials [12,36–40]: mitochondrial functionality, cell apoptosis, measurements of the DNA synthesis and fragmentation, cell proliferation, cell membrane integrity tests, cell cycle analysis, protein expression, binding assays, protein opsonization assays in human plasma, protein-nanoparticle interactions, immunoassays, blood compatibility analysis (haemolysis), reactive oxygen and nitric oxide species analysis (ROS and RNOS) – cell stress, cell morphology, cellular inflammation and cell–biomaterial association/uptake studies. However, not all assays are suitable for testing with mesoporous materials. For example, the 3-(4,5-dimethylthiazol-2-yl)-2,5-diphenyltetrazolium bromide (MTT) assay (or its variants) widely used in the evaluation of cytotoxicity cannot be used to study the cytotoxicity of certain PSi materials, because these materials might act as a reducing agents of the product assay, preventing the correct prediction of the *in vitro* toxicity [41]. In addition, to test the effect of direct contact with cells, an indirect assessment of the exposure to drugs, residual solvents, degradation products, etc., should also be considered.

A thorough *in vitro* analysis is the first screening for the understanding of the mechanisms resulting from the cell–biomaterial interactions and the first assessment before moving to *in vivo* tests. This is because *in vitro* experiments are cheaper and faster, less time-consuming, and more advantageous and ethical than animal experimentation. Recently, there has been increasing concern about the toxicity of biomaterials, in particular regarding nanoparticulate systems that can enter the cells and induce cell damage or immune response. The particle toxicity is size-dependent and can result from the interaction of the cell membrane and the biomaterial or be due to the internalization of the particles by the cells. Most of the cells cannot take up large microparticles (e.g. sizes above 25 μm) and the resulting toxicity is mainly due to membrane interactions. Nano- and submicron particles can be engulfed by macrophages, whereas nanoparticles up to few hundred nanometers in diameter can also be endocytosed by other common cells, which may induce cytotoxicity [12, 36, 39, 40].

After the *in vitro* assessment, *in vivo* studies also play an important role in further deepening the understanding of the biocompatibility of the materials, because they take into account the biological environment contrary to the *in vitro* cell-based models. The biomaterial may not be directly cytotoxic, but may yet induce a reaction that is destructive in the body. For example, bioinert materials can cause inflammatory (tissue) reactions which cannot be predicted *in vitro*. The fast or the prolonged/slower release of a drug may also be harmful or cause local toxicity. Therefore, the biocompatibility involves more than just simple interactions between the material and one cell type; it involves also possible degradation products, and separate effects by the drug and the carrier. The information regarding the biocompatibility *in vitro* and *in vivo*, the biodistribution, retention and clearance assessments of the mesoporous materials *in vivo* are, therefore, very valuable. Despite the great potential of porous materials in biomedical applications, our knowledge of the toxicological effects *in vivo* is still very limited. This is because many features of the materials, such as concentration, particle size, surface area, chemical composition, surface structure, solubility, particle shape and aggregation, can act as potential toxic triggers.

15.4.1 *In Vitro* Studies

In 1995, PSi materials were reported to be biocompatible for the first time [42], opening up an avenue for further development of PSi-based materials for biomedical purposes. Several *in vitro* studies have demonstrated that PSi-based materials could promote calcification for tissue and bone engineering [43, 44], with the possibility of cell adhesion and culturing [45], growth of neural networks [46], as well as being suitable for protein adsorption [47]. The adhesion of human blood-derived monocytes and primary rat hepatocytes to PSi surfaces (nanoporous, nanochannelled and flat) and to PSi hydrosilylated with carboxylic acid species (undecanoic acid modified), respectively, resulted in non-immunogenic, biocompatible and stable responses similar to those observed in tissue culture polystyrene without the formation of significant ROS and inflammatory reactions [48, 49]. Naturally, the physicochemical and surface properties of the PSi particles affect the cyto-adhesion. For example, human lens epithelial cells poorly adhered to non-treated PSi (20 and 60 μm in size; ~68–70% porosity) causing complete cell death as a result of the ROS formation induced by the cell culture medium [50]. On the other hand, oxidized PSi microparticles of the same size (with and without a medium containing red phenol) completely mitigated any cytotoxic effect. The biocompatibility, stabilization and wettability of the PSi materials may also be further improved by coating, for example, with self-assembling biofilms of hydrophobin proteins [51], which gives them beneficial features for cyto-adhesion as implantable devices.

In order to efficiently and safely use the PSi-based particles in oral drug delivery, their toxicity at the intestinal level should be minimized. The most significant parameters of several PSi-based microparticles of different size fractions and surface chemistries (Table 15.1) affecting the cytotoxicity of Caco-2 intestinal cell models (human epithelial colorectal adenocarcinoma cell line) were reported to be the particle dose and size, and the surface chemistry of the particles [36]. The cytotoxic mechanism was described as a result of the mitochondrial disruption, ATP depletion, increase in the ROS production and cell apoptosis. For shorter exposure times similar to the intestinal transit time (3–8 h) those effects were minimal in contrast to longer exposure times (11–24 h), in particular for particle sizes below 25 μm. Table 15.1 shows that among the tested surface chemistries, oxidized PSi was the most favorable surface regarding the toxicity on the Caco-2 intestinal cells. Therefore, when developing PSi-based formulations for oral drug delivery, the particle dose may be an issue for drug efficiency, and the non-toxic threshold doses should be determined beforehand. For the hydrocarbonized and carbonized PSi particles of sizes above 25 μm, doses below 2 and 4 mg/ml, respectively, are the possible starting points for the development of the drug formulations encompassing PSi-based materials. The corresponding value for oxidized PSi particles would be below 4 mg/ml.

In addition to intestinal cells, the biomaterials may interact with other kind of cells when administrated to humans or animals. Therefore, for the representative cytotoxicity screening a large number of different cell lines should be tested. For example, cultured murine macrophages (RAW 264.7) were exposed to non-porous silicon particles of various sizes in the range of 3–3000 nm for 24 and 48 h. The cytotoxicity and inflammatory responses induced were reported to be size- and dose- dependent [52]. The nanoparticles were better tolerated at concentrations up to 20 μg/ml. Nano- and microparticles at concentrations above 20 and 200 μg/ml, respectively, were more critical to the cells. In case of silicon microparticles, a dose-related increase in cytokine production was observed, whereas increasing the doses of nanoparticles affected the cytokine production less.

A further example of the cytotoxic profile of PSi nano- and microparticles is shown in Figure 15.5. Here, Caco-2 and RAW 264.7 macrophage cells exposed to hydrocarbonized (hydrophobic) and oxidized (hydrophilic) PSi particles (the sizes ranging from 97 nm to 25 μm; zeta-potential ca. –30 mV) showed results consistent with the dose- and size-dependence effect, as well as with the influence of the surface chemistry on cytotoxicity [12, 40]. Nevertheless, the induced toxicity, oxidative stress (ROS and RNOS) and inflammatory responses (e.g. tumor necrosis factor, TNF-α) did not follow a linear relationship between the cell cultures and the particle size. Generally, based on these results, hydrocarbonized and oxidized PSi particles can be considered more critical to macrophages than to Caco-2 cells. An optimal particle size showing dose independent bioinertness was found to be ca. 142–164 nm for both hydrocarbonized PSi and oxidized PSi particles.

Table 15.1 *Physical characteristics of the PSi-based microparticles (average pore diameter, D, pore volume, V, porosity and specific surface area, A) and their effects on the in vitro cytotoxicity and ROS production as a function of the particle size and surface modification: (−) minimal effect, (+) moderated effect, (+ +) strong effect. Reprinted with permission from [36].*

	Particle size fraction/μm				Assay/determination				
						Flow cytometry		Intracellular ROS	
	D (nm)	V (cm^3/g)	Porosity (%)	A, (m^2/g)	Luminescence ATP content	Apoptosis	Necrosis	H_2O_2	Superoxides
THCPSi									
1.2–25	12.8	1.03	71.2	456	−/+	+	+	−	−/+
25–38	12.7	1.03	71.2	450	−/+	+	+	−	−/+
38–53	13.6	1.16	73.6	461	−	+	+	−	−
53–75	12.9	1.27	75.3	476	−	−/+	−/+	−	−
TCPSi									
1.2–25	13.9	0.92	68.7	302	+ +	+	+	+ +	+ +
25–38	14.9	0.99	70.3	289	+ +	+	+	+	+ +
38–53	12.8	0.96	69.7	328	−/+	+	+	+	+
53–75	12.5	0.97	70.0	343	−/+	+	+	+	+
TOPSi									
1.2–25	12.2	0.79	65.5	331	−	−/+	−/+	−	−/+
25–38	11.6	0.76	64.5	319	−	−	−	−	−/+
38–53	10.7	0.84	66.8	361	−	−	−	−	−
53–75	10.8	0.81	66.0	341	−	−	−	−	−

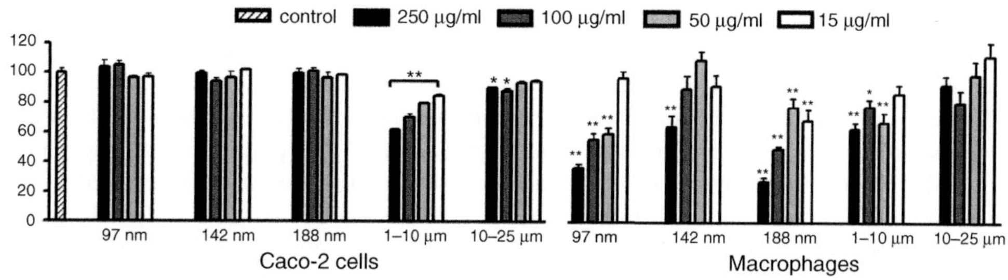

Figure 15.5 Viability of Caco-2 and RAW 264.7 macrophage cells after 24 h incubation with different concentrations (μg/ml) of hydrocarbonized PSi particles, assessed by an ATP luminescent-based assay. Reprinted with permission from [12].

Modification of the particle surface with fluorescent-based probes, such as fluorescein isothiocyanate (FITC), allows tracking of the particles in contact with the cells, for example, by confocal fluorescence microscopy. In this context, the FITC-modified hydrocarbonized PSi nanoparticles showed a strong association with Caco-2 intestinal cell monolayers without permeating across the cell membranes regardless of the particle dose [12]. Similar behavior was also observed with oxidized PSi nanoparticles, but in this case the particles were more readily phagocytized by the macrophages and also dissolved faster in fasted-state simulated intestinal fluid (FaSSIF, pH 6.5, with over 80% of the nanoparticles (164 nm) dissolved in silicic acid in ca. 6 days) than in gastric simulated fluid (SGF, pH 1.2) (Figure 15.6).

The degradation (or dissolution) kinetics of the PSi particles can be further controlled by changing the surface properties of the material. For example, by covalently attaching PEGs to the PSi surface (pore sizes of 30–50 nm) the degradation kinetics of PSi structures can be tuned [53]. The crafting of PSi with PEGs with different molecular weights (245–5000 Da) slowed down the degradation of the particles

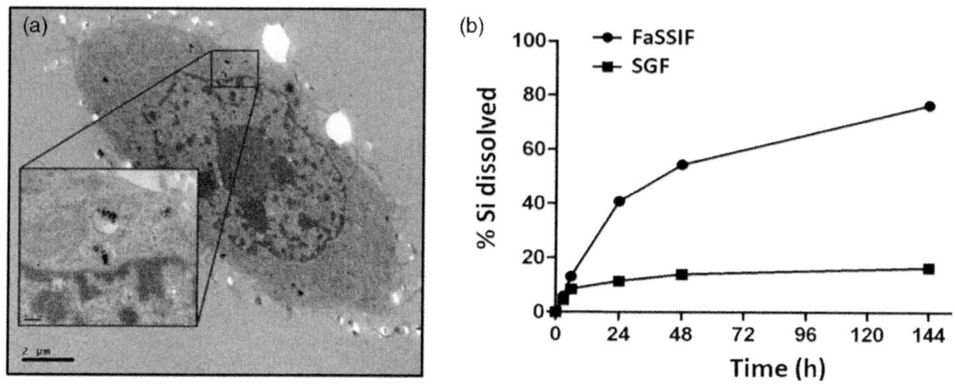

Figure 15.6 (a) TEM pictures of flat embedded ultrathin sections of oxidized PSi nanoparticles (164 nm, 50 μg/ml) inside endosomal compartments of RAW 264.7 macrophages; (b) Dissolution profiles of oxidized PSi nanoparticles (164 nm, 1 mg/ml) incubated in blank FaSSIF and SGF at 37°C as a function of time.

in phosphate buffer solution (pH 7.2) and serum at 37°C. The degradation was hindered with increased molecular weight of PEG and no extensive release of pro-inflammatory cytokines in THP1 human macrophages (human acute monocytic leukemia cell line) was induced. The longest PEG chains inhibited the degradation of the PSi particles down to ca. 4 days for the total degradation of the particles. The mechanism of the PSi degradation and dissolution in physiological solutions is associated with the erosion of the pores and with more irregular and less smooth particle surface formed during the degradation process. The biodegradation of PSi in simulated body fluids [54,55] is also dependent on its porosity. Usually, materials with porosity above 70% dissolve practically in all the simulated body fluids (except gastric fluids, see Figure 15.6B), whereas PSi with porosity below 70% degrades more slowly. Thus, regarding various drug delivery applications PSi-based materials have the advantage that the biodegradation can be controlled by the overall porosity, pore size, particle size and surface properties, which in turn are controlled by the choice of the fabrication parameters of the material. The great advantage is that, in the body, PSi is degraded completely into non-toxic and resorbable silicic acid ($Si(OH)_4$), which is the most natural form of Si in the environment and non-toxic. Silic acid is also important in the human physiology [56].

15.4.2 *In Vivo* and *Ex Vivo* Studies

Contrary to *in vitro* tests, *in vivo* tests on PSi-based materials are still in their infancy. Several studies have been reported in the past few years, but there is still lack of understanding on the effect of PSi materials in an *in vivo* situation. Regulatory bodies in Singapore, the UK and the USA (the Food and Drug Administration) have approved PSi for clinical trials in brachytherapy (use of radioactive sources brought into close contact with the target tissues) [57]. For example, Brachysil™ (an active radiologic medical device comprising radioactive phosphorus ^{32}P and BioSilicon microparticles of 20 μm size) of pSivida Ltd. is now in phase II clinical trials for inoperable primary liver cancer and also for pancreatic cancer [58]. Brachysil™ is an active implantable (radiologic) device containing highly pure silicon, doped with phosphorus, which is irradiated to form ^{32}P for the treatment of unresectable solid tumours. By combining PSi with ^{32}P, the product can be delivered directly to a tumour and due to the porosity of the particles they can be immobilized within the tissues and deliver a restricted and targeted dose of beta radiation without significant leakages.

Other systems have been developed to allow a multistage drug delivery (MDS). The idea of this delivery system is to enable the loaded therapeutic nanoparticles to sequentially overcome biological barriers and reach their target. The MDS is composed of PSi particles constituting the first stage particles (S1MP) with the dimensions of several hundred of nm to a few microns, which accommodate second stage therapeutic nanoparticles (S2NP) in their porous structure of up to 100 nm in size [59]. For example, the biocompatibility of S1MP particles (diameter of 1.6 μm and pore size of 36 nm; zeta potential of −32 mV and 8.7 mV) after intravenous administration to immunocompetent mice in acute single doses (10^7, 10^8, 5×10^8 PSi particles/animal) and subchronic multiple doses (10^8 PSi particles/animal/week for 4 weeks) showed an accumulation (histopathology) of the particles in the liver and spleen without significant inflammatory responses (both innate and adaptive immunity) [60]. The lactate dehydrogenase levels of renal and hepatic biomarkers and the response of

23 different plasma cytokines remained unchanged upon administration, which shows that single and sub-chronic multiple administrations of those particles are well tolerated in mice after intravenous administration. Corresponding results on plasma cytokine concentrations after single subcutaneous administration of hydrocarbonized PSi microparticles have also been reported with mice [61].

Besides the therapeutic purposes, PSi-based materials can serve as platforms and diagnostic probes for biomedical imaging applications *in vivo*, such as for tumour and inflammation imaging. Imaging can be employed by following the fluorescently labelled or intrinsically photoluminescent particles circulating in the body. Taking advantage of the photoluminescent property of PSi, Park and co-workers prepared PSi nanoparticles (126 nm in size; pore diameter of 5–10 nm) and loaded the particles with anticancer doxorubicin drug [62]. They showed that the particles slowly release the drug in physiological conditions, reaching a maximum within 8 h, without inducing significant toxicological effects to HeLa cells (immortal cell line derived from cervical cancer cells). At pH 7.4 and 37°C the particles also dissolved into silicic acid and simultaneously lost their luminescence. It was demonstrated that the particles passively accumulated in the tumour, liver and spleen, followed by renal clearance from the body within 4 weeks, without significant inflammatory responses observed from the degradation product of Si in the animal tissues.

Despite the above-mentioned advantages of PSi-based materials, *in vivo* fluorescence imaging itself in the animal tissues/organs is not an easy task and has its own limitations. For example, it is not fully quantitative due to signal attenuation in tissue, especially when an object is located deep in the body. Therefore, the *in vivo* optical imaging is limited to small laboratory animals. As an alternative, radio-labelled PSi materials can be used for further imaging and pre-diagnostics *in vivo*. In this context, Bimbo and co-workers developed hydrocarbonized PSi nanoparticles (~142 nm) labelled with ^{18}F radionuclide and used the system to assess the biodistribution of PSi particles after enteral and parenteral administration in rats [12]. Positron emission tomography (PET) is fully translational and qualitative *in vivo* imaging modality, which is widely used in clinical diagnostics and drug development. Radionuclide ^{18}F is a PET compliant isotope with high sensitivity, good temporal resolution and it has 110 min. half-life ideally suited for dynamic imaging of particle distribution in the body. ^{18}F-hydrocarbonized PSi nanoparticles can be administrated orally, subcutaneously and intravenously, rendering the possibility of monitoring the biodistribution of the particles as they travel through, for example, the gastrointestinal (GI) tract of the animals (Figure 15.7) or their accumulation in the different organs. The described radio-labelled nanosystem exhibited excellent *in vivo* stability, which further renders the PSi-based material a flexible and multifunctional platform for various drug delivery applications.

PSi has also been used as a drug delivery platform for intraocular diseases. In these studies, various PSi particulate systems were used [63, 64] with positive biocompatibility results: oxidized PSi microparticles did not cause any significant toxicity when dispersed *in vitro* into the rabbit and rat eye, and aminosilanized PSi membranes did not erode the adjacent tissue and did not cause marked accumulation of inflammatory cells around the PSi membrane.

The *in vitro* and *in vivo* assessments are essential steps in the development of biomaterials for drug delivery applications. In the future, it is expected that there will be an exponential increase in the amount of toxicological and biocompatibility data related to PSi-based

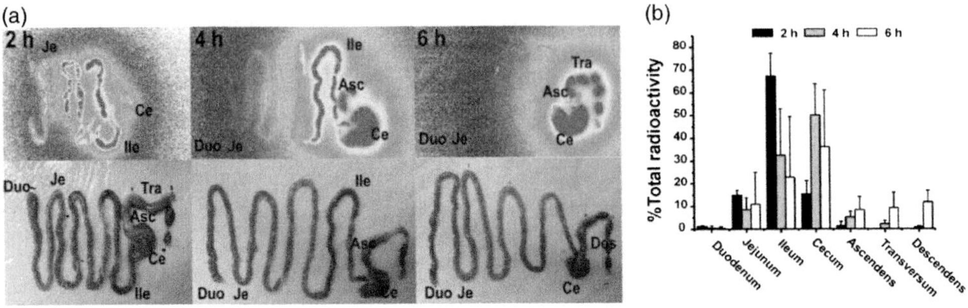

Figure 15.7 *(a) Ex vivo macroautoradiographs and respective photographs of the rat GI tracts at 2, 4, and 6 h after oral administration of* ^{18}*F-hydrocarbonized PSi nanoparticles (abbreviations: Duo, duodenum; Je, jejunum; Ile, ileum; Ce, cecum; Asc, ascendens; Tra, transversum; Des, descendens). (b) Radioactive distribution of* ^{18}*F-hydrocarbonized PSi nanoparticles after oral administration in the lower GI tract as quantified from autoradiography. Reprinted with permission from [12].* (See colour plate section.)

materials both *in vitro* and *in vivo*, as well as a further improvement in the chemical versatility of PSi to allow better blood circulation times, adsorption-specificity, biodegradability, clearance from the animal/human body, specific targetting, etc. In addition, it is also envisioned that further modifications and functionalizations of the PSi surface will allow the application of these materials as theranostic nanovehicles in clinical trials.

15.5 Nanostructured Silicon Materials as DDS

15.5.1 Drug-Loading Procedures

Drug loading into a carrier material can be performed in several ways, but the fabrication method of the carrier may limit the alternatives. In many cases, the drug need to be added during the synthesis of the carrier material. This may restrict the usable drug concentrations and lead to relative low drug loads. Also, the chemical environment in the synthesis can be too harsh for sensitive compounds, such as peptides and proteins. In the case of mesoporous carrier materials, it is advantageous that the drug loading can be performed separately after the fabrication of the carrier.

Although there is a wide variety of loading methods, they can be classified mainly into two groups [5]: immersion and impregnation methods. In an immersion method, the mesoporous particles are immersed in the drug solution, in which a certain amount of the drug is dissolved in a desired solvent and the volume of the solution is substantially higher than the pore volume of the particles. After the adsorption equilibrium is reached, the particles are separated from the solution by filtration or centrifugation and dried to obtain the dry powder of microparticles. In the impregnation or incipient wetness method, a controlled amount of drug solution is added to the particles and allowed to infuse through capillary forces into the pores. The excess solvent is then removed by evaporation, while the drug remains in the pores. The impregnation method is especially preferred when loading

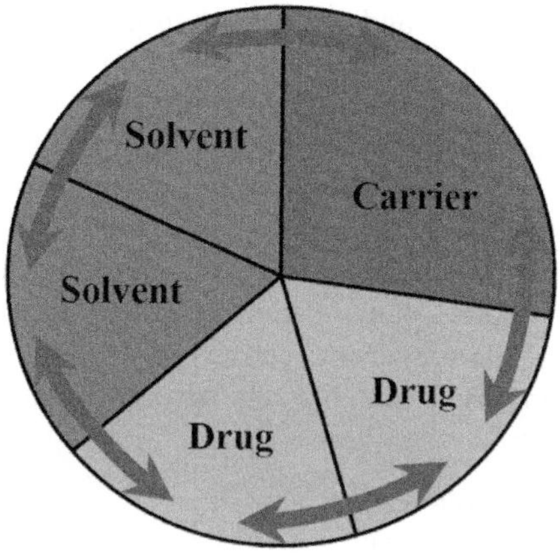

Figure 15.8 Pie chart illustration of different interactions (indicated with the arrows) in the drug loading processes.

expensive drugs or when only small amounts of the drug are available, because practically all of the drug will be eventually located in the end product. In addition, the amount of the drug in the loaded material is easy to determine in advance. However, the uniformity of the loading and the possible crystalline fraction on the external surface of the particles are more difficult to control than in the immersion method.

There are also many variations of these two methods described above. For example, a melt method, in which a physical mixture of drug and carrier is heated above the melting point of the drug, can be seen as a special case of the impregnation method [65, 66]. It should be noted that the melt method can be applied in special cases only. Most of the drugs cannot stand melting without decomposition and the viscosity of the melted drug rarely enables a successful loading. In addition, the melting temperature of poorly soluble drugs is usually relatively high, which may lead to chemical reactions between drug molecules and the carrier surface. Mellaerts and co-workers combined these methods using an excess amount of the initially dilute drug solution as in the immersion method, but they allowed the mixture to evaporate as in the impregnation method [66]. This approach also allows the use of more practical loading methods for large-scale processes such as fluid bed and rotavapor methods [67].

Despite its apparent simplicity, drug loading by immersion has a complex set of interactions, which all affect the final product. In Figure 15.8, the main interactions are illustrated. The most important interactions are carrier–drug, drug–solvent and carrier–solvent, but also drug–drug and solvent–solvent interactions may play an important role in the loading processes. The interactions between the drug and the solvent have an important role also in the selection of the proper solvent, which is important for the successful loading. However, in the case of very poorly soluble drugs, the number of possible solvents is usually limited, preventing the possibility of controlling the loading process by the solvent selection.

The drug-loading process begins with a monolayer adsorption on the surface of the carrier, which is determined mainly by drug–carrier, drug–solvent and carrier–solvent interactions. For efficient monolayer adsorption, the affinity of the drug molecule to the carrier surface should be dominant and stronger than drug–solvent interactions. Strong carrier–solvent interaction may have a hindering effect on the adsorption process. In the subsequent multilayer adsorption, the drug–drug interaction becomes significant, whereas the effect of carrier–drug and carrier–solvent interactions diminishes due to coverage of the carrier surface. It is of crucial importance to know whether the affinity of the solute drug molecule to the adsorbed molecule is stronger than the drug–solvent interaction, and to understand the adsorption mechanism of the monolayer molecules.

Adsorption is a general term for a phenomenon where the adsorbate (the dissolved drug) is attached to the surface of the adsorbent (carrier) in a liquid or gaseous phase due to physical or chemical interactions. Three different adsorption mechanisms can affect drug loading into porous materials:

1. physical adsorption (physisorption);
2. chemical adsorption (chemisorption);
3. physical trapping of the payload.

The primary mechanism by which adsorption takes place has a significant effect on the loading process and especially on the drug release. Physisorption refers to an adsorption process in which the molecule is adsorbed on the surface with weak physical interactions and it is based on three different interactions:

1. Attractive electrostatic forces between permanent charges like dipoles, quadrupoles or higher order multipoles (Keesom forces).
2. Attractive polarization forces in which a permanent multipole (charge) induces multipole on another (nonpolar) compound (Debye forces).
3. Attractive force between two instantaneously induced dipoles arising from a transient fluctuation of electron density in molecules (London forces).

Physisorption is the most favourable loading mechanism, if the aim is to enhance the solubility of the payload compound, because physisorption is reversible and allows the quick release of the loaded compound with minimal activation. Chemisorption provides a strong chemical interaction (bonding) between the drug molecule and the surface of the carrier. This is favourable when the aim is to have sustained release and makes it possible to control the release rate of the drug by tailoring the dissolution rate of the carrier material itself. Wu and co-workers used this approach to delivery ophthalmic drug daunorubicin-loaded PSi into the eye of the rabbit [68]. Daunorubicin was covalently attached to the undecylenic acid-treated PSi and its release rate was dependent on the dissolution rate of the carrier material providing almost constant therapeutical level of the drug for 30 days. However, there are also some problems in the covalent bonding approach. Only the monolayer of the adsorbate can be chemisorbed onto the carrier, limiting the maximum payload, which means that the surface area of the carrier determines the maximal loading capacity. In addition, the release mechanism of the attached drug molecule should be known in order to ensure that the therapeutic activity is maintained and no chemical degradation of the payload molecules takes place.

Physical trapping of payloads is a rarely used procedure, but it can be applied basically using two different methods. Gu and co-workers trapped magnetic iron oxide nanoparticles in the PSi by loading the nanoparticles into the pores and subsequently reducing the pore size by thermal oxidation [69]. The other method is to adsorb large compounds (proteins or polymers) on the external surface and block the pore opening with them [70, 71]. The trapping may also include some release-triggering properties, for example, based on a change of pH.

The most important physicochemical characteristics of the drug loaded into mesoporous materials are the amount and integrity of the loaded drug, and the fractions of the drug inside the pores and on the external surface of PSi. Although there are several methods to quantify the amount of the loaded drug, none of them can give adequate information alone.

There are four possible physical states in which the drug compound loaded in porous materials can be found: (a) a thin liquid-like layer next to the pore wall, which does not undergo any phase transitions due to the strong interaction with the pore wall (i.e., a delta layer); (b) the amorphous drug in the middle of the pores; (c) the nanocrystalline drug in the middle of the pores; and (d) the bulk crystalline fraction on the external surface of the particles. At least part of the drug always exists in the liquid-like state, but the thickness of this delta layer is dependent on the strength of the surface–drug interactions, while the existence of the other states depends more on the diameter of pores, the size of the drug molecules, the arrangement of the drug molecules in the pores and on the filling factor of the pores (i.e., how full the pores are).

As the system might be complex, a combination of different analyzing methods is preferable. Chromatographic and spectroscopy methods can be used to determine the amount of the released drug as well as its chemical purity. A combination of thermoanalytical methods, such as thermogravimetry (TGA) and differential scanning calorimetry (DSC) can give adequate information about the amount of the drug load and also its crystallinity [5]. However, a critical judgment of the results is needed to avoid erroneous conclusions. For example, due to confined space in the pores, crystallization can take several weeks, which should be taken into account in the characterization of the loaded PSi materials [72].

15.5.2 Enhanced Drug Release

The prerequisites for a feasible drug delivery system are biocompatibility and non-toxicity of the carrier applied, no matter if the system is intended for targeted drug delivery or for oral administration of poorly soluble drugs. It is estimated that more than 90% of new drug candidates suffer from poor bioavailability due to low solubility/dissolution rate in the stomach and intestinal lumen, poor permeability properties in the GI tract and high intestinal and/or hepatic first pass metabolism. Poor solubility alone is an issue for 40% of the new drug candidates. As the issue related to poor solubility is a limiting factor for drug efficacy, several technologies have been developed to improve the situation, PSi being one of the potential choices. As reviewed in the previous paragraphs, PSi is an acceptable biomaterial and it possesses several features advantageous as an efficient drug carrier for poorly soluble drugs, such as:

- The confinement of the drug molecules in the pores produces a disordered structure of the payload drug, which hinders the crystallization and stabilizes the disordered structure against crystallization or reduces the crystal size.
- The confinement of the payload molecules protects the molecules from the harsh conditions of the body environment and from enzymatic degradation.

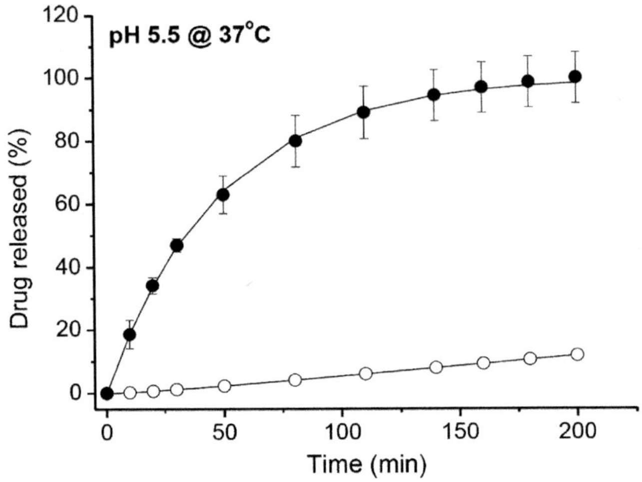

Figure 15.9 *Dissolution profiles of indomethacin-loaded in carbonized PSi microparticles (solid symbols) and of the bulk crystalline drug (open symbols). The dissolution medium is phosphate buffered saline (pH 5.5) at 37°C. The drug-loading degree is 16 w-%. Indomethacin is a drug with low solubility but high permeability (Biopharmaceutical Classification System, BCS, classified as a Class II drug).*

- The release kinetics can be controlled with the pore size and the interaction between the payload molecule and the pore wall.
- The wettability can be improved by utilizing hydrophilic PSi particles.
- High loading degrees up to 60 w-% can be obtained.
- The physicochemical properties (surface chemistry, particle size and pore size, shape and morphology) of PSi carriers can easily be controlled by the fabrication parameters.

A typical dissolution profile for a poorly soluble drug loaded in PSi microparticles is presented in Figure 15.9 and compared with the corresponding profile obtained with the crystalline drug. When the intention is to enhance the drug dissolution behavior, the pore surface of the PSi carrier needs to be chosen so that only physisorption takes place. To improve the loading degree, further adsorption of extra drug molecules in the PSi particles is assisted by the capillary effect. As an optimal situation, the PSi pores are fully filled with the drug with no crystalline drug material on the external surface of the carrier. In this case the pore volume limits the loading degree. The improvement of the drug dissolution behavior is based on the fact that in the disordered structure the lattice energy of the drug compound is reduced and the release of the molecules in the solution phase is enhanced when compared with the bulk crystalline drug. As the pores of PSi are large enough for the fast diffusion of the molecules out of the pores, the overall fast release is obtained and typically obeys the first order kinetics. However, depending on the pore diameter and morphology, also the diffusion of the drug molecules might be the limiting step of the drug release [73]. Considering the versatile fabrication of PSi and the critical physicochemical parameters of the carrier to control the release kinetics, PSi is a potential material to optimize the performance of the carrier and to obtain high bioavailability of poorly soluble drugs.

In oral delivery applications, PSi microparticles are usually preferable due to their high drug load capacity and easy handling [74, 75]. In the very first report demonstrating PSi as drug carrier, insulin was delivered across Caco-2 cell monolayers [76]. After that, several drug molecules have been successfully loaded into PSi microparticles, for example, ibuprofen [77], steroid dexamethasone [78], doxorubicin [79], and cisplatin [80]. The drug-loading degree can be optimized by choosing the desired material properties (hydrophilicity/hydrophobicity, pore size and volume, surface chemistry, etc.) and the loading parameters in relation to the drug molecule. The amount of the drug confined in the pores of mesoporous materials is usually quantified by a combination of different techniques such as thermogravimetry, differential scanning calorimetry, nitrogen adsorption/desorption and high-performance liquid chromatography [72, 75, 81]. By utilizing thermoanalytical methods it is also possible to distinguish different thermodynamic states of drugs when loaded into PSi microparticles with various surface chemistries and pore diameters as discussed earlier.

The chemical stability of the payload drug needs to be verified after the loading and the loading parameters need to be changed accordingly if problems rise. Another important aspect of the PSi carrier is that the loaded drug must be chemically and physically stable for long periods of time after loading. For this purpose, the physicochemical properties of the drug need to be monitored for the samples stored under stressed conditions of temperature and humidity. Also here, the surface chemistry of the carrier plays an essential role and it has been reported that hydrophilic carbonized PSi microparticles are the most stable and induce fast ibuprofen dissolution, with no changes in the drug release profiles even after 3 months of storage [77]. The surface chemistry of the particles is also a key parameter that dictates the behavior of the drug release profiles besides the pore size. For example, fast release of dexamethasone can be obtained from freshly etched, hydrogen-terminated PSi films due to the chemical instability of PSi at physiologic pH [78], but the chemical modification of the PSi films by thermal hydrosilylation with dodecene improves the stability of the film and slows down the release. Also, doxorubicin loaded into a two-layer PSi matrix with pores sizes of 2 μm and a 200 nm nanoporous coverage exhibited a time-dependent drug release profile that was modulated by the thickness of the biodegradable nanoporous silicon layer [82].

As a versatile material, PSi-based materials can be prepared to meet the needs of oral drug delivery, i.e., improved fast drug dissolution, enhanced drug permeation across intestinal barriers and reduction of the drug pH-dependency. As the interaction between the pore surface and the payload molecules is essential for the loading, five different model compounds were loaded in both oxidized and carbonized PSi microparticles, and also the effect of the loading solution was screened [75], varying the drug-loading degrees between 9 w-% and 45 w-%. Generally, by loading poorly soluble drugs in PSi microparticles with pores of ca. 9 nm in diameter, faster dissolution rates are obtained and also the pH-dependency of the dissolution is reduced (Figure 15.10).

Caco-2 cell confluent monolayers on Transwell® insert filters were used [40, 83] to study simultaneously the drug dissolution and permeation behavior of the drug loaded in carbonized PSi microparticles (furosemide) and oxidized PSi nanoparticles (griseofulvin). Caco-2 cell monolayers can be used to study several active and passive permeability properties simultaneously, while more simplified *in vitro* methods only provide information on the specific factors they are designed to model. Caco-2 cells are immortalized, widely

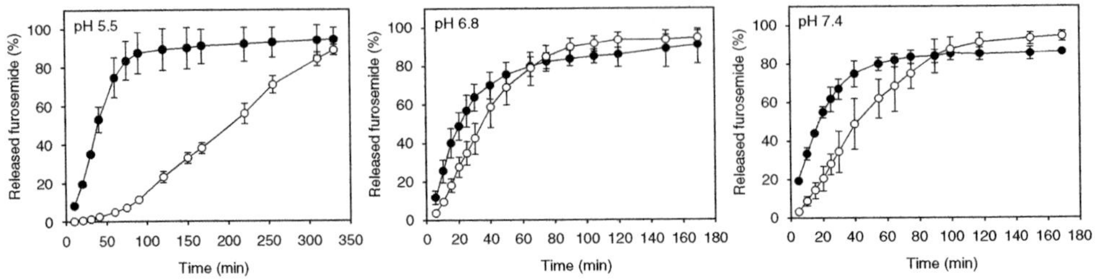

Figure 15.10 Dissolution profiles of furosemide loaded in carbonized PSi microparticles (solid symbols) and pure drug (open symbols) at pH values of 5.5, 6.8 and 7.4, and at 37°C. Reprinted with permission from [75].

used and well-differentiated cell lines derived from human colorectal adenocarcinoma. The permeability across Caco-2 cell monolayers is considered to model the intestinal absorption, because the cells represent many of the characteristics and functions of the epithelium of the small intestine. During the permeation study the electrical resistance of the cell monolayers (transepithelial electrical resistance, TEER) is often tested to evaluate the tight junction formation. Usually, a TEER value of 300–500 Ωcm^2 indicates reasonably tight junctions that prevent leakage and, thus, high-background basal drug concentration in the receiver compartment. Basically, the permeability coefficient of a drug compound is calculated according to $P_{app} = (dM/dt)/C_0A$, where M is the drug amount transported from the donor to the receiver side of the insert at the time point t, C_0 is the initial drug concentration, and A is the surface area of the Caco-2 cell membrane on the insert. Following this procedure, the permeability and flux (μg/min) of furosemide loaded in carbonized PSi microparticles yielded ca. 1.5- and 5.2-fold improvement in the drug permeability, respectively, across Caco-2 cell monolayers from pH 5.5 (donor compartment) to 7.4 (receiver compartment) when compared with pre-dissolved drug solution (Table 15.2). When the pH of the medium increased, the permeability of the dissolved furosemide decreased, but the corresponding P_{app} value of the furosemide loaded in carbonized PSi microparticles increased ca. 4.7-fold at pH 7.4, where the permeability of furosemide is the poorest. The TEER values did not decrease due to the microparticles showing that the integrity of the monolayer was not compromised by the TCPSi microparticles. Similarly, also oxidized PSi nanoparticles (170 nm) were able to enhance permeation of griseofulvin across Caco-2 monolayers regardless

Table 15.2 Apparent permeability (P_{app}) of pre-dissolved furosemide (control solutions) and furosemide-loaded carbonized PSi microparticles across Caco-2 cell monolayers. Reprinted with permission from [83].

	Control solutions		TCPSi microparticles	
pH	Concentration (μg/ml)	$P_{app} \times 10^6$ (cm/s)	Concentration (μg/ml)	$P_{app} \times 10^6$ (cm/s)
7.4	650	0.30 ± 0.01	670	1.38 ± 0.05
6.8	600	0.93 ± 0.38	840	4.1 ± 0.5
5.5	150	12.2 ±1.0	520	18.0 ± 1.3

Table 15.3 *Pharmacokinetic parameters of pure indomethacin and indomethacin formulations after intravenous and oral administration in rats. Reprinted with permission from [84].*

Parameter	Pure indomethacin		Indocid (oral)	Indomethacin-loaded oxidized PSi (oral)
	Intravenous	Oral		
Dose, mmol	0.005	0.01	0.01	0.01
T_{max}, h	0	2.75 ± 0.65	2.00 ± 0.00	0.56 ± 0.31
C_{max}, μg/ml		2.48 ± 0.38	3.49 ± 0.54	6.46 ± 0.52
$AUC_{0\text{-}24h}$, h μg/ml/mg (area under the curve)	66.92 ± 6.73	35.83 ± 1.95	51.82 ± 2.33	66.98 ± 2.62
F (absolute bioavailability), %		53.54 ± 2.91	77.43 ± 3.9	100.1 ± 3.9

of the pH conditions [40]. All these results showed improvement of the physicochemical properties (solubility/dissolution and permeability) of the drug molecules after being loaded into PSi particles and suggest that the high local concentrations provided by the enhanced dissolution properties of drug-loaded particles may prove beneficial for drug absorption in an *in vivo* situation.

Currently, major challenges faced in the formulation of poorly soluble drugs are the solid state stability and the establishment of a reliable *in vitro-in vivo* correlation (IVIVC). In the case of indomethacin, the improved dissolution behavior obtained by loading the drug in a mesoporous carrier was also reflected in the improved pharmacokinetic results [84]. The advantage of utilizing oxidized PSi microparticles as the carrier is the fast drug dissolution at pH 7.2 similar to the commercial formulation Indocid® compared with pure crystalline indomethacin. Improved *in vivo* pharmacokinetics (increase of the maximum plasma concentration, C_{max}, increase in bioavailability and decrease of T_{max}, which is the time at which C_{max} is reached) is achieved when indomethacin is loaded into oxidized PSi microparticles compared with the pure indomethacin or Indocid® formulation. The plasma concentration of indomethacin in fasted rat models was significantly increased when the drug was formulated into oxidized PSi microparticles with C_{max} ~6.5 μg/mL and bioavailability ~100% and with a remarkable decrease in T_{max} (~0.6 h) compared to the free indomethacin and the Indocid® formulation (Table 15.3). In addition, no adverse health effects on the rats after administration of the indomethacin-loaded oxidized PSi microparticles were reported. Most importantly, the linear IVIVC relationship (level A) obtained for both the indomethacin-loaded oxidized PSi microparticles and crystalline indomethacin samples (in good agreement with the BCS classification of a Class II drug) shows that the *in vivo* absorption of indomethacin-loaded oxidized PSi microparticles can be predicted from *in vitro* dissolution profiles.

For sustained drug delivery applications, the drug release rate should be prolonged and this can be effectively achieved by covalently attaching the drug molecules on the pore walls of the inorganic mesoporous carrier. With this approach the drug release can be triggered and fine-tuned by the incorporation of responsive polymers creating a so-called 'capping' or 'gatekeeper' concept. This kind of approach can be attained by employing a pH-responsive, chitosan-based hydrogel film to cap the pores of the PSi layer after being loaded with insulin [71]. The porous layer is formed by oxidized PSi particles and the hydrogel film composed of chitosan and glycidoxypropyltrimethoxysilane. By switching the pH from 7.4 to 6.0 an immediate release of insulin from the oxidized PSi films into the solution is obtained as a result of a volume phase transition in the hydrogel. In the same line, doxorubicin can be physically adsorbed to the PSi surface (30–50 μm particle size; 20–30 nm pore diameter) or covalently attached to a 10-undecenoic acid linker by thermal hydrosilylation and grafted to the PSi surface [79]. In the case of the drug physically adsorbed on oxidized PSi a rapid and complete drug release is observed within 24 h, whereas the particles containing a combination of covalently attached and physisorbed doxorubicin released the drug slowly over a period of 24 h. The doxorubicin release is extensively prolonged when attached covalently to the particles for over 5 days. The release is a result of the breaking of the covalent bonds followed by a two-step mechanism involving both oxidation and subsequent dissolution of the PSi matrix.

The bioavailability of peptides can also be increased by using PSi-based materials. In order to avoid multiple injections, sustained release of peptides for longer periods of time is desirable. Peptides studied *in vitro* and *in vivo* include insulin [76] and ghrelin antagonist [61].

15.5.3 Intracellular Uptake

The studies on the intracellular uptake of PSi materials and their payloads are still in their infancy and there are only few reports about the efficient internalization of PSi particles for intracellular delivery using several different cell lines. The cellular internalization of the mesoporous carriers is controlled by different factors, such as size and morphology of the carriers, electrostatic interactions between the biomaterials and the cell membrane, as well as the surface functionalization of the materials.

By combining the drug carrier and cell targeting moieties it is possible to enhance the intracellular uptake of the carriers and the delivery of the payloads inside the cells. This is particularly important in cancer therapies due to the lack of target specificity of the antitumor drugs alone, which results in adverse side effects and limited effectiveness. Thus, targeted and controlled drug delivery improves the pharmacological and therapeutic effectiveness of the drugs with minimized adverse effects. For example, the endothelium has recently been explored as one of the barriers for tissue-specific targeting. PSi-based MDS loaded with one or more types of second stage therapeutic nanoparticles (S2NPs) is an alternative for targeting delivery [59, 85]. This strategy has been extensively applied for intravascular targeting by administering first stage particles (S1MPs) to endothelial cells in the tumour vasculature [86]. S1MPs carry and release S2NPs after intravenous injection through the bloodstream triggered by environmental pH or proteases, or as a result of the degradation of the S1MP system, reaching the lesion by increasing the permeability through the endothelial fenestrations or by transcytosis.

Interestingly, hemispherical-shaped particles are more beneficial to surmount some biological barriers such as the cellular membranes and to escape from the MPS uptake compared to the spherical ones. They also migrate better towards the vasculature wall in the circulation [86]. S2NPs (e.g. quantum-dots and PEG-FITC-single wall carbon nanotubes) loaded into S1MPs can enter the cytosol and accumulate inside distinct vesicles in the cells [59]. PSi particles (diameters of 600 nm–3.2 μm) consisting of quasi-hemispherical and discoidal shaped particles are successfully phagocytosed by vascular endothelial cells [85, 86, 87]. In addition, the modification of the PSi surface with non-specific antibody-labels and the size of S1MP do not significantly alter the nature of the intracellular vesicle housing of S1MP [85]. Oxidized and modified-PSi microparticles induce a faster internalization by vascular endothelial cells than PEG-modified PSi throughout an actin-dependent mechanism involving both phagocytosis and macropinocytosis. Further modification of PSi microparticles with mPEG-5000 even suppresses the internalization of the microparticles in human umbilical vein endothelial cells [86].

The intracellular trafficking of APTES-modified PSi particles loaded with amino-PEG superparamagnetic iron oxide nanoparticles (15 nm) in human microvascular endothelial cells showed that the cell membranes surrounded each particle types predominately as a tight-fitting enclosure [88]. In the presence of 3.2 μm PSi particles loaded with 150 nM nocodazole, the membrane-encapsulated PSi particles migrate to the perinuclear region of the cell by a microtubule-driven mechanism within 50 min. The surface charge, shape (discoidal and hemispherical) and size (1.6 and 3.2 μm; pore size 26 and 51 nm, respectively) of the particles do not affect the rate of migration extensively, but the alteration of the microtubule network may change the motility of individual microparticle-laden phagosomes, and thus, reduce their accumulation in the perinuclear region by 80%. Phagosomes containing particles localized in the perinuclear region are still able to mature even in presence of the particles without significant arrest of the movement or localization of the endo-/phagosomes. Interestingly, the payloads of the PSi carriers are further able to be sorted into multi-vesicular bodies that dissociate from the phagosome into novel membrane-bound compartments.

All together, the above-mentioned studies reinforced the concept of PSi as multifunctional nanocarriers, in which different particles are able to perform specific tasks, in order to deliver single- or multi-component payloads to specific sub-cellular compartments. One of the most difficult and rather challenging in targeted delivery is the cancer therapy by small RNA interference (siRNA). This is a versatile approach to silence genes with a known sequence. By employing the above-mentioned PSi MDS, Tanaka and co-workers demonstrated the possibility of loading and sustaining the release of S2NPs composed of dioleoyl phosphatidylcholine (DOPC) nanoliposomal siRNA (siRNA-DOPC) [89]. *In vivo* studies performed in two orthotopic mouse models of human ovarian cancer and treated either twice weekly with siRNA-DOPC over 3 weeks, or by a single injection with S1MP-EphA2-siRNA-DOPC (EphA2 is a tyrosine kinase receptor in the Ephrin family overexpressed in >70% of human cancers, such as ovarian, breast, lung, and melanoma) showed that a single injection of siRNA-DOPC (serum half-life ~2.7 h) was efficient for sustained gene silencing for 3 weeks without observable toxicity. This combinatorial PSi MDS approach further induced a substantial reduction in the tumour weight and burden, improving significantly the pharmacokinetics of the system, and decreasing the angiogenesis and cellular proliferation.

In summary, the studies reported above show that further biofunctionalization of PSi-based nanoparticles can be targeted for enhanced cancer therapy and drug delivery inside of the cells. However, there are still many difficulties to overcome in order to create an ideal nanocarrier. Some of the most important challenges are the conjugation of the nanocarriers with the host molecules (because they usually exhibit different chemical properties), multiple drug loading as well as the nanocarrier's biodegradability and potential nanotoxicity after administration.

15.6 Conclusion

To improve the dissolution behavior of poorly water soluble drugs, PSi is applicable to most drugs since the loading of the drug molecules is intended to take place by physisorption and subsequent pore filling. Therefore, the loaded drug is in a stabilized disordered or nanocrystalline state due to the confined space of the pore without any particular requirements for the chemistry of the drug molecules. Naturally, the loading parameters need to be optimized and it must be ascertained that the drug molecules are nonreactive to the PSi carrier.

Even though there are several in *vitro* studies to improve the dissolution behavior by means of PSi, there are only few *in vivo* studies on the topic. Thus, all the features of PSi critical for oral delivery of drugs have not yet been thoroughly considered – the carrier can improve not only the dissolution behavior of the drugs but also their permeation/absorption properties. As PSi materials can be fabricated also as nanoparticles and the external surface of the particles can be modified to be, for example, mucoadhesive, the bioavailability can still be increased even though the nanoparticles themselves do not penetrate into the systemic circulation from the gastrointestinal tract. Basically, the PSi material is easy to produce on a small scale, but regarding the commercial intention the up-scaling of the production to produce appropriate PSi materials lucratively in large amounts might be rather challenging. Another issue is the safety of the material especially in case of nanoparticles, which has not yet been fully evaluated. Nevertheless, so far, the PSi research results have been promising and it is well argued to put more efforts in the studies to develop PSi as a versatile drug delivery platform.

References

1. K.W. Kolasinski, Silicon nanostructures from electroless electrochemical etching, *Curr Opin Solid St M*, **9**, 73–83 (2005).
2. G. Korotcenkov, B.K. Cho, Silicon porosification: State of the art, *Crit Rev Solid State Mater Sci*, **35**, 153–260 (2010).
3. A. Loni, D. Barwick, L. Batchelor, J. Tunbridge, Y. Han, Z.Y. Li, L.T. Canham, Extremely high surface area metallurgical-grade porous silicon powder prepared by metal-assisted etching, *Electrochem Solid-State Lett*, **14**, K25–K27 (2011).
4. K.W. Kolasinski, J.D. Hartline, B.T. Kelly, J. Yadlovskiy, Dynamics of porous silicon formation by etching in HF + V_2O_5 solutions, *Mol Phys*, **108**, 1033–1043 (2010).
5. J. Salonen, V.P. Lehto, Fabrication and chemical surface modification of mesoporous silicon for biomedical applications, *Chem Eng J*, **137**, 162–172 (2008).
6. K.W. Kolasinski, Etching of silicon in fluoride solutions, *Surf Sci*, **603**, 1904–1911 (2009).
7. V. Lehmann, R. Stengl, A. Luigart, On the morphology and the electrochemical formation mechanism of mesoporous silicon, *Mat Sci Eng B*, **69**, 11–22 (2000).

8. K. Takemoto, Y. Nakamura, O. Nittono, Microstructure and crystallinity of N-Type porous silicon, *Jpn J Appl Phys*, **33**, 6432–6436 (1994).
9. Zhang, G. (ed.), *Electrochemistry of Silicon and Its Oxide*, New York: Kluwer Academic/Plenum Publishers (2004).
10. J. Salonen, E. Mäkilä, J. Riikonen, T. Heikkilä,V.P. Lehto, Controlled enlargement of pores by annealing of porous silicon, *Phys Status Solidi A*, **206**, 1313–1317 (2009).
11. J.L. Heinrich, C.L. Curtis, G.M. Credo, K.L. Kavanagh, M.J. Sailor, Luminescent colloidal silicon suspensions from porous silicon, *Science*, **255**, 66–68 (1992).
12. L.M. Bimbo, M. Sarparanta, H.A. Santos, A.J. Airaksinen, E. Mäkilä, T. Laaksonen, L. Peltonen, V.P. Lehto, J. Hirvonen, J. Salonen, Biocompatibility of thermally hydrocarbonized porous silicon nanoparticles and their biodistribution in rats. *ACS Nano*, **4**, 3023–3032 (2010).
13. J.M. Buriak, Organometallic chemistry on silicon and germanium surfaces. *Chem Rev*, **102**, 1271–1308 (2002).
14. J. Salonen, V.P. Lehto, E. Laine, The room temperature oxidation of porous silicon, *Appl Surf Sci*, **120**, 191–198 (1997).
15. J.E. Bateman, R.D. Eagling, B.R. Horrocks, A. Houlton, D.R. Worrall, Rôle for organic molecules in the oxidation of porous silicon. *Chem Commun*, 2275–2276 (1997).
16. A. Janshoff, K.P.S. Dancil, C. Steinem, D.P. Greiner, V.S.Y. Lin, C. Gurtner, K. Motesharei, M.J. Sailor, M. Reza Ghadiri, Macroporous p-type silicon Fabry-Perot layers. Fabrication, characterization, and applications in biosensing, *J Am Chem Soc*, **120**, 12108–12116 (1998).
17. M.P. Stewart, J.M. Buriak, New approaches toward the formation of silicon-carbon bonds on porous silicon, *Comment Inorg Chem*, **23**, 179–203 (2002).
18. E. Pastor, E. Matveeva, V. Parkhutik, J. Curiel-Esparza, M.C. Millan, Influence of porous silicon oxidation on its behavior in simulated body fluid, *Phys Status Solidi C*, **4**, 2136–2140 (2007).
19. J. Salonen, V.P. Lehto, E. Laine, Thermal oxidation of free-standing porous silicon films, *Appl Phys Lett*, **70**, 637–639 (1997).
20. D.B. Mawhinney, J.A. Glass, J.T. Yates, FTIR study of the oxidation of porous silicon, *J 20 Phys Chem B*, **101**, 1202–1206 (1997).
21. L. Debarge, J.P. Stoquert, A. Slaoui, L. Stalmans, J. Poortmans, Rapid thermal oxidation of porous silicon for surface passivation, *Mater Sci Semicond Process*, **1**, 281–285 (1998).
22. R. Czaputa, R. Fritzl, A. Popitsch, Anomalous luminescence degradation behavior of chemically oxidized porous silicon, *Thin Solid Films*, **255**, 212–215 (1995).
23. B.V. Rama Mohana Rao, P.K. Basu, J.C. Biswas, S.K. Lahiri, S. Ghosh, D.N. Bose, Large enhancement of photoluminescence from porous silicon films by post-anodization treatment in boiling hydrogen peroxide, *Solid State Commun*, **97**, 417–418 (1996).
24. Y. Fukuda, Y. Yingda, W. Zhou, K. Furuya, H. Suzuki, Quenching of porous silicon photoluminescence by ammonia hydrogen peroxide mixture, *J Electrochem Soc*, **147**, 3914–3916 (2000).
25. D. Xu, L. Sun, H. Li, L. Zhang, G. Guo, X. Zhao, L. Gui, Hydrolysis and silanization of the hydrosilicon surface of freshly prepared porous silicon by an amine catalytic reaction, *New J Chem*, **27**, 300–306 (2003).

26. R. Boukherroub, S. Morin, D.D.M. Wayner, D.J. Lockwood, Thermal route for chemical modification and photoluminescence stabilization of porous silicon. *Phys Status Solidi A*, **182**, 117–121 (2000).
27. B. Gelloz, H. Sano, R. Boukherroub, D.D.M. Wayner, D.J. Lockwood, N. Koshida, Stable electroluminescence from passivated nano-crystalline porous silicon using undecylenic acid, *Phys Status Solidi C*, **2**, 3273–3277 (2005).
28. R. Boukherroub, D.J. Lockwood, D.D.M. Wayner, L.T. Canham, Thermal functionalization of hydrogen-terminated porous silicon surfaces with terminal alkenes and aldehydes. *Mater Res Soc Symp Proc*, **638**, F1141–F1146 (2001).
29. R. Boukherroub, S. Morin, D.D.M. Wayner, F. Bensebaa, G.I. Sproule, J.M. Baribeau, D.J. Lockwood, Ideal passivation of luminescent porous silicon by thermal, noncatalytic reaction with alkenes and aldehydes. *Chem Mater*, **13**, 2002–2011 (2001).
30. J. Salonen, E. Laine, L. Niinistö, Thermal carbonization of porous silicon surface by acetylene, *J Appl Phys*, **91**, 456–461 (2002).
31. M.J. Sweetman, F.J. Harding, S.D. Graney, N.H. Voelcker, Effect of oligoethylene glycol moieties in porous silicon surface functionalisation on protein adsorption and cell attachment, *Appl Surf Sci*, **257**, 6768–6775 (2011).
32. B. Sciacca, S.D. Alvarez, F. Geobaldo, M.J. Sailor, Bioconjugate functionalization of thermally carbonized porous silicon using a radical coupling reaction, *Dalton Trans*, **39**, 10847–10853 (2010).
33. M. Hiraoui, M. Guendouz, N. Lorrain, A. Moadhen, L. Haji, M. Oueslati, Spectroscopy studies of functionalized oxidized porous silicon surface for biosensing applications, *Mater Chem Phys*, **128**, 151–156 (2011).
34. T. Böcking, K.A. Kilian, K. Gaus, J.J. Gooding, Modifying porous silicon with self-assembled monolayers for biomedical applications: The influence of surface coverage on stability and biomolecule coupling. *Adv Funct Mater*, **18**, 3827–3833 (2008).
35. D.S. Kohane, R. Langer, Biocompatibility and drug delivery systems, *Chem Sci*, **1**, 441–446 (2010).
36. H.A. Santos, J. Riikonen, J. Salonen, E. Mäkilä, T. Heikkilä, T. Laaksonen, L. Peltonen, V.P. Lehto, J. Hirvonen, In vitro cytotoxicity of porous silicon microparticles: Effect of the particle concentration, surface chemistry and size. *Acta Biomater*, **6**, 2721–2731 (2010).
37. H.A. Santos, L.M. Bimbo, V.P. Lehto, A.J. Airaksinen, J. Salonen, J. Hirvonen, Multifunctional porous silicon for therapeutic drug delivery and imaging, *Curr Drug Discov Technol*, **8**, 228–249 (2011).
38. H.A. Santos, J. Salonen, L.M. Bimbo, V.-. Lehto, L. Peltonen, J. Hirvonen, Mesoporous materials as controlled drug delivery formulations, *J Drug Deliv Sci*, **21**, 139–155 (2011).
39. T. Heikkilä, H.A. Santos, N. Kumar, D.Y. Murzin, J. Salonen, T. Laaksonen, L. Peltonen, J. Hirvonen, V.P. Lehto, Cytotoxicity study of ordered mesoporous silica MCM-41 and SBA-15 microparticles on Caco-2 cells, *Eur J Pharm Biopharm*, **74**, 483–494 (2010).
40. L.M. Bimbo, E. Mäkilä, T. Laaksonen, V.P. Lehto, J. Salonen, J. Hirvonen, H.A. Santos, Drug permeation across intestinal epithelial cells using porous silicon nanoparticles. *Biomaterials*, **32**, 2625–2633 (2011).

41. T. Laaksonen, Santos, H. Vihola, J. Salonen, J. Riikonen, T. Heikkila, L. Peltonen, N. Kumar, D.Y. Murzin, V.P. Lehto and J. Hirvonen, Failure of MTT as a toxicity testing agent for mesoporous silicon microparticles, *Chem Res Toxicol*, **20**, 1913–1918 (2007).
42. L.T. Canham, Bioactive silicon structure fabrication through nanoetching techniques. *Adv Mater*, **7**, 1033–1037 (1995).
43. L.T. Canham, C.L. Reeves, D.O. King, P.J. Branfield, J.G. Crabb, M.C.L. Ward, Bioactive polycrystalline silicon. *Adv Mater*, **8**, 850–852 (1996).
44. V.V. Seregin, J.L. Coffer, Bias-assisted in vitro calcification of calcium disilicide growth layers on spark-processed silicon, *Biomaterials*, **27**, 3726–3737 (2006).
45. S.P. Low, K.A. Williams, L.T. Canham, N.H. Voelcker, Evaluation of mammalian cell adhesion on surface-modified porous silicon, *Biomaterials*, **27**, 4538–4546 (2006).
46. A.V. Sapelkin, S.C. Bayliss, B. Unal, A. Charalambou, Interaction of B50 rat hippocampal cells with stain-etched porous silicon, *Biomaterials*, **27**, 842–846 (2006).
47. L.M. Karlsson, P. Tengvall, I. Lundström, H. Arwin, Penetration and loading of human serum albumin in porous silicon layers with different pore sizes and thicknesses, *J Colloid Interface Sci*, **266**, 40–47 (2003).
48. K.M. Ainslie, S.L. Tao, K.C. Popat, T.A. Desai, In vitro immunogenicity of silicon-based micro- and nanostructured surfaces. *ACS Nano*, **2**, 1076–1084 (2008).
49. S.D. Alvarez, A.M. Derfus, M.P. Schwartz, S.N. Bhatia, M.J. Sailor, The compatibility of hepatocytes with chemically modified porous silicon with reference to in vitro biosensors. *Biomaterials*, **30**, 26–34 (2009).
50. S.P. Low, K.A. Williams, L.T. Canham, N.H. Voelcker, Generation of reactive oxygen species from porous silicon microparticles in cell culture medium, *J Biomed Mater Res, Part A*, **93**, 1124–1131 (2010).
51. L. De Stefano, I. Rea, A. Armenante, P. Giardina, M. Giocondo, I. Rendina, Self-assembled biofilm of hydrophobins protects the silicon surface in the KOH wet etch process, *Langmuir*, **23**, 7920–7922 (2007).
52. J. Choi, Q. Zhang, V. Reipa, N.S. Wang, M.E. Stratmeyer, V.M. Hitchins, P.L. Goering, Comparison of cytotoxic and inflammatory responses of photoluminescent silicon nanoparticles with silicon micron-sized particles in RAW 264.7 macrophages, *J Appl Toxicol*, **29**, 52–60 (2009).
53. B. Godin, J. Gu, R.E. Serda, R. Bhavane, E. Tasciotti, C. Chiappini, X. Liu, T. Tanaka, P. Decuzzi, M. Ferrari, Tailoring the degradation kinetics of mesoporous silicon structures through PEGylation, *J Biomed Mater Res, Part A*, **94**, 1236–1243 (2010).
54. L.T. Canham, M.P. Stewart, J.M. Buriak, C.L. Reeves, M. Anderson, E.K. Squire, P. Allcock, P.A. Snow, Derivatized porous silicon mirrors: Implantable optical components with slow resorbability. *Phys Status Solidi A*, **182**, 521–525 (2000).
55. S.H.C. Anderson, H. Elliott, D.J. Wallis, L.T. Canham, J.J. Powell, Dissolution of different forms of partially porous silicon wafers under simulated physiological conditions. *Phys Status Solidi A*, **197**, 331–335 (2003).
56. D.M. Reffitt, R. Jugdaohsingh, R.P.H. Thompson, J.J. Powell, Silicic acid: Its gastrointestinal uptake and urinary excretion in man and effects on aluminium excretion, *J Inorg Biochem*, **76**, 141–147 (1999).
57. A.S.W. Goh, A.Y.F. Chung, R.H.G. Lo, T.N. Lau, S.W.K. Yu, M. Chng, S. Satchithanantham, S.L.E. Loong, D.C.E. Ng, B.C. Lim, S. Connor, P.K.H. Chow, A novel approach

to brachytherapy in hepatocellular carcinoma using a phosphorous32, (32P) brachytherapy delivery device: a first-in-man study, *Int J Radiat Oncol*, **67**, 786–792 (2007).
58. C.A. Prestidge, T.J. Barnes, C.H. Lau, C. Barnett, A. Loni, L. Canham, Mesoporous silicon: A platform for the delivery of therapeutics, *Expert Opin Drug Deliv*, **4**, 101–110 (2007).
59. E. Tasciotti, X. Liu, R. Bhavane, K. Plant, A.D. Leonard, B.K. Price, M.M.C. Cheng, P. Decuzzi, J.M. Tour, F. Robertson, M. Ferrari, Mesoporous silicon particles as a multistage delivery system for imaging and therapeutic applications, *Nat Nanotechnol*, **3**, 151–157 (2008).
60. T. Tanaka, B. Godin, R. Bhavane, R. Nieves-Alicea, J. Gu, X. Liu, C. Chiappini, J.R. Fakhoury, S. Amra, A. Ewing, Q. Li, I.J. Fidler, M. Ferrari, In vivo evaluation of safety of nanoporous silicon carriers following single and multiple dose intravenous administrations in mice, *Int J Pharm*, **402**, 190–197 (2010).
61. M. Kilpeläinen, J. Riikonen, M.A. Vlasova, A. Huotari, V.P. Lehto, J. Salonen, K.H. Herzig, K. Järvinen, In vivo delivery of a peptide, ghrelin antagonist, with mesoporous silicon microparticles, *J Controlled Release*, **137**, 166–170 (2009).
62. J.H. Park, L. Gu, G. Von Maltzahn, E. Ruoslahti, S.N. Bhatia, M.J. Sailor, Biodegradable luminescent porous silicon nanoparticles for in vivo applications, *Nat Mater*, **8**, 331–336 (2009).
63. L. Cheng, E. Anglin, F. Cunin, D. Kim, M.J. Sailor, I. Falkenstein, A. Tammewar, W.R. Freeman, Intravitreal properties of porous silicon photonic crystals: A potential self-reporting intraocular drug-delivery vehicle, *Br J Ophthalmol*, **92**, 705–711 (2008).
64. S.P. Low, N.H. Voelcker, L.T. Canham, K.A. Williams, The biocompatibility of porous silicon in tissues of the eye, *Biomaterials*, **30**, 2873–2880 (2009).
65. M. Kinoshita, K. Baba, A. Nagayasu, K. Yamabe, T. Shimooka, Y. Takeichi, M. Azuma, H. Houchi, K. Minakuchi, Improvement of solubility and oral bioavailability of a poorly water-soluble drug, TAS-301, by its melt-adsorption on a porous calcium silicate, *J Pharm Sci*, **91**, 362–370 (2002).
66. R. Mellaerts, J.A.G. Jammaer, M. Van Speybroeck, H. Chen, J. Van Humbeeck, P. Augustijns, G. Van den Mooter, J.A. Martens, Physical state of poorly water soluble therapeutic molecules loaded into SBA-15 ordered mesoporous silica carriers: A case study with itraconazole and ibuprofen, *Langmuir*, **24**, 8651–8659 (2008).
67. T. Limnell, H.A. Santos, E. Mäkilä, T. Heikkilä, J. Salonen, D.Y. Murzin, N. Kumar, T. Laaksonen, L. Peltonen, J. Hirvonen, Drug delivery formulations of ordered and nonordered mesoporous silica: Comparison of three drug loading methods, *J Pharm Sci*, **100**, 3294–3306 (2011).
68. E.C. Wu, J.S. Andrew, L. Cheng, W.R. Freeman, L. Pearson, M.J. Sailor, Real-time monitoring of sustained drug release using the optical properties of porous silicon photonic crystal particles, *Biomaterials*, **32**, 1957–1966 (2011).
69. L. Gu, J.H. Park, K.H. Duong, E. Ruoslahti, M.J. Sailor, Magnetic luminescent porous silicon microparticles for localized delivery of molecular drug payloads, *Small*, **6**, 2546–2552 (2010).
70. L.A. Perelman, C. Pacholski, Y.Y. Li, M.S. VanNieuwenhz, M.J. Sailor, pH-triggered release of vancomycin from protein-capped porous silicon films, *Nanomedicine*, **3**, 31–43 (2008).

71. J. Wu, M.J. Sailor, Chitosan hydrogel-capped porous SiO_2 as a pH responsive nanovalve for triggered release of insulin, *Adv Funct Mater*, **19**, 733–741 (2009).
72. J. Riikonen, E. Mäkilä, J. Salonen, V.P. Lehto, Determination of physical state of drug molecules in mesoporous silicon with different surface chemistries, *Langmuir*, **25**, 6137–6142 (2009).
73. T. Heikkilä, J. Salonen, J. Tuura, N. Kumar, T. Salmi, D.Y. Murzin, M.S. Hamdy, G. Mul, L. Laitinen, A. Kaukonen, J. Hirvonen, V.P. Lehto, Evaluation of mesoporous TCPSi, MCM-41, SBA-15, and TUD-1 materials as API carriers for oral drug delivery, *Drug Deliv*, **15**, 337–347 (2007).
74. J. Salonen, A. Kaukonen, J. Hirvonen, V.P. Lehto, Mesoporous silicon in drug delivery applications, *J Pharm Sci*, **97**, 632–653 (2008).
75. J. Salonen, L. Laitinen, A. Kaukonen, J. Tuura, M. Björkqvist, T. Heikkilä, K. Vähä-Heikkilä, J. Hirvonen, V.P. Lehto, Mesoporous silicon microparticles for oral drug delivery: Loading and release of five model drugs, *J Controlled Release*, **108**, 362–374 (2005).
76. A.B. Foraker, R.J. Walczak, M.H. Cohen, T.A. Boiarski, C.F. Grove, P.W. Swaan, Microfabricated porous silicon particles enhance paracellular delivery of insulin across intestinal caco-2 cell monolayers, *Pharm Res*, **20**, 110–116 (2003).
77. T. Limnell, J. Riikonen, J. Salonen, A. Kaukonen, L. Laitinen, J. Hirvonen, V.P. Lehto, Surface chemistry and pore size affect carrier properties of mesoporous silicon microparticles, *Int J Pharm*, **343**, 141–147 (2007).
78. E.J. Anglin, M.P. Schwartz, V.P. Ng, L.A. Perelman,M.J. Sailor, Engineering the chemistry and nanostructure of porous silicon fabry-pérot films for loading and release of a steroid. *Langmuir*, **20**, 11264–11269 (2004).
79. E.C. Wu, J.H. Park, J. Park, E. Segal, F. Cunin, M.J. Sailor, Oxidation-triggered release of fluorescent molecules or drugs from mesoporous Si microparticles, *ACS Nano*, **2**, 2401–2409 (2008).
80. J.L. Coffer, J.L. Montchamp, J.B. Aimone, R.P. Weis, Routes to calcified porous silicon: Implications for drug delivery and biosensing, *Phys Status Solidi A*, **197**, 336–339 (2003).
81. V.P. Lehto, K. Vähä-Heikkilä, J. Paski, J. Salonen, Use of thermoanalytical methods in quantification of drug load in mesoporous silicon microparticles, *J Therm Anal Calorim*, **80**, 393–397 (2005).
82. L. Vaccari, D. Canton, N. Zaffaroni, R. Villa, M. Tormen, E. di Fabrizio, Porous silicon as drug carrier for controlled delivery of doxorubicin anticancer agent, *Microelectron Eng*, **83**, 1598–1601 (2006).
83. A.M. Kaukonen, L. Laitinen, J. Salonen, J. Tuura, T. Heikkilä, T. Limnell, J. Hirvonen, V.P. Lehto, Enhanced in vitro permeation of furosemide loaded into thermally carbonized mesoporous silicon (TCPSi) microparticles, *Eur J Pharm Biopharm*, **66**, 348–356 (2007).
84. F. Wang, H. Hui, T.J. Barnes, C. Barnett, C.A. Prestidge, Oxidized mesoporous silicon microparticles for improved oral delivery of poorly soluble drugs, *Mol Pharmaceutics*, **7**, 227–236 (2010).
85. R.E. Serda, A. MacK, M. Pulikkathara, A.M. Zaske, C. Chiappini, J.R. Fakhoury, D. Webb, B. Godin, J.L. Conyers, X.W. Liu, J.A. Bankson, M. Ferrari, Cellular association and assembly of a multistage delivery system, *Small*, **6**, 1329–1340 (2010).

86. R.E. Serda, J. Gu, R.C. Bhavane, X. Liu, C. Chiappini, P. Decuzzi, M. Ferrari, The association of silicon microparticles with endothelial cells in drug delivery to the vasculature, *Biomaterials*, **30**, 2440–2448 (2009).
87. R.E. Serda, J. Gu, J.K. Burks, K. Ferrari, C. Ferrari, M. Ferrari, Quantitative mechanics of endothelial phagocytosis of silicon microparticles, *Cytometry A*, **75**, 752–760 (2009).
88. S. Ferrati, A. Mack, C. Chiappini, X. Liu, A.J. Bean, M. Ferrari, R.E. Serda, Intracellular trafficking of silicon particles and logic-embedded vectors, *Nanoscale*, **2**, 1512–1520 (2010).
89. T. Tanaka, L.S. Mangala, P.E. Vivas-Mejia, R. Nieves-Alicea, A.P. Mann, E. Mora, H.D. Han, M.M.K. Shahzad, X. Liu, R. Bhavane, J. Gu, J.R. Fakhoury, C. Chiappini, C. Lu, K. Matsuo, B. Godin, R.L. Stone, A.M. Nick, G. Lopez-Berestein, A.K. Sood, M. Ferrari, Sustained small interfering RNA delivery by mesoporous silicon particles, *Cancer Res*, **70**, 3687–3696 (2010).

16

Micro- and Nanosizing of Poorly Soluble Drugs by Grinding Techniques

Stefan Scheler

16.1 Introduction

The bioavailability of an orally administered drug substance is the result of both its solubility in the gastrointestinal fluids and its transepithelial permeability. Apart from rare exceptions, a drug passes the intestinal mucosa in the dissolved state. If the permeability is not the stage that limits the rate, the entry into the systemic circulation is determined by the release of the drug from the dosage form and, particularly in the case of poorly soluble compounds, by its dissolution rate. Thus, for many drugs, especially those belonging to Class II of the Biopharmaceutics Classification System (BCS) but also some of BCS Class IV, the dissolution rate governs the absorption process. As soon the drug has reached the central compartment, it becomes subject to different mechanisms of metabolism and elimination. Significant biopharmaceutical parameters like the maximum plasma concentration (c_{max}) and the area under the curve (AUC) are determined by the ratio of the absorption rate to the elimination rate. Because the latter usually cannot be influenced therapeutically, the absorption rate, and thus the dissolution of the drug, are the principal parameters to control the plasma concentration profile and the intensity and duration of the action. One-tenth of the drugs currently on the market have to be considered as problematic with respect to their solubility behavior. However, the number of such compounds will rise in the future, for more than one-third of drugs in the research pipelines and nearly two-thirds of newly synthesized and screened drug candidates have low solubilities (<0.1 mg/ml) [1].

Drug Delivery Strategies for Poorly Water-Soluble Drugs, First Edition. Edited by Dennis Douroumis and Alfred Fahr.

16.2 Kinetics of Drug Dissolution

As described by Nernst and Brunner, the dissolution rate *dc/dt* of solid drugs increases in proportion to the surface area *S* and thus inversely to the particle size [2, 3].

$$\frac{dc}{dt} = \frac{DS}{Vh}(c_s - c) \tag{16.1}$$

with *D*, *V* and *h* being the diffusion coefficient, the solvent volume and the thickness of the diffusion layer, respectively. c_s is the saturation concentration and *c* is the concentration at time *t*. When approaching nanometer dimensions, both the saturation solubility c_s and the layer thickness *h*, which can be considered to be constants in the case of larger particles, become also dependent on the particle size. The relationship between c_s and the particle radius *r* is given by the Freundlich–Ostwald equation:

$$c_s = c_s^{\infty} \exp\left(\frac{2\gamma M}{r\rho RT}\right) \tag{16.2}$$

where c_s^{∞} is the saturation solubility of an infinitely large drug crystal, γ is the crystal-medium interfacial tension, *M* is the molecular weight of the drug, ρ is the density, *R* is the gas constant and *T* is the temperature [4]. The thickness of the diffusion layer can be considered as constant for larger particles, however, it becomes size dependent if the particle radius decreases below 30 μm [5–7]. In the case of spherical particles, the layer thickness *h* can be calculated as

$$h = \frac{2r}{Sh} \tag{16.3}$$

The Sherwood number *Sh* is defined by the semi-empirical Ranz-Marshall equation

$$Sh = 2 + 0.6Re^{1/2}Sc^{1/3} \tag{16.4}$$

with *Re* being the Reynolds number and *Sc* the Schmidt number. In slowly agitated media, where the laminar flow can be assumed, the Reynolds number becomes negligible and thus the thickness of the diffusion layer equals the particle radius *r* [5, 7]:

$$\begin{cases} h = r & r < 30\ \mu\text{m} \\ h = 30\ \mu\text{m} & r > 30\ \mu\text{m} \end{cases} \tag{16.5}$$

These relationships show that both the dissolution rate and the saturation solubility depend on the particle radius. Size reduction is therefore an effective method to increase both parameters and hence to enhance the absorption of poorly soluble drugs.

16.3 Micronization and Nanosizing of Drugs

16.3.1 Dissolution Enhancement by Micronization and Nanonization

A common approach used for many years in pharmaceutical development to increase the oral bioavailability of poorly soluble drugs is micronization, which means size reduction in the range between typically 1–10 μm. One of the oldest and best-known examples of a drug whose bioavailability could be significantly enhanced by micronization is griseofulvin.

Upon administration of 500 mg of the micronized substance, identical plasma levels were achieved as with 1000 mg doses of non-micronized griseofulvin ($c_s = 10$ μg/ml) [8]. Also in the case of digoxin ($c_s = 64.8$ μg/ml) the maximum plasma levels (c_{max}) could be doubled by reducing the average particle size from 22 μm to 3.7 μm [9]. Further drugs which were brought on the market in micronized form are spironolactone ($c_s = 22$ μg/ml), progesterone ($c_s = 7$ μg/ml), glibenclamide ($c_s = 24.8$ μg/ml), and fenofibrate ($c_s = 0.8$ μg/ml).

However, an increasing number of newly developed drugs have such low solubilities that micronization is no longer enough. The maximum possible enhancement of the surface area, and thus consequently of the dissolution rate, by size reduction within the micrometer range is still too low to overcome the bioavailability problems of very poorly soluble drugs of the BCS Class II. The logical next step was to move from micronization to nanonization [10].

16.3.2 Dry and Wet Milling Technologies

One of the oldest and most common techniques for size reduction is mechanical milling. It is the most important method for decreasing the particle size of crystalline drug substances in order to adapt their solubility properties to the desired requirements. Milling processes are comparatively cheap, fast and easy to scale up. Dry milling, e.g. with hammer-mills, rod-mills or jet mills, is the most frequently applied milling technique in pharmaceutical applications. However, dry mills are barely able to reduce the particle size below 1–10 μm [11–14] and fail to produce significant yields at the sub-micron level. In addition, due to the strong heat generation of most dry milling processes, temperature-sensitive drugs are likely to be prone to degradation [15, 16], but also phase transitions or melting may occur. However, the main problem in the production of particles below 1 μm is the aggregation of such high surface area materials.

Slurry or wet milling is able to overcome these problems as the suspension liquid dissipates the heat generated in the process and deagglomerates the particles. Also in the case of highly potent drugs, if dry milling is not optimal because of the potential emission of toxic dust, or in the case of compounds which are susceptible to amorphization, wet milling is often preferred. Two main principles of wet milling are applied in pharmaceutical processes: rotor stator technologies and media (bead, pearl or ball) milling.

Rotor stator mills are used for the comminution of solids but also for emulsification and deagglomeration. Devices for the former purpose are mostly designed as disk attrition mills. They consist of two profiled or perforated disks made of steel, sintered material or corundum, one of them acting as a rotor, the other as a stator. In between is an adjustable annular gap of the order of 30–500 μm in which the particles are ground during their passage under the action of centrifugal forces. These types of mills are widely used for slurry milling of pharmaceuticals and chemicals to achieve particles in the 10–30 μm range [17, 18].

Ball mills have been in use since the late 1800s. The oldest type is the tumbling mill, which consists of a horizontally rotating drum filled with the milling product and a grinding material such as iron balls. Upon rotation the balls either roll against each other or, above a certain speed limit, they are lifted and fall. It is known that in a tumbling mill, a material can be ground faster and more efficiently when it is suspended in water than in the dry state [19, 20]. As a well-known fact, the size limit of particles produced by conventional ball milling is several micrometers. Since the 1980s smaller beads of 0.5 to 5 mm are available which even allow materials to be crushed into fine particles at the submicron level [21, 22].

16.3.3 NanoCrystal® Technology

Further size reduction of the grinding beads and application of high-energy agitated media mills allowed the generation of drug particle formulations with typical sizes of the order of less than 200–600 nm [23, 24] and even as small as 10 nm [25]. The application of this technique (NanoCrystal® technology) for milling of drug substances was patented in 1992 by Liversidge *et al.*, formerly owned by Sterling Drug Inc., and later acquired by Elan Drug Technologies (Monksland, Athlone, Ireland) [26, 27]. They used both conventional ball mills as well as high-shear media mills. While process times of 4–5 days were necessary in the first case, with high-energy mills it was possible to generate nanosized drug particles within a few hours or even less.

Nanomilling is an effective method for producing nanosized drug materials and thus improving the solution rate of poorly soluble drugs. It is economic, time-efficient and easy to scale up. Furthermore, chemical degradation may be reduced because most of the drug is in the solid state (usually crystalline). Consequently it is less accessible to and thus less susceptible to destabilizing factors such as light or oxygen [28–30]. Most nanocrystalline drug products which have been brought to the market so far are produced by the NanoCrystal® wet-milling technology. Five products for oral administration (all of them drugs of BCS Classes II or IV) and one for parenteral use are commercially available (Table 16.1). Annual in-market sales exceed US$1.9 billion [31]. Further products are in the research pipelines of pharmaceutical companies, with an estimate of about 20 drugs being in clinical studies, some of them already in phase III [23].

16.4 Theory of Grinding Operations

16.4.1 Fraction under Compressive Stress

The theoretical strength of a crystalline material is the result of interatomic attractive and repulsive forces. In experimental settings, however, the material strength is found to be many times smaller than what is calculated on this theoretical basis. The difference is caused by imperfections of the crystal structure which occur in the form of fissures or irregularities of the lattice. Thus, the strength of a material depends on its microstructure, especially on the size distribution of flaws and in particular on the number of critical-sized flaws. They cause a local increase in stress (force per area) when a load is applied because stress is concentrated around the tips of such flaws. Consequently, these are the points where inelastic deformation and fracturing start. Once this process is activated, the crack propagates rapidly through the material, thereby encountering other flaws which may, in turn, trigger the growth of secondary cracks. The strength of a material therefore depends on the random distribution of flaws and can be considered a statistical quantity.

For most materials, their strength is greater in compression than in tension. This means that their failure typically occurs under tension. It is therefore unfortunate that tensile stresses cannot be utilized directly for milling operations. Most comminution methods impose compressive stresses on the particles. They do not cause failure in a direct manner

Table 16.1 *Commercialized drug products based on the NanoCrystal® technology.*

	Product name	Drug	Manufacturer	Therapeutic class	BCS class	FDA approval
Oral NanoCrystal® formulations	Rapamune	Sirolimus	Wyeth	immunosuppressant	BCS II	2000
	Emend	Aprepitant	Merck	antiemetic	BCS IV	2003
	TriCor	Fenofibrate	Abbott	hypolipidemic agent	BCS II	2004
	Megace	Megestrol acetate	Par Pharmaceutical	antineoplastic agent, hormone	BCS II	2005
Parenteral NanoCrystal® formulations	Invega Sustenna	Paliperidone palmitate	Janssen	antipsychotic		2009

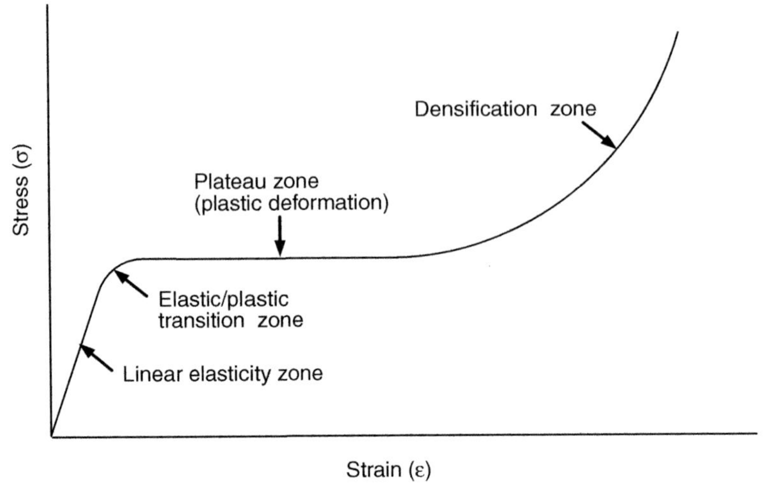

Figure 16.1 *Typical stress–strain plot in compression mode.*

but generate, by distortion, sufficient tensile or shear stress to form a crack tip in a region away from the point of primary stress application. This mechanism is not as efficient as a tensile load but it is the only technical way to subject small particles to fracturing forces [32].

16.4.2 Brittle-Ductile Transition and Grinding Limit

Solids under stress are strained and deformed before fracturing. A stress–strain curve is a graphical representation of the relationship between the stress, derived from measuring the load applied on the sample, and the strain, derived from measuring the thus induced deformation. The stress–strain curves for materials in compression differ from those measured under tension. Because the primary stresses in milling processes are of a compressive nature, only compression stress–strain plots are discussed below. A typical stress–strain curve in compression is characterized by a linear elastic zone, an elastic/plastic transition zone, a plastic deformation plateau zone and a densification zone in which the curve increases again (Figure 16.1). The total area under the curve represents the strain energy stored in the material and is an approximate measure of its impact strength. As this energy is not uniformly distributed throughout the material, there are a lot of weaker regions where particle failure can originate. When the application of a force causes permanent deformation, energy is expended and in the case of brittle materials, cracks may be initiated. The furthest extension of a crack is the point of greatest stress. However, most of these cracks do not propagate very far after initiating because they encounter internal boundaries which arrest them. In the stress–strain diagram this progressive formation of small fissures or microcracks can be observed as a saw-tooth pattern of the plastic deformation zone which is preceded by a relatively short elastic zone up to only a small strain. Each stress drop represents a sudden drop in the material's load-carrying capacity. Finally, a strain is reached at which complete failure occurs and the specimen is crushed down to dust. Up to this point the solid has become riddled with microcracks which suddenly coalesce when they have reached a certain degree of spreading within the material [33].

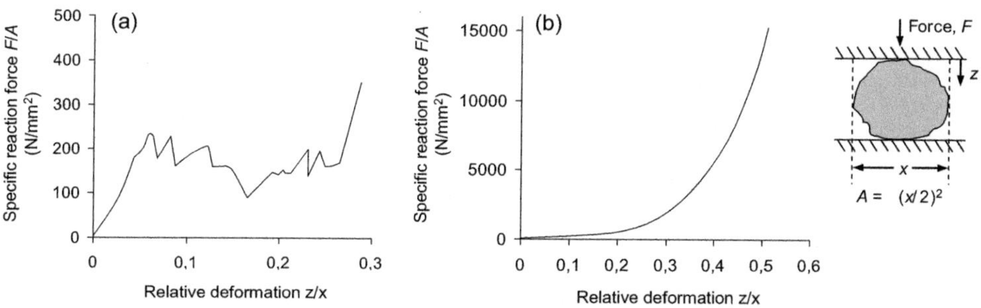

Figure 16.2 *Loading diagrams of cement clinker particles with different diameters (A: 16.1 μm, B: 2.5 μm). Reproduced from [17] with permission from Wiley-VCH Verlag GmbH & Co. KgaA.*

If an ideal crystal is charged with an increasing pressure, the force is distributed uniformly throughout the volume until the crystal disintegrates into smaller units [20]. However, as already mentioned, most crystals do not have a perfect structure since they contain lattice voids, interstices, and other imperfections that act as starting points for the formation of microcracks. The applied stress is concentrated around the tips of those cracks or flaws. This stress concentration can be quantified using the parameter K which is defined as local stress per mean stress in a particle and is calculated according to

$$K = 1 + 2(LR)^{1/2} \tag{16.6}$$

where L is half the crack length and R is the radius of the crack tip [34]. As can be seen from Equation 16.6, the stress around a crack tip depends on the length of the crack and increases as the crack propagates. In small particles, the maximum crack length is limited by the particles' dimensions. With decreasing particle size, the smaller crack lengths cause reduced stress concentration, and with the achievement of more and more faultless domains in smaller particles, the material strength increases until it reaches the theoretical strength. As a result, the material exhibits increasingly plastic behavior, which means that it undergoes substantial plastic deformation with high energy absorption before fracture. Figure 16.2 shows loading diagrams, being analog to stress–strain plots, of cement clinker particles with different diameters (16.1 μm and 2.5 μm) [17]. Multiple peaks and the serrated pattern of the larger particles' curve indicate fracturing and brittleness, whereas the smooth curve obtained with the smaller particles represents a plastic type of deformation. Because smaller particles contain only small flaws, much larger stresses are required to reach the critical yield stress which is necessary to initiate fracturing.

Every macroscopically brittle material becomes ductile below a characteristic particle size. Kendall [35] suggested that there is a critical diameter d_{crit} below which the particles start to behave in a ductile manner:

$$d_{crit} = (AK_{Ic}/\sigma_y)[2] \tag{16.7}$$

Table 16.2 Critical particle size for brittle-ductile transition of selected materials.

Substance	Critical particle size for brittle-ductile transition	Reference
Ibuprofen	125–854 μm	37,38
α-Lactose monohydrate	45 μm	39
Sucrose	12–20 μm	40
Quartz	1.7–6.8 μm	35, 41, 42

where K_{Ic} is the critical stress intensity factor, also referred to as fracture toughness, and σ_y is the compressive yield stress. A is a constant whose value depends on the geometry of the test specimen and the stress application. For compression of rectangular samples with large cracks, $A = (32/3)^{0.5}$ [35]. For other geometries, A varies between 50 and 1 [36].

Below d_{crit} is becomes difficult to comminute particles by compression or impact because they undergo plastic deformation instead of brittle fracture and the only effect of compressive forces is a change of the particle shape. It must be noted that the critical diameter provides only a rough guide because effects such as temperature, stress rate, and crack initiation are not considered [37]. The critical particle size of the brittle-ductile transition varies widely but for most materials it is in the lower micrometer range, as Table 16.2 shows.

16.4.3 Milling Beyond the Brittle-Ductile Transition Limit

For a long time it did not seem possible to achieve smaller particles by crushing and grinding. However, high-shear media milling has been shown to overcome this limit and to further comminute also submicron particle fragments which should already have lost their brittle properties and are assumed to deform plastically rather than fracture.

Theories which could explain submicron grinding mechanisms were developed mainly in the field of petrology and geophysics. A first explanation can be found if it is considered that the compressive yield stress σ_y is not a non-influenced material constant but increases with decreasing temperature and increasing loading rate. According to Ashby and Verral, for each substance an upper threshold limit for compressive yield stress σ_y^{max} can be calculated, which is not exceeded at any temperature and at any strain rate [43,44]. The critical particle size calculated with this maximum possible stress value is much smaller than the respective particle diameter in Table 16.2 and is well within the nanometer range. In the case of quartz, for example, a minimum particle size of 86 nm should be theoretically achievable under compression at a sufficiently high strain rate and low temperature. Dry grinding at low temperatures is a method already used for the grinding of soft materials, for example, for pharmaceuticals, polymers and food ingredients. Due to the embrittlement at low temperatures, the transition to the ductile state takes place at smaller particle sizes and thus a finer product can be obtained [45]. Most recently, also wet milling using liquid nitrogen as a cryogenic fluid has been reported as a method for producing nanocrystals [46]. However, there are also methods to overcome the grinding limit at ambient temperatures caused by plastication. With regard to the strain rate, the threshold size of plastic deformation can be reduced to a minimum by means of shock loading. In this case, stresses are as much as 50 times higher than the flow stress in quasi-statical measurement [47], which, according to

Equation 16.7, can decrease the minimum particle size by a factor of 2500, well below the nanometer scale [44].

16.4.4 Fatigue Fracture

Some authors suggest a fatigue fracture mechanism for particle breakage in the nanometer range [17, 48, 49]. They suppose that below a certain particle size the transferred stress intensity from the milling beads is no longer sufficient to achieve a brittle fracture with a single impact and further breakage can only occur if the particles are stressed repeatedly. This leads to an accumulation of lattice imperfections and to an enhancement of the total amount of elastic energy stored in the crystallites until the energy balance is fulfilled and fracturing is able to traverse the entire particle [17]. The smaller the particles are ground, the more energy is required for further breakage. This behavior explains the reduced kinetics of particle breakage in the nanometer range. The comminution process can only be continued if further defects are generated by repeated stressing events until the stored energy combined with the energy immediately provided by the milling media is large enough to enable fracturing. Thus, a 'true' grinding limit is not reached before the crystallites become so small that nearly no defects can be stored in the lattice any longer [50].

16.4.5 Agglomeration

So far, only particle breakage has been addressed, ignoring the fact that another counteracting mechanism becomes increasingly important with diminishing particle size. According to Karagedov, particle grinding can be described as a four-step procedure in which particle growth by different mechanisms takes place simultaneously with comminution in all stages except the initial one (Figure 16.3) [51].

A graphical illustration of the size changes during a milling process is given in Figure 16.4, which shows the alteration of the particle size as a function of the grinding time at constant intensity. However, similar-shaped curves can be also obtained as a function of the mechanical energy input (by varying the grinding intensity) during a fixed time of milling. Curve A describes the change of the particle size if no agglomeration occurs. Because the energy needed for particle comminution increases with decreasing particle size, the curve tends to a size limit when the energy necessary for further fragmentation approaches the total amount of mechanical energy introduced into the system. If the time is kept constant and the grinding intensity is gradually increased, fragmentation ceases when a particle size is reached at which the energy of brittle fracture exceeds the value necessary for plastic deformation (brittle-ductile transition).

The main mechanism of particle fracture in stirred media mills was demonstrated to be the stressing of particles between two grinding beads [57]. Particle–particle collisions, however, are regarded as causative for agglomeration. According to Annapragada and Adjei, the collision rate depends on the number concentration, the linear velocity and the radius of the particles [58] and can be shown to increase if the particle size of a given amount of mill charge gradually becomes smaller. Although the radius decreases in this process, its lowering effect on the collision rate is outweighed by the increasing particle number. In the case of sub-micron particles, the collision rate increases also due to the effect of Brownian motion [48]. At a particle size below about 1 μm, attractive dispersion or van der Waals forces become more and more dominant, causing aggregation

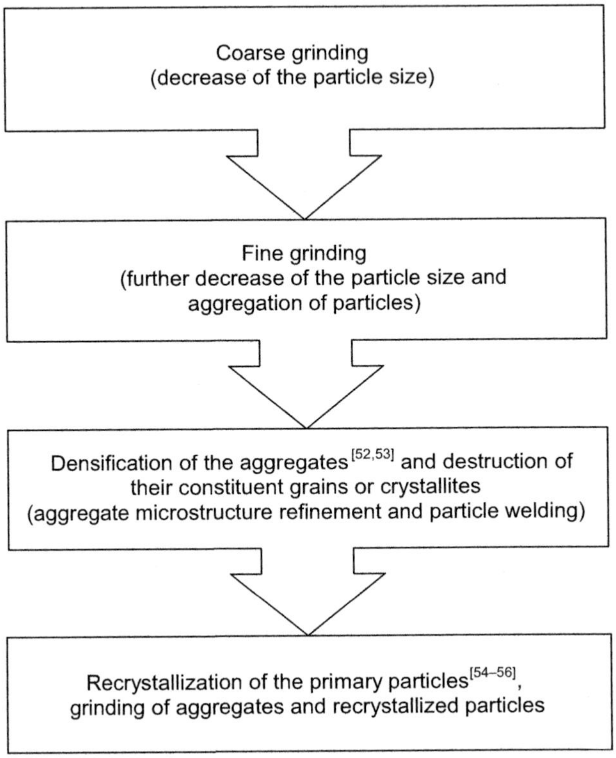

Figure 16.3 *Stages in wet grinding [51–56].*

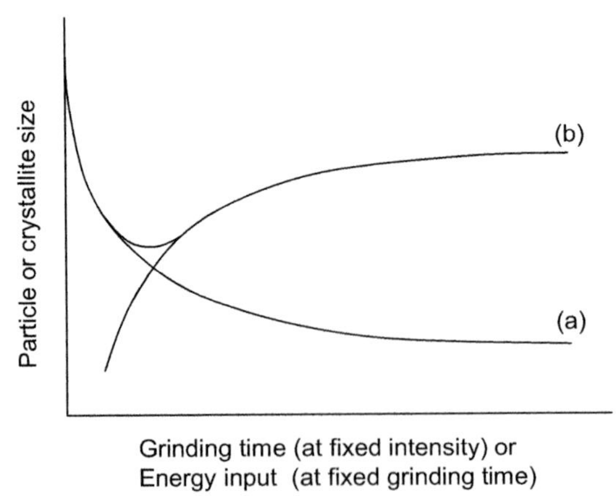

Figure 16.4 *Particle size as a function of the grinding time or of the energy input [51].*

and subsequently agglomeration. Aggregation is characterized by weak, reversible cohesion of particles due to dispersive forces. It appears already in the early stages of the grinding process or at low intensities [59]. When grinding progresses or with increasing input of energy, the particles are subjected more and more to plastic deformation. Consequently the enhancing contact geometry promotes further van der Waals interactions and finally dense aggregates are formed by cold welding. Some researchers have proposed that also fracture-induced surface modifications are involved [60, 61]. Particularly in the case of organic materials which behave more plastically than inorganic solids, the temperature, which rises with increasing agitation, also favours reagglomeration. In Figure 16.4, this process is characterized by curve b. The particle size increases until a dynamic equilibrium is achieved between agglomeration and deagglomeration due to gravitation and shear forces. Besides deagglomeration, the larger particles and aggregates are also subjected to brittle fracture. Before the stage is achieved in which comminution and aggregation counterbalance each other, the particle size passes through a minimum which marks the best achievable grinding result (combined curve a–b). This explains why in some cases an extension of the grinding time may even increase the size and heterogeneity of the particles, while a decrease of the milling speed is able to reduce the particle size [25].

Inkyo *et al.* studied the process of deagglomeration and subsequent reagglomeration with TiO_2 particles in a stirred media mill [62]. Using beads of different sizes they found that the time course of grinding is remarkably influenced by the diameter of the beads. With all tested beads larger than 15 μm (30 μm, 50 μm, 100 μm), the particle size distribution first shifted to smaller sizes within 2 hours of grinding and then gradually shifted back to larger sizes until the process was stopped after a total time of 4 or 5 hours. This indicates an initial break-up of agglomerates, followed by reagglomeration as milling is continued. The 100 and 50 μm beads were not capable of completely breaking up the nanoparticle agglomerates and could not provide size distributions in the range of the primary TiO_2 particles. However, the 30 and 15 μm beads succeeded in completely dispersing the particles and with the latter no reagglomeration could be observed. Using much larger beads (300–400 μm and 400–600 μm) also Bel Fadhel *et al.* reported similar results which confirms that agglomeration is favoured by high energy input and larger bead sizes and can be reduced or avoided by choosing fine grinding media [63].

An established assumption in nanomilling is that the rate of particle breakage is proportional to the rate of collisions between the particles and the milling beads. It was often demonstrated that smaller beads cause a higher degree of size reduction, an observation that is attributed to an increased collision frequency due to the larger number of beads in the case of smaller grinding media [25]. On the other hand, larger beads have greater dispersing and impulsive power per bead, but lower number concentrations in the milling chamber [62]. Smaller beads are able to perform the same work as larger ones but with a lower energy input. Similar results have been reported also by Kwade [64] who correlated the comminution efficiency with the number of stress events in the grinding chamber and with the intensity acting at these events.

16.4.6 Amorphization

Upon continuous input of mechanical energy, particle surfaces are activated and crystal defects are generated due to the disordering of the lattice structures [65, 66]. When the number of defects reaches a critical concentration, the amorphous state becomes more stable than the disordered crystals and transformation into an amorphous state occurs [25, 67, 68]. The transformation starts with the formation of local amorphous regions in the crystalline material which increases the surface energy of the particles. With increasing input of grinding energy the amorphous fraction of the material rises at the expense of crystallinity [69]. It should be pointed out here that the term 'drug nanocrystals' implies a crystalline state of the discrete particles, but depending on the production method, they can also be partially or completely amorphous [10]. Although the amorphous state seems advantageous for rapid dissolution and high saturation solubility, it is often undesirable with respect to an impaired physical and chemical stability of the finished product during storage. Amorphous regions in crystalline nanoparticles are able to favour chemical degradation or particle aggregation. Moreover, in many cases the amount and location of the amorphous content are only poorly predictable [70]. Smaller grinding beads are less effective with respect to amorphization than larger ones for they transmit less impact energy [62]. Thus the use of small grinding media is highly beneficial for the preparation of stable products which obtain their improved solubility mainly by crystal fineness rather than by amorphicity. However, wet milling *per se* often favours a high degree of crystallinity. Water is able to inhibit amorphization for it reduces the glass-transition temperature [25, 70].

16.5 Influence of the Stabilizer

16.5.1 Effects of Stabilization

As discussed above, agglomeration becomes a major problem if particles are ground to the nanometer range. One reason is that further comminution is impeded and smaller particles may coalesce into larger particles again. The other reason is the occurrence of flocculation which increases the viscosity of the suspension if particles become smaller. According to the semi-empirical Krieger–Dougherty equation, the viscosity ratio between the suspension and the liquid phase η_r can be calculated from the volume fraction of the suspended solids.

$$\eta_r = \frac{\eta}{\eta_s} = \left(1 - \frac{\phi}{\phi_{\max}}\right)^{-[\eta]\phi_{\max}} \tag{16.8}$$

with η being the viscosity of the suspension and η_s the viscosity of the suspending fluid. $[\eta]$ is the intrinsic viscosity (value for hard spheres $= 2.5$), ϕ is the volume fraction of the particles and $\phi_{\max}$ is the maximum packing fraction. The latter is 0.74 for the closest packing of equally sized spheres but for log-normal distributed spherical particles under real conditions, a range between 0.64 and 0.74 can be calculated [71]. In the case of bulky agglomerates, however, $\phi_{\max}$ becomes even smaller, thus significantly increasing the viscosity of the suspension. The smaller the particles are ground and the more their surface energy rises, the more they tend to agglomerate with the result of considerable rheological thickening of the slurry and ultimately the development of a yield stress. The occurrence of a yield value impedes the production of small particles, limits the solid concentration and reduces the throughput [72]. Equation 16.8 shows that a high fraction of solids ϕ increases the viscosity but the effect can be overcompensated by an effective deagglomeration, thus increasing also ϕ_{max}. Greenwood *et al.* demonstrated that without any dispersant, the typical maximum percentage of solids in a suspension is restricted to approximately 50% (wt) for ultrafine grinding in stirred media mills while an upper limit of solid concentration of 80% can be achieved in the presence of an optimum dispersant [73]. The use of dispersants is necessary in nanomilling to avoid particle agglomeration and flocculation by creating repulsive forces between the particles. Dispersants, also referred to as stabilizers, can be surfactants or polymers, both of them ionic or non-ionic.

16.5.2 Steric and Electrostatic Stabilization

Two different mechanisms of stabilization have to be distinguished: steric stabilization, which is accomplished by uncharged molecules, and electrostatic stabilization which is achieved by ionic substances. Another type of stabilizer combines both principles within the same molecule to achieve both steric and electrostatic effects, which is referred to as electrosteric stabilization. The same principle of action can also be achieved by the combined use of a steric and an electrostatic stabilizer. Suspension stabilizers are usually surfactants or polymers. Nonionic compounds from these substance groups act by steric effects while ionic surfactants and polymers stabilize the system by electrostatic repulsion or, in case of higher molecular weight or chain length, by electrosteric action [25] (Figure 16.5).

- *Electrostatic stabilizers:* Ionic stabilizers are effective in an aqueous environment, but upon drying they may lose effectiveness because the charged state is not maintained on dry surfaces [74]. Due to their ionic nature, electrostatic stabilizers are sensitive to changes in pH and ionic strength. The zeta potential is the usual measure for the surface charge of the particles. A physically stable nanosuspension which is stabilized only by electrostatic repulsion requires a zeta potential of at least 30 mV [26]. As the zeta potential strongly depends on the pH, it is important that sufficiently high values are achieved not only under artificial measurement conditions but at the genuine pH of the suspension. Also the presence of charged species, particularly multivalent ions, dissolved in the dispersion medium can lower the surface charge of the particles, hence reducing their interparticle distance and thus favouring rapid agglomeration [75].
- *Steric stabilizers:* Nonionic stabilizers do not have any charged functional groups and accordingly, as predicted by the DLVO theory, they cannot provide any significant long-range repulsion. Instead, their stabilizing action is based on steric effects: depending on the size and structure of the stabilizer molecules, they form more or less pronounced layers on the particle surfaces which prevent the particles from approaching close enough to each other and prevent the van der Waals attractive forces from dominating [25]. Steric stabilization is more robust than electrostatic stabilization, resulting in a higher insensitivity towards electrolytes. Furthermore, steric stabilization is effective not only in aqueous but also in nonaqueous environments, and is able to provide lower viscous systems even in the case of high solid concentrations. The high viscosity of the suspension enhances the stability of the finished product, however, during the milling process it may prolong the process time. A disadvantage of steric stabilization is its greater sensitivity to temperature fluctuations. This may pose problems if the product needs to be autoclaved. The

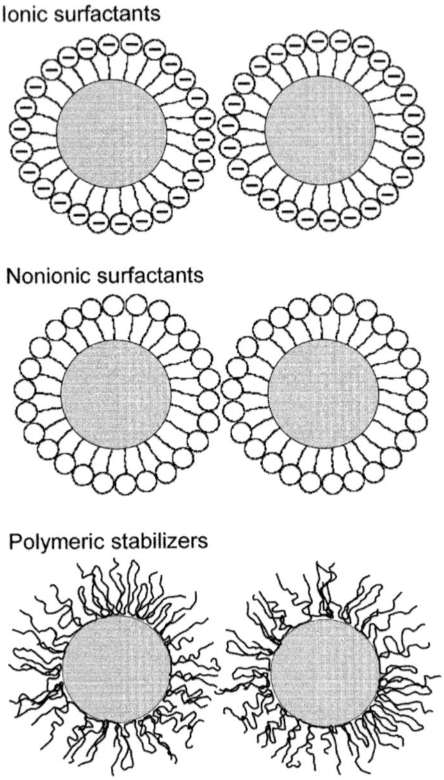

Figure 16.5 *Types of stabilizers.*

effectiveness of a steric stabilizer depends mainly on its molecular structure, particularly the chain length, flexibility and branching of the molecules. Formerly it was considered that longer chains would provide better stabilization due to the thickness of the steric barrier, however, studies have shown that longer ethylene oxide or alkyl chains reduce the stability of dispersions [76]. The reason is the higher packing density of longer chains which causes the surface layer to lose conformational entropy [76, 77]. To counteract this entropy loss, surfactant mixtures can be used to increase the conformational disorder of the stabilizer molecules and hence to achieve better protection against agglomeration [25, 78].

An essential prerequisite for the effectiveness of the stabilizer molecules is their ability to adsorb onto the particles' surface. The adsorption should be strong enough and robust enough to last for a sufficient time. Various binding forces, such as hydrogen bonds, van der Waals, polar, ionic or ion-dipole interactions may contribute to the adsorption of the stabilizer [79]. From a kinetic point of view, high adsorption and low desorption rates are desirable. As the stabilizer has to cover the particles' surface, the amount necessary for proper stabilization is directly related to the surface area and thus to the particle size. Though one could assume that effective stabilization needs full coverage of the particle surface, there is evidence against this view. The adsorption characteristics of a stabilizer can vary between the different crystal faces of the same material. Depending on the raw material and the parameters of the milling process, it is very probable that different crystal faces are formed, revealing different solubility properties [80–83]. Also the adsorption of excipients may be preferred or even may only be possible on certain faces of the crystal, depending on the affinity to the newly formed surfaces and their surface energies [84]. With microcrystals it has been observed that no complete surface coverage by the stabilizer is necessary to exert a marked effect on the particle properties [85]. In drug nanomilling, physically stable formulations are obtained with a weight ratio of drug to stabilizer from 20:1 to 2:1 (w/w) which corresponds to a stabilizer fraction of 5–33% [24]. If the amount of the stabilizer is too low, particles tend to aggregate, while concentrations exceeding the optimum may promote Ostwald ripening which means that larger crystals grow at the expense of smaller crystals [25].

A plethora of different substances have been used and tested as stabilizers for technical applications of wet grinding, many of them toxic or harmful compounds. For pharmaceutical purposes stabilizers – surfactants as well as polymers – are required to be classified as GRAS (Generally Recognized As Safe) by the U.S. Food and Drug Administration. Table 16.3 gives an overview of stabilizers used for the preparation of nanocrystals by wet milling.

16.5.3 Surfactants

In a screening study of surface stabilization during the production of drug nanocrystals, van Eerdenbrugh *et al.* compared 13 stabilizers by testing them for milling of nine structurally different drugs. The study included both polymers (synthetic and semisynthetic) and surfactants. Their stabilizing performance increased from semi-synthetic polymers to linear synthetic polymers to synthetic copolymers. Best results were achieved with the surfactants. This was explained as being caused by their low viscosity and high surface activity. Furthermore, it was found that the surface hydrophobicity was the decisive factor determining the agglomeration tendency of the particles and hence the ease of nanosuspension stabilization [92]. Poorly soluble drugs are usually hydrophobic and thus are highly prone to agglomeration if not sufficiently stabilized. It was shown with a series of different types of nonionic stabilizers (Brij, Span, Symperonic, Tween) that the degree of stabilization increases as the HLB number decreases [93]. This was explained by the number of surfactant molecules present on the particle surface. Hydrophobic surfactants with lower HLB values escape more readily from the aqueous phase to the particle surfaces than hydrophilic compounds with higher HLB values [25].

Surfactants typically employed for nanomilling include, for example, Tween 80, Span 80, sodium lauryl sulphate, poloxamer, and D-α-tocopherol polyethylene glycol 1000 succinate. When selecting a stabililizer, subsequent process steps should also be considered. For example, if drying of the particles is intended after milling, a low melting point of the stabilizer may limit the selection of a suitable method. Thus, in the case of particles stabilized with Tween or poloxamer, spray drying may cause aggregation. In steam sterilization, the temperature usually exceeds the cloud point of the stabilizing surfactant, posing the risk of an increased particle size. Aggregation can also occur as a consequence of irradiation which has been noted to decrease the zeta potential [25, 94].

Table 16.3 *Stabilizers employed in nanomilling.*

1st stabilizer	% (w/w)	2nd stabilizer	% (w/w)	Drug	% (w/w)	Mean (nm)	Part. prop.	Stab.	Ref.
HPMC	1.25			Megestrol acet.	5	314	a	N/A	86
HPMC	1.25			Megestrol acet.	5	185	a	e	86
HPMC	2			Raloxifene HCl	5	211			87
HPC-SL	1.25			Megestrol acet.	5	623	a	N/A	86
HPC-SL	1.25			Megestrol acet.	5	176	a	e	86
HPC-SL	2			Raloxifene HCl	5	198			87
HPC-SL	2			Finasteride	5	292		N/A	88
PVP K-12	1			Meloxicam	5	98			89
PVP K-17	1			Meloxicam	5	98			89
PVP K29/32	1.25			Megestrol acet.	5	24816	a	N/A	86
PVP K29-32	2			Raloxifene HCl	5	186			87
Copovidone K25-34	1.25			Megestrol acet.	5	292	a	N/A	86
Copovidone K25-34	2			Raloxifene HCl	5	225			87
Poloxamer 188	1			Meloxicam	5	133			89
Poloxamer 308	1.25			Megestrol acet.	5	459	a	N/A	86
Poloxamer 308	1.5			Raloxifene HCl	5	215		g	87
Poloxamer 308	1.5			Finasteride	5	1343		f	88
Poloxamer 308	2			Tacrolimus	10	237			90
Poloxamer 388	1			Meloxicam	5	129			89
Poloxamer 407	1			Megestrol acet.	5	228		f	86
Poloxamer 407	1.5			Raloxifene HCl	5	209		g	87
Lysozyme	1			Meloxicam	5	95	a		89
Lysozyme	1.25			Megestrol acet.	5	179		N/A	86
Lysozyme	1.25			Megestrol acet.	5	209		d	86
Lysozyme	4			Megestrol acet.	10	108			86
		PS 80	1	Meloxicam	5	227			89
		PS 80	1.25	Megestrol acet.	5	210		N/A	86
		PS 80	1.25	Megestrol acet.	5	157		f	86
		PS 80	1.5	Raloxifene HCl	5	513		f	87
		PS 80	1.5	Finasteride	5	-	b	N/A	88

		PS 80	2	Tacrolimus	10	208		g	90
		SLS	0.1	Raloxifene HCl	5	186		e	87
		DOSS	0.1	Raloxifene HCl	5	206			87
		Tyloxapol	1.25	Megestrol acet.	5	214		N/A	86
		Tyloxapol	1.25	Megestrol acet.	5	208		f	86
		Tyloxapol	1.25	Finasteride	5	376	(a)	N/A	88
		SDC	1	Meloxicam	5	119			89
		Lecithin	1	Meloxicam	5	190	(a)		89
HPMC	1	SLS	0.05	Megestrol acet.	5	148			86
HPMC	1	SLS	0.05	Megestrol acet.	5	152		d	86
HPMC	1.25	SLS	0.0625	Megestrol acet.	5	161			86
HPMC	2	SLS	0.1	Megestrol acet.	10	146			86
HPMC	1	DOSS	0.05	Megestrol acet.	5	156			86
HPMC	1.25	DOSS	0.05	Raloxifene HCl	5	213			87
HPMC	1.25	DOSS	0.05	Finasteride	5	585		N/A	88
HPMC	2	DOSS	0.05	Tacrolimus	10	215			90
HPMC	2	DOSS	0.1	Megestrol acet.	10	150			86
HPMC	6.5	DOSS	0.325	Megestrol acet.	32.5	160			86
HPC-LF	1.25	SLS	0.05	Miconazole	20	156*		(f)	91
HPC-LF	5	SLS	0.05	Miconazole	5	150*		f	91
HPC-LF	5	SLS	0.05	Miconazole	20	140*		(f)	91
HPC-LF	5	SLS	0.2	Miconazole	20	152*		(f)	91
HPC-SL	1.25	SLS	0.05	Megestrol acet.	5	167			86
HPC-SL	1	DOSS	0.05	Megestrol acet.	5	167			86
HPC-SL	1	DOSS	0.15	Tacrolimus	5	169			90
HPC-SL	1.25	DOSS	0.05	Raloxifene HCl	5	192			87
HPC-SL	1.25	DOSS	0.05	Finasteride	5	386	a	N/A	88
HPC-SL	2	DOSS	0.1	Tacrolimus	10	389			90
HPC-SL	1	SDC	0.1	Tacrolimus	5	1780	c		90
PVP	1	DOSS	0.05	Megestrol acet.	5	167			86
PVP K12	2	SDC	0.15	Tacrolimus	10	329			90
PVP K15.5-17.5	1.25	SDC	0.05	Raloxifene HCl	5	169			87

(continued)

Table 16.3 *(Continued)*

1st stabilizer	% (w/w)	2nd stabilizer	% (w/w)	Drug	% (w/w)	Mean (nm)	Part. prop.	Stab.	Ref.
PVP K15.5-17.5	1.25	SDC	0.05	Finasteride	5	444		N/A	88
PVP K17	1.25	BAC	0.05	Raloxifene HCl	5	195			87
PVP K17	1.25	BAC	0.05	Finasteride	5	1218		N/A	88
PVP K29-32	1.5	SLS	0.05	Raloxifene HCl	5	250			87
PVP K29/32	1.25	SLS	0.05	Finasteride	5	1608			88
PVP K29/32	2	DOSS	0.05	Tacrolimus	10	192			90
Copovidone K25-34	1.25	SLS	0.05	Finasteride	5	354		N/A	88
Copovidone K25-34	1.5	SLS	0.05	Raloxifene HCl	5	178		e	87
Copovidone K25-34	1	DOSS	0.05	Megestrol acet.	5	164			86
Copovidone K25-34	3	DOSS	0.05	Tacrolimus	20	171			90
Poloxamer 188	1.25	DOSS	0.05	Raloxifene HCl	5	283		g	87
Poloxamer 188	1.25	DOSS	0.05	Finasteride	5	397	a	N/A	88
Poloxamer 308	1.0	PS 80	1.0	Raloxifene HCl	5	180		f	87
Poloxamer 338	1	PS 80	1	Finasteride	5	211	a	N/A	88

Notes:

BAC	Benzalkonium chloride
DOSS	Dioctyl sodium sulfosuccinate
HPC	Hydroxypropyl cellulose
HPMC	Hydroxypropylmethyl cellulose
Megestrol acet.	Megestrol acetate
PS 80	Polysorbate 80
PVP	Polyvidone (Polyvinylpyrrolidone)
SDC	Sodium deoxycholate
SLS	Sodium lauryl sulphate
a	Aggregation
b	No particulate product
c	Unsuccessful
d	Aggregation upon dilution
e	Aggregation upon storage
f	Crystal growth upon storage
g	Instable upon storage
N/A	No stability upon storage investigated
*)	d_{50} instead of mean diameter

As already mentioned, the amount of stabilizer is a critical factor, as it affects the stability of the suspension and needs to be tailored to the appropriate particle size. High amounts of surfactants are able to solubilize the drug by micelle formation, thus destroying the crystal structure of the particles. On the other hand, micelles formed by excess surfactants may develop an adhesive effect and promote the formation of aggregates [93].

16.5.4 Polymers

The molecular weight of a polymeric stabilizer has an important influence on its action. The chains should be long enough to form a layer of an optimum thickness on the particles' surface to overcome the van der Waals forces of attraction. It is easy to understand that below a certain chain length the steric barrier becomes too thin, thus promoting particle aggregation. However, also exceeding an optimum molecular weight is unfavourable as thick polymer brush layers on the surfaces may cause particle bridging and thus may also lead to agglomeration. Usually a molecular weight between approximately 5000–25000 g/mol is in an optimum range for steric repulsion and thus particle stabilization [95]. Polymers usually have a lower adsorption potential than nonionic surfactants with equal chain lengths [93]. In the context of molecular weight also, its effect on the viscosity has to be considered. As polymers enhance the viscosity of the suspension medium and reduce the diffusivity depending on their molecular weight, their adsorption to the particle surfaces may be retarded, creating the need for longer process times [96]. However, in a study with four poorly soluble anti-cancer agents, the most effective stabilizers turned out to be the higher molecular weight types of each polymer [97]. Polymer adsorption and the related steric stabilization may be abolished if the drug crystals have a low melting point and the surface structure of the particles disintegrates upon exceeding this temperature during the process. This can even occur at higher temperatures than expected for the melting point is size dependent and dependent on changes in the nanometer range [25, 98].

The surface energy of drugs and polymers was found to be an important factor, in stabilizing a nanosuspension, although it cannot solely explain all experimental observations. Specific interactions, such as hydrogen bonding between the drug and the polymer, are able to interfere with the stabilization mechanism, and may hinder deagglomeration. For example, in an experimental setting with seven different drugs, the stabilizing activity of hydroxypropyl cellulose did not prove sufficient for those compounds with hydroxyl groups, as the latter are also part of the polymer structure [99]. However, not only hydrogen bonds but also ionic attractions can act between drug and polymer and impede stabilization. Such ion-pairing, for example, was reported to occur between the lysine component of amphiphilic amino acid copolymers used as stabilizers and the carboxylic acid group of naproxen [25, 95].

16.6 Milling Equipment and Technology

16.6.1 Grinding Beads

The technique of nanomilling in media mills is characterized by the application of high shear forces and small grinding beads. The typical size of such pearls is between 1000 μm and down to 15 μm. An optimum utilization of power input is achieved with spherical grinding media. Irregularly shaped particles revealed a worse grinding performance and that a higher specific energy is necessary to achieve a certain particle size [48]. Another essential requirement for high energy utilization is the low plastic deformability of the beads because the energy expended for plastic deformation of the milling media is not available for the comminution of the drug particles [100]. For this reason ceramic materials are preferred to steel. Most commonly used are grinding beads made of yttrium-stabilized zirconia (95% ZrO_2 and 5% Y_2O_3) which has the highest fracture-toughness among all oxide ceramics. The material has a high density (6.05 g/cm^3) and a high wear resistance. Becker and Schwedes found yttrium-stabilized zirconia beads to generate the fewest wear of all tested grinding media (among the others: Al_2O_3, magnesium-stabilized zirconia and silicon nitride) [100]. Bead breakage results in fragments which cannot be removed from the nanoparticle slurry and would end up contaminating the product. Inkyo *et al.* studied the contamination caused by bead fragments in correlation to the media size using zirconia beads. A nanoparticle slurry was dispersed using 100 μm and 15 μm beads, respectively. After a process time of 15 minutes an amount of 1400 ppm could be detected in the suspension milled with the larger beads and 220 ppm was produced by the 15 μm pearls. This result implies that the larger beads were more likely to break during the milling process. A similar result was also found by Yoden and Itoh [101]. As larger beads have higher impact energies, they are more likely to fragment during collisions [62]. Although zirconia is a highly inert substance with a low degree of product contamination, it is reported that under certain circumstances the suspension stabilizer was able to cause extraction of yttrium from the milling beads into the solution [102]. In the late 1990s, researchers at the Eastman Kodak Company worked on media milling methods for the large-scale production of pigment nanoparticles. They developed small polymeric beads made of very hard, highly cross-linked polystyrene and found them to be extremely effective for this purpose. In a series of patents it was claimed that these polymeric beads do not produce any significant wear when used in a stirred media mill for producing pigment nanoparticles [103–106]. Merisko-Liversidge *et al.* used polystyrene beads also for milling drugs and reported values of less than 50 ppm insolubles (based on the drug concentration of the dispersion) shed from the milling media and less than 50 ppb residual monomers [24].

16.6.2 Types of Media Mills

16.6.2.1 Conventional Ball Mills

At the laboratory scale different types of mills can be used to agitate a mixture of suspended drug and grinding beads. Low and high energy mills are distinguished. Conventional ball mills (tumbling mills, roller mills, jar mills) belong to the former type. They are usually operated at 60–85% of the critical speed which is reached if the centrifugal forces are equal to the weight of the beads and the beads just begin to centrifuge with the mill. The critical speed (n_c [rpm]) can be calculated from the inner diameter of the milling jar (d [m]) by Equation 16.9:

$$n_c = \frac{42.29}{\sqrt{d}} \tag{16.9}$$

Hence, if laboratory tumbling mills with drum diameters of 15–20 cm are used, they have to be rotated typically between 80 and 90 rpm. In wet milling a somewhat lower speed is used than in dry milling [20]. The drum should be 30–50% (v/v) filled with grinding media with respect to the bulk volume of the beads (about 20–35% (v/v) with respect to the true volume). For optimum results and shorter process times, the upper end of this range is preferred. The void volume between the grinding media represents 10–20% of the total volume of the mill. Grinding occurs fastest if the interparticulate voids are just filled with the drug/stabilizer slurry and the grinding media are slightly covered. This is usually the case if the slurry fills about 20–30% of the grinding chamber. The level of the suspension should be well above the surface of the grinding media because excessive wear might develop if dry beads move in close contact with each other. Using sub-micrometer grinding media and appropriate stabilizers, even nanosizing of drug crystals is possible with low energy mills. However, process times of at least 3 to 5 days are necessary [107–109]. In a tumbling ball mill the media velocity and thus the maximum energy are limited by the occurrence of centrifugal forces which cause the beads to revolve with the jar instead of rolling and cascading. Hence, planetary and vibrating ball mills are employed to facilitate media milling on a more energy-intensive level. In a planetary ball mill, the grinding beads are accelerated by Coriolis forces generated by superimposed rotational movements. These forces are much higher than gravitational forces acting in tumbling mills and thus reduce the particle size to the nanometer range within process times of about 24 hours [92]. Optimum bead loads are the same as in the case of tumbling ball mills. A vibratory mill is driven by an eccentric-activated mechanism which vibrates the mill 1500–2500 cpm with an amplitude of several millimeters. The beads are moved by short and sharp impacts which cause a much greater acceleration than is possible due to gravity. Usually the media charge is approximately 80% of the milling chamber volume, however, also bead loads of 16% are described [20, 96]. Within a process time of only 30 min., a variety of different drugs could be ground to particle sizes between 150 and 450 nm [96].

16.6.2.2 Stirred Media Mills

In media milling, two basic principles of agitation can be distinguished. Either the milling medium is stirred by an agitator, or the complete container is moved in a more or less complex pattern and sets the grinding beads in motion. If the grinding chamber is completely filled with milling material (for which a volume fraction of about 64–74% solids can be assumed), it is hardly possible to achieve agitation of the milling beads by movement of the complete container [110]. Above the critical speed the beads are pushed against the wall and revolve with the drum. High rotational speeds up to 5000 rpm as typically used in high shear milling can only be transformed into shear forces if the grinding media are moved by agitators instead of rotating the drum. Thus stirred media mills became established with a stationary milling chamber and a rotating agitator.

As a modification of existing ball mills with rotating milling drums, the first stirred media mill was developed in 1928, consisting of a vertical stationary grinding chamber in which spherical grinding media were moved by a slowly rotating agitator. Mainly in the ceramic and mineral industry, such low-speed stirred media mills, also called 'attritors' are still in use. They exist in a variety of different sizes and construction types with capacities up to 10 t/h. Also a combined agitation by stirrer and rotating body is known. A common feature is a circumferential stirrer speed usually not above 7 m/s. The achievable product fineness is in the higher micrometer range but also particle sizes of 5–10 μm can be obtained. In parallel to further improvement of such mills, high-speed stirred media mills were developed, working with circumferential stirrer speeds of 8–20 m/s (typically 5–15 m/s) and with smaller grinding media. The first high-speed stirred media mills were constructed in a vertical grinding chamber arrangement with an open top covered by a sieve cage. Today, however, mainly closed-type mills have become established. These are available in two different designs with either a vertically or a horizontally mounted vessel.

The grinding chamber can be lined with ceramic, tungsten carbide or polyurethane and is usually equipped with a water-jacket for cooling purposes. Usually the bulk volume of the grinding media occupies 70% and up to 90% of the free grinding chamber volume. The size and material of the beads are chosen based on type and particle size of the feed material but also depending on the desired degree of comminution (usually between 200 and 1000 μm). By agitating the grinding media via stirring at a circumferential speed of up to 20 m/s, centrifugal accelerations are achieved which are more than 50 times those attainable by means of gravity. Stirred media mills are available in a broad range of different sizes from small-scale laboratory equipment with milling chambers of 10 ml up to production machines with holding capacities of more than 1 m^3 and with drive powers of more than 1 MW.

A huge variety of different types of stirred media mills have been developed particularly since the 1990s. As well as devices for technical purposes, also special mill types for use in drug production have been constructed, meeting GMP requirements with regard to materials and design. The main differences between the various types can be found in their chamber and stirrer geometry and in their separation devices. The latter is responsible for a free discharge of the product while retaining the grinding beads in the milling chamber. Usually, it is designed as a screen (e.g. flat screens, screen cartridges, rotating or stationary cylindrical screens) or as a rotating gap. For an optimum separating function, the width of the rotating gap or the mesh size of the screen should be less than half the diameter of the grinding media. Due to the multiple advantages of small milling beads, as discussed above, increasingly smaller grinding media are employed. In order not to unduly constrict the discharge of the suspension, increasingly high shear mills are equipped with screens instead of rotating gaps. The advantage of gap separators is that most of them are easily adjustable to be adapted to different sizes of grinding media. Sieve separators are often constructed as ring screen assemblies which are either positioned in the cylindrical wall of the milling chamber or integrated into the rotating stirrer shaft. In both cases they have to be designed and fixed in a manner, that they cause no dead zones in the grinding chamber. Especially with outlet screens positioned within the hollow agitator shaft, a high separation performance can be achieved by taking advantage of the centrifugal forces acting inside the mill which direct the motion of the grinding media radially outwards and thus away from the screen. Highly effective is also the use of a special kind of rejector wheel classifier upstream of the outlet of the vertically arranged grinding chamber. Concerning the chamber and stirrer geometry, three construction types of stirred media mills can be distinguished, which are shown in Figure 16.6:

- stirred media mill with disc stirrer;
- stirred media mill with pin stirrer or pin-counter-pin stirrer;
- annular gap mills.

The simplest agitator type is the disc stirrer. From regularly shaped unbaffled discs the energy is transferred to the feed particles and grinding media mainly by friction. However, many agitator discs are built of special designs, concentrically or eccentrically fixed, equipped with grooves, holes or slits, so that displacement forces become effective

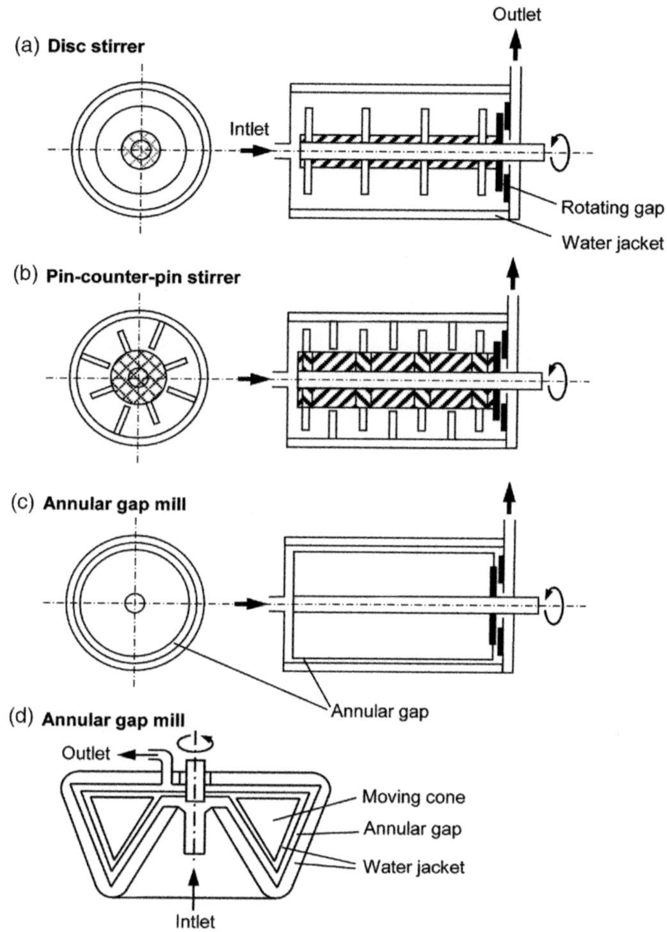

Figure 16.6 *Types of stirred media mills. (a–c) Reproduced with permission from [64], (d) Reproduced from [34] with permission from John Wiley and Sons.*

and additional energy can be transferred to the beads and particles of the stirred mixture. A pin-stirrer moves the slurry mainly by displacement forces. Especially if the milling chamber is equipped with counter-pins, larger power densities can be achieved than with disc stirrers at the same circumferential speed. The highest energy transfer from the rotor to the beads and thus the largest power densities can be obtained in annular gap mills. The width of this gap between the rotor and the stator is usually no larger than about 4 to 10 times the diameter of the grinding beads. If the mills are equipped with a smooth rotor and a smooth chamber wall, the energy is transferred only by adhesion forces. Not least because of the smaller volume of the grinding chamber and its mainly two-dimensional geometry, the power density is very evenly distributed and the residence times of the particles in each circulation cycle are less variable than in case of disc or pin-stirrers. Some annular gap mills are also equipped with baffles, in order to increase the energy transferred to the beads and particles. In the past few years, annular gap mills have become widely used and now a variety of different models and construction types are available from many manufacturers. They are designed with a single or double conical or cylindrical annular gap, or with a flat-shaped rotor disc. In some types of annular gap mills, the grinding media are internally circulated in order to reduce the risk of accumulation and blocking of the beads in front of the separator [64].

The grinding process can be performed in either batch or recirculation mode. In batch milling the suspension is processed within the confines of the grinding container and the batch size is limited to the grinding chamber volume. The recirculation mode, by contrast, offers the possibility of increasing the batch size beyond this limit by circulating the feed suspension between a storage container and the grinding chamber while retaining the grinding media in the mill. Besides this, also a continuous operation is possible in which the feed material is passed only once through the mill. However, far less comminution efficiency has to be expected in this mode.

During the milling process a considerable amount of mechanical energy is introduced into the system and most of it is transformed into thermal energy. Because the heating of the feed material can cause a variety of negative effects, as discussed above, effective cooling of the milling chamber is necessary. For this purpose, the mill and in the recirculation mode also the storage container are jacketed in order to control the temperature. The inlet suspension is around 5 to 10°C while the outlet product usually has a temperature between 15 to 30°C, or is even warmer, depending on the mill speed and the product viscosity [111]. It should be noted that most of the materials used as anti-abrasive linings of the grinding chamber (e.g. zirconia) are poor heat conductors which could impede the cooling efficiency. Table 16.4 compares the technical data of different types of stirred media mills from the small laboratory scale to large production equipment.

16.6.3 Process Parameters

The feed material for nanomilling processes typically should have a particle size between a few microns and a hundred microns. For small-scale trial batches also larger-sized starting material can be employed. However, at the larger scale, it runs the risk of clogging the mill. If only coarsely crystalline material is available, a pre-milling step may be necessary [111]. This can be done by dry milling or, alternatively, in a high-shear media mill using larger grinding beads. In the second case, after pre-milling, the process can be continued in the same equipment after changing the grinding media.

The most critical process parameters in nanomilling are the amounts of drug and grinding media, the size of the grinding beads, the agitator speed, the process time and the temperature. In the circulation mode also the flow rate has to be considered as another factor of influence. The optimum settings vary within broad limits. In typical processes the amount of drug in the drug/stabilizer slurry ranges between 0.1 to 40% (w/v) [24, 86–92, 96, 99, 108, 109]. The process times and milling speeds vary considerably and depend mainly on the mill type employed. Nanocrystals can be obtained either by low milling speeds (80–90 rpm) and long milling times (1–5 days) [92, 107–109, 112] or high milling speeds (1800–6000 rpm) and short milling times (30–60 min) [58, 96, 99, 113]. Thus, depending on the mill type and energy input, the nanomilling process may last from hours to several days

Table 16.4 *Technical data of stirred media mills for wet grinding (selection).*

	Operation mode	Batch volume (l) or capacity range (l/h)	Size of grinding chamber (l)	Lining of chamber and agitator	Separator	Bead size (μm)	Pharma version available
DYNO®-MILL RL	cont.	≥ 0.1 l	0.075	container: 5, 8 agitator: 3, 5	flat slit sieve	50–1000	x
DYNO®-MILL ML	discont. cont.	0.085 l / 0.170 l ≥ 0.5 l 1-50 l/h	0.15 / 0.3 0.15 – 1.4	container: 1, 4 agitator: 1, 2, 5, 7, 10, 11 container: 1, 2, 5, 8 agitator: 1, 2, 3, 5, 7, 10, 11, 12	without separator dynamic gap separator	100–2000	x x
DYNO®-MILL ECM	cont.	10-6000 l/h	1.5-60	container: 1, 2, 5, 8 agitator: 3, 5	slit sieve cartridge	200–2000	x
DYNO®-MILL KD	cont.	20-8000 l/h	6-612	container: 1, 2, 5, 8, 10 agitator: 3, 5	sieve or gap	300–2500	x
DYNO®-MILL NPM	cont.	10-100 l/h	2	container: 5, 8 agitator: 3, 5	dynamic separation system	50–200	x
NETZSCH MicroSeries	discont. cont.	0.075-0.080 l 0.140-0.500 l	0.08	container: 1, 5, 9, 10 agitator: 1, 5, 9, 10	centrifugal separation system	50–800	x
NETZSCH MiniSeries	discont. cont.	0.135-0.140 l 0.250-0.500 l	0.14	container: 1, 5, 9, 10 agitator: 1, 5, 9, 10	centrifugal separation system	50–2000	x
NETZSCH LMZ 2 - LMZ 150	cont.	50 - > 4000 l	1.6-150	container: 1, 2, 5, 10 agitator: 1, 2, 5, 10	centrifugal separation system	90–3000	x

(*continued*)

Table 16.4 *(Continued)*

	Opera-tion mode	Batch volume (l) or capacity range (l/h)	Size of grinding chamber (l)	Lining of chamber and agitator	Separator	Bead size (μm)	Pharma version available
HOSOKAWA ALPINE 50 AHM (Pico*liq*)	discont. cont.	≥ 0.008 l ≤ 0.5 l ≤ 20 l/h	0.012, 0.025, 0.090	agitator: 1, 5, 6, 10, 11 container: 1, 5, 6, 8, 10, 11	dynamically pressure compensated slit sieve	50–1000	-
HOSOKAWA ALPINE 90 AHM	discont. cont.	0,5 - 10 l 20 - 100 l/h	0.25, 1.1	agitator: 1, 5, 10, 11 container: 1, 5, 8, 10, 11	dynamically pressure compensated slit sieve	50–1500	x
HOSOKAWA ALPINE 132 AHM -800 AHM	discont. cont.	≥ 10 l 100-12000 l/h	5-1056	agitator: 1, 3, 5, 10 container: 1, 3, 5, 8, 10	dynamically pressure compensated slit sieve	200–3000	x
Kotobuki Industries Co., Ltd., ULTRA APEX MILL			0.15-10		centrifugal separation system	15–50	

Notes: Lining materials of grinding containers and agitators:
1 stainless steel
2 hardened steel
3 hardened chrome alloy
4 glass
5 zirconium oxide
6 aluminium oxide
7 tungsten carbide
8 silicon carbide
9 silicon carbide-silicon nitride (SiC/Si_3N_4)
10 polyurethane
11 polyamide
12 polyethylene

and up to a week, however, high shear media mills allow for time-efficient high throughput production cycles [25].

16.7 Process Development from Laboratory to Commercial Scale

16.7.1 Early Development

When in the early stages of development the active pharmaceutical ingredient (API) is still in short supply and there are often no more than a few hundred milligrams available, special laboratory-scale equipment is necessary for milling. As Table 16.4 shows, the smallest batches which can be processed in such lab-scale mills are the size of 75–85 ml in the case of the Dyno®-Mill Multi Lab and the Netzsch MicroSeries line, whereas the NanoMill-01™ from Elan and the Picoliq from Hosokawa Alpine, which are designed for even smaller working capacities, have smallest grinding chamber volumes of only 4 and 12 ml, respectively. Because the drug concentration of the feed slurry usually ranges from 1 to 400 mg/ml, milling of only a few milligrams of API is possible with such equipment, providing yields up to 85 [113] or even more than 95% [111]. According to Kesisoglou *et al.*, the success rate for reaching a particle size in the 100–250 nm range and consequently enhancing the oral bioavailability of the drug is around 80–90%. At a mill speed of about 5000 rpm, a batch is usually obtained within no more than a few hours. Thus, this way of small-scale processing is suitable for feasibility studies and serial tests, for example, in compound screening. Considering that so far 30% of all discovered chemical entities are not tested in animal or toxicity studies but discarded from further evaluation because of formulation problems arising from poor water solubility, nanomilling could provide a simple and fast method for the preparation of sufficiently bioavailable test formulations and thus increase the chances of success [114]. At this time, only short-term physical stability needs to be tested, mainly to cover the duration of the animal study.

16.7.2 Toxicological Studies

If nanosuspensions are prepared for toxicology studies, which are typically the next step in development, usually an amount of a few hundred to a few thousand grams of drug has to be milled. In contrast to clinical studies, for which solid formulations are preferred, it is acceptable to use liquid nanosuspensions without drying for the purpose of toxicological evaluation. As an early-phase toxicology study typically spans 2 weeks to 3 months, the formulation needs to be stable over this extended period of time and the choice of type and amount of stabilizer become increasingly important. In order to find a suitable stabilizer and to define an optimum concentration, a series of formulations have to be produced and evaluated in short time stability tests. These formulations can be prepared on a small scale (<1 g) as described above. Based on a drug concentration of 100–200 mg/ml and doses between 100–1000 mg/kg for a typical toxicology study (in dogs or rats), suspension volumes of more than 5 l are often required. Larger types of laboratory mills are used in this stage. If the capacity of the grinding chamber is smaller than the intended batch size, the mill can be operated in the circulation mode. As the time frame of a toxicological study can be up to 3 months, physical and microbiological stability are important concerns. It can be seen as an advantage, that the bioburden in nanosuspensions is quite low due to the shear-induced rupture of the microbial cells. Of course, the grinding procedure alone does not guarantee sterility. The best storage conditions with respect to both physical and microbiological stability are obtained in the dry or in the frozen state. Drying, however, needs additional effort for development and often involves time-consuming processes. Thus, freezing may be the more straightforward alternative. In order to obtain redispersible products, usually the addition of a cryo-protectant is necessary. In many cases it is sufficient to add sugar (e.g. sucrose), in equal weights to the drug.

16.7.3 Clinical Studies

When the development process reaches the stage of clinical studies, a variety of challenges arise. The preferred formulation for the market will mostly be a solid dosage form. This requires additional processing steps for drying and consequently additional development efforts. Every subsequent treatment of the grinded suspension runs the risk of changing its properties and biopharmaceutical behavior. Though early phase study formulations could be applied also as liquid suspensions, at least the registration batches must be designed to reflect the performance of the marketed product. In order to avoid agglomeration upon contact with water or gastric fluids, good redispersibility has to be ensured.

16.7.4 Drying

Many conventional pharmaceutical drying operations, such as fluidbed- or spray-drying are also applicable to nanosuspensions. In case of injectable formulations, freeze-drying will be the preferred method. If milling is performed with highly concentrated suspensions, to maximize the throughput, it can be necessary to dilute them before drying in order to decrease the viscosity. For reasons of dispersibility, a suitable stabilizer is added prior to the drying step. The most frequently used redispersants for freeze-drying are sugars and sugar alcohols like trehalose, sucrose, lactose, glucose and mannitol. They form a glassy matrix in the dry state in which the nanoparticles are immobilized. Basically, the redispersibility depends primarily on the choice of these cryoprotectants, but can sometimes additionally be influenced by the polymeric or surfactant stabilizers.

16.7.5 Further Processing of Drug Nanoparticles

Nanocrystals for parenteral administration need to be produced under aseptic conditions. If the particles are small enough, sterile filtration may be a suitable method. Otherwise, if the drug is sufficiently stable, autoclaving or gamma irradiation may be possible alternatives [10, 108]. If nanocrystals are going to be compressed into tablets, it has to be ensured that the primary particle size is maintained and the particles are released from the tablet in the form of an ultrafine nanosuspension. In the case when non-dispersible aggregates are formed, the dissolution rate is reduced, thus decreasing the oral bioavailability. It can be demonstrated that there is an upper limit of the drug concentration within the excipient mixture above which the nanocrystals come into contact with each other and might fuse to larger crystals under the compression force. Low-dose formulations, such as Rapamune (with 0.5–2 mg sirolimus per tablet) do not face this problem. A nanoparticulate drug fraction of less than 1% is well below the critical percentage. Based on the percolation theory, the critical amount of drug can be expected at about 30%. If the drug load exceeds

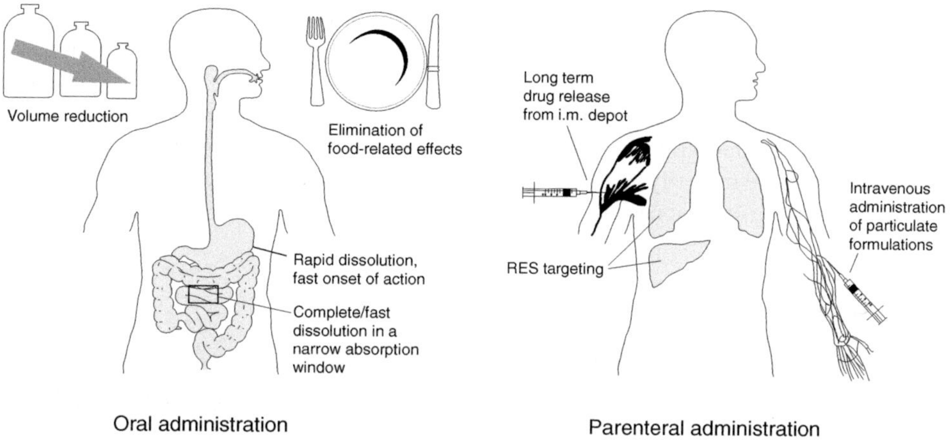

Figure 16.7 *Therapeutic benefits of nanocrystalline drug particles.*

this limit, the nanocrystals approach each other sufficiently closely and are likely to fuse at the contact points [110]. For this reason, the preparation of well-dispersible tablets with high concentrations of nanoparticles is not a trivial problem. However, special methods have been developed which allow for drug loads up to 90% [115].

16.8 Application and Biopharmaceutical Properties

One of the foremost objectives of nanosizing in pharmaceutical applications is the enhancement of drug dissolution and thus the improvement of bioavailability. Moreover, a variety of other biopharmaceutical effects can be achieved by nanosized drugs (Figure 16.7). As is commonly known, reducing the dimensions to the nanoscale causes fundamental changes in the material properties and behavior. Nanosizing of drug substances does not mean only a further gradual increase of alterations which already can be detected during comminution to the micrometer scale. Particularly with respect to biopharmaceutical interactions at the cellular level, new properties emerge if the particle size enters the nanometer range. For example, organ or membrane barriers can be passed by the particles as such and they become able to be taken up by phagocytic cells. However, not least, the dramatically increased dissolution rate of nanosized drug crystals offers a great potential for various applications, most notably for the oral route of administration.

The oral route of administration is the most common and convenient method of drug delivery due to ease of ingestion, good compliance, and cost effectiveness. When the decision has to be made which formulation should be chosen for a new substance, pharmaceutical developers generally prefer oral dosage forms. Currently, oral products account for more than 60% of the marketed drugs. By the introduction of combinatorial chemistry and high-throughput screening, the synthesis of new drug candidate compounds has accelerated immensely. Many of these substances, however, are poorly soluble, poorly absorbed or both. Such compounds represent an increasing proportion of newly synthesized candidate drugs

[22, 116, 117]. While currently about 40% of the drugs in use belong to BCS Classes II and IV (~30% Class II, ~10% Class IV), it was estimated that up to 90% of all new molecular entities are Class II and IV drugs (~70% Class II, ~20% Class IV) [118, 119]. Hence, the improvement of the bioavailability will be one of the most important challenges in the development of oral dosage forms. Nanomilling was shown to be an extremely promising technique. So far, the majority of the commercially available oral dosage forms containing nanoparticulate drug substances are manufactured by the NanoCrystal® technology.

16.8.1 Oral Drug Delivery

16.8.1.1 Increase of Bioavailability

Rapamune® tablets were the first nanocrystalline particle product approved by the FDA. As an immunosuppressant containing nanomilled sirolimus, it was introduced to the market in July 2001. The tablets are made from a nanosuspension of the drug, stabilized with poloxamer 188, which is incorporated in a sugar coating surrounding an inert tablet core [120]. Compared to the Rapamune® solution, these tablets reveal a 27% increased bioavailability [121].

Two years later, in 2003, Emend® capsules, containing a nanomilled powder of the antiemetic drug aprepitant, was launched as the second NanoCrystal® product. EMEND is presented in the form of gelatine hard capsules filled with inert microcrystalline cellulose pellets, which are coated with 80 or 125 mg of aprepitant. The nanocrystals are stabilized with 4% hydroxypropylcellulose (HPC-SL) and 0.08% sodium lauryl sulphate. After premilling a suspension of the drug in a hydroxypropylcellulose solution, sodium lauryl sulphate is added and the slurry is media milled to form a nanocrystalline dispersion. This nanosuspension is mixed with an aqueous sucrose dispersion and sprayed onto microcrystalline cellulose beads, which are then blended with micronized sodium lauryl sulphate and encapsulated [122]. By reducing the average particle size to 120 nm, the surface area of the drug crystals could be increased to 41.5 times the surface area of the 5 μm particles in the standard suspension. In beagle dogs administration of aprepitant nanocrystals resulted in an approximately four-fold relative bioavailability compared to a conventional suspension (5.49 μm) [123]. The absorption of aprepitant is limited to the upper gastrointestinal tract [110]. Due to their rapid dissolution, nanoparticulate drug crystals dissolve already in the stomach and in the upper duodenum instead of delaying the drug release until a major portion of unsolved particles has passed the absorption window. As aprepitant, which belongs to BCS Class IV, clearly shows, the benefits of nanosizing in oral drug therapy are not confined to high permeability compounds. Dissolution enhancement by size reduction is also able to increase the bioavailability of low permeable drugs, provided dissolution is the rate-determining step in the drug delivery process.

16.8.1.2 Elimination of Food Effects

Particularly in the case of BSC Class II to IV drugs, dissolution in the gastrointestinal fluids and absorption from the gut are often strongly influenced by food intake. Many examples have shown that nanosizing of the drug crystals not only increases the bioavailability but is also a good means to reduce the influence of food-related effects.

One of these examples is Tricor® (in some countries marketed as Lipidil®), an oral formulation of nanomilled fenofibrate, available with 48 or 145 mg drug per tablet. Fenofibrate is a lipid-lowering drug belonging to BCS Class II. The bioavailability of fenofibrate from conventional formulations depends strongly on whether the patient is in the fed or nonfed state. The drug is absorbed up to 35% better in fed patients than in nonfed patients. After administration of the nanocrystalline formulation, plasma levels in fed and fasting conditions are about the same. Fenofibrate is a lipophilic compound and practically insoluble in water. Having no dissociable functionalities, the solubility of fenofibrate is not influenced by food-related changes of the gastrointestinal pH. Instead fenofibrate is solubilized by lipids and surface active-active components of the digested food (e.g. cholesterol) as well as by bile salts secreted in response to the food stimulation. By nanonizing the drug, the solubility is enhanced even in the fasted state, thus causing the food-induced solubilization to have less effect on the drug concentration and consequently on absorption and bioavailability [110].

Also Megace ES® (ES = Enhanced Stability) was designed to reduce the effects of food on absorption. As a BCS Class II drug, Megestrol acetate, the active ingredient of this formulation, is well absorbed in the gastrointestinal tract but is subject to food-related variations. This could be demonstrated in a bioavailability study comparing the peak plasma concentration and extent of absorption of Megace ES and megestrol acetate oral suspension [124]. While the absorption of the oral suspension depended substantially on food intake, the bioavailability of Megace ES® was only minimally reduced in unfed patients. Also in this case, size reduction to the nanometer range improved the dissolution behavior and thus equalized the absorption in the fasted and fed state. Another benefit of nanoparticulate formulations was realized in Megace ES®: Because the drug substance is concentrated in solid particles, less liquid is necessary to disperse the drug. In comparison to the oral solution, the volume of a single dose could be reduced by the factor four by using the nanosuspension (daily dose 625 mg of megestrol in 5 ml of fluid). The nanoparticulate size of the drug inhibits rapid sedimentation of the crystals and alleviates the need for viscosity enhancement. All these benefits – reduced volume, low viscosity, and enhanced bioavailability – improve the patients' compliance and enable a more individualized dose adjustment [110, 111].

Although not yet on the market as a nanocrystalline formulation, also cilastatin was shown to benefit from nanosizing. Jinno *et al.* compared the biopharmaceutical properties of hammer-milled, jet-milled, and nanocrystalline cilastatine [125]. *In vitro* dissolution studies with simulated intestinal fluids in the fasted state (FaSSIF) and the fed state (FeSSIF) (USP Apparatus 2, 50 rpm, 37°C) revealed on average about 25% dissolution of the hammer-milled, about 80% dissolution of the jet-milled, and 100% dissolution of the nanocrystalline drug after 15 min. with the concentrations measured in FeSSIF being about 10% and 20% higher with hammer-milled and jet-milled cilastatine, respectively. As Table 16.5 shows, also *in vivo* bioavailability parameters were increased with decreasing particle size. In the cases of hammer-milled and jet-milled drug, the administration in the fed state resulted in AUC and c_{max} values up to 3.7-fold higher than in the fasted state. By contrast, the nanocrystal formulation showed even a slight but not statistically significant decrease of AUC and c_{max} in the fed state. As cilostazol is classified in BCS Class II, the dissolution process could be the rate-limiting step in systemic uptake. Therefore, it is an obvious conclusion that the plasma levels could directly reflect the *in vivo* dissolution behavior

Table 16.5 *Biopharmaceutical properties of hammer-milled, jet-milled, and nanocrystalline cilastatine [125].*

	AUC fasted (ng h/ml)	AUC fed (ng h/ml)	c_{max} fasted (ng h/ml)	c_{max} fed (ng h/ml)
Hammer-milled (100 mg/body)	2722 ± 803	4694 ± 1612	582 ± 154	1152 ± 221
Jet-milled (100 mg/body)	2875 ± 587	10669 ± 3417	1029 ± 218	2901 ± 314
NanoCrystal® (100 mg/body)	17832 ± 4994	13589 ± 3895	5371 ± 1173	4872 ± 1112

and indicate that the food intake enhanced the dissolution of cilostazol in the cases of the hammer-milled and the jet-milled suspensions. On the other hand, it can be assumed that in the case of the nanocrystalline suspension the dissolution was already maximized so that no further enhancement by food effects was possible. However, it is conceivable that the slightly negative effect observed was caused by partial adsorption of the nanoparticles to food components. Though food intake was shown to increase the total degree of drug absorption in the case of non-nanosized particles, it did not however, enhance the apparent initial absorption rate. This may be due to a counteracting influence of the delayed gastric emptying after the food intake. As gastric emptying is another possible rate-limiting step for drug absorption, its retardation could offset dissolution-enhancing effects.

16.8.1.3 Accelerated Onset of Action

As a result of the increased dissolution rate, nanoparticulate drug crystals are able to enter the systemic circulation faster than conventional formulations, leading to a rapid onset of action. Fast-acting formulations are often desirable, for example, in the case of analgesics like naproxen. This BCS class II drug has a high permeability and shows nearly complete absorption. As a weak acid, the compound exhibits a high solubility in the intestinal pH range but dissolves only slowly in the stomach. Merisko-Liversidge *et al.* compared the pharmacokinetics of nanocrystalline naproxen (mean particle size 147 nm) with those of two marketed products (Naprosyn® suspension and Anaprox® tablets) [24]. The time to reach maximal drug concentrations in the plasma was approximately 50% (1.5 h) less for the nanodispersion and during the first hour of the study the nanoparticles achieved a 2.5–4.5-fold increase in the area under the curve (AUC). Besides the faster onset of action, the naproxen nanosuspension was found also to show a reduced gastric irritancy [10, 97].

16.8.2 Parenteral Drug Delivery

16.8.2.1 Sustained Release

While in oral drug delivery nanocrystals have the potential to accelerate the onset of action, in parenteral therapy they can be used for long-term controlled drug release. So far the only parenteral product on the market manufactured by the NanoCrystal® technology is Invega® Sustenna® (Ortho-McNeil-Janssen Pharmaceuticals, Inc., Titusville, NJ, USA), an extended-release injectable suspension for i.m. administration. The active compound, paliperidone palmitate, is the palmitoyl ester of paliperidone, an active metabolite of risperidone. The practically insoluble substance ($\ll$ 10 μg/ml in water [118]) is nanomilled to obtain a particle size which facilitates a very slow dissolution of the drug, thus enabling monthly dose intervals.

16.8.2.2 Intravenous Administration

The main advantage of nanoparticles in parenteral therapy is their potential to be administered also intravenously rather than only intramuscularly, the latter being also an acceptable administration route for microparticulate formulations. As discussed above, nanomilling is a simple and fast method to provide intravenously injectable formulations for toxicological studies of compounds which are not sufficiently soluble in any biocompatible solvent. As in many conventional formulation strategies harsh solvents or co-solvents are used to facilitate clear solutions of poorly soluble drugs, it has to be considered that maximum doses might be limited by the toxicity of these excipients. Nanocrystalline formulations are free of organic solvents and if parenteral use is intended, one can resort to a series of stabilizers which are well established and proven safe for i.v. and i.m. use. This enhances the efficacy by exceeding the maximum dose determined by the solvents' toxicity. Paclitaxel, a poorly water-soluble antineoplastic agent, is a popular example of a drug which is solubilized using a highly concentrated solvent/surfactant mixture (ethanol, Cremophor EL). Merisko-Liversidge *et al.* compared the efficacy of the marketed paclitaxel product (Taxol®), based on an ethanolic solution, with a nanoparticulate suspension in a mouse xenograft tumour model [24]. The maximum tolerated dose of the nanoparticulate paclitaxel was demonstrated to be three times higher than that of the ethanol/surfactant solution, leading to a significantly improved efficacy of the nanoparticle formulation. As shown also with other poorly water-soluble drugs like camphtothecin and etoposide, in general, nanoparticulate formulations can be administered intravenously in higher doses than solutions. The fate of nanoparticles after i.v. administration is determined by the solubility of the drug in the plasma [126]. Upon entering a blood vessel the drug particles are instantaneously exposed to sink conditions. Due to their extremely large surface area, nanoparticulate drugs with relatively high solubilities in the mg/ml range or ionizable compounds will rapidly dissolve in the bloodstream after intravenous injection. In this case, the biodistribution profile of the nanoparticulate drug will be indistinguishable from that of a solution. However, if the solubility is of the order of μg/ml, the particles may not dissolve when administered in larger doses. Nevertheless in low doses, nanoparticles of even less soluble drugs were shown to quickly dissolve in the bloodstream. Comparing a nanoparticulate suspension of itraconazole (mean diameter about 300 nm) with a solution of a hydroypropyl-β-cyclodextrin complex of the same drug, Mouton *et al.* [127] found similar pharmacokinetic profiles of both formulations in healthy human subjects.

16.8.2.3 RES Targeting

A different pharmacokinetic behavior can be observed if the injected particles dissolve slowly [128]. The circulating particles of crystalline drug are captured by the reticuloendothelial system (RES) which may act as a deep compartment, prolonging the half-life at higher doses [127–129]. Generally, nanoparticles larger than 150 nm undergo rapid hepatic clearance. Within 5–10 minutes, 60–90% of them are taken up by Kupffer cells. Another 2–20% of the drug is deposited in the spleen and smaller fractions in the lungs and in the bone marrow (0.005–1%) [131, 139]. Itraconazole particles about twice as large as those mentioned above (mean diameter about 580 nm) were investigated by Rabinow *et al.* versus a hydroxypropyl-β-cyclodextrin complex [128]. Pharmacokinetic studies in rats revealed a reduced c_{max} and a prolonged $t_{\frac{1}{2}}$ of the nanocrystalline suspension compared to the solution of the drug-cyclodextrin complex. Uptake of drug nanoparticles by the RES followed by a sustained release in these cells and tissues allows for passive targeting to lung, liver and spleen. As a consequence, this may reduce the toxicity and increase the tolerability of the drug, thus enabling higher doses to be administered. However, the biological effects of nanoparticles on the cellular level, immunological actions, and long-term consequences are still not well understood. For this reason, attention should be paid to the potential toxicity of chronic exposure of nanoparticles to phagocytic organs [28]. Nevertheless, based on their ability for passive targeting to the reticuloendothelial system, nanoformulations were considered to have a great potential in the therapy of infections with facultative or obligate intracellular pathogens. Such organisms like Mycobacterium avium or Leishmania spp. invade macrophages and are able to survive in these cells largely protected from the effects of common anti-infectives. Nanocrystals and other nanoparticulate drug carriers are also internalized by phagocytic cells and thus can release the drug in closest proximity to the pathogenes. Peters *et al.* compared a nanocrystalline with a liposomal formulation of clofazimine for the treatment of artificially induced Mycobacterium avium infections in mice [131]. Doses of 20 mg clofazimine/kg bodyweight were administered intravenously twice-weekly for three weeks. Both formulations achieved comparable drug concentrations in lung, liver, and spleen and also a comparable therapeutic response. Kayser demonstrated with a nanosuspension of aphidicolin an about 140-fold increase in anti-Leishmania activity relative to the effect achieved with a solution of the drug in dimethyl sulfoxide. This indicates that in the case of intra-macrophage protozoa and bacteria, cellular drug delivery is able to improve the activity of anti-infectives considerably [132].

16.8.3 Extracorporal Therapy

Another application of nanocrystals is currently in phase I/II clinical development. It is an extracorporal photodynamic therapy (ECP) with 4,5-dibromorhodamine methyl ester (TH9402, Celmed BioSciences Inc, Saint-Laurent, QC, Canada) which is evaluated for the prevention of graft-versus-host disease (GvHD), the treatment of autoimmune diseases, and the eradication of cancerous cells from bone marrow transplants in non-Hodgkin's lymphoma (NHL). Because of the poor solubility of TH9402, the compound is nanomilled using the NanoCrystal® technology from Elan Nanosystems. TH9402 is a non-toxic molecule that passively diffuses into all cells and accumulates in the mitochondria. However, only in actively growing cells in which the P-glycoprotein (Pgp) pump is inactive, is the drug retained. Upon exposition of the cells to visible light using Celmed's Theralux device, the drug is photodynamically activated, causing the death of the drug-enriched cells. In order to minimize side-effects and toxicity, the process is performed *ex vivo* (outside the body). Once the blood has been purged of the pathologically altered cells, it is reinfused into the circulation, after the patient has undergone high-dose chemotherapy [110].

16.9 Conclusion

Particle size reduction by grinding is a well-known method to increase the dissolution rate and thus to improve the bioavailability of poorly water-soluble drugs. However, despite being successfully applied for a very long time, conventional grinding techniques increasingly face their limits due to the growing number of drugs with extremely low aqueous solubilities far below 1 μg/ml. Grinding to the nanometer size range is able to overcome this problem by strongly enhancing the dissolution rate and therefore increasing the ratio between the residence time within the absorption window and the dissolution time. Because plasticizing and agglomeration tendencies emerge if the particle size is reduced below the micrometer scale, special wet milling techniques and the sophisticated use of stabilizers are necessary to obtain nanoparticles in high yield. Media milling is the method of choice and is effective even in the sub-micrometer range if the size of the grinding beads is reduced well below 1 mm. Smaller beads were shown to perform the same work as larger ones but with a lower energy expenditure. As they transmit less impact energy to the drug particles, amorphization is impeded, thus favouring crystalline particles which benefit from a high storage stability. The stabilizer plays a key role in this technique by inhibiting reagglomeration of separated particles which is the main process counteracting particle comminution. The right choice regarding type and concentration of the stabilizer is most critical to the success of milling. Nanomilling can be performed in conventional ball mills or in high-shear stirred media mills. Depending on batch size, equipment and energy input process times reach from a few weeks to a few minutes. The technology can be easily transferred to the commercial production scale. Currently the main field of application is in oral drug delivery where it has proven to be an effective method for enhancing the bioavailability of poorly soluble drugs. Besides this, nanomilling is also able to reduce food-related effects on drug dissolution and absorption and accelerates the onset of action. In parenteral therapy nanoparticulate formulations are currently used for sustained drug release. Because they pass the lung capillaries, they have the potential to be administered also intravenously, thus being an alternative to drug solutions containing ethanol or other organic solvents. Many other benefits and applications of nanoparticles prepared by the grinding of drug crystals have been described. Being the first technology used for the manufacturing of a marketed nanosized drug solid product, nanomilling has a great potential to become established as a standard method in the formulation of poorly soluble drugs.

References

1. C.H. Dubin, Formulation strategies for poorly soluble drugs. *Drug Del Technol*, **6**, 34–38 (2006).
2. W. Nernst, Theorie der Reaktionsgeschwindigkeit in heterogenen Systemen. *Z Phys Chem*, **47**, 52–55 (1904).
3. E. Brunner, Reaktionsgeschwindigkeit in heterogenen Systemen. *Z Phys Chem*, **47**, 56–102 (1904).
4. R.H. Muller, K. Peters, Nanosuspensions for the formulation of poorly soluble drugs – I. Preparation by a size-reduction technique. *Int J Pharm*, **160**, 229–237 (1998).

5. R.J. Hintz, K.C. Johnson, The effect of particle size distribution on dissolution rate and oral absorption. *Int J Pharm*, **51**, 9–17 (1989).
6. P. Harriott, Mass transfer to particles: Part 1. Suspended inagitated tanks, *AIChE J*, **8**, 93–102 (1962).
7. A. Okazaki, T. Mano, K. Sugano, Theoretical dissolution model of poly-disperse drug particles in biorelevant media. *J Pharm Sci*, **97**, 1843–1852 (2008).
8. M. Kraml, J. Dubuc, R. Gaudry, Gastrointestinal absorption of griseofulvin – II. Influence of particle size in man. *Antibiot Chemother*, **12**, 239–242 (1962).
9. T.R. Shaw, J.E. Carless, The effect of particle size on the absorption of digoxin. *Eur J Clin Pharmacol*, **7**, 269–273 (1974).
10. S. Katteboinaa, V.S.R. Chandrasekhar, S. Balaji, Drug nanocrystals: A novel formulation approach for poorly soluble drugs. *International Journal of Pharm Tech Research*, **1**, 682–694 (2009).
11. J. Chingunpitak, S. Puttipipatkhachorn, Y. Tozuka, K. Moribe, K. Yamamoto, Micronization of dihydroartemisinin by rapid expansion of supercritical solutions. *Drug Dev Ind Pharm*, **34**, 609–617 (2008).
12. V. Joshi, S. Dwivedi, G.H. Ward, Increase in the specific surface area of budesonide during storage postmicronization. *Pharm Res*, **19**, 7–12 (2002).
13. T. Van Nijlen, K. Brennan, G. Van den Mooter, N. Blaton, R. Kinget, P. Augustijns, Improvement of the dissolution rate of artemisinin by means of supercritical fluid technology and solid dispersions. *Int J Pharm*, **254**, 173–181 (2003).
14. C.C. Lee, L.W. Chan, P.W. Heng, Use of a fluidized bed hammer mill for size reduction and classification: Effects of process variables and starting materials on the particle size distribution of milled lactose batches. *Pharm Dev Technol*, **8**, 431–442 (2003).
15. A. Saleki-Gerhardt, C. Ahlneck, G. Zografi, Assessment of disorder in crystalline solids. *Int J Pharm*, **101**, 237–247 (1994).
16. Y. Tanaka, M. Inkyo, R. Yumoto, J. Nagai, M. Takano, S. Nagata, Nanoparticulation of poorly water soluble drugs using a wet-mill process and physicochemical properties of the nanopowders. *Chem Pharm Bull*, **57**, 1050–1057 (2009).
17. S. Bernotat, K. Schönert, Size reduction. In *Ullmann's Encyclopedia of Industrial Chemistry*. New York: John Wiley & Sons, Inc., 2010.
18. I. Lee, N. Variankaval, C. Lindemann, C. Starbuck, Rotor-stator milling of APIs – empirical scale-up parameters and theoretical relationships between the morphology and breakage of crystals. *Amer Pharm Review*, **7**, 120–123 (2004).
19. C. Tangsathitkulchai, Acceleration of particle breakage rates in wet batch ball milling. *Powder Technology*, **124**, 67–75 (2002).
20. E.L. Parrott, Milling of pharmaceutical solids. *J Pharm Sci*, **63**, 813–829 (1974).
21. Y. Kondo, Y. Hashizuka, M. Nakahara, K. Yokota, K. Ishizaki, Influence of combination of ball diameters and rotation speed on grinding performance of aluminia by ball milling. *Journal of the Ceramic Society of Japan, Int Edition*, **101**, 797–801 (1993).
22. K. Tanaka, I. Inada, K. Wakino, I. Uei, Submicron grinding of low-soda alumina by ball milling. *Zairyo*, **37**, 1319–1325 (1988).
23. C. Keck, S. Kobierski, R. Mauludin, R. Müller, Second generation of drug nanocrystals for delivery of poorly soluble drugs: SmartCrystals technology. *Dosis*, **24**, 124–128 (2008).

24. E. Merisko-Liversidge, G.G. Liversidge, E.R. Cooper, Nanosizing: A formulation approach for poorly-water-soluble compounds. *Eur J Pharm Sci*, **18**, 113–120 (2003).
25. L. Peltonen, J. Hirvonen, Pharmaceutical nanocrystals by nanomilling: Critical process parameters, particle fracturing and stabilization methods. *J Pharm Pharmacol*, **62**, 1569–1579 (2010).
26. J. Hu, K.P. Johnston, R.O. Williams, Nanoparticle engineering processes for enhancing the dissolution rates of poorly water soluble drugs. *Drug Dev Ind Pharm*, **30**, 233–245 (2004).
27. G.G. Liversidge, K.C. Cundy, J.F. Bishop, D.A. Czekai, Surface Modified Drug Nanoparticles. *US Patent 5, 145, 684*, 1992.
28. Y. Shi, W. Porter, T. Merdan, L.C. Li, Recent advances in intravenous delivery of poorly water-soluble compounds. *Expert Opin Drug Deliv*, **6**, 1261–1282 (2009).
29. B.E. Rabinow, Nanosuspensions in drug delivery. *Nat Rev Drug Discov*, **3**, 785–796 (2004).
30. J. Moschwitzer, G. Achleitner, H. Pomper, R.H. Müller, Development of an intravenously injectable chemically stable aqueous omeprazole formulation using nanosuspension technology. *Eur J Pharm Biopharm*, **58**, 615–619 (2004).
31. http://www.elandrugtechnologies.com/technology [Accessed 31 May 2011].
32. A.J. Hickey, D. Ganderton, Pharmaceutical process engineering. In *Drugs and the Pharmaceutical Sciences*, Vol. 112. New York: Marcel Dekker, Inc. (2001).
33. Y. Rabin, P.S. Steif, M.J. Taylor, T.B. Julian, N. Wolmark, An experimental study of the mechanical response of frozen biological tissues at cryogenic temperatures. *Cryobiology*, **33**, 472–482 (1996).
34. M. Prior, Size reduction. In *Kirk-Othmer Encyclopedia of Chemical Technology*. New York: John Wiley & Sons (2000).
35. K. Kendall, The impossibility of comminuting small particles by compression. *Nature*, **272**, 710–711 (1978).
36. K.E. Puttick, The correlation of fracture transitions. *J Phys D: Appl Phys*, **13**, 2249–2262 (1980).
37. I. Larsson, H.G. Kristensen, Comminution of a brittle/ductile material in a Micros Ring Mill. *Powder Technology*, **107**, 175–178 (2000).
38. H. Leuenberger, The compressibility and compactibility of powder systems. *Int J Pharm*, **12**, 41–55 (1982).
39. R.C. Rowe, R.J. Roberts, Mechanical properties. In *Pharmaceutical Powder Compaction*, G. Alderborn and C. Nystrom (eds). New York: Marcel Dekker (1996).
40. R.J. Roberts, R.C Rowe, Brittle-ductile transitions in sucrose and the influence of lateral stresses during compaction. *J Pharm Pharmacol*, **52**, 147–150 (2000).
41. C. Prasher, *Crushing and Grinding Process Handbook*. New York: John Wiley & Sons Limited (1987).
42. J. Hadizadeh, W.K. Johnson, Estimating local strain due to comminution in experimental cataclastic textures. *Journal of Structural Geology*, **25**, 1973–1979 (2003).
43. M.F. Ashby, R.A. Verrall, Micromechanisms of flow and fracture, and their relevance to the rheology of the upper mantle. *Philos Trans R Soc London Ser A*, **288**, 59–93 (1978).
44. C.G. Sammis, Y. Ben-Zion, Mechanics of grain-size reduction in fault zones. *J Geophys Res*, **113**, B02306 doi:10.1029/2006JB004892 (2008).

45. M. Junhke, R. Weichert, Nanoparticles of soft materials by high-energy milling at low temperatures. In *Congress Manuscripts of the 7th World Congress of Chemical Engineering*, 10–14. July 2005, Glasgow, Scotland.
46. T. Niwa, Y. Nakanishi, K. Danjo, One-step preparation of pharmaceutical nanocrystals using ultra cryomilling technique in liquid nitrogen, *Eur J Pharm Sci*, **41**, 78–85 (2010).
47. M. Meyers, *Dynamic Behavior of Materials*, New York: Wiley-Interscience (1994).
48. C. Knieke, C. Steinborn, S. Romeis, W. Peukert, S. Breitung-Faes, A. Kwade, Nanoparticle production with stirred media mills: Opportunities and limits. *Chemical Engineering & Technology*, **33**, 1401–1411 (2010).
49. C. Knieke, S. Romeis, W. Peukert, Influence of process parameters on breakage kinetics and grinding limit at the nanoscale. *AIChE J*, **57** (2011), DOI: 10.1002/aic.12408.
50. C. Knieke, W. Peukert, Particle breakage in the nanometer range. In *Proceedings of the AIChE 2008 Annual Meeting*. Philadelphia (2008).
51. G.R. Karagedov, N.Z. Lyakhov, Mechanochemical grinding of inorganic oxides. *KONA*, **21**, 76–87 (2003).
52. G. Jimbo, O.O. Zhao, T. Yokoyama, Y. Taniyama, The grinding limit and the negative grinding phenomenon. *Proc 2nd World Congress Particle Technology, Kyoto*, Japan, 305–312 (1990).
53. B.B. Bokhonov, Y.T. Pavlukhin, A.I. Rykov, S.M. Paramzin, V.V. Boldyrev, Crystal growth during mechanical activation of zinc ferrite. *J Mater Synth Process*, **1**, 341–346 (1993).
54. V.V. Alekseenko, G.R. Karagedov, Interaction of solids subjected to mechanical stress. *Journal of Mechanochemistry and Mechanical Alloying*, **1**, 1–6 (1994).
55. B.B. Bokhonov, I.G. Konstanchuk, V.V. Boldyrev, Structural and morphological changes during the mechanical activation of nano-size particles. *Materials Research Bulletin*, **30**, 1277–1284 (1995).
56. V.V. Alekseenko, G.R. Karagedov, Interaction of crystals under conditions modelling mechanical activation. *Neorganicheskie Materialy*, **33**, 817–821 (1997).
57. A. Kwade, Determination of the most important grinding mechanism in stirred media mills by calculating stress intensity and stress number. *Powder Technology*, **105**, 382–388 (1999).
58. A. Annapragada, A. Adjei, Numerical simulation of milling processes as an aid to process design. *Int J Pharm*, **136**, 1–11 (1996).
59. S. Fadda, A. Cincotti, A. Concas, M. Pisu, G. Cao, Modelling breakage and reagglomeration during fine dry grinding in ball milling devices. *Powder Technology*, **194**, 207–216 (2009).
60. F. Müller, W. Peukert, R. Polke, F. Stenger, Dispersing nanoparticles in liquids. *Int J Miner Process*, **74**, S31–S41 (2004).
61. F. Müller, R.F. Polke, From the product and process requirements to the milling facility. *Powder Technol*, **105**, 2–13 (1999).
62. M. Inkyo, T. Tahara, T. Iwaki, F. Iskandar, C.J. Hogan, K. Okuyama, Experimental investigation of nanoparticle dispersion by beads milling with centrifugal bead separation. *J Colloid and Interface Sci*, **304**, 535–540 (2006).
63. H. Bel Fadhel, C. Frances, A. Mamourian, Investigations on ultra-fine grinding of titanium dioxide in a stirred media mill. *Powder Technology*, **105**, 362–373 (1999).

64. A. Kwade, Wet comminution in stirred media mills – research and its practical application. *Powder Technology*, **105**, 14–20 (1999).
65. K. Brodka-Pfeiffer, P. Langguth, P. Grass, H. Häusler, Influence of mechanical activation on the physical stability of salbutamol sulphate. *Eur J Pharm Biopharm*, **56**, 393–400 (2003).
66. R. Hüttenrauch, S. Fricke, P. Zielke, Mechanical activation of pharmaceutical systems. *Pharm Res*, **2**, 302–306 (1985).
67. K.J. Crowley, G. Zografi, Cryogenic grinding of indomethacin polymorphs and solvates: Assessment of amorphous phase formation and amorphous phase physical stability. *J Pharm Sci*, **91**, 492–507 (2002).
68. H.J. Fecht, Defect-induced melting and solid-state amorphization, *Nature*, **356**, 133–135 (1992).
69. P. Pourghahramani, E. Altin, M.R. Mallembakam, W. Peukert, E. Forssberg, Microstructural characterization of hematite during wet and dry millings using Rietveld and XRD line profile analyses. *Powder Technology*, **186**, 9–21 (2008).
70. P. Sharma, W.A. Denny, S. Garg, Effect of wet milling process on the solid state of indomethacin and simvastatin. *Int J Pharm*, **380**, 40–48 (2009).
71. R.S. Farr, R.D. Groot, Close packing density of polydisperse hard spheres. *J Chem Phys*, **131**, 244104 (2009).
72. R.R. Klimpel, The selection of wet grinding chemical additives based on slurry rheology control. *Powder Technology*, **105**, 430–435 (1999).
73. R. Greenwood, N. Rowson, S. Kingman, G. Brown, A new method for determining the optimum dispersant concentration in aqueous grinding. *Powder Technology*, **123**, 199–207 (2002).
74. S. Farrokhpay, A review of polymeric dispersant stabilisation of titania pigment. *Adv Colloid Interface Sci*, **151**, 24–32 (2009).
75. B.J. Palla, D.O. Shah, Stabilization of high ionic strength slurries using the synergistic effects of a mixed surfactant system. *J Colloid Interface Sci*, **223**, 102–111 (2000).
76. A. Badia, L. Cuccia, L. Demers, F. Morin, R.B. Lennox, Structure and dynamics in alkanethiolate monolayers self-assembled on gold nanoparticles: A DSC, FT-IR, and deuterium NMR study. *J Am Chem Soc*, **119**, 2682–2692 (1997).
77. L. Peltonen, J. Yliruusi, Surface pressure, hysteresis, interfacial tension, and CMC of four sorbitan monesters at water-air, water-hexane, and hexane-air interfaces. *J Colloid Interface Sci*, **227**, 1–6 (2000).
78. L. Piao, K.H. Leeb, W.J. Kwonc, S.-H. Kimd and S. Yoon, The simple and facile methods to improve dispersion stability of nanoparticles: Different chain length alkylcarboxylate mixtures. *J Colloid Interface Sci*, **334**, 208–211 (2009).
79. S.L. Law, J.P. Kayes, Adsorption of non-ionic water-soluble cellulose polymers at the solid-water interface and their effect on suspension stability. *Int J Pharm*, **15**, 251–260 (1983).
80. J.Y. Heng, F. Thielmann, D.R. Williams, The effects of milling on the surface properties of form I paracetamol crystals. *Pharm Res*, **23**, 1918–1927 (2006).
81. P. York, M.D. Ticehurst, J.C. Osborn, R.J. Roberts, R.C. Rowe, Characterization of the surface energies of milled DL-propanolol hydrochloride using inverse gas chromatography and molecular modelling. *Int J Pharm*, **174**, 179–186 (1998).

82. K.V.R. Prasad, R.I. Ristic, D.B. Sheen, J.N. Sherwood, Dissolution kinetics of paracetamol single crystals. *Int J Pharm*, **238**, 29–41 (2002).
83. A. Watanabe, Y. Yamaoka, K. Takada, Crystal habits and dissolution behaviour of aspirin. *Chem Pharm Bull*, **30**, 2958–2963 (1982).
84. N. Rasenack, H. Hartenhauer, B.W. Müller, Microcrystals for dissolution rate enhancement of poorly water-soluble drugs. *Int J Pharm*, **254**, 137–145 (2003).
85. A. Zimmermann, A. Millqvist-Fureby, M.R. Elema, T. Hansen, A. Müllertz, L. Hovgaard, Adsorption of pharmaceutical excipients onto microcrystals of siramesine hydrochloride: Effects on physicochemical properties. *Eur J Pharm Biopharm*, **71**, 109–116 (2009).
86. D. Hovey, J. Pruitt, T. Ryde, Nanoparticulate megestrol formulations. *US Patent 0233001 A1*, 2005.
87. G. Liversidge, S. Jenkins, Nanoparticulate benzothiophene formulations. *US Patent 0035366 A1*, 2009.
88. G. Liversidge, S. Jenkins, Formulations of a nanoparticulate finasteride, dutasteride or tamsulosin hydrochloride, and mixtures thereof. *US Patent 0204588 A1*, 2006.
89. E.R. Cooper, T. Ryde, J. Pruitt, L. Kline, Nanoparticulate meloxicam formulations. *US Patent 0297252 A1*, 2010.
90. S. Jenkins, G. Liverdidge, E. Liversidge, Nanoparticulate tacrolimus formulations. *US Patent 0152720 A1*, 2008.
91. A.M. Cerdeira, M. Mazzotti, B. Gander, Miconazole nanosuspensions: Influence of formulation variables on particle size reduction and physical stability. *Int J Pharm*, **396**, 210–218 (2010).
92. B. Van Eerdenbrugh, J. Vermant, J.A. Martens, L. Froyen, J. Van Humbeeck, P. Augustijns, G. Van den Mooter, A screening study of surface stabilization during the production of drug nanocrystals. *J Pharm Sci*, **98**, 2091–2103 (2009).
93. B.J. Palla, D.O. Shah, Stabilization of high ionic strength slurries using surfactant mixtures: Molecular factors that determine optimal stability. *J Colloid Interface Sci*, **256**, 143–152 (2002).
94. R.H. Müller, J.-U. Junghanns, Drug nanocrystals/nanosuspensions for the delivery of poorly soluble drugs. In *Nanoparticulates as Drug Carriers*, V. Torchilin (ed.), London: Imperial College Press (2006).
95. J. Lee, S.J. Lee, J.Y. Choi, J.Y. Yoo, C.H. Ahn, Amphiphilic amino acid copolymers as stabilizers for the preparation of nanocrystal dispersion. *Eur J Pharm Sci*, **24**, 441–449 (2005).
96. J. Lee, J.Y. Choi, C.H. Park, Characteristics of polymers enabling nanocomminution of water-insoluble drugs. *Int J Pharm*, **355**, 328–336 (2008).
97. G.G. Liversidge, P. Conzentino, Drug particle size reduction for decreasing gastric irritancy and enhancing absorption of naproxen in rats. *Int J Pharm*, **125**, 309–313 (1995).
98. X. Liu, P. Yang, Q. Jiang, Size effect on melting temperature of nanostructured drugs. *Mat Chem Phys*, **103**, 1–4 (2007).
99. J.Y. Choi, J.Y. Yoo, H.S. Kwak, B.U. Nam, J. Lee, Role of polymeric stabilizers for drug nanocrystal dispersions. *Curr Appl Phys*, **5**, 472–474 (2005).
100. M. Becker, J. Schwedes, Comminution of ceramics in stirred media mills and wear of grinding beads. *Powder Technology*, **105**, 374–381 (1999).

101. H. Yoden, J. Itoh, Effect of small beads on dispersion of nanometer-sized silica particle by wet beads mill process. *J Soc Powder Technol Jpn*, **41**, 457–464 (2004).
102. M. He, Y. Wang, E. Forssberg, Slurry rheology in wet ultrafine grinding of industrial minerals: A review. *Powder Technology*, **147**, 94–112 (2004).
103. J.F. Bishop, D.A. Czekai, Ink jet inks containing nanoparticles of organic pigments. *US Patent 5679138* (1997).
104. D.A. Czekai, L.P. Seaman, Comminution with small particle milling media. *US Patent 5500331* (1996).
105. C.E. Romano, D.E. Bugner, R.C. VanHanehem, J.R. Bennett, D.E. Smith, R.A. Guistina, Process for preparing an ink jet ink. *US Patent 6210474* (2001).
106. E. Bilgili, R. Hamey, B. Scarlett, Nano-milling of pigment agglomerates using a wet stirred media mill: Elucidation of the kinetics and breakage mechanisms. *Chem Eng Sci*, **61**, 149–157 (2006).
107. E. Merisko-Liversidge, P. Sarpotdar, J. Bruno, S. Hajj, L. Wei, N. Peltier, J. Rake, J.M. Shaw, S. Pugh, L. Polin, J. Jones, T. Corbett, E. Cooper, G.G. Liversidge, Formulation and antitumor activity evaluation of nanocrystalline suspensions of poorly soluble anticancer drugs. *Pharm Res*, **13**, 272–278 (1996).
108. G.C. Na, J. Stevens, B.O. Yuan, N. Rajagopalan, Physical stability of ethyl diatrizoate nanocrystalline suspension in steam sterilization. *Pharm Res*, **16**, 569–574 (1999).
109. J. Lee, Drug nano- and microparticles processed into solid dosage forms: Physical properties, *J Pharm Sci*, **92**, 2057–2068 (2003).
110. J.-U. Junghanns, R.H. Müller, Nanocrystal technology, drug delivery and clinical applications, *International Journal of Nanomedicine*, **3**, 295–309 (2008).
111. F. Kesisoglou, S. Panmai, Y. Wu, Nanosizing: Oral formulation development and biopharmaceutical evaluation. *Adv Drug Deliv Rev*, **59**, 631–644 (2007).
112. G.G. Liversidge, K.C. Cundy, Particle size reduction for improvement of oral bioavailability of hydrophobic drugs: I. Absolute oral bioavailability of nanocrystalline danazol in beagle dogs. *Int J Pharm*, **125**, 91–97 (1995).
113. Z. Deng, S. Xu, S. Li, Understanding a relaxation behavior in a nanoparticle suspension for drug delivery applications. *Int J Pharm*, **351**, 236–243 (2008).
114. E.P. Tefft, M.M. Simkin, Application of nanotechnology: A case study in the pharmaceutical arena, Foley & Lardner LLP, online available from: http://www.foley.com/files/tbl_s31Publications/FileUpload137/2007/1180354_1.pdf [Accessed 15 May 2011].
115. J. Möschwitzer, Method for producing ultrafine submicronic suspensions, *US Patent Application US020080279928A1*, 2008.
116. C.A. Lipinski, Avoiding investment in doomed drugs: Is solubility an industry-wide problem? *Curr Drug Discov*, 17–19 (2001).
117. C.A. Lipinski, Poor aqueous solubility: An industry-wide problem in drug delivery. *Am Pharm Rev*, **5**, 82–85 (2002).
118. S. Stokbroekx, EUFEPS and COST B25 Conference on Bioavailability and Bioequivalence: Focus on Physiological Factors and Variability, Athens, Greece,1–2 October 2007, online available from: http://www.eufeps.org/document/BA_BE_presentations/Stokbroekx_Sigrid.pdf [Accessed 22 May 2011].
119. L. Benet, European drug absorption network, Annual EDAN meeting at Leuven University, Belgium, 18–20 March, 2007.

120. European Medicines Agency, Rapamune: EPAR – Scientific Discussion, 08/08/2006, online available from: http://www.ema.europa.eu/docs/en_GB/document_library/EPAR_-_Scientific_Discussion/human/000273/WC500046434.pdf.
121. Wyeth Research Drug Information, Rapamune (Sirolimus) Oral Solutions and Tablets. Company Communications (2004).
122. European Medicines Agency, Emend: EPAR – Scientific Discussion, 05/09/2006 online available from: http://www.ema.europa.eu/docs/en_GB/document_library/EPAR_-_Scientific_Discussion/human/000527/WC500026534.pdf.
123. Y. Wu, A. Loper, E. Landis, L. Hettrick, L. Novak, K. Lynn, C. Chen, K. Thompson, R. Higgins, U. Batra, S. Shelukar, G. Kwei, D. Storey, The role of biopharmaceutics in the development of a clinical nanoparticle formulation of MK-0869: A beagle dog model predicts improved bioavailability and diminished food effect on absorption in humans. *Int J Pharm*, **285**, 135–146 (2004).
124. R. Femia, Megestrol acetate nanocrystal: Results of dose-escalating studies under fed and fasting conditions. Poster presented at AMFAR's 17th National HIV/AIDS Update Conference, California, USA, 2005.
125. J. Jinno, N. Kamada, M. Miyake, K. Yamada, T. Mukai, M. Odomi, H. Toguchi, G.G. Liversidge, K. Higaki, T. Kimura, Effect of particle size reduction on dissolution and oral absorption of a poorly water-soluble drug, cilostazol, in beagle dogs. *J Contr Release*, **111**, 56–64 (2006).
126. E.M. Merisko-Liversidge, G.G. Liversidge, Drug nanoparticles: Formulating poorly water-soluble compounds. *Toxicol Path*, **36**, 43–48 (2008).
127. J.W. Mouton, A. Van Peer, K. De Beule, A. Van Vliet, J.P. Donnelly, P.A. Soons, Pharmacokinetics of itraconazole and hydroxyitraconazole in healthy subjects after single and multiple doses of a novel formulation. *Antimicrob Agents Chemother*, **50**, 4096–4102 (2006).
128. B. Rabinow, J. Kipp, P. Papadopoulos, J. Wong, J. Glosson, J. Gass, C.S. Sun, T. Wielgos, R. White, C. Cook, K. Barker, K. Wood, Itraconazole IV nanosuspension enhances efficacy through altered pharmacokinetics in the rat. *Int J Pharm*, **339**, 251–260 (2007).
129. S.M. Moghimi, A.C. Hunter, J.C. Murray, Long-circulating and target-specific nanoparticles: Theory to practice. *Pharmacol Rev*, **53**, 283–318 (2001).
130. J. Kreuter, Drug targeting with nanoparticles. *Eur J Drug Metabol Pharmacokinet*, **3**, 253–256 (1994).
131. K. Peters, S. Leitzke, J.E. Diederichs, K. Borner, H. Hahn, R.H. Müller, S. Ehlers, Preparation of a clofazimine nanosuspension for intravenous use and evaluation of its therapeutic efficacy in murine Mycobacterium avium infection. *J Antimicrob Chemother*, **45**, 77–83 (2000).
132. O. Kayser, Nanosuspensions for the formulation of aphidicolin to improve drug targeting effects against Leishmania infected macrophages. *Int J Pharm*, **196**, 253–256 (2000).

17

Enhanced Solubility of Poorly Soluble Drugs Via Spray Drying

Cordin Arpagaus, David Rütti and Marco Meuri

17.1 Introduction

It is estimated that approximately 40% of all new drugs in the pharmaceutical market display poor aqueous solubility [1–6]. According to the Biopharmaceutics Classification Scheme, such drug compounds (low solubility, high permeability) are categorized as Class II drugs, whereby the dissolution step is the rate-limiting process of drug absorption [7].

Several formulation strategies have been applied by pharmaceutical scientists to enhance the solubility of poorly water-soluble drugs. These include:

- particle size reduction;
- drug dispersion in a suitable carrier or matrix;
- modifications to the crystal habit;
- microemulsions drug delivery systems.

Physical methods such as milling [4, 8, 9], high pressure homogenization [10, 11], and super-critical precipitation techniques [12] have all been successfully applied for particle size reduction.

Thereafter, techniques such as freeze drying, spray drying, centrifugation and ultra filtration are often employed to concentrate, dry, and stabilize the formulated drugs to dosage forms such as tablets or capsules without compromising their physical, chemical, and pharmaceutical properties.

Drug dispersion in a suitable carrier or matrix in a solid state is another approach. The main aim is to stabilize and protect the drug during formulation. Various carrier materials

Drug Delivery Strategies for Poorly Water-Soluble Drugs, First Edition. Edited by Dennis Douroumis and Alfred Fahr.

are used, including acids, carbohydrates (e.g. lactose [13], mannitol, sucrose and trehalose, polymeric materials, such as PLGA [14], chitosan [15], and cyclodextrins or surfactants).

Nanoparticle formulations are another promising way to enhance the dissolution rates of poorly water-soluble drugs [1, 10]. Thanks to their reduced size, and hence larger specific surface area, drug nanoparticles enable new developments for targeted drug delivery systems for cancer therapy, diabetes, asthma, and so forth [12, 16–20]. Micro-emulsion drug delivery systems are of great interest to the pharmaceutical scientist because of their considerable potential to act as drug delivery vehicles by incorporating a wide range of drug molecules [1]. These systems are essentially mixes of surfactant, oil, and drug that form emulsions on mixing with water with little or no energy input [19]. Self-emulsifying drug delivery systems [21] have received particular attention as a means of enhancing the oral bioavailability of poorly absorbed drugs.

A popular way to overcome the drawback of thermodynamic instability is by the transformation of emulsions into dry powdered emulsions by removal of the aqueous phase. Typically, the processes applied are spray drying, lyophilization, or solvent evaporation [1, 21].

The final conversion of drug dispersions into dry powders is often performed by spray drying [5, 10, 11]. Spray drying imparts key principles to enhance drug solubility in one process, such as particle size reduction (e.g. size and morphology), crystal habit modification (e.g. changing the morphological form of the solid from a crystalline structure to amorphous states by the rapid removal of the solvent), and encapsulation (e.g. stabilization and protection) of the drug in a suitable carrier or matrix. Figure 17.1 exemplifies

Figure 17.1 *Colorized SEM images of spray-dried cyclosporin-loaded PLGA particles [43].* (See colour plate section.)

Table 17.1 *Advantages of spray drying compared to other methods for enhancing drug solubility.*

Advantages of spray drying
1. Control of particle size, shape, morphology, porosity, and density depending on spray drying conditions.
2. One-step process to directly transform various liquids (solutions, emulsions, dispersions, slurries, pastes or event melts) into dry formulations.
3. Highly dispersible and spherical particles with narrow particle size distribution.
4. Process simplicity, cost-effectiveness, and scale-up capability.

the morphological capabilities of spray-dried nanosuspensions. The colorized SEM picture illustrates spray-dried cyclosporin nanoparticles encapsulated within PLGA as an example of slightly water soluble immune system-suppressant drug [43]. The generated aggregated particles revealed a highly porous structure (micron particles consisting of nanosized drug particles) and narrow size distribution.

17.2 Advantages of Spray Drying

Compared to other methods used in drug delivery applications, spray drying is a continuous process to directly transform various liquids (solutions, emulsions, dispersions, slurries, pastes or event melts) into solid particles with adjustable size, distribution, shape, porosity, density, and chemical composition.

Spray drying is a simple, rapid, reproducible, and scalable drying technique [22], allowing mild temperature conditions suited to heat-sensitive biopharmaceutical compounds (Table 17.1). The spray-dried particles are generally free-flowing and spherical, requiring no additional processing (such as dry granulation) before compaction into tables or capsules.

The main applications are found in the food, chemical, and materials industries to enhance ingredient conservation, particle properties, powder handling, and storage. However, there are numerous application examples available in the formulation of pharmaceuticals for drug delivery, such as:

- particles with increased dissolution rates [23], controlled drug release profiles [24], better pharmacokinetic performance [8], and enhanced bioavailability of poorly soluble drugs [18, 25];
- formulated powders for inhalation (pulmonary drug delivery) to treat asthma, cystic fibrosis, chronic pulmonary infections or lung cancers [13, 14, 22, 26, 27];
- powders with increased storage stability [21, 23, 24, 28].

17.3 Principles and Instrumentation of Spray Drying Processes

17.3.1 Principal Function of a Spray Dryer

The complete spray drying process generally consists of a sequence of four steps:

1. feed preparation;
2. atomization/droplet formation, and hot gas contact;

3. evaporation, particle shape formation, and drying;
4. separation of the dried product from the gas.

Briefly, the liquid stream is atomized into fine droplets via an appropriate device, is then subjected to drying in a drying chamber to yield the solid particles, and is finally separated and collected via a suitable dry powder collector [17].

A comprehensive discussion of the general spray drying process is available in the books, *Spray Drying in Practice* [29] and *Guide to Industrial Drying* [30]. Pharmaceutical applications were extensively reviewed by Broadhead *et al.* [31], Wendel and Celik [32], and Vehring *et al.* [33].

17.3.1.1 Feed Preparation

The first requirement is for the feed to be pumpable, homogenized, and free from impurities. A typical feed can be an emulsion, suspension, or solution made up of a solvent and the substance requiring spray drying.

Feed preparation has a significant influence on particle shape, morphology, and coating. Typically, a small percentage of additives, such as lubricants, anti-foaming, or other special agents are mixed into the feed to provide particle shape control and higher solids concentrations in the feed. Organic binders, for example, PVA, maltodextrin, arabic gum, or gelatin are used to hold the material together after pressing the powder into tablets or capsules. Glycerine or ethylene glycol are used to soften the particles. Increasing the concentration of feed solids not only improves the thermal efficiency and plant capacity of the spray drying process, but also has a strong influence on the bulk density of the final powder.

17.3.1.2 Atomization/Droplet Formation, and Hot Gas Contact

Different kinds of energy are used to disperse the liquid into fine droplets in a spray dryer. Designs typically used are based on two-fluid nozzles, pressure nozzles, rotary disk atomizers, and ultrasonic nozzles (Figure 17.2). The selection of the atomizer type depends upon the nature and rate of the feed, and upon the desired characteristics of the dried product. The higher the energy for the dispersion, the smaller the generated droplets are. In addition, the size of the droplets is also controlled by the surface tension, viscosity, and density of the fluid.

A specific characteristic of any spray drying process is the drying over a large liquid surface area enabling high heat transfer coefficients and drying within seconds. As an

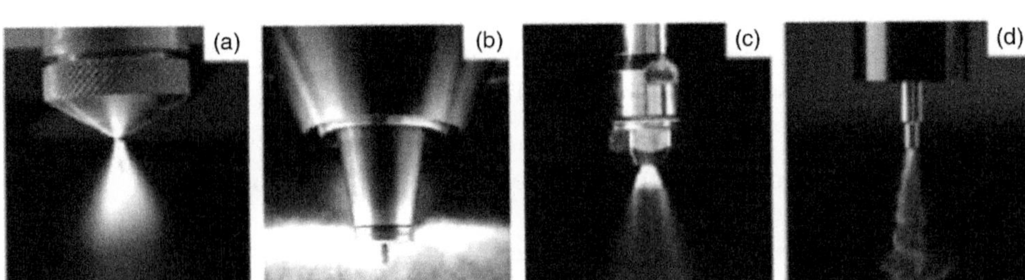

Figure 17.2 *Atomizing principles: (a) two-fluid nozzle; (b) rotary atomizer; (c) pressure nozzle; (d) ultrasonic nozzle [35].*

example, 100 mL of a solution sprayed into 25 μm size droplets has a total number of about 800 million droplets. This results in approximately 12 m^2 of surface area. Hence, it is possible to dry at moderate temperatures, which enables even heat-sensitive materials (e.g. enzymes, proteins, antibiotics, etc.) to remain active after spray drying without major loss of activity.

Two-fluid nozzles provide the energy for atomization by the rapid expansion of a spray gas mixed with the liquid feed at its nozzle tip. Atomization with two-fluid nozzles is considered suitable for laboratory-scale spray dryers due to its relatively low pressure consumption and low particle velocity, and the resulting shorter drying time requirement in the spray chamber. Dryers utilizing two-fluid nozzles generally achieve smaller droplets in comparison to other atomization techniques.

Rotary disk atomizers are the most frequently used industrial-scale atomizing devices, and provide high feed throughputs of the order of tons per hour. They are typically used for the production of particles in the size range of 80–120 μm. Chambers used are generally larger than for two-fluid or pressure nozzles. The droplet size is determined by the rotational speed of the disk. The higher the speed, the smaller the droplets.

Pressure nozzles are the preferred choice if large droplets in the range of 150–350 μm are required. Their basic function is to convert the pressure energy supplied by high-pressure pumps into kinetic energy in the liquid. The droplet size produced varies with the feed rate and feed viscosity. Pressure nozzles are applied for high capacity multi-nozzle systems.

Ultrasonic atomizers show droplet size dependence in the sprays produced that is governed by the frequency at which the nozzle vibrates and by the surface tension and density of the liquid being atomized. However, the frequency is the predominant factor. The higher the frequency, the smaller the median drop size. Ultrasonic atomizers have the advantage of a more uniform particle size distribution than, for example, two-fluid nozzles.

17.3.1.3 Evaporation, Particle Shape Formation, and Drying

As soon as the droplets make contact with the hot drying gas, evaporation takes place at the saturated vapor film that is quickly established on the droplet surface. The drying forces are the temperature and moisture gradients present between the droplets and the drying air.

The evaporation process of a droplet can be modeled by the dimensionless number *Peclet* (*Pe*), which is the ratio between the droplet evaporation rate and the diffusion motion of the solutes inside the droplet [33]. The rate of evaporation is determined by the balance of energy required to vaporize the solvent, and the energy transported to the surface of the droplet.

For a small value of *Pe* ($Pe < 1$), drying of a sprayed droplet is sufficiently slow that the solute has enough time to evenly distribute inside the evaporating droplet by diffusion. Because the surface enrichment is small, the chances are high that the solute will precipitate instantly throughout the droplet, leading to dense dried particles [33].

In contrast, for high *Pe* number conditions, as discussed by Vehring *et al.* [33], hollow-like structures are formed. Drying of the droplet is quick, and the solute in the droplet does not have sufficient time to diffuse from the surface to the center. This leads to accumulation of the solute at the surface of the droplet. Solute at the surface is saturated, and can precipitate to a shell or skin, impeding further shrinking of the droplet. In the final drying stage, solute in the center is accumulated in the shell layer due to slow diffusion through the shell and transportation of the solute (Figure 17.3).

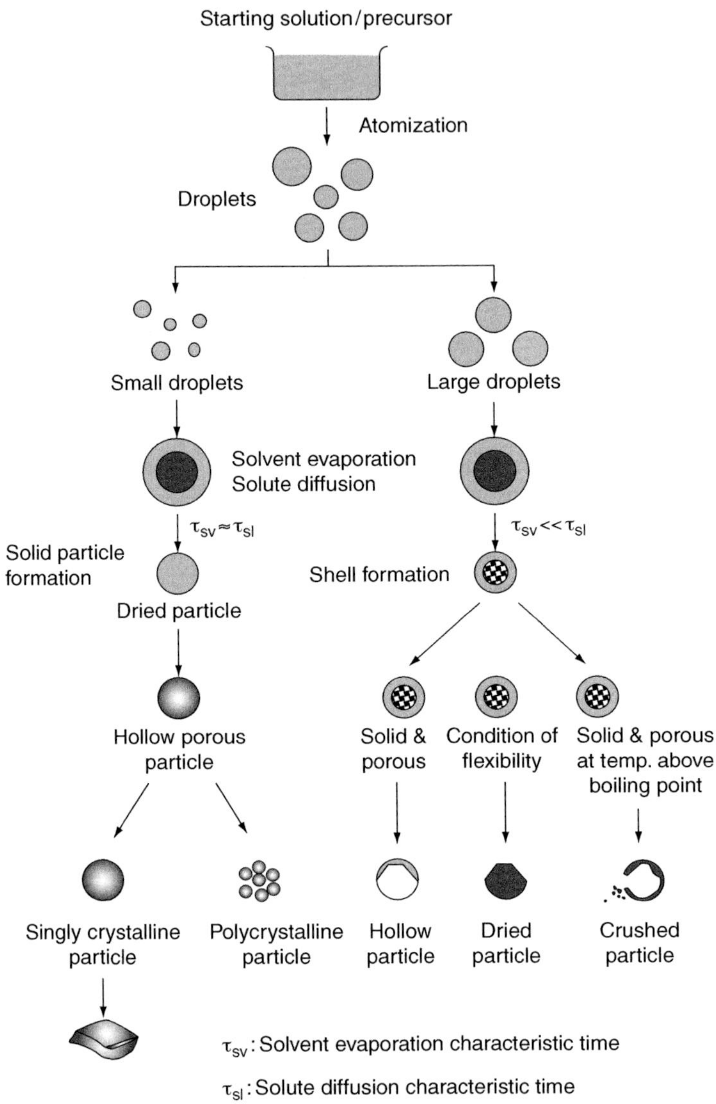

Figure 17.3 *Morphology of particles prepared by spray drying. Reproduced with permission from [34].*

The properties of the solvent, the composition of the sample, the flow rate of the drying gas (i.e. residence time of heating or the solvent evaporation), and the temperature are the main parameters affecting the morphology of particles generated by spray drying. Concerning the evolution of the particle morphology during spray drying, in addition to solid (dense) and spherical particles, hollow or fragmented particles are often formed.

Typically, the hot drying gas is air. However, sensitive materials such as pharmaceuticals, and solvents such as ethanol, require oxygen-free drying. In this case, nitrogen gas is used instead (e.g. in closed mode operation and inertization).

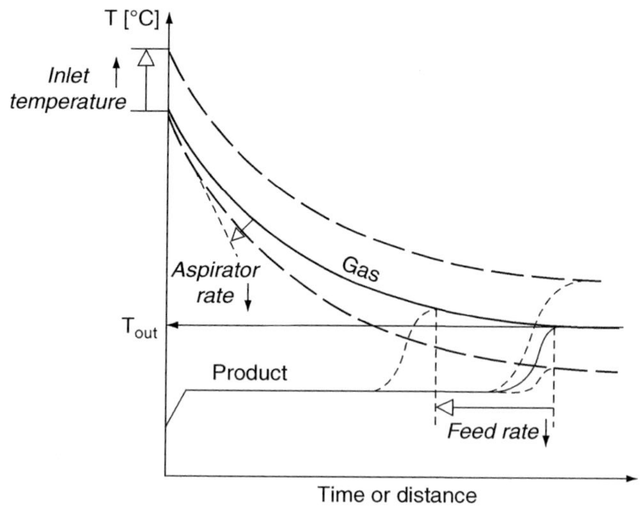

Figure 17.4 *Temperature profiles of the product and gas during spray drying in co-current mode [35].*

Figure 17.4 illustrates the temperature profile of the drying gas and the droplet (product) in the spray chamber in a co-current flow. The gas temperature decreases with time (or more precisely, the drying time). At the start of drying, the droplet temperature increases to the cooling limit temperature (the wet-bulb temperature). It then remains constant during evaporation. The droplets can shrink, agglomerate, or lose sphericity as moisture is evaporated.

At the end of drying, the dried particle surface temperature (product curve) approximates the temperature of the surrounding gas at the dryer outlet, which determines the final outlet temperature (T out). The outlet temperature is the result of the applied drying gas flow (aspirator rate), inlet drying gas temperature, and the liquid feed rate.

A higher inlet temperature moves the outlet temperature to higher values. A lower liquid feed rate increases the outlet temperature due to the fact that less liquid has to be evaporated over the same time period. A lower aspirator rate decreases the slope of the gas temperature profile and leads to a lower outlet temperature.

The manner in which the sprayed liquid droplets contact the drying gas is an important factor in spray drying design. It has a great influence on the droplets' behavior and the properties of the dried product. The commonly used flow configurations in spray dryers are co-current flow, counter-current flow, and combined mode (also called mixed mode or fountain mode). These are displayed in Figure 17.5.

In *co-current flow* the material is sprayed in the same downwards direction as the flow of the hot gas. The droplets come into contact with the hot gas at the top, and exit the spray chamber at the bottom where the gas temperature is the lowest. By this means, the product is treated with care. Co-current flow is therefore widely used and preferred for the spray drying of heat-sensitive products.

In *counter-current flow*, the material is sprayed in the opposite direction to the hot gas. The hot gas enters at the bottom and flows upwards, while the product falls through increasingly hot gas into the spray chamber. The residual moisture is evaporated, and the product becomes hotter at the end than in the co-current mode. This method is

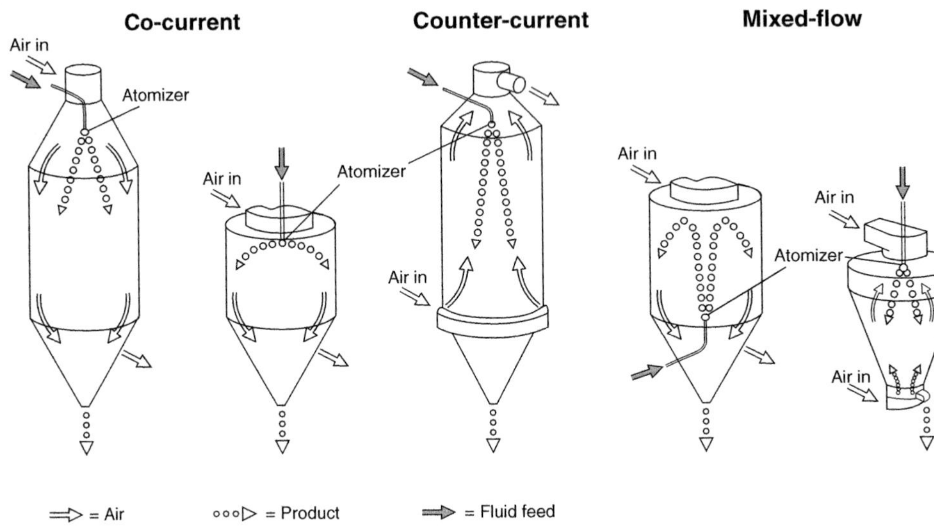

Figure 17.5 Flow configurations of spray dryers [35].

suitable only for thermally stable products, for example, inorganic materials such as ceramic powders.

In *mixed flow* the advantages of both spraying methods (co-current flow and counter-current flow) are applied. This method increases the residence time of the droplets in the drying gas, allowing drying of larger droplets than the co-current or counter-current modes. Gravity pulls the product back into the cooler zone again.

Closed-cycle layouts are mainly used for non-aqueous feedstock (i.e., organic solvents) which require inert gas as the drying medium. They are employed for flammable and explosive products. Oxygen-sensitive products can also require closed-cycle layouts.

17.3.1.4 Separation of the Dried Product from the Gas

The primary separation of the dried particles from the drying gas takes place at the base of the spray drying chamber. The finer particles can be recovered from the drying gas by means of a separation device such as a cyclone. Based on inertial forces, the particles flow to the cyclone wall and separate from the gas as a downwards strain (Figure 17.6). Other systems for particle separation are electrostatic precipitators (Figure 17.7), bag (textile) filters, and wet collectors such as scrubbers.

Typical cyclones used in commercial laboratory spray dryers are unable to collect small particles below 2 μm efficiently [12, 17, 36]. Therefore, the only feasible options for nano-particles are filter bags and electrostatic precipitators [16]. Electrostatic precipitators have the advantages of being able to collect particles with smaller sizes and have the ease of uncontaminated powder recovery.

17.3.2 Traditional Spray Dryers

The Mini Spray Dryer B-290 is the commonest traditional laboratory-scale spray dryer worldwide and works according to the principle of the co-current flow. Figure 17.8 shows the

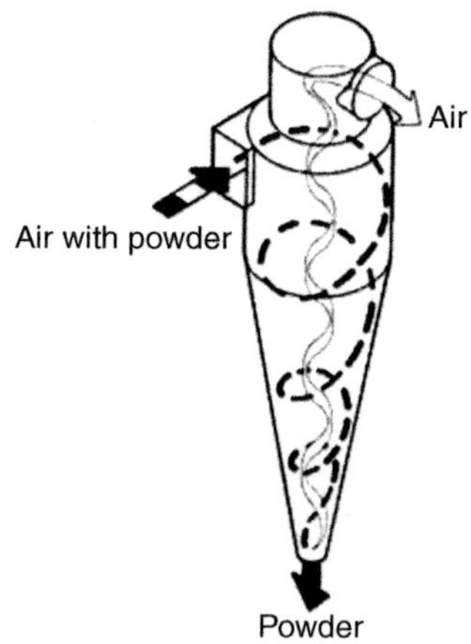

Figure 17.6 *Functional principle of a cyclone separator [35].*

spray dryer model with its process diagram and process parameters. Table 17.2 summarizes the main features and benefits.

The Mini Spray Dryer B-290 is designed to evaporate up to 1 liter of water or organic solvent per hour. This is particularly valuable for pharmaceutical R&D applications, as they can obtain small product volumes within a short time. The speed of the process and the consequently short drying time enable the drying of even temperature-sensitive products without degradation.

All the important parts – the drying chamber, cyclone, and product collection vessels – are manufactured from glassware. This enables the supervision of the complete drying process from the two-fluid nozzle down to the powder collection vessel. The peristaltic pump feeds the liquid sample from the feed solution container to the nozzle. The integrated

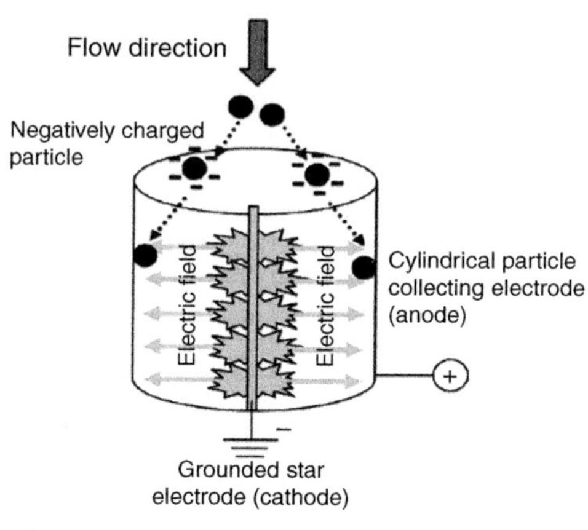

Figure 17.7 *Functional principle of an electrostatic particle collector. Reproduced with permission from [18].*

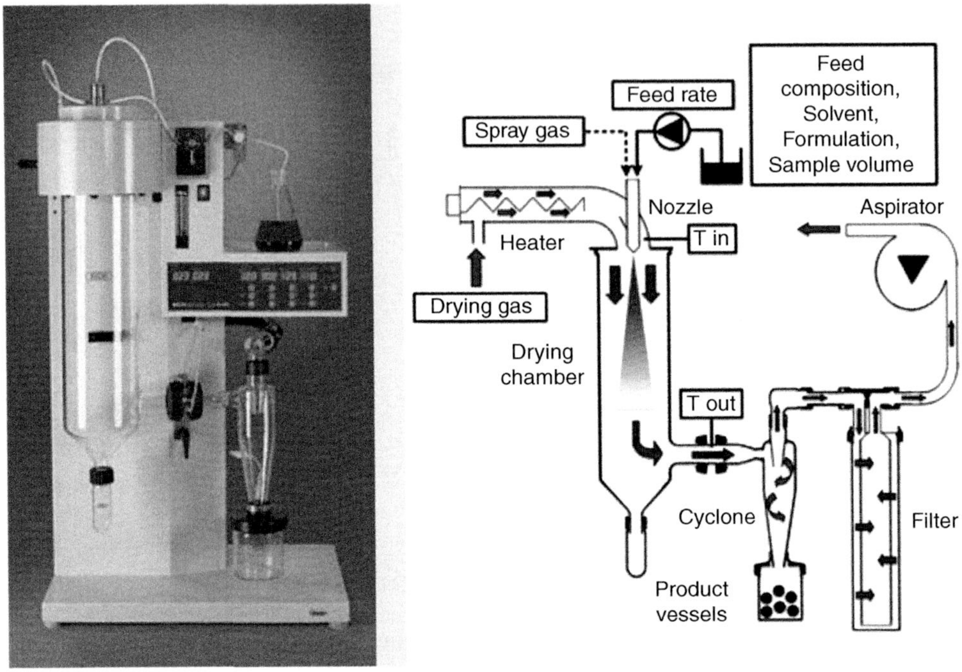

Figure 17.8 Mini Spray Dryer B-290 with process diagram and process parameters [13].

Table 17.2 Main features and benefits of the Mini Spray Dryer B-290 and Nano Spray Dryer B-90 [13, 16].

	Mini Spray Dryer B-290	Nano Spray Dryer B-90
Main benefit	for traditional spray drying, established process	for small quantities, finest particles, highest yields
Max. inlet temperature	220°C	120°C
Water evaporation	1.0 kg/h, higher for solvents	max. 0.2 kg/h
Nozzle type	two-fluid nozzle, three-fluid nozzle	piezoelectric driven vibrating mesh
Particle size	2–25 μm (from solutions)	300 nm–5 μm
Particle separation technique	cyclone	electrostatic particle collector
Typical yield	typically around 50%–70%	up to 90%
Min. sample volume	30 mL	3 mL
Max. sample viscosity	300 cps (viscous samples and juices possible)	10 cps (diluted samples)
Scale-up	possible to scale-up to kg- and tons-scale	limited by spray head and electrical particle collector

two-fluid nozzle dispenses the fluid feed with compressed gas (e.g. air or nitrogen) into a drying chamber.

The inlet drying gas (e.g. air or nitrogen) is heated by an electrical heater, is blown through the system by an aspirator, and is used as the drying gas. The spray gas is used to disperse the liquid feed into fine droplets (size adjustment), which are subsequently dried in the spray chamber. The cyclone separates the particles from the gas stream, and the dried particles fall into a product collection vessel. The outlet filter captures very fine particles, preventing them from leaving the system.

The residence time of the drying air within the spray chamber is about 1.5 seconds. The powder collection is provided by a glass cyclone separator, which is internally coated with a nanometer thin antistatic film to reduce powder adhesion to the glass wall. The separation works by centrifugal forces by virtue of the inertia of the solid particles.

17.3.3 Recent Developments in Spray Drying

The recent rise of nanomedicine and nanotherapeutics has further increased the pressure on existing spray dryer systems to generate nanoparticles with a good yield and a narrow size distribution [8, 17]. BÜCHI Labortechnik AG is well known for the development of spray drying technology, notably for its laboratory-scale devices and the development of the innovative Nano Spray Dryer B-90 (see Figure 17.9 and main features and benefits in Table 17.2). This laboratory spray dryer is based on a fundamentally new concept in

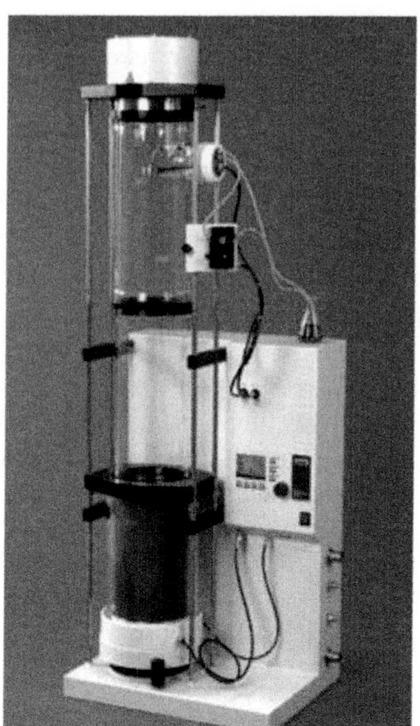

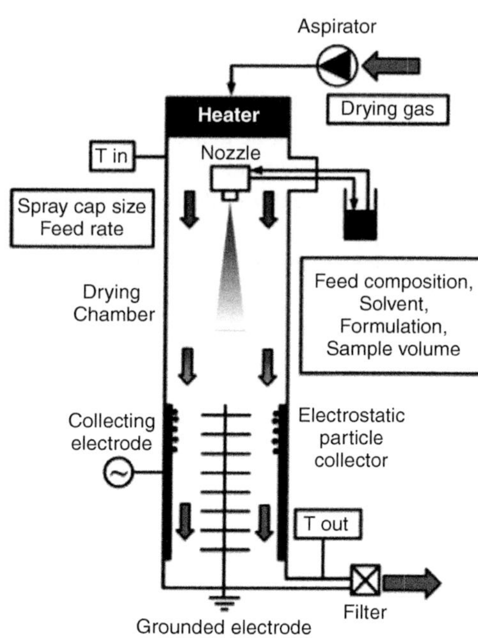

Figure 17.9 Nano Spray Dryer B-90 with process diagram and process parameters [37].

spray drying technology which allows the generation and collection of submicron particles directly from a solution [16, 17, 19].

This new system has been developed specifically to generate particle sizes ranging from 300 nm to 5 μm for milligram sample quantities at high yields, and is especially suited for the early stages of product development in the pharmaceutical industry [37]. The technological novelty of this spray dryer lies in the gentle laminar drying flow, the vibrating mesh spray technology, and the highly efficient electrostatic particle collector [37, 38].

The droplet generation is based on a piezoelectric crystal-driven actuator vibrating a thin stainless steel mesh containing extremely small holes. The actuator operates at an ultrasonic frequency and thereby creates millions of precisely sized droplets in a range of 3–15 μm (median diameter 5–7 μm, depending on the mesh size) with narrow droplet size distribution and span values of approximately 1.2 [36]. It is possible to spray diluted liquids with viscosities of up to 10 cPa s [16]. The sample is fed through a peristaltic pump in the form of an organic or aqueous solution or even as a dispersion.

The drying gas (typically air in open mode or nitrogen gas in closed mode) enters the apparatus from its top through a compact porous metal heater which provides optimal energy transfer and short heating-up rates of up to 120°C inlet temperature setting. The heater enables a laminar gas flow in the drying section and very short droplet drying times in the order of 11.3 ms (calculated with an initial droplet of 7.0 μm, 75°C drying temperature and 100 L/min drying air flow rate) [39]. The drying gas exits the spray dryer, and the outlet temperature is measured while the gas is filtered.

The fine droplets generated are gently dried into solid particles. Spray-dried particles as small as 350 nm [19], 460 nm [18], and 500 nm [20] can be created. These dried particles are electrostatically charged and are deposited at the collecting electrode surface. The electrostatic collection mechanism is independent of the particle mass – in contrast to cyclones – which enables the capture of nanoparticles at excellent particle recovery rates (e.g. yields). Uniquely high yields can be obtained: 72% [18], 78% [20, 36], 94% [40], and up to 94.5% [19] for small batch sizes in a range of 30–500 mg. This is something which most likely cannot be achieved with traditional cyclone separators [36].

The electrostatic precipitator collects even thin-walled particles without breaking them [39]. The powder is gently collected from the internal wall of the electrode cylinder with a particle scraper. Due to the modular glass assembly of the spray cylinder, the set-up is simple, the cleaning is easy, and considerable process time can be saved in daily lab work.

Recent applicative results derived on a Nano Spray Dryer B-90 suggest that this instrument has the ability to effectively formulate temperature-sensitive compounds, for example, proteins [18, 36], enzymes [40], and amino acids [39], at the submicron scale. This makes the Nano Spray Dryer B-90 highly desirable for drug delivery applications for research and early development studies. Published applications can be found in the areas of inhalation drug delivery [13, 18, 26, 40], nanotherapeutics [18], the encapsulation of nanoemulsions and the formulation of nanocrystals [19], the formulation of novel nanoparticulate pharmaceutical excipients [20, 36], and the generation of polymeric nanoparticles [19].

Increased customer demand for the laboratory model, coupled with promising new applications, is expected to eventually encourage and stimulate the development of further industrial-relevant spray dryer models for nanoparticle applications [17].

17.4 Optimizing Spray Drying Process Parameters

Spray drying is a method in which the properties of the resulting product are highly dependent upon the correlation and interdependency of the process parameters. Thus, the instrument settings are:

- drying gas flow rate (aspirator level);
- drying gas humidity;
- inlet temperature;
- spray gas flow;
- feed rate;
- solid concentration;
- organic solvent (instead of water).

Combined with the equipment choices (e.g. nozzle diameter, design), these all influence the properties of the final product, such as:

- outlet temperature (thermal load);
- particle size and shape;
- humidity in the final product (moisture content);
- yield;
- flow properties.

The optimization of these parameters usually follows a trial and error procedure. Figure 17.10 gives an overview of the spray drying parameters and their influence on the final product properties. The influence of spray drying process parameters on outlet temperature, particle size, humidity, and yield is discussed in the following. Some guidelines for quick process optimization are identified.

17.4.1 Drying Gas Flow Rate (Aspirator Rate)

The drying gas is sucked or blown through the device by the aspirator motor, creating a slight under- or over-pressure in the system. By regulating the aspirator rate, the amount of heated drying gas is increased or decreased, which has a significant effect on the drying performance.

The optimum setting must be determined experimentally using the guidelines:

- A higher aspirator rate leads to a higher degree of separation in the cyclone and increases the outlet gas temperature.
- A lower aspirator rate results in higher residual moisture content in the product.

17.4.2 Drying Gas Humidity

Higher drying gas humidity means less vapor uptake capability in the gas stream. Saturation occurs easily, leading to a higher outlet temperature. This can increase the humidity of the final product and result in moist particles, which could adhere to the glassware, thereby decreasing the yield.

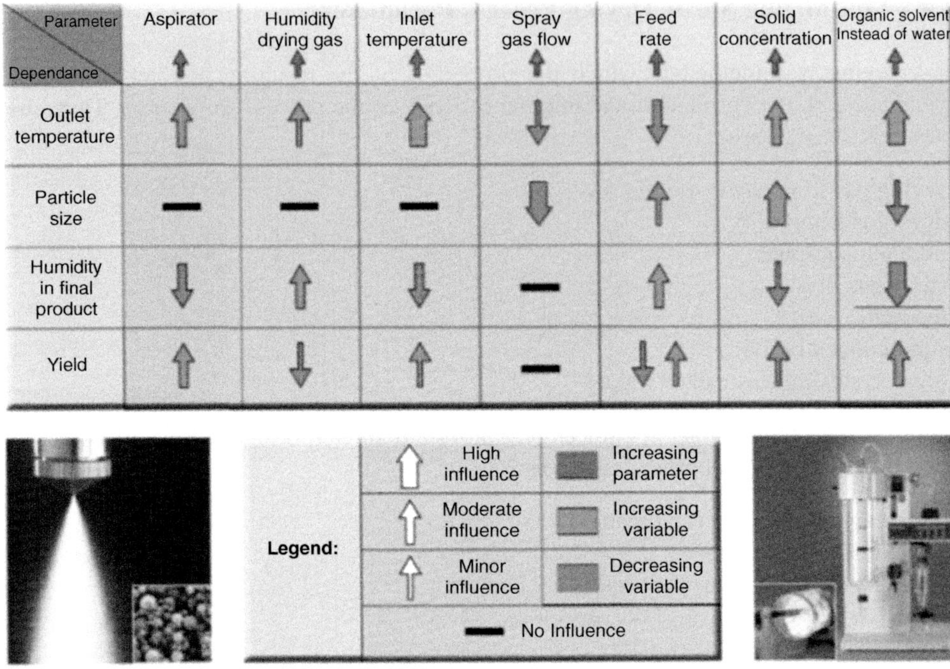

Figure 17.10 *Influence of spray drying process parameters on the product characteristics [35].*

17.4.3 Inlet Temperature

The inlet temperature is understood as being the temperature of the heated drying gas. The drying gas is sucked or blown into the system by the aspirator motor and is heated up before it enters the spray chamber. The inlet temperature is measured at the inlet of the drying chamber. The inlet temperature is set to a value at which the temperature gradient between the wet droplet surface and the hot saturated gas leads to evaporation of the solvent.

The following can be considered a guideline. Higher inlet temperatures increase outlet temperatures proportionally. They reduce the relative humidity of the drying gas, producing drier powder. This offers a dryer, less sticky product, and increases the yield.

The outlet temperature is determined by the heat and mass balance in the drying cylinder, and is influenced by the following parameters (shown in Figure 17.10):

- inlet temperature;
- aspirator flow rate (quantity of drying gas, e.g. air or nitrogen);
- feed pump setting;
- solid concentration of the sample material.

As a rule of thumb, the maximum temperature the product can experience is the outlet temperature.

One of the most important considerations is the choice of the temperature difference between the inlet and outlet. Other points are product-specific factors, such as the melting point, the glass transition temperature [41] or the degradation temperature. In spite of this, there is still some room for parameter adjustment and optimization.

To achieve a final product with a very small amount of residual moisture, the inlet temperature must be set as high as possible and the temperature difference must be as small as possible.

17.4.4 Spray Gas Flow

The gas spray flow is the amount of compressed gas needed to disperse the liquid solution, emulsion, or suspension into a well-formed spray cone. In a closed mode configuration, compressed nitrogen is used instead of air.

As a guideline, the spray gas flow is considered to be the key process parameter for adjusting the droplet size:

- A higher spray gas flow leads to smaller droplets and to smaller dried particles.

Figure 17.11 illustrates this relationship, showing the shift of the particle and droplet size distribution towards smaller droplets and particles with higher gas spray flow.

17.4.5 Feed Concentration

This influence is obviously due to the pure mass balance of the input and output of solid material. The solid concentration depends strongly on the application. Increasing the solid concentration of the feed means less liquid is available to vaporize, and leads to increased outlet temperatures. It also produces bigger particles.

17.4.6 Feed Rate

The pump feeds the spray solution to the nozzle. The feed rate affects the temperature difference between the inlet and the outlet. The pump rate corresponds directly to the inlet mass. The higher the throughput, the more energy is needed to evaporate the droplets into solid particles. Thus, the outlet temperature decreases.

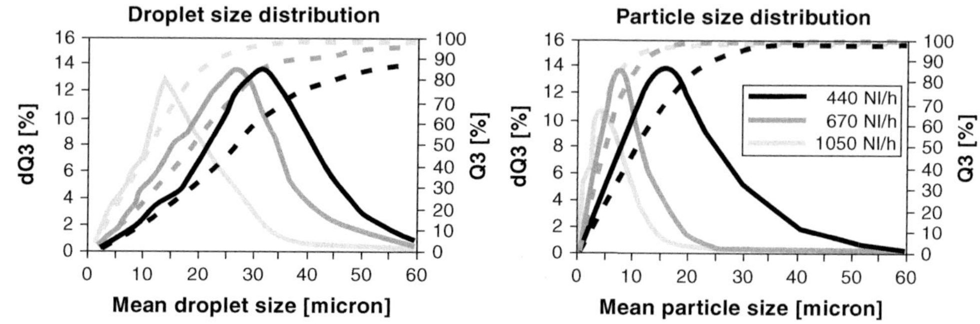

Figure 17.11 Droplet and particle size distributions as a function of the spray gas flow in a two-fluid nozzle (20% maltodextrin solution, 100% aspirator rate, 140°C inlet temperature) [35].

The selection of too high a pump rate can lead to wet and sticky particles, which adhere to the spray chamber wall. The optimal pump rate depends also upon factors such as the viscosity of the spray solution and the diameter of the feed tubing.

A general guideline for the feed pump rate is:

- An increase of the feed pump rate lowers the outlet temperature and increases the temperature difference between the inlet and the outlet. Reducing the pump rate, while holding the inlet temperature and aspirator rate constant, leads to a drier final product.

17.4.7 Organic Solvent Instead of Water

In pharmaceutical applications, mainly aqueous solvents are used. However, organic solvents, typically ethanol, ethyl acetate, acetone, and dichloromethane are also applied [14,22,42].

Organic solvents offer a liquid requiring less energy to vaporize, and the outlet temperature of the spray dryer increases. Conversion factors (compared to water) are dependent mainly on the heat of evaporation. Note that most organic solvents are highly flammable and if sprayed into air, they constitute an explosion risk. Inertization and drying in closed mode operation are applied.

Organic solvents typically result in slightly smaller particles, due to the lower surface tension.

17.5 Spray Drying of Water-Insoluble Drugs: Case Studies

17.5.1 Nanosuspensions

The use of nanoparticles of poorly soluble drugs has enormously widened the window of achievable pharmacokinetic performance [18]. For the successful development of this technology, it is essential to understand the characteristics of nanoparticle suspensions. A suspension (D) is created from the product to be treated (A), a matrix (B), and water with additional filmogen (C). This suspension is then sprayed into small droplets (E). The matrix and/or filmogen lead to an agglomeration or coating of the suspended particles (F) (Figure 17.12).

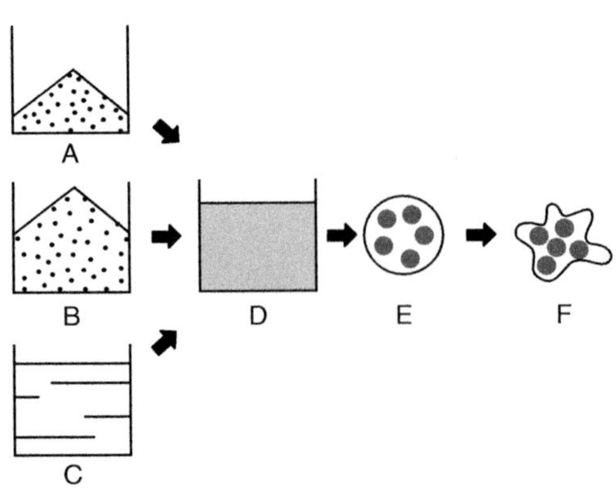

Figure 17.12 Microencapsulation by spray drying from a (nano)suspension.

In order to prepare an effective solid dosage form for a nanosuspension by spray drying, it is important that the spray-dried nanoparticles return to their original particle size when reconstituted in an aqueous system.

Chaubal and Popescu [10] examined the impact of various formulation and processing parameters on the redispersibility of spray-dried itraconalzole (antifungal drug) nanoparticles. First, a crystalline nanosuspension of itraconazole was prepared using a high-pressure homogenization process. Then spray drying was conducted with lactose and mannitol as a matrix to prepare aggregated powders with good flowability which redispersed back to the original submicron size upon reconstitution in an aqueous buffer. Spray-dried powders containing lactose typically showed 4–6% moisture content, whereas mannitol-based formulations contained less than 2%. The dissolution profiles of the spray-dried powders were comparable to the original nanosuspension.

Schaffazick *et al.* [24] prepared nanocapsules by interfacial deposition of a preformed polymer (e.g. Eudragit® S100) containing melatonin (a hormone regulating the sleep–wake cycle), using an organic phase (e.g. oil, polymer, drug and surfactant in acetone). This organic phase was added with magnetic stirring into an aqueous solution containing surfactant. After nanoprecipitation, acetone and some water were removed by rotary evaporation under reduced pressure. In addition to the suspension of nanocapsules, colloidal silicone dioxide was added and the mixture was spray dried. This powder improved the physical stability of the product compared to the original aqueous nanocapsule suspension. The melatonin-loaded powder showed a controlled release profile in comparison to the dissolution of the pure melatonin. An encapsulation efficiency of up to 93% was achieved with a moisture content lower than 2%. No morphological alterations were observed after 12 months of storage.

Schafroth *et al.* [43] optimized spray drying processing parameters to generate PLGA nanoparticles with cyclosporine A (an immune system-suppressant drug) and dexamethasone (an anti-inflammatory steroid drug) encapsulated within a biodegradable poly(DL-lactide-co-glycolide) core. In order to achieve the controlled release of the encapsulated drugs, two PLGA grades with different molecular weights were applied. To obtain drug-loaded nanoparticles, the drugs and the biodegradable PLGA polymer were dissolved in dichloromethane/ethanol organic solution. As seen in Figure 17.13, the spray-dried

Figure 17.13 *Spray-dried cyclosporine-loaded PLGA particles. Reproduced with permission from [43].*

nanoparticles showed spherical shape, narrow size distribution and highly porous surface. The particle morphology appeared to be governed by the PLGA grade and the solvent composition. X-ray diffraction analysis suggested that cyclosporine existed in an amorphous form in the PLGA particles.

17.5.2 Solid Lipid Nanoparticles

Solid lipid nanoparticles are colloidal drug delivery systems derived from oil-in-water emulsions by replacing the liquid lipid (oil) by a solid lipid [1]. Solid lipid nanoparticles open a broad field of applications including oral, dermal, and pulmonary administration. The conversion of the liquid dispersion into a dry product is useful and often necessary.

Freitas and Müller [11] report on their investigations of solid lipid nanoparticles dispersions from compritol, a lubricant found in oral solid dosage formulations and cosmetics. Aqueous dispersions of solid lipid nanoparticles were prepared by high-pressure homogenization and then converted by spray drying into dry, non-sticking, fine-grained powders which could be stored over long periods. An identical size distribution of 300 nm to 900 nm before and after the spray drying process, followed by subsequent redispersion, was still acceptable for use with respect to the particle size distribution and toxicity.

Coalescence of molten lipid droplets was prevented by spray drying alcoholic, instead of aqueous, dispersions. The addition of carbohydrates, such as mannitol, lactose, and trehalose reduced the extent of melting by building a sugar layer around the particles. Surfactant addition helped to increase the stability and entrapment efficiency.

For lung delivery applications, microparticles containing solid lipid nanoparticles also received increasing attention as carrier systems. Microparticles offer the potential of reaching the deep lung and of enabling effective redispersion of the nanoparticles in the presence of water-soluble excipients.

Li *et al.* [44] prepared thymopentin-loaded solid lipid nanoparticles using a double emulsion technique. With the addition of mannitol and leucine, microparticles were prepared by spray drying that showed the necessary dispersibility for redispersion. It was concluded that the bioavailability and therapeutic efficacy of thymopentin were remarkably improved by the pulmonary administration of the microparticles, as compared to the thymopentin solution. Drug release studies from the microparticles displayed a sustained drug release pattern without any burst effects.

17.5.3 Silica-Lipid Hybrid Microcapsules

Silica-lipid hybrid microcapsules are a novel oral delivery system for poorly soluble drugs. Tan *et al.* [5] reported on it for the first time while combining the solubilizing effect of lipids and the stabilizing effect of silica nanoparticles to produce a novel silica-lipid hybrid microcapsule system to enhance the dissolution kinetics and pharmacokinetic properties of BCS Class II drug compounds.

The preparation of silica-lipid hybrid microcapsules follows a two-step process, i.e., high pressure homogenization followed by the spray drying of the silica-stabilized emulsions. Tan *et al.* [5] prepared silica-lipid hybrid microcapsules containing the model drug celecoxib. Celecoxib is a COX-2 inhibitor classified as a BCS Class II drug, with an aqueous solubility of only 3–7 μg/ml (at pH 7.0, 40°C), which is practically insoluble at

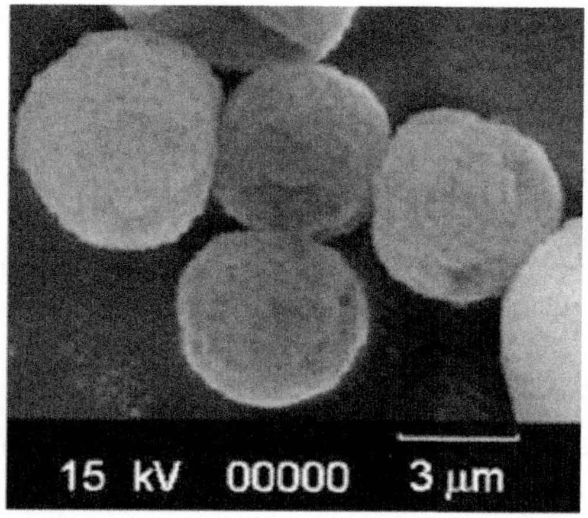

Figure 17.14 Silica-lipid hybrid microcapsules containing celecoxib. Reproduced with permission from [5].

gastrointestinal pH 1 to pH 9. The spray-dried solid lipid hybrid microcapsules consist of well-separated spherical particles with diameters of 1–5 μm and porous matrix structures (pore sizes 100–500 nm). The microcapsules were readily redispersible with an excellent physicochemical stability at room temperature for at least 6 months. The powder exhibited improved bioavailability, oil-loading levels of 42–51%, and oil entrapment efficiencies of 92–100%.

Tan *et al.* [5] (Figure 17.14) and Simovic *et al.* [45] (Figure 17.15) encapsulated the lipophilic drug compound indomethacin into microcapsules and demonstrated that dry lipid capsules stabilized by silica nanoparticles could be engineered as a complex porous nanostructure material with specific properties that enhance the delivery of poorly soluble drugs in comparison with equivalent emulsion droplets. *In vivo* studies in orally dosed rats

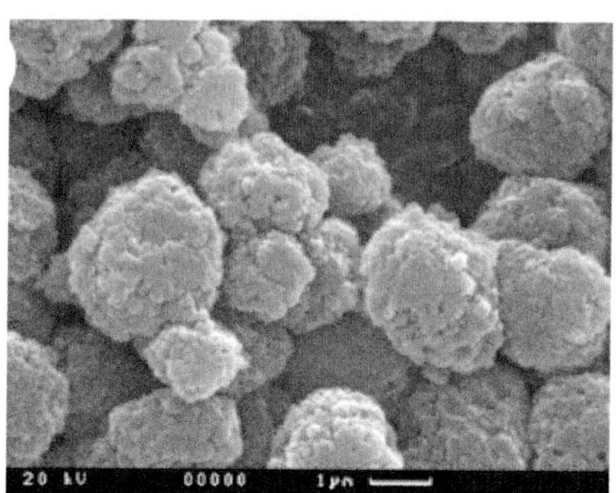

Figure 17.15 Silica-lipid hybrid microcapsules containing indomethacin. Reproduced with permission from [45].

confirmed statistically higher ($p < 0.05$) bioavailability and highest plasma concentration in comparison with o/w submicron emulsions and an aqueous drug suspension.

17.5.4 Milled Nanoparticles

As discussed, the dissolution of poorly water soluble drugs correlates with the surface area of the drug particles. Particle size reduction methods, such as wet milling technology to formulate poorly water-soluble compounds, have significantly advanced in the past decades as a viable approach capable of resolving many issues with developing and commercializing poorly water-soluble molecules [4]. Spray drying is commonly used to then convert such nanosuspensions into dry powders [9, 18].

Lee *et al.* [8] prepared nanosuspensions (100–200 nm) of an API substance (original size of 8–10 μm) in water by ball milling for 5 days, and spray dried the nanosuspension into micro aggregates using a small amount of hydroxypropyl cellulose as a stabilizer (Figure 17.16). During the spray drying process, nanoparticles easily formed aggregates yet could be slowly redispersed into nanoparticles in water over hours.

The fact that the micro aggregates were processed into compacts and were redispersed slowly in water over hours, is promising for applications such as controlled release and intravenous delivery systems. The aggregates were able to disintegrate into nanoparticles in water with time.

Traditional Chinese medicine (TCM) is also hard to absorb effectively into the body due to its large particle size. Peiyan *et al.* [9] prepared several types of TCM herbs in nanosize using a high speed centrifugal sheering pulverizer (a newly developed fine mill to obtain submicron-sized particles within a short time). As an example, the widely applied herbal drug *Radix salvia miltiorrhiza* (which prevents blood stagnation) was finely ground to 133.5 nm mean particles and then spray dried using gelatin and a CMC-Na solution as wall materials to microcapsules with rough surfaces but improved medicinal effect (e.g. dissolution rate) (Figure 17.17).

Figure 17.16 *Particles made of hydroxypropyl cellulose and API nanosuspensions, scale bar 10 μm. Reproduced from [8] with permission from Wiley-Blackwell.*

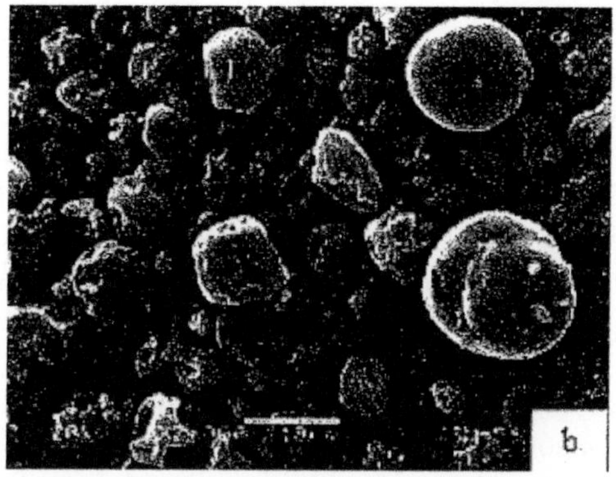

Figure 17.17 *Encapsulated Radix salvia miltiorrhiza nanoparticles in gelatin and CMC-Na. Reproduced from [9] with permission from Bentham Science Publishers.*

17.5.5 Inhalation Dosage Forms

A trend in pulmonary drug delivery is a move from liquid or pressurized delivery to dry powder inhalation formulations to treat asthma, cystic fibrosis, chronic pulmonary infections and lung cancers [13, 14, 22, 27]. This, in part, is due to the advantages of dry powder systems, including breath-actuated inhalation, limited coordination requirements, no propellant requirement, and short treatment time [13].

For effective inhalation dosage forms, the drug must be deposited in the lower airway. Generally, the aerosol particles must be between about 1–5 μm aerodynamic diameter to be effectively deposited in the lower respiratory tract [32]. Spray drying has the ability to generate highly dispersible powders in this size range. The feasibility of spray drying for aerosol delivery was extensively investigated in the inhalable research field. Several authors have prepared drug-loaded micro particles by spray drying.

Sham *et al.* [27] formulated and characterized spray dried lactose powders (size about 3 μm) containing polybutyl-cyanoacrylate and gelatine nanoparticles. Depending on the carrier type, the particles were found to be hollow while others had a continuous matrix. The nanoparticles remained in the nanosize range after spray drying.

Hadinoto *et al.* [26] demonstrated that large hollow nanoparticulate aggregated polyacrylate can be formulated in a straightforward manner using spray drying. Such aggregates were capable of serving as carriers for nanoparticulate drugs for delivery by dry powder inhalers. They offer high flowability and therapeutic efficiency.

Li *et al.* [44] prepared thymopentin-loaded (immunomodulator) particles with mannitol and leucine and investigated their aerosolization properties, as well as pharmacokinetics and -dynamics. Spherical and hollow particles of about 4.1 μm size were generated with high aerosolization efficiency (e.g. 98% emitted dose, 51% respirable fraction).

Bürki *et al.* [40] studied the effects of spray mesh size, inlet temperature, and ethanol concentration on the enzyme activity, particle size, yield, and shelf life of the model enzyme ß-galactosidase, with trehalose as stabilizer. Spherical and smooth particles of 1–5 μm size with full activity after 3 weeks and high yields of up to about 90% were achieved at a low inlet temperature of 80°C.

17.5.6 Porous Products

Porous microparticles are known to be beneficial for drug delivery to the respiratory tract by inhalation, as they have reduced interparticulate attractive forces and improved flow characteristics compared to a micronized drug [46].

The increased porosity associated with porous microparticles is reflected in an increased powder surface area, which in turn is likely to result in an increased dissolution rate. Several authors have used the spray drying technique to prepare porous particles.

Healy *et al.* [46] prepared excipient-free nanoporous microparticles of the model drug bendroflumethiazide (a thiazide diuretic used to treat hypertension) (Figure 17.18). The process involved spray drying of the active ingredient in an alcoholic solution containing ammonium carbonate as a pore-forming agent.

This process involves spray drying of solutions rather than surfactant-stabilized emulsions, and provides excipient-free microparticles with desirable physical properties such as an amorphous solid state, a porosity between 20 and 500 nm, and low bulk density. The median particles sizes measured were less than 3 μm. In short, this new engineering technology has potential in the design of other therapeutic agents such as those used in pulmonary delivery.

Nolan *et al.* [25] continued this work for the budesonide drug used for asthma treatments (Figure 17.19). The nanoporous microparticles (1–5 μm size, 10.5 m^2/g surface area) showed improved *in vitro* deposition properties compared to micronized or non-porous spray dried material. This could be attributable to improved aerosolization due to reduced interparticulate interaction/adhesion in the porous powder. The excipient-free porous microparticles showed good physical stability for over one year.

17.5.7 Microemulsions

Spray drying of emulsions has been an intensive area of research in food science, specifically regarding the drying of dairy products [33]. For pharmaceutical applications, microemulsions are attracting great interest due their considerable potential to act as drug delivery vehicles by incorporating a wide range of drug molecules [1].

Figure 17.18 *BFMT/ammonium carbonate 85:15 system spray dried from 80% v/v methanol, 1 to 5 μm nanoporous particles. Reproduced with permission from [46].*

Figure 17.19 *Nanoporous microparticles of budesonide spray dried from 80%/20% ethanol/water solution. Reproduced with permission from [25].*

These systems are essentially mixes of surfactant, oil, and drug, that form emulsions on mixing with water, and require little or no energy input [19]. Self-emulsifying drug delivery systems have received particular attention as a means of enhancing oral bioavailability for poorly absorbed drugs [21].

Emulsions are known for their physical and thermodynamic instability [1,21]. A popular way to overcome such a drawback is by their transformation into dry powdered emulsions (e.g. solid state emulsions) by removal of the aqueous phase. One widely used technique to prepare a dry emulsion is spray drying [1].

In brief, an emulsion (D) is typically created from the liquid product to be treated (A), a carrier substance (B) such as maltodextrin, and a filmogen solution (C) such as chitosan in acidic water. This emulsion is then sprayed into small droplets (E). The solvent evaporates leading to a solid matrix around the dispersed second phase (F). The result is that the small droplets of the product (A) are stored in the carrier substance (B) and embedded in the filmogen (C) (Figure 17.20).

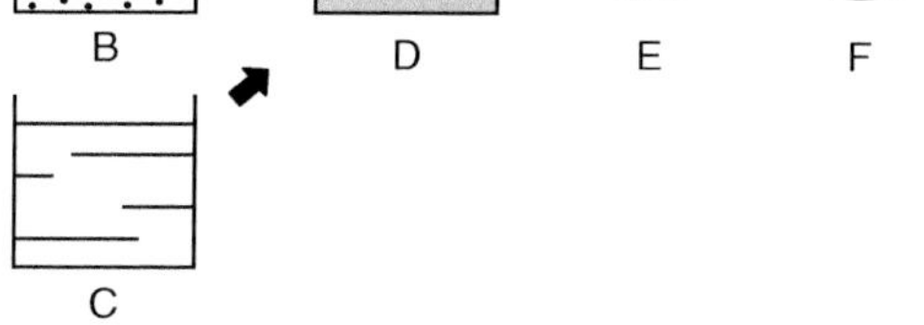

Figure 17.20 *Microencapsulation by spray drying from an emulsion [35].*

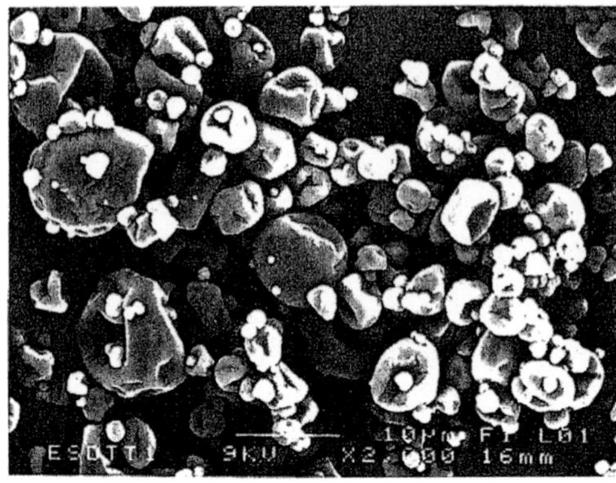

Figure 17.21 Spray-dried emulsion containing lipophilic model drug [21].
Source: Reproduced by permission of Elsevier.

A physically stabilized dry emulsion dosage form of the highly lipophilic model drug 5-phenyl-1,2-dithiole-3-thione (5-PDTT) has been developed by Dollo *et al.* [21] (Figure 17.21). An oil-in-water emulsion containing maltodextrin (as the carrier) and sodium caseinate (as the emulsifying agent) was easily spray dried into dry solid-state emulsions.

Different oil-to-water as well as maltodextrin-to-water ratios were tested to find the optimum formulation. The resulting dry emulsion exhibited good physical stability after 6 months of storage, good fluidity, a spray drying yield of nearly 70%, and a satisfactory drug incorporation efficiency of 90%. Bioavailability of the drug could be improved as demonstrated by an *in-vivo* rabbit model. The spray-dried powder consisted of well-separated 2–15 μm particles with spherical morphology and relatively deep dents.

De Jaeghere *et al.* [23] prepared RR01-loaded microparticles by the emulsification-diffusion method (oil-in-water emulsion) and spray dried it in a methanol solution containing pH-dependent dissolving poly(methacrylic acid-*co*-ethylacrylate) copolymer (Eudragit® L100-55). The emulsification-diffusion and spray drying methods allowed the preparation of nanoparticles and microparticles respectively, with optimal drug entrapment efficiency of nearly 100%. The production yields ranged around 95% for nanoparticles and around 70% for microparticles. The spray-dried microparticles were about 10 μm in size and exhibited shriveled surfaces. The absence of recrystallization, even after 12 months of storage, indicated the physical stability of the drug within the polymeric matrices.

Li *et al.* [19] used a low-energy nanoemulsion, i.e., simply formed by spontaneous emulsification, as a model to investigate the potential of the novel Nano Spray Dryer B-90 [37] for lipid encapsulation (Figure 17.22). Vitamin E acetate was chosen as a model oily phase mixed with the nonionic surfactant and aqueous phase, spontaneously generating nanoemulsions.

According to the data obtained, the Nano Spray Dryer B-90 was able to encapsulate emulsions smaller than 100 nm into separated (i.e. not aggregated) submicron solid particles. Redissolving the spray dried powder in water resulted in a redispersion of the encapsulated nano-emulsions without any significant degradation and/or size increase. The preliminary results of the encapsulated nano-emulsions appear to be of extreme interest for API formulation and offer promising perspectives for new pharmaceutical applications using spray drying [19].

Figure 17.22 Whey protein particles with encapsulated low-energy nano-emulsion of Vitamin E (bar = 10 μm). Reproduced with permission from [19].

Feng *et al.* [39] prepared various types of paclitaxel-loaded biodegradable PLGA particles (paclitaxel is a natural anti-cancer drug with very low aqueous solubility) by a single emulsion technique using dichloromethane as the solvent (Figure 17.23). For the aqueous phase, additives such as polyvinyl alcohol, gelatin, phospholipids, and cholesterol were applied to investigate their effects on the encapsulation efficiency.

Differential scanning calorimetric data revealed that paclitaxel-formulated dry emulsions were likely to be in an amorphous form after spray drying. Nanoparticles in the biopolymer sized from 550 to 850 nm were encapsulated with high drug encapsulation efficiencies of up to 100% and with the desired drug release kinetics. Moreover, the nanoparticles showed high cell uptake.

17.5.8 Application Examples: Summary

Table 17.3 outlines selected application examples for the spray drying of a poorly soluble drug using a laboratory-scale spray dryer from BÜCHI Labortechnik AG.

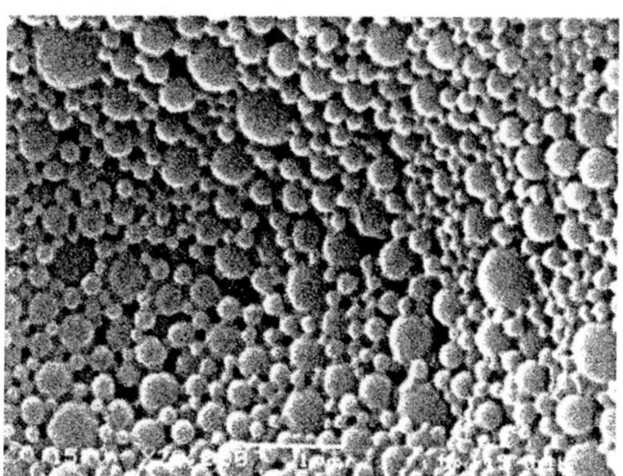

Figure 17.23 Paclitaxel loaded PLGA particles (bar = 1 μm). Reproduced from [39] with permission from Bentham Science Publishers.

Table 17.3 *Applications and process parameters for spray dried poorly soluble drugs using BÜCHI spray dryers.*

Application	Carrier, concentration, solvent	Spray drying process parameters	Particle size, morphology, yield, powder amount	Literature Reference
Galactosidase in trehalose (model protein)	5% aqueous concentration with 1:2 (w/w) enzyme to trehalose ratio	Nano Spray Dryer B-90, spray cap 4.0 μm, T in 80°C, T out 36–53°C, drying air flow 100–110 L/min, spray rate 100%	1–5 μm (respirable) spherical and smooth particles, residual enzyme activity 75% to 100% (after 3 weeks), yields 60% to 94%, sample amounts 500 mg	[40]
Itraconazole (antifungal drug)	Crystalline nanosuspenion of itraconazole by high-pressure homogenization (456 nm size), 2–10% sugars (lactose, sucrose, mannitol) in suspension, including Poloxamer 188 and sodium deoxycholate as surfactants	Mini Spray Dryer B-290, T in 90–100°C, T out 40–60°C, pump rate 10–30%, feed volume 100 ml	10–40 μm spherical aggregates, redispersible powder, mannitol was the most favorable carrier	[10]
RR01 (highly lipophilic drug) in poly-methacrylic acid-co-ethylacrylate copolymer	RR01-loaded nanoparticles prepared as oil-in-water emulsion, methanol solution containing Eudragit L100-55, drug:polymer ratio 1:19 and 1:1	Mini Spray Dryer B-190, 0.5 mm two-fluid nozzle, T in 50°C, T out 37–42°C, spray gas 500 NL/h, aspirator 15, feed rate 4 mL/min	Nanoparticles 292–297 nm, spray-dried particles 9.9–10.9 μm, shriveled surfaces, yield 68%, drug loading 50%, excellent batch-to-batch reproducibility	[23]
5-phenyl-1,2-dithiole-3-thione (5-PDTT) lipophilic model drug as o/w emulsion	Maltodextrin as carrier and sodium caseinate as emulsifying agent	Mini Spray Dryer B-191, T in 110°C, T out 75–85°C, aspirator 100%, spray gas 500 NL/h, feed rate 7 ml/min	2–15 μm particles, spherical with dents, yield 44–74%	[21]

Paclitaxel (anticancer drug)	Paclitaxel-loaded PLGA nanoparticles by single emulsion technique, polymeric materials with paclitaxel and additives dissolved in DCM (PVA, gelatin, DPPC and cholesterol)	Mini Spray Dryer B-191, T in 50°C, T out 42°C, spray air 700 NL/h (6 bar), aspirator 100%, pump rate 4.5–5.0 mL/min	550–850 nm nanoparticles in biopolymer, spherical and smooth surface, drug encapsulation of up to 100%, desired drug release kinetics, high cell uptake of nanoparticles	[39]
L-leucine and trehalose dispersing agent for inhalable drugs)	20 to 35 mg/mL total feed concentration, L-leucine to trehalose ratio varied from 0–100%	Nano Spray Dryer B-90, spray cap 4.0 μm, T in 75°C, T out 45°C, drying air flow 100 L/min, droplet drying time 11.3 ms	2.1–5.4 μm aerodynamic diameter solid, spherical and hollow, pure trehalose amorphous, with >25% leucine content crystalline, low-density, well dispersing sample amounts <80 mg	[47]
Compritol (lubricant in oral solid dosage formulations and cosmetics)	Solid lipid nanoparticles (cetylpalmitate, compritol, synchrowax) prepared by high pressure homogenization	Mini Spray Dryer B-190, 0.5 mm two-fluid nozzle, T in 90–110°C, T out 50–60°C, spray air 550 NL/h, aspirator 10, pump rate 5 mL/min	300–900 nm solid lipid nanoparticles with mannitol, lactose or trehalose, dry powders, non-sticking, fine-grained	[11]
Polyacrylate and silica nanoparticles	Polyacrylate nanoparticles via a solvent replacement technique, nanoparticulate suspensions of silica prepared by adding silica nanoparticles to an ethanol-water solution	Mini Spray Dryer B-290 with Inert Loop B-295, T in 110°C, T out 60°C, spray gas N2 350 L/h, drying gas flow 40 m3/h, pump rate 4 mL/min	2.5 ± 1 μm hollow nanoparticulate aggregates, tap density 0.15 g/cm3, high flowability, high terapeutic efficiency	[26]
Bendroflumethiazide (high blood pressure treatment) in nanoporous microparticles	2.5% solution of bendroflumethiazide in ethanol/water (80% ethanol) or methanol/water (80% methanol) solutions with ammonium carbonate (15 w% of total solids) as pore forming agent	Mini Spray Dryer B-290 with Inert Loop B-295, T in 78°C, T out 53°C (ethanol), T in 110°C, T out 74°C (methanol), spray air 670 NL/h, aspirator 100%, pump setting 30%	1–5 μm particles, spherical and nanoporous, 0.24–0.32 g/cm3 tap density	[46]

(continued)

Table 17.3 *(Continued).*

Application	Carrier, concentration, solvent	Spray drying process parameters	Particle size, morphology, yield, powder amount	Literature Reference
API drug (unknown)	Nanosuspension of API particles mixed in water with hydroxypropyl cellulose (HPC) and yittria-stabilized zirconium beads produced by ball milling	Mini Spray Dryer B-191, T in 100°C, T out 63°C, spray air 650 L/h, pump rate 1.0 kg/h	Nano- and microparticles of API in HPC as nanoparticle aggregates, redispersible in water over hours	[8]
Bovine serum albumin (model protein)	0.1% aqueous solution with 0.05% surfactant polyoxyethylene sorbitan monoleate (Tween 80)	Nano Spray Dryer B-90, spray cap 4.0 μm, T in 120°C, T out 51–55°C, T in 100°C, T out 42–45°C, T in 80°C, T out 36–40°C, drying air flow 150 L/min	460 ± 10 nm (span 1.03) to 2.6 μm in all 18 runs, median particle size < 5 μm, smooth spherical particles, yield 72 ± 4%	[18]
Polymeric wall materials Furosemide (diuretic drug) Vitamin E acetate nanoemulsion	0.1, 1 and 10 w% polymeric wall materials (arabic gum, whey protein, polyvinyl alcohol, modified starch, maltodextrin), 1.25 w% furosemide in acetone, Vitamin E acetate nanoemulsion in wall materials at 1:4 ratio	Nano Spray Dryer B-90, spray cap 4.0 μm, T in 100°C, T out 38–60°C, drying air flow 100 L/min, feed rate 3–25 mL/h, closed loop for organic solvents (N2 drying gas, O2 < 4%)	460–730 nm distributions, mainly < 1 μm, yields 43–94.5% for 30–300 mg powder amounts, encapsulated oil-in-water nanoemulsions with oil droplets below 100 nm	[19]
Thymopentin (synthetic pentapeptide, immunomodulator)	Thymopentin-loaded solid lipid nanoparticles by double emulsion technique, microparticles prepared by spray drying the solid lipid nanoparticles with mannitol (nontoxic, nonhygroscopic) and leucine (dispersibility enhancer)	Mini Spray Dryer B-290, T in 110°C, T out 52°C, spray air 600 L/h, aspirator 100%, pump rate 4.0 ml/min	4.1 μm, spherical and hollow microparticles, tap density 0.175 g/cm3, high aerosolization efficiency, emitted dose of 98%, respirable fraction 51%, after spray drying the mean particle size increased from 147 nm to 218 nm	[44]

Budesonide (glucocorticoid steroid for asthma treatment)	Budesonide with and without ammonium carbonate from ethanol/water or methanol/water solutions	Mini Spray Dryer B-290 with Inert Loop B-295, T in 78°C, T out 50°C (open air mode, ethanol/water), T in 78°C, T out 53°C (closed N2 mode, 80% ethanol/20% water) T in 70°C, T out 45–48°C (closed N2 mode, 80% methanol/20% water), spray gas 6	1–5 μm excipient-free porous microparticles, surface area 10.5 m^2/g, amorphous, low density suitable for pulmonary drug delivery, good physical stability for over 1 year	[25]
Nano *Radix Salvia Miltiorrhiza* (herbal Traditional Chinese Medicine)	5.5 % w/v *radix salvia miltiorrhiza* nanoparticles suspension in distilled water, gelatin solution (30 g/L) and CMC-Na solution (5 g/L) mixed 1:2 and stirred to form an aqueous solution, by stirring to form a homogeneous emulsion	Mini Spray Dryer B-290, T in 80°C, spray air 357 L/h, pump setting 6%	133.5 nm mean particles size after ultrafinely high speed grinding, spherical capsules with rough surfaces, improved medicinal effect by nano technology (dissolution rate)	[9]
Cyclosporin A (immune system suppressant) Dexamethasone (anti-inflammatory steroid drug) in PLGA	100 mg Cyclosporin A or Dexamethasone encapsulated in 500 mg of biodegradable PLGA, 0.5–2% solid content dissolved in DCM/ethanol (70/30 vol%)	Nano Spray Dryer B-90 with Inert Loop B-295, spray meshes 4.0 and 5.5 μm, T in 29–32°C, T out 28–32°C, drying gas flow 102–130 L/min, feed rate 10–16 mL/h	Spherical shape with porous surface, sustained release profiles over several days with low MW PLGA grades, particle size 0.9–2.2 μm	[43]
Melatonin (hormone to regulate the sleep-wake cycle)	Nanocapsule suspensions (204–280 nm) containing melatonin by interfacial deposition method, organic phase: caprylic/capric triglyceride or mineral oil, poly(ε-caprolactone, poly(lactide), Eudragit, melatonin and surfactant in acetone, colloidal silicon dioxide added	Mini Spray Dryer B-190, T in 150ºC, T out 107ºC, pump rate 3 mL/min	2.4–10.3 μm, spherical, narrow distribution (span 1.6), encapsulation efficiency 93%, yield 74%, moisture content < 2%, no morphological changes after 12 months	[24]

(continued)

Table 17.3 (Continued).

Application	Carrier, concentration, solvent	Spray drying process parameters	Particle size, morphology, yield, powder amount	Literature Reference
Trehalose (protein stabilizer)	α,α-trehalose-dihydrate as 1% and 0.1% aqueous solutions, either with or without 0.05% polysorbate 20	Spray cap 7 μm or, 3 μm, T in 60, 80 and 100°C T out 30–45°C, sample volumes 5–20 mL, drying air flow 115 L/min	600 nm and span 1.6 at 0.1%, 1.2 μm and span 0.8 at 1%, smallest powder amounts down to 10 mg	[36]
Griseovulvin (antifungal drug) with mannitol and polysorbate 20	0.44% griseofulvin in methanol/acetone, 1% salicylic acid in ethyl acetate, 5% salicylic acid in acetone, 1% benzocaine in ethanol, 0.1% disodium phosphate, 1% mannitol or trehalose with 0.05% polysorbate 20	Nano Spray Dryer B-90 and Inert Loop B-295, spray cap 4.0 μm, T in 50–120°C, T out 28–59°C sample amounts 10 mL, drying air flow 120 L/min in open mode for aqueous solutions, N2 drying gas, O2 < 4%) closed loop for organic solvents	500 nm disodium phosphate, 800 nm trehalose, down to 4.8 μm mean droplet sizes (span approx. 1.2), disodium phosphate, yield 75%, mannitol yield 65–78%, trehalose yield 50–75%, sample amounts 50 mg	[20]
Polybutyl-cyanoacrylate and gelatine nanoparticles in lactose (pulmonary drug delivery)	Lactose in water mixed with either gelatin nanoparticles (protein model) or polybutylcyanoacrylate nanoparticles (model for synthetic particles)	Mini Spray Dryer B-191, T in 170–180°C, spray air 700 l/h (80 psi), aspirator 40 m3/h, pump rate 2 ml/min	3.0 ± 0.2 μm respirable lactose carrier particles, nanoparticles remain in nanosize after spray drying	[27]
Indomethacin (lipophilic non-steroidal anti-inflammatory drug)	Silica stabilized Pickering o/w emulsions by high pressure homogenization, oil phase: meduim chain tryglicerides with dissolved lipophilic emulsifiers, lecithin or oleylamine, aqueous phase: hydrophilic silica nanoparticles	Mini Spray Dryer B-190, T in 160°C, T out 85°C, aspirator setting 10, spray air, 0.6 m3/min, pump rate 5 mL/min	1–5 μm microcapsules with internal porous structure (20–100 nm), 50–85% oil content, excellent storage stability, high bioavailability, enhanced drug release	[28]

Indomethacin in porous silica-lipid hybrid microcapsules (non-steroidal anti-inflammatory drug)	Simultaneous emulsification of Pickering emulsions, oleylamine dissolved in oil, followed by addition of indomethacin, silica aqueous dispersion (fumed silica nanoparticles, Aerosil 380) added as the continuous phase, the coarse o/w emulsions were homogenised and then spray dried	Mini Spray Dryer B-290, T in 160°C, T out 85°C, spray air 0.6 m3/min, aspirator setting 10, pump rate 5 ml/min	1–5 μm spherical microcapsules with nanoparticle layers on surface, internal porous structure (25–100 nm), oil content 60–63%, oil entrapment efficiencies 81–83%, stable structural integrity after in-vitro release	[45]
Celecoxib (rheumatoid) in silica-lipid hybrid microcapsules	Spray-dried o/w emulsions (lecithin in oil and celecoxib), stabilized with hydrophilic silica nanoparticles (average size 7 nm), medium chain triglyceride as lipid matrix, soybean lecithin as emulsifier	Mini Spray Dryer B-290, T in 160°C, T out 85°C, aspirator100%, spray air 0.6 m3/min, pump rate 5 ml/min	Novel silica-lipid hybrid microcapsules 1–5 μm, spherical, readily redispersible, with internal porous matrix (pore sizes 100–500 nm), stable for at least 6 months, improved bioavailability, oil loading 42–51%, oil entrapment 92–100%	[5]
COX-2 inhibitor BMS-347070 (non-steroidal anti-inflammatory drug)	BMS-347070 (COX-2 inhibitor, particle size between 5 and 10 μm) and Pluronic F127 NF	Mini Spray Dryer B-191, two cyclones in series, T in and T out at ambient temperature, spray air 400–700 NL/h, backpressure -28 mbar, pump setting 5 to 6 (500–800 mL/h)	Smooth and spherical morphology, microparticles containing nanocrystalline drug <100 nm, faster dissolution profiles compared to micronized bulk drug	[6]

17.6 Conclusion

Spray drying is a continuous process that is well suited to the generation of bulk drug substances and excipients. Using this process, the physical properties of the resulting product, such as particle size and shape, moisture content, and flow properties can be controlled through the selection of equipment design and optimization of process parameters. Thus, the final spray-dried powder may not need further processing (wet or dry granulation) before compaction into tablets or capsules.

Spray drying processes offer several advantages when the solid-state properties of drug substances need to be modified. Using this process, solubility and dissolution rates of poorly soluble materials can be increased several fold, and the stability of the amorphous form of the materials can be improved significantly.

The key benefits of this technology are the opportunities to control the size and morphology of the particles while using a relatively gentle processing method. Indeed, spray drying has been proven for the drying of heat-sensitive materials such as, for example, protein-based drugs. Process parameters are discussed for the creation of spherical particles with a smooth or structured surface and porous morphology.

The feasibility of the traditional Mini Spray Dryer models in lab-scale research has been demonstrated by several scientific literature studies. The collected data show the feasibility of encapsulating various drugs within biodegradable polymers for novel drug delivery systems. While the traditional bench-top spray dryers have been shown to be capable tools for the generation of respiratory-sized particles, the area of process technology is ever-evolving.

The Nano Spray Dryer B-90 offers new possibilities in the field of laboratory-scale spray drying, including increased recovery (up to 90%), small powder quantity (100 mg amounts), and highly-definable particle size ranges (300 nm to 5 μm).

Because of its initial inherent costs, spray drying is not always considered a processing option for many conventional formulations, especially for small batch size operations. However, when a specialized particle type is required by the active ingredient or dosage form, spray drying can become a feasible alternative to more conventional manufacturing processes. Such particle types include microcapsules, controlled release particles, nanoparticles, and microemulsions.

The literature review presented shows that spray drying is an attractive method for encapsulation of poorly soluble drugs. The advantage of spray drying is that it is a one-step method enabling the fast processing of small batches with reasonable yields. Spray dried microparticles have suitable characteristics (e.g. size and shape) for inhalation and other applications.

Pharmacological effects can be changed significantly when the medical material size is decreased to a micro-nano size. The particle sizes of materials are an important physical property that affects such pharmaceutical behavior as dissolution, chemical stability, and the bioavailability of solid dosage forms.

Finally, nanosuspensions of poorly soluble drugs can be spray dried to obtain flowable powders with high surface areas that can be easily redispersed. These optimized powders also showed significantly improved dissolution rates when compared to the micronized drug, or to unoptimized nanosuspensions.

References

1. A. Fahr, X. Liu. Drug delivery strategies for poorly water-soluble drugs. *Expert Opinion Drug Delivery*, **4**, 403–416 (2007).
2. C. Lipinski, Poor aqueous solubility: An industry-wide problem in drug discovery. *American Pharmaceutical Review*, **5**, 82–85 (2002).
3. M. Martinez, L. Augsburger, T. Johnston, W.W. Jones, Applying the biopharmaceutics classification system to veterinary pharmaceutical products. Part I. Bio-pharmaceutics and formulation considerations. *Advanced Drug Delivery Reviews*, **54**, 805–824 (2002).
4. E. Merisko-Liversidge, G.G. Liversidge, E.R. Cooper, Nanosizing: a formulation approach for poorly-water-soluble compounds. *European Journal of Pharmaceutical Sciences*, **18**, 113–120 (2003).
5. A. Tan, S. Simovic, A.K. Davey, T. Rades, C.A. Prestidge, Silica-lipid hybrid (SLH) microcapsules: A novel oral delivery system for poorly soluble drugs. *Journal of Controlled Release*, **134**, 62–70 (2009).
6. S.X. Yin, M. Franchini, J. Chen, A. Hsieh, S. Jen, T. Lee, M. Hussain, R. Smith, Bioavailability enhancement of a COX-2 inhibitor, BMS-347070, from a nanocrystalline dispersion prepared by spray-drying. *Journal of Pharmaceutical Sciences*, **94**, 7, 1598–1607 (2005).
7. G.L. Amidon, H. Lunnernas, V.P. Shah, J.R. Crison. A theoretical basis for a biopharmaceutic drug classification: The correlation of *in vitro* drug product dissolution and *in vivo* bioavailability. *Pharm. Res.*, **12**, 413–420 (1995).
8. J. Lee, Drug nano- and microparticles processed into solid dosage forms: Physical properties. *Journal of Pharmaceutical Sciences*, **92**, 10, 2057–2068 (2003).
9. M. Peiyan, M.F. Zhengyi, S. Yanli, W. Weimin, W. Hao, Z. Jinyong, W. Yucheng and Z. Qingjie, Fabrication and physicochemical and medicinal characterization of nano traditional Chinese medicine. *Recent Patents on Engineering*, **2**, 104–113 (2008).
10. M.V. Chaubal, C. Popescu, Conversion of nanosuspensions into dry powders by spray drying: A case study. *Pharmaceutical Research*, **25**, 10, 2302–2308 (2008).
11. C. Freitas, H. Müller, Spray drying of solid lipid nanoparticles. *European Journal of Pharmaceutics and Biopharmaceutics*, **46**, 145–151 (1998).
12. H.-K. Chan, P.C.L. Kwok, Production methods for nanodrug particles using the bottom-up approach. *Advanced Drug Delivery Reviews*, **63**, 406–416 (2011).
13. C. Arpagaus, N. Schafroth, M. Meuri, Laboratory-scale spray drying of inhalable drugs: A review. *BUCHI Brochure*, best@buchi 59/2010 en (2010).
14. C. Arpagaus, N. Schafroth, M. Meuri, Laboratory-scale spray drying of lactose: A review. *BUCHI Brochure*, best@buchi 57/2010 en (2010).
15. S. Gautier, C. Arpagaus, N. Schafroth, M. Meuri, A. Deschamps, V. Maquet, Very fine chitosan microparticles with narrow and controlled size distribution using spray-drying technologies. *Drug Delivery Technology*, **10**, 8, 30–37 (2010).
16. C. Arpagaus, Nano Spray Dryer B-90: Literature review and applications. *BUCHI Brochure*, best@buchi 63/2011 en (2011).
17. D. Heng, S.H. Lee, W.K. Ng, R.B.H. Tan, The Nano Spray Dryer B-90. *Expert Opinion Drug Delivery*, **8**, 7, 965–972 (2011).

18. S.H. Lee, D. Heng, W.K. Ng, H.-K. Chan, R.B.H. Tan, Nano spray drying: A novel method for preparing protein nanoparticles for protein therapy. *International Journal of Pharmaceutics*, **403**, 192–200 (2011).
19. X. Li, N. Anton, C. Arpagaus, F. Belleteix, T.F. Vandamme, Nanoparticles by spray drying using innovative new technology: The BUCHI Nano Spray Dryer B-90. *Journal of Controlled Release*, **147**, 304–310 (2010).
20. K. Schmid, C. Arpagaus, W. Friess, Evaluation of the Nano Spray Dryer B-90 for pharmaceutical applications. *Pharmaceutical Development and Technology*, **16**, 4, 287–294 (2011).
21. G. Dollo, P. Le Corre, A. Guérin, F. Chevanne, J.L. Burgot, R. Leverge, Spray-dried redispersible oil-in-water emulsion to improve oral bioavailability of poorly soluble drugs. *European Journal of Pharmaceutical Sciences*, **19**, 273–280 (2003).
22. C. Arpagaus, H. Schwartzbach, Scale-up from the Büchi Mini Spray Dryer B-290 to the Niro MOBILE MINOR^{TM}. *BUCHI Brochure*, best@buchi 52/2008 en (2007).
23. F. De Jaeghere, pH-dependent dissolving nano- and microparticles for improved peroral delivery of a highly lipophilic compound in dogs. *AAPS Pharmsci*, www.pharmsci.org, 3, 1, article 8 (2001).
24. S.R. Schaffazick, A.R. Pohlmann, G. Mezzalira, S.S. Guterres, Development of nanocapsule suspensions and nanocapsule spray-dried powders containing melatonin. *Journal of Brazilian Chemical Society*, **17**, **3**, 562–569 (2009).
25. L.M. Nolan, L. Tajber, B.F. McDonald, A.S. Barham, O.I. Corrigan, A.M. Healy, Excipient-free nanoporous microparticles of budesonide for pulmonary delivery. *European Journal of Pharmaceutical Sciences*, **37**, 593–602 (2009).
26. K. Hadinoto, P. Phanapavudhikul, Z. Kewu, R.B.H. Tan, Novel formulation of large hollow nanoparticles aggregates as potential carriers in inhaled delivery of nanoparticulate drugs. *Industrial Engineering Chemical Research*, **45**, 3697–3706 (2006).
27. J.O.-H. Sham, Y. Zhang, W.H. Finlay, W.H. Roa, R. Löbenberg, Formulation and characterization of spray-dried powders containing nanoparticles for aerosol delivery to the lung. *International Journal of Pharmaceutics*, **269**, 457–467 (2004).
28. S. Simovic, P. Heard, H. Hui, Y. Song, F. Peddie, A.K. Davey, A. Lewis, T. Rades, C.A. Prestidge, Dry hybrid lipid-silica microcapsules engineered from submicron lipid droplets and nanoparticles as a novel delivery system for poorly soluble drugs. *Molecular Pharmaceutics*, **6**, 3, 861–872 (2009).
29. K. Masters, *Spray Drying in Practice*. SprayDryConsult, Denmark (2002).
30. A.S. Mujumdar, *Guide to Industrial Drying: Principles, Equipment and New Development*. IDS2008, Hyderabad, India (2008).
31. J. Broadhead, S.K. Edmond Rouan, C.T. Rhodes, The spray drying of pharmaceuticals. *Drug Development and Industrial Pharmacy*, **18**, 11–12, 1169–1206 (1992).
32. S.C. Wendel, M. Celik, Spray drying and pharmaceutical applications. In *Handbook of Pharmaceutical Granulation Technology*, D.M. Parikh (ed.), 3rd edn. New York: Taylor & Francis Group, LLC, 2005.
33. R. Vehring, Pharmaceutical particle engineering via spray drying. *Pharmaceutical Research*, **25**, 5, 999–1022 (2008).
34. K. Okuyama, I.W. Lenggoro, Preparation of nanoparticles via spray route. *Chemical Engineering Science*, **58**, 537–547 (2003).

35. C. Portmann, S. Flückiger, Th. Ziolko, A. Tauer, A. Wieprächtiger, J. Müller, S. Kleinhans, C. Arpagaus, G. Schönenberger, *The Laboratory Assistant*. BÜCHI Labortechnik AG, Flawil, Switzerland, Third edition (2007).
36. K. Schmid, C. Arpagaus, W. Friess, Evaluation of a vibrating mesh spray dryer for preparation of submicron particles. *Respiratory Drug Delivery Europe*, **30**, 323–326 (2009).
37. BÜCHI Labortechnik AG, *Nano Spray Dryer B-90 Commercial Brochure*. Flawil, Switzerland (2009).
38. M. Schön, R. Baumgartner, Heating, method for heating and laminating, electrostatic separator, spray drier, separating device and method for separating particles. *Patent*, EP 2056037 (A1) (2009).
39. S.-S. Feng, L. Mu, K.Y. Win, G. Huang, Nanoparticles of biodegradable polymers for clinical administration of paclitaxel. *Current Medicinal Chemistry*, **11**, 413–424 (2004).
40. K. Bürki, I. Jeon, C. Arpagaus, G. Betz, New insights into respirable protein powder preparation using a nano spray dryer. *International Journal of Pharmaceutics*, **48**, 248–256 (2011).
41. B.R. Bhandari, N. Datta, T. Howes, Problems associated with spray drying of sugar-rich foods. *Drying Technology*, **15**, 671-684 (1997).
42. C. Arpagaus, N. Schafroth, Spray dried biodegradable polymers as target material for controlled drug delivery. *BUCHI Brochure*, best@buchi 46/2007 en (2007).
43. N. Schafroth, C. Arpagaus, U.Y. Jadhav, S. Makne, D. Douroumis, Nano and microparticle engineering of water insoluble drugs using a novel spray-drying process. *Colloids and Surfaces B: Biointerfaces*, **90**, 8–15 (2012).
44. Y.-Z. Li, X. Sun, T. Gong, J. Liu, J. Zuo, Z.-R. Zhang, Inhalable microparticles as carriers for pulmonary delivery of thymopentin-loaded solid lipid nanoparticles. *Pharmaceutical Research*, **27**, 1977–1986 (2010).
45. S. Simovic, H. Hui, Y. Song, A.K. Davey, T. Rades, C.A. Prestidge, An oral delivery system for indomethicin engineered from cationic lipid emulsions and silica nanoparticles. *Journal of Controlled Release*, **143**, 367–373 (2010).
46. A.M. Healy, B.F. McDonald, L. Taijber, O.I. Corrigan, Characterisation of excipient-free nanoporous microparticles (NPMPs) of bendroflumethiazide. *European Journal of Pharmaceutics and Biopharmaceutics*, **69**, 1182–1186 (2008).
47. A.L. Feng, M.A. Boraey, M.A. Gwin, P.R. Finlay, P.J. Kuehl, R. Vehring, Mechanistic models facilitate efficient development of leucine containing microparticles for pulmonary drug delivery. *International Journal of Pharmaceutics*, **409**, 156–163 (2011).

Index

Drug Delivery Strategies for Poorly Water-Soluble Drugs, First Edition. Edited by Dennis Douroumis and Alfred Fahr.

Index

Printed and bound by CPI Group (UK) Ltd, Croydon, CR0 4YY
16/09/2026
14963384-0001